FIFTH EDITION

LINEAR ALGEBRA

WITH APPLICATIONS

W. Keith Nicholson
University of Calgary

Toronto   Montréal   Boston   Burr Ridge, IL   Dubuque, IA   Madison, WI
New York   San Francisco   St. Louis   Bangkok   Bogotá   Caracas
Kuala Lumpur   Lisbon   London   Madrid   Mexico City   Milan   New Delhi
Santiago   Seoul   Singapore   Sydney   Taipei

The McGraw-Hill Companies

McGraw-Hill Ryerson

Linear Algebra with Applications
Fifth Edition

Copyright © 2006, 2003 by McGraw-Hill Ryerson Limited, a Subsidiary of The McGraw-Hill Companies. All rights reserved. Copyright © 1995 by PWS Publishing Company. Copyright © 1990 by PWS-KENT Publishing Company. Copyright © 1986 by PWS Publishers. All rights reserved. No part of this publication may be reproduced or transmitted in any form or by any means, or stored in a data base or retrieval system, without the prior written permission of McGraw-Hill Ryerson Limited, or in the case of photocopying or other reprographic copying, a licence from The Canadian Copyright Licensing Agency (Access Copyright). For an Access Copyright licence, visit www.accesscopyright.ca or call toll free to 1-800-893-5777.

ISBN 0-07-092277-2

2 3 4 5 6 7 8 9 10 TCP 0 9 8 7 6

PRINTED AND BOUND IN CANADA

Care has been taken to trace ownership of copyright material contained in this text; however, the publisher will welcome any information that enables them to rectify any reference or credit for subsequent editions.

Publisher, Higher Education: Lynn Fisher
Sponsoring Editor: Leanna MacLean
Developmental Editor: Jennifer Matyczak
Senior Marketing Manager: Suzanne Beaudoin
Senior Supervising Editor: Margaret Henderson
Copy Editor: Edie Franks
Production Coordinator: Paula Brown
Composition: S R Nova Pvt Ltd, Bangalore, India
Cover Design: Greg Devitt
Cover Image: Karin Slade/Gettyone
Printer: Transcontinental Printing Company

Library and Archives Canada Cataloguing in Publication

Nicholson, W. Keith
  Linear algebra with applications / W. Keith Nicholson. — 5th ed.

Includes bibliographical references and index.
ISBN 0-07-092277-2

  1. Algebras, Linear—Textbooks. I. Title.

QA184.N53 2006   512'.5   C2005-906366-1

Contents

Preface

This textbook is a basic introduction to the ideas and techniques of linear algebra for first- or second-year students who have a working knowledge of high school algebra. Its aim is to achieve a balance among the computational skills, theory, and applications of linear algebra, while keeping the level suitable for beginning students. The contents are arranged to permit enough flexibility to allow the presentation of a traditional introduction to the subject, or to allow a more applied course. Calculus is not a prerequisite; places where it is mentioned may be omitted.

Linear algebra has wide application to the mathematical and natural sciences, to engineering, to computer science, and (increasingly) to management and the social sciences. As a rule, students of linear algebra learn the subject by studying examples and solving problems. More than 370 solved examples are included in the text, many of a computational nature, together with a wide variety of exercises. In addition, a number of sections are devoted to applications and to the computational side of the subject. The applications are optional, but they are included at the end of the relevant chapters (rather than at the end of the book) to encourage students to browse.

The examples are also used to motivate (and illustrate) concepts and theorems, carrying the student from concrete to abstract; however, most proofs are presented at a level appropriate to the student. This means that the book can be used to give a course emphasizing computation and examples (and omitting many proofs) or to give a more rigorous treatment (a few longer proofs are deferred to the end of the chapter).

Two innovations that have proven to be very successful are retained in the fifth edition:

First, as requested by engineers and scientists, *early treatment of diagonalization* in Chapter 3 (using only determinants and matrix inverses) opens up a wealth of applications early in the text, helping to answer the constant student query: "Why are we studying this stuff?"

Second, the "bridging" chapter (Chapter 5) allows difficult concepts like subspaces, spanning, independence, and dimension to be assimilated first in the concrete context of $\mathbb{R}^n$, which mitigates students "hitting the wall" when abstract vector spaces are introduced in Chapter 6.

The fifth edition maintains a balance between the abstract theory on one hand and matrix computations and applications on the other. Both early diagonalization and the "bridge" to abstraction have been retained and refined, and more examples have been included. A new feature is the early introduction of matrix transformations in Section 2.5, immediately after matrix multiplication and inversion are introduced. This helps students to visualize the effect of these operations, and is part of a general effort (in response to user suggestions) to introduce linear transformations earlier in the book.

New in the Fifth Edition

Reorganization of Topics

- The fifth edition has been reorganized into 10 chapters. Change of Basis is now covered in a chapter of its own (Chapter 9), while Linear Transformations is introduced earlier (Chapter 7), right after abstract spaces are defined.
- **Matrix transformations are introduced much earlier** (in Section 2.5). This provides students with a geometric picture of matrix multiplication in $\mathbb{R}^2$ and $\mathbb{R}^3$ and, in response to comments from readers, introduces the concept of a linear transformation of $\mathbb{R}^n$ at an early stage. It also paves the way for new examples of geometric linear transformations in Section 4.4. This will particularly appeal to visual learners and to instructors with a geometrical approach to the subject.
- **Increased emphasis on linear transformations**. As already mentioned, matrix transformations are introduced early in Chapter 2. In addition, the general theory of linear transformations is introduced in (a new) Chapter 7 immediately following abstract vector spaces, where kernels, images, and the dimension theorem are presented. However, the discussion of change of basis remains in (the new) Chapter 9.
- Section 2.4 on elementary matrices has been completely rewritten, as has the discussion of the Smith normal form.
- Vectors in $\mathbb{R}^2$ and $\mathbb{R}^3$ are defined intrinsically in terms of length and direction, and (matrix) coordinates are introduced. Vector operations are now defined using matrices and *then* interpreted to give the usual intrinsic definitions.
- The cross product is introduced early in Section 4.2 with just enough detail to apply it to planes. The details are in Section 4.3, which may be omitted with no loss of continuity.

New Topics and Expanded Coverage

- The "bridging" chapter (Chapter 5), has been expanded to cover concepts like subspaces, spanning, and independence in **two sections**. Most reviewers acknowledge that this is the most difficult part of the course for students and so welcome additional discussion of these concepts in Chapter 6. In addition, several reviewers pointed out that Chapter 5 is the end of the study of linear algebra for their students, and so they welcome the chance to introduce these topics in $\mathbb{R}^n$.
- Some material on **orthogonality is now presented earlier** (in Section 5.3), including the Cauchy inequality, orthogonal bases, the expansion theorem, and Pythagoras' theorem. This too met with reviewer approval and fits with the view that Chapter 5 is the end of Linear Algebra for some students. The full treatment of orthogonality in Chapter 8 is unchanged except that a section on finite fields and linear codes has been added.
- The introduction to the section on isometries (now Section 10.4) has been rewritten to focus first on the distance preserving operators on $\mathbb{R}^n$, and then on showing that those that fix zero are linear (the isometries). Hence, the distance preserving operators are just the composites of a translation and an isometry.
- Section 4.4 contains new material on matrix transformations in $\mathbb{R}^3$, discussing rotations, reflections, and projections, and presenting details on how volumes and areas are related to the determinant.
- The discussion of the rank theorem (now in Section 5.4) has been rewritten. If A is an $m \times n$ matrix, the characterizations when rank A is m or n have been expanded, with simpler proofs.
- A compact proof of the Jordan canonical form has been added in Section 9.5.
- At the request of several instructors (and students), a short discussion of what constitutes a proof is presented in a new Appendix B, found at the back of the book.

- A short proof of Pythagoras' theorem is presented in Section 4.1, using only similar triangles.
- Brief introductions have been added to several chapters and sections, often reflecting the history of the subject. To make the history more real to students, some images of mathematicians prominent in the development of linear algebra have been included.

New Applications

- A number of applications (including 5 new) are presented, showing how linear algebra clarifies and solves problems, and providing relevance for students. To encourage students to browse, these are placed at the end of the chapter containing the necessary techniques.
- A new application in Section 5.6 reveals how to represent data samples in $\mathbb{R}^n$, and how to use the dot product to calculate means, variances, and covariances. A nice feature is how the Cauchy inequality shows that the correlation coefficient lies between 1 and -1.
- A **new application to linear codes** is presented in Section 8.8. The field $\mathbb{Z}_p$ of integers modulo p is constructed informally for any prime p and used to describe UPC and ISBN codes, as well as other check-digit codes. Then codes are defined in general over any finite field, the Hamming distance is introduced, and nearest neighbour decoding is described. Linear codes are presented in terms of generator and parity-check matrices, and syndrome decoding is discussed. The section closes with an investigation of orthogonal codes (requiring an interesting look at the dot product over a finite field).
- Several other new applications have been included:
    Directed graphs in Section 2.2.
    Google PageRank in Section 3.3.
    Computer graphics in Section 4.5.

Other Features

- Presentation of techniques in examples, with emphasis on concrete computations and on the algorithmic nature of some procedures, allows students to readily master new skills. The text has over 370 solved examples that cover the basic techniques, illustrate the central ideas, and are keyed to the exercises in each section.
- A wide variety of exercises (over 1200, many with multiple parts) are included, keyed to the examples, starting with routine computational problems, and progressing to more theoretical exercises that help students develop skills in an appropriate, logically paced fashion.
- Exercises with a ♦ symbol have an answer at the end of the book, enabling students to check the accuracy of their computation immediately.

Ancillary Materials

LILA: Lyryx Interactive Linear Algebra

http://lila.lyryx.com

LILA—Lyryx Interactive Linear Algebra is a Web-based teaching/learning tool that has captured the attention of post-secondary institutions across the country. Offering significant benefits not only to the student but *also* to the instructor, LILA has instant appeal because it parallels the classroom environment.

Each chapter is broken down into several *Lessons*, condensed versions of the material in the textbook. Lessons are animated with voice over and act as the student's own personal tutor.

LYRYX LEARNING INC
Online Learning and Assessment
lyryx.com

Lessons are supported by highly interactive *Explorations* with graphical and computational capabilities well-suited as a demonstration tool in the class or for individual use. Concepts can be explored by performing "what if" analysis and students can immediately see the resulting impact, allowing them to fully appreciate complex concepts.

Each section contains an Assessment component, referred to as a *Lab*. Algorithmically generated and automatically graded, students get instant grades and feedback—no need to wait until the next class to find out how well you did!

Grades are instantly recorded in a grade book that the student can view. Instructors can view the grades of all students, which means less time spent marking, but with the pedagogical advantages of marked assignments—a dream come true!

Students are motivated to do their LILA Labs for two reasons: first, because it can be tied to assessment, and second, because they can try the Lab as many times as they wish prior to the due date with only their best grade being recorded. Instructors know from experience that if students are doing their linear algebra homework, they will be successful in the course. Recent research has shown that when Labs are tied to assessment, even if worth only a small percentage of the total grade for the course, students WILL do their homework—and MORE THAN ONCE!

For the Instructor

Your **Integrated Learning Sales Specialist** is a McGraw-Hill Ryerson representative who has the experience, product knowledge, training, and support to help you assess and integrate any of the products, technology, and services noted below into your course for optimum teaching and learning performance. Whether it's helping your students improve their grades or putting your entire course online, your *i*Learning Sales Specialist is there to help you do it. Contact your local *i*Learning Sales Specialist today to learn how to maximize all of McGraw-Hill Ryerson's resources!

McGraw-Hill Ryerson offers a unique *i***Learning Services Program** designed for Canadian faculty. Our mission is to equip providers of higher education with superior tools and resources required for excellence in teaching. For additional information visit www.mcgrawhill.ca/highereducation/iservices.

The **Instructor's Online Learning Centre (OLC)** at www.mcgrawhill.ca/college/nicholson offers downloadable instructor's supplements in a password-protected environment, including an Instructor's Solution Manual, Additional Problems, a section on Linear Programming, and PageOut, the McGraw-Hill Ryerson course management system.

The **Instructor's Solution Manual**, available on the OLC, contains answers or solutions to all the exercises found in the book (in PDF format).

A **Test Bank** of over 100 problems with complete solutions covers the entire course and is suitable for exams.

For the Student

The **Partial Student Solutions Manual** (0-07-092702-2) contains detailed solutions to all exercises with an answer in the text (marked with a ♦). The Partial Student Solutions Manual provides complete worked solutions. It also contains solutions to other selected exercises (often theoretical ones).

Chapter Summaries

Chapter 1: Systems of Linear Equations. A standard treatment of gaussian elimination is given. The rank of a matrix is introduced via the row-echelon form. Applications to network flows, electrical networks, and chemical reactions are provided.

Chapter 2: Matrix Algebra. Matrix operations (including transposition) are defined, matrix multiplication is related to directed graphs, and matrix inverses are characterized. Elementary matrices are discussed, and the Smith normal form is proved. Matrix transformations in the plane (and in space) are introduced and used to reveal the geometrical meaning of matrix multiplication and inverses. Linear transformations of $\mathbb{R}^n$ are defined and shown to be matrix transformations. The relationship of matrix algebra to linear equations is described, and block multiplication is introduced emphasizing those cases needed later in the book. LU-factorization is introduced and applications to economic models and Markov chains are given.

Chapter 3: Determinants and Diagonalization. The cofactor expansion is stated (proved by induction later) and used to define determinants inductively and to deduce the basic rules. The product and adjugate theorems are proved. Then, motivated by an example about radioactive decay, the diagonalization algorithm is presented, leading to the idea of a discrete dynamical system and to applications to linear recurrences and population growth (popular with students). Note that diagonalization is carried out using only determinants and matrix inverses, avoiding any appeal to subspaces and dimension. A brief discussion of Google PageRank is included.

Chapter 4: Vector Geometry. Vectors are defined intrinsically (that is, in terms of length and direction), and related to matrices via coordinates. Then, in contrast to the fourth edition, vector operations are defined using matrices and then interpreted to give the usual intrinsic definitions. Next, dot products and projections are introduced and used to solve (primarily geometric) problems about lines and planes. This leads to the cross product. Matrix transformations are introduced in $\mathbb{R}^3$ and lead to the calculation of areas and volumes using determinants. The chapter closes with an application to computer graphics.

Chapter 5: The Vector Space $\mathbb{R}^n$. Subspaces, spanning, independence, and dimension are introduced in the context of $\mathbb{R}^n$ in the first two sections (expanded from the terse, one-section treatment in the fourth edition). As in the fourth edition, this is a "bridging" chapter, easing the transition to abstract spaces. In addition, orthogonal bases are introduced and used to derive the expansion theorem. After a rigorous study of rank and diagonalization, the chapter closes with a (new) application to correlation and variance, and an application to least squares approximation.

Introducing this material in Chapter 5 is important because: (1) for many students Chapter 5 is the end of the course, and instructors welcome the chance to introduce this material; (2) in many universities Chapter 5 is the end of the first half of a two-semester course; and (3) this is the most difficult part of the course, so many students welcome a repeated discussion of concepts like subspaces, independence, and spanning, albeit in the abstract setting.

Chapter 6: Vector Spaces. Building on the work on $\mathbb{R}^n$ in Chapter 5, the basic theory of abstract finite dimensional vector spaces is developed emphasizing examples like matrices, polynomials, and functions as well as $\mathbb{R}^n$. This is the first acquaintance most students have had with an abstract system, so not having to deal with spanning, independence, and dimension in the general context eases the transition to abstract thinking. Applications to polynomials and to differential equations are included.

Chapter 7: Linear Transformations. This is a new chapter in the fifth edition, inserted to bring in the general theory of linear transformations as early as possible (in keeping with the earlier treatment of matrix transformations in Section 2.5). The chapter consists primarily of Sections 8.1 to 8.3 in the fourth edition. General linear transformations are introduced, motivated by many examples from geometry, matrix theory, and calculus. Then kernels and images are defined, the dimension theorem is proved, and isomorphisms are discussed. The chapter ends with an application to linear recurrences.

Chapter 8: Orthogonality. The study of orthogonality in $\mathbb{R}^n$, begun in Chapter 5, is continued. Orthogonal complements and projections are defined and used to study orthogonal diagonalization (obtaining the principal axis theorem), positive definite matrices (the Cholesky factorization), the QR-factorization, and best approximation and least squares. The theory is extended to $\mathbb{C}^n$ in Section 8.6 where hermitian and unitary matrices are discussed, culminating in Schur's theorem and the spectral theorem (a new short proof of the Cayley–Hamilton theorem is also given).

In a new section (Section 8.8) the field of integers modulo p is constructed informally for any prime p, and used to describe UPC and ISBN codes, as well as other check-digit codes. Then codes are defined over any finite field, the Hamming distance is introduced, and nearest neighbour decoding is described. Next, linear codes are described in terms of generator and parity-check matrices, syndrome decoding is discussed, and orthogonal codes are investigated.

The chapter concludes with applications to quadratic forms and systems of differential equations.

Chapter 9: Change of Basis. The matrix of a general linear transformation is defined and studied. In the case of an operator, the relationship between basis changes and similarity is revealed. Invariant subspaces and direct sums are introduced and used to derive the block triangular form. That, in turn, is used to give a compact proof of the Jordan canonical form (new in the fifth edition).

Chapter 10: Inner Product Spaces. General inner products are introduced and distance, norms, and the Schwarz inequality are discussed. The Gram-Schmidt algorithm is given, projections are defined, and the approximation theorem is proved (with an application to Fourier approximation). Finally, isometries are characterized, and distance-preserving operators are shown to be composites of translations and isometries.

Appendices. In Appendix A, complex arithmetic is developed far enough to find nth roots. In Appendix B (new in the fifth edition), methods of proof are discussed. In Appendix C, mathematical induction is presented.

Section Dependencies

The following suggestions are made for alternative sequences through the text. Each section requires that sections above it have been completed.

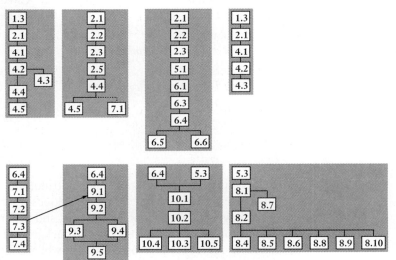

List of Applications

- Network Flow (Section 1.4)
- Electrical Networks (Section 1.5)
- Chemical Reactions (Section 1.6)
- Directed Graphs (Section 2.2)

- Input-Output Economic Models (Section 2.7)
- Markov Chains (Section 2.8)
- Polynomial Interpolation (Section 3.2)
- Radioactive Decay (Examples 1 and 10, Section 3.3)
- Google PageRank (Section 3.3)
- Linear Recurrences (Section 3.4; see also Section 7.4)
- Population Growth (Section 3.5)
- Computer Graphics (Section 4.5)
- Correlation and Variance (Section 5.6)
- Least Squares Approximation (Section 5.7)
- Polynomials (Section 6.5)
- Differential Equations (Section 6.6; see also Section 8.10)
- Linear Recurrences (Section 7.4)
- UPC, ISBN, and other check-digit codes (Section 8.8)
- Error Correcting Codes (Section 8.8)
- Quadratic Forms (Section 8.9)
- Systems of Differential Equations (Section 8.10)
- Fourier Approximation (Section 10.5)

Acknowledgments

Comments and suggestions that have been invaluable to the development of this edition were provided by a variety of reviewers, and I thank the following instructors:

Natalia Cheredeko
University of Toronto

John MacDonald
University of British Columbia

Thi Dinh
University of Calgary

Arturo Pianzola
University of Alberta

Nelly Faycal
Algonquin College

Dorette Pronk
Dalhousie University

Deirdre Haskell
McMaster University

Clive M. Reis
University of Victoria

Ulrich Haussmann
University of British Columbia

F. C. Y. Tang
University of Waterloo

Lev V. Idels
Malaspina University College

Walter Tholen
York University

Colin Ingalls
University of New Brunswick

Ross Willard
University of Waterloo

Yuanlin Li
Brock University

Peter Zizler
Mount Royal College

It is a pleasure to recognize the contributions of several people to this book. First, thanks go to Leanna MacLean, Sponsoring Editor, for her enthusiasm and effort in getting the project under way, and to Jennifer Matyczak, Developmental Editor, for her work on the editorial background to the book. Next, I would like to thank Margaret Henderson, Paula Brown, and Lynn Fisher at McGraw-Hill Ryerson for their part in the project, as well as Kelly Dickson and the rest of the production staff. Special thanks go to Jason Nicholson in Calgary for his help in preparing the Solution Manual and for the technical check for the text, and to Edie Franks for copy editing. Finally, I want to thank my wife Kathleen without whose understanding and cooperation this book would not exist.

W. Keith Nicholson

1 Systems of Linear Equations

SECTION 1.1   Solutions and Elementary Operations

Practical problems in many fields of study—such as biology, business, chemistry, computer science, economics, electronics, engineering, and the social sciences—can often be reduced to solving a system of linear equations. Linear algebra arose from attempts to find systematic methods for solving these systems, so it is natural to begin this book by studying linear equations. Here is an example.

Example 1

A biologist wants to create a diet from fish and meal that contains 183 grams of protein and 93 grams of carbohydrate per day. If fish contains 70% protein and 10% carbohydrate, and meal contains 30% protein and 60% carbohydrate, how much of each food is required each day?

Solution

If x grams of fish and y grams of meal are given each day, the amount of protein and carbohydrate in the diet are represented by these equations:

$$0.7x + 0.3y = 183$$
$$0.1x + 0.6y = 93$$

Multiplying the second equation by 7 gives

$$0.7x + 0.3y = 183$$
$$0.7x + 4.2y = 651$$

Subtracting the top equation from the bottom equation gives $3.9y = 468$; then dividing by 3.9 gives $y = 120$. Substituting this into the equation $0.1x + 0.6y = 93$ gives $x = 210$. Hence, there is exactly one satisfactory diet: It consists of 210 grams of fish and 120 grams of meal daily.

If a, b, and c are real numbers, the graph of an equation of the form

$$ax + by = c$$

is a straight line (if a and b are not both zero), so such an equation is called a linear equation in the variables x and y. However, it is often convenient to write the variables as $x_1, x_2, \ldots, x_n$, particularly when more than two variables are involved. An equation of the form

$$a_1 x_1 + a_2 x_2 + \cdots + a_n x_n = b$$

is called a **linear equation** in the n variables $x_1, x_2, \ldots, x_n$. Here $a_1, a_2, \ldots, a_n$ denote real numbers (called the **coefficients** of $x_1, x_2, \ldots, x_n$, respectively) and b is also a number (called the **constant term** of the equation). A finite collection of linear equations in the variables $x_1, x_2, \ldots, x_n$ is called a **system of linear equations** in these variables. Hence,

$$2x_1 - 3x_2 + 5x_3 = 7$$

is a linear equation; the coefficients of x_1, x_2, and x_3 are 2, -3, and 5, and the constant term is 7. Note that each variable in a linear equation occurs to the first power only.

Given a linear equation $a_1 x_1 + a_2 x_2 + \cdots + a_n x_n = b$, a sequence $s_1, s_2, \ldots, s_n$ of n numbers is called a **solution** to the equation if

$$a_1 s_1 + a_2 s_2 + \cdots + a_n s_n = b$$

that is, if the equation is satisfied when the substitutions $x_1 = s_1, x_2 = s_2, \ldots, x_n = s_n$ are made. A sequence of numbers is called **a solution to a system** of equations if it is a solution to every equation in the system.

For example, $x = -2, y = 5, z = 0$ and $x = 0, y = 4, z = -1$ are both solutions to the system

$$\begin{aligned} x + y + z &= 3 \\ 2x + y + 3z &= 1 \end{aligned}$$

A system may have no solution at all, or it may have an infinite family of solutions. For instance, the system $x + y = 2, x + y = 3$ has no solution because the sum of two numbers cannot be 2 and 3 simultaneously. A system that has no solution is called **inconsistent**; a system with at least one solution is called **consistent**. The system in Example 2 below has infinitely many solutions.

Example 2

Show that, for arbitrary values of s and t,

$$\begin{aligned} x_1 &= t - s + 1 \\ x_2 &= t + s + 2 \\ x_3 &= s \\ x_4 &= t \end{aligned}$$

is a solution to the system

$$\begin{aligned} x_1 - 2x_2 + 3x_3 + x_4 &= -3 \\ 2x_1 - x_2 + 3x_3 - x_4 &= 0 \end{aligned}$$

Solution   Simply substitute these values of $x_1, x_2, x_3,$ and x_4 in each equation.

$$\begin{aligned} x_1 - 2x_2 + 3x_3 + x_4 &= (t - s + 1) - 2(t + s + 2) + 3s + t = -3 \\ 2x_1 - x_2 + 3x_3 - x_4 &= 2(t - s + 1) - (t + s + 2) + 3s - t = 0 \end{aligned}$$

Because both equations are satisfied, it is a solution for all s and t.

The quantities s and t in Example 2 are called **parameters**, and the set of solutions, described in this way, is said to be given in **parametric form** and is called the **general solution** to the system. It turns out that the solutions to *every* system of equations (if there *are* solutions) can be given in parametric form (that is, the x's are given in terms of new independent variables s, t, etc.). The following example shows how this happens in the simplest systems where only one equation is present.

Example 3

Describe all solutions to $3x - y + 2z = 6$ in parametric form.

Solution   Solving the equation for y in terms of x and z, we get $y = 3x + 2z - 6$. If s and t are arbitrary then, setting $x = s$, $z = t$, we get solutions

$$x = s$$
$$y = 3s + 2t - 6 \qquad s \text{ and } t \text{ arbitrary}$$
$$z = t$$

Of course we could have solved for x : $x = \frac{1}{3}(y - 2z + 6)$. Then, if we take $y = p$, $z = q$, the solutions are represented as follows:

$$x = \frac{1}{3}(p - 2q + 6)$$
$$y = p \qquad p \text{ and } q \text{ arbitrary}$$
$$z = q$$

The same family of solutions can "look" quite different!

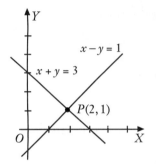

(a) Unique solution
($x = 2$, $y = 1$)

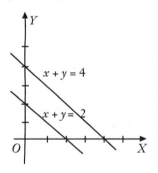

(b) No solution

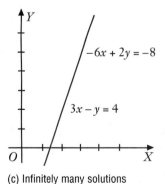

(c) Infinitely many solutions
($x = t$, $y = 3t - 4$)

Figure 1.1

When only two variables are involved, the solutions to systems of linear equations can be described geometrically because the graph of a linear equation $ax + by = c$ is a straight line if a and b are not both zero. Moreover, a point $P(s, t)$ with coordinates s and t lies on the line if and only if $as + bt = c$—that is, when $x = s$, $y = t$ is a solution to the equation. Hence the solutions to a *system* of linear equations correspond to the points $P(s, t)$ that lie on *all* the lines in question.

In particular, if the system consists of just one equation, there must be infinitely many solutions because there are infinitely many points on a line. If the system has two equations, there are three possibilities for the corresponding straight lines:

1. The lines intersect in a single point. Then the system has a *unique solution* corresponding to that point.
2. The lines are parallel (and distinct) and so do not intersect. Then the system has *no solution*.
3. The lines are identical. Then the system has *infinitely many solutions*—one for each point on the (common) line.

These three situations are illustrated in Figure 1.1. In each case the graphs of two specific lines are plotted and the corresponding equations indicated. In the last case, the equations are $3x - y = 4$ and $-6x + 2y = -8$, which have identical graphs.

When three variables are present, the graph of an equation $ax + by + cz = d$ can be shown to be a plane and so again provides a "picture" of the set of solutions. However, this graphical method has its limitations: When more than three variables are involved, no physical image of the graphs (called hyperplanes) is possible. It is necessary to turn to a more "algebraic" method of solution.

Before describing the method, we introduce a concept that simplifies the computations involved. Consider the following system

$$
\begin{aligned}
3x_1 + 2x_2 - x_3 + x_4 &= -1 \\
2x_1 \quad\quad - x_3 + 2x_4 &= 0 \\
3x_1 + x_2 + 2x_3 + 5x_4 &= 2
\end{aligned}
$$

of three equations in four variables. The array of numbers[1]

$$
\left[\begin{array}{cccc|c}
3 & 2 & -1 & 1 & -1 \\
2 & 0 & -1 & 2 & 0 \\
3 & 1 & 2 & 5 & 2
\end{array}\right]
$$

occurring in the system is called the **augmented matrix** of the system. Each row of the matrix consists of the coefficients of the variables (in order) from the corresponding equation, together with the constant term. For clarity, the constants are separated by a vertical line. The augmented matrix is just a different way of describing the system of equations. The array of coefficients of the variables

$$
\left[\begin{array}{cccc}
3 & 2 & -1 & 1 \\
2 & 0 & -1 & 2 \\
3 & 1 & 2 & 5
\end{array}\right]
$$

is called the **coefficient matrix** of the system and $\left[\begin{array}{c} -1 \\ 0 \\ 2 \end{array}\right]$ is called the **constant matrix** of the system.

Elementary Operations

The algebraic method for solving systems of linear equations is described as follows. Two such systems are said to be **equivalent** if they have the same set of solutions. A system is solved by writing a series of systems, one after the other, each equivalent to the previous system. Each of these systems has the same set of solutions as the original one; the aim is to end up with a system that is easy to solve. Each system in the series is obtained from the preceding system by a simple manipulation chosen so that it does not change the set of solutions.

As an illustration, we solve the system $x + 2y = -2$, $2x + y = 7$ in this manner. At each stage, the corresponding augmented matrix is displayed. The original system is

$$
\begin{aligned}
x + 2y &= -2 \\
2x + y &= 7
\end{aligned}
\qquad
\left[\begin{array}{cc|c}
1 & 2 & -2 \\
2 & 1 & 7
\end{array}\right]
$$

First, subtract twice the first equation from the second. The resulting system is

$$
\begin{aligned}
x + 2y &= -2 \\
- 3y &= 11
\end{aligned}
\qquad
\left[\begin{array}{cc|c}
1 & 2 & -2 \\
0 & -3 & 11
\end{array}\right]
$$

---

1 A rectangular array of numbers is called a **matrix**. Matrices will be discussed in more detail in Chapter 2.

which is equivalent to the original (see Theorem 1). At this stage we obtain $y = -\frac{11}{3}$ by multiplying the second equation by $-\frac{1}{3}$. The result is the equivalent system

$$
\begin{aligned}
x + 2y &= -2 \\
y &= -\tfrac{11}{3}
\end{aligned}
\qquad
\begin{bmatrix} 1 & 2 & \big| & -2 \\ 0 & 1 & \big| & -\tfrac{11}{3} \end{bmatrix}
$$

Finally, we subtract twice the second equation from the first to get another equivalent system.

$$
\begin{aligned}
x &= \tfrac{16}{3} \\
y &= -\tfrac{11}{3}
\end{aligned}
\qquad
\begin{bmatrix} 1 & 0 & \big| & \tfrac{16}{3} \\ 0 & 1 & \big| & -\tfrac{11}{3} \end{bmatrix}
$$

Now *this* system is easy to solve! And because it is equivalent to the original system, it provides the solution to that system.

Observe that, at each stage, a certain operation is performed on the system (and thus on the augmented matrix) to produce an equivalent system. The following operations, called **elementary operations**, can routinely be performed on systems of linear equations to produce equivalent systems.

    I. Interchange two equations.

    II. Multiply one equation by a nonzero number.

    III. Add a multiple of one equation to a different equation.

Theorem 1

Suppose that an elementary operation is performed on a system of linear equations. Then the resulting system has the same set of solutions as the original, so the two systems are equivalent.

PROOF

If an elementary operation ρ is applied to a system of equations, we must show that the new system has the same solutions as the original system. Observe that both adding or subtracting equations and multiplying an equation by a constant result in another equation. Since ρ is an elementary operation, this means that every solution of the original system is a solution of the new system. So it is enough to show that we can carry the new system *back* to the original system by a "reverse" elementary operation ρ'. In fact we can choose ρ' to be of the same type as ρ:

    1. If ρ interchanges two equations, then ρ' also interchanges them (that is $\rho' = \rho$).

    2. If ρ multiplies equation p by $k \neq 0$, then ρ' multiplies equation p by $1/k$.

    3. If ρ adds k times equation p to equation q where $q \neq p$, then ρ' *subtracts* k times equation p from equation q. (Notice that $q \neq p$ is essential here.)

Hence ρ' is elementary and carries the new system back to the original, as required.

The fact that each elementary row operation has a "reverse" means that, if we can carry system A to system B by row operations then we can carry B back to A by doing the reverse operations. In this case we say that A and B are **row equivalent**. We will have more to say about this in Section 2.4.

Elementary operations performed on a system of equations produce corresponding manipulations of the *rows* of the augmented matrix. Thus, multiplying a row of a matrix by a number *k* means multiplying *every entry* of the row by *k*. Adding one row to another row means adding *each entry* of that row to the corresponding entry of the other row. Subtracting two rows is done similarly.

In hand calculations (and in computer programs) we usually manipulate the rows of the augmented matrix rather than the equations. For this reason we restate these elementary operations for matrices.

The following are called **elementary row operations** on a matrix.

I. Interchange two rows.

II. Multiply one row by a nonzero number.

III. Add a multiple of one row to a different row.

In the illustration preceding Theorem 1 these operations led to a matrix of the form

$$\left[\begin{array}{cc|c} 1 & 0 & * \\ 0 & 1 & * \end{array}\right]$$

where the asterisks represent arbitrary numbers. In the case of three equations in three variables, the goal is to produce a matrix of the form

$$\left[\begin{array}{ccc|c} 1 & 0 & 0 & * \\ 0 & 1 & 0 & * \\ 0 & 0 & 1 & * \end{array}\right]$$

This does not always happen, as we will see in the next section. Here is an example in which it does happen.

Example 4

Find all solutions to the following system of equations.

$$
\begin{array}{rcrcrcr}
3x & + & 4y & + & z & = & 1 \\
2x & + & 3y & & & = & 0 \\
4x & + & 3y & - & z & = & -2
\end{array}
$$

Solution   The augmented matrix of the original system is

$$\left[\begin{array}{ccc|c} 3 & 4 & 1 & 1 \\ 2 & 3 & 0 & 0 \\ 4 & 3 & -1 & -2 \end{array}\right]$$

To create a 1 in the upper left corner we could multiply row 1 through by $\frac{1}{3}$. However, the 1 can be obtained without introducing fractions by subtracting row 2 from row 1. The result is

$$\left[\begin{array}{ccc|c} 1 & 1 & 1 & 1 \\ 2 & 3 & 0 & 0 \\ 4 & 3 & -1 & -2 \end{array}\right]$$

The upper left 1 is now used to "clean up" the first column, that is create zeros in the other positions in the column. First subtract 2 times row 1 from row 2 to obtain

$$\begin{bmatrix} 1 & 1 & 1 & | & 1 \\ 0 & 1 & -2 & | & -2 \\ 4 & 3 & -1 & | & -2 \end{bmatrix}$$

Next subtract 4 times row 1 from row 3. The result is

$$\begin{bmatrix} 1 & 1 & 1 & | & 1 \\ 0 & 1 & -2 & | & -2 \\ 0 & -1 & -5 & | & -6 \end{bmatrix}$$

This completes the work on column 1. We now use the 1 in the second position of the second row to clean up the second column by subtracting row 2 from row 1 and adding row 2 to row 3. For convenience, both row operations are done in one step. The result is

$$\begin{bmatrix} 1 & 0 & 3 & | & 3 \\ 0 & 1 & -2 & | & -2 \\ 0 & 0 & -7 & | & -8 \end{bmatrix}$$

Note that these manipulations *did not affect* the first column (the second row has a zero there), so our previous effort there has not been undermined. Finally we clean up the third column. Begin by multiplying row 3 by $-\frac{1}{7}$ to obtain

$$\begin{bmatrix} 1 & 0 & 3 & | & 3 \\ 0 & 1 & -2 & | & -2 \\ 0 & 0 & 1 & | & \frac{8}{7} \end{bmatrix}$$

Now subtract 3 times row 3 from row 1, and add 2 times row 3 to row 2 to get

$$\begin{bmatrix} 1 & 0 & 0 & | & -\frac{3}{7} \\ 0 & 1 & 0 & | & \frac{2}{7} \\ 0 & 0 & 1 & | & \frac{8}{7} \end{bmatrix}$$

The corresponding equations are $x = -\frac{3}{7}$, $y = \frac{2}{7}$, and $z = \frac{8}{7}$, which give the (unique) solution.

Exercises 1.1

1. In each case verify that the following are solutions for all values of s and t.

   (a) $x = 19t - 35$
   $y = 25 - 13t$
   $z = t$
   is a solution of
   $2x + 3y + z = 5$
   $5x + 7y - 4z = 0$

   (b) $x_1 = 2s + 12t + 13$
   $x_2 = s$
   $x_3 = -s - 3t - 3$
   $x_4 = t$
   is a solution of
   $2x_1 + 5x_2 + 9x_3 + 3x_4 = -1$
   $x_1 + 2x_2 + 4x_3 \quad\quad = 1$

2. Find all solutions to the following in parametric form in two ways.

   (a) $3x + y = 2$ ♦^{2}(b) $2x + 3y = 1$
   (c) $3x - y + 2z = 5$ ♦ (d) $x - 2y + 5z = 1$

3. Regarding $2x = 5$ as the equation $2x + 0y = 5$ in two variables, find all solutions in parametric form.

♦4. Regarding $4x - 2y = 3$ as the equation $4x - 2y + 0z = 3$ in three variables, find all solutions in parametric form.

♦5. Find all solutions to the general system $ax = b$ of one equation in one variable (a) when $a = 0$ and (b) when $a \neq 0$.

6. Show that a system consisting of exactly one linear equation can have no solution, one solution, or infinitely many solutions. Give examples.

7. Write the augmented matrix for each of the following systems of linear equations.

   (a) $x - 3y = 5$ ♦(b) $x + 2y = 0$
       $2x + y = 1$ $y = 1$
   (c) $x - y + z = 2$ ♦(d) $x + y = 1$
       $x - z = 1$ $y + z = 0$
       $y + 2x = 0$ $z - x = 2$

8. Write a system of linear equations that has each of the following augmented matrices.

   (a) $\begin{bmatrix} 1 & -1 & 6 & | & 0 \\ 0 & 1 & 0 & | & 3 \\ 2 & -1 & 0 & | & 1 \end{bmatrix}$ ♦(b) $\begin{bmatrix} 2 & -1 & 0 & | & -1 \\ -3 & 2 & 1 & | & 0 \\ 0 & 1 & 1 & | & 3 \end{bmatrix}$

9. Find the solution of each of the following systems of linear equations using augmented matrices.

   (a) $x - 3y = 1$ ♦(b) $x + 2y = 1$
       $2x - 7y = 3$ $3x + 4y = -1$
   (c) $2x + 3y = -1$ ♦(d) $3x + 4y = 1$
       $3x + 4y = 2$ $4x + 5y = -3$

10. Find the solution of each of the following systems of linear equations using augmented matrices.

   (a) $x + y + 2z = -1$ ♦(b) $2x + y + z = -1$
       $2x + y + 3z = 0$ $x + 2y + z = 0$
       $-2y + z = 2$ $3x - 2z = 5$

11. Find all solutions (if any) of the following systems of linear equations.

   (a) $3x - 2y = 5$ ♦(b) $3x - 2y = 5$
       $-12x + 8y = -20$ $-12x + 8y = 16$

12. Show that $\begin{cases} x + 2y - z = a \\ 2x + y + 3z = b \\ x - 4y + 9z = c \end{cases}$ is inconsistent unless $c = 2b - 3a$.

13. By examining the possible positions of lines in the plane, show that three equations in two variables can have zero, one, or infinitely many solutions.

14. In each case either show that the statement is true, or give an example^3 showing it is false.

   (a) If a linear system has n variables and m equations, then the augmented matrix has n rows.

   ♦(b) A consistent linear system must have infinitely many solutions.

   (c) If a row operation is done to a consistent linear system, the resulting system must be consistent.

   ♦(d) If a series of row operations on a linear system results in an inconsistent system, the original system is inconsistent.

   (e) If A is row equivalent to B and B is row equivalent to C, then A is row equivalent to C.

15. Find a quadratic $a + bx + cx^2$ such that the graph of $y = a + bx + cx^2$ contains each of the points $(-1, 6)$, $(2, 0)$, and $(3, 2)$.

♦16. Solve the system $\begin{matrix} 3x + 2y = 5 \\ 7x + 5y = 1 \end{matrix}$ by changing variables $\begin{matrix} x = 5x' - 2y' \\ y = -7x' + 3y' \end{matrix}$ and solving the resulting equations for x' and y'.

♦17. Find a, b, and c such that

$$\frac{x^2 - x + 3}{(x^2 + 2)(2x - 1)} = \frac{ax + b}{x^2 + 2} + \frac{c}{2x - 1}$$

[*Hint:* Multiply through by $(x^2 + 2)(2x - 1)$ and equate coefficients of powers of x.]

---

2 A ♦ indicates that the exercise has an answer at the end of the book.

3 Such an example is called a **counterexample**. For example, if the statement is that "all philosophers have beards", the existence of a non-bearded philosopher would be a counterexample proving that the statement is false. This is discussed again in Appendix B.

18. A zookeeper wants to give an animal 42 mg of vitamin A and 65 mg of vitamin D per day. He has two supplements: the first contains 10% vitamin A and 25% vitamin D; the second contains 20% vitamin A and 25% vitamin D. How much of each supplement should he give the animal each day?

♦19. Workmen John and Joe earn a total of $24.60 when John works 2 hours and Joe works 3 hours. If John works 3 hours and Joe works 2 hours, they get $23.90. Find their hourly rates.

SECTION 1.2 Gaussian Elimination

The algebraic method introduced in the preceding section can be summarized as follows: Given a system of linear equations, use a series of elementary row operations to carry the augmented matrix to a "nice" matrix (meaning that the corresponding equations are easy to solve). In Example 4 §1.1,[4] this nice matrix took the form

$$\begin{bmatrix} 1 & 0 & 0 & | & * \\ 0 & 1 & 0 & | & * \\ 0 & 0 & 1 & | & * \end{bmatrix}$$

The following definitions identify the nice matrices that arise in this process.

A matrix is said to be in **row-echelon form** (and will be called a **row-echelon matrix**) if it satisfies the following three conditions:

1. All **zero rows** (consisting entirely of zeros) are at the bottom.

2. The first nonzero entry from the left in each nonzero row is a 1, called the **leading 1** for that row.

3. Each leading 1 is to the right of all leading 1's in the rows above it.

A row-echelon matrix is said to be in **reduced row-echelon form** (and will be called a **reduced row-echelon matrix**) if, in addition, it satisfies the following condition:

4. Each leading 1 is the only nonzero entry in its column.

The row-echelon matrices have a "staircase" form, as indicated by the following example (the asterisks indicate arbitrary numbers).

$$\begin{bmatrix} 0 & 1 & * & * & * & * & * \\ 0 & 0 & 0 & 1 & * & * & * \\ 0 & 0 & 0 & 0 & 1 & * & * \\ 0 & 0 & 0 & 0 & 0 & 0 & 1 \\ 0 & 0 & 0 & 0 & 0 & 0 & 0 \end{bmatrix}$$

The leading 1's proceed "down and to the right" through the matrix. Entries above and to the right of the leading 1's are arbitrary, but all entries below and to the left of them are zero. Hence, a matrix in row-echelon form is in reduced form if, in addition, the entries directly above each leading 1 are all zero. Note that a matrix in row-echelon form can, with a few more row operations, be carried to reduced form (create zeros above each leading one in succession, beginning from the right).

---

4 This means Example 4 in Section 1.1.

Example 1

The following matrices are in row-echelon form (for any choice of numbers in ∗-positions).

$$
\begin{bmatrix} 1 & * & * \\ 0 & 0 & 1 \end{bmatrix}
\begin{bmatrix} 0 & 1 & * & * \\ 0 & 0 & 1 & * \\ 0 & 0 & 0 & 0 \end{bmatrix}
\begin{bmatrix} 1 & * & * & * \\ 0 & 1 & * & * \\ 0 & 0 & 0 & 1 \end{bmatrix}
\begin{bmatrix} 1 & * & * \\ 0 & 1 & * \\ 0 & 0 & 1 \end{bmatrix}
$$

The following, on the other hand, are in reduced row-echelon form.

$$
\begin{bmatrix} 1 & * & 0 \\ 0 & 0 & 1 \end{bmatrix}
\begin{bmatrix} 0 & 1 & 0 & * \\ 0 & 0 & 1 & * \\ 0 & 0 & 0 & 0 \end{bmatrix}
\begin{bmatrix} 1 & 0 & * & 0 \\ 0 & 1 & * & 0 \\ 0 & 0 & 0 & 1 \end{bmatrix}
\begin{bmatrix} 1 & 0 & 0 \\ 0 & 1 & 0 \\ 0 & 0 & 1 \end{bmatrix}
$$

Clearly the choice of the positions for the leading 1's determines the (reduced) row-echelon form (apart from the numbers in ∗-positions).

The importance of row-echelon matrices comes from the following theorem.

Theorem 1

Every matrix can be brought to (reduced) row-echelon form by a series of elementary row operations.

In fact we can give a step-by-step procedure for actually finding the row-echelon matrix. Observe that there are many sequences of row operations that will bring a matrix to row-echelon form; this one is systematic and is easy to program on a computer. Note that the algorithm deals with matrices in general, possibly with columns of zeros.

Carl Friedrich Gauss.
Photo © Corbis.

Gaussian Algorithm[5]

Step 1. If the matrix consists entirely of zeros, stop—it is *already* in row-echelon form.

Step 2. Otherwise, find the first column from the left containing a nonzero entry (call it *a*), and move the row containing that entry to the top position.

Step 3. Now multiply that row by $1/a$ to create a leading 1.

Step 4. By subtracting multiples of that row from rows below it, make each entry below the leading 1 zero.

This completes the first row, and all further row operations are carried out on the other rows.

Step 5. Repeat steps 1–4 on the matrix consisting of the remaining rows.

The process stops when either no rows remain at step 5 or the remaining rows consist of zeros.

---

5   Carl Friedrich Gauss (1777–1855) ranks with Archimedes and Newton as one of the three greatest mathematicians of all time. He was a child prodigy and, at the age of 21, he gave the first proof that every polynomial has a complex root. In 1801 he published a timeless masterpiece, *Disquisitiones Arithmeticae*, in which he founded modern number theory. He went on to make ground-breaking contributions to nearly every branch of mathematics, often well before others rediscovered and published the results. In addition, he did fundamental work in both physics and astronomy. Gauss is said to have been the last mathematician to know everything in his subject; it is no wonder he is called "the prince of mathematicians."

Observe that the gaussian algorithm is recursive: When the first leading 1 has been obtained, the procedure is repeated on the remaining rows of the matrix. This makes the algorithm easy to use on a computer.

Returning to systems of linear equations, observe that the solution of Example 4 §1.1 actually used the gaussian algorithm to carry the augmented matrix to reduced row-echelon form. In hand calculations it is sometimes more convenient to vary the procedure a bit. However, the general pattern is usually used: Create the leading ones from left to right, using each in turn to create zeros below it. Here are two more examples.

Example 2

Solve the following system of equations.

$$
\begin{aligned}
x \quad\quad + 10z &= 5 \\
3x + y - 4z &= -1 \\
4x + y + 6z &= 1
\end{aligned}
$$

Solution   We manipulate the augmented matrix

$$
\begin{bmatrix}
1 & 0 & 10 & 5 \\
3 & 1 & -4 & -1 \\
4 & 1 & 6 & 1
\end{bmatrix}
$$

The first leading one is in place, so we create zeros below it by subtracting 3 times row 1 from row 2 and subtracting 4 times row 1 from row 3. The result is

$$
\begin{bmatrix}
1 & 0 & 10 & 5 \\
0 & 1 & -34 & -16 \\
0 & 1 & -34 & -19
\end{bmatrix}
$$

Now subtract row 2 from row 3 to obtain

$$
\begin{bmatrix}
1 & 0 & 10 & 5 \\
0 & 1 & -34 & -16 \\
0 & 0 & 0 & -3
\end{bmatrix}
$$

This means that the following system of equations

$$
\begin{aligned}
x \quad + 10z &= 5 \\
y - 34z &= -16 \\
0 &= -3
\end{aligned}
$$

is equivalent to the original system. In other words, the two have the *same* solutions. But this last system clearly has *no* solution (the last equation requires that x, y and z satisfy $0x + 0y + 0z = -3$, and no such numbers exist). Hence the original system has *no* solution.

Example 3

Solve the following system of equations.

$$
\begin{aligned}
x_1 - 2x_2 - x_3 + 3x_4 &= 1 \\
2x_1 - 4x_2 + x_3 \quad\quad &= 5 \\
x_1 - 2x_2 + 2x_3 - 3x_4 &= 4
\end{aligned}
$$

Solution   The augmented matrix is

$$\begin{bmatrix} 1 & -2 & -1 & 3 & | & 1 \\ 2 & -4 & 1 & 0 & | & 5 \\ 1 & -2 & 2 & -3 & | & 4 \end{bmatrix}$$

Subtracting twice row 1 from row 2 and subtracting row 1 from row 3 gives

$$\begin{bmatrix} 1 & -2 & -1 & 3 & | & 1 \\ 0 & 0 & 3 & -6 & | & 3 \\ 0 & 0 & 3 & -6 & | & 3 \end{bmatrix}$$

Now subtract row 2 from row 3 and multiply row 2 by $\frac{1}{3}$ to get

$$\begin{bmatrix} 1 & -2 & -1 & 3 & | & 1 \\ 0 & 0 & 1 & -2 & | & 1 \\ 0 & 0 & 0 & 0 & | & 0 \end{bmatrix}$$

This is in row-echelon form, and we take it to reduced form by adding row 2 to row 1:

$$\begin{bmatrix} 1 & -2 & 0 & 1 & | & 2 \\ 0 & 0 & 1 & -2 & | & 1 \\ 0 & 0 & 0 & 0 & | & 0 \end{bmatrix}$$

The corresponding system of equations is

$$\begin{aligned} x_1 - 2x_2 \quad + \quad x_4 &= 2 \\ x_3 - 2x_4 &= 1 \\ 0 &= 0 \end{aligned}$$

The leading ones are in columns 1 and 3 here, so the corresponding variables x_1 and x_3 are called leading variables. Because the matrix is in reduced row-echelon form, these equations can be used to solve for the leading variables in terms of the nonleading variables x_2 and x_4. More precisely, in the present example we set $x_2 = s$ and $x_4 = t$ where s and t are arbitrary, so these equations become

$$x_1 - 2s + t = 2 \quad \text{and} \quad x_3 - 2t = 1.$$

Finally the solutions are given by

$$\begin{aligned} x_1 &= 2 + 2s - t \\ x_2 &= s \\ x_3 &= 1 + 2t \\ x_4 &= t \end{aligned}$$

where s and t are arbitrary.

The solution of Example 3 is typical of the general case. To solve a linear system, the augmented matrix is carried to reduced row-echelon form, and the variables corresponding to the leading ones are called **leading variables**. Because the matrix is in reduced form, each leading variable occurs in exactly one equation, so that equation can be solved to give a formula for the leading variable in terms of the nonleading variables. It is customary to label the nonleading variables by new variables s, t, ... , called **parameters**. Hence, as in Example 3, every variable x_i is given by a formula in terms of the parameters s, t, ... , every such choice leads to a solution to the system, and every solution arises in this way. This procedure works in general, and has come to be called

Gauss–Jordan Elimination[6]

To solve a system of linear equations proceed as follows:
1. Carry the augmented matrix to a reduced row-echelon matrix using elementary row operations.
2. If a row $[0 \ 0 \ 0 \ \cdots \ 0 \ 1]$ occurs, the system is inconsistent.
3. Otherwise, assign the nonleading variables (if any) as parameters, and use the equations corresponding to the reduced row-echelon matrix to solve for the leading variables in terms of the parameters.

There is a variant of this procedure, called **Gaussian Elimination**, wherein the augmented matrix is carried only to row-echelon form. The nonleading variables are assigned as parameters as before. Then the last equation (corresponding to the row-echelon form) is used to solve for the last leading variable in terms of the parameters. This last leading variable is then substituted into all the preceding equations. Next, the second last equation yields the second last leading variable, which is also substituted back. The process continues to give the general solution. This procedure is called **back-substitution**. The motivation for using it at all is that it can be shown to be numerically more efficient than Gauss-Jordan elimination, and so is important when solving very large systems.[7]

Example 4

Find a condition on the numbers a, b, and c such that the following system of equations is consistent. When that condition is satisfied, find all solutions (in terms of a, b, and c).

$$\begin{aligned} x_1 + 3x_2 + x_3 &= a \\ -x_1 - 2x_2 + x_3 &= b \\ 3x_1 + 7x_2 - x_3 &= c \end{aligned}$$

Solution We use Gauss-Jordan elimination except that now the augmented matrix

$$\begin{bmatrix} 1 & 3 & 1 & a \\ -1 & -2 & 1 & b \\ 3 & 7 & -1 & c \end{bmatrix}$$

has entries a, b, and c as well as known numbers. The first leading one is in place, so we create zeros below it in column 1:

$$\begin{bmatrix} 1 & 3 & 1 & a \\ 0 & 1 & 2 & a + b \\ 0 & -2 & -4 & c - 3a \end{bmatrix}$$

---

6 The second name honours Wilhelm Jordan who used the method in 1888 in his popular textbook on geodesy. He should not be confused with the famous French mathematician Camille Jordan.

7 With n equations where n is large, Gauss–Jordan elimination requires roughly $n^3/3$ multiplications and divisions, whereas this number is roughly $n^3/2$ if gaussian elimination is used.

The second leading one has appeared, so use it to create zeros in the rest of column 2:

$$\begin{bmatrix} 1 & 0 & -5 & -2a-3b \\ 0 & 1 & 2 & a+b \\ 0 & 0 & 0 & c-a+2b \end{bmatrix}$$

Now the whole solution depends on the number $c-a+2b = c-(a-2b)$. The last row corresponds to an equation $0 = c-(a-2b)$. If $c \neq a-2b$, there is *no* solution (just as in Example 2). Hence:

The system is consistent if and only if $c = a - 2b$.

In this case the last matrix becomes

$$\begin{bmatrix} 1 & 0 & -5 & -2a-3b \\ 0 & 1 & 2 & a+b \\ 0 & 0 & 0 & 0 \end{bmatrix}$$

Thus, if $c = a - 2b$, taking $x_3 = t$ where t is a parameter gives the solutions

$$x_1 = 5t - (2a+3b) \qquad x_2 = (a+b) - 2t \qquad x_3 = t.$$

Rank

It can be proven that the *reduced* row-echelon form of a matrix A is uniquely determined by A. That is, no matter which series of row operations is used to carry A to a reduced row-echelon matrix, the result will always be the same matrix. (See Supplementary Exercise 7 in Chapter 6.) By contrast, this is not true for row-echelon matrices: Different series of row operations can carry the same matrix A to *different* row-echelon matrices. Indeed, the matrix $A = \begin{bmatrix} 1 & -1 & 4 \\ 2 & -1 & 2 \end{bmatrix}$ can be carried (by one row operation) to the row-echelon matrix $\begin{bmatrix} 1 & -1 & 4 \\ 0 & 1 & -6 \end{bmatrix}$, and then by another row operation to the (reduced) row-echelon matrix $\begin{bmatrix} 1 & 0 & -2 \\ 0 & 1 & -6 \end{bmatrix}$.

However, it *is* true that the number r of leading 1's must be the same in each of these row-echelon matrices (this will be proved in Chapter 5). Hence, the number r depends only on A and not on the way in which A is carried to row-echelon form, and r is called the **rank** of the matrix A, written $r = \text{rank } A$.

Example 5

Compute the rank of $A = \begin{bmatrix} 1 & 1 & -1 & 4 \\ 2 & 1 & 3 & 0 \\ 0 & 1 & -5 & 8 \end{bmatrix}$.

Solution   The reduction of A to row-echelon form is

$$A = \begin{bmatrix} 1 & 1 & -1 & 4 \\ 2 & 1 & 3 & 0 \\ 0 & 1 & -5 & 8 \end{bmatrix} \rightarrow \begin{bmatrix} 1 & 1 & -1 & 4 \\ 0 & -1 & 5 & -8 \\ 0 & 1 & -5 & 8 \end{bmatrix} \rightarrow \begin{bmatrix} 1 & 1 & -1 & 4 \\ 0 & 1 & -5 & 8 \\ 0 & 0 & 0 & 0 \end{bmatrix}$$

Because this row-echelon matrix has two leading 1's, rank $A = 2$.

The notion of rank of a matrix has a useful application to equations.

Theorem 2

Suppose a system of m equations in n variables has a solution. If the rank of the augmented matrix is r, the set of solutions involves exactly $n - r$ parameters.

PROOF

The fact that the rank of the augmented matrix is r means there are exactly r leading variables, and hence exactly $n - r$ nonleading variables. These nonleading variables are all assigned as parameters in the gaussian algorithm, so the set of solutions involves exactly $n - r$ parameters.

In particular, this shows that, for any system of linear equations, exactly three possibilities exist:

1. *No solution.* This occurs when a row $[0 \ 0 \ \cdots \ 0 \ 1]$ occurs in the row-echelon form.

2. *Unique solution.* This occurs when *every* variable is a leading variable.

3. *Infinitely many solutions.* This occurs when there is at least one nonleading variable, so a parameter is involved.

Example 6

Suppose the matrix A in Example 5 is the augmented matrix of a system of $m = 3$ linear equations in $n = 3$ variables. As rank $A = r = 2$, the set of solutions will have $n - r = 1$ parameter. The reader can verify this fact directly.

Exercises 1.2

1. Which of the following matrices are in reduced row-echelon form? Which are in row-echelon form?

(a) $\begin{bmatrix} 1 & -1 & 2 \\ 0 & 0 & 0 \\ 0 & 0 & 1 \end{bmatrix}$

♦(b) $\begin{bmatrix} 2 & 1 & -1 & 3 \\ 0 & 0 & 0 & 0 \end{bmatrix}$

(c) $\begin{bmatrix} 1 & -2 & 3 & 5 \\ 0 & 0 & 0 & 1 \end{bmatrix}$

♦(d) $\begin{bmatrix} 1 & 0 & 0 & 3 & 1 \\ 0 & 0 & 0 & 1 & 1 \\ 0 & 0 & 0 & 0 & 1 \end{bmatrix}$

(e) $\begin{bmatrix} 1 & 1 \\ 0 & 1 \end{bmatrix}$

♦(f) $\begin{bmatrix} 0 & 0 & 1 \\ 0 & 0 & 1 \\ 0 & 0 & 1 \end{bmatrix}$

2. Carry each of the following matrices to reduced row-echelon form.

(a) $\begin{bmatrix} 0 & -1 & 2 & 1 & 2 & 1 & -1 \\ 0 & 1 & -2 & 2 & 7 & 2 & 4 \\ 0 & -2 & 4 & 3 & 7 & 1 & 0 \\ 0 & 3 & -6 & 1 & 6 & 4 & 1 \end{bmatrix}$

♦(b) $\begin{bmatrix} 0 & -1 & 3 & 1 & 3 & 2 & 1 \\ 0 & -2 & 6 & 1 & -5 & 0 & -1 \\ 0 & 3 & -9 & 2 & 4 & 1 & -1 \\ 0 & 1 & -3 & -1 & 3 & 0 & 1 \end{bmatrix}$

3. The augmented matrix of a system of linear equations has been carried to the following by row operations. In each case solve the system.

(a) $\begin{bmatrix} 1 & 2 & 0 & 3 & 1 & 0 & | & -1 \\ 0 & 0 & 1 & -1 & 1 & 0 & | & 2 \\ 0 & 0 & 0 & 0 & 0 & 1 & | & 3 \\ 0 & 0 & 0 & 0 & 0 & 0 & | & 0 \end{bmatrix}$

◆(b) $\begin{bmatrix} 1 & -2 & 0 & 2 & 0 & 1 & | & 1 \\ 0 & 0 & 1 & 5 & 0 & -3 & | & -1 \\ 0 & 0 & 0 & 0 & 1 & 6 & | & 1 \\ 0 & 0 & 0 & 0 & 0 & 0 & | & 0 \end{bmatrix}$

(c) $\begin{bmatrix} 1 & 2 & 1 & 3 & 1 & | & 1 \\ 0 & 1 & -1 & 0 & 1 & | & 1 \\ 0 & 0 & 0 & 1 & -1 & | & 0 \\ 0 & 0 & 0 & 0 & 0 & | & 0 \end{bmatrix}$

◆(d) $\begin{bmatrix} 1 & -1 & 2 & 4 & 6 & | & 2 \\ 0 & 1 & 2 & 1 & -1 & | & -1 \\ 0 & 0 & 0 & 1 & 0 & | & 1 \\ 0 & 0 & 0 & 0 & 0 & | & 0 \end{bmatrix}$

4. Find all solutions (if any) to each of the following systems of linear equations.

(a) $x - 2y = 1$
$4y - x = -2$

◆(b) $3x - y = 0$
$2x - 3y = 1$

(c) $2x + y = 5$
$3x + 2y = 6$

◆(d) $3x - y = 2$
$2y - 6x = -4$

(e) $3x - y = 4$
$2y - 6x = 1$

◆(f) $2x - 3y = 5$
$3y - 2x = 2$

5. Find all solutions (if any) to each of the following systems of linear equations.

(a) $x + y + 2z = 8$
$3x - y + z = 0$
$-x + 3y + 4z = -4$

◆(b) $-2x + 3y + 3z = -9$
$3x - 4y + z = 5$
$-5x + 7y + 2z = -14$

(c) $x + y - z = 10$
$-x + 4y + 5z = -5$
$x + 6y + 3z = 15$

◆(d) $x + 2y - z = 2$
$2x + 5y - 3z = 1$
$x + 4y - 3z = 3$

(e) $5x + y = 2$
$3x - y + 2z = 1$
$x + y - z = 5$

◆(f) $3x - 2y + z = -2$
$x - y + 3z = 5$
$-x + y + z = -1$

(g) $x + y + z = 2$
$x + z = 1$
$2x + 5y + 2z = 7$

◆(h) $x + 2y - 4z = 10$
$2x - y + 2z = 5$
$x + y - 2z = 7$

6. Express the last equation of each system as a sum of multiples of the first two equations. [*Hint:* Label the equations, use the gaussian algorithm.]

(a) $x_1 + x_2 + x_3 = 1$
$2x_1 - x_2 + 3x_3 = 3$
$x_1 - 2x_2 + 2x_3 = 2$

◆(b) $x_1 + 2x_2 - 3x_3 = -3$
$x_1 + 3x_2 - 5x_3 = 5$
$x_1 - 2x_2 + 5x_3 = -35$

7. Find all solutions to the following systems.

(a) $3x_1 + 8x_2 - 3x_3 - 14x_4 = 2$
$2x_1 + 3x_2 - x_3 - 2x_4 = 1$
$x_1 - 2x_2 + x_3 + 10x_4 = 0$
$x_1 + 5x_2 - 2x_3 - 12x_4 = 1$

◆(b) $x_1 - x_2 + x_3 - x_4 = 0$
$-x_1 + x_2 + x_3 + x_4 = 0$
$x_1 + x_2 - x_3 + x_4 = 0$
$x_1 + x_2 + x_3 + x_4 = 0$

(c) $x_1 - x_2 + x_3 - 2x_4 = 1$
$-x_1 + x_2 + x_3 + x_4 = -1$
$-x_1 + 2x_2 + 3x_3 - x_4 = 2$
$x_1 - x_2 + 2x_3 + x_4 = 1$

◆(d) $x_1 + x_2 + 2x_3 - x_4 = 4$
$3x_2 - x_3 + 4x_4 = 2$
$x_1 + 2x_2 - 3x_3 + 5x_4 = 0$
$x_1 + x_2 - 5x_3 + 6x_4 = -3$

8. In each of the following, find conditions on a and b such that the system has no solution, one solution, and infinitely many solutions.

(a) $x - 2y = 1$
$ax + by = 5$

◆(b) $x + by = -1$
$ax + 2y = 5$

(c) $x - by = -1$
$x + ay = 3$

◆(d) $ax + y = 1$
$2x + y = b$

9. In each of the following, find (if possible) conditions on a, b, and c such that the system has no solution, one solution, or infinitely many solutions.

(a) $3x + y - z = a$
$x - y + 2z = b$
$5x + 3y - 4z = c$

◆(b) $2x + y - z = a$
$2y + 3z = b$
$x - z = c$

(c) $-x + 3y + 2z = -8$
$x + z = 2$
$3x + 3y + az = b$

◆(d) $x + ay = 0$
$y + bz = 0$
$z + cx = 0$

(e) $3x - y + 2z = 3$
  $x + y - z = 2$
  $2x - 2y + 3z = b$

♦(f) $x + ay - z = 1$
  $-x + (a-2)y + z = -1$
  $2x + 2y + (a-2)z = 1$

10. Find the rank of each of the matrices in Exercise 1.

11. Find the rank of each of the following matrices.

(a) $\begin{bmatrix} 1 & 1 & 2 \\ 3 & -1 & 1 \\ -1 & 3 & 4 \end{bmatrix}$ ♦(b) $\begin{bmatrix} -2 & 3 & 3 \\ 3 & -4 & 1 \\ -5 & 7 & 2 \end{bmatrix}$

(c) $\begin{bmatrix} 1 & 1 & -1 & 3 \\ -1 & 4 & 5 & -2 \\ 1 & 6 & 3 & 4 \end{bmatrix}$ ♦(d) $\begin{bmatrix} 3 & -2 & 1 & -2 \\ 1 & -1 & 3 & 5 \\ -1 & 1 & 1 & -1 \end{bmatrix}$

(e) $\begin{bmatrix} 1 & 2 & -1 & 0 \\ 0 & a & 1-a & a^2+1 \\ 1 & 2-a & -1 & -2a^2 \end{bmatrix}$

♦(f) $\begin{bmatrix} 1 & 1 & 2 & a^2 \\ 1 & 1-a & 2 & 0 \\ 2 & 2-a & 6-a & 4 \end{bmatrix}$

12. Consider a system of linear equations with augmented matrix A and coefficient matrix C. In each case either prove the statement or give an example showing that it is false.

(a) If there is more than one solution, A has a row of zeros.

♦(b) If A has a row of zeros, there is more than one solution.

(c) If there is no solution, the row-echelon form of C has a row of zeros.

♦(d) If the row-echelon form of C has a row of zeros, there is no solution.

(e) There is no system that is inconsistent for every choice of constants.

♦(f) If the system is consistent for some choice of constants, it is consistent for every choice of constants.

Now assume that the augmented matrix A has 3 rows and 5 columns.

(g) If the system is consistent, there is more than one solution.

♦(h) The rank of A is at most 3.

(i) If rank $A = 3$, the system is consistent.

(j) If rank $C = 3$, the system is consistent.

13. Find a sequence of row operations carrying

$$\begin{bmatrix} b_1+c_1 & b_2+c_2 & b_3+c_3 \\ c_1+a_1 & c_2+a_2 & c_3+a_3 \\ a_1+b_1 & a_2+b_2 & a_3+b_3 \end{bmatrix} \text{ to } \begin{bmatrix} a_1 & a_2 & a_3 \\ b_1 & b_2 & b_3 \\ c_1 & c_2 & c_3 \end{bmatrix}$$

14. In each case, show that the reduced row-echelon form is as given.

(a) $\begin{bmatrix} p & 0 & a \\ b & 0 & 0 \\ q & c & r \end{bmatrix}$ with $abc \neq 0$; $\begin{bmatrix} 1 & 0 & 0 \\ 0 & 1 & 0 \\ 0 & 0 & 1 \end{bmatrix}$

(b) $\begin{bmatrix} 1 & a & b+c \\ 1 & b & c+a \\ 1 & c & a+b \end{bmatrix}$ where $c \neq a$ or $b \neq a$; $\begin{bmatrix} 1 & 0 & * \\ 0 & 1 & * \\ 0 & 0 & 0 \end{bmatrix}$

15. Show that $\begin{cases} ax + by + cz = 0 \\ a_1x + b_1y + c_1z = 0 \end{cases}$ always has a solution other than $x = 0$, $y = 0$, $z = 0$.

16. Find the circle $x^2 + y^2 + ax + by + c = 0$ passing through the following points.

(a) $(-2, 1)$, $(5, 0)$, and $(4, 1)$

♦(b) $(1, 1)$, $(5, -3)$, and $(-3, -3)$

17. Three Nissans, two Fords, and four Chevrolets can be rented for $106 per day. At the same rates two Nissans, four Fords, and three Chevrolets cost $107 per day, whereas four Nissans, three Fords, and two Chevrolets cost $102 per day. Find the rental rates for all three kinds of cars.

♦18. A school has three clubs and each student is required to belong to exactly one club. One year the students switched club membership as follows:

Club A. $\frac{4}{10}$ remain in A, $\frac{1}{10}$ switch to B, $\frac{5}{10}$ switch to C.

Club B. $\frac{7}{10}$ remain in B, $\frac{2}{10}$ switch to A, $\frac{1}{10}$ switch to C.

Club C. $\frac{6}{10}$ remain in C, $\frac{2}{10}$ switch to A, $\frac{2}{10}$ switch to B.

If the fraction of the student population in each club is unchanged, find each of these fractions.

19. Given points (p_1, q_1), (p_2, q_2), and (p_3, q_3) in the plane with p_1, p_2, and p_3 distinct, show that they lie on some curve with equation $y = a + bx + cx^2$. [*Hint:* Solve for a, b, and c.]

20. The scores of three players in a tournament have been lost. The only information available is the total of the scores for players 1 and 2, the total for players 2 and 3, and the total for players 3 and 1.

(a) Show that the individual scores can be rediscovered.

(b) Is this possible with four players (knowing the totals for players 1 and 2, 2 and 3, 3 and 4, and 4 and 1)?

21. A boy finds $1.05 in dimes, nickels, and pennies. If there are 17 coins in all, how many coins of each type can he have?

22. If a consistent system has more variables than equations, show that it has infinitely many solutions. [*Hint:* Use Theorem 2.]

SECTION 1.3　Homogeneous Equations

A system of equations in the variables $x_1, x_2, \ldots, x_n$ is called **homogeneous** if all the constant terms are zero—that is, if each equation of the system has the form

$$a_1 x_1 + a_2 x_2 + \cdots + a_n x_n = 0$$

Clearly $x_1 = 0, x_2 = 0, \ldots, x_n = 0$ is a solution to such a system; it is called the **trivial solution**. Any solution in which at least one variable has a nonzero value is called a **nontrivial solution**. Our chief goal in this section is to give a useful condition for a homogeneous system to have nontrivial solutions. The following example is instructive.

Example 1

Show that the following homogeneous system has nontrivial solutions.

$$\begin{aligned} x_1 - x_2 + 2x_3 + x_4 &= 0 \\ 2x_1 + 2x_2 \quad\quad - x_4 &= 0 \\ 3x_1 + x_2 + 2x_3 + x_4 &= 0 \end{aligned}$$

Solution　The reduction of the augmented matrix to reduced row-echelon form is outlined below.

$$\begin{bmatrix} 1 & -1 & 2 & 1 & | & 0 \\ 2 & 2 & 0 & -1 & | & 0 \\ 3 & 1 & 2 & 1 & | & 0 \end{bmatrix} \rightarrow \begin{bmatrix} 1 & -1 & 2 & 1 & | & 0 \\ 0 & 4 & -4 & -3 & | & 0 \\ 0 & 4 & -4 & -2 & | & 0 \end{bmatrix} \rightarrow \begin{bmatrix} 1 & 0 & 1 & 0 & | & 0 \\ 0 & 1 & -1 & 0 & | & 0 \\ 0 & 0 & 0 & 1 & | & 0 \end{bmatrix}$$

The leading variables are x_1, x_2, and x_4, so x_3 is assigned as a parameter—say $x_3 = t$. Then the general solution is $x_1 = -t$, $x_2 = t$, $x_3 = t$, $x_4 = 0$. Hence, taking $t = 1$ (say), we get a nontrivial solution: $x_1 = -1$, $x_2 = 1$, $x_3 = 1$, $x_4 = 0$.

The existence of a nontrivial solution in Example 1 is ensured by the presence of a parameter in the solution. This is due to the fact that there is a *nonleading* variable (x_3 in this case). But there *must* be a nonleading variable here because there are four variables and only three equations (and hence at *most* three leading variables). This discussion generalizes to a proof of the following useful theorem.

Theorem 1

If a homogeneous system of linear equations has more variables than equations, then it has a nontrivial solution (in fact, infinitely many).

PROOF

Suppose there are m equations in n variables where $n > m$, and let R denote the reduced row-echelon form of the augmented matrix. If there are r leading variables, there are $n - r$ nonleading variables, and so $n - r$ parameters. Hence, it suffices to show that $r < n$. But $r \le m$ because R has r leading 1's and m rows, and $m < n$ by hypothesis.

Note that the converse of Theorem 1 is not true: if a homogeneous system has nontrivial solutions, it need not have more variables than equations (Exercise 3(b)).

Surprisingly enough, Theorem 1 is useful in applications. The next example provides an illustration from geometry.

Example 2

We call the graph of an equation $ax^2 + bxy + cy^2 + dx + ey + f = 0$ a **conic** if the numbers a, b, and c are not all zero. Show that there is at least one conic through any five points in the plane that are not all on a line.

Solution   Let the coordinates of the five points be (p_1, q_1), (p_2, q_2), (p_3, q_3), (p_4, q_4), and (p_5, q_5). The graph of $ax^2 + bxy + cy^2 + dx + ey + f = 0$ passes through (p_i, q_i) if

$$ap_i^2 + bp_iq_i + cq_i^2 + dp_i + eq_i + f = 0$$

This gives five equations, linear in the six variables a, b, c, d, e, and f. Hence, there is a nontrivial solution by Theorem 1. If $a = b = c = 0$, the five points all lie on the line $dx + ey + f = 0$, contrary to assumption. Hence, one of a, b, c is nonzero.

Exercises 1.3

1. Consider the following statements about a system of linear equations with augmented matrix A. In each case either prove the statement or give an example for which it is false.

   (a) If the system is homogeneous, every solution is trivial.

   ◆(b) If the system has a nontrivial solution, it cannot be homogeneous. $\begin{bmatrix} 1 & 0 & 0 & 0 \\ 0 & 1 & -1 & 0 \end{bmatrix}$

   (c) If there exists a trivial solution, the system is homogeneous.

   ◆(d) If the system is consistent, it must be homogeneous. $\begin{bmatrix} 1 & 0 & 0 \\ 0 & 1 & 0 \end{bmatrix}$

   Now assume that the system is homogeneous.

   (e) If there exists a nontrivial solution, there is no trivial solution.

   ◆(f) If there exists a solution, there are infinitely many solutions.

   (g) If there exist nontrivial solutions, the row-echelon form of A has a row of zeros.

   ◆(h) If the row-echelon form of A has a row of zeros, there exist nontrivial solutions.

   (i) If a row operation is applied to the system, the new system is also homogeneous.

2. In each of the following, find all values of a for which the system has nontrivial solutions, and determine all solutions in each case.

   (a) $\begin{aligned} x - 2y + z &= 0 \\ x + ay - 3z &= 0 \\ -x + 6y - 5z &= 0 \end{aligned}$ ◆(b) $\begin{aligned} x + 2y + z &= 0 \\ x + 3y + 6z &= 0 \\ 2x + 3y + az &= 0 \end{aligned}$

   (c) $\begin{aligned} x + y - z &= 0 \\ ay - z &= 0 \\ x + y + az &= 0 \end{aligned}$ ◆(d) $\begin{aligned} ax + y + z &= 0 \\ x + y - z &= 0 \\ x + y + az &= 0 \end{aligned}$

3. (a) Does Theorem 1 imply that the system
$\begin{aligned} -x + 3y &= 0 \\ 2x - 6y &= 0 \end{aligned}$ has nontrivial solutions?
Explain.

   (b) Show that the converse to Theorem 1 is not true. That is, show that the existence of nontrivial solutions does *not* imply that there are more variables than equations.

4. In each case determine how many solutions (and how many parameters) are possible for a homogeneous system of four linear equations in six variables with augmented matrix A. Assume that A has nonzero entries. Give all possibilities.

   (a) Rank $A = 2$.

   ◆(b) Rank $A = 1$.

   (c) A has a row of zeros.

   ◆(d) The row-echelon form of A has a row of zeros.

5. The graph of an equation $ax + by + cz = 0$ is a plane through the origin (provided that not all of a, b, and c are zero). Use Theorem 1 to

show that two planes through the origin have a point in common other than the origin $(0, 0, 0)$.

6. (a) Show that there is a line through any pair of points in the plane. [*Hint:* Every line has equation $ax + by + c = 0$, where a, b, and c are not all zero.]

   (b) Generalize and show that there is a plane $ax + by + cz + d = 0$ through any three points in space.

7. The graph of $a(x^2 + y^2) + bx + cy + d = 0$ is a circle if $a \neq 0$. Show that there is a circle through any three points in the plane that are not all on a line.

8. Consider a homogeneous system of linear equations in n variables, and suppose that the augmented matrix has rank r. Show that the system has nontrivial solutions if and only if $n > r$.[8]

9. If a consistent (possibly nonhomogeneous) system of linear equations has more variables than equations, prove that it has more than one solution.

SECTION 1.4   An Application to Network Flow

There are many types of problems that concern a network of conductors along which some sort of flow is observed. Examples of these include an irrigation network and a network of streets or freeways. There are often points in the system at which a net flow either enters or leaves the system. The basic principle behind the analysis of such systems is that the total flow into the system must equal the total flow out. In fact, we apply this principle at every junction in the system.

Junction Rule

At each of the junctions in the network, the total flow into that junction must equal the total flow out.

This requirement gives a linear equation relating the flows in conductors emanating from the junction.

Example 1

A network of one-way streets is shown in the accompanying diagram. The rate of flow of cars into intersection A is 500 cars per hour, and 400 and 100 cars per hour emerge from B and C, respectively. Find the possible flows along each street.

---

8 If p and q are statements, we say that p implies q if q is true whenever p is true. Then "p if and only if q" means that both p implies q and q implies p. See Appendix B for more on this.

Solution

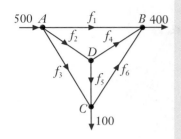

Suppose the flows along the streets are $f_1, f_2, f_3, f_4, f_5,$ and f_6 cars per hour in the directions shown. Then, equating the flow in with the flow out at each intersection, we get

Intersection A	$500 = f_1 + f_2 + f_3$
Intersection B	$f_1 + f_4 + f_6 = 400$
Intersection C	$f_3 + f_5 = f_6 + 100$
Intersection D	$f_2 = f_4 + f_5$

These give four equations in the six variables $f_1, f_2, \ldots, f_6$.

$$\begin{aligned} f_1 + f_2 + f_3 \qquad\qquad\qquad &= 500 \\ f_1 \qquad\quad + f_4 \qquad + f_6 &= 400 \\ f_3 \qquad + f_5 - f_6 &= 100 \\ f_2 \qquad - f_4 - f_5 \qquad &= 0 \end{aligned}$$

The reduction of the augmented matrix is

$$\left[\begin{array}{cccccc|c} 1 & 1 & 1 & 0 & 0 & 0 & 500 \\ 1 & 0 & 0 & 1 & 0 & 1 & 400 \\ 0 & 0 & 1 & 0 & 1 & -1 & 100 \\ 0 & 1 & 0 & -1 & -1 & 0 & 0 \end{array}\right] \rightarrow \left[\begin{array}{cccccc|c} 1 & 0 & 0 & 1 & 0 & 1 & 400 \\ 0 & 1 & 0 & -1 & -1 & 0 & 0 \\ 0 & 0 & 1 & 0 & 1 & -1 & 100 \\ 0 & 0 & 0 & 0 & 0 & 0 & 0 \end{array}\right]$$

Hence, when we use $f_4, f_5,$ and f_6 as parameters, the general solution is

$$\begin{aligned} f_1 &= 400 - f_4 - f_6 \\ f_2 &= f_4 + f_5 \\ f_3 &= 100 - f_5 + f_6 \end{aligned}$$

This gives all solutions to the system of equations and hence all the possible flows.

Of course, not all these solutions may be acceptable in the real situation. For example, the flows $f_1, f_2, \ldots, f_6$ are all *positive* in the present context (if one came out negative, it would mean traffic flowed in the opposite direction). This imposes constraints on the flows: $f_1 \geq 0$ and $f_3 \geq 0$ become

$$\begin{aligned} f_4 + f_6 &\leq 400 \\ f_5 - f_6 &\leq 100 \end{aligned}$$

Further constraints might be imposed by insisting on maximum values on the flow in each street.

Exercises 1.4

1. Find the possible flows in each of the following networks of pipes.

(a)

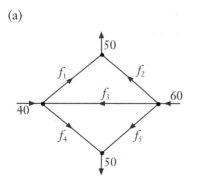

◆(b)

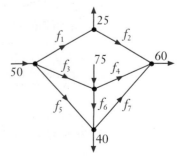

2. A proposed network of irrigation canals is described in the accompanying diagram.

At peak demand, the flows at interchanges A, B, C, and D are as shown.

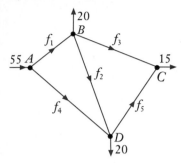

(a) Find the possible flows.

♦(b) If canal BC is closed, what range of flow on AD must be maintained so that no canal carries a flow of more than 30?

3. A traffic circle has five one-way streets, and vehicles enter and leave as shown in the accompanying diagram.

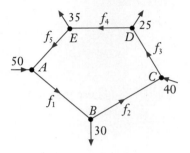

(a) Compute the possible flows.

♦(b) Which road has the heaviest flow?

SECTION 1.5   An Application to Electrical Networks[9]

In an electrical network it is often necessary to find the current in amperes (A) flowing in various parts of the network. These networks usually contain resistors that retard the current. The resistors are indicated by a symbol ⎍⎍⎍, and the resistance is measured in ohms (Ω). Also, the current is increased at various points by voltage sources (for example, a battery). The voltage of these sources is measured in volts (V), and they are represented by the symbol ⫣⊢. We assume these voltage sources have no resistance. The flow of current is governed by the following principles.

Ohm's Law

The current I and the voltage drop V across a resistance R are related by the equation $V = RI$.

Kirchhoff's Laws

1. (Junction Rule) The current flow into a junction equals the current flow out of that junction.
2. (Circuit Rule) The algebraic sum of the voltage drops (due to resistances) around any closed circuit of the network must equal the sum of the voltage increases around the circuit.

When applying rule 2, select a direction (clockwise or counterclockwise) around the closed circuit and then consider all voltages and currents positive when in this direction and negative when in the opposite direction. This is why the term *algebraic sum* is used in rule 2. Here is an example.

---

9 This section is independent of Section 1.4.

Example 1

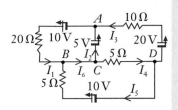

Find the various currents in the circuit shown.

Solution First apply the junction rule at junctions A, B, C, and D to obtain

$$
\begin{aligned}
\text{Junction } A \quad & I_1 = I_2 + I_3 \\
\text{Junction } B \quad & I_6 = I_1 + I_5 \\
\text{Junction } C \quad & I_2 + I_4 = I_6 \\
\text{Junction } D \quad & I_3 + I_5 = I_4
\end{aligned}
$$

Note that these equations are not independent (in fact, the third is an easy consequence of the other three).

Next, the circuit rule insists that the sum of the voltage increases (due to the sources) around a closed circuit must equal the sum of the voltage drops (due to resistances). By Ohm's law, the voltage loss across a resistance R (in the direction of the current I) is RI. Going counterclockwise around three closed circuits yields

$$
\begin{aligned}
\text{Upper left} \quad & 10 + 5 = 20I_1 \\
\text{Upper right} \quad & -5 + 20 = 10I_3 + 5I_4 \\
\text{Lower} \quad & {-10} = -5I_5 - 5I_4
\end{aligned}
$$

Hence, disregarding the redundant equation obtained at junction C, we have six equations in the six unknowns $I_1, \ldots, I_6$. The solution is

$$
\begin{aligned}
I_1 = \tfrac{15}{20} \qquad & I_4 = \tfrac{28}{20} \\
I_2 = \tfrac{-1}{20} \qquad & I_5 = \tfrac{12}{20} \\
I_3 = \tfrac{16}{20} \qquad & I_6 = \tfrac{27}{20}
\end{aligned}
$$

The fact that I_2 is negative means, of course, that this current is in the opposite direction, with a magnitude of $\tfrac{1}{20}$ amperes.

Exercises 1.5

In Exercises 1–4, find the currents in the circuits.

1.

2.

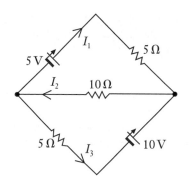

3.

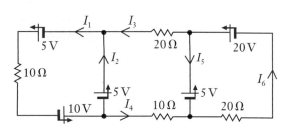

◆4. All resistances are 10 Ω.

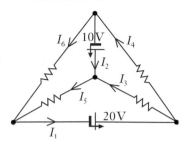

5. Find the voltage x such that the current $I_1 = 0$.

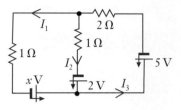

SECTION 1.6   An Application to Chemical Reactions

When a chemical reaction takes place a number of molecules combine to produce new molecules. Hence, when hydrogen H_2 and oxygen O_2 molecules combine, the result is water H_2O. We express this as

$$H_2 + O_2 \rightarrow H_2O$$

Individual atoms are neither created nor destroyed, so the number of hydrogen and oxygen atoms going into the reaction must equal the number coming out (in the form of water). In this case the reaction is said to be *balanced*. Note that each hydrogen molecule H_2 consists of two atoms as does each oxygen molecule O_2, while a water molecule H_2O consists of two hydrogen atoms and one oxygen atom. In the above reaction, this requires that twice as many hydrogen molecules enter the reaction; we express this as follows:

$$2H_2 + O_2 \rightarrow 2H_2O$$

This is now balanced because there are 4 hydrogen atoms and 2 oxygen atoms on each side of the reaction.

Example 1

Balance the following reaction for burning octane C_8H_{18} in oxygen O_2:

$$C_8H_{18} + O_2 \rightarrow CO_2 + H_2O$$

where CO_2 represents carbon dioxide. We must find positive integers $x, y, z,$ and w such that

$$x\,C_8H_{18} + y\,O_2 \rightarrow z\,CO_2 + w\,H_2O$$

Equating the number of carbon, hydrogen, and oxygen atoms on each side gives $8x = z$, $18x = 2w$ and $2y = 2z + w$, respectively. These can be written as a homogeneous linear system

$$\begin{aligned} 8x \quad\quad - z \quad\quad &= 0 \\ 18x \quad\quad\quad - 2w &= 0 \\ 2y - 2z - w &= 0 \end{aligned}$$

which can be solved by gaussian elimination. In larger systems this is necessary but, in such a simple situation, it is easier to solve directly. Set $w = t$, so that $x = \frac{1}{9}t$, $z = \frac{8}{9}t$, $2y = \frac{16}{9}t + t = \frac{25}{9}t$. But $x, y, z,$ and w must be positive integers, so the smallest value of t that eliminates fractions is 18. Hence, $x = 2$, $y = 25$, $z = 16$, and $w = 18$, and the balanced reaction is

$$2C_8H_{18} + 25O_2 \rightarrow 16CO_2 + 18H_2O$$

The reader can verify that this is indeed balanced.

It is worth noting that this problem introduces a new element into the theory of linear equations: the insistence that the solution must consist of positive integers.

Exercises 1.6

In each case balance the chemical reaction.

1. $CH_4 + O_2 \rightarrow CO_2 + H_2O$. This is the burning of methane CH_4.

♦2. $NH_3 + CuO \rightarrow N_2 + Cu + H_2O$. Here NH_3 is ammonia, CuO is copper oxide, Cu is copper, and N_2 is nitrogen.

3. $CO_2 + H_2O \rightarrow C_6H_{12}O_6 + O_2$. This is called the photosynthesis reaction—$C_6H_{12}O_6$ is glucose.

♦4. $Pb(N_3)_2 + Cr(MnO_4)_2 \rightarrow Cr_2O_3 + MnO_2 + Pb_3O_4 + NO$.

Supplementary Exercises for Chapter 1

1. We show in Chapter 4 that the graph of an equation $ax + by + cz = d$ is a plane in space when not all of a, b, and c are zero.

    (a) By examining the possible positions of planes in space, show that three equations in three variables can have zero, one, or infinitely many solutions.

    ♦(b) Can two equations in three variables have a unique solution? Give reasons for your answer.

2. Find all solutions to the following systems of linear equations.

    (a)
    $$\begin{aligned} x_1 + x_2 + x_3 - x_4 &= 3 \\ 3x_1 + 5x_2 - 2x_3 + x_4 &= 1 \\ -3x_1 - 7x_2 + 7x_3 - 5x_4 &= 7 \\ x_1 + 3x_2 - 4x_3 + 3x_4 &= -5 \end{aligned}$$

    ♦(b)
    $$\begin{aligned} x_1 + 4x_2 - x_3 + x_4 &= 2 \\ 3x_1 + 2x_2 + x_3 + 2x_4 &= 5 \\ x_1 - 6x_2 + 3x_3 &= 1 \\ x_1 + 14x_2 - 5x_3 + 2x_4 &= 3 \end{aligned}$$

3. In each case find (if possible) conditions on a, b, and c such that the system has zero, one, or infinitely many solutions.

    (a)
    $$\begin{aligned} x + 2y - 4z &= 4 \\ 3x - y + 13z &= 2 \\ 4x + y + a^2z &= a + 3 \end{aligned}$$

    ♦(b)
    $$\begin{aligned} x + y + 3z &= a \\ ax + y + 5z &= 4 \\ x + ay + 4z &= a \end{aligned}$$

♦4. Show that any two rows of a matrix can be interchanged by elementary row transformations of the other two types.

5. If $ad \neq bc$, show that $\begin{bmatrix} a & b \\ c & d \end{bmatrix}$ has reduced row-echelon form $\begin{bmatrix} 1 & 0 \\ 0 & 1 \end{bmatrix}$.

♦6. Find a, b, and c so that the system
    $$\begin{aligned} x + ay + cz &= 0 \\ bx + cy - 3z &= 1 \\ ax + 2y + bz &= 5 \end{aligned}$$
    has the solution $x = 3, y = -1, z = 2$.

7. Solve the system
    $$\begin{aligned} x + 2y + 2z &= -3 \\ 2x + y + z &= 0 \\ x - y - iz &= i \end{aligned}$$
    where $i^2 = -1$. [See Appendix A.]

8. Show that the *real* system
    $$\begin{cases} x + y + z = 5 \\ 2x - y - z = 1 \\ -3x + 2y + 2z = 0 \end{cases}$$
    has a *complex* solution: $x = 2, y = i, z = 3 - i$ where $i^2 = -1$. Explain. What happens when such a real system has a unique solution?

9. A man is ordered by his doctor to take 5 units of vitamin A, 13 units of vitamin B, and 23 units of vitamin C each day. Three brands of vitamin pills are available, and the number of units of each vitamin per pill are shown in the accompanying table.

Brand	Vitamin		
	A	B	C
1	1	2	4
2	1	1	3
3	0	1	1

    (a) Find all combinations of pills that provide exactly the required amount of vitamins (no partial pills allowed).

◆(b) If brands 1, 2, and 3 cost 3¢, 2¢, and 5¢ per pill, respectively, find the least expensive treatment.

10. A restaurant owner plans to use x tables seating 4, y tables seating 6, and z tables seating 8, for a total of 20 tables. When fully occupied, the tables seat 108 customers. If only half of the x tables, half of the y tables, and one-fourth of the z tables are used, each fully occupied, then 46 customers will be seated. Find x, y, and z.

11. (a) Show that a matrix with two rows and two columns that is in reduced row-echelon form must have one of the following forms:

$$\begin{bmatrix} 1 & 0 \\ 0 & 1 \end{bmatrix} \quad \begin{bmatrix} 0 & 1 \\ 0 & 0 \end{bmatrix} \quad \begin{bmatrix} 0 & 0 \\ 0 & 0 \end{bmatrix} \quad \begin{bmatrix} 1 & * \\ 0 & 0 \end{bmatrix}$$

[*Hint:* The leading 1 in the first row must be in column 1 or 2 or not exist.]

(b) List the seven reduced row-echelon forms for matrices with two rows and three columns.

(c) List the four reduced row-echelon forms for matrices with three rows and two columns.

12. An amusement park charges $7 for adults, $2 for youths, and $0.50 for children. If 150 people enter and pay a total of $100, find the numbers of adults, youths, and children.
[*Hint:* These numbers are nonnegative *integers*.]

13. Solve the following system of equations for x and y.

$$x^2 + xy - y^2 = 1$$
$$2x^2 - xy + 3y^2 = 13$$
$$x^2 + 3xy + 2y^2 = 0$$

[*Hint:* These equations are linear in the new variables $x_1 = x^2$, $x_2 = xy$, and $x_3 = y^2$.]

2 Matrix Algebra

In the study of systems of linear equations in Chapter 1, we found it convenient to manipulate the augmented matrix of the system. Our aim was to reduce it to row-echelon form (using elementary row operations) and hence to write down all solutions to the system. In the present chapter we consider matrices for their own sake. While some of the motivation comes from linear equations, it turns out that matrices can be multiplied and added and so form an algebraic system somewhat analogous to the real numbers. This "matrix algebra" is useful in ways that are quite different from the study of linear equations. For example, the geometrical transformations obtained by rotating the euclidean plane about the origin can be viewed as multiplications by certain 2×2 matrices. These "matrix transformations" are an important tool in geometry and, in turn, the geometry provides a "picture" of the matrices. Furthermore, matrix algebra has many other applications, some of which are will be explored in this chapter. This subject is quite old and was first studied systematically in 1858 by Arthur Cayley.[1]

Arthur Cayley. Photo
© Corbis.

SECTION 2.1 Matrix Addition, Scalar Multiplication, and Transposition

A rectangular array of numbers is called a **matrix** (the plural is **matrices**), and the numbers are called the **entries** of the matrix. Matrices are usually denoted by uppercase letters: A, B, C, and so on. Hence,

$$A = \begin{bmatrix} 1 & 2 & -1 \\ 0 & 5 & 6 \end{bmatrix} \qquad B = \begin{bmatrix} 1 & -1 \\ 0 & 2 \end{bmatrix} \qquad C = \begin{bmatrix} 1 \\ 3 \\ 2 \end{bmatrix}$$

are matrices. Clearly matrices come in various shapes depending on the number of **rows** and **columns**. For example, the matrix A shown has 2 rows and 3 columns. In general, a matrix with m rows and n columns is referred to as an $m \times n$ **matrix** or as having **size $m \times n$**. Thus matrices A, B, and C above have sizes 2×3, 2×2, and 3×1, respectively. A matrix of size $1 \times n$ is called a **row matrix**, whereas one of size $n \times 1$ is called a **column matrix**. Matrices of size $n \times n$ for some n are called **square** matrices.

---

1 Arthur Cayley (1821–1895) showed his mathematical talent early and graduated from Cambridge in 1842 as senior wrangler. With no employment in mathematics in view, he took legal training and worked as a lawyer while continuing to do mathematics, publishing nearly 300 papers in fourteen years. Finally, in 1863, he accepted the Sadlerian professorship at Cambridge and remained there for the rest of his life, valued for his administrative and teaching skills as well as for his scholarship. His mathematical achievements were of the first rank. In addition to originating matrix theory and the theory of determinants, he did fundamental work in group theory, in higher-dimensional geometry, and in the theory of invariants. He was one of the most prolific mathematicians of all time and produced 966 papers, filling thirteen volumes of 600 pages each.

Each entry of a matrix is identified by the row and column in which it lies. The rows are numbered from the top down, and the columns are numbered from left to right. Then the (i, j)-**entry** of a matrix is the number lying simultaneously in row i and column j. For example,

$$\text{The } (1, 2)\text{-entry of } \begin{bmatrix} 1 & -1 \\ 0 & 1 \end{bmatrix} \text{ is } -1$$

$$\text{The } (2, 3)\text{-entry of } \begin{bmatrix} 1 & 2 & -1 \\ 0 & 5 & 6 \end{bmatrix} \text{ is } 6$$

A special notation has been devised for the entries of a matrix. If A is an $m \times n$ matrix, and if the (i, j)-entry of A is denoted as a_{ij}, then A is displayed as follows:

$$A = \begin{bmatrix} a_{11} & a_{12} & a_{13} & \cdots & a_{1n} \\ a_{21} & a_{22} & a_{23} & \cdots & a_{2n} \\ \vdots & \vdots & \vdots & & \vdots \\ a_{m1} & a_{m2} & a_{m3} & \cdots & a_{mn} \end{bmatrix}$$

This is usually denoted simply as $A = [a_{ij}]$. Thus a_{ij} is the entry in row i and column j of A. For example, a 3×4 matrix in this notation is written

$$A = \begin{bmatrix} a_{11} & a_{12} & a_{13} & a_{14} \\ a_{21} & a_{22} & a_{23} & a_{24} \\ a_{31} & a_{32} & a_{33} & a_{34} \end{bmatrix}$$

It is worth pointing out a convention regarding rows and columns: *Rows are mentioned before columns.* For example:

- If a matrix has size $m \times n$, it has m rows and n columns.
- If we speak of the (i, j)-entry of a matrix, it lies in row i and column j.
- If an entry is denoted a_{ij}, the first subscript i refers to the row and the second subscript j to the column in which a_{ij} lies.

Two points (x_1, y_1) and (x_2, y_2) in the plane are equal exactly when they have the same coordinates, that is $x_1 = x_2$ and $y_1 = y_2$. Similarly, two matrices A and B are called **equal** (written $A = B$) if and only if:

1. They have the same size.
2. Corresponding entries are equal.

If the entries of A and B are written in the form $A = [a_{ij}]$, $B = [b_{ij}]$, described earlier, then the second condition takes the following form:

$$[a_{ij}] = [b_{ij}] \quad \text{means} \quad a_{ij} = b_{ij} \text{ for all } i \text{ and } j$$

Example 1

Given $A = \begin{bmatrix} a & b \\ c & d \end{bmatrix}$, $B = \begin{bmatrix} 1 & 2 & -1 \\ 3 & 0 & 1 \end{bmatrix}$, and $C = \begin{bmatrix} 1 & 0 \\ -1 & 2 \end{bmatrix}$, discuss the possibility that $A = B$, $B = C$, $A = C$.

Solution

$A = B$ is impossible because A and B are of different sizes: A is 2×2 whereas B is 2×3. Similarly, $B = C$ is impossible. But $A = C$ is possible provided that corresponding entries are equal: $\begin{bmatrix} a & b \\ c & d \end{bmatrix} = \begin{bmatrix} 1 & 0 \\ -1 & 2 \end{bmatrix}$ means $a = 1$, $b = 0$, $c = -1$, and $d = 2$.

Matrix Addition

If A and B are matrices of the same size, their **sum** $A + B$ is the matrix formed by adding corresponding entries. If $A = [a_{ij}]$ and $B = [b_{ij}]$, this takes the form

$$A + B = [a_{ij} + b_{ij}]$$

Note that addition is *not* defined for matrices of different sizes.

Example 2

If $A = \begin{bmatrix} 2 & 1 & 3 \\ -1 & 2 & 0 \end{bmatrix}$ and $B = \begin{bmatrix} 1 & 1 & -1 \\ 2 & 0 & 6 \end{bmatrix}$, compute $A + B$.

Solution $A + B = \begin{bmatrix} 2+1 & 1+1 & 3-1 \\ -1+2 & 2+0 & 0+6 \end{bmatrix} = \begin{bmatrix} 3 & 2 & 2 \\ 1 & 2 & 6 \end{bmatrix}$.

Example 3

Find a, b, and c if $[a \ \ b \ \ c] + [c \ \ a \ \ b] = [3 \ \ 2 \ \ -1]$.

Solution   Add the matrices on the left side to obtain

$$[a+c \ \ b+a \ \ c+b] = [3 \ \ 2 \ \ -1]$$

Because corresponding entries must be equal, this gives three equations: $a + c = 3$, $b + a = 2$, and $c + b = -1$. Solving these yields $a = 3$, $b = -1$, $c = 0$.

If A, B, and C are any matrices *of the same size*, then

$$A + B = B + A \qquad \text{(commutative law)}$$
$$A + (B + C) = (A + B) + C \qquad \text{(associative law)}$$

In fact, if $A = [a_{ij}]$ and $B = [b_{ij}]$, then the (i, j)-entries of $A + B$ and $B + A$ are, respectively, $a_{ij} + b_{ij}$ and $b_{ij} + a_{ij}$. Since these are equal for all i and j, we get

$$A + B = [a_{ij} + b_{ij}] = [b_{ij} + a_{ij}] = B + A$$

The associative law is verified similarly.

The $m \times n$ matrix in which every entry is zero is called the **zero matrix** and is denoted as 0 (or 0_{mn} if it is important to emphasize the size). Hence,

$$0 + X = X$$

holds for all $m \times n$ matrices X. The **negative** of an $m \times n$ matrix A (written $-A$) is defined to be the $m \times n$ matrix obtained by multiplying each entry of A by -1. If $A = [a_{ij}]$, this becomes $-A = [-a_{ij}]$. Hence,

$$A + (-A) = 0$$

holds for all matrices A where, of course, 0 is the zero matrix of the same size as A.

A closely related notion is that of subtracting matrices. If A and B are two $m \times n$ matrices, their **difference** $A - B$ is defined by

$$A - B = A + (-B)$$

Note that if $A = [a_{ij}]$ and $B = [b_{ij}]$, then

$$A - B = [a_{ij}] + [-b_{ij}] = [a_{ij} - b_{ij}]$$

is the $m \times n$ matrix formed by *subtracting* corresponding entries.

Example 4

Let $A = \begin{bmatrix} 3 & -1 & 0 \\ 1 & 2 & -4 \end{bmatrix}$, $B = \begin{bmatrix} 1 & -1 & 1 \\ -2 & 0 & 6 \end{bmatrix}$, and $C = \begin{bmatrix} 1 & 0 & -2 \\ 3 & 1 & 1 \end{bmatrix}$. Compute $-A$, $A - B$, and $A + B - C$.

Solution

$$-A = \begin{bmatrix} -3 & 1 & 0 \\ -1 & -2 & 4 \end{bmatrix}$$

$$A - B = \begin{bmatrix} 3 - 1 & -1 - (-1) & 0 - 1 \\ 1 - (-2) & 2 - 0 & -4 - 6 \end{bmatrix} = \begin{bmatrix} 2 & 0 & -1 \\ 3 & 2 & -10 \end{bmatrix}$$

$$A + B - C = \begin{bmatrix} 3 + 1 - 1 & -1 - 1 - 0 & 0 + 1 - (-2) \\ 1 - 2 - 3 & 2 + 0 - 1 & -4 + 6 - 1 \end{bmatrix} = \begin{bmatrix} 3 & -2 & 3 \\ -4 & 1 & 1 \end{bmatrix}$$

Example 5

Solve $\begin{bmatrix} 3 & 2 \\ -1 & 1 \end{bmatrix} + X = \begin{bmatrix} 1 & 0 \\ -1 & 2 \end{bmatrix}$, where X is a matrix.

Solution

We solve a numerical equation $a + x = b$ by subtracting the number a from both sides to obtain $x = b - a$. This also works for matrices. To solve $\begin{bmatrix} 3 & 2 \\ -1 & 1 \end{bmatrix} + X = \begin{bmatrix} 1 & 0 \\ -1 & 2 \end{bmatrix}$, simply subtract the matrix $\begin{bmatrix} 3 & 2 \\ -1 & 1 \end{bmatrix}$ from both sides to get

$$X = \begin{bmatrix} 1 & 0 \\ -1 & 2 \end{bmatrix} - \begin{bmatrix} 3 & 2 \\ -1 & 1 \end{bmatrix} = \begin{bmatrix} 1 - 3 & 0 - 2 \\ -1 - (-1) & 2 - 1 \end{bmatrix} = \begin{bmatrix} -2 & -2 \\ 0 & 1 \end{bmatrix}$$

The reader should verify that this matrix X does indeed satisfy the original equation.

The solution in Example 5 solves the single matrix equation $A + X = B$ directly via matrix subtraction: $X = B - A$. This ability to work with matrices as entities lies at the heart of matrix algebra.

It is important to note that the sizes of matrices involved in some calculations are often determined by the context. For example, if

$$A + C = \begin{bmatrix} 1 & 3 & -1 \\ 2 & 0 & 1 \end{bmatrix}$$

then A and C must be the same size (so that $A + C$ makes sense), and that size must be 2×3 (so that the sum is 2×3). For simplicity we shall often omit reference to such facts when they are clear from the context.

Scalar Multiplication

In gaussian elimination, multiplying a row of a matrix by a number k means multiplying *every* entry of that row by k. More generally, if A is any matrix and k is any number, the **scalar multiple** kA is the matrix obtained from A by multiplying each entry of A by k. If $A = [a_{ij}]$, this is

$$kA = [ka_{ij}]$$

The term *scalar* arises here because the set of numbers from which the entries are drawn is usually referred to as the set of scalars. We have been using real numbers as scalars, but we could equally well have been using complex numbers.

Example 6

If $A = \begin{bmatrix} 3 & -1 & 4 \\ 2 & 0 & 6 \end{bmatrix}$ and $B = \begin{bmatrix} 1 & 2 & -1 \\ 0 & 3 & 2 \end{bmatrix}$, compute $5A$, $\frac{1}{2}B$, and $3A - 2B$.

Solution

$$5A = \begin{bmatrix} 15 & -5 & 20 \\ 10 & 0 & 30 \end{bmatrix}, \qquad \frac{1}{2}B = \begin{bmatrix} \frac{1}{2} & 1 & -\frac{1}{2} \\ 0 & \frac{3}{2} & 1 \end{bmatrix}$$

$$3A - 2B = \begin{bmatrix} 9 & -3 & 12 \\ 6 & 0 & 18 \end{bmatrix} - \begin{bmatrix} 2 & 4 & -2 \\ 0 & 6 & 4 \end{bmatrix} = \begin{bmatrix} 7 & -7 & 14 \\ 6 & -6 & 14 \end{bmatrix}$$

If A is any matrix, note that kA is the same size as A for all scalars k. We also have

$$0A = 0 \quad \text{and} \quad k0 = 0$$

because the zero matrix has every entry zero. In other words, $kA = 0$ if either $k = 0$ or $A = 0$. The converse of these properties is also true, as Example 7 shows.

Example 7

If $kA = 0$, show that either $k = 0$ or $A = 0$.

Solution   Write $A = [a_{ij}]$ so that $kA = 0$ means $ka_{ij} = 0$ for all i and j. If $k = 0$, there is nothing to do. If $k \neq 0$, then $ka_{ij} = 0$ implies that $a_{ij} = 0$ for all i and j; that is, $A = 0$.

For future reference, the basic properties of matrix addition and scalar multiplication are listed in Theorem 1.

Theorem 1

Let A, B, and C denote arbitrary $m \times n$ matrices where m and n are fixed. Let k and p denote arbitrary real numbers. Then
  1. $A + B = B + A$.
  2. $A + (B + C) = (A + B) + C$.
  3. There is an $m \times n$ matrix 0, such that $0 + A = A$ for each A.
  4. For each A there is an $m \times n$ matrix, $-A$, such that $A + (-A) = 0$.
  5. $k(A + B) = kA + kB$.
  6. $(k + p)A = kA + pA$.
  7. $(kp)A = k(pA)$.
  8. $1A = A$.

PROOF

Properties 1–4 were given previously. To check property 5, let $A = [a_{ij}]$ and $B = [b_{ij}]$ denote matrices of the same size. Then $A + B = [a_{ij} + b_{ij}]$, as before, so the (i, j)-entry of $k(A + B)$ is

$$k(a_{ij} + b_{ij}) = ka_{ij} + kb_{ij}$$

But this is just the (i, j)-entry of $kA + kB$, and it follows that $k(A + B) = kA + kB$. The other properties can be similarly verified; the details are left to the reader.

The properties in Theorem 1 enable us to do calculations with matrices in much the same way that numerical calculations are carried out. To begin, property 2 implies that the sum $(A + B) + C = A + (B + C)$ is the same no matter how it is formed and so is written as $A + B + C$. Similarly, the sum $A + B + C + D$ is independent of how it is formed; for example, it equals both $(A + B) + (C + D)$ and $A + [B + (C + D)]$. Furthermore, property 1 ensures that, for example, $B + D + A + C = A + B + C + D$. In other words, the *order* in which the matrices are added does not matter. A similar remark applies to sums of five (or more) matrices.

Properties 5 and 6 in Theorem 1 are called distributive laws for scalar multiplication, and they extend to sums of more than two terms. For example,

$$k(A + B + C) = kA + kB + kC$$
$$(k + p + m)A = kA + pA + mA$$

Similar observations hold for more than three summands. These facts, together with properties 7 and 8, enable us to simplify expressions by collecting like terms, expanding, and taking common factors in exactly the same way that algebraic expressions involving variables and real numbers are manipulated. The following example illustrates these techniques.

Example 8

Simplify $2(A + 3C) - 3(2C - B) - 3[2(2A + B - 4C) - 4(A - 2C)]$ where A, B, and C are all matrices of the same size.

Solution The reduction proceeds as though A, B, and C were variables.

$$2(A + 3C) - 3(2C - B) - 3[2(2A + B - 4C) - 4(A - 2C)]$$
$$= 2A + 6C - 6C + 3B - 3[4A + 2B - 8C - 4A + 8C]$$
$$= 2A + 3B - 3[2B]$$
$$= 2A - 3B$$

Transpose

Many results about a matrix A involve the *rows* of A, and the corresponding result for columns is derived in an analogous way, essentially by replacing the word *row* by the word *column* throughout. The following definition is made with such applications in mind.

If A is an $m \times n$ matrix, the **transpose** of A, written A^T, is the $n \times m$ matrix whose rows are just the columns of A in the same order. In other words, the first row of A^T is the first column of A, the second row of A^T is the second column of A, and so on.

Example 9

Write down the transpose of each of the following matrices.

$$A = \begin{bmatrix} 1 \\ 3 \\ 2 \end{bmatrix} \quad B = [5\ 2\ 6] \quad C = \begin{bmatrix} 1 & 2 \\ 3 & 4 \\ 5 & 6 \end{bmatrix} \quad D = \begin{bmatrix} 3 & 1 & -1 \\ 1 & 3 & 2 \\ -1 & 2 & 1 \end{bmatrix}$$

Solution $A^T = [1\ 3\ 2]$, $B^T = \begin{bmatrix} 5 \\ 2 \\ 6 \end{bmatrix}$, $C^T = \begin{bmatrix} 1 & 3 & 5 \\ 2 & 4 & 6 \end{bmatrix}$, and $D^T = D$.

If $A = [a_{ij}]$ is a matrix, write $A^T = [b_{ij}]$. Then b_{ij} is the jth element of the ith row of A^T and so is the jth element of the ith *column* of A. This means $b_{ij} = a_{ji}$, so the definition of A^T can be stated as follows:

$$\text{If } A = [a_{ij}], \text{ then } A^T = [a_{ji}] \qquad (*)$$

This is useful in verifying the following properties of transposition.

> ### Theorem 2
>
> Let A and B denote matrices of the same size, and let k denote a scalar.
> 1. If A is an $m \times n$ matrix, then A^T is an $n \times m$ matrix.
> 2. $(A^T)^T = A$.
> 3. $(kA)^T = kA^T$.
> 4. $(A + B)^T = A^T + B^T$.

PROOF

Property 1 is part of the definition of A^T, and property 2 follows from $(*)$. As to property 3: If $A = [a_{ij}]$, then $kA = [ka_{ij}]$, so $(*)$ gives

$$(kA)^T = [ka_{ji}] = k[a_{ji}] = kA^T$$

Finally, if $B = [b_{ij}]$, then $A + B = [c_{ij}]$ where $c_{ij} = a_{ij} + b_{ij}$. Then $(*)$ gives property 4:

$$(A + B)^T = [c_{ij}]^T = [c_{ji}] = [a_{ji} + b_{ji}] = [a_{ji}] + [b_{ji}] = A^T + B^T$$

There is another useful way to think of transposition. If $A = [a_{ij}]$ is an $m \times n$ matrix, the elements $a_{11}, a_{22}, a_{33}, \ldots$ are called the **main diagonal** of A. Hence the main diagonal extends down and to the right from the upper left corner of the matrix A; it is shaded in the following examples:

$$\begin{bmatrix} a_{11} & a_{12} \\ a_{21} & a_{22} \\ a_{31} & a_{32} \end{bmatrix} \qquad \begin{bmatrix} a_{11} & a_{12} & a_{13} \\ a_{21} & a_{22} & a_{23} \end{bmatrix} \qquad \begin{bmatrix} a_{11} & a_{12} & a_{13} \\ a_{21} & a_{22} & a_{23} \\ a_{31} & a_{32} & a_{33} \end{bmatrix} \qquad \begin{bmatrix} a_{11} \\ a_{21} \end{bmatrix}$$

Thus forming the transpose of a matrix A can be viewed as "flipping" A about its main diagonal, or as "rotating" A through $180°$ about the line containing the main diagonal. This makes property 2 in Theorem 2 transparent.

Example 10

Solve for A if $\left[2A^T - 3\begin{bmatrix} 1 & 2 \\ -1 & 1 \end{bmatrix} \right]^T = \begin{bmatrix} 2 & 3 \\ -1 & 2 \end{bmatrix}$.

Solution Using Theorem 2, the left side of the equation is

$$\left[2A^T - 3\begin{bmatrix} 1 & 2 \\ -1 & 1 \end{bmatrix} \right]^T = 2(A^T)^T - 3\begin{bmatrix} 1 & 2 \\ -1 & 1 \end{bmatrix}^T = 2A - 3\begin{bmatrix} 1 & -1 \\ 2 & 1 \end{bmatrix}$$

Hence the equation becomes

$$2A - 3\begin{bmatrix} 1 & -1 \\ 2 & 1 \end{bmatrix} = \begin{bmatrix} 2 & 3 \\ -1 & 2 \end{bmatrix}$$

Thus $2A = \begin{bmatrix} 2 & 3 \\ -1 & 2 \end{bmatrix} + 3\begin{bmatrix} 1 & -1 \\ 2 & 1 \end{bmatrix} = \begin{bmatrix} 5 & 0 \\ 5 & 5 \end{bmatrix}$, so finally $A = \frac{1}{2}\begin{bmatrix} 5 & 0 \\ 5 & 5 \end{bmatrix} = \frac{5}{2}\begin{bmatrix} 1 & 0 \\ 1 & 1 \end{bmatrix}$.

Note that this exercise also can be solved by first transposing both sides, then solving for A^T, and so obtaining $A = (A^T)^T$. The reader should do this.

The matrix D in Example 9 has the property that $D = D^T$. Such matrices are important; a matrix A is called **symmetric** if $A = A^T$. A symmetric matrix A is necessarily square (if A is $m \times n$, then A^T is $n \times m$, so $A = A^T$ forces $n = m$). The name comes from the fact that these matrices exhibit a symmetry about the main diagonal. That is, entries that are directly across the main diagonal from each other are equal.

For example, $\begin{bmatrix} a & b & c \\ b' & d & e \\ c' & e' & f \end{bmatrix}$ is symmetric when $b = b'$, $c = c'$, and $e = e'$.

Example 11

If A and B are symmetric $n \times n$ matrices, show that $A + B$ is symmetric.

Solution   We have $A^T = A$ and $B^T = B$, so, by Theorem 2, $(A + B)^T = A^T + B^T = A + B$. Hence $A + B$ is symmetric.

Example 12

Suppose a square matrix A satisfies $A = 2A^T$. Show that necessarily $A = 0$.

Solution   If we iterate the given equation, Theorem 2 gives

$$A = 2A^T = 2[2A^T]^T = 2[2(A^T)^T] = 4A$$

Subtracting A from both sides gives $3A = 0$, so $A = \frac{1}{3}(3A) = \frac{1}{3}(0) = 0$.

Exercises 2.1

1. Find a, b, c, and d if
   (a) $\begin{bmatrix} a & b \\ c & d \end{bmatrix} = \begin{bmatrix} c-3d & -d \\ 2a+d & a+b \end{bmatrix}$
   ◆(b) $\begin{bmatrix} a-b & b-c \\ c-d & d-a \end{bmatrix} = 2\begin{bmatrix} 1 & 1 \\ -3 & 1 \end{bmatrix}$
   (c) $3\begin{bmatrix} a \\ b \end{bmatrix} + 2\begin{bmatrix} b \\ a \end{bmatrix} = \begin{bmatrix} 1 \\ 2 \end{bmatrix}$ ◆(d) $\begin{bmatrix} a & b \\ c & d \end{bmatrix} = \begin{bmatrix} b & c \\ d & a \end{bmatrix}$

2. Compute the following:
   (a) $\begin{bmatrix} 3 & 2 & 1 \\ 5 & 1 & 0 \end{bmatrix} - 5\begin{bmatrix} 3 & 0 & -2 \\ 1 & -1 & 2 \end{bmatrix}$

◆(b) $3\begin{bmatrix} 3 \\ -1 \end{bmatrix} - 5\begin{bmatrix} 6 \\ 2 \end{bmatrix} + 7\begin{bmatrix} 1 \\ -1 \end{bmatrix}$
   (c) $\begin{bmatrix} -2 & 1 \\ 3 & 2 \end{bmatrix} - 4\begin{bmatrix} 1 & -2 \\ 0 & -1 \end{bmatrix} + 3\begin{bmatrix} 2 & -3 \\ -1 & -2 \end{bmatrix}$
   ◆(d) $[3\ -1\ 2] - 2[9\ 3\ 4] + [3\ 11\ -6]$
   (e) $\begin{bmatrix} 1 & -5 & 4 & 0 \\ 2 & 1 & 0 & 6 \end{bmatrix}^T$ ◆(f) $\begin{bmatrix} 0 & -1 & 2 \\ 1 & 0 & -4 \\ -2 & 4 & 0 \end{bmatrix}^T$
   (g) $\begin{bmatrix} 3 & -1 \\ 2 & 1 \end{bmatrix} - 2\begin{bmatrix} 1 & -2 \\ 1 & 1 \end{bmatrix}^T$ ◆(h) $3\begin{bmatrix} 2 & 1 \\ -1 & 0 \end{bmatrix}^T - 2\begin{bmatrix} 1 & -1 \\ 2 & 3 \end{bmatrix}$

3. Let $A = \begin{bmatrix} 2 & 1 \\ 0 & -1 \end{bmatrix}$, $B = \begin{bmatrix} 3 & -1 & 2 \\ 0 & 1 & 4 \end{bmatrix}$, $C = \begin{bmatrix} 3 & -1 \\ 2 & 0 \end{bmatrix}$,

$D = \begin{bmatrix} 1 & 3 \\ -1 & 0 \\ 1 & 4 \end{bmatrix}$, and $E = \begin{bmatrix} 1 & 0 & 1 \\ 0 & 1 & 0 \end{bmatrix}$. Compute the

following (where possible).

(a) $3A - 2B$ ◆(b) $5C$ (c) $3E^T$
◆(d) $B + D$ (e) $4A^T - 3C$ ◆(f) $(A + C)^T$
(g) $2B - 3E$ ◆(h) $A - D$ (i) $(B - 2E)^T$

4. Find A if:

(a) $5A - \begin{bmatrix} 1 & 0 \\ 2 & 3 \end{bmatrix} = 3A - \begin{bmatrix} 5 & 2 \\ 6 & 1 \end{bmatrix}$

◆(b) $3A + \begin{bmatrix} 2 \\ 1 \end{bmatrix} = 5A - 2\begin{bmatrix} 3 \\ 0 \end{bmatrix}$

5. Find A in terms of B if:

(a) $A + B = 3A + 2B$
◆(b) $2A - B = 5(A + 2B)$

6. If X, Y, A, and B are matrices of the same size, solve the following equations to obtain X and Y in terms of A and B.

(a) $5X + 3Y = A$ ◆(b) $4X + 3Y = A$
   $2X + Y = B$ $5X + 4Y = B$

7. Find all matrices X and Y such that:

(a) $3X - 2Y = [3 \ \ -1]$ ◆(b) $2X - 5Y = [1 \ \ 2]$

8. Simplify the following expressions where A, B, and C are matrices.

(a) $2[9(A - B) + 7(2B - A)]$
   $-2[3(2B + A) - 2(A + 3B) - 5(A + B)]$
◆(b) $5[3(A - B + 2C) - 2(3C - B) - A]$
   $+2[3(3A - B + C) + 2(B - 2A) - 2C]$

9. If A is any 2×2 matrix, show that:

(a) $A = a\begin{bmatrix} 1 & 0 \\ 0 & 0 \end{bmatrix} + b\begin{bmatrix} 0 & 1 \\ 0 & 0 \end{bmatrix} + c\begin{bmatrix} 0 & 0 \\ 1 & 0 \end{bmatrix} + d\begin{bmatrix} 0 & 0 \\ 0 & 1 \end{bmatrix}$

for some numbers a, b, c, and d.

◆(b) $A = p\begin{bmatrix} 1 & 0 \\ 0 & 1 \end{bmatrix} + q\begin{bmatrix} 1 & 1 \\ 0 & 0 \end{bmatrix} + r\begin{bmatrix} 1 & 0 \\ 1 & 0 \end{bmatrix} + s\begin{bmatrix} 0 & 1 \\ 1 & 0 \end{bmatrix}$

for some numbers p, q, r, and s.

10. Let $A = [1 \ 1 \ -1]$, $B = [0 \ 1 \ 2]$, and $C = [3 \ 0 \ 1]$. If $rA + sB + tC = 0$ for some scalars r, s, and t, show that necessarily $r = s = t = 0$.

11. (a) If $Q + A = A$ holds for every $m \times n$ matrix A, show that $Q = 0_{mn}$.

◆(b) If A is an $m \times n$ matrix and $A + A' = 0_{mn}$, show that $A' = -A$.

12. If A denotes an $m \times n$ matrix, show that $A = -A$ if and only if $A = 0$.

13. A square matrix is called a **diagonal** matrix if all the entries off the main diagonal are zero. If A and B are diagonal matrices, show that the following matrices are also diagonal.

(a) $A + B$ (b) $A - B$ (c) kA for any number k

14. In each case determine all s and t such that the given matrix is symmetric:

(a) $\begin{bmatrix} 1 & s \\ -2 & t \end{bmatrix}$ ◆(b) $\begin{bmatrix} s & t \\ st & 1 \end{bmatrix}$

(c) $\begin{bmatrix} s & 2s & st \\ t & -1 & s \\ t & s^2 & s \end{bmatrix}$ ◆(d) $\begin{bmatrix} 2 & s & t \\ 2s & 0 & s+t \\ 3 & 3 & t \end{bmatrix}$

15. In each case find the matrix A.

(a) $\left(A + 3\begin{bmatrix} 1 & -1 & 0 \\ 1 & 2 & 4 \end{bmatrix} \right)^T = \begin{bmatrix} 2 & 1 \\ 0 & 5 \\ 3 & 8 \end{bmatrix}$

◆(b) $\left(3A^T + 2\begin{bmatrix} 1 & 0 \\ 0 & 2 \end{bmatrix} \right)^T = \begin{bmatrix} 8 & 0 \\ 3 & 1 \end{bmatrix}$

(c) $(2A - 3[1 \ 2 \ 0])^T = 3A^T + [2 \ 1 \ -1]^T$

◆(d) $\left(2A^T - 5\begin{bmatrix} 1 & 0 \\ -1 & 2 \end{bmatrix} \right)^T = 4A - 9\begin{bmatrix} 1 & 1 \\ -1 & 0 \end{bmatrix}$

16. Let A and B be symmetric (of the same size). Show that each of the following is symmetric.

(a) $(A - B)$ ◆(b) kA for any scalar k

17. Show that $A + A^T$ is symmetric for *any* square matrix A.

18. If A is a square matrix and $A = kA^T$, show that $A = 0$ if $k \neq \pm 1$.

19. In each case either show that the statement is true or give an example showing it is false.

(a) If $A + B = A + C$, then B and C have the same size.
◆(b) If $A + B = 0$, then $B = 0$.
(c) If the $(3, 1)$-entry of A is 5, then the $(1, 3)$-entry of A^T is -5.
◆(d) A and A^T have the same main diagonal for every matrix A.

(e) If $A^T = 3I$, then $A = 3I$.

♦(f) If A and B are symmetric, then $kA + mB$ is symmetric for any scalars k and m.

20. A square matrix W is called **skew-symmetric** if $W^T = -W$. Let A be any square matrix.

   (a) Show that $A - A^T$ is skew-symmetric.

   (b) Find a symmetric matrix S and a skew-symmetric matrix W such that $A = S + W$.

   ♦(c) Show that S and W in part (b) are uniquely determined by A.

21. If W is skew-symmetric (Exercise 20), show that the entries on the main diagonal are zero.

22. Prove the following parts of Theorem 1.

   (a) $(k + p)A = kA + pA$ ♦(b) $(kp)A = k(pA)$

23. Let $A, A_1, A_2, \ldots, A_n$ denote matrices of the same size. Use induction on n to verify the following extensions of properties 5 and 6 of Theorem 1.

   (a) $k(A_1 + A_2 + \cdots + A_n) = kA_1 + kA_2 + \cdots + kA_n$ for any number k

   (b) $(k_1 + k_2 + \cdots + k_n)A = k_1A + k_2A + \cdots + k_nA$ for any numbers $k_1, k_2, \ldots, k_n$

24. Let A be a square matrix. If $A = pB^T$ and $B = qA^T$ for some matrix B and numbers p and q, show that either $A = 0 = B$ or $pq = 1$. [*Hint:* Example 7.]

SECTION 2.2   Matrix Multiplication

Matrices can sometimes be multiplied to yield another matrix. This is useful in the study of linear equations, but the following example illustrates other applications.

Example 1

A farmer produces corn and barley as animal feed, and uses three chemicals 1, 2, and 3 to control insects and weeds. Matrix A below gives the number of milligrams of each chemical absorbed by a kilogram of each grain, and matrix B gives the number of kilograms of each grain eaten (per month) by cows and hogs.

$$
\begin{array}{c}
\\
\text{Chemical 1} \\
\text{Chemical 2} \\
\text{Chemical 3}
\end{array}
\begin{array}{c}
\text{Corn}\ \ \text{Barley} \\
\begin{bmatrix} 1 & 2 \\ 2 & 1 \\ 3 & 2 \end{bmatrix}
\end{array} = A
\qquad
\begin{array}{c}
\\
\text{Corn} \\
\text{Barley}
\end{array}
\begin{array}{c}
\text{Cows}\ \ \text{Hogs} \\
\begin{bmatrix} 27 & 15 \\ 15 & 5 \end{bmatrix}
\end{array} = B
$$

The owner of the market garden wants to know how much of each chemical is injested each month by cows and hogs. The result is matrix C below.

$$
\begin{array}{c}
\\
\text{Chemical 1} \\
\text{Chemical 2} \\
\text{Chemical 3}
\end{array}
\begin{array}{c}
\text{Cows}\ \ \text{Hogs} \\
\begin{bmatrix} 57 & 25 \\ 69 & 35 \\ 111 & 55 \end{bmatrix}
\end{array} = C
$$

The entries of matrix C can be computed directly from the entries of A and B. For example, the amount (57 mg) of chemical 1 injested by a cow each month comes as follows: 1 mg from each of the 27 kg of corn it eats, and 2 mg from each of the 15 kg of barley it eats. Hence the total is $(1\text{ mg})27 + (2\text{ mg})15 = 57$ mg. Observe that this can be calculated by multiplying each entry of the first row $[1\ 2]$ of A by the corresponding entry of the first column $\begin{bmatrix} 27 \\ 15 \end{bmatrix}$ of B and adding the results. This is called taking the dot product of row 1 of A and column 1 of B, and it is easily verified that, for each i and j, the (i, j)-entry of matrix C is the dot product of row i of A and row j of B.

In general, the **dot product** of a row $R = [r_1 \ r_2 \ \cdots \ r_n]$ and a column $C = [c_1 \ c_2 \ \cdots \ c_n]^T$ is defined to be the number

$$r_1c_1 + r_2c_2 + \cdots + r_nc_n$$

formed by multiplying corresponding entries and adding the results. With this we can define matrix multiplication:

Definition

The **product** AB of an $m \times n$ matrix A and an $n \times k$ matrix B is defined to be the $m \times k$ matrix AB whose (i, j)-entry is the dot product of row i of A and column j of B.

Thus computing the (i, j)-entry of AB means going across *row i* of A, down *column j* of B, multiplying corresponding entries, and adding the results. Note that this requires that the rows of A must be the same length as the columns of B. The following rule is useful for remembering this and deciding the size of the product matrix AB.

Rule

If A is $m \times n$ and B is $n' \times k$, then AB can be formed only if $n = n'$. In this case AB has size $m \times k$, and we say that the product AB is **defined**.

$$
\begin{array}{cc}
A & B \\
m \times n & n' \times k
\end{array}
$$

Example 2

Compute the $(1, 3)$- and $(2, 4)$-entries of AB where

$$A = \begin{bmatrix} 3 & -1 & 2 \\ 0 & 1 & 4 \end{bmatrix} \quad \text{and} \quad B = \begin{bmatrix} 2 & 1 & 6 & 0 \\ 0 & 2 & 3 & 4 \\ -1 & 0 & 5 & 8 \end{bmatrix}$$

Then compute AB.

Solution

The $(1, 3)$-entry of AB is the dot product of row 1 of A and column 3 of B (highlighted in the following display), computed by multiplying corresponding entries and adding the results.

$$\begin{bmatrix} 3 & -1 & 2 \\ 0 & 1 & 4 \end{bmatrix}\begin{bmatrix} 2 & 1 & 6 & 0 \\ 0 & 2 & 3 & 4 \\ -1 & 0 & 5 & 8 \end{bmatrix} \quad (1, 3)\text{-entry} = 3 \cdot 6 + (-1) \cdot 3 + 2 \cdot 5 = 25$$

Similarly, the $(2, 4)$ entry of AB involves row 2 of A and column 4 of B.

$$\begin{bmatrix} 3 & -1 & 2 \\ 0 & 1 & 4 \end{bmatrix}\begin{bmatrix} 2 & 1 & 6 & 0 \\ 0 & 2 & 3 & 4 \\ -1 & 0 & 5 & 8 \end{bmatrix} \quad (2, 4)\text{-entry} = 0 \cdot 0 + 1 \cdot 4 + 4 \cdot 8 = 36$$

Since A is 2×3 and B is 3×4, the product is 2×4.

$$\begin{bmatrix} 3 & -1 & 2 \\ 0 & 1 & 4 \end{bmatrix}\begin{bmatrix} 2 & 1 & 6 & 0 \\ 0 & 2 & 3 & 4 \\ -1 & 0 & 5 & 8 \end{bmatrix} = \begin{bmatrix} 4 & 1 & 25 & 12 \\ -4 & 2 & 23 & 36 \end{bmatrix}$$

Example 3

If $A = [1 \ 3 \ 2]$ and $B = \begin{bmatrix} 5 \\ 6 \\ 4 \end{bmatrix}$, compute A^2, AB, BA, and B^2 when they are defined.

Solution   Here, A is a 1×3 matrix and B is a 3×1 matrix, so A^2 and B^2 are not defined. However, the rule reads

$$\begin{matrix} A & B \\ 1 \times 3 & 3 \times 1 \end{matrix} \quad \text{and} \quad \begin{matrix} B & A \\ 3 \times 1 & 1 \times 3 \end{matrix}$$

so both AB and BA can be formed and these are 1×1 and 3×3 matrices, respectively.

$$AB = [1 \ 3 \ 2] \begin{bmatrix} 5 \\ 6 \\ 4 \end{bmatrix} = [1 \cdot 5 + 3 \cdot 6 + 2 \cdot 4] = [31]$$

$$BA = \begin{bmatrix} 5 \\ 6 \\ 4 \end{bmatrix} [1 \ 3 \ 2] = \begin{bmatrix} 5 \cdot 1 & 5 \cdot 3 & 5 \cdot 2 \\ 6 \cdot 1 & 6 \cdot 3 & 6 \cdot 2 \\ 4 \cdot 1 & 4 \cdot 3 & 4 \cdot 2 \end{bmatrix} = \begin{bmatrix} 5 & 15 & 10 \\ 6 & 18 & 12 \\ 4 & 12 & 8 \end{bmatrix}$$

Unlike numerical multiplication, matrix products AB and BA *need not be equal*. In fact they need not even be the same size, as Example 3 shows. It turns out to be rare that $AB = BA$ (although it is by no means impossible). A and B are said to **commute** when this happens.

Example 4

Let $A = \begin{bmatrix} 6 & 9 \\ -4 & -6 \end{bmatrix}$ and $B = \begin{bmatrix} 1 & 2 \\ -1 & 0 \end{bmatrix}$. Compute A^2, AB, and BA.

Solution $A^2 = \begin{bmatrix} 6 & 9 \\ -4 & -6 \end{bmatrix} \begin{bmatrix} 6 & 9 \\ -4 & -6 \end{bmatrix} = \begin{bmatrix} 0 & 0 \\ 0 & 0 \end{bmatrix}$, so $A^2 = 0$ can occur even if $A \neq 0$. Next,

$$AB = \begin{bmatrix} 6 & 9 \\ -4 & -6 \end{bmatrix} \begin{bmatrix} 1 & 2 \\ -1 & 0 \end{bmatrix} = \begin{bmatrix} -3 & 12 \\ 2 & -8 \end{bmatrix}$$

$$BA = \begin{bmatrix} 1 & 2 \\ -1 & 0 \end{bmatrix} \begin{bmatrix} 6 & 9 \\ -4 & -6 \end{bmatrix} = \begin{bmatrix} -2 & -3 \\ -6 & -9 \end{bmatrix}$$

Hence $AB \neq BA$, even though AB and BA are the same size.

The number 1 plays a neutral role in numerical multiplication in the sense that $1 \cdot a = a$ and $a \cdot 1 = a$ for all numbers a. An analogous role for matrix multiplication is played by square matrices of the following types:

$$\begin{bmatrix} 1 & 0 \\ 0 & 1 \end{bmatrix}, \quad \begin{bmatrix} 1 & 0 & 0 \\ 0 & 1 & 0 \\ 0 & 0 & 1 \end{bmatrix}, \quad \begin{bmatrix} 1 & 0 & 0 & 0 \\ 0 & 1 & 0 & 0 \\ 0 & 0 & 1 & 0 \\ 0 & 0 & 0 & 1 \end{bmatrix}, \quad \text{and so on.}$$

In general, an **identity matrix** I is a *square matrix with 1's on the main diagonal and zeros elsewhere*. If it is important to stress the size of an $n \times n$ identity matrix, we shall denote it by I_n; however, these matrices are usually written simply as I. Identity matrices play a neutral role with respect to matrix multiplication in the sense that

$$AI = A \quad \text{and} \quad IB = B$$

whenever the products are defined. These and other properties of matrix multiplication are collected in the following theorem. The proof is at the end of this section.

Theorem 1

Assume that k is an arbitrary scalar and that A, B, and C are matrices of sizes such that the indicated operations can be performed.

1. $IA = A$, $BI = B$.
2. $A(BC) = (AB)C$.
3. $A(B + C) = AB + AC$; $A(B - C) = AB - AC$.
4. $(B + C)A = BA + CA$; $(B - C)A = BA - CA$.
5. $k(AB) = (kA)B = A(kB)$.
6. $(AB)^T = B^T A^T$.

Property 2 in Theorem 1 asserts that the **associative law** $A(BC) = (AB)C$ holds for all matrices (if the products are defined). Hence, the product is the same no matter how it is formed and so is simply written ABC. This extends: The product $ABCD$ of four matrices can be formed several ways—for example, $(AB)(CD)$, $[A(BC)]D$, and $A[B(CD)]$—but property 2 implies that they are all equal and so are written simply as $ABCD$. A similar remark applies in general: Matrix products can be written unambiguously with no parentheses.

However, a note of caution about matrix multiplication *is* in order. The fact that AB and BA need *not* be equal means that the *order* of the factors is important in a product of matrices. For example, $ABCD$ and $ADCB$ may *not* be equal.

Warning

If the order of the factors in a product of matrices is changed, the product matrix may change (or may not exist).

Ignoring this warning is a source of many errors by students of linear algebra!

Properties 3 and 4 in Theorem 1 are called the **distributive laws**, and they extend to more than two terms. For example,

$$A(B - C + D - E) = AB - AC + AD - AE$$
$$(A + C - D)B = AB + CB - DB$$

Note again that the warning is in effect: For example, $A(B - C)$ need *not* equal $AB - CA$. Together with property 5 of Theorem 1, the distributive laws make possible a lot of simplification of matrix expressions.

Example 5

Simplify the expression $A(BC - CD) + A(C - B)D - AB(C - D)$.

Solution

$$A(BC - CD) + A(C - B)D - AB(C - D)$$
$$= ABC - ACD + (AC - AB)D - ABC + ABD$$
$$= ABC - ACD + ACD - ABD - ABC + ABD$$
$$= 0$$

Examples 6 and 7 show how we can use the properties in Theorem 1 to deduce facts about matrix multiplication.

Example 6

> Suppose that A, B, and C are $n \times n$ matrices and that both A and B commute with C; that is, $AC = CA$ and $BC = CB$. Show that AB commutes with C.
>
> *Solution*   Showing that AB commutes with C means verifying that $(AB)C = C(AB)$. The computation uses property 2 of Theorem 1 several times, as well as the given facts that $AC = CA$ and $BC = CB$.
>
> $$(AB)C = A(BC) = A(CB) = (AC)B = (CA)B = C(AB)$$

Example 7

> Show that $AB = BA$ if and only if $(A - B)(A + B) = A^2 - B^2$.
>
> *Solution*   Theorem 1 shows that the following always holds:
>
> $$(A - B)(A + B) = A(A + B) - B(A + B) = A^2 + AB - BA - B^2 \qquad (*)$$
>
> Hence if $AB = BA$, then $(A - B)(A + B) = A^2 - B^2$ follows. Conversely, if this last equation holds, then equation $(*)$ becomes
>
> $$A^2 - B^2 = A^2 + AB - BA - B^2$$
>
> This gives $0 = AB - BA$, and $AB = BA$ follows.

Matrices and Linear Equations

One of the most important motivations for matrix multiplication results from the fact that it allows a system of linear equations to be written as a single matrix equation. To see how, consider the following system of linear equations:

$$\begin{aligned} 3x_1 - 2x_2 + x_3 &= b_1 \\ 2x_1 + x_2 - x_3 &= b_2 \end{aligned}$$

By the definition of matrix equality, this system is the same as the matrix equation

$$\begin{bmatrix} 3x_1 - 2x_2 + x_3 \\ 2x_1 + x_2 - x_3 \end{bmatrix} = \begin{bmatrix} b_1 \\ b_2 \end{bmatrix}$$

Now observe that the matrix on the left can be factored as a product of matrices

$$\begin{bmatrix} 3 & -2 & 1 \\ 2 & 1 & -1 \end{bmatrix} \begin{bmatrix} x_1 \\ x_2 \\ x_3 \end{bmatrix} = \begin{bmatrix} b_1 \\ b_2 \end{bmatrix}$$

If we write $A = \begin{bmatrix} 3 & -2 & 1 \\ 2 & 1 & -1 \end{bmatrix}$, $X = \begin{bmatrix} x_1 \\ x_2 \\ x_3 \end{bmatrix}$ and $B = \begin{bmatrix} b_1 \\ b_2 \end{bmatrix}$, the above system of equations becomes the single matrix equation

$$AX = B$$

In the same way, consider any system of linear equations

$$
\begin{aligned}
a_{11}x_1 + a_{12}x_2 + \cdots + a_{1n}x_n &= b_1 \\
a_{21}x_1 + a_{22}x_2 + \cdots + a_{2n}x_n &= b_2 \\
\vdots \qquad \vdots \qquad\qquad \vdots \quad &\;\; \vdots \\
a_{m1}x_1 + a_{m2}x_2 + \cdots + a_{mn}x_n &= b_m
\end{aligned}
$$

If $A = \begin{bmatrix} a_{11} & a_{12} & \cdots & a_{1n} \\ a_{21} & a_{22} & \cdots & a_{2n} \\ \vdots & \vdots & & \vdots \\ a_{m1} & a_{m2} & \cdots & a_{mn} \end{bmatrix}$, $X = \begin{bmatrix} x_1 \\ x_2 \\ \vdots \\ x_n \end{bmatrix}$ and $B = \begin{bmatrix} b_1 \\ b_2 \\ \vdots \\ b_m \end{bmatrix}$, these equations become the

single matrix equation

$$AX = B$$

This is called the **matrix form** of the system of equations, and B is called the **constant matrix**. As in Section 1.1, A is called the **coefficient matrix** of the system, and a column X_1 is called a **solution** to the system if $AX_1 = B$.

The matrix form is useful for formulating results about solutions of systems of linear equations. Given a system $AX = B$ there is a related system

$$AX = 0$$

called the **associated homogeneous system**. If X_1 is a solution to $AX = B$ and if X_0 is a solution to $AX = 0$, then $X_1 + X_0$ is a solution to $AX = B$. Indeed, $AX_1 = B$ and $AX_0 = 0$, so

$$A(X_1 + X_0) = AX_1 + AX_0 = B + 0 = B$$

This observation has a useful converse.

Theorem 2

Suppose X_1 is a particular solution to the system $AX = B$ of linear equations. Then every solution X_2 to $AX = B$ has the form

$$X_2 = X_0 + X_1$$

for some solution X_0 of the associated homogeneous system $AX = 0$.

PROOF

Suppose that X_2 is *any* solution to $AX = B$ so that $AX_2 = B$. Write $X_0 = X_2 - X_1$. Then $X_2 = X_0 + X_1$, and we compute:

$$AX_0 = A(X_2 - X_1) = AX_2 - AX_1 = B - B = 0$$

Thus X_0 is a solution to the associated homogeneous system $AX = 0$.

The importance of Theorem 2 lies in the fact that sometimes a particular solution X_1 is easily found, and so the problem of finding *all* solutions is reduced to solving the associated homogeneous system.

Example 8

Express every solution to the following system as the sum of a specific solution plus a solution to the associated homogeneous system.

$$\begin{aligned} x - y - z &= 2 \\ 2x - y - 3z &= 6 \\ x \quad\quad - 2z &= 4 \end{aligned}$$

Solution   Gaussian elimination gives $x = 4 + 2t$, $y = 2 + t$, $z = t$, where t is arbitrary. Hence the general solution can be written

$$X = \begin{bmatrix} x \\ y \\ z \end{bmatrix} = \begin{bmatrix} 4 + 2t \\ 2 + t \\ t \end{bmatrix} = \begin{bmatrix} 4 \\ 2 \\ 0 \end{bmatrix} + t \begin{bmatrix} 2 \\ 1 \\ 1 \end{bmatrix}$$

Thus $X_0 = \begin{bmatrix} 4 \\ 2 \\ 0 \end{bmatrix}$ is a specific solution, and $X_1 = t \begin{bmatrix} 2 \\ 1 \\ 1 \end{bmatrix}$ gives *all* solutions to the associated homogeneous system (do the gaussian elimination with all the constants zero).

Theorem 2 focuses attention on homogeneous systems. In that case there is a convenient matrix form for the solutions that will be needed later. Example 9 provides an illustration.

Example 9

Solve the homogeneous system $AX = 0$ where

$Ax = B$

$$A = \begin{bmatrix} 1 & -2 & 3 & -2 \\ -3 & 6 & 1 & 0 \\ -2 & 4 & 4 & -2 \end{bmatrix}$$

Solution   The reduction of the augmented matrix to reduced form is

$$\begin{bmatrix} 1 & -2 & 3 & -2 & | & 0 \\ -3 & 6 & 1 & 0 & | & 0 \\ -2 & 4 & 4 & -2 & | & 0 \end{bmatrix} \rightarrow \begin{bmatrix} 1 & -2 & 0 & -\frac{1}{5} & | & 0 \\ 0 & 0 & 1 & -\frac{3}{5} & | & 0 \\ 0 & 0 & 0 & 0 & | & 0 \end{bmatrix}$$

so the solutions are $x_1 = 2s + \frac{1}{5}t$, $x_2 = s$, $x_3 = \frac{3}{5}t$, and $x_4 = t$ by gaussian elimination. Hence we can write the general solution X in the matrix form

$$X = \begin{bmatrix} x_1 \\ x_2 \\ x_3 \\ x_4 \end{bmatrix} = \begin{bmatrix} 2s + \frac{1}{5}t \\ s \\ \frac{3}{5}t \\ t \end{bmatrix} = s \begin{bmatrix} 2 \\ 1 \\ 0 \\ 0 \end{bmatrix} + t \begin{bmatrix} \frac{1}{5} \\ 0 \\ \frac{3}{5} \\ 1 \end{bmatrix} = sX_1 + tX_2$$

where $X_1 = [2\ 1\ 0\ 0]^T$ and $X_2 = [\frac{1}{5}\ 0\ \frac{3}{5}\ 1]^T$ are particular solutions determined by the gaussian algorithm.

The solutions X_1 and X_2 in Example 9 are called **basic solutions** to the homogeneous system, and a solution of the form $sX_1 + tX_2$ is called a **linear combination** of X_1 and X_2. In Example 9 the general solution X becomes

$$X = s \begin{bmatrix} 2 \\ 1 \\ 0 \\ 0 \end{bmatrix} + t \begin{bmatrix} \frac{1}{5} \\ 0 \\ \frac{3}{5} \\ 1 \end{bmatrix} = s \begin{bmatrix} 2 \\ 1 \\ 0 \\ 0 \end{bmatrix} + \frac{t}{5} \begin{bmatrix} 1 \\ 0 \\ 3 \\ 5 \end{bmatrix}$$

Hence by introducing a new parameter $r = t/5$ we can multiply the original basic solution $X_2 = [\frac{1}{5} \ 0 \ \frac{3}{5} \ 1]^T$ by 5 and so eliminate fractions. Thus

Definition   Any nonzero scalar multiple of a basic solution will be called a basic solution.

In the same way, the gaussian algorithm produces basic solutions to *every* homogeneous system $AX = 0$, one for each parameter (there are no basic solutions if the system has only the trivial solution). Moreover every solution is given by the algorithm as a linear combination of these basic solutions (as in Example 9). If A has rank r, Theorem 2 §1.2 shows that there are exactly $n - r$ parameters, and so $n - r$ basic solutions. This proves:

Theorem 3

Let A be an $m \times n$ matrix of rank r, and consider the homogeneous system $AX = 0$ in n variables. Then:
1. The system has exactly $n - r$ basic solutions, one for each parameter.
2. Every solution is a linear combination of these basic solutions.

Example 10

Find basic solutions of the system $AX = 0$ and express every solution as a linear combination of the basic solutions, where

$$A = \begin{bmatrix} 1 & -3 & 0 & 2 & 2 \\ -2 & 6 & 1 & 2 & -5 \\ 3 & -9 & -1 & 0 & 7 \\ -3 & 9 & 2 & 6 & -8 \end{bmatrix}$$

Solution   The reduction of the augmented matrix to reduced row-echelon form is

$$\begin{bmatrix} 1 & -3 & 0 & 2 & 2 & | & 0 \\ -2 & 6 & 1 & 2 & -5 & | & 0 \\ 3 & -9 & -1 & 0 & 7 & | & 0 \\ -3 & 9 & 2 & 6 & -8 & | & 0 \end{bmatrix} \rightarrow \begin{bmatrix} 1 & -3 & 0 & 2 & 2 & | & 0 \\ 0 & 0 & 1 & 6 & -1 & | & 0 \\ 0 & 0 & 0 & 0 & 0 & | & 0 \\ 0 & 0 & 0 & 0 & 0 & | & 0 \end{bmatrix}$$

so the general solution is $x_1 = 3r - 2s - 2t$, $x_2 = r$, $x_3 = -6s + t$, $x_4 = s$, and $x_5 = t$ where r, s, and t are parameters. In matrix form this is

$$X = \begin{bmatrix} x_1 \\ x_2 \\ x_3 \\ x_4 \\ x_5 \end{bmatrix} = \begin{bmatrix} 3r - 2s - 2t \\ r \\ -6s + t \\ s \\ t \end{bmatrix} = r \begin{bmatrix} 3 \\ 1 \\ 0 \\ 0 \\ 0 \end{bmatrix} + s \begin{bmatrix} -2 \\ 0 \\ -6 \\ 1 \\ 0 \end{bmatrix} + t \begin{bmatrix} -2 \\ 0 \\ 1 \\ 0 \\ 1 \end{bmatrix}$$

Hence basic solutions are $X_1 = [3 \ 1 \ 0 \ 0 \ 0]^T$, $X_2 = [-2 \ 0 \ -6 \ 1 \ 0]^T$, and $X_3 = [-2 \ 0 \ 1 \ 0 \ 1]^T$.

Block Multiplication

When forming a matrix product BA it is often convenient to view the matrix as a row of columns. If A is $m \times n$ and $C_1, C_2, \ldots, C_n$ are the columns of A, we write

$$A = [C_1 \ C_2 \ \cdots \ C_n]$$

Now observe that column j of BA is BC_j by the definition of matrix multiplication (verify). Hence we obtain the matrix BA in terms of its columns:

$$BA = B[C_1 \ C_2 \ \cdots \ C_n] = [BC_1 \ BC_2 \ \cdots \ BC_n]$$

This is an example of block multiplication. As another illustration, let $A = \begin{bmatrix} a_1 & a_2 & a_3 \\ b_1 & b_2 & b_3 \end{bmatrix}$ be 2×3, so that $C_1 = \begin{bmatrix} a_2 \\ b_2 \end{bmatrix}$, $C_2 = \begin{bmatrix} a_2 \\ b_2 \end{bmatrix}$ and $C_3 = \begin{bmatrix} a_3 \\ b_3 \end{bmatrix}$. If $X = \begin{bmatrix} x_1 \\ x_2 \\ x_3 \end{bmatrix}$ and we claim that

$$AX = [C_1 \ C_2 \ C_3] \begin{bmatrix} x_1 \\ x_2 \\ x_3 \end{bmatrix} = x_1 C_1 + x_2 C_2 + x_3 C_3$$

Indeed:

$$AX = \begin{bmatrix} a_1 & a_2 & a_3 \\ b_1 & b_2 & b_3 \end{bmatrix} \begin{bmatrix} x_1 \\ x_2 \\ x_3 \end{bmatrix} = \begin{bmatrix} a_1 x_1 + a_2 x_2 + a_3 x_3 \\ b_1 x_1 + b_2 x_2 + b_3 x_3 \end{bmatrix}$$

$$= x_1 \begin{bmatrix} a_1 \\ b_1 \end{bmatrix} + x_2 \begin{bmatrix} a_2 \\ b_2 \end{bmatrix} + x_3 \begin{bmatrix} a_3 \\ b_3 \end{bmatrix} = x_1 C_1 + x_2 C_2 + x_3 C_3$$

Hence $AX = [C_1 \ C_2 \ C_3] \begin{bmatrix} x_1 \\ x_2 \\ x_3 \end{bmatrix} = x_1 C_1 + x_2 C_2 + x_3 C_3$ can be computed using ordinary matrix multiplication except that we view A as a 1×3 matrix with the columns C_j as entries. This holds in general and is recorded in the following theorem. The result will be used several times in this book.

Theorem 4

Let $A = [C_1 \ C_2 \ \cdots \ C_n]$ be an $m \times n$ matrix with columns $C_1, C_2, \ldots, C_n$. If $X = [x_1 \ x_2 \ \cdots \ x_n]^T$ is any column, then

$$AX = [C_1 \ C_2 \ \cdots \ C_n] \begin{bmatrix} x_1 \\ x_2 \\ \vdots \\ x_n \end{bmatrix} = x_1 C_1 + x_2 C_2 + \cdots + x_n C_n$$

These are special cases of a more general way of looking at matrices that, among its other uses, can greatly simplify matrix multiplications. The idea is to partition a matrix A into smaller matrices (called **blocks**) by inserting vertical lines between the columns and horizontal lines between the rows.

As an example, consider the matrices

$$
A = \begin{bmatrix} 1 & 0 & 0 & 0 & 0 \\ 0 & 1 & 0 & 0 & 0 \\ 2 & -1 & 4 & 2 & 1 \\ 3 & 1 & -1 & 7 & 5 \end{bmatrix} = \begin{bmatrix} I_2 & 0_{23} \\ P & Q \end{bmatrix} \quad \text{and} \quad B = \begin{bmatrix} 4 & -2 \\ 5 & 6 \\ 7 & 3 \\ -1 & 0 \\ 1 & 6 \end{bmatrix} = \begin{bmatrix} X \\ Y \end{bmatrix}
$$

where the blocks have been labelled as indicated. This is a natural way to think of A in view of the blocks I_2 and 0_{23} that occur. This notation is particularly useful when we are multiplying the matrices A and B because the product AB can be computed in block form as follows:

$$
AB = \begin{bmatrix} I & 0 \\ P & Q \end{bmatrix}\begin{bmatrix} X \\ Y \end{bmatrix} = \begin{bmatrix} IX + 0Y \\ PX + QY \end{bmatrix} = \begin{bmatrix} X \\ PX + QY \end{bmatrix} = \begin{bmatrix} 4 & -2 \\ 5 & 6 \\ 30 & 8 \\ 8 & 27 \end{bmatrix}
$$

This is easily checked to be the product AB, computed in the conventional manner.

In other words, *we can compute the product AB by ordinary matrix multiplication, using blocks as entries.* The only requirement is that the blocks be **compatible**. That is, *the sizes of the blocks must be such that all (matrix) products of blocks that occur make sense.* This means that the number of columns in each block of A must equal the number of rows in the corresponding block of B.

Block Multiplication

If matrices A and B are partitioned compatibly into blocks, the product AB can be computed by matrix multiplication using blocks as entries.

We omit the proof and instead give one more example of block multiplication that will be used below.

Theorem 5

Suppose that matrices $A = \begin{bmatrix} B & X \\ 0 & C \end{bmatrix}$ and $A_1 = \begin{bmatrix} B_1 & X_1 \\ 0 & C_1 \end{bmatrix}$ are partitioned as shown ,

where B and B_1 are square matrices of the same size, and C and C_1 are square of the same size. These are compatible partitionings and block multiplication gives

$$
AA_1 = \begin{bmatrix} B & X \\ 0 & C \end{bmatrix}\begin{bmatrix} B_1 & X_1 \\ 0 & C_1 \end{bmatrix} = \begin{bmatrix} BB_1 & BX_1 + XC_1 \\ 0 & CC_1 \end{bmatrix}
$$

Block multiplication is useful in computing products of matrices in a computer with limited memory capacity. The matrices are partitioned into blocks in such a way that each product of blocks can be handled. Then the blocks are stored in auxiliary memory (on tape, for example), and their products are computed one by one.

Directed Graphs

The study of directed graphs gives another illustration of how matrix multiplication arises in ways other than the study of linear equations.

A directed graph consists of a set of points (called **vertices**) connected by arrows (called **edges**). For example, the vertices could represent cities and the edges available flights. If the graph has n vertices $v_1, v_2, \ldots, v_n$, the **adjacency** matrix $A = [a_{ij}]$ is the $n \times n$ matrix whose (i, j)-entry a_{ij} is 1 if there is an edge from v_j to v_i, and zero otherwise. For example, the adjacency matrix of the directed graph shown is

$A = \begin{bmatrix} 1 & 1 & 0 \\ 1 & 0 & 1 \\ 1 & 0 & 0 \end{bmatrix}$. A **path of length** r (or an r-path) from vertex j to vertex i is a

sequence of r edges leading from v_j to v_i. Thus $v_1 \to v_2 \to v_1 \to v_1 \to v_3$ is a 4-path from v_1 to v_3. The edges are just the paths of length 1, so the (i, j)-entry a_{ij} of the adjacency matrix A is the number of 1-paths from v_j to v_i. This observation has an important extension:

Theorem 6

If A is the adjacency matrix of a directed graph with n vertices, then the (i, j)-entry of A^r is the number of r-paths $v_j \to v_i$.

As an illustration, consider the adjacency matrix A in the graph shown. Then

$$A = \begin{bmatrix} 1 & 1 & 0 \\ 1 & 0 & 1 \\ 1 & 0 & 0 \end{bmatrix}, \quad A^2 = \begin{bmatrix} 2 & 1 & 1 \\ 2 & 1 & 0 \\ 1 & 1 & 0 \end{bmatrix} \quad \text{and} \quad A^3 = \begin{bmatrix} 4 & 2 & 1 \\ 3 & 2 & 1 \\ 2 & 1 & 1 \end{bmatrix}.$$

Hence, since the (2, 1)-entry of A^2 is 2, there are two 2-paths $v_1 \to v_2$ (in fact $v_1 \to v_1 \to v_2$ and $v_1 \to v_3 \to v_2$). Similarly, the (2, 3)-entry of A^2 is zero, so there are *no* 2-paths $v_3 \to v_2$, as the reader can verify. The fact that no entry of A^3 is zero shows that it is possible to go from any vertex to any other vertex in exactly three steps.

Before giving the proof of Theorem 6, we must state the definition of matrix multiplication more formally. If $A = [a_{ij}]$ is $m \times n$ and $B = [b_{ij}]$ is $n \times p$, row i of A and column j of B are, respectively,

$$[a_{i1} \ a_{i2} \ \cdots \ a_{in}] \quad \text{and} \quad [b_{1j} \ b_{2j} \ \cdots \ b_{nj}]^T.$$

Hence the (i, j)-entry of the product matrix AB is the dot product

$$a_{i1}b_{1j} + a_{i2}b_{2j} + \cdots + a_{in}b_{nj} = \sum_{k=1}^{n} a_{ik}b_{kj} \tag{$*$}$$

where summation notation has been introduced for convenience.[2]

PROOF OF THEOREM 6

Write $A = [a_{ij}]$. Theorem is clear if $r = 1$, so assume inductively that it is true for $r \ge 1$. Write $A^r = [b_{ij}]$ so that there are b_{ij} r-paths $v_j \to v_i$. We must show that the

---

2 Summation notation is a convenient shorthand way to write sums of similar expressions. For example $a_1 + a_2 + a_3 + a_4 = \sum_{i=1}^{4} a_i$, $a_5b_5 + a_6b_6 + a_7b_7 + a_8b_8 = \sum_{k=5}^{8} a_kb_k$, and $1^2 + 2^2 + 3^2 + 4^2 + 5^2 = \sum_{j=1}^{5} j^2$.

(i, j)-entry of A^{r+1} is the number of ($r + 1$)-paths from v_j to v_i. But every such path is the result of an r-path $v_j \to v_k$ for some k, followed by a 1-path $v_k \to v_i$, and there are b_{kj} of the former and a_{ik} of the latter. Hence there are $b_{kj}a_{ik} = a_{ik}b_{kj}$ paths of length $r + 1$ from v_j to v_i via v_k. Summing over k, this shows that there are $a_{i1}b_{1j} + a_{i2}b_{2j} + \cdots + a_{in}b_{nj}$ such paths in all. Since this sum is the (i, j)-entry of $AA^r = A^{r+1}$, this completes the proof.

PROOF OF THEOREM 1

Property 1 is routine—try it for 2×2 and 3×3 matrices, and you will see how to do the general case. *Property 5* is a routine consequence of ($*$).

Property 2. Assume that C is of size $p \times q$ and write $C = [c_{ij}]$ and $BC = [d_{ij}]$. Then the (i, j)-entry of $A(BC)$ is $\sum_{k=1}^{n} a_{ik}d_{kj}$ by ($*$). Now $d_{kj} = \sum_{l=1}^{p} b_{kl}c_{lj}$, so the ($i$, j)-entry of $A(BC)$ is

$$\sum_{k=1}^{n} a_{ik}d_{kj} = \sum_{k=1}^{n} a_{ik}\left(\sum_{l=1}^{p} b_{kl}c_{lj}\right) = \sum_{k=1}^{n}\sum_{l=1}^{p} a_{ik}b_{kl}c_{lj}.$$

A similar argument shows that this double sum is the (i, j)-entry of $(AB)C$.

Properties 3 and 4. We prove only property 3; property 4 is analogous.

Write $A = [a_{ij}]$, $B = [b_{ij}]$, and $C = [c_{ij}]$ and assume that A is $m \times n$ and that B and C are $n \times p$. Then $B + C = [b_{ij} + c_{ij}]$, so the ($i$, j)-entry of $A(B + C)$ is

$$\sum_{k=1}^{n} a_{ik}(b_{kj} + c_{kj}) = \sum_{k=1}^{n}(a_{ik}b_{kj} + a_{ik}c_{kj}) = \sum_{k=1}^{n} a_{ik}b_{kj} + \sum_{k=1}^{n} a_{ik}c_{kj}$$

This is the (i, j)-entry of $AB + AC$ because the sums on the right are the (i, j)-entries of AB and AC, respectively. Hence $A(B + C) = AB + AC$; the other equation in property 3 is proved similarly.

Property 6. Write $A^T = [a'_{ij}]$ and $B^T = [b'_{ij}]$, where $a'_{ij} = a_{ji}$ and $b'_{ij} = b_{ji}$. If B^T and A^T are $p \times n$ and $n \times m$, respectively, the (i, j)-entry of $B^T A^T$ is

$$\sum_{k=1}^{n} b'_{ik}a'_{kj} = \sum_{k=1}^{n} b_{ki}a_{jk} = \sum_{k=1}^{n} a_{jk}b_{ki}$$

This is the (j, i)-entry of AB—that is, the (i, j)-entry of $(AB)^T$. Hence $B^T A^T = (AB)^T$.

Exercises 2.2

1. Compute the following matrix products.

(a) $\begin{bmatrix} 1 & 3 \\ 0 & -2 \end{bmatrix}\begin{bmatrix} 2 & -1 \\ 0 & 1 \end{bmatrix}$

♦(b) $\begin{bmatrix} 1 & -1 & 2 \\ 2 & 0 & 4 \end{bmatrix}\begin{bmatrix} 2 & 3 & 1 \\ 1 & 9 & 7 \\ -1 & 0 & 2 \end{bmatrix}$

(c) $\begin{bmatrix} 5 & 0 & -7 \\ 1 & 5 & 9 \end{bmatrix}\begin{bmatrix} 3 \\ 1 \\ -1 \end{bmatrix}$

♦(d) $\begin{bmatrix} 1 & 3 & -3 \end{bmatrix}\begin{bmatrix} 3 & 0 \\ -2 & 1 \\ 0 & 6 \end{bmatrix}$

(e) $\begin{bmatrix} 1 & 0 & 0 \\ 0 & 1 & 0 \\ 0 & 0 & 1 \end{bmatrix}\begin{bmatrix} 3 & -2 \\ 5 & -7 \\ 9 & 7 \end{bmatrix}$

♦(f) $\begin{bmatrix} 1 & -1 & 3 \end{bmatrix}\begin{bmatrix} 2 \\ 1 \\ -8 \end{bmatrix}$

(g) $\begin{bmatrix} 2 \\ 1 \\ -7 \end{bmatrix}\begin{bmatrix} 1 & -1 & 3 \end{bmatrix}$

♦(h) $\begin{bmatrix} 3 & 1 \\ 5 & 2 \end{bmatrix}\begin{bmatrix} 2 & -1 \\ -5 & 3 \end{bmatrix}$

(i) $\begin{bmatrix} 2 & 3 & 1 \\ 5 & 7 & 4 \end{bmatrix} \begin{bmatrix} a & 0 & 0 \\ 0 & b & 0 \\ 0 & 0 & c \end{bmatrix}$ ◆(j) $\begin{bmatrix} a & 0 & 0 \\ 0 & b & 0 \\ 0 & 0 & c \end{bmatrix} \begin{bmatrix} a' & 0 & 0 \\ 0 & b' & 0 \\ 0 & 0 & c' \end{bmatrix}$

2. In each of the following cases, find all possible products A^2, AB, AC, and so on.

(a) $A = \begin{bmatrix} 1 & 2 & 3 \\ -1 & 0 & 0 \end{bmatrix}$, $B = \begin{bmatrix} 1 & -2 \\ \frac{1}{2} & 3 \end{bmatrix}$, $C = \begin{bmatrix} -1 & 0 \\ 2 & 5 \\ 0 & 3 \end{bmatrix}$

◆(b) $A = \begin{bmatrix} 1 & 2 & 4 \\ 0 & 1 & -1 \end{bmatrix}$, $B = \begin{bmatrix} -1 & 6 \\ 1 & 0 \end{bmatrix}$, $C = \begin{bmatrix} 2 & 0 \\ -1 & 1 \\ 1 & 2 \end{bmatrix}$

3. Find a, b, a_1, and b_1 if:

(a) $\begin{bmatrix} a & b \\ a_1 & b_1 \end{bmatrix} \begin{bmatrix} 3 & -5 \\ -1 & 2 \end{bmatrix} = \begin{bmatrix} 1 & -1 \\ 2 & 0 \end{bmatrix}$

◆(b) $\begin{bmatrix} 2 & 1 \\ -1 & 2 \end{bmatrix} \begin{bmatrix} a & b \\ a_1 & b_1 \end{bmatrix} = \begin{bmatrix} 7 & 2 \\ -1 & 4 \end{bmatrix}$

4. Verify that $A^2 - A - 6I = 0$ if:

(a) $A = \begin{bmatrix} 3 & -1 \\ 0 & -2 \end{bmatrix}$ ◆(b) $A = \begin{bmatrix} 2 & 2 \\ 2 & -1 \end{bmatrix}$

5. Given $A = \begin{bmatrix} 1 & -1 \\ 0 & 1 \end{bmatrix}$, $B = \begin{bmatrix} 1 & 0 & -2 \\ 3 & 1 & 0 \end{bmatrix}$, $C = \begin{bmatrix} 1 & 0 \\ 2 & 1 \\ 5 & 8 \end{bmatrix}$,

and $D = \begin{bmatrix} 3 & -1 & 2 \\ 1 & 0 & 5 \end{bmatrix}$, verify the following facts

from Theorem 1.

(a) $A(B - D) = AB - AD$ ◆(b) $A(BC) = (AB)C$

(c) $(CD)^T = D^T C^T$

6. Let A be a 2×2 matrix.

(a) If A commutes with $\begin{bmatrix} 0 & 1 \\ 0 & 0 \end{bmatrix}$, show that $A = \begin{bmatrix} a & b \\ 0 & a \end{bmatrix}$ for some a and b.

(b) If A commutes with $\begin{bmatrix} 0 & 0 \\ 1 & 0 \end{bmatrix}$, show, that $A = \begin{bmatrix} a & 0 \\ c & a \end{bmatrix}$ for some a and c.

(c) Show that A commutes with *every* 2×2 matrix if and only if $A = \begin{bmatrix} a & 0 \\ 0 & a \end{bmatrix}$ for some a.

7. Write each of the following systems of linear equations in matrix form.

(a) $3x_1 + 2x_2 - x_3 + x_4 = 1$
$x_1 - x_2 + 3x_4 = 0$
$2x_1 - x_2 - x_3 = 5$

◆(b) $-x_1 + 2x_2 - x_3 + x_4 = 6$
$2x_1 + x_2 - x_3 + 2x_4 = 1$
$3x_1 - 2x_2 + x_4 = 0$

8. In each case, express every solution of the system as a sum of a specific solution plus a solution of the associated homogeneous system.

(a) $x + y + z = 2$
$2x + y = 3$
$x - y - 3z = 0$

◆(b) $x - y - 4z = -4$
$x + 2y + 5z = 2$
$x + y + 2z = 0$

(c) $x_1 + x_2 - x_3 - 5x_5 = 2$
$x_2 + x_3 - 4x_5 = -1$
$x_2 + x_3 + x_4 - x_5 = -1$
$2x_1 - 4x_3 + x_4 + x_5 = 6$

◆(d) $2x_1 + x_2 - x_3 - x_4 = -1$
$3x_1 + x_2 + x_3 - 2x_4 = -2$
$-x_1 - x_2 + 2x_3 + x_4 = 2$
$-2x_1 - x_2 + 2x_4 = 3$

9. If X_0 and X_1 are solutions to the homogeneous system of equations $AX = 0$, use Theorem 1 to show that $sX_0 + tX_1$ is also a solution for any scalars s and t (called a **linear combination** of X_0 and X_1).

10. In each of the following, find the basic solutions, and write the general solution as a linear combination of the basic solutions.

(a) $x_1 + 2x_2 - x_3 + 2x_4 + x_5 = 0$
$x_1 + 2x_2 + 2x_3 + x_5 = 0$
$2x_1 + 4x_2 - 2x_3 + 3x_4 + x_5 = 0$

◆(b) $x_1 + x_2 - 2x_3 + 3x_4 + 2x_5 = 0$
$2x_1 - x_2 + 3x_3 + 4x_4 + x_5 = 0$
$-x_1 - 2x_2 + 3x_3 + x_4 = 0$
$3x_1 + x_3 + 7x_4 + 2x_5 = 0$

11. Assume that $A \begin{bmatrix} 1 \\ -1 \\ 2 \end{bmatrix} = 0 = A \begin{bmatrix} 2 \\ 0 \\ 3 \end{bmatrix}$, and that

$AX = B$ has a solution $X_0 = \begin{bmatrix} 2 \\ -1 \\ 3 \end{bmatrix}$ Find a

two-parameter family of solutions to $AX = B$.

12. (a) If A^2 can be formed, what can be said about the size of A?

   ♦(b) If AB and BA can both be formed, describe the sizes of A and B.

   (c) If ABC can be formed, A is 3×3, and C is 5×5, what size is B?

13. (a) Find two 2×2 matrices A such that $A^2 = 0$.

   ♦(b) Find three 2×2 matrices A such that
   (i) $A^2 = I$; (ii) $A^2 = A$.

   (c) Find 2×2 matrices A and B such that $AB = 0$ but $BA \neq 0$.

14. Write $P = \begin{bmatrix} 1 & 0 & 0 \\ 0 & 0 & 1 \\ 0 & 1 & 0 \end{bmatrix}$, and let A be $3 \times n$ and B be $m \times 3$.

   (a) Describe PA in terms of the rows of A.

   (b) Describe BP in terms of the columns of B.

15. Let A, B, and C be as in Exercise 5. Find the $(3, 1)$-entry of CAB using exactly six numerical multiplications.

16. (a) Compute AB, using the indicated block partitioning.

$$A = \begin{bmatrix} 2 & -1 & 3 & 1 \\ 1 & 0 & 1 & 2 \\ \hline 0 & 0 & 1 & 0 \\ 0 & 0 & 0 & 1 \end{bmatrix} \quad B = \begin{bmatrix} 1 & 2 & 0 \\ -1 & 0 & 0 \\ \hline 0 & 5 & 1 \\ 1 & -1 & 0 \end{bmatrix}$$

   ♦(b) Partition A and B in part (a) differently and compute AB again.

   (c) Find A^2 using the partitioning in part (a) and then again using a different partitioning.

17. In each case give formulas for all powers A, A^2, A^3, ... of A using the block decomposition indicated.

   (a) $A = \begin{bmatrix} 1 & 0 & 0 \\ \hline 1 & 1 & -1 \\ 1 & -1 & 1 \end{bmatrix}$ ♦(b) $A = \begin{bmatrix} 1 & -1 & 2 & -1 \\ 0 & 1 & 0 & 0 \\ \hline 0 & 0 & -1 & 1 \\ 0 & 0 & 0 & 1 \end{bmatrix}$

18. Compute the following using block multiplication (all blocks are $k \times k$).

   (a) $\begin{bmatrix} I & X \\ -Y & I \end{bmatrix}\begin{bmatrix} I & 0 \\ Y & I \end{bmatrix}$ ♦(b) $\begin{bmatrix} I & X \\ 0 & I \end{bmatrix}\begin{bmatrix} I & -X \\ 0 & I \end{bmatrix}$

   (c) $[I \ X][I \ X]^T$ ♦(d) $[I \ X^T][-X \ I]^T$

   (e) $\begin{bmatrix} I & X \\ 0 & -I \end{bmatrix}^n$, any $n \geq 1$ ♦(f) $\begin{bmatrix} 0 & X \\ I & 0 \end{bmatrix}^n$, any $n \geq 1$

19. (a) If A has a row of zeros, show that the same is true of AB for any B.

   (b) If B has a column of zeros, show that the same is true of AB for any A.

20. Let A denote an $m \times n$ matrix.

   (a) If $AX = 0$ for every $n \times 1$ matrix X, show that $A = 0$.

   ♦(b) If $YA = 0$ for every $1 \times m$ matrix Y, show that $A = 0$.

21. (a) If $U = \begin{bmatrix} 1 & 2 \\ 0 & -1 \end{bmatrix}$, and $AU = 0$, show that $A = 0$.

   (b) Let U be such that $AU = 0$ implies that $A = 0$. If $PU = QU$, show that $P = Q$.

22. Simplify the following expressions where A, B, and C represent matrices.

   (a) $A(3B - C) + (A - 2B)C + 2B(C + 2A)$

   ♦(b) $A(B + C - D) + B(C - A + D) - (A + B)C + (A - B)D$

   (c) $AB(BC - CB) + (CA - AB)BC + CA(A - B)C$

   ♦(d) $(A - B)(C - A) + (C - B)(A - C) + (C - A)^2$

23. If $A = \begin{bmatrix} a & b \\ c & d \end{bmatrix}$ where $a \neq 0$, show that A factors in the form $A = \begin{bmatrix} 1 & 0 \\ x & 1 \end{bmatrix}\begin{bmatrix} y & z \\ 0 & w \end{bmatrix}$.

24. If A and B commute with C, show that the same is true of:

   (a) $A + B$ ♦(b) kA, k any scalar

25. If A is any matrix, show that AA^T and A^TA are symmetric.

♦26. If A and B are symmetric, show that AB is symmetric if and only if $AB = BA$.

27. If A is a 2×2 matrix, show that $A^TA = AA^T$ if and only if A is symmetric or $A = \begin{bmatrix} a & b \\ -b & a \end{bmatrix}$ for some a and b.

28. (a) Find all symmetric 2×2 matrices A such that $A^2 = 0$.

   (b) Repeat (a) if A is 3×3.

   (c) Repeat (a) if A is $n \times n$.

29. Show that there exist no 2×2 matrices A and B such that $AB - BA = I$. [*Hint:* Examine the $(1, 1)$- and $(2, 2)$-entries.]

♦30. Let B be an $n \times n$ matrix. Suppose $AB = 0$ for some nonzero $m \times n$ matrix A. Show that no $n \times n$ matrix C exists such that $BC = I$.

31. An autoparts manufacturer makes fenders, doors, and hoods. Each requires assembly and packaging carried out at factories: Plant 1, Plant 2, and Plant 3. Matrix A below gives the number of hours for assembly and packaging, and matrix B gives the hourly rates at the three plants. Explain the meaning of the $(3, 2)$-entry in the matrix AB. Which plant is the most economical to operate? Give reasons.

$$
\begin{array}{c}
\quad\quad\quad \text{Assembly} \quad \text{Packaging} \\
\begin{array}{c} \text{Fenders} \\ \text{Doors} \\ \text{Hoods} \end{array}
\left[\begin{array}{cc} 12 & 2 \\ 21 & 3 \\ 10 & 2 \end{array} \right] = A
\end{array}
$$

$$
\begin{array}{c}
\quad\quad \text{Plant 1} \quad \text{Plant 2} \quad \text{Plant 3} \\
\begin{array}{c} \text{Assembly} \\ \text{Packaging} \end{array}
\left[\begin{array}{ccc} 21 & 18 & 20 \\ 14 & 10 & 13 \end{array} \right] = B
\end{array}
$$

♦32. For the directed graph at the right, find the adjacency matrix A, compute A^3, and determine the number of paths of length 3 from v_1 to v_4 and from v_2 to v_3.

33. In each case either show the statement is true, or give an example showing that it is false.

(a) If A is $m \times n$ where $m < n$, then $AX = B$ has a solution for every column B.

♦(b) If $AX = B$ has a solution for some column B, then it has a solution for every column B.

(c) If A is square, then $(A^T)^3 = (A^3)^T$.

♦(d) If X_1 and X_2 are solutions to $AX = B$, then $X_1 - X_2$ is a solution to $AX = 0$.

(e) If $AB = AC$ and $A \neq 0$, then $B = C$.

♦(f) If $A \neq 0$, then $A^2 \neq 0$.

(g) If A has a row of zeros, so also does BA for all B.

♦(h) If A commutes with $A + B$, then A commutes with B.

34. (a) If A and B are 2×2 matrices whose rows sum to 1, show that the rows of AB also sum to 1.

(b) Repeat part (a) for the case where A and B are $n \times n$.

35. Let A and B be $n \times n$ matrices for which the systems of equations $AX = 0$ and $BX = 0$ each have only the trivial solution $X = 0$. Show that the system $(AB)X = 0$ has only the trivial solution.

36. The **trace** of a square matrix A, denoted tr A, is the sum of the elements on the main diagonal of A. Show that, if A and B are $n \times n$ matrices:

(a) $\operatorname{tr}(A + B) = \operatorname{tr} A + \operatorname{tr} B$.

♦(b) $\operatorname{tr}(kA) = k\operatorname{tr}(A)$ for any number k.

(c) $\operatorname{tr}(A^T) = \operatorname{tr}(A)$.

(d) $\operatorname{tr}(AB) = \operatorname{tr}(BA)$.

♦(e) $\operatorname{tr}(AA^T)$ is the sum of the squares of all entries of A.

37. Show that $AB - BA = I$ is impossible. [*Hint:* See the preceding exercise.]

38. A square matrix P is called an **idempotent** if $P^2 = P$. Show that:

(a) 0 and I are idempotents.

(b) $\begin{bmatrix} 1 & 1 \\ 0 & 0 \end{bmatrix}$, $\begin{bmatrix} 1 & 0 \\ 1 & 0 \end{bmatrix}$, and $\frac{1}{\sqrt{2}}\begin{bmatrix} 1 & 1 \\ 1 & 1 \end{bmatrix}$ are idempotents.

(c) If P is an idempotent, so is $I - P$. Show further that $P(I - P) = 0$.

(d) If P is an idempotent, so is P^T.

♦(e) If P is an idempotent, so is $Q = P + AP - PAP$ for any square matrix A (of the same size as P).

(f) If A is $n \times m$ and B is $m \times n$, and if $AB = I_n$, then BA is an idempotent.

39. Let A and B be $n \times n$ **diagonal matrices** (all entries off the main diagonal are zero).

(a) Show that AB is diagonal and $AB = BA$.

(b) Formulate a rule for calculating XA if X is $m \times n$.

(c) Formulate a rule for calculating AY if Y is $n \times k$.

40. If A and B are $n \times n$ matrices, show that:

(a) $AB = BA$ if and only if
$$(A + B)^2 = A^2 + 2AB + B^2.$$

♦(b) $AB = BA$ if and only if
$$(A + B)(A - B) = (A - B)(A + B).$$

♦41. (V. Camillo) Show that the product of two reduced row-echelon matrices is also reduced row-echelon.

SECTION 2.3   Matrix Inverses

Three basic operations on matrices, addition, multiplication, and subtraction, are analogs for matrices of the same operations for numbers. In this section we introduce the matrix analog of numerical division.

To begin, consider how a numerical equation

$$ax = b$$

is solved when a and b are known numbers. If $a = 0$, there is no solution (unless $b = 0$). But if $a \neq 0$, we can multiply both sides by the inverse a^{-1} to obtain the solution $x = a^{-1}b$. This multiplication by a^{-1} is commonly called dividing by a, and the property of a^{-1} that makes this work is that $a^{-1}a = 1$. Moreover, we saw in Section 2.2 that the role that 1 plays in arithmetic is played in matrix algebra by the identity matrix I.

This suggests the following definition. If A is a square matrix, a matrix B is called an **inverse** of A if and only if

$$AB = I \quad \text{and} \quad BA = I$$

A matrix A that has an inverse is called an **invertible matrix**.[3]

Example 1

Show that $B = \begin{bmatrix} -1 & 1 \\ 1 & 0 \end{bmatrix}$ is an inverse of $A = \begin{bmatrix} 0 & 1 \\ 1 & 1 \end{bmatrix}$.

Solution   Compute AB and BA.

$$AB = \begin{bmatrix} 0 & 1 \\ 1 & 1 \end{bmatrix}\begin{bmatrix} -1 & 1 \\ 1 & 0 \end{bmatrix} = \begin{bmatrix} 1 & 0 \\ 0 & 1 \end{bmatrix} \qquad BA = \begin{bmatrix} -1 & 1 \\ 1 & 0 \end{bmatrix}\begin{bmatrix} 0 & 1 \\ 1 & 1 \end{bmatrix} = \begin{bmatrix} 1 & 0 \\ 0 & 1 \end{bmatrix}$$

Hence $AB = I = BA$, so B is indeed an inverse of A.

Example 2

Show that $A = \begin{bmatrix} 0 & 0 \\ 1 & 3 \end{bmatrix}$ has no inverse.

Solution   Let $B = \begin{bmatrix} a & b \\ c & d \end{bmatrix}$ denote an arbitrary 2×2 matrix. Then

$$AB = \begin{bmatrix} 0 & 0 \\ 1 & 3 \end{bmatrix}\begin{bmatrix} a & b \\ c & d \end{bmatrix} = \begin{bmatrix} 0 & 0 \\ a + 3c & b + 3d \end{bmatrix}$$

so AB has a row of zeros. Hence AB cannot equal I for any B.

The argument in Example 2 shows that no zero matrix has an inverse. But Example 2 also shows that, unlike arithmetic, *it is possible for a nonzero matrix to have no inverse*. However, if a matrix *does* have an inverse, it has only one.

Theorem 1

If B and C are both inverses of A, then $B = C$.

PROOF

Since B and C are both inverses of A, we have $CA = I = AB$.
Hence $B = IB = (CA)B = C(AB) = CI = C$.

---

[3] Only square matrices have inverses. Even though it is plausible that nonsquare matrices A and B exist such that $AB = I_m$ and $BA = I_n$, where A is $m \times n$ and B is $n \times m$, we claim that this forces $n = m$. Indeed, if $m < n$ there exists a nonzero column X such that $AX = 0$ (by Theorem 1 §1.3), so $X = I_nX = (BA)X = B(AX) = B(0) = 0$, a contradiction. Hence $m \geq n$. Similarly, the condition $AB = I_m$ implies that $n \geq m$.

If A is an invertible matrix, the (unique) inverse of A is denoted A^{-1}. Hence A^{-1} (when it exists) is a square matrix of the same size as A with the property that

$$AA^{-1} = I \quad \text{and} \quad A^{-1}A = I$$

These equations characterize A^{-1} in the following sense: If somehow a matrix B can be found such that $AB = I = BA$, then A is invertible and B is the inverse of A; in symbols, $B = A^{-1}$. This gives us a way of verifying that the inverse of a matrix exists. Examples 3 and 4 offer illustrations.

Example 3

If $A = \begin{bmatrix} 0 & -1 \\ 1 & -1 \end{bmatrix}$, show that $A^3 = I$ and so find A^{-1}.

Solution   We have $A^2 = \begin{bmatrix} 0 & -1 \\ 1 & -1 \end{bmatrix}\begin{bmatrix} 0 & -1 \\ 1 & -1 \end{bmatrix} = \begin{bmatrix} -1 & 1 \\ -1 & 0 \end{bmatrix}$, and so

$$A^3 = A^2A = \begin{bmatrix} -1 & 1 \\ -1 & 0 \end{bmatrix}\begin{bmatrix} 0 & -1 \\ 1 & -1 \end{bmatrix} = \begin{bmatrix} 1 & 0 \\ 0 & 1 \end{bmatrix} = I$$

Hence $A^3 = I$, as asserted. This can be written as $A^2A = I = AA^2$, so it shows that A^2 is the inverse of A. That is, $A^{-1} = A^2 = \begin{bmatrix} -1 & 1 \\ -1 & 0 \end{bmatrix}$.

The next example presents a useful formula for the inverse of a 2×2 matrix $A = \begin{bmatrix} a & b \\ c & d \end{bmatrix}$. To state it, we define the **determinant** $\det A$ and the **adjugate** $\text{adj } A$ of the matrix A as follows:

$$\det \begin{bmatrix} a & b \\ c & d \end{bmatrix} = ad - bc, \quad \text{and} \quad \text{adj} \begin{bmatrix} a & b \\ c & d \end{bmatrix} = \begin{bmatrix} d & -b \\ -c & a \end{bmatrix}$$

Example 4

If $A = \begin{bmatrix} a & b \\ c & d \end{bmatrix}$, show that A has an inverse if and only if $\det A \neq 0$, and in this case

$$A^{-1} = \frac{1}{\det A}\text{adj } A$$

Solution   For convenience, write $e = \det A = ad - bc$ and $B = \text{adj } A = \begin{bmatrix} d & -b \\ -c & a \end{bmatrix}$. Then $AB = eI = BA$ as the reader can verify. So if $e \neq 0$, scalar multiplication by $1/e$ gives $A(\frac{1}{e}B) = I = (\frac{1}{e}B)A$. Hence A is invertible and $A^{-1} = \frac{1}{e}B$. Thus it remains only to show that if A^{-1} exists, then $e \neq 0$.

We prove this by showing that assuming $e = 0$ leads to a contradiction. In fact, if $e = 0$, then $AB = eI = 0$, so left multiplication by A^{-1} gives $A^{-1}AB = A^{-1}0$, that is $IB = 0$, so $B = 0$. But this implies that a, b, c, and d are all zero, so $A = 0$, contrary to the assumption that A^{-1} exists.

As an illustration, if $A = \begin{bmatrix} 2 & 4 \\ -3 & 8 \end{bmatrix}$ then $\det A = 2 \cdot 8 - 4 \cdot (-3) = 28 \neq 0$. Hence A is invertible and $A^{-1} = \dfrac{1}{\det A}\operatorname{adj} A = \frac{1}{28}\begin{bmatrix} 8 & -4 \\ 3 & 2 \end{bmatrix}$, as the reader is invited to verify.

The determinant and adjugate will be defined in Chapter 3 for any $n \times n$ matrix, and the conclusions in Example 4 will be proved in full generality.

Inverses and Linear Systems

Matrix inverses can be used to solve certain systems of linear equations. Recall that a *system* of linear equations can be written as a *single* matrix equation

$$AX = B$$

where A and B are known matrices and X is to be determined. If A is invertible, we multiply each side of the equation on the left by A^{-1} to get

$$A^{-1}AX = A^{-1}B$$
$$IX = A^{-1}B$$
$$X = A^{-1}B$$

This gives the solution to the system of equations (the reader should verify that $X = A^{-1}B$ really does satisfy $AX = B$). Furthermore, the argument shows that if X is *any* solution, then necessarily $X = A^{-1}B$, so the solution is unique. Of course the technique works only when the coefficient matrix A has an inverse. This proves Theorem 2.

Theorem 2

Suppose a system of n equations in n variables is written in matrix form as
$$AX = B$$
If the $n \times n$ coefficient matrix A is invertible, the system has the unique solution
$$X = A^{-1}B$$

Example 5

Use Example 4 to solve the system $\begin{cases} 5x_1 - 3x_2 = -4 \\ 7x_1 + 4x_2 = \ \ 8 \end{cases}$.

Solution   In matrix form this is $AX = B$ where $A = \begin{bmatrix} 5 & -3 \\ 7 & 4 \end{bmatrix}$, $X = \begin{bmatrix} x_1 \\ x_2 \end{bmatrix}$, and $B = \begin{bmatrix} -4 \\ 8 \end{bmatrix}$.

Then $\det A = 5 \cdot 4 - (-3) \cdot 7 = 41$, so A is invertible and $A^{-1} = \frac{1}{41}\begin{bmatrix} 4 & 3 \\ -7 & 5 \end{bmatrix}$ by Example 4. Thus Theorem 2 gives

$$X = A^{-1}B = \frac{1}{41}\begin{bmatrix} 4 & 3 \\ -7 & 5 \end{bmatrix}\begin{bmatrix} -4 \\ 8 \end{bmatrix} = \frac{1}{41}\begin{bmatrix} 8 \\ 68 \end{bmatrix},$$

so the solution is $x_1 = \frac{8}{41}$ and $x_2 = \frac{68}{41}$.

An Inversion Method

Given a particular $n \times n$ matrix A, it is desirable to have an efficient technique to determine whether A has an inverse and, if so, to find that inverse. For simplicity, we shall derive the technique for 2×2 matrices; the $n \times n$ case is entirely analogous.

Given the invertible 2×2 matrix A, we determine A^{-1} from the equation $AA^{-1} = I$. Write

$$A^{-1} = \begin{bmatrix} x_1 & x_2 \\ y_1 & y_2 \end{bmatrix}$$

where $x_1, y_1, x_2,$ and y_2 are to be determined. Equating columns in the equation $AA^{-1} = I$ gives

$$A\begin{bmatrix} x_1 \\ y_1 \end{bmatrix} = \begin{bmatrix} 1 \\ 0 \end{bmatrix} \quad \text{and} \quad A\begin{bmatrix} x_2 \\ y_2 \end{bmatrix} = \begin{bmatrix} 0 \\ 1 \end{bmatrix}$$

These are systems of linear equations, each with A as coefficient matrix. Since A is invertible, each system has a unique solution by Theorem 2. But this means that the reduced row-echelon form R of A cannot have a row of zeros, and so is the identity matrix (R is square). Hence, there is a sequence of elementary row operations carrying A to the 2×2 identity matrix I. This sequence carries the augmented matrices of both systems to reduced row-echelon form and so solves the systems:

$$\begin{bmatrix} A & \Big| & 1 \\ & & 0 \end{bmatrix} \to \begin{bmatrix} I & \Big| & x_1 \\ & & y_1 \end{bmatrix} \quad \text{and} \quad \begin{bmatrix} A & \Big| & 0 \\ & & 1 \end{bmatrix} \to \begin{bmatrix} I & \Big| & x_2 \\ & & y_2 \end{bmatrix}$$

Hence, we can do *both* calculations simultaneously by applying the elementary row operations to the matrix.

$$\begin{bmatrix} A & \Big| & 1 & 0 \\ & & 0 & 1 \end{bmatrix} \to \begin{bmatrix} I & \Big| & x_1 & x_2 \\ & & y_1 & y_2 \end{bmatrix}$$

This can be written more compactly as follows:

$$[A \ \ I] \to [I \ \ A^{-1}]$$

In other words, the sequence of row operations that carries A to I also carries I to A^{-1}. This is the desired algorithm.

Matrix Inversion Algorithm

If A is a (square) invertible matrix, there exists a sequence of elementary row operations that carry A to the identity matrix I of the same size, written $A \to I$. This same series of row operations carries I to A^{-1}; that is, $I \to A^{-1}$. The algorithm can be summarized as follows:

$$[A \ \ I] \to [I \ \ A^{-1}]$$

where the row operations on A and I are carried out simultaneously.

Example 6

Use the inversion algorithm to find the inverse of the matrix

$$A = \begin{bmatrix} 2 & 7 & 1 \\ 1 & 4 & -1 \\ 1 & 3 & 0 \end{bmatrix}$$

Solution   Apply elementary row operations to the double matrix

$$[A \ I] = \begin{bmatrix} 2 & 7 & 1 & | & 1 & 0 & 0 \\ 1 & 4 & -1 & | & 0 & 1 & 0 \\ 1 & 3 & 0 & | & 0 & 0 & 1 \end{bmatrix}$$

so as to carry A to I. First interchange rows 1 and 2.

$$\begin{bmatrix} 1 & 4 & -1 & | & 0 & 1 & 0 \\ 2 & 7 & 1 & | & 1 & 0 & 0 \\ 1 & 3 & 0 & | & 0 & 0 & 1 \end{bmatrix}$$

Next subtract 2 times row 1 from row 2, and subtract row 1 from row 3.

$$\begin{bmatrix} 1 & 4 & -1 & | & 0 & 1 & 0 \\ 0 & -1 & 3 & | & 1 & -2 & 0 \\ 0 & -1 & 1 & | & 0 & -1 & 1 \end{bmatrix}$$

Continue to reduced row-echelon form.

$$\begin{bmatrix} 1 & 0 & 11 & | & 4 & -7 & 0 \\ 0 & 1 & -3 & | & -1 & 2 & 0 \\ 0 & 0 & -2 & | & -1 & 1 & 1 \end{bmatrix}$$

$$\begin{bmatrix} 1 & 0 & 0 & | & \frac{-3}{2} & \frac{-3}{2} & \frac{11}{2} \\ 0 & 1 & 0 & | & \frac{1}{2} & \frac{1}{2} & \frac{-3}{2} \\ 0 & 0 & 1 & | & \frac{1}{2} & \frac{-1}{2} & \frac{-1}{2} \end{bmatrix}$$

Hence $A^{-1} = \frac{1}{2} \begin{bmatrix} -3 & -3 & 11 \\ 1 & 1 & -3 \\ 1 & -1 & -1 \end{bmatrix}$, as is readily verified.

Given any $n \times n$ matrix A, Theorem 1 §1.2 shows that A can be carried by elementary row operations to a matrix R in reduced row-echelon form. If $R = I$, the matrix A is invertible (this will be proved in the next section), so the algorithm produces A^{-1}. If $R \neq I$, then R has a row of zeros (it is square), so no system of linear equations $AX = B$ can have a unique solution. But then A is not invertible by Theorem 2. Hence, the algorithm is effective in the sense conveyed in Theorem 3 below.

Theorem 3

If A is an $n \times n$ matrix, either A can be reduced to I by elementary row operations or it cannot. In the first case, the algorithm produces A^{-1}; in the second case, A^{-1} does not exist.

Properties of Inverses

Sometimes the inverse of a matrix is given by a formula. Example 4 is one illustration; Examples 7 and 8 provide two more. Given a square matrix A, recall that if a matrix B can be found such that $AB = I = BA$, then A is invertible and $A^{-1} = B$.

Example 7

If A is an invertible matrix, show that the transpose A^T is also invertible. Show further that the inverse of A^T is just the transpose of A^{-1}; in symbols, $(A^T)^{-1} = (A^{-1})^T$.

Solution　A^{-1} exists (by assumption). Its transpose $(A^{-1})^T$ is the candidate proposed for the inverse of A^T. Using Theorem 1(6) in §2.2, we test it as follows:

$$A^T(A^{-1})^T = (A^{-1}A)^T = I^T = I$$
$$(A^{-1})^T A^T = (AA^{-1})^T = I^T = I$$

Hence $(A^{-1})^T$ is indeed the inverse of A^T; that is, $(A^T)^{-1} = (A^{-1})^T$.

Example 8

If A and B are invertible $n \times n$ matrices, show that their product AB is also invertible and $(AB)^{-1} = B^{-1}A^{-1}$.

Solution　We are given a candidate for the inverse of AB, namely $B^{-1}A^{-1}$. We test it as follows:

$$(B^{-1}A^{-1})(AB) = B^{-1}(A^{-1}A)B = B^{-1}IB = B^{-1}B = I$$
$$(AB)(B^{-1}A^{-1}) = A(BB^{-1})A^{-1} = AIA^{-1} = AA^{-1} = I$$

Hence $B^{-1}A^{-1}$ is the inverse of AB; in symbols, $(AB)^{-1} = B^{-1}A^{-1}$.

We now collect several basic properties of matrix inverses for reference.

Theorem 4

All the following matrices are square matrices of the same size.

1. I is invertible and $I^{-1} = I$.
2. If A is invertible, so is A^{-1}, and $(A^{-1})^{-1} = A$.
3. If A and B are invertible, so is AB, and $(AB)^{-1} = B^{-1}A^{-1}$.
4. If $A_1, A_2, \ldots, A_k$ are all invertible, so is their product $A_1 A_2 \cdots A_k$, and $(A_1 A_2 \cdots A_k)^{-1} = A_k^{-1} \cdots A_2^{-1} A_1^{-1}$.
5. If A is invertible, so is A^k for $k \geq 1$, and $(A^k)^{-1} = (A^{-1})^k$.
6. If A is invertible and $a \neq 0$ is a number, then aA is invertible and $(aA)^{-1} = \frac{1}{a} A^{-1}$.
7. If A is invertible, so is its transpose A^T, and $(A^T)^{-1} = (A^{-1})^T$.

PROOF

1. This is an immediate consequence of the formula $I^2 = I$.
2. The equations $AA^{-1} = I = A^{-1}A$ show that A is the inverse of A^{-1}; in symbols, $(A^{-1})^{-1} = A$.
3. This is Example 8.
4. Use induction on k. If $k = 1$, there is nothing to prove, and if $k = 2$, the result is property 3. If $k > 2$, assume inductively that

$(A_1 A_2 \cdots A_{k-1})^{-1} = A_{k-1}^{-1} \cdots A_2^{-1} A_1^{-1}$. We apply this fact together with property 3 as follows:

$$[A_1 A_2 \cdots A_{k-1} A_k]^{-1} = [(A_1 A_2 \cdots A_{k-1}) A_k]^{-1}$$
$$= A_k^{-1} (A_1 A_2 \cdots A_{k-1})^{-1}$$
$$= A_k^{-1} (A_{k-1}^{-1} \cdots A_2^{-1} A_1^{-1})$$

So the proof by induction is complete.

5. This is property 4 with $A_1 = A_2 = \cdots = A_k = A$.
6. This is left as Exercise 28.
7. This is Example 7.

Part 7 of Theorem 4 together with the fact that $(A^T)^T = A$ give

Corollary

A square matrix A is invertible if and only if A^T is invertible.

Example 9

Find A if $(A^T - 2I)^{-1} = \begin{bmatrix} 2 & 1 \\ -1 & 0 \end{bmatrix}$.

Solution   By Theorem 4(2) and Example 4

$$(A^T - 2I) = [(A^T - 2I)^{-1}]^{-1} = \begin{bmatrix} 2 & 1 \\ -1 & 0 \end{bmatrix}^{-1} = \begin{bmatrix} 0 & -1 \\ 1 & 2 \end{bmatrix}$$

Hence $A^T = 2I + \begin{bmatrix} 0 & -1 \\ 1 & 2 \end{bmatrix} = \begin{bmatrix} 2 & -1 \\ 1 & 4 \end{bmatrix}$, so $A = \begin{bmatrix} 2 & 1 \\ -1 & 4 \end{bmatrix}$.

The reversal of the order of the inverses in properties 3 and 4 of Theorem 4 is a consequence of the fact that matrix multiplication is not commutative. Another manifestation of this comes when matrix equations are dealt with. If a matrix equation $B = C$ is given, it can be *left-multiplied* by a matrix A to yield $AB = AC$. Similarly, *right-multiplication* gives $BA = CA$. However, we cannot mix the two: If $B = C$, it need *not* be the case that $AB = CA$.

We conclude this section with an important theorem that collects a number of conditions all equivalent[4] to invertibility. It will be referred to frequently below.

Theorem 5

The following conditions are equivalent for an $n \times n$ matrix A:
1. A is invertible.
2. The homogeneous system $AX = 0$ has only the trivial solution $X = 0$.
3. A can be carried to the identity matrix I_n by elementary row operations.
4. The system $AX = B$ has at least one solution X for every choice of column B.
5. There exists an $n \times n$ matrix C such that $AC = I_n$.

---

4 If p and q are statements, we say that p **implies** q (written $p \Rightarrow q$) if q is true whenever p is true. The statements are called **equivalent** if both $p \Rightarrow q$ and $q \Rightarrow p$ (written $p \Leftrightarrow q$, spoken "p if and only if q"). See Appendix B.

PROOF

We show that each of these conditions implies the next, and that (5) implies (1).

(1) $\Rightarrow$ (2). If A^{-1} exists, then $AX = 0$ gives $X = I_n X = A^{-1}AX = A^{-1}0 = 0$.

(2) $\Rightarrow$ (3). Assume that (2) is true. Certainly $A \to R$ by row operations where R is a reduced, row-echelon matrix; we show that $R = I_n$. Suppose on the contrary that $R \neq I_n$. Then R has a row of zeros (being square), and we consider the augmented matrix $[A \ \ 0]$ of the system $AX = 0$. Then $[A \ \ 0] \to [R \ \ 0]$ is the row-echelon form and $[R \ \ 0]$ also has a row of zeros. Since A is square, this means that there is at least one nonleading variable, and hence at least one parameter. Thus $AX = 0$ has infinitely many solutions, contrary to (2). So $R = I_n$ after all.

(3) $\Rightarrow$ (4). Consider the augmented matrix $[A \ \ B]$ of the system. Using (3), let $A \to I$ by a sequence of row operations. Then these same operations carry $[A \ \ B] \to [I \ \ C]$ for some column C. Hence the system $AX = B$ has a solution (in fact unique) by gaussian elimination.

(4) $\Rightarrow$ (5). Write $I_n = [E_1 \ \ E_2 \ \cdots \ E_n]$ where $E_1, E_2, \ldots, E_n$ are the columns of I_n. For each $j = 1, 2, \ldots, n$, the system $AX = E_j$ has a solution C_j by (4), so $AC_j = E_j$. Now let $C = [C_1 \ \ C_2 \ \cdots \ C_n]$ be the $n \times n$ matrix with these matrices C_j as its columns. Then the definition of matrix multiplication gives (5):

$$AC = A[C_1 \ \ C_2 \ \cdots \ C_n] = [AC_1 \ \ AC_2 \ \cdots \ AC_n] = [E_1 \ \ E_2 \ \cdots \ E_n] = I_n$$

(5) $\Rightarrow$ (1). Assume that (5) is true so that $AC = I_n$ for some matrix C. Then $CX = 0$ implies $X = 0$ (because $X = I_n X = ACX = A0 = 0$). Thus condition (2) holds for the matrix C rather than A. Hence the argument above that (2) $\Rightarrow$ (3) $\Rightarrow$ (4) $\Rightarrow$ (5) (with A replaced by C) shows that a matrix C' exists such that $CC' = I$. But then

$$A = AI_n = A(CC') = (AC)C' = IC' = C'$$

Thus $CA = CC' = I$ which, together with $AC = I$, shows that C is the inverse of A. This proves (1).

The proof of (5) $\Rightarrow$ (1) in Theorem 5 shows that if $AC = I$ for square matrices, then necessarily $CA = I$, and hence that C and A are inverses of each other. We record this important fact for reference.

Corollary

If A and C are square matrices such that $AC = I$, then also $CA = I$. In particular, both A and C are invertible, $C = A^{-1}$, and $A = C^{-1}$.

Observe that the Corollary is false if A and C are not square matrices. For example, we have

$$\begin{bmatrix} 1 & 2 & 1 \\ 1 & 1 & 1 \end{bmatrix} \begin{bmatrix} -1 & 1 \\ 1 & -1 \\ 0 & 1 \end{bmatrix} = I_2 \quad \text{but} \quad \begin{bmatrix} -1 & 1 \\ 1 & -1 \\ 0 & 1 \end{bmatrix} \begin{bmatrix} 1 & 2 & 1 \\ 1 & 1 & 1 \end{bmatrix} \neq I_3$$

In fact, it is verified in the footnote on page 51 that if $AB = I_m$ and $BA = I_n$, where A is $m \times n$ and B is $n \times m$, then $m = n$ and A and B are (square) inverses of each other.

Example 10

Show using Theorem 5 that $A = \begin{bmatrix} 6 & 8 \\ 15 & 20 \end{bmatrix}$ has no inverse.

Solution Observe that $AX = 0$ where $X = \begin{bmatrix} 4 \\ -3 \end{bmatrix}$. Hence A has no inverse by Part (2) of Theorem 5. Note that we do not need Theorem 5 for this: If A^{-1} exists then left-multiplying $AX = 0$ by A^{-1} gives $A^{-1}AX = A^{-1}0$, that is $IX = 0$. This means that $X = 0$, which is not the case. So A^{-1} does not exist.

Exercises 2.3

1. In each case, show that the matrices are inverses of each other.

(a) $\begin{bmatrix} 3 & 5 \\ 1 & 2 \end{bmatrix}$, $\begin{bmatrix} 2 & -5 \\ -1 & 3 \end{bmatrix}$ (b) $\begin{bmatrix} 3 & 0 \\ 1 & -4 \end{bmatrix}$, $\frac{1}{12}\begin{bmatrix} 4 & 0 \\ 1 & -3 \end{bmatrix}$

(c) $\begin{bmatrix} 1 & 2 & 0 \\ 0 & 2 & 3 \\ 1 & 3 & 1 \end{bmatrix}$, $\begin{bmatrix} 7 & 2 & -6 \\ -3 & -1 & 3 \\ 2 & 1 & -2 \end{bmatrix}$ (d) $\begin{bmatrix} 3 & 0 \\ 0 & 5 \end{bmatrix}$, $\begin{bmatrix} \frac{1}{3} & 0 \\ 0 & \frac{1}{5} \end{bmatrix}$

2. Find the inverse of each of the following matrices.

(a) $\begin{bmatrix} 1 & -1 \\ -1 & 3 \end{bmatrix}$ ◆(b) $\begin{bmatrix} 4 & 1 \\ 3 & 2 \end{bmatrix}$

(c) $\begin{bmatrix} 1 & 0 & -1 \\ 3 & 2 & 0 \\ -1 & -1 & 0 \end{bmatrix}$ ◆(d) $\begin{bmatrix} 1 & -1 & 2 \\ -5 & 7 & -11 \\ -2 & 3 & -5 \end{bmatrix}$

(e) $\begin{bmatrix} 3 & 5 & 0 \\ 3 & 7 & 1 \\ 1 & 2 & 1 \end{bmatrix}$ ◆(f) $\begin{bmatrix} 3 & 1 & -1 \\ 2 & 1 & 0 \\ 1 & 5 & -1 \end{bmatrix}$

(g) $\begin{bmatrix} 2 & 4 & 1 \\ 3 & 3 & 2 \\ 4 & 1 & 4 \end{bmatrix}$ ◆(h) $\begin{bmatrix} 3 & 1 & -1 \\ 5 & 2 & 0 \\ 1 & 1 & -1 \end{bmatrix}$

(i) $\begin{bmatrix} 3 & 1 & 2 \\ 1 & -1 & 3 \\ 1 & 2 & 4 \end{bmatrix}$ ◆(j) $\begin{bmatrix} -1 & 4 & 5 & 2 \\ 0 & 0 & 0 & -1 \\ 1 & -2 & -2 & 0 \\ 0 & -1 & -1 & 0 \end{bmatrix}$

(k) $\begin{bmatrix} 1 & 0 & 7 & 5 \\ 0 & 1 & 3 & 6 \\ 1 & -1 & 5 & 2 \\ 1 & -1 & 5 & 1 \end{bmatrix}$ ◆(l) $\begin{bmatrix} 1 & 2 & 0 & 0 & 0 \\ 0 & 1 & 3 & 0 & 0 \\ 0 & 0 & 1 & 5 & 0 \\ 0 & 0 & 0 & 1 & 7 \\ 0 & 0 & 0 & 0 & 1 \end{bmatrix}$

3. In each case, solve the systems of equations by finding the inverse of the coefficient matrix.

(a) $3x - y = 5$
$2x + 3y = 1$

◆(b) $2x - 3y = 0$
$x - 4y = 1$

(c) $x + y + 2z = 5$
$x + y + z = 0$
$x + 2y + 4z = -2$

◆(d) $x + 4y + 2z = 1$
$2x + 3y + 3z = -1$
$4x + y + 4z = 0$

(e) $x + y \quad\quad - w = 1$
$-x + y - z \quad\quad = -1$
$y + z + w = 0$
$x \quad\quad - z + w = 1$

◆(f) $x + y + z + w = 1$
$x + y \quad\quad\quad = 0$
$y \quad\quad + w = -1$
$x \quad\quad\quad + w = 2$

4. Given $A^{-1} = \begin{bmatrix} 1 & -1 & 3 \\ 2 & 0 & 5 \\ -1 & 1 & 0 \end{bmatrix}$:

(a) Solve the system of equations $AX = \begin{bmatrix} 1 \\ -1 \\ 3 \end{bmatrix}$.

◆(b) Find a matrix B such that $AB = \begin{bmatrix} 1 & -1 & 2 \\ 0 & 1 & 1 \\ 1 & 0 & 0 \end{bmatrix}$.

(c) Find a matrix C such that $CA = \begin{bmatrix} 1 & 2 & -1 \\ 3 & 1 & 1 \end{bmatrix}$.

5. Find A when

(a) $(3A)^{-1} = \begin{bmatrix} 1 & -1 \\ 0 & 1 \end{bmatrix}$ ◆(b) $(2A)^T = \begin{bmatrix} 1 & -1 \\ 2 & 3 \end{bmatrix}^{-1}$

(c) $(I + 3A)^{-1} = \begin{bmatrix} 2 & 0 \\ 1 & -1 \end{bmatrix}$

◆(d) $(I - 2A^T)^{-1} = \begin{bmatrix} 2 & 1 \\ 1 & 1 \end{bmatrix}$

(e) $\left(A\begin{bmatrix} 1 & -1 \\ 0 & 1 \end{bmatrix} \right)^{-1} = \begin{bmatrix} 2 & 3 \\ 1 & 1 \end{bmatrix}$

♦(f) $\left(\begin{bmatrix} 1 & 0 \\ 2 & 1 \end{bmatrix} A \right)^{-1} = \begin{bmatrix} 1 & 0 \\ 2 & 2 \end{bmatrix}$

(g) $(A^T - 2I)^{-1} = 2\begin{bmatrix} 1 & 1 \\ 2 & 3 \end{bmatrix}$

♦(h) $(A^{-1} - 2I)^T = -2\begin{bmatrix} 1 & 1 \\ 1 & 0 \end{bmatrix}$

6. Find A when:

(a) $A^{-1} = \begin{bmatrix} 1 & -1 & 3 \\ 2 & 1 & 1 \\ 0 & 2 & -2 \end{bmatrix}$ ♦(b) $A^{-1} = \begin{bmatrix} 0 & 1 & -1 \\ 1 & 2 & 1 \\ 1 & 0 & 1 \end{bmatrix}$

7. Given $\begin{bmatrix} x_1 \\ x_2 \\ x_3 \end{bmatrix} = \begin{bmatrix} 3 & -1 & 2 \\ 1 & 0 & 4 \\ 2 & 1 & 0 \end{bmatrix}\begin{bmatrix} y_1 \\ y_2 \\ y_3 \end{bmatrix}$ and

$\begin{bmatrix} z_1 \\ z_2 \\ z_3 \end{bmatrix} = \begin{bmatrix} 1 & -1 & 1 \\ 2 & -3 & 0 \\ -1 & 1 & -2 \end{bmatrix}\begin{bmatrix} y_1 \\ y_2 \\ y_3 \end{bmatrix}$, express the variables x_1, x_2, and x_3 in terms of z_1, z_2, and z_3.

8. (a) In the system $\begin{array}{l} 3x + 4y = 7 \\ 4x + 5y = 1 \end{array}$, substitute the new variables x' and y' given by $\begin{array}{l} x = -5x' + 4y' \\ y = 4x' - 3y' \end{array}$.

Then find x and y.

♦(b) Explain part (a) by writing the equations as $A\begin{bmatrix} x \\ y \end{bmatrix} = \begin{bmatrix} 7 \\ 1 \end{bmatrix}$ and $\begin{bmatrix} x \\ y \end{bmatrix} = B\begin{bmatrix} x' \\ y' \end{bmatrix}$. What is the relationship between A and B? Generalize.

9. In each case either prove the assertion or give an example showing that it is false.

(a) If $A \neq 0$ is a square matrix, then A is invertible.

✓ ♦(b) If A and B are both invertible, then $A + B$ is invertible.

(c) If A and B are both invertible, then $(A^{-1}B)^T$ is invertible.

◗(d) If $A^4 = 3I$, then A is invertible.

(e) If $A^2 = A$ and $A \neq 0$, then A is invertible.

♦(f) If $AB = B$ for some $B \neq 0$, then A is invertible.

(g) If A is invertible and skew symmetric $(A^T = -A)$, the same is true of A^{-1}.

♦(h) If A^2 is invertible, then A is invertible.

(i) If $AB = I$, then A and B commute.

♦10. (a) If A, B, and C are square matrices and $AB = I = CA$, show that A is invertible and $B = C = A^{-1}$.

♦(b) If $C^{-1} = A$, find the inverse of C^T in terms of A.

11. Suppose $CA = I_m$, where C is $m \times n$ and A is $n \times m$. Consider the system $AX = B$ of n equations in m variables.

(a) Show that this system has a unique solution CB if it is consistent.

♦(b) If $C = \begin{bmatrix} 0 & -5 & 1 \\ 3 & 0 & -1 \end{bmatrix}$ and $A = \begin{bmatrix} 2 & -3 \\ 1 & -2 \\ 6 & -10 \end{bmatrix}$,

find X (if it exists) when (i) $B = \begin{bmatrix} 1 \\ 0 \\ 3 \end{bmatrix}$; and

(ii) $B = \begin{bmatrix} 7 \\ 4 \\ 22 \end{bmatrix}$.

12. Verify that $A = \begin{bmatrix} 1 & -1 \\ 0 & 2 \end{bmatrix}$ satisfies $A^2 - 3A + 2I = 0$, and use this fact to show that $A^{-1} = \frac{1}{2}(3I - A)$.

13. Let $Q = \begin{bmatrix} a & -b & -c & -d \\ b & a & -d & c \\ c & d & a & -b \\ d & -c & b & a \end{bmatrix}$. Compute QQ^T and so find Q^{-1}.

14. Let $U = \begin{bmatrix} 0 & 1 \\ 1 & 0 \end{bmatrix}$. Show that each of U, $-U$, and $-I_2$ is its own inverse and that the product of any two of these is the third.

15. Consider $A = \begin{bmatrix} 1 & 1 \\ -1 & 0 \end{bmatrix}$, $B = \begin{bmatrix} 0 & -1 \\ 1 & 0 \end{bmatrix}$, $C = \begin{bmatrix} 0 & 1 & 0 \\ 0 & 0 & 1 \\ 5 & 0 & 0 \end{bmatrix}$.

Find the inverses by computing (a) A^6; ♦(b) B^4; and (c) C^3.

♦16. Find the inverse of $\begin{bmatrix} 1 & 0 & 1 \\ c & 1 & c \\ 3 & c & 2 \end{bmatrix}$ in terms of c.

17. If $c \neq 0$, find the inverse of $\begin{bmatrix} 1 & -1 & 1 \\ 2 & -1 & 2 \\ 0 & 2 & c \end{bmatrix}$ in terms of c.

◆18. Find the inverse of $\begin{bmatrix} \sin\theta & \cos\theta \\ -\cos\theta & \sin\theta \end{bmatrix}$ for any real number θ.

19. Show that A has no inverse when
    (a) A has a row of zeros.
    ◆(b) A has a column of zeros.
    (c) each row of A sums to 0. [*Hint:* Theorem 5(2).]
    ◆(d) each column of A sums to 0. [*Hint:* Corollary, Theorem 4.]

20. Let A denote a square matrix.
    (a) Let $YA = 0$ for some matrix $Y \neq 0$. Show that A has no inverse. [*Hint:* Corollary, Theorem 4.]
    (b) Use part (a) to show that (i) $\begin{bmatrix} 1 & -1 & 1 \\ 0 & 1 & 1 \\ 1 & 0 & 2 \end{bmatrix}$; and
    ◆(ii) $\begin{bmatrix} 2 & 1 & -1 \\ 1 & 1 & 0 \\ 1 & 0 & -1 \end{bmatrix}$ have no inverse.
    [*Hint:* For part (ii) compare row 3 with the difference between row 1 and row 2.]

21. If A is invertible, show that
    (a) $A^2 \neq 0$.
    (b) $A^k \neq 0$ for all $k = 1, 2, \ldots$.
    (c) $AX = AY$ implies $X = Y$.
    ◆(d) $PA = QA$ implies $P = Q$.

22. Suppose $AB = 0$, where A and B are square matrices. Show that:
    (a) If one of A and B has an inverse, the other is zero.
    (b) It is impossible for both A and B to have inverses.
    (c) $(BA)^2 = 0$.

23. (a) Show that $\begin{bmatrix} a & 0 \\ 0 & b \end{bmatrix}$ is invertible if and only if $a \neq 0$ and $b \neq 0$. Describe the inverse.
    (b) Show that a diagonal matrix is invertible if and only if all the main diagonal entries are nonzero. Describe the inverse.
    (c) If A and B are square matrices, show that
    (i) the block matrix $\begin{bmatrix} A & 0 \\ 0 & B \end{bmatrix}$ is invertible if

and only if A and B are both invertible; and
    (ii) $\begin{bmatrix} A & 0 \\ 0 & B \end{bmatrix}^{-1} = \begin{bmatrix} A^{-1} & 0 \\ 0 & B^{-1} \end{bmatrix}$.
    (d) Use part (c) to find the inverses of:
    (i) $\begin{bmatrix} 1 & 0 & 0 \\ 0 & 2 & -1 \\ 0 & 1 & -1 \end{bmatrix}$ ◆(ii) $\begin{bmatrix} 3 & 1 & 0 \\ 5 & 2 & 0 \\ 0 & 0 & -1 \end{bmatrix}$
    (iii) $\begin{bmatrix} 2 & 1 & 0 & 0 \\ 1 & 1 & 0 & 0 \\ 0 & 0 & 1 & -1 \\ 0 & 0 & 1 & -2 \end{bmatrix}$ ◆(iv) $\begin{bmatrix} 3 & 4 & 0 & 0 \\ 2 & 3 & 0 & 0 \\ 0 & 0 & 1 & 3 \\ 0 & 0 & 0 & -1 \end{bmatrix}$
    (e) Extend part (c) to **block diagonal matrices** —that is, matrices with square blocks down the main diagonal and zero blocks elsewhere.

24. (a) Show that $\begin{bmatrix} a & x \\ 0 & b \end{bmatrix}$ is invertible if and only if $a \neq 0$ and $b \neq 0$.
    (b) If A and B are square and invertible, show that (i) the block matrix $\begin{bmatrix} A & X \\ 0 & B \end{bmatrix}$ is invertible for any X; and
    ◆(ii) $\begin{bmatrix} A & X \\ 0 & B \end{bmatrix}^{-1} = \begin{bmatrix} A^{-1} & -A^{-1}XB^{-1} \\ 0 & B^{-1} \end{bmatrix}$.
    (c) If $\begin{bmatrix} A & X \\ 0 & B \end{bmatrix}$ is invertible, show that A and B are invertible. [*Hint:* Write the inverse as $\begin{bmatrix} A_1 & Y \\ Z & B_1 \end{bmatrix}$ where A_1 and B_1 are the same size as A and B. Use block multiplication.]

25. If A and B are invertible symmetric matrices such that $AB = BA$, show that A^{-1}, AB, AB^{-1}, and $A^{-1}B^{-1}$ are also invertible and symmetric.

26. (a) Let $A = \begin{bmatrix} 1 & 1 \\ 0 & 1 \end{bmatrix}$, $B = \begin{bmatrix} 0 & 0 \\ 1 & 2 \end{bmatrix}$, and $C = \begin{bmatrix} 1 & 1 \\ 1 & 1 \end{bmatrix}$. Verify that $AB = CA$, A is invertible, but $B \neq C$. (See Exercise 21, (c) and (d).)
    ◆(b) Find 2×2 matrices P, Q, and R such that $PQ = PR$, P is not invertible, and $Q \neq R$. (See Exercise 21.)

27. Let A be an $n \times n$ matrix and let I be the $n \times n$ identity matrix.
    (a) If $A^2 = 0$, verify that $(I - A)^{-1} = I + A$.

(b) If $A^3 = 0$, verify that $(I - A)^{-1} = I + A + A^2$.

(c) Find the inverse of $\begin{bmatrix} 1 & 2 & -1 \\ 0 & 1 & 3 \\ 0 & 0 & 1 \end{bmatrix}$.

♦(d) If $A^n = 0$, find the formula for $(I - A)^{-1}$.

28. Prove property 6 of Theorem 4: If A is invertible and $a \neq 0$, then aA is invertible and $(aA)^{-1} = \frac{1}{a}A^{-1}$.

29. Let A, B, and C denote $n \times n$ matrices. Using only Theorem 4, show that:

   (a) If A and AB are both invertible, B is invertible.

   ♦(b) If AB and BA are both invertible, A and B are both invertible.

   (c) If A, C, and ABC are all invertible, B is invertible.

30. Let A and B denote invertible $n \times n$ matrices.

   (a) If $A^{-1} = B^{-1}$, does it mean that $A = B$? Explain.

   (b) Show that $A = B$ if and only if $A^{-1}B = I$.

31. Let A, B, and C be $n \times n$ matrices, with A and B invertible. Show that

   ♦(a) If A commutes with C, then A^{-1} commutes with C.

   (b) If A commutes with B, then A^{-1} commutes with B^{-1}.

32. Let A and B be square matrices of the same size.

   (a) Show that $(AB)^2 = A^2B^2$ if $AB = BA$.

   (b) If A and B are invertible and $(AB)^2 = A^2B^2$, show that $AB = BA$.

   (c) If $A = \begin{bmatrix} 1 & 0 \\ 0 & C \end{bmatrix}$ and $B = \begin{bmatrix} 1 & 1 \\ 0 & 0 \end{bmatrix}$, show that

   $(AB)^2 = A^2B^2$ but $AB \neq BA$.

♦33. Let A and B be $n \times n$ matrices for which AB is invertible. Show that A and B are both invertible.

34. Consider $A = \begin{bmatrix} 1 & 3 & -1 \\ 2 & 1 & 5 \\ 1 & -7 & 13 \end{bmatrix}$, $B = \begin{bmatrix} 1 & 1 & 2 \\ 3 & 0 & -3 \\ -2 & 5 & 17 \end{bmatrix}$.

   (a) Show that A is not invertible by finding a nonzero 1×3 matrix Y such that $YA = 0$.
   [*Hint:* Row 3 of A equals 2(row 2) – 3(row 1).]

   ♦(b) Show that B is not invertible.
   [*Hint:* Column 3 = 3(column 2) – column 1.]

35. Show that a square matrix A is invertible if and only if it can be left-cancelled: $AB = AC$ implies $B = C$.

36. If $U^2 = I$, show that $I + U$ is not invertible unless $U = I$.

37. (a) If J is the 4×4 matrix with every entry 1, show that $I - \frac{1}{2}J$ is self-inverse and symmetric.

   (b) If X is $n \times m$ and satisfies $X^TX = I_m$, show that $I_n - 2XX^T$ is self-inverse and symmetric.

38. An $n \times n$ matrix P is called an idempotent if $P^2 = P$. Show that:

   (a) I is the only invertible idempotent.

   ♦(b) P is an idempotent if and only if $I - 2P$ is self-inverse.

   (c) U is self-inverse if and only if $U = I - 2P$ for some idempotent P.

   (d) $I - aP$ is invertible for any $a \neq 1$, and
   $$(I - aP)^{-1} = I + \left(\frac{a}{1-a}\right)P.$$

39. If $A^2 = kA$, where $k \neq 0$, show that A is invertible if and only if $A = kI$.

40. Let A and B denote $n \times n$ invertible matrices.

   (a) Show that $A^{-1} + B^{-1} = A^{-1}(A + B)B^{-1}$.

   ♦(b) If $A + B$ is also invertible, show that $A^{-1} + B^{-1}$ is invertible and find a formula for $(A^{-1} + B^{-1})^{-1}$.

41. Let A and B be $n \times n$ matrices, and let I be the $n \times n$ identity matrix.

   (a) Verify that $A(I + BA) = (I + AB)A$ and that $(I + BA)B = B(I + AB)$.

   (b) If $I + AB$ is invertible, verify that $I + BA$ is also invertible and that $(I + BA)^{-1} = I - B(I + AB)^{-1}A$.

SECTION 2.4   Elementary Matrices

It is now clear that elementary row operations are important in linear algebra. It turns out that they can be performed by left multiplying by certain invertible matrices. These matrices are the subject of this section.

An $n \times n$ matrix E is called an **elementary matrix** if it can be obtained from the identity matrix I_n by a single elementary row operation (called the operation **corresponding** to E). We say that E is of type I, II, or III if the operation is of that type (see page 6). Hence

$$E_1 = \begin{bmatrix} 0 & 1 \\ 1 & 0 \end{bmatrix}, \quad E_2 = \begin{bmatrix} 1 & 0 \\ 0 & 9 \end{bmatrix}, \quad \text{and} \quad E_3 = \begin{bmatrix} 1 & 5 \\ 0 & 1 \end{bmatrix}$$

are elementary of types I, II, and III, respectively, obtained from the 2×2 identity matrix by interchanging rows 1 and 2, multiplying row 2 by 9, and adding 5 times row 2 to row 1.

Suppose now that a matrix $A = \begin{bmatrix} a & b & c \\ p & q & r \end{bmatrix}$ is left multiplied by the elementary matrices E_1, E_2, and E_3. The results are:

$$E_1 A = \begin{bmatrix} 0 & 1 \\ 1 & 0 \end{bmatrix}\begin{bmatrix} a & b & c \\ p & q & r \end{bmatrix} = \begin{bmatrix} p & q & r \\ a & b & c \end{bmatrix}$$

$$E_2 A = \begin{bmatrix} 1 & 0 \\ 0 & 9 \end{bmatrix}\begin{bmatrix} a & b & c \\ p & q & r \end{bmatrix} = \begin{bmatrix} a & b & c \\ 9p & 9q & 9r \end{bmatrix}$$

$$E_3 A = \begin{bmatrix} 1 & 5 \\ 0 & 1 \end{bmatrix}\begin{bmatrix} a & b & c \\ p & q & r \end{bmatrix} = \begin{bmatrix} a + 5p & b + 5q & c + 5r \\ p & q & r \end{bmatrix}$$

In each case, left multiplying A by the elementary matrix has the effect of doing the corresponding row operation to A. This works is general.

Lemma 1[5]

If an elementary row operation is performed on an $m \times n$ matrix A, the result is EA where E is the elementary matrix obtained by performing the operation on the $m \times m$ identity matrix.

PROOF

Suppose the operation is of type III, say adding k times row p to row q. Let $K_1, K_2, \ldots, K_m$ denote the rows of I_m. Then row i of E is K_i if $i \neq q$, and it is $K_q + kK_p$ if $i = q$. Hence:

If $i \neq q$, then row i of $EA = K_i A = $ row i of A;

If $i = q$, then row i of $EA = (K_q + kK_p)A = K_q A + kK_p A = $ row q of A plus k(row p of A).

This proves the Lemma for type III operations; the proofs for types I and II are left as exercises.

---

5 A *lemma* is an auxiliary theorem used in the proof of other theorems.

The effect of an elementary row operation can be reversed by another such operation (called its inverse) which is also elementary of the same type (see the proof of Theorem 1 §1.1). It follows that each elementary matrix E is invertible. In fact, if a row operation on I produces E, then the inverse operation carries E back to I. By Lemma 1, this means that $FE = 1$ where F is the elementary matrix corresponding to the inverse operation. Thus $F = E^{-1}$, and we have proved

Lemma 2

Every elementary matrix E is invertible, and E^{-1} is also a elementary matrix (of the same type). Moreover, E^{-1} corresponds to the inverse of the row operation that produces E.

The following table gives the inverse of each type of elementary row operation:

Type	Operation	Inverse Operation
I	Interchange rows p and q	Interchange rows p and q
II	Multiply row p by $k \neq 0$	Multiply row p by $1/k$
III	Add k times row p to row $q \neq p$	Subtract k times row p from row q

Note that elementary matrices of type I are self-inverse.

Example 1

Find the inverse of each of the elementary matrices

$$E_1 = \begin{bmatrix} 0 & 1 & 0 \\ 1 & 0 & 0 \\ 0 & 0 & 1 \end{bmatrix}, \quad E_2 = \begin{bmatrix} 1 & 0 & 0 \\ 0 & 1 & 0 \\ 0 & 0 & 9 \end{bmatrix}, \quad \text{and} \quad E_3 = \begin{bmatrix} 1 & 0 & 5 \\ 0 & 1 & 0 \\ 0 & 0 & 1 \end{bmatrix}.$$

Solution E_1, E_2, and E_3 are of Type I, II, and III respectively, so the table gives

$$E_1^{-1} = \begin{bmatrix} 0 & 1 & 0 \\ 1 & 0 & 0 \\ 0 & 0 & 1 \end{bmatrix} = E_1, \quad E_2^{-1} = \begin{bmatrix} 1 & 0 & 0 \\ 0 & 1 & 0 \\ 0 & 0 & \frac{1}{9} \end{bmatrix}, \quad \text{and} \quad E_3^{-1} = \begin{bmatrix} 1 & 0 & -5 \\ 0 & 1 & 0 \\ 0 & 0 & 1 \end{bmatrix}.$$

Inverses and Rank

Suppose that an $m \times n$ matrix A is carried to a matrix B (written $A \to B$) by a series of k elementary row operations. Let $E_1, E_2, \ldots, E_k$ denote the corresponding elementary matrices. By Lemma 1, the reduction becomes

$$A \to E_1 A \to E_2 E_1 A \to E_3 E_2 E_1 A \to \cdots \to E_k E_{k-1} \cdots E_2 E_1 A = B.$$

In other words,

$$A \to UA = B \quad \text{where } U = E_k E_{k-1} \cdots E_2 E_1.$$

The matrix $U = E_k E_{k-1} \cdots E_2 E_1$ is invertible by Lemma 2 and Theorem 4 §2.3. Moreover, U can be computed without finding the E_i as follows: If the above series of operations carrying $A \to B$ is performed on I_m in place of A, the result is

$I_m \to UI_m = U$. Hence this series of operations carries the block matrix $[A\ I_m] \to [B\ U]$. This, together with the above discussion, proves

---

Theorem 1

Suppose A is $m \times n$ and $A \to B$ by elementary row operations.
1. $B = UA$ where U is an $m \times m$ invertible matrix.
2. U can be computed by $[A\ I_m] \to [B\ U]$ using the operations carrying $A \to B$.
3. $U = E_k E_{k-1} \cdots E_2 E_1$ where $E_1, E_2, \ldots, E_k$ are the elementary matrices corresponding (in order) to the elementary row operations carrying A to B.

Example 2

If $A = \begin{bmatrix} 2 & 3 & 1 \\ 1 & 2 & 1 \end{bmatrix}$, express the reduced row-echelon form R of A as $R = UA$ where U is invertible.

Solution Reduce the double matrix $[A\ I] \to [R\ U]$ as follows:

$$[A\ I] = \begin{bmatrix} 2 & 3 & 1 & | & 1 & 0 \\ 1 & 2 & 1 & | & 0 & 1 \end{bmatrix} \to \begin{bmatrix} 1 & 2 & 1 & | & 0 & 1 \\ 2 & 3 & 1 & | & 1 & 0 \end{bmatrix} \to \begin{bmatrix} 1 & 2 & 1 & | & 0 & 1 \\ 0 & -1 & -1 & | & 1 & -2 \end{bmatrix}$$

$$\to \begin{bmatrix} 1 & 0 & -1 & | & 2 & -3 \\ 0 & 1 & 1 & | & -1 & 2 \end{bmatrix}$$

Hence $R = \begin{bmatrix} 1 & 0 & -1 \\ 0 & 1 & 1 \end{bmatrix}$ and $U = \begin{bmatrix} 2 & -3 \\ -1 & 2 \end{bmatrix}$.

Now suppose that A is invertible. We know that $A \to I$ by Theorem 5 §2.3, so taking $B = I$ in Theorem 1 gives $[A\ I] \to [I\ U]$ where $I = UA$. Thus $U = A^{-1}$, so we have $[A\ I] \to [I\ A^{-1}]$. This is the matrix inversion algorithm, derived (in another way) in Section 2.3. However, more is true: Theorem 1 gives $A^{-1} = U = E_k E_{k-1} \cdots E_2 E_1$ where $E_1, E_2, \ldots, E_k$ are the elementary matrices corresponding (in order) to the row operations carrying $A \to I$. Hence

$$A = (A^{-1})^{-1} = (E_k E_{k-1} \cdots E_2 E_1)^{-1} = E_1^{-1} E_2^{-1} \cdots E_{k-1}^{-1} E_k^{-1}. \tag{*}$$

By Lemma 2, this shows that every invertible matrix A is a product of elementary matrices. Since elementary matrices are invertible (again by Lemma 2), this proves the following important characterization of invertible matrices.

Theorem 2

A square matrix is invertible if and only if it is a product of elementary matrices.

Example 3

Express $A = \begin{bmatrix} -2 & 3 \\ 1 & 0 \end{bmatrix}$ as a product of elementary matrices.

Solution   The reduction of $A \to I$ is as follows:

$$A = \begin{bmatrix} -2 & 3 \\ 1 & 0 \end{bmatrix} \xrightarrow{E_1} \begin{bmatrix} 1 & 0 \\ -2 & 3 \end{bmatrix} \xrightarrow{E_2} \begin{bmatrix} 1 & 0 \\ 0 & 3 \end{bmatrix} \xrightarrow{E_3} \begin{bmatrix} 1 & 0 \\ 0 & 1 \end{bmatrix}$$

where the corresponding elementary matrices are

$$E_1 = \begin{bmatrix} 0 & 1 \\ 1 & 0 \end{bmatrix}, \quad E_2 = \begin{bmatrix} 1 & 0 \\ 2 & 1 \end{bmatrix}, \quad E_3 = \begin{bmatrix} 1 & 0 \\ 0 & \frac{1}{3} \end{bmatrix}.$$

Hence equation (∗) above gives the required factorization of A:

$$A = (E_3 E_2 E_1)^{-1} = E_1^{-1} E_2^{-1} E_3^{-1} = \begin{bmatrix} 0 & 1 \\ 1 & 0 \end{bmatrix}\begin{bmatrix} 1 & 0 \\ -2 & 1 \end{bmatrix}\begin{bmatrix} 1 & 0 \\ 0 & 3 \end{bmatrix}.$$

Smith Normal Form

Let A be an $m \times n$ matrix of rank r, and let R be the reduced row-echelon form of A. Theorem 1 shows that $R = UA$ where U is invertible, and that U can be found from $[A \ I_m] \to [R \ U]$.

The matrix R has r leading ones (since *rank* $A = r$), so the $n \times m$ matrix R^T contains each row of I_r in the first r columns. Thus row operations will carry $R^T \to \begin{bmatrix} I_r & 0 \\ 0 & 0 \end{bmatrix}_{n \times m}$. Hence Theorem 1 (again) shows that $\begin{bmatrix} I_r & 0 \\ 0 & 0 \end{bmatrix}_{n \times m} = U_1 R^T$ where U_1 is an $n \times n$ invertible matrix. Writing $V = U_1^T$, we obtain

$$UAV = RV = RU_1^T = (U_1 R^T)^T = \left(\begin{bmatrix} I_r & 0 \\ 0 & 0 \end{bmatrix}_{n \times m} \right)^T = \begin{bmatrix} I_r & 0 \\ 0 & 0 \end{bmatrix}_{m \times n}.$$

Moreover, the matrix $U_1 = V^T$ can be computed by $[R^T \ I_n] \to \left[\begin{bmatrix} I_r & 0 \\ 0 & 0 \end{bmatrix}_{n \times m} \ V^T \right]$. This proves

Theorem 3

Let A be an $m \times n$ matrix of rank r. There exist invertible matrices U and V of size $m \times m$ and $n \times n$, respectively, such that

$$UAV = \begin{bmatrix} I_r & 0 \\ 0 & 0 \end{bmatrix}_{m \times n}.$$

Moreover, if R is the reduced row-echelon form of A, then:

1. U can be computed by $[A \ I_m] \to [R \ U]$;

2. V can be computed by $[R^T \ I_n] \to \left[\begin{bmatrix} I_r & 0 \\ 0 & 0 \end{bmatrix}_{n \times m} \ V^T \right]$.

If A is an $m \times n$ matrix of rank r, the matrix $\begin{bmatrix} I_r & 0 \\ 0 & 0 \end{bmatrix}$ is called the **Smith normal form**[6] of A. Whereas the reduced row-echelon form of A is the "nicest" matrix to which A can be carried by row operations, the Smith canonical form is the "nicest"

---

6 Named after Henry John Stephen Smith (1826–83).

matrix to which A can be carried by row *and column* operations. This is because doing row operations to R^T amounts to doing *column* operations to R and then transposing.

Example 4

Given $A = \begin{bmatrix} 1 & -1 & 1 & 2 \\ 2 & -2 & 1 & -1 \\ -1 & 1 & 0 & 3 \end{bmatrix}$ find invertible matrices U and V such that

$UAV = \begin{bmatrix} I_r & 0 \\ 0 & 0 \end{bmatrix}$, where $r = \text{rank } A$.

Solution The matrix U and the reduced row-echelon form R of A are computed by the row reduction $[A \ I_3] \to [R \ U]$.

$$\begin{bmatrix} 1 & -1 & 1 & 2 & | & 1 & 0 & 0 \\ 2 & -2 & 1 & -1 & | & 0 & 1 & 0 \\ -1 & 1 & 0 & 3 & | & 0 & 0 & 1 \end{bmatrix} \to \begin{bmatrix} 1 & -1 & 0 & -3 & | & -1 & 1 & 0 \\ 0 & 0 & 1 & 5 & | & 2 & -1 & 0 \\ 0 & 0 & 0 & 0 & | & -1 & 1 & 1 \end{bmatrix}$$

Hence

$$R = \begin{bmatrix} 1 & -1 & 0 & -3 \\ 0 & 0 & 1 & 5 \\ 0 & 0 & 0 & 0 \end{bmatrix} \quad \text{and} \quad U = \begin{bmatrix} -1 & 1 & 0 \\ 2 & -1 & 0 \\ -1 & 1 & 1 \end{bmatrix}$$

In particular, $r = \text{rank } R = 2$. Now row-reduce $[R^T \ I_4] \to \left[\begin{bmatrix} I_r & 0 \\ 0 & 0 \end{bmatrix} V^T\right]$:

$$\begin{bmatrix} 1 & 0 & 0 & | & 1 & 0 & 0 & 0 \\ -1 & 0 & 0 & | & 0 & 1 & 0 & 0 \\ 0 & 1 & 0 & | & 0 & 0 & 1 & 0 \\ -3 & 5 & 0 & | & 0 & 0 & 0 & 1 \end{bmatrix} \to \begin{bmatrix} 1 & 0 & 0 & | & 1 & 0 & 0 & 0 \\ 0 & 1 & 0 & | & 0 & 0 & 1 & 0 \\ 0 & 0 & 0 & | & 1 & 1 & 0 & 0 \\ 0 & 0 & 0 & | & 3 & 0 & -5 & 1 \end{bmatrix}$$

whence

$$V^T = \begin{bmatrix} 1 & 0 & 0 & 0 \\ 0 & 0 & 1 & 0 \\ 1 & 1 & 0 & 0 \\ 3 & 0 & -5 & 1 \end{bmatrix} \quad V = \begin{bmatrix} 1 & 0 & 1 & 3 \\ 0 & 0 & 1 & 0 \\ 0 & 1 & 0 & -5 \\ 0 & 0 & 0 & 1 \end{bmatrix}$$

Then $UAV = \begin{bmatrix} I_2 & 0 \\ 0 & 0 \end{bmatrix}$ as is easily verified.

Uniqueness of the Reduced Row-echelon Form

In this short subsection, Theorem 1 is used to prove the following important theorem. The result is not needed in the sequel and so may be omitted.

Theorem 4

If a matrix A is carried to reduced row-echelon matrices R and S by row operations, then $R = S$.

PROOF

Observe first that $UR = S$ for some invertible matrix U (by Theorem 1 there exist invertible matrices P and Q such that $R = PA$ and $S = QA$; take $U = QP^{-1}$). We show that $R = S$ by induction on the number m of rows of R and S. The case $m = 1$ is left to the reader. If R_j and S_j denote column j in R and S respectively, the fact that $UR = S$ gives

$$UR_j = S_j \quad \text{for each } j. \tag{$*$}$$

Since U is invertible, this shows that R and S have the same zero columns. Hence, by passing to the matrices obtained by deleting the zero columns from R and S, we may assume that R and S have no zero columns.

But then the first column of R and S is the first column of I_m, so $(*)$ shows that the first column of U is column 1 of I_m. Now write U, R, and S in block form as follows.

$$U = \begin{bmatrix} 1 & X \\ 0 & V \end{bmatrix}, \quad R = \begin{bmatrix} 1 & X \\ 0 & R' \end{bmatrix}, \quad \text{and} \quad S = \begin{bmatrix} 1 & Z \\ 0 & S' \end{bmatrix}.$$

Since $UR = S$, block multiplication gives $VR' = S'$ so, since V is invertible (U is invertible) and both R' and S' are reduced row-echelon, we obtain $R' = S'$ by induction.

It follows that R and S have leading ones in the same columns, say r of them. Applying $(*)$ to these columns shows that the first r columns of U are the first r columns of I_m. Hence we can write U, R, and S in block form as follows:

$$U = \begin{bmatrix} I_r & M \\ 0 & W \end{bmatrix}, \quad R = \begin{bmatrix} R_1 & R_2 \\ 0 & 0 \end{bmatrix}, \quad \text{and} \quad S = \begin{bmatrix} S_1 & S_2 \\ 0 & 0 \end{bmatrix}$$

where R_1 and S_1 are $r \times r$. Then block multiplication gives $UR = R$; that is, $S = R$. This completes the proof.

Exercises 2.4

1. For each of the following elementary matrices, describe the corresponding elementary row operation and write the inverse.

(a) $E = \begin{bmatrix} 1 & 0 & 3 \\ 0 & 1 & 0 \\ 0 & 0 & 1 \end{bmatrix}$ ◆(b) $E = \begin{bmatrix} 0 & 0 & 1 \\ 0 & 1 & 0 \\ 1 & 0 & 0 \end{bmatrix}$

(c) $E = \begin{bmatrix} 1 & 0 & 0 \\ 0 & \frac{1}{2} & 0 \\ 0 & 0 & 1 \end{bmatrix}$ ◆(d) $E = \begin{bmatrix} 1 & 0 & 0 \\ -2 & 1 & 0 \\ 0 & 0 & 1 \end{bmatrix}$

(e) $E = \begin{bmatrix} 0 & 1 & 0 \\ 1 & 0 & 0 \\ 0 & 0 & 1 \end{bmatrix}$ ◆(f) $E = \begin{bmatrix} 1 & 0 & 0 \\ 0 & 1 & 0 \\ 0 & 0 & 5 \end{bmatrix}$

2. In each case find an elementary matrix E such that $B = EA$.

(a) $A = \begin{bmatrix} 2 & 1 \\ 3 & -1 \end{bmatrix}$, $B = \begin{bmatrix} 2 & 1 \\ 1 & -2 \end{bmatrix}$

◆(b) $A = \begin{bmatrix} -1 & 2 \\ 0 & 1 \end{bmatrix}$, $B = \begin{bmatrix} 1 & -2 \\ 0 & 1 \end{bmatrix}$

(c) $A = \begin{bmatrix} 1 & 1 \\ -1 & 2 \end{bmatrix}$, $B = \begin{bmatrix} -1 & 2 \\ 1 & 1 \end{bmatrix}$

◆(d) $A = \begin{bmatrix} 4 & 1 \\ 3 & 2 \end{bmatrix}$, $B = \begin{bmatrix} 1 & -1 \\ 3 & 2 \end{bmatrix}$

(e) $A = \begin{bmatrix} -1 & 1 \\ 1 & -1 \end{bmatrix}$, $B = \begin{bmatrix} -1 & 1 \\ -1 & 1 \end{bmatrix}$

$\blacklozenge$(f) $A = \begin{bmatrix} 2 & 1 \\ -1 & 3 \end{bmatrix}$, $\quad B = \begin{bmatrix} -1 & 3 \\ 2 & 1 \end{bmatrix}$

3. Let $A = \begin{bmatrix} 1 & 2 \\ -1 & 1 \end{bmatrix}$ and $C = \begin{bmatrix} -1 & 1 \\ 2 & 1 \end{bmatrix}$.

    (a) Find elementary matrices E_1 and E_2 such that $C = E_2 E_1 A$.

    $\blacklozenge$(b) Show that there is *no* elementary matrix E such that $C = EA$.

4. If E is elementary, show that A and EA differ in at most two rows.

5. (a) Is I an elementary matrix? Explain.

    $\blacklozenge$(b) Is 0 an elementary matrix? Explain.

6. In each case find an invertible matrix U such that $UA = R$ is in reduced row-echelon form, and express U as a product of elementary matrices.

    (a) $A = \begin{bmatrix} 1 & -1 & 2 \\ -2 & 1 & 0 \end{bmatrix}$ $\blacklozenge$(b) $A = \begin{bmatrix} 1 & 2 & 1 \\ 5 & 12 & -1 \end{bmatrix}$

    (c) $A = \begin{bmatrix} 1 & 2 & -1 & 0 \\ 3 & 1 & 1 & 2 \\ 1 & -3 & 3 & 2 \end{bmatrix}$ $\blacklozenge$(d) $A = \begin{bmatrix} 2 & 1 & -1 & 0 \\ 3 & -1 & 2 & 1 \\ 1 & -2 & 3 & 1 \end{bmatrix}$

7. In each case find an invertible matrix U such that $UA = B$, and express U as a product of elementary matrices.

    (a) $A = \begin{bmatrix} 2 & 1 & 3 \\ -1 & 1 & 2 \end{bmatrix}$, $\quad B = \begin{bmatrix} 1 & -1 & -2 \\ 3 & 0 & 1 \end{bmatrix}$

    $\blacklozenge$(b) $A = \begin{bmatrix} 2 & -1 & 0 \\ 1 & 1 & 1 \end{bmatrix}$, $\quad B = \begin{bmatrix} 3 & 0 & 1 \\ 2 & -1 & 0 \end{bmatrix}$

8. In each case factor A as a product of elementary matrices.

    (a) $A = \begin{bmatrix} 1 & 1 \\ 2 & 1 \end{bmatrix}$ $\blacklozenge$(b) $A = \begin{bmatrix} 2 & 3 \\ 1 & 2 \end{bmatrix}$

    (c) $A = \begin{bmatrix} 1 & 0 & 2 \\ 0 & 1 & 1 \\ 2 & 1 & 6 \end{bmatrix}$ $\blacklozenge$(d) $A = \begin{bmatrix} 1 & 0 & -3 \\ 0 & 1 & 4 \\ -2 & 2 & 15 \end{bmatrix}$

9. Let E be an elementary matrix.

    (a) Show that E^T is also elementary of the same type.

    (b) Show that $E^T = E$ if E is of type I or II.

$\blacklozenge$10. Show that every matrix A can be factored as $A = UR$ where U is invertible and R is in reduced row-echelon form.

11. If $A = \begin{bmatrix} 1 & 2 \\ 1 & -3 \end{bmatrix}$ and $B = \begin{bmatrix} 5 & 2 \\ -5 & -3 \end{bmatrix}$, find an elementary matrix F such that $AF = B$. [*Hint:* See Exercise 9.]

12. In each case find invertible U and V such that $UAV = \begin{bmatrix} I_r & 0 \\ 0 & 0 \end{bmatrix}$, where $r = \operatorname{rank} A$.

    (a) $A = \begin{bmatrix} 1 & 1 & -1 \\ -2 & -2 & 4 \end{bmatrix}$ $\blacklozenge$(b) $A = \begin{bmatrix} 3 & 2 \\ 2 & 1 \end{bmatrix}$

    (c) $A = \begin{bmatrix} 1 & -1 & 2 & 1 \\ 2 & -1 & 0 & 3 \\ 0 & 1 & -4 & 1 \end{bmatrix}$ $\blacklozenge$(d) $A = \begin{bmatrix} 1 & 1 & 0 & -1 \\ 3 & 2 & 1 & 1 \\ 1 & 0 & 1 & 3 \end{bmatrix}$

13. Prove Lemma 1 for elementary matrices of: (a) type I; (b) type II.

14. While trying to invert A, $[A \; I]$ is carried to $[P \; Q]$ by row operations. Show that $P = QA$.

15. If A and B are $n \times n$ matrices and AB is a product of elementary matrices, show that the same is true of A.

16. If U is invertible, show that the reduced row-echelon form of a matrix $[U \; A]$ is $[I \; U^{-1}A]$.

17. Two matrices A and B are called **row-equivalent** (written $A \stackrel{r}{\sim} B$) if there is a sequence of elementary row operations carrying A to B.

    (a) Show that $A \stackrel{r}{\sim} B$ if and only if $A = UB$ for some invertible matrix U.

    $\blacklozenge$(b) Show that:

        (i) $A \stackrel{r}{\sim} A$ for all matrices A.

        (ii) If $A \stackrel{r}{\sim} B$, then $B \stackrel{r}{\sim} A$.

        (iii) If $A \stackrel{r}{\sim} B$ and $B \stackrel{r}{\sim} C$, then $A \stackrel{r}{\sim} C$.

    (c) Show that, if A and B are both row-equivalent to some third matrix, then $A \stackrel{r}{\sim} B$.

    (d) Show that $\begin{bmatrix} 1 & -1 & 3 & 2 \\ 0 & 1 & 4 & 1 \\ 1 & 0 & 8 & 6 \end{bmatrix}$ and $\begin{bmatrix} 1 & -1 & 4 & 5 \\ -2 & 1 & -11 & -8 \\ -1 & 2 & 2 & 2 \end{bmatrix}$ are row-equivalent. [*Hint:* Consider (c) and Theorem 1 §1.2.]

18. If U and V are invertible $n \times n$ matrices, show that $U \stackrel{r}{\sim} V$. (See Exercise 17.)

19. (See Exercise 17.) Find all matrices that are row-equivalent to:

    (a) $\begin{bmatrix} 0 & 0 & 0 \\ 0 & 0 & 0 \end{bmatrix}$ ◆(b) $\begin{bmatrix} 0 & 0 & 0 \\ 0 & 0 & 1 \end{bmatrix}$

    (c) $\begin{bmatrix} 1 & 0 & 0 \\ 0 & 1 & 0 \end{bmatrix}$ (d) $\begin{bmatrix} 1 & 2 & 0 \\ 0 & 0 & 1 \end{bmatrix}$

20. Let A and B be $m \times n$ and $n \times m$ matrices, respectively. If $m > n$, show that AB is not invertible. [*Hint:* Use Theorem 1 §1.3 to find $X \neq 0$ with $BX = 0$.]

21. Define an *elementary column operation* on a matrix to be one of the following: (I) Interchange two columns. (II) Multiply a column by a nonzero scalar. (III) Add a multiple of a column to another column. Show that:

    (a) If an elementary column operation is done to an $m \times n$ matrix A, the result is AF, where F is an $n \times n$ elementary matrix.

    (b) Given any $m \times n$ matrix A, there exist $m \times m$ elementary matrices $E_1, \ldots, E_k$ and $n \times n$ elementary matrices $F_1, \ldots, F_p$ such that, in block form,

$$E_k \cdots E_1 A F_1 \cdots F_p = \begin{bmatrix} I_r & 0 \\ 0 & 0 \end{bmatrix}.$$

22. Suppose B is obtained from A by:

    (a) interchanging rows i and j;

    ◆(b) multiplying row i by $k \neq 0$;

    (c) adding k times row i to row j $(i \neq j)$.

    In each case describe how to obtain B^{-1} from A^{-1}. [*Hint:* See part (a) of the preceding exercise.]

23. Two $m \times n$ matrices A and B are called **equivalent** (written $A \stackrel{e}{\sim} B$ if there exist invertible matrices U and V (sizes $m \times m$ and $n \times n$) such that $A = UBV$.

    (a) Prove the following the properties of equivalence.

        (i) $A \stackrel{e}{\sim} A$ for all $m \times n$ matrices A.

        (ii) If $A \stackrel{e}{\sim} B$, then $B \stackrel{e}{\sim} A$.

        (iii) If $A \stackrel{e}{\sim} B$ and $B \stackrel{e}{\sim} C$, then $A \stackrel{e}{\sim} C$.

    (b) Prove that two $m \times n$ matrices are equivalent if they have the same rank. [*Hint:* Use part (a) and Theorem 3.]

SECTION 2.5   Matrix Transformations

Up to now, our treatment of matrices has been algebraic, focusing on matrix addition, multiplication, inverses, etc. In this section we view matrices as functions (called transformations) and so reveal a whole new geometrical interpretation of matrix multiplication.

Let $\mathbb{R}$ denote the set of real numbers. If $n \geq 1$, an ordered sequence

$$(a_1, a_2, \ldots, a_n)$$

of n real numbers is called an **ordered n-tuple**. The word "ordered" is used here because we insist that two ordered n-tuples are equal if and only if corresponding entries are equal; that is, if and only if they are equal as matrices. The set of all ordered n-tuples has a special notation:

$$\mathbb{R}^n \text{ denotes the set of all ordered } n\text{-tuples.}$$

The n-tuples in $\mathbb{R}^n$ will be called **vectors** or **n-vectors**, and we will usually write them as column matrices (sometimes as rows, depending on the situation).

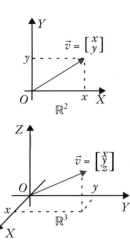

For reasons that will become clear in Chapter 4, we adopt a special notation in $\mathbb{R}^2$ and $\mathbb{R}^3$. We identify $\mathbb{R}^2$ with the euclidean plane as follows: The point $P(x, y)$ with coordinates (x, y) will be written as $\vec{v} = \begin{bmatrix} x \\ y \end{bmatrix}$ in $\mathbb{R}^2$, and denoted graphically by an arrow from the origin to P (see the first diagram). Similarly we identify $\mathbb{R}^3$ with euclidean space by writing a point $P(x, y, z)$ in $\mathbb{R}^3$ as $\vec{v} = \begin{bmatrix} x \\ y \\ z \end{bmatrix} = [x \ y \ z]^T$, again represented by an arrow in the second diagram. Thus the terms "point" and "vector" mean the same thing in the plane and in space. The reason for denoting them as arrows from the origin will become clear in Chapter 4. Finally, in the same way, the real line is identified with $\mathbb{R}^1$, and we write $\mathbb{R}^1 = \mathbb{R}$.

Transformations

We begin by describing a geometric transformation of the plane $\mathbb{R}^2$.

Example 1

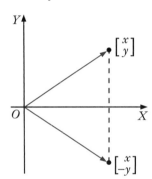

Consider the geometrical transformation of $\mathbb{R}^2$ given by *reflection* in the X axis. This operation carries the vector $\begin{bmatrix} x \\ y \end{bmatrix}$ to its reflection $\begin{bmatrix} x \\ -y \end{bmatrix}$, and so can be described geometrically as in the diagram. Now observe that

$$\begin{bmatrix} x \\ -y \end{bmatrix} = \begin{bmatrix} 1 & 0 \\ 0 & -1 \end{bmatrix}\begin{bmatrix} x \\ y \end{bmatrix}$$ so reflecting $\begin{bmatrix} x \\ y \end{bmatrix}$ in the X axis can be achieved by

multiplying by the matrix $\begin{bmatrix} 1 & 0 \\ 0 & -1 \end{bmatrix}$.

Reflection in the X axis *transforms* every vector $\vec{v}$ in $\mathbb{R}^2$ into another vector $A\vec{v}$ in $\mathbb{R}^2$. It is thus an example of a function $T : \mathbb{R}^2 \to \mathbb{R}^2$ where $T(\vec{v}) = A\vec{v}$, and as such is a generalization of the familiar functions $f : \mathbb{R} \to \mathbb{R}$ that carry each real number x to $f(x)$.

More generally, a **transformation** T from $\mathbb{R}^n$ to $\mathbb{R}^m$ is a rule that assigns to every vector X in $\mathbb{R}^n$ a uniquely determined vector $T(X)$ in $\mathbb{R}^m$. When this is the case we write

$$T : \mathbb{R}^n \to \mathbb{R}^m \quad \text{or} \quad \mathbb{R}^n \xrightarrow{T} \mathbb{R}^m$$

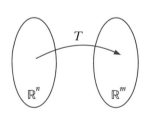

and call $\mathbb{R}^n$ and $\mathbb{R}^m$ the **domain** and **codomain** of T, respectively. The transformation T can be visualized as in the diagram.

To describe such a transformation T we must specify the vector $T(X)$ for every vector X in $\mathbb{R}^n$. This is referred to as **defining** T, or as specifying the **action** of T. Saying that the action *defines* the transformation T means that we regard two transformations $S : \mathbb{R}^n \to \mathbb{R}^m$ and $T : \mathbb{R}^n \to \mathbb{R}^m$ as **equal** if they have the same action; more formally

$$S = T \quad \text{if and only if} \quad S(X) = T(X) \quad \text{for all } X \text{ in } \mathbb{R}^n$$

Transformations are often defined by a formula.

Example 2

The formula $T[x_1 \ x_2 \ x_3 \ x_4]^T = [x_1 + x_2 \ x_2 + x_3 \ x_3 + x_4]^T$ defines a transformation $T : \mathbb{R}^4 \to \mathbb{R}^3$.

As in Example 1, matrix multiplication provides an important way of defining transformations $\mathbb{R}^n \to \mathbb{R}^m$. If A is any $m \times n$ matrix, multiplication by A gives a transformation

$$T : \mathbb{R}^n \to \mathbb{R}^m \quad \text{defined by} \quad T(X) = AX \text{ for every } X \text{ in } \mathbb{R}^n.$$

This is called the **matrix transformation induced** by A. Thus reflection in the X axis is the matrix transformation $\mathbb{R}^2 \to \mathbb{R}^2$ induced by $A = \begin{bmatrix} 1 & 0 \\ 0 & -1 \end{bmatrix}$, and the transformation $\mathbb{R}^4 \to \mathbb{R}^3$ in Example 2 is the matrix transformation induced by $A = \begin{bmatrix} 1 & 1 & 0 & 0 \\ 0 & 1 & 1 & 0 \\ 0 & 0 & 1 & 1 \end{bmatrix}$ because $\begin{bmatrix} 1 & 1 & 0 & 0 \\ 0 & 1 & 1 & 0 \\ 0 & 0 & 1 & 1 \end{bmatrix} \begin{bmatrix} x_1 \\ x_2 \\ x_3 \\ x_4 \end{bmatrix} = \begin{bmatrix} x_1 + x_2 \\ x_2 + x_3 \\ x_3 + x_4 \end{bmatrix}$.

If $A = 0$ is the zero matrix, the corresponding matrix transformation T is given by $T(X) = AX = 0$ for every X in $\mathbb{R}^n$. This is called the **zero transformation** and is denoted $T = 0$. Another example is the **identity transformation** $1_{\mathbb{R}^n}$ induced by the identity matrix I_n:

$$1_{\mathbb{R}^n} : \mathbb{R}^n \to \mathbb{R}^n \quad \text{defined by} \quad 1_{\mathbb{R}^n}(X) = X \text{ for all } X \text{ in } \mathbb{R}^n.$$

Example 3

If $a > 0$, the matrix transformation $T\begin{bmatrix} x \\ y \end{bmatrix} = \begin{bmatrix} ax \\ y \end{bmatrix}$ induced by the matrix $A = \begin{bmatrix} a & 0 \\ 0 & 1 \end{bmatrix}$ is called an **X-expansion** of $\mathbb{R}^2$ if $a > 1$, and an **X-compression** if $0 < a < 1$. The reason for the name is clear in the diagram below. Similarly, if $b > 0$ the matrix $\begin{bmatrix} 1 & 0 \\ 0 & b \end{bmatrix}$ gives rise to **Y-expansions** and **Y-compressions**.

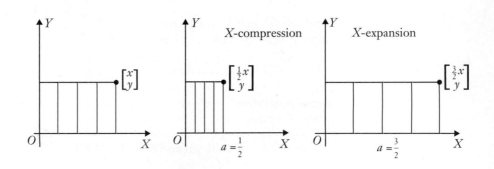

Example 4

If a is a number, the matrix transformation $T\begin{bmatrix} x \\ y \end{bmatrix} = \begin{bmatrix} x + ay \\ y \end{bmatrix}$ induced by the

matrix $A = \begin{bmatrix} 1 & a \\ 0 & 1 \end{bmatrix}$ is called an **X-shear** of $\mathbb{R}^2$ (**positive** if $a > 0$ and **negative** if

$a < 0$). Its effect is illustrated below when $a = \frac{1}{3}$ and $a = -\frac{1}{3}$.

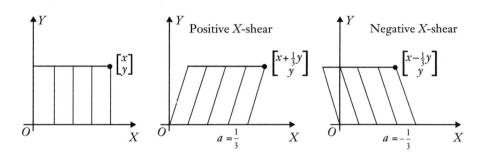

Example 5

If θ is any angle, let R_θ denote the transformation that rotates $\mathbb{R}^2$ counterclockwise about the origin through the angle θ. Then R_θ is the matrix transformation induced by the matrix $\begin{bmatrix} \cos\theta & -\sin\theta \\ \sin\theta & \cos\theta \end{bmatrix}$.

Solution

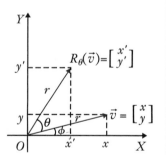

Given $\vec{v} = \begin{bmatrix} x \\ y \end{bmatrix}$, let r denote the distance from $\vec{v}$ to the origin and let ϕ be the

angle between the positive X axis and the line from the origin to $\vec{v}$ (see the

diagram). Write $R_\theta(\vec{v}) = \begin{bmatrix} x' \\ y' \end{bmatrix}$. Then the double angle formula from

trigonometry gives

$$\begin{aligned} x' &= r\cos(\theta + \phi) = r(\cos\theta\cos\phi - \sin\theta\sin\phi) \\ &= \cos\theta(r\cos\phi) - \sin\theta(r\sin\phi) \\ &= x\cos\theta - y\sin\theta \end{aligned}$$

Similarly, $y' = \sin(\theta + \phi) = x\sin\theta + y\cos\theta$. Hence

$$R_\theta(\vec{x}) = \begin{bmatrix} x' \\ y' \end{bmatrix} = \begin{bmatrix} x\cos\theta - y\sin\theta \\ x\sin\theta + y\cos\theta \end{bmatrix} = \begin{bmatrix} \cos\theta & -\sin\theta \\ \sin\theta & \cos\theta \end{bmatrix}\begin{bmatrix} x \\ y \end{bmatrix} = \begin{bmatrix} \cos\theta & -\sin\theta \\ \sin\theta & \cos\theta \end{bmatrix}\vec{x}.$$

Since this holds for every $\vec{x}$ in $\mathbb{R}^n$, we are done.

Example 6

Find the matrix of the rotation of $\mathbb{R}^3$ about the Z axis through an angle θ from the X axis to the Y axis.

Solution

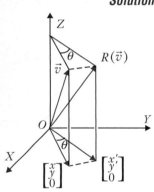

Denote the rotation by $R : \mathbb{R}^3 \to \mathbb{R}^3$. Given a point $\vec{v} = [x \ y \ z]^T$ in $\mathbb{R}^3$, we have $R(\vec{v}) = [x' \ y' \ z]^T$ where $[x' \ y' \ 0]^T$ is the result of rotating $[x \ y \ 0]^T$ through the angle θ (see the diagram). Hence $x' = x \cos\theta - y \sin\theta$ and $y' = x \sin\theta + y \cos\theta$ by Example 5. Hence

$$R(\vec{v}) = \begin{bmatrix} x \cos\theta - y \sin\theta \\ x \sin\theta + y \cos\theta \\ z \end{bmatrix} = \begin{bmatrix} \cos\theta & -\sin\theta & 0 \\ \sin\theta & \cos\theta & 0 \\ 0 & 0 & 1 \end{bmatrix} \begin{bmatrix} x \\ y \\ z \end{bmatrix} = A\vec{v}$$

so the required matrix is $A = \begin{bmatrix} \cos\theta & -\sin\theta & 0 \\ \sin\theta & \cos\theta & 0 \\ 0 & 0 & 1 \end{bmatrix}$.

We hasten to note that there are important transformations that are *not* matrix transformations. For example, if Y is a fixed column in $\mathbb{R}^n$, define the transformation $T_Y : \mathbb{R}^n \to \mathbb{R}^n$ by

$$T_Y(X) = X + Y \quad \text{for all } X \text{ in } \mathbb{R}^n.$$

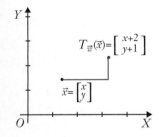

Then T_Y is called **translation** by Y. In particular, if $Y = \vec{w} = \begin{bmatrix} 2 \\ 1 \end{bmatrix}$ in $\mathbb{R}^2$, the effect of $T_{\vec{w}}$ on $\vec{v}$ is to translate $\vec{v}$ two units to the right and one unit up (see the diagram).

The translation $T_{\vec{w}}$ is not a matrix transformation unless $\vec{w} = \vec{0}$. Indeed, if $T_{\vec{w}}$ were induced by a matrix A, then $\vec{v} + \vec{w} = T_{\vec{w}}(\vec{v}) = A\vec{v}$ would hold for every $\vec{v}$ in $\mathbb{R}^n$. In particular, taking $\vec{v} = \vec{0}$ gives $\vec{w} = A\vec{0} = \vec{0}$.

Linear Transformations

Matrix transformations can be characterized by two simple properties which are embodied in the following definition. A transformation $T : \mathbb{R}^n \to \mathbb{R}^m$ is called a **linear transformation** if it satisfies the following two properties for all vectors X and Y in $\mathbb{R}^n$ and all scalars a:

T1. $T(X + Y) = T(X) + T(Y)$

T2. $T(aX) = aT(X)$

Of course, $X + Y$ and aX here are matrix addition and scalar multiplication. We say that *T preserves addition* if T1 holds, and that *T preserves scalar multiplication* if T2 holds. Moreover, taking $a = 0$ and $a = -1$ in T2 gives

$$T(0) = 0 \quad \text{and} \quad T(-X) = -T(X)$$

Hence T preserves the zero vector and negatives. Even more is true.

A vector Y in $\mathbb{R}^n$ is called a **linear combination** of vectors $X_1, X_2, \ldots, X_k$ if Y has the form

$$Y = a_1 X_1 + a_2 X_2 + \cdots + a_k X_k$$

for some scalars $a_1, a_2, \ldots, a_k$. Conditions T1 and T2 combine to show that every linear transformation T *preserves linear combinations* in the sense of the following theorem.

Theorem 1

If $T : \mathbb{R}^n \to \mathbb{R}^m$ is a linear transformation, then
$$T(a_1 X_1 + a_2 X_2 + \cdots + a_k X_k) = a_1 T(X_1) + a_2 T(X_2) + \cdots + a_k T(X_k)$$
for all scalars a_i and all vectors X_i in $\mathbb{R}^n$.

PROOF

We use induction on k. If $k = 1$, it is T2. If it holds for some $k > 1$, then

$$
\begin{aligned}
T(a_1 X_1 &+ a_2 X_2 + \cdots + a_k X_k + a_{k+1} X_{k+1}) \\
&= T[(a_1 X_1 + a_2 X_2 + \cdots + a_k X_k) + a_{k+1} X_{k+1}] \\
&= T(a_1 X_1 + a_2 X_2 + \cdots + a_k X_k) + T(a_{k+1} X_{k+1}) \\
&= [a_1 T(X_1) + a_2 T(X_2) + \cdots + a_k T(X_k)] + a_{k+1} T(X_{k+1})
\end{aligned}
$$

This proves the result for $k + 1$, and so completes the induction.

Example 7

If a linear transformation $T : \mathbb{R}^2 \to \mathbb{R}^2$ satisfies $T[1 \; 1]^T = [2 \; -3]^T$ and $T[1 \; -2]^T = [5 \; 1]^T$, find $T[4 \; 3]^T$.

Solution Write $Z = [4 \; 3]^T$, $X = [1 \; 1]^T$, and $Y = [1 \; -2]^T$ for convenience. Then we know $T(X)$ and $T(Y)$ and we want $T(Z)$, so it is enough by Theorem 1 to express Z as a linear combination of X and Y. That is, we want to find numbers a and b such that $Z = aX + bY$. Equating entries gives two equations $4 = a + b$ and $3 = a - 2b$. Hence, $a = \frac{11}{3}$ and $b = \frac{1}{3}$, so $Z = \frac{11}{3} X + \frac{1}{3} Y$. Thus Theorem 1 gives

$$T(Z) = \tfrac{11}{3} T(X) + \tfrac{1}{3} T(Y) = \tfrac{11}{3}\begin{bmatrix} 2 \\ -3 \end{bmatrix} + \tfrac{1}{3}\begin{bmatrix} 5 \\ 1 \end{bmatrix} = \tfrac{1}{3}\begin{bmatrix} 27 \\ -32 \end{bmatrix}$$

This is what we wanted.

Example 8

Every matrix transformation $T : \mathbb{R}^n \to \mathbb{R}^m$ is a linear transformation.

Solution If T is induced by the $m \times n$ matrix A, then $T(X) = AX$ for all X in $\mathbb{R}^n$. Hence the basics of matrix multiplication show that

$$T(X + Y) = A(X + Y) = AX + AY = T(X) + T(Y), \quad \text{and}$$
$$T(aX) = A(aX) = a(AX) = aT(X)$$

hold for all X and Y in $\mathbb{R}^n$ and all scalars a. Hence T satisfies T1 and T2, and so is linear.

The remarkable thing is that the *converse* of Example 8 is true: Every linear transformation $T : \mathbb{R}^n \to \mathbb{R}^m$ is actually a matrix transformation. To prove this, we define the **standard basis** of $\mathbb{R}^n$ to be the columns

$$E_1, E_2, \ldots, E_n$$

of the identity matrix I_n. Then each E_i is in $\mathbb{R}^n$ and every vector $X = [v_1\ v_2\ \cdots\ v_n]^T$ in $\mathbb{R}^n$ is a linear combination of the E_i. In fact:

$$X = [v_1\ v_2\ \cdots\ v_n]^T = v_1E_1 + v_2E_2 + \cdots + v_nE_n$$

as the reader can verify. Hence Theorem 1 shows that

$$T(X) = T(v_1E_1 + v_2E_2 + \cdots + v_nE_n) = v_1T(E_1) + v_2T(E_2) + \cdots + v_nT(E_n).$$

Now observe that each $T(E_i)$ is a column in $\mathbb{R}^m$, so the matrix $A = [T(E_1)\ T(E_2)\ \cdots\ T(E_n)]$ is an $m \times n$ matrix. Hence we can apply Theorem 4 §2.2 to get

$$T(X) = v_1T(E_1) + v_2T(E_2) + \cdots + v_nT(E_n)$$

$$= [T(E_1)\ T(E_2) \cdots T(E_n)]\begin{bmatrix} v_1 \\ v_2 \\ \vdots \\ v_n \end{bmatrix}$$

$$= AX.$$

Since this holds for every X in $\mathbb{R}^n$, it shows that T is the matrix transformation induced by A, and so proves most of the following theorem.

Theorem 2

Let $T : \mathbb{R}^n \to \mathbb{R}^m$ be a transformation.
1. T is linear if and only if it is a matrix transformation.
2. If T is linear, then T is induced by a unique matrix A, given in terms of its columns by
$$A = [T(E_1)\ T(E_2)\ \cdots\ T(E_n)]$$
where $\{E_1, E_2, \ldots, E_n\}$ is the standard basis of $\mathbb{R}^n$.

PROOF

It remains to verify that the matrix A is unique. If T is induced by another matrix B, then $BX = T(X) = AX$ holds for all X in $\mathbb{R}^n$. Hence

$$B = BI_n = B[E_1\ \cdots\ E_n] = [BE_1\ \cdots\ BE_n] = [AE_1\ \cdots\ AE_n] = AI_n = A$$

This is what we wanted.

Hence we can speak of *the* matrix of a linear transformation. Because of Theorem 2, we may (and will) use the phrases "linear transformation" and "matrix transformation" interchangeably.

The argument in the proof of Theorem 2 also proves the following.

Lemma 1

Let A and B be $m \times n$ matrices. If $AX = BX$ for every column X in $\mathbb{R}^n$, then $A = B$.

As an application, consider the rotation R_θ of $\mathbb{R}^2$ counterclockwise through the angle θ. This transformation is linear by Example 5 so we can use Theorem 2 to

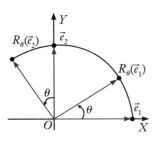

obtain its matrix. Indeed, if $\vec{e}_1 = \begin{bmatrix} 1 \\ 0 \end{bmatrix}$ and $\vec{e}_2 = \begin{bmatrix} 0 \\ 1 \end{bmatrix}$, then the diagram shows that

$$R_\theta(\vec{e}_1) = \begin{bmatrix} \cos\theta \\ \sin\theta \end{bmatrix} \quad \text{and} \quad R_\theta(\vec{e}_2) = \begin{bmatrix} -\sin\theta \\ \cos\theta \end{bmatrix},$$

so the matrix is $[R_\theta(\vec{e}_1) \quad R_\theta(\vec{e}_2)] = \begin{bmatrix} \cos\theta & -\sin\theta \\ \sin\theta & \cos\theta \end{bmatrix}$, as in Example 5.

Of course, we must know that R_θ is linear to apply Theorem 2.

Composition

Sometimes two transformations $T : \mathbb{R}^n \to \mathbb{R}^m$ and $S : \mathbb{R}^m \to \mathbb{R}^k$ link together because the codomain of T is the same as the domain of S. This is illustrated graphically if we write them as follows:

$$\mathbb{R}^n \xrightarrow{T} \mathbb{R}^m \xrightarrow{S} \mathbb{R}^k.$$

Hence we can apply T first and then apply S, and the result is a new transformation

$$\mathbb{R}^n \xrightarrow{S \circ T} \mathbb{R}^k$$

called the **composite** of S and T, defined by

$$(S \circ T)(X) = S[T(X)] \quad \text{for all } X \text{ in } \mathbb{R}^n.$$

The action of $S \circ T$ can be described as "first T and then S" (note the order!)[7]. This transformation is described in the diagram.

A natural question is whether the composite of two linear transformations is again linear. The answer is "yes" and even more is true.

Theorem 3

Let $T : \mathbb{R}^n \to \mathbb{R}^m$ and $S : \mathbb{R}^m \to \mathbb{R}^k$ be linear transformations. Then $S \circ T : \mathbb{R}^n \to \mathbb{R}^k$ is also linear. Indeed:

If S has matrix A and T has matrix B, then $S \circ T$ has matrix AB.

PROOF

If X is any vector in $\mathbb{R}^n$, then $S(X) = AX$ and $T(X) = BX$, so

$$(S \circ T)(X) = S[T(X)] = S[BX] = A[BX] = (AB)X$$

by the associativity of matrix multiplication. Hence $S \circ T$ is the matrix transformation with matrix AB, and so is linear.

Theorem 3 enables the action of the composite of two geometric transformations to be computed algebraically using matrix multiplication. But it also provides a very useful geometric interpretation of matrix multiplication: The product matrix AB

---

7 When reading the notation $S \circ T$ we read S first and then T, even though the action is "first T and then S". This annoying state of affairs results because we write $T(X)$ for the effect of the transformation T on X, with T on the left. If we wrote this as $(X)T$ this confusion would not occur. However the notation $T(X)$ is well established.

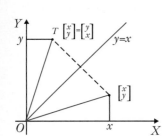

corresponds to the transformation resulting from first applying B and then applying A (again note the order). Thus the study of matrices can cast light on geometrical transformations, and vice versa.

Before giving an example, let $T : \mathbb{R}^2 \to \mathbb{R}^2$ denote reflection in the line $y = x$. Then the diagram shows that $T\begin{bmatrix} x \\ y \end{bmatrix} = \begin{bmatrix} y \\ x \end{bmatrix}$. Hence $T\begin{bmatrix} x \\ y \end{bmatrix} = \begin{bmatrix} 0 & 1 \\ 1 & 0 \end{bmatrix}\begin{bmatrix} y \\ x \end{bmatrix}$, so T is the matrix transformation induced by the matrix $A = \begin{bmatrix} 0 & 1 \\ 1 & 0 \end{bmatrix}$. We use this in the following example.

Example 9

Show that reflection in the X axis followed by rotation through $\pi/2$ is reflection in the line $y = x$.

Solution

The composite in question is $R_{\pi/2} \circ T$ where T is reflection in the X axis and $R_{\pi/2}$ is rotation through $\pi/2$. Since $R_{\pi/2}$ has matrix $A = \begin{bmatrix} 0 & -1 \\ 1 & 0 \end{bmatrix}$ (by Example 1), and T has matrix $B = \begin{bmatrix} 1 & 0 \\ 0 & -1 \end{bmatrix}$, the matrix of $R_{\pi/2} \circ T$ is

$$AB = \begin{bmatrix} 0 & -1 \\ 1 & 0 \end{bmatrix}\begin{bmatrix} 1 & 0 \\ 0 & -1 \end{bmatrix} = \begin{bmatrix} 0 & 1 \\ 1 & 0 \end{bmatrix}$$ by Theorem 3, which is the matrix of reflection

in the line $y = x$ by the above discussion.

Theorem 4

Let R, S, and T be linear transformations.
  1. If $\mathbb{R}^n \xrightarrow{T} \mathbb{R}^m$, then $T \circ 1_{\mathbb{R}^n} = T$ and $1_{\mathbb{R}^m} \circ T = T$.
  2. If $\mathbb{R}^n \xrightarrow{T} \mathbb{R}^m \xrightarrow{S} \mathbb{R}^k \xrightarrow{R} \mathbb{R}^l$, then $R \circ (S \circ T) = (R \circ S) \circ T$.

PROOF

1. We have $(T \circ 1_{\mathbb{R}^n})(X) = T[1_{\mathbb{R}^n}(X)] = T[X]$ for all X in $\mathbb{R}^n$, so $T \circ 1_{\mathbb{R}^n} = T$. A similar computation shows that $1_{\mathbb{R}^m} \circ T = T$.
2. Here
$$\{R \circ (S \circ T)\}(X) = R\{(S \circ T)(X)\} = R\{S[T(X)]\} \quad \text{for all } X \text{ in } \mathbb{R}^n.$$

Similarly $\{(R \circ S) \circ T\}(X) = R\{S[T(X)]\}$, and (2) follows.

If R, S, and T have matrices A, B, and C respectively, it follows from Theorem 3 that $R \circ (S \circ T)$ has matrix $A(BC)$, and $(R \circ S) \circ T$ has matrix $(AB)C$. Thus $A(BC)$ and $(AB)C$ are both matrices of the same transformation $R \circ (S \circ T) = (R \circ S) \circ T$, and so $A(BC) = (AB)C$ by the *uniqueness* in Theorem 2. Of course we knew this in Section 2.2, but it also follows from Theorems 3 and 4.[8]

---

8 Note that this does not *prove* that $A(BC) = (AB)C$ because we *used* this fact to establish Theorem 3!

Example 10

Let L denote the line through the origin in $\mathbb{R}^2$ that makes an angle θ with the positive X axis. If $Q : \mathbb{R}^2 \to \mathbb{R}^2$ is reflection in L, show that Q is linear with matrix $\begin{bmatrix} \cos(2\theta) & \sin(2\theta) \\ \sin(2\theta) & -\cos(2\theta) \end{bmatrix}$.

Solution   Let Q_X denote reflection in the X axis. The transformation Q can be accomplished by first rotating the line through $-\theta$, then reflecting in the X axis, and finally rotating back through θ. In other words,

$$Q = R_\theta \circ Q_X \circ R_{-\theta}.$$

Hence Theorem 3 shows that Q is linear. Moreover, since R_θ and Q_X have matrices $\begin{bmatrix} \cos\theta & -\sin\theta \\ \sin\theta & \cos\theta \end{bmatrix}$ and $\begin{bmatrix} 1 & 0 \\ 0 & -1 \end{bmatrix}$ respectively, T has matrix

$$\begin{bmatrix} \cos\theta & -\sin\theta \\ \sin\theta & \cos\theta \end{bmatrix}\begin{bmatrix} 1 & 0 \\ 0 & -1 \end{bmatrix}\begin{bmatrix} \cos(-\theta) & -\sin(-\theta) \\ \sin(-\theta) & \cos(-\theta) \end{bmatrix} = \begin{bmatrix} \cos\theta & \sin\theta \\ \sin\theta & -\cos\theta \end{bmatrix}\begin{bmatrix} \cos\theta & \sin\theta \\ -\sin\theta & \cos\theta \end{bmatrix}$$

$$= \begin{bmatrix} \cos^2\theta - \sin^2\theta & 2\cos\theta\sin\theta \\ 2\cos\theta\sin\theta & \sin^2\theta - \cos^2\theta \end{bmatrix}$$

$$= \begin{bmatrix} \cos(2\theta) & \sin(2\theta) \\ \sin(2\theta) & -\cos(2\theta) \end{bmatrix}$$

Inverses

Let $T : \mathbb{R}^n \to \mathbb{R}^n$ denote a linear transformation, and let A denote its matrix. Since A is square, it may very well be invertible, and this leads to the question:

What does it mean geometrically for T that A is invertible?

To answer this, let $T' : \mathbb{R}^n \to \mathbb{R}^n$ denote the transformation induced by A^{-1}. Then

$$\begin{aligned} T'[T(X)] &= A^{-1}[AX] = IX = X \\ T[T'(X)] &= A[A^{-1}X] = IX = X \end{aligned} \quad \text{for all } X \text{ in } \mathbb{R}^n \qquad (*)$$

The first of these equations asserts that, if T carries X to a vector $T(X)$, then T' carries $T(X)$ right back to X; that is T' "reverses" the action of T. Similarly T "reverses" the action of T'. The conditions in $(*)$ can be stated compactly in terms of composition:

$$T' \circ T = 1_{\mathbb{R}^n} \quad \text{and} \quad T \circ T' = 1_{\mathbb{R}^n} \qquad (**)$$

When these conditions hold, we say that the linear transformation T' is an **inverse** of T, and we have shown that if the matrix A of T is invertible, then T has an inverse (induced by A^{-1}).

The converse is also true: If T has an inverse, then its matrix A must be invertible. Indeed, suppose $S : \mathbb{R}^n \to \mathbb{R}^n$ is any inverse of T, so that $S \circ T = 1_{\mathbb{R}^n}$ and $T \circ S = 1_{\mathbb{R}^n}$. If B is the matrix of S, we have

$$BAX = S[T(X)] = (S \circ T)(X) = 1_{\mathbb{R}^n}(X) = X \quad \text{for all } X \text{ in } \mathbb{R}^n$$

It follows by Lemma 1 that $BA = I_n$, and a similar argument shows that $AB = I_n$. Hence A is invertible with $A^{-1} = B$. Furthermore, the inverse transformation S has matrix A^{-1}, so $S = T'$ using the earlier notation. This proves the following important theorem.

Theorem 5

Let $T : \mathbb{R}^n \to \mathbb{R}^n$ denote a linear transformation with matrix A. Then

$$A \text{ is invertible if and only if } T \text{ has an inverse.}$$

In this case, T has exactly one inverse (which we denote as T^{-1}), and $T^{-1} : \mathbb{R}^n \to \mathbb{R}^n$ is the transformation induced by the matrix A^{-1}.

The geometrical relationship between T and T^{-1} is embodied in equations $(*)$ above:

$$\begin{array}{l} T^{-1}[T(X)] = X \\ T[T^{-1}(X)] = X \end{array} \quad \text{for all } X \text{ in } \mathbb{R}^n$$

These equations are called the **fundamental identities** relating T and T^{-1}. Loosely speaking, they assert that each of T and T^{-1} "reverses" or "undoes" the action of the other.

This geometric view of the inverse of a linear transformation provides a new way to find the inverse of a matrix A. More precisely, if A is an invertible matrix, we proceed as follows:

1. Let T be the linear transformation induced by A.
2. Obtain the linear transformation T^{-1} which "reverses" the action of T.
3. Then A^{-1} is the matrix of T^{-1}.

Here is an example.

Example 11

If θ is any angle, find A^{-1} if $A = \begin{bmatrix} \cos\theta & -\sin\theta \\ \sin\theta & \cos\theta \end{bmatrix}$.

Solution   By Example 5, A is the matrix of the rotation R_θ through the angle θ. Hence A^{-1} will be the matrix of the inverse transformation $(R_\theta)^{-1}$. But $(R_\theta)^{-1}$ must be the rotation $R_{-\theta}$ through the angle $-\theta$ because $R_{-\theta}$ "reverses" the action of R_θ. Hence A^{-1} is the matrix of $R_{-\theta}$, so (by Example 5 again with θ replaced by $-\theta$)

$$A^{-1} = \begin{bmatrix} \cos(-\theta) & -\sin(-\theta) \\ \sin(-\theta) & \cos(-\theta) \end{bmatrix} = \begin{bmatrix} \cos\theta & \sin\theta \\ -\sin\theta & \cos\theta \end{bmatrix}$$

Of course, this can be verified directly.

Exercises 2.5

1. Give the matrix of the transformation T in each case:

   (a) $T : \mathbb{R}^2 \to \mathbb{R}^2$ is reflection in the Y axis.

   ♦(b) $T : \mathbb{R}^2 \to \mathbb{R}^2$ is reflection in the line $y = -x$.

   (c) $T : \mathbb{R}^2 \to \mathbb{R}^2$ is rotation through $\pi/4$.

   ♦(d) $T : \mathbb{R}^2 \to \mathbb{R}^2$ is rotation through $-\pi/2$.

   (e) $T : \mathbb{R}^3 \to \mathbb{R}^3$ is reflection in the X-Y plane.

   ♦(f) $T : \mathbb{R}^3 \to \mathbb{R}^3$ is reflection in the X-Z plane.

   (g) $T : \mathbb{R}^3 \to \mathbb{R}^3$ is rotation through θ about the X axis (from the Y axis to the Z axis).

   ♦(h) $T : \mathbb{R}^3 \to \mathbb{R}^3$ is rotation through θ about the Y axis (from the X axis to the Z axis).

2. In each case show that $T : \mathbb{R}^2 \to \mathbb{R}^2$ is not a linear transformation:

   (a) $T([x \ y]^T) = [xy \ 0]^T$.

   ♦(b) $T([x \ y]^T) = [0 \ y^2]^T$.

3. If $T: \mathbb{R}^2 \to \mathbb{R}^2$ is as in (a), ◆(b), (c), and ◆(d) of Exercise 1, find $T([1\ 1]^T)$ and $T([2\ -1]^T)$.

4. If $a > 0$ is a fixed real number, define $T: \mathbb{R}^n \to \mathbb{R}^n$ by $T(X) = aX$ for all X in $\mathbb{R}^n$. Show that T is a linear transformation, and find its matrix. [T is called a **dilation** if $a > 0$, and a **contraction** if $a < 0$.]

5. The transformation $T: \mathbb{R}^3 \to \mathbb{R}^2$ given by $T([x\ y\ z]^T) = [x\ y]^T$ for all $[x\ y\ z]^T$ in $\mathbb{R}^3$ is called a projection of $\mathbb{R}^3$ onto $\mathbb{R}^2$. Show that T is linear and find its matrix.

◆6. Let $C_1, C_2, \ldots, C_n$ be fixed columns in $\mathbb{R}^m$, and define $T: \mathbb{R}^n \to \mathbb{R}^m$ by $T([x_1\ x_2\ \cdots\ x_n]^T) = x_1 C_1 + x_2 C_2 + \cdots + x_n C_n$ for all $[x_1\ x_2\ \cdots\ x_n]^T$ in $\mathbb{R}^n$. Show that T is a linear transformation and find the matrix of T.

7. Let R and S be matrix transformations $\mathbb{R}^n \to \mathbb{R}^m$ induced by matrices A and B respectively. In each case, show that T is a matrix transformation and describe its matrix in terms of A and B.
   (a) $T(X) = R(X) + S(X)$ for all X in $\mathbb{R}^n$.
   ◆(b) $T(X) = aR(X)$ for all X in $\mathbb{R}^n$ (where a is a fixed real number).

8. Show that the following hold for all linear transformations $T: \mathbb{R}^n \to \mathbb{R}^m$:
   (a) $T(0) = 0$.
   ◆(b) $T(-X) = -T(X)$ for all X in $\mathbb{R}^n$.

9. Let $T: \mathbb{R}^4 \to \mathbb{R}^3$ be a linear transformation. In each case determine $T(X)$. See Example 7.
   (a) $T([1\ 1\ 0\ -2]^T) = [2\ 3\ -1]^T$ and $T([0\ -1\ 1\ 1]^T) = [5\ 0\ 1]^T$; $X = [1\ 3\ -2\ -4]^T$.
   ◆(b) $T([1\ 1\ 1\ 1]^T) = [5\ 1\ -3]^T$ and $T([-1\ 1\ 0\ 2]^T) = [2\ 0\ 1]^T$; $X = [5\ -1\ 2\ -4]^T$.

10. Express reflection in the line $y = -x$ [Exercise 1(b)] as the composition of a rotation followed by a reflection in the line $y = x$.

11. Show that $R_\theta \circ R_\phi = R_{\theta+\phi}$ in two ways: (a) Geometrically; (b) Using matrices.

12. In each case find a rotation or reflection that equals the given transformation.

(a) Reflection in the Y axis followed by rotation through $\pi/2$.
◆(b) Rotation through π followed by reflection in the X axis.
(c) Rotation through $\pi/2$ followed by reflection in the line $y = x$.
◆(d) Reflection in the X axis followed by rotation through $\pi/2$.
(e) Reflection in the line $y = x$ followed by reflection in the X axis.
◆(f) Reflection in the X axis followed by reflection in the line $y = x$.

13. Find the inverse of the X-expansion in Example 3 and describe it geometrically.

14. Find the inverse of the shear transformation in Example 4 and describe it geometrically.

15. The transformation $T: \mathbb{R}^n \to \mathbb{R}^m$ defined by $T(X) = 0$ for all X in $\mathbb{R}^n$ is called the **zero transformation**.
   (a) Show that the zero transformation is linear and find its matrix.
   (b) Let $E_1, E_2, \ldots, E_n$ denote the columns of the $n \times n$ identity matrix. If $T: \mathbb{R}^n \to \mathbb{R}^m$ is linear and $T(E_i) = 0$ for each i, show that T is the zero transformation. [*Hint:* Theorem 1.]

16. Write the elements of $\mathbb{R}^n$ and $\mathbb{R}^m$ as rows. If A is an $m \times n$ matrix, define $T: \mathbb{R}^m \to \mathbb{R}^n$ by $T(Y) = YA$ for all rows Y in $\mathbb{R}^m$.
   (a) Show that T is a linear transformation.
   (b) Show that the rows of A are $T(F_1), T(F_2), \ldots, T(F_m)$ where F_i denotes row i of I_m.

17. Let $S: \mathbb{R}^n \to \mathbb{R}^n$ and $T: \mathbb{R}^n \to \mathbb{R}^n$ be linear transformations with matrices A and B respectively. [*Hint:* Lemma 1.]
   (a) Show that $B^2 = B$ if and only if $T^2 = T$ (where T^2 means $T \circ T$).
   ◆(b) Show that $B^2 = I$ if and only if $T^2 = 1_{\mathbb{R}^n}$.
   (c) Show that $AB = BA$ if and only if $S \circ T = T \circ S$.

18. Let L denote the line through the origin in $\mathbb{R}^2$ with slope m. Show that reflection in L has

matrix $A = \dfrac{1}{1+m^2} \begin{bmatrix} 1 - m^2 & 2m \\ 2m & m^2 - 1 \end{bmatrix}$.

[*Hint:* See Example 10 where $m = \tan \theta$.]

19. Given a matrix $A = \begin{bmatrix} a & b \\ c & d \end{bmatrix}$, use it to define a real

    valued function f_A by $f_A(x) = \dfrac{ax + b}{cx + d}$.

(a) If B is 2×2 show that $f_B \circ f_A = f_{BA}$.

(b) If A is invertible and $f_A(x) = f_A(y)$, show that $x = y$.

SECTION 2.6   LU-Factorization[9]

The solution to a system $AX = B$ of linear equations can be solved quickly if A can be factored as $A = LU$ where L and U are of a particularly nice form. In this section we show that gaussian elimination can be used to find such factorizations.

Triangular Matrices

As for square matrices, if $A = [a_{ij}]$ is an $m \times n$ matrix, the set of elements $a_{11}, a_{22}, a_{33}, \ldots$ form the **main diagonal** of A. Then A is called **upper triangular** if every entry below and to the left of the main diagonal is zero. Every row-echelon matrix is upper triangular, as are the matrices

$$\begin{bmatrix} 1 & -1 & 0 & 3 \\ 0 & 2 & 1 & 1 \\ 0 & 0 & -3 & 0 \end{bmatrix} \quad \begin{bmatrix} 0 & 2 & 1 & 0 & 5 \\ 0 & 0 & 0 & 3 & 1 \\ 0 & 0 & 1 & 0 & 1 \end{bmatrix} \quad \begin{bmatrix} 1 & 1 & 1 \\ 0 & -1 & 1 \\ 0 & 0 & 0 \\ 0 & 0 & 0 \end{bmatrix}$$

By analogy, a matrix A is called **lower triangular** if its transpose is upper triangular, that is if each entry above and to the right of the main diagonal is zero. A matrix is called **triangular** if it is upper or lower triangular.

Example 1

Solve the system

$$\begin{aligned} x_1 + 2x_2 - 3x_3 - x_4 + 5x_5 &= 3 \\ 5x_3 + x_4 + x_5 &= 8 \\ 2x_5 &= 6 \end{aligned}$$

where the coefficient matrix is upper triangular.

Solution   As for a row-echelon matrix, let $x_2 = s$ and $x_4 = t$. Then solve for x_5, x_3, and x_1 in that order as follows. The last equation gives

$$x_5 = \tfrac{6}{2} = 3$$

Substitution into the second last equation gives

$$x_3 = 1 - \tfrac{1}{5}t$$

Finally, substitution of both x_5 and x_3 into the first equation gives

$$x_1 = -9 - 2s + \tfrac{2}{5}t.$$

The method used in Example 1 is called **back substitution** because later variables are substituted into earlier equations. It works because the coefficient matrix is

---

9 This section is not used later and so may be omitted with no loss of continuity.

upper triangular. Similarly, if the coefficient matrix is lower triangular the system can be solved by **forward substitution** where earlier variables are substituted into later equations. As observed in Section 1.2, these procedures are more efficient than Gauss–Jordan elimination.

Now consider a system $AX = B$ where A can be factored as $A = LU$ where L is lower triangular and U is upper triangular. Then the system $AX = B$ can be solved in two stages as follows:

1. First solve $LY = B$ for Y by forward substitution.
2. Then solve $UX = Y$ for X by back substitution.

Then X is a solution to $AX = B$ because $AX = LUX = LY = B$. Moreover, every solution X arises this way (take $Y = UX$). Furthermore the method adapts easily for use in a computer.

This focuses attention on efficiently obtaining such factorizations $A = LU$. The following result will be needed; the proof is straightforward and is left as Exercises 7 and 8.

Lemma 1

Let A and B denote matrices.
1. If A and B are both lower (upper) triangular, the same is true of AB.
2. If A is $n \times n$ and lower (upper) triangular, then A is invertible if and only if no main diagonal entry is zero. In this case A^{-1} is also lower (upper) triangular.

LU-Factorization

Let A be an $m \times n$ matrix. Then A can be carried to a row-echelon matrix U (that is, upper triangular). As in Section 2.4, the reduction is

$$A \to E_1 A \to E_2 E_1 A \to E_3 E_2 E_1 A \to \cdots \to E_k E_{k-1} \cdots E_2 E_1 A = U$$

where $E_1, E_2, \ldots, E_k$ are elementary matrices corresponding to the row operations used. Hence

$$A = LU$$

where $L = (E_k E_{k-1} \cdots E_2 E_1)^{-1} = E_1^{-1} E_2^{-1} \cdots E_{k-1}^{-1} E_k^{-1}$. If we do not insist that U is reduced then, except for row interchanges, none of these row operations involve adding a row to a row *above* it. Thus, if no row interchanges are used, all the E_i are *lower* triangular, and so L is lower triangular (and invertible) by Lemma 1. This proves the following theorem. For convenience, let us say that A can be **lower reduced** if it can be carried to row-echelon form using no row interchanges.

Theorem 1

If A can be lower reduced to a row-echelon matrix U, then

$$A = LU$$

where L is lower triangular and invertible and U is upper triangular and row-echelon.

A factorization $A = LU$ as in Theorem 1 is called an **LU-factorization** of A. Such a factorization may not exist (Exercise 4) because A cannot be carried to row-echelon form using no row interchange. A procedure for dealing with this situation will be outlined later. However, if an LU-factorization $A = LU$ does exist, then the gaussian algorithm gives U and also leads to a procedure for finding L. Example 2 provides an illustration. For convenience, the first nonzero column from the left in a matrix A is called the **leading column** of A.

Example 2

Find an LU-factorization of $A = \begin{bmatrix} 0 & 2 & -6 & -2 & 4 \\ 0 & -1 & 3 & 3 & 2 \\ 0 & -1 & 3 & 7 & 10 \end{bmatrix}$.

Solution   We lower reduce A to row-echelon form as follows:

$$A = \begin{bmatrix} 0 & \boxed{2} & -6 & -2 & 4 \\ 0 & -1 & 3 & 3 & 2 \\ 0 & -1 & 3 & 7 & 10 \end{bmatrix} \rightarrow \begin{bmatrix} 0 & 1 & -3 & -1 & 2 \\ 0 & 0 & 0 & \boxed{2} & 4 \\ 0 & 0 & 0 & \boxed{6} & 12 \end{bmatrix} \rightarrow \begin{bmatrix} 0 & 1 & -3 & -1 & 2 \\ 0 & 0 & 0 & 1 & 2 \\ 0 & 0 & 0 & 0 & 0 \end{bmatrix} = U$$

The circled columns are determined as follows: The first is the leading column of A, and is used (by lower reduction) to create the first leading 1 and create zeros below it. This completes the work on row 1, and we repeat the procedure on the matrix consisting of the remaining rows. Thus the second circled column is the leading column of this smaller matrix, which we use to create the second leading 1 and the zeros below it. As the remaining row is zero here, we are finished. Then $A = LU$ where

$$L = \begin{bmatrix} 2 & 0 & 0 \\ -1 & 2 & 0 \\ -1 & 6 & 1 \end{bmatrix}.$$

This matrix L is obtained from I_3 by replacing the bottom of the first two columns by the circled columns in the reduction. Note that the rank of A is 2 here, and this is the number of circled columns.

The calculation in Example 2 works in general. There is no need to calculate the elementary matrices E_i, and the method is suitable for use in a computer because the circled columns can be stored in memory as they are created. The procedure can be formally stated as follows:

LU-Algorithm

Let A be an $m \times n$ matrix of rank r, and suppose that A can be lower reduced to a row-echelon matrix U. Then $A = LU$ where the lower triangular, invertible matrix L is constructed as follows:

1. If $A = 0$, take $L = I_m$ and $U = 0$.
2. If $A \neq 0$, write $A_1 = A$ and let C_1 be the leading column of A_1. Use C_1 to create the first leading 1 and create zeros below it (using lower reduction). When this is completed, let A_2 denote the matrix consisting of rows 2 to m of the matrix just created.
3. If $A_2 \neq 0$, let C_2 be the leading column of A_2 and repeat Step 2 on A_2 to create A_3.
4. Continue in this way until U is reached, where all rows below the last leading 1 consist of zeros. This will happen after r steps.
5. Create L by placing $C_1, C_2, \dots, C_r$ at the bottom of the first r columns of I_m.

A proof of the LU-algorithm is given at the end of this section.

LU-factorization is particularly important if, as often happens in business and industry, a series of equations $AX = B_1$, $AX = B_2, \ldots$, $AX = B_k$ must be solved, each with the same coefficient matrix A. It is very efficient to solve the first system by gaussian elimination, simultaneously creating an LU-factorization of A, and then using the factorization to solve the remaining systems by forward and back substitution.

Example 3

Find an LU-factorization for $A = \begin{bmatrix} 5 & -5 & 10 & 0 & 5 \\ -3 & 3 & 2 & 2 & 1 \\ -2 & 2 & 0 & -1 & 0 \\ 1 & -1 & 10 & 2 & 5 \end{bmatrix}$.

Solution The reduction to row-echelon form is

$$\begin{bmatrix} 5 & -5 & 10 & 0 & 5 \\ -3 & 3 & 2 & 2 & 1 \\ -2 & 2 & 0 & -1 & 0 \\ 1 & -1 & 10 & 2 & 5 \end{bmatrix} \rightarrow \begin{bmatrix} 1 & -1 & 2 & 0 & 1 \\ 0 & 0 & 8 & 2 & 4 \\ 0 & 0 & 4 & -1 & 2 \\ 0 & 0 & 8 & 2 & 4 \end{bmatrix}$$

$$\rightarrow \begin{bmatrix} 1 & -1 & 2 & 0 & 1 \\ 0 & 0 & 1 & \frac{1}{4} & \frac{1}{2} \\ 0 & 0 & 0 & -2 & 0 \\ 0 & 0 & 0 & 0 & 0 \end{bmatrix}$$

$$\rightarrow \begin{bmatrix} 1 & -1 & 2 & 0 & 1 \\ 0 & 0 & 1 & \frac{1}{4} & \frac{1}{2} \\ 0 & 0 & 0 & 1 & 0 \\ 0 & 0 & 0 & 0 & 0 \end{bmatrix}$$

If U denotes this row-echelon matrix, then $A = LU$, where

$$L = \begin{bmatrix} 5 & 0 & 0 & 0 \\ -3 & 8 & 0 & 0 \\ -2 & 4 & -2 & 0 \\ 1 & 8 & 0 & 1 \end{bmatrix}$$

The next example deals with a case where no row of zeros is present in U (in fact, A is invertible).

Example 4

Find an LU-factorization for $A = \begin{bmatrix} 2 & 4 & 2 \\ 1 & 1 & 2 \\ -1 & 0 & 2 \end{bmatrix}$.

Solution The reduction to row-echelon form is

$$\begin{bmatrix} 2 & 4 & 2 \\ 1 & 1 & 2 \\ -1 & 0 & 2 \end{bmatrix} \rightarrow \begin{bmatrix} 1 & 2 & 1 \\ 0 & -1 & 1 \\ 0 & 2 & 3 \end{bmatrix} \rightarrow \begin{bmatrix} 1 & 2 & 1 \\ 0 & 1 & -1 \\ 0 & 0 & 5 \end{bmatrix} \rightarrow \begin{bmatrix} 1 & 2 & 1 \\ 0 & 1 & -1 \\ 0 & 0 & 1 \end{bmatrix} = U$$

Hence $A = LU$ where $L = \begin{bmatrix} 2 & 0 & 0 \\ 1 & -1 & 0 \\ -1 & 2 & 5 \end{bmatrix}$.

There are matrices (for example $\begin{bmatrix} 0 & 1 \\ 1 & 0 \end{bmatrix}$) that have no LU-factorization and so require at least one row interchange when being carried to row-echelon form via the gaussian algorithm. However, it turns out that, if all the row interchanges encountered in the algorithm are carried out first, the resulting matrix requires no interchanges and so has an LU-factorization. Here is the precise result.

Theorem 2

Suppose an $m \times n$ matrix A is carried to a row-echelon matrix U via the gaussian algorithm. Let $P_1, P_2, \ldots, P_s$ be the elementary matrices corresponding (in order) to the row interchanges used, and write $P = P_s \cdots P_2 P_1$. (If no interchanges are used take $P = I_m$.) Then:

1. PA is the matrix obtained from A by doing these interchanges (in order) to A.
2. PA has an LU-factorization.

The proof is given at the end of this section.

A matrix P that is the product of elementary matrices corresponding to row interchanges is called a **permutation matrix**. Such a matrix is obtained from the identity matrix by arranging the rows in a different order, so it has exactly one 1 in each row and each column, and has zeros elsewhere. We regard the identity matrix as a permutation matrix. The elementary permutation matrices are those obtained from I by a single row interchange, and every permutation matrix is a product of elementary ones.

Example 5

If $A = \begin{bmatrix} 0 & 0 & -1 & 2 \\ -1 & -1 & 1 & 2 \\ 2 & 1 & -3 & 6 \\ 0 & 1 & -1 & 4 \end{bmatrix}$, find a permutation matrix P such that PA has an LU-factorization, and then find the factorization.

Solution   Apply the gaussian algorithm to A:

$$A \overset{*}{\to} \begin{bmatrix} -1 & -1 & 1 & 2 \\ 0 & 0 & -1 & 2 \\ 2 & 1 & -3 & 6 \\ 0 & 1 & -1 & 4 \end{bmatrix} \to \begin{bmatrix} 1 & 1 & -1 & -2 \\ 0 & 0 & -1 & 2 \\ 0 & -1 & -1 & 10 \\ 0 & 1 & -1 & 4 \end{bmatrix} \overset{*}{\to} \begin{bmatrix} 1 & 1 & -1 & -2 \\ 0 & -1 & -1 & 10 \\ 0 & 0 & -1 & 2 \\ 0 & 1 & -1 & 4 \end{bmatrix}$$

$$\to \begin{bmatrix} 1 & 1 & -1 & -2 \\ 0 & 1 & 1 & -10 \\ 0 & 0 & -1 & 2 \\ 0 & 0 & -2 & 14 \end{bmatrix} \to \begin{bmatrix} 1 & 1 & -1 & -2 \\ 0 & 1 & 1 & -10 \\ 0 & 0 & 1 & -2 \\ 0 & 0 & 0 & 10 \end{bmatrix}$$

Two row interchanges were needed (marked with $*$), first rows 1 and 2 and then rows 2 and 3. Hence, as in Theorem 2,

$$P = \begin{bmatrix} 1 & 0 & 0 & 0 \\ 0 & 0 & 1 & 0 \\ 0 & 1 & 0 & 0 \\ 0 & 0 & 0 & 1 \end{bmatrix} \begin{bmatrix} 0 & 1 & 0 & 0 \\ 1 & 0 & 0 & 0 \\ 0 & 0 & 1 & 0 \\ 0 & 0 & 0 & 1 \end{bmatrix} = \begin{bmatrix} 0 & 1 & 0 & 0 \\ 0 & 0 & 1 & 0 \\ 1 & 0 & 0 & 0 \\ 0 & 0 & 0 & 1 \end{bmatrix}$$

If we do these interchanges (in order) to A, the result is PA. Now apply the LU-algorithm to PA:

$$PA = \begin{bmatrix} -1 & -1 & 1 & 2 \\ 2 & 1 & -3 & 6 \\ 0 & 0 & -1 & 2 \\ 0 & 1 & -1 & 4 \end{bmatrix} \rightarrow \begin{bmatrix} 1 & 1 & -1 & -2 \\ 0 & -1 & -1 & 10 \\ 0 & 0 & -1 & 2 \\ 0 & 1 & -1 & 4 \end{bmatrix} \rightarrow \begin{bmatrix} 1 & 1 & -1 & -2 \\ 0 & 1 & 1 & -10 \\ 0 & 0 & -1 & 2 \\ 0 & 0 & -2 & 14 \end{bmatrix}$$

$$\rightarrow \begin{bmatrix} 1 & 1 & -1 & -2 \\ 0 & 1 & 1 & -10 \\ 0 & 0 & 1 & -2 \\ 0 & 0 & 0 & 10 \end{bmatrix} \rightarrow \begin{bmatrix} 1 & 1 & -1 & -2 \\ 0 & 1 & 1 & -10 \\ 0 & 0 & 1 & -2 \\ 0 & 0 & 0 & 1 \end{bmatrix} = U$$

Hence, $PA = LU$, where $L = \begin{bmatrix} -1 & 0 & 0 & 0 \\ 2 & -1 & 0 & 0 \\ 0 & 0 & -1 & 0 \\ 0 & 1 & -2 & 10 \end{bmatrix}$ and $U = \begin{bmatrix} 1 & 1 & -1 & -2 \\ 0 & 1 & 1 & -10 \\ 0 & 0 & 1 & -2 \\ 0 & 0 & 0 & 1 \end{bmatrix}$.

Theorem 2 provides an important general factorization theorem for matrices. If A is any $m \times n$ matrix, it asserts that there exists a permutation matrix P and an LU-factorization $PA = LU$. Moreover, it shows that either $P = I$ or $P = P_s \cdots P_2 P_1$, where $P_1, P_2, \ldots, P_s$ are the elementary permutation matrices arising in the reduction of A to row-echelon form. Now observe that $P_i^{-1} = P_i$ for each i (they are elementary row interchanges). Thus, $P^{-1} = P_1 P_2 \cdots P_s$, so the matrix A can be factored as

$$A = P^{-1}LU$$

where P^{-1} is a permutation matrix, L is lower triangular and invertible, and U is a row-echelon matrix. This is called a **PLU-factorization** of A.

The LU-factorization in Theorem 1 is not unique. For example,

$$\begin{bmatrix} 1 & 0 \\ 3 & 2 \end{bmatrix}\begin{bmatrix} 1 & -2 & 3 \\ 0 & 0 & 0 \end{bmatrix} = \begin{bmatrix} 1 & 0 \\ 3 & 1 \end{bmatrix}\begin{bmatrix} 1 & -2 & 3 \\ 0 & 0 & 0 \end{bmatrix}$$

However, it is necessary here that the row-echelon matrix has a row of zeros. Recall that the rank of a matrix A is the number of nonzero rows in any row-echelon matrix U to which A can be carried by row operations. Thus, if A is $m \times n$, the matrix U has no row of zeros if and only if A has rank m.

Theorem 3

Let A be an $m \times n$ matrix that has an LU-factorization

$$A = LU$$

If A has rank m (that is, U has no row of zeros), then L and U are uniquely determined by A.

PROOF

Suppose $A = MV$ is another LU-factorization of A, so M is lower triangular and invertible and V is row-echelon. Hence $LU = MV$, and we must show that $L = M$ and $U = V$. We write $N = M^{-1}L$. Then N is lower triangular and invertible (Lemma 1) and $NU = V$, so it suffices to prove that $N = I$. If N is $m \times m$, we use induction on m. The case $m = 1$ is left to the reader. If $m > 1$, observe first that column 1 of V is N times column 1 of U. Thus if either column is zero, so is the other (N is invertible). Hence, we can assume (by deleting zero columns) that the $(1, 1)$-entry is 1 in both U and V.

Now we write $N = \begin{bmatrix} a & 0 \\ X & N_1 \end{bmatrix}$, $U = \begin{bmatrix} 1 & Y \\ 0 & U_1 \end{bmatrix}$, and $V = \begin{bmatrix} 1 & Z \\ 0 & V_1 \end{bmatrix}$ in block form.

Then $NU = V$ becomes $\begin{bmatrix} a & aY \\ X & XY + N_1U_1 \end{bmatrix} = \begin{bmatrix} 1 & Z \\ 0 & V_1 \end{bmatrix}$. Hence $a = 1$, $Y = Z$, $X = 0$, and $N_1U_1 = V_1$. But $N_1U_1 = V_1$ implies $N_1 = I$ by induction, whence $N = I$.

If A is an $m \times m$ invertible matrix, then A has rank m by Theorem 5 §2.3. Hence, we get the following important special case of Theorem 3.

Corollary

If an invertible matrix A has an LU-factorization $A = LU$, then L and U are uniquely determined by A.

Of course, in this case U is an upper triangular matrix with 1s along the main diagonal.

Proofs of Theorems

PROOF OF THE LU-ALGORITHM

If $C_1, C_2, \ldots, C_r$ are columns of lengths $m, m - 1, \ldots, m - r + 1$, respectively, write $L^{(m)}(C_1, C_2, \ldots, C_r)$ for the lower triangular $m \times m$ matrix obtained from I_m by placing $C_1, C_2, \ldots, C_r$ at the bottom of the first r columns of I_m.

Proceed by induction on n. If $A = 0$ or $n = 1$, it is left to the reader. If $n > 1$, let C_1 denote the leading column of A and let K_1 denote the first column of the $m \times m$ identity matrix. There exist elementary matrices $E_1, \ldots, E_k$ such that, in block form,

$$(E_k \cdots E_2E_1)A = \left[\; 0 \;\middle|\; K_1 \;\middle|\; \frac{X_1}{A_1} \right] \quad \text{where } (E_k \cdots E_2E_1)C_1 = K_1.$$

Moreover, each E_j can be taken to be lower triangular (by assumption). Write

$$G = (E_k \cdots E_2E_1)^{-1} = E_1^{-1}E_2^{-1} \cdots E_k^{-1}$$

Then G is lower triangular, and $GK_1 = C_1$. Also, each E_j (and so each E_j^{-1}) is the result of either multiplying row 1 of I_m by a constant or adding a multiple of row 1 to another row. Hence,

$$G = (E_1^{-1}E_2^{-1} \cdots E_k^{-1})I_m = \left[\; C_1 \;\middle|\; \frac{0}{I_{m-1}} \right]$$

in block form. Now, by induction, let $A_1 = L_1U_1$ be an LU-factorization of A_1, where $L_1 = L^{(m-1)}[C_2, \ldots, C_r]$ and U_1 is row-echelon. Then block multiplication gives

$$G^{-1}A = \left[\; 0 \;\middle|\; K_1 \;\middle|\; \frac{X_1}{L_1U_1} \right] = \begin{bmatrix} 1 & 0 \\ 0 & L_1 \end{bmatrix}\left[\begin{array}{cc|c} 0 & 1 & X_1 \\ 0 & 0 & U_1 \end{array}\right]$$

Hence $A = LU$, where $U = \left[\begin{array}{cc|c} 0 & 1 & X_1 \\ 0 & 0 & U_1 \end{array}\right]$ is row-echelon and

$$L = \left[\; C_1 \;\middle|\; \frac{0}{I_{m-1}} \right]\begin{bmatrix} 1 & 0 \\ 0 & L_1 \end{bmatrix} = \left[\; C_1 \;\middle|\; \frac{0}{L_1} \right] = L^{(m)}[C_1, C_2, \ldots, C_r].$$

This completes the proof.

PROOF OF THEOREM 2

Let A be a nonzero $m \times n$ matrix and let K_j denote column j of I_m. There is a permutation matrix P_1 (where either P_1 is elementary or $P_1 = I_m$) such that the first nonzero column C_1 of $P_1 A$ has a nonzero entry on top. Hence, as in the LU-algorithm,

$$L^{(m)}[C_1]^{-1} \cdot P_1 \cdot A = \left[\begin{array}{c|c|c} 0 & 1 & X_1 \\ \hline 0 & 0 & A_1 \end{array}\right]$$

in block form. Then let P_2 be a permutation matrix (either elementary or I_m) such that

$$P_2 \cdot L^{(m)}[C_1]^{-1} \cdot P_1 \cdot A = \left[\begin{array}{c|c|c} 0 & 1 & X_1 \\ \hline 0 & 0 & A_1' \end{array}\right]$$

and the first nonzero column C_2 of A_1' has a nonzero entry on top. Thus,

$$L^{(m)}[K_1, C_2]^{-1} \cdot P_2 \cdot L^{(m)}[C_1]^{-1} \cdot P_1 \cdot A = \left[\begin{array}{cc|c} 0 & 1 & X_1 \\ \hline 0 & 0 & \begin{array}{c|c|c} 0 & 1 & X_2 \\ \hline 0 & 0 & A_2 \end{array} \end{array}\right]$$

in block form. Continue to obtain elementary permutation matrices $P_1, P_2, \ldots, P_r$ and columns $C_1, C_2, \ldots, C_r$ of lengths $m, m-1, \ldots$, such that

$$(L_r P_r L_{r-1} P_{r-1} \cdots L_2 P_2 L_1 P_1) A = U$$

where U is a row-echelon matrix and $L_j = L^{(m)}[K_1, \ldots, K_{j-1}, C_j]^{-1}$ for each j, where the notation means the first $j-1$ columns are those of I_m. It is not hard to verify that each L_j has the form $L_j = L^{(m)}[K_1, \ldots, K_{j-1}, C_j']$ where C_j' is a column of length $m-j+1$. We now claim that each permutation matrix P_k can be "moved past" each matrix L_j to the right of it, in the sense that

$$P_k L_j = L_j' P_k$$

where $L_j' = L^{(m)}[K_1, \ldots, K_{j-1}, C_j'']$ for some column C_j'' of length $m-j+1$. Given that this is true, we obtain a factorization of the form

$$(L_r L_{r-1}' \cdots L_2' L_1')(P_r P_{r-1} \cdots P_2 P_1) A = U$$

If we write $P = P_r P_{r-1} \cdots P_2 P_1$, this shows that PA has an LU-factorization because $L_r L_{r-1}' \cdots L_2' L_1'$ is lower triangular and invertible. All that remains is to prove the following rather technical result.

Lemma 2

Let P_k result from interchanging row k of I_m with a row below it. If $j < k$, let C_j be a column of length $m-j+1$. Then there is another column C_j' of length $m-j+1$ such that

$$P_k \cdot L^{(m)}[K_1, \ldots, K_{j-1}, C_j] = L^{(m)}[K_1, \ldots, K_{j-1}, C_j'] \cdot P_k$$

The proof is left as Exercise 11.

Exercises 2.6

1. Find an LU-factorization of the following matrices.

(a) $\begin{bmatrix} 2 & 6 & -2 & 0 & 2 \\ 3 & 9 & -3 & 3 & 1 \\ -1 & -3 & 1 & -3 & 1 \end{bmatrix}$

♦(b) $\begin{bmatrix} 2 & 4 & 2 \\ 1 & -1 & 3 \\ -1 & 7 & -7 \end{bmatrix}$

(c) $\begin{bmatrix} 2 & 6 & -2 & 0 & 2 \\ 1 & 5 & -1 & 2 & 5 \\ 3 & 7 & -3 & -2 & 5 \\ -1 & -1 & 1 & 2 & 3 \end{bmatrix}$

◆(d) $\begin{bmatrix} -1 & -3 & 1 & 0 & -1 \\ 1 & 4 & 1 & 1 & 1 \\ 1 & 2 & -3 & -1 & 1 \\ 0 & -2 & -4 & -2 & 0 \end{bmatrix}$

(e) $\begin{bmatrix} 2 & 2 & 4 & 6 & 0 & 2 \\ 1 & -1 & 2 & 1 & 3 & 1 \\ -2 & 2 & -4 & -1 & 1 & 6 \\ 0 & 2 & 0 & 3 & 4 & 8 \\ -2 & 4 & -4 & 1 & -2 & 6 \end{bmatrix}$ ◆(f) $\begin{bmatrix} 2 & 2 & -2 & 4 & 2 \\ 1 & -1 & 0 & 2 & 1 \\ 3 & 1 & -2 & 6 & 3 \\ 1 & 3 & -2 & 2 & 1 \end{bmatrix}$

2. Find a permutation matrix P and an LU-factorization of PA if A is:

(a) $\begin{bmatrix} 0 & 0 & 2 \\ 0 & -1 & 4 \\ 3 & 5 & 1 \end{bmatrix}$ ◆(b) $\begin{bmatrix} 0 & -1 & 2 \\ 0 & 0 & 4 \\ -1 & 2 & 1 \end{bmatrix}$

(c) $\begin{bmatrix} 0 & -1 & 2 & 1 & 3 \\ -1 & 1 & 3 & 1 & 4 \\ 1 & -1 & -3 & 6 & 2 \\ 2 & -2 & -4 & 1 & 0 \end{bmatrix}$ ◆(d) $\begin{bmatrix} -1 & -2 & 3 & 0 \\ 2 & 4 & -6 & 5 \\ 1 & 1 & -1 & 3 \\ 2 & 5 & -10 & 1 \end{bmatrix}$

3. In each case use the given LU-decomposition of A to solve the system $AX = B$ by finding Y such that $LY = B$, and then X such that $UX = Y$:

(a) $A = \begin{bmatrix} 2 & 0 & 0 \\ 0 & -1 & 0 \\ 1 & 1 & 3 \end{bmatrix}\begin{bmatrix} 1 & 0 & 0 & 1 \\ 0 & 0 & 1 & 2 \\ 0 & 0 & 0 & 1 \end{bmatrix}$; $B = \begin{bmatrix} 1 \\ -1 \\ 2 \end{bmatrix}$

◆(b) $A = \begin{bmatrix} 2 & 0 & 0 \\ 1 & 3 & 0 \\ -1 & 2 & 1 \end{bmatrix}\begin{bmatrix} 1 & 1 & 0 & -1 \\ 0 & 1 & 0 & 1 \\ 0 & 0 & 0 & 0 \end{bmatrix}$; $B = \begin{bmatrix} -2 \\ -1 \\ 1 \end{bmatrix}$

(c) $A = \begin{bmatrix} -2 & 0 & 0 & 0 \\ 1 & -1 & 0 & 0 \\ -1 & 0 & 2 & 0 \\ 0 & 1 & 0 & 2 \end{bmatrix}\begin{bmatrix} 1 & -1 & 2 & -1 \\ 0 & 1 & 1 & -4 \\ 0 & 0 & 1 & -\frac{1}{2} \\ 0 & 0 & 0 & 1 \end{bmatrix}$; $B = \begin{bmatrix} 1 \\ -1 \\ 2 \\ 0 \end{bmatrix}$

◆(d) $A = \begin{bmatrix} 2 & 0 & 0 & 0 \\ 1 & -1 & 0 & 0 \\ -1 & 1 & 2 & 0 \\ 3 & 0 & 1 & -1 \end{bmatrix}\begin{bmatrix} 1 & -1 & 0 & 1 \\ 0 & 1 & -2 & -1 \\ 0 & 0 & 1 & 1 \\ 0 & 0 & 0 & 0 \end{bmatrix}$; $B = \begin{bmatrix} 4 \\ -6 \\ 4 \\ 5 \end{bmatrix}$

4. Show that $\begin{bmatrix} 0 & 1 \\ 1 & 0 \end{bmatrix} = LU$ is impossible where L is lower triangular and U is upper triangular.

◆5. Show that we can accomplish any row interchange by using only row operations of other types.

6. (a) Let L and L_1 be invertible lower triangular matrices, and let U and U_1 be invertible upper triangular matrices. Show that $LU = L_1U_1$ if and only if there exists an invertible diagonal matrix D such that $L_1 = LD$ and $U_1 = D^{-1}U$. [*Hint:* Scrutinize $L^{-1}L_1 = UU_1^{-1}$.]

◆(b) Use part (a) to prove Theorem 3 in the case that A is invertible.

◆7. Prove Lemma 1(1). [*Hint:* Use block multiplication and induction.]

8. Prove Lemma 1(2). [*Hint:* Use block multiplication and induction.]

9. A triangular matrix is called **unit triangular** if it is square and every main diagonal element is a 1.

(a) If A can be carried by the gaussian algorithm to row-echelon form using no row interchanges, show that $A = LU$ where L is unit lower triangular and U is upper triangular.

(b) Show that the factorization in (a) is unique.

10. Let $C_1, C_2, \ldots, C_r$ be columns of lengths $m, m-1, \ldots, m-r+1$. If K_j denotes column j of I_m, show that $L^{(m)}[C_1, C_2, \ldots, C_r] = L^{(m)}[C_1]L^{(m)}[K_1, C_2]L^{(m)}[K_1, K_2, C_3]\cdots L^{(m)}[K_1, K_2, \ldots, K_{r-1}, C_r]$. The notation is as in the proof of Theorem 2. [*Hint:* Use induction on m and block multiplication.]

11. Prove Lemma 2. [*Hint:* $P_k^{-1} = P_k$. Write $P_k = \begin{bmatrix} I_k & 0 \\ 0 & P_0 \end{bmatrix}$ in block form where P_0 is an $(m-k) \times (m-k)$ permutation matrix.]

SECTION 2.7   An Application to Input-Output Economic Models[10]

In 1973 Wassily Leontief was awarded the Nobel prize in economics for his work on mathematical models.[11] Roughly speaking, an economic system in this model

---

10 The applications in this section and the next are independent and may be taken in any order.

11 See W. W. Leontief, "The world economy of the year 2000," *Scientific American*, Sept. 1980.

consists of several industries, each of which produces a product and each of which uses some of the production of the other industries. The following example is typical.

Example 1

A primitive society has three basic needs: food, shelter, and clothing. There are thus three industries in the society—the farming, housing, and garment industries—that produce these commodities. Each of these industries consumes a certain proportion of the total output of each commodity according to the following table.

		OUTPUT		
		Farming	**Housing**	**Garment**
	Farming	0.4	0.2	0.3
CONSUMPTION	**Housing**	0.2	0.6	0.4
	Garment	0.4	0.2	0.3

Find the annual prices that each industry must charge for its income to equal its expenditures.

Solution   Let p_1, p_2, and p_3 be the prices charged per year by the farming, housing, and garment industries, respectively, for their total output. To see how these prices are determined, consider the farming industry. It receives p_1 for its production in any year. But it *consumes* products from all these industries in the following amounts (from row 1 of the table): 40% of the food, 20% of the housing, and 30% of the clothing. Hence, the expenditures of the farming industry are $0.4p_1 + 0.2p_2 + 0.3p_3$, so

$$0.4p_1 + 0.2p_2 + 0.3p_3 = p_1$$

A similar analysis of the other two industries leads to the following system of equations.

$$0.4p_1 + 0.2p_2 + 0.3p_3 = p_1$$
$$0.2p_1 + 0.6p_2 + 0.4p_3 = p_2$$
$$0.4p_1 + 0.2p_2 + 0.3p_3 = p_3$$

This has the matrix form $EP = P$, where

$$E = \begin{bmatrix} 0.4 & 0.2 & 0.3 \\ 0.2 & 0.6 & 0.4 \\ 0.4 & 0.2 & 0.3 \end{bmatrix} \quad \text{and} \quad P = \begin{bmatrix} p_1 \\ p_2 \\ p_3 \end{bmatrix}$$

The equations can be written as the homogeneous system

$$(I - E)P = 0$$

where I is the 3×3 identity matrix, and the solutions are

$$P = \begin{bmatrix} 2t \\ 3t \\ 2t \end{bmatrix}$$

where t is a parameter. Thus, the pricing must be such that the total output of the farming industry has the same value as the total output of the garment industry, whereas the total value of the housing industry must be $\frac{3}{2}$ as much.

In general, suppose an economy has n industries, each of which uses some (possibly none) of the production of every industry. We assume first that the economy is **closed** (that is, no product is exported or imported) and that all product is used. Given two industries i and j, let e_{ij} denote the proportion of the total annual output of industry j that is consumed by industry i. Then $E = [e_{ij}]$ is called the **input-output** matrix for the economy. Clearly,

$$0 \le e_{ij} \le 1 \quad \text{for all } i \text{ and } j \tag{1}$$

Moreover, all the output from industry j is used by *some* industry (the model is closed), so

$$e_{1j} + e_{2j} + \cdots + e_{nj} = 1 \quad \text{for each } j \tag{2}$$

This condition asserts that each column of E sums to 1. Matrices satisfying conditions 1 and 2 are called **stochastic matrices**.

As in Example 1, let p_i denote the price of the total annual production of industry i. Then p_i is the annual revenue of industry i. On the other hand, industry i spends $e_{i1}p_1 + e_{i2}p_2 + \cdots + e_{in}p_n$ annually for the product it uses ($e_{ij}p_j$ is the cost for product from industry j). The closed economic system is said to be in **equilibrium** if the annual expenditure equals the annual revenue for each industry—that is, if

$$e_{i1}p_1 + e_{i2}p_2 + \cdots + e_{in}p_n = p_i \quad \text{for each } i = 1, 2, \ldots, n$$

If we write $P = \begin{bmatrix} p_1 \\ p_2 \\ \vdots \\ p_n \end{bmatrix}$, these equations can be written as the matrix equation

$$EP = P$$

This is called the **equilibrium condition**, and the solutions P are called **equilibrium price structures**. The equilibrium condition can be written as

$$(I - E)P = 0$$

which is a system of homogeneous equations for P. Moreover, there is always a non-trivial solution P. Indeed, the column sums of $I - E$ are all 0 (because E is stochastic), so the row-echelon form of $I - E$ has a row of zeros. In fact, more is true:

Theorem 1

Let E be any $n \times n$ stochastic matrix. Then there is a nonzero $n \times 1$ matrix P with nonnegative entries such that $EP = P$. If all the entries of E are positive, the matrix P can be chosen with all entries positive.

Theorem 1 guarantees the existence of an equilibrium price structure for any closed input-output system of the type discussed here. The proof is beyond the scope of this book.[12]

Example 2

Find the equilibrium price structures for four industries if the input-output matrix is

$$E = \begin{bmatrix} .6 & .2 & .1 & .1 \\ .3 & .4 & .2 & 0 \\ .1 & .3 & .5 & .2 \\ 0 & .1 & .2 & .7 \end{bmatrix}$$

Find the prices if the total value of business is \$1000.

---

12 The interested reader is referred to P. Lancaster's *Theory of Matrices* (New York: Academic Press, 1969) or to E. Seneta's *Non-negative Matrices* (New York: Wiley, 1973).

Solution   If $P = \begin{bmatrix} p_1 \\ p_2 \\ p_3 \\ p_4 \end{bmatrix}$ is the equilibrium price structure, then the equilibrium condition

is $EP = P$. When we write this as $(I - E)P = 0$, the methods of Chapter 1 yield the following family of solutions:

$$P = \begin{bmatrix} 44t \\ 39t \\ 51t \\ 47t \end{bmatrix}$$

where t is a parameter. If we insist that $p_1 + p_2 + p_3 + p_4 = 1000$, then $t = 5.525$ (to four figures). Hence

$$P = \begin{bmatrix} 243.09 \\ 215.47 \\ 281.76 \\ 259.67 \end{bmatrix}$$

to five figures.

The Open Model

We now assume that there is a demand for products in the **open sector** of the economy, which is the part of the economy other than the producing industries (for example, consumers). Let d_i denote the total value of the demand for product i in the open sector. If p_i and e_{ij} are as before, the value of the annual demand for product i by the producing industries themselves is $e_{i1}p_1 + e_{i2}p_2 + \cdots + e_{in}p_n$, so the total annual revenue p_i of industry i breaks down as follows:

$$p_i = (e_{i1}p_1 + e_{i2}p_2 + \cdots + e_{in}p_n) + d_i \quad \text{for each } i = 1, 2, \ldots, n$$

The column $D = \begin{bmatrix} d_1 \\ \vdots \\ d_n \end{bmatrix}$ is called the **demand matrix**, and this gives a matrix equation

$$P = EP + D$$

or

$$(I - E)P = D \tag{*}$$

This is a system of linear equations for P, and we ask for a solution P with every entry nonnegative. Note that every entry of E is between 0 and 1, but the column sums of E need not equal 1 as in the closed model.

Before proceeding, it is convenient to introduce a useful notation. If $A = [a_{ij}]$ and $B = [b_{ij}]$ are matrices of the same size, we write $A > B$ if $a_{ij} > b_{ij}$ for all i and j, and we write $A \geq B$ if $a_{ij} \geq b_{ij}$ for all i and j. Thus $P \geq 0$ means that every entry of P is nonnegative. Note that $A \geq 0$ and $B \geq 0$ implies that $AB \geq 0$.

Now, given a demand matrix $D \geq 0$, we look for a production matrix $P \geq 0$ satisfying equation (*). This certainly exists if $I - E$ is invertible and $(I - E)^{-1} \geq 0$. On the other hand, the fact that $D \geq 0$ means any solution P to equation (*) satisfies $P \geq EP$. Hence, the following theorem is not too surprising.

Theorem 2

Let $E \geq 0$ be a square matrix. Then $I - E$ is invertible and $(I - E)^{-1} \geq 0$ if and only if there exists a column $P > 0$ such that $P > EP$.

HEURISTIC PROOF

If $(I - E)^{-1} \geq 0$, the existence of $P > 0$ with $P > EP$ is left as Exercise 11. Conversely, suppose such a column P exists. Observe that

$$(I - E)(I + E + E^2 + \cdots + E^{k-1}) = I - E^k$$

holds for all $k \geq 2$. If we can show that every entry of E^k approaches 0 as k becomes large then, intuitively, the infinite matrix sum

$$U = I + E + E^2 + \cdots$$

exists and $(I - E)U = I$. Since $U \geq 0$, this does it. To show that E^k approaches 0, it suffices to show that $EP < \mu P$ for some number μ with $0 < \mu < 1$ (then $E^k P < \mu^k P$ for all $k \geq 1$ by induction). The existence of μ is left as Exercise 12.

The condition $P > EP$ in Theorem 2 has a simple economic interpretation. If P is a production matrix, entry i of EP is the total value of all product used by industry i in a year. Hence, the condition $P > EP$ means that, for each i, the value of product produced by industry i exceeds the value of the product it uses. In other words, each industry runs at a profit.

Example 3

If $E = \begin{bmatrix} 0.6 & 0.2 & 0.3 \\ 0.1 & 0.4 & 0.2 \\ 0.2 & 0.5 & 0.1 \end{bmatrix}$, show that $I - E$ is invertible and $(I - E)^{-1} \geq 0$.

Solution Use $P = [3 \ 2 \ 2]^T$ in Theorem 2.

If $P_0 = [1 \ 1 \ \cdots \ 1]^T$, the entries of EP_0 are the row sums of E. Hence $P_0 > EP_0$ holds if the row sums of E are all less than 1. This proves the first of the following useful facts (the second is Exercise 10).

Corollary

Let $E \geq 0$ be a square matrix. In each of the following cases, $I - E$ is invertible and $(I - E)^{-1} \geq 0$:
1. All row sums of E are less than 1.
2. All column sums of E are less than 1.

Exercises 2.7

1. Find the possible equilibrium price structures when the input-output matrices are:

(a) $\begin{bmatrix} 0.1 & 0.2 & 0.3 \\ 0.6 & 0.2 & 0.3 \\ 0.3 & 0.6 & 0.4 \end{bmatrix}$
♦(b) $\begin{bmatrix} 0.5 & 0 & 0.5 \\ 0.1 & 0.9 & 0.2 \\ 0.4 & 0.1 & 0.3 \end{bmatrix}$
(c) $\begin{bmatrix} .3 & .1 & .1 & .2 \\ .2 & .3 & .1 & 0 \\ .3 & .3 & .2 & .3 \\ .2 & .3 & .6 & .5 \end{bmatrix}$
♦(d) $\begin{bmatrix} .5 & 0 & .1 & .1 \\ .2 & .7 & 0 & .1 \\ .1 & .2 & .8 & .2 \\ .2 & .1 & .1 & .6 \end{bmatrix}$

♦2. Three industries A, B, and C are such that all the output of A is used by B, all the output of B is used by C, and all the output of C is used by A. Find the possible equilibrium price structures.

3. Find the possible equilibrium price structures for three industries where the input-output matrix is $\begin{bmatrix} 1 & 0 & 0 \\ 0 & 0 & 1 \\ 0 & 1 & 0 \end{bmatrix}$. Discuss why there are two parameters here.

♦4. Prove Theorem 1 for a 2×2 stochastic matrix E by first writing it in the form $E = \begin{bmatrix} a & b \\ 1-a & 1-b \end{bmatrix}$, where $0 \le a \le 1$ and $0 \le b \le 1$.

5. If E is an $n \times n$ stochastic matrix and C is an $n \times 1$ matrix, show that the sum of the entries of C equals the sum of the entries of the $n \times 1$ matrix EC.

6. Let $W = [1\ 1\ 1\ \cdots\ 1]$. Let E and F denote $n \times n$ matrices with nonnegative entries.

   (a) Show that E is a stochastic matrix if and only if $WE = W$.

   (b) Use part (a) to deduce that, if E and F are both stochastic matrices, then EF is also stochastic.

7. Find a 2×2 matrix E with entries between 0 and 1 such that:

   (a) $I - E$ has no inverse.

   ♦(b) $I - E$ has an inverse but not all entries of $(I - E)^{-1}$ are nonnegative.

8. If E is a 2×2 matrix with entries between 0 and 1, show that $I - E$ is invertible and $(I - E)^{-1} \ge 0$ if and only if tr $E < 1 + \det E$. Here, if $E = \begin{bmatrix} a & b \\ c & d \end{bmatrix}$, then tr $E = a + d$ and $\det E = ad - bc$.

9. In each case show that $I - E$ is invertible and $(I - E)^{-1} \ge 0$.

   (a) $\begin{bmatrix} 0.6 & 0.5 & 0.1 \\ 0.1 & 0.3 & 0.3 \\ 0.2 & 0.1 & 0.4 \end{bmatrix}$ ♦(b) $\begin{bmatrix} 0.7 & 0.1 & 0.3 \\ 0.2 & 0.5 & 0.2 \\ 0.1 & 0.1 & 0.4 \end{bmatrix}$

   (c) $\begin{bmatrix} 0.6 & 0.2 & 0.1 \\ 0.3 & 0.4 & 0.2 \\ 0.2 & 0.5 & 0.1 \end{bmatrix}$ (d) $\begin{bmatrix} 0.8 & 0.1 & 0.1 \\ 0.3 & 0.1 & 0.2 \\ 0.3 & 0.3 & 0.2 \end{bmatrix}$

10. Prove that (1) implies (2) in the Corollary to Theorem 2.

11. If $(I - E)^{-1} \ge 0$, find $P > 0$ such that $P > EP$.

12. If $EP < P$ where $E \ge 0$ and $P > 0$, find a number μ such that $EP < \mu P$ and $0 < \mu < 1$. [*Hint:* If $EP = [q_1, \ldots, q_n]^T$ and $P = [p_1, \ldots, p_n]^T$, take any number μ such that $\max\left\{\frac{q_1}{p_1}, \ldots, \frac{q_n}{p_n}\right\} < \mu < 1$.]

SECTION 2.8   An Application to Markov Chains

Many natural phenomena progress through various stages and can be in a variety of states at each stage. For example, the weather in a given city progresses day by day and, on any given day, may be sunny or rainy. Here the states are "sun" and "rain," and the weather progresses from one state to another in daily stages. Another example might be a football team: The stages of its evolution are the games it plays, and the possible states are "win," "draw," and "loss."

The general setup is as follows: A "system" evolves through a series of "stages," and at any stage it can be in any one of a finite number of "states." At any given stage, the state to which it will go at the next stage depends on the past and present history of the system—that is, on the sequence of states it has occupied to date. A **Markov chain** is such an evolving system wherein the state to which it will go next depends *only* on its *present* state and does not depend on the earlier history of the system.[13]

Even in the case of a Markov chain, the state the system will occupy at any stage is determined only in terms of probabilities. In other words, chance plays a role. For example, if a football team wins a particular game, we do not know whether it will win, draw, or lose the next game. On the other hand, we may know that the team tends to persist in winning streaks; for example, if it wins one game it may win the next game $\frac{1}{2}$ of the time, lose $\frac{4}{10}$ of the time, and draw $\frac{1}{10}$ of the time. These

---

13 The name honours Andrei Andreyevich Markov (1856–1922) who was a professor at the university in St. Petersburg, Russia.

fractions are called the **probabilities** of these various possibilities. Similarly, if the team loses, it may lose the next game with probability $\frac{1}{2}$ (that is, half the time), win with probability $\frac{1}{4}$, and draw with probability $\frac{1}{4}$. The probabilities of the various outcomes after a drawn game will also be known.

We shall treat probabilities informally here: *The probability that a given event will occur is the long-run proportion of the time that the event does indeed occur.* Hence, all probabilities are numbers between 0 and 1. A probability of 0 means the event is impossible and never occurs; events with probability 1 are certain to occur.

If a Markov chain is in a particular state, the probabilities that it goes to the various states at the next stage of its evolution are called the **transition probabilities** for the chain, and they are assumed to be known quantities. To motivate the general conditions that follow, consider the following simple example. Here the system is a man, the stages are his successive lunches, and the states are the two restaurants he chooses.

Example 1

A man always eats lunch at one of two restaurants, A and B. He never eats at A twice in a row. However, if he eats at B, he is three times as likely to eat at B next time as at A. Initially, he is equally likely to eat at either restaurant.

(a) What is the probability that he eats at A on the third day after the initial one?

(b) What proportion of his lunches does he eat at A?

Solution   The table of transition probabilities follows. The A column indicates that if he eats at A on one day, he never eats there again on the next day and so is certain to go to B.

		PRESENT LUNCH	
		A	**B**
NEXT LUNCH	**A**	0	0.25
	B	1	0.75

The B column shows that, if he eats at B on one day, he will eat there on the next day $\frac{3}{4}$ of the time and switches to A only $\frac{1}{4}$ of the time.

The restaurant he visits on a given day is not determined. The most that we can expect is to know the probability that he will visit A or B on that day. Let $S_m = \begin{bmatrix} s_1^{(m)} \\ s_2^{(m)} \end{bmatrix}$ denote the *state vector* for day m. Here $s_1^{(m)}$ denotes the probability that he eats at A on day m, and $s_2^{(m)}$ is the probability that he eats at B on day m. It is convenient to let S_0 correspond to the initial day. Because he is equally likely to eat at A or B on that initial day, $s_1^{(0)} = 0.5$ and $s_2^{(0)} = 0.5$, so $S_0 = \begin{bmatrix} 0.5 \\ 0.5 \end{bmatrix}$.

Now let

$$P = \begin{bmatrix} 0 & 0.25 \\ 1 & 0.75 \end{bmatrix}$$

denote the *transition matrix*. We claim that the relationship

$$S_{m+1} = PS_m$$

holds for all integers $m \geq 0$. This will be derived later; for now, we use it as follows to successively compute $S_1, S_2, S_3, \ldots$.

$$S_1 = PS_0 = \begin{bmatrix} 0 & 0.25 \\ 1 & 0.75 \end{bmatrix} \begin{bmatrix} 0.5 \\ 0.5 \end{bmatrix} = \begin{bmatrix} 0.125 \\ 0.875 \end{bmatrix}$$

$$S_2 = PS_1 = \begin{bmatrix} 0 & 0.25 \\ 1 & 0.75 \end{bmatrix} \begin{bmatrix} 0.125 \\ 0.875 \end{bmatrix} = \begin{bmatrix} 0.218\,75 \\ 0.781\,25 \end{bmatrix}$$

$$S_3 = PS_2 = \begin{bmatrix} 0 & 0.25 \\ 1 & 0.75 \end{bmatrix} \begin{bmatrix} 0.218\,75 \\ 0.781\,25 \end{bmatrix} = \begin{bmatrix} 0.1\,953\,125 \\ 0.8\,046\,875 \end{bmatrix}$$

Hence, the probability that his third lunch (after the initial one) is at A is approximately 0.195, whereas the probability that it is at B is 0.805.

If we carry these calculations on, the next state vectors are (to five figures)

$$S_4 = \begin{bmatrix} 0.201\,17 \\ 0.798\,83 \end{bmatrix} \qquad S_5 = \begin{bmatrix} 0.199\,71 \\ 0.800\,29 \end{bmatrix}$$

$$S_6 = \begin{bmatrix} 0.200\,07 \\ 0.799\,93 \end{bmatrix} \qquad S_7 = \begin{bmatrix} 0.199\,98 \\ 0.800\,02 \end{bmatrix}$$

Moreover, as m increases the entries of S_m get closer and closer to the corresponding entries of $\begin{bmatrix} 0.2 \\ 0.8 \end{bmatrix}$. Hence, in the long run, he eats 20% of his lunches at A and 80% at B.

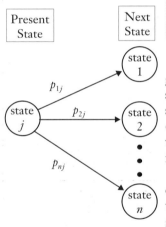

Present State		Next State

Example 1 incorporates most of the essential features of all Markov chains. The general model is as follows: The system evolves through various stages and at each stage can be in exactly one of n distinct states. It progresses through a sequence of states as time goes on. If a Markov chain is in state j at a particular stage of its development, the probability p_{ij} that it goes to state i at the next stage is called the **transition probability**. The $n \times n$ matrix $P = [p_{ij}]$ is called the **transition matrix** for the Markov chain. The situation is depicted graphically in the diagram.

We make one important assumption about the transition matrix $P = [p_{ij}]$: It does *not* depend on which stage the process is in. This assumption means that the transition probabilities are *independent of time*—that is, they do not change as time goes on. It is this assumption that distinguishes Markov chains in the literature of this subject.

Example 2

Suppose the transition matrix of a three-state Markov chain is

Present state

$$P = \begin{bmatrix} p_{11} & p_{12} & p_{13} \\ p_{21} & p_{22} & p_{23} \\ p_{31} & p_{32} & p_{33} \end{bmatrix} = \begin{bmatrix} 0.3 & 0.1 & 0.6 \\ 0.5 & 0.9 & 0.2 \\ 0.2 & 0 & 0.2 \end{bmatrix} \begin{matrix} 1 \\ 2 \\ 3 \end{matrix} \quad \text{Next state}$$

with columns labeled $1 \quad 2 \quad 3$.

If, for example, the system is in state 2, column 2 lists the probabilities of where it goes next. Thus, the probability is $p_{12} = 0.1$ that it goes from state 2 to state 1, and the probability is $p_{22} = 0.9$ that it goes from state 2 to state 2. The fact that $p_{32} = 0$ means that it is impossible for it to go from state 2 to state 3 at the next stage.

Consider the *j*th column of the transition matrix *P*.

$$\begin{bmatrix} p_{1j} \\ p_{2j} \\ \vdots \\ p_{nj} \end{bmatrix}$$

If the system is in state *j* at some stage of its evolution, the transition probabilities $p_{1j}, p_{2j}, \ldots, p_{nj}$ represent the fraction of the time that the system will move to state 1, state 2, ... , state *n*, respectively, at the next stage. We assume that it has to go to *some* state at each transition, so the sum of these probabilities equals 1:

$$p_{1j} + p_{2j} + \cdots + p_{nj} = 1 \quad \text{for each } j$$

Thus, the columns of *P* all sum to 1 and the entries of *P* lie between 0 and 1, so *P* is a **stochastic matrix**.

As in Example 1, we introduce the following notation: Let $s_i^{(m)}$ denote the probability that the system is in state *i* after *m* transitions. Then $n \times 1$ matrices

$$S_m = \begin{bmatrix} s_1^{(m)} \\ s_2^{(m)} \\ \vdots \\ s_n^{(m)} \end{bmatrix} \quad m = 0, 1, 2, \ldots$$

are called the **state vectors** for the Markov chain. Note that the sum of the entries of S_m must equal 1 because the system must be in *some* state after *m* transitions. The matrix S_0 is called the **initial state vector** for the Markov chain and is given as part of the data of the particular chain. For example, if the chain has only two states, then an initial vector $S_0 = \begin{bmatrix} 1 \\ 0 \end{bmatrix}$ means that it started in state 1. If it started in state 2, the initial vector would be $S_0 = \begin{bmatrix} 0 \\ 1 \end{bmatrix}$. If $S_0 = \begin{bmatrix} 0.5 \\ 0.5 \end{bmatrix}$, it is equally likely that the system started in state 1 or in state 2.

Theorem 1

Let *P* be the transition matrix for an *n*-state Markov chain. If S_m is the state vector at stage *m*, then

$$S_{m+1} = PS_m$$

for each $m = 0, 1, 2, \ldots$.

HEURISTIC PROOF

Suppose that the Markov chain has been run *N* times, each time starting with the same initial state vector. Recall that p_{ij} is the proportion of the time the system goes from state *j* at some stage to state *i* at the next stage, whereas $s_i^{(m)}$ is the proportion of the time it is in state *i* at stage *m*. Hence

$$s_i^{(m+1)} N$$

is (approximately) the number of times the system is in state *i* at stage $m + 1$. We are going to calculate this number another way. The system got to state *i* at stage $m + 1$ through *some* other state (say state j) at stage m. The number of times it was *in* state j at that stage is (approximately) $s_j^{(m)}N$, so the number of times it got to state i via state j is $p_{ij}(s_j^{(m)}N)$. Summing over j gives the number of times the system is in state i (at stage $m + 1$). This is the number we calculated before, so

$$s_i^{(m+1)}N = p_{i1}s_1^{(m)}N + p_{i2}s_2^{(m)}N + \cdots + p_{in}s_n^{(m)}N$$

Dividing by N gives $s_i^{(m+1)} = p_{i1}s_1^{(m)} + p_{i2}s_2^{(m)} + \cdots + p_{in}s_n^{(m)}$ for each i, and this can be expressed as the matrix equation $S_{m+1} = PS_m$.

If the initial probability vector S_0 and the transition matrix P are given, Theorem 1 gives $S_1, S_2, S_3, \ldots$, one after the other, as follows:

$$S_1 = PS_0$$
$$S_2 = PS_1$$
$$S_3 = PS_2$$
$$\vdots$$

Hence, the state vector S_m is completely determined for each $m = 0, 1, 2, \ldots$ by P and S_0.

Example 3

A wolf pack always hunts in one of three regions R_1, R_2, and R_3. Its hunting habits are as follows:

1. If it hunts in some region one day, it is as likely as not to hunt there again the next day.
2. If it hunts in R_1, it never hunts in R_2 the next day.
3. If it hunts in R_2 or R_3, it is equally likely to hunt in each of the other regions the next day.

If the pack hunts in R_1 on Monday, find the probability that it hunts there on Thursday.

Solution The stages of this process are the successive days; the states are the three regions. The transition matrix P is determined as follows (see the table): The first habit asserts that $p_{11} = p_{22} = p_{33} = \frac{1}{2}$. Now column 1 displays what happens when the pack starts in R_1: It never goes to state 2, so $p_{21} = 0$ and, because the column must sum to 1, $p_{31} = \frac{1}{2}$. Column 2 describes what happens if it starts in R_2: $p_{22} = \frac{1}{2}$ and p_{12} and p_{32} are equal (by habit 3), so $p_{12} = p_{32} = \frac{1}{4}$ because the column sum must equal 1. Column 3 is filled in a similar way.

Now let Monday be the initial stage. Then $S_0 = \begin{bmatrix} 1 \\ 0 \\ 0 \end{bmatrix}$ because the pack hunts in R_1 on that day. Then S_1, S_2, and S_3 describe Tuesday, Wednesday, and Thursday, respectively, and we compute them using Theorem 1.

	R_1	R_2	R_3
R_1	$\frac{1}{2}$	$\frac{1}{4}$	$\frac{1}{4}$
R_2	0	$\frac{1}{2}$	$\frac{1}{4}$
R_3	$\frac{1}{2}$	$\frac{1}{4}$	$\frac{1}{2}$

$$S_1 = PS_0 = \begin{bmatrix} \frac{1}{2} \\ 0 \\ \frac{1}{2} \end{bmatrix} \qquad S_2 = PS_1 = \begin{bmatrix} \frac{3}{8} \\ \frac{1}{8} \\ \frac{4}{8} \end{bmatrix} \qquad S_3 = PS_2 = \begin{bmatrix} \frac{11}{32} \\ \frac{6}{32} \\ \frac{15}{32} \end{bmatrix}$$

Hence, the probability that the pack hunts in Region R_1 on Thursday is $\frac{11}{32}$.

Another phenomenon that was observed in Example 1 can be expressed in general terms. The state vectors $S_0, S_1, S_2, \ldots$ were calculated in that example and were found to "approach" $S = \begin{bmatrix} 0.2 \\ 0.8 \end{bmatrix}$. This means that the first component of S_m becomes and remains very close to 0.2 as m becomes large, whereas the second component gets close to 0.8 as m increases. When this is the case, we say that S_m **converges** to S. For large m, then, there is very little error in taking $S_m = S$, so the long-term probability that the system is in state 1 is 0.2, whereas the probability that it is in state 2 is 0.8. In Example 1, enough state vectors were computed for the limiting vector S to be apparent. However, there is a better way to do this that works in most cases.

Suppose P is the transition matrix of a Markov chain, and assume that the state vectors S_m converge to a limiting vector S. Then S_m is very close to S for sufficiently large m, so S_{m+1} is also very close to S. Thus, the equation $S_{m+1} = PS_m$ from Theorem 1 is closely approximated by

$$S = PS$$

so it is not surprising that S should be a solution to this matrix equation. Moreover, it is easily solved because it can be written as a system of linear equations

$$(I - P)S = 0$$

with the entries of S as variables.

In Example 1, where $P = \begin{bmatrix} 0 & 0.25 \\ 1 & 0.75 \end{bmatrix}$, the general solution to $(I - P)S = 0$ is $S = \begin{bmatrix} t \\ 4t \end{bmatrix}$, where t is a parameter. But if we insist that the entries of S sum to 1 (as must be true of all state vectors), we find $t = 0.2$ and so $S = \begin{bmatrix} 0.2 \\ 0.8 \end{bmatrix}$ as before.

All this is predicated on the existence of a limiting vector for the sequence of state vectors of the Markov chain, and such a vector may not always exist. However, it does exist in one commonly occurring situation. A stochastic matrix P is called **regular** if some power P^m of P has every entry greater than zero. The matrix $P = \begin{bmatrix} 0 & 0.25 \\ 1 & 0.75 \end{bmatrix}$ of Example 1 is regular (in this case, each entry of P^2 is positive), and the general theorem is as follows:

Theorem 2

Let P be the transition matrix of a Markov chain and assume that P is regular. Then there is a unique column matrix S satisfying the following conditions:
1. $PS = S$.
2. The entries of S are positive and sum to 1.

Moreover, condition 1 can be written as

$$(I - P)S = 0$$

and so gives a homogeneous system of linear equations for S. Finally, the sequence of state vectors $S_0, S_1, S_2, \ldots$ converges to S in the sense that if m is large enough, each entry of S_m is closely approximated by the corresponding entry of S.

This theorem will not be proved here.[14]

---

14 The interested reader can find an elementary proof in J. Kemeny, H. Mirkil, J. Snell, and G. Thompson, *Finite Mathematical Structures* (Englewood Cliffs, N.J.: Prentice-Hall, 1958).

If P is the regular transition matrix of a Markov chain, the column S satisfying conditions 1 and 2 of Theorem 2 is called the **steady-state vector** for the Markov chain. The entries of S are the long-term probabilities that the chain will be in each of the various states.

Example 4

A man eats one of three soups—beef, chicken, and vegetable—each day. He never eats the same soup two days in a row. If he eats beef soup on a certain day, he is equally likely to eat each of the others the next day; if he does not eat beef soup, he is twice as likely to eat it the next day as the alternative.

   (a) If he has beef soup one day, what is the probability that he has it again two days later?
   (b) What are the long-run probabilities that he eats each of the three soups?

Solution   The states here are B, C, and V, the three soups. The transition matrix P is given in the table. (Recall that, for each state, the corresponding column lists the probabilities for the next state.) If he has beef soup initially, then the initial state vector is

	B	C	V
B	0	$\frac{2}{3}$	$\frac{2}{3}$
C	$\frac{1}{2}$	0	$\frac{1}{3}$
V	$\frac{1}{2}$	$\frac{1}{3}$	0

$$S_0 = \begin{bmatrix} 1 \\ 0 \\ 0 \end{bmatrix}$$

Then two days later the state vector is S_2. If P is the transition matrix, then

$$S_1 = PS_0 = \frac{1}{2}\begin{bmatrix} 0 \\ 1 \\ 1 \end{bmatrix}, \qquad S_2 = PS_1 = \frac{1}{6}\begin{bmatrix} 4 \\ 1 \\ 1 \end{bmatrix}$$

so he eats beef soup two days later with probability $\frac{2}{3}$. This answers (a) and also shows that he eats chicken and vegetable soup each with probability $\frac{1}{6}$.

To find the long-run probabilities, we must find the steady-state vector S. Theorem 2 applies because P is regular (P^2 has positive entries), so S satisfies $PS = S$. That is, $(I - P)S = 0$ where

$$I - P = \frac{1}{6}\begin{bmatrix} 6 & -4 & -4 \\ -3 & 6 & -2 \\ -3 & -2 & 6 \end{bmatrix}$$

The solution is $S = \begin{bmatrix} 4t \\ 3t \\ 3t \end{bmatrix}$, where t is a parameter, and we use $S = \begin{bmatrix} 0.4 \\ 0.3 \\ 0.3 \end{bmatrix}$ because the entries of S must sum to 1. Hence, in the long run, he eats beef soup 40% of the time and eats chicken soup and vegetable soup each 30% of the time.

Exercises 2.8

1. Which of the following stochastic matrices is regular?

   (a) $\begin{bmatrix} 0 & 0 & \frac{1}{2} \\ 1 & 0 & \frac{1}{2} \\ 0 & 1 & 0 \end{bmatrix}$ ♦(b) $\begin{bmatrix} \frac{1}{2} & 0 & \frac{1}{3} \\ \frac{1}{4} & 1 & \frac{1}{3} \\ \frac{1}{4} & 0 & \frac{1}{3} \end{bmatrix}$

2. In each case find the steady-state vector and, assuming that it starts in state 1, find the probability that it is in state 2 after 3 transitions.

   (a) $\begin{bmatrix} 0.5 & 0.3 \\ 0.5 & 0.7 \end{bmatrix}$ ♦(b) $\begin{bmatrix} \frac{1}{2} & 1 \\ \frac{1}{2} & 0 \end{bmatrix}$

(c) $\begin{bmatrix} 0 & \frac{1}{2} & \frac{1}{4} \\ 1 & 0 & \frac{1}{4} \\ 0 & \frac{1}{2} & \frac{1}{2} \end{bmatrix}$ ◆(d) $\begin{bmatrix} 0.4 & 0.1 & 0.5 \\ 0.2 & 0.6 & 0.2 \\ 0.4 & 0.3 & 0.3 \end{bmatrix}$

(e) $\begin{bmatrix} 0.8 & 0 & 0.2 \\ 0.1 & 0.6 & 0.1 \\ 0.1 & 0.4 & 0.7 \end{bmatrix}$ ◆(f) $\begin{bmatrix} 0.1 & 0.3 & 0.3 \\ 0.3 & 0.1 & 0.6 \\ 0.6 & 0.6 & 0.1 \end{bmatrix}$

3. A fox hunts in three territories A, B, and C. He never hunts in the same territory on two successive days. If he hunts in A, then he hunts in C the next day. If he hunts in B or C, he is twice as likely to hunt in A the next day as in the other territory.

   (a) What proportion of his time does he spend in A, in B, and in C?

   (b) If he hunts in A on Monday (C on Monday), what is the probability that he will hunt in B on Thursday?

4. Assume that there are three classes—upper, middle, and lower—and that social mobility behaves as follows:

   1. Of the children of upper-class parents, 70% remain upper-class, whereas 10% become middle-class and 20% become lower-class.

   2. Of the children of middle-class parents, 80% remain middle-class, whereas the others are evenly split between the upper class and the lower class.

   3. For the children of lower-class parents, 60% remain lower-class, whereas 30% become middle-class and 10% upper-class.

   (a) Find the probability that the grandchild of lower-class parents becomes upper-class.

   ◆(b) Find the long-term breakdown of society into classes.

5. The Prime Minister says she will call an election. This gossip is passed from person to person with a probability $p \neq 0$ that the information is passed incorrectly at any stage. Assume that when a person hears the gossip he or she passes it to one person who does not know. Find the long-term probability that a person will hear that there is going to be an election.

◆6. John makes it to work on time one Monday out of four. On other work days his behaviour is as follows: If he is late one day, he is twice as likely to come to work on time the next day as to be late. If he is on time one day, he is as likely to be late as not the next day. Find the probability of his being late and that of his being on time Wednesdays.

7. Suppose you have 1¢ and match coins with a friend. At each match you either win or lose 1¢ with equal probability. If you go broke or ever get 4¢, you quit. Assume your friend never quits. If the states are 0, 1, 2, 3, and 4 representing your wealth, show that the corresponding transition matrix P is not regular. Find the probability that you will go broke after 3 matches.

◆8. A mouse is put into a maze of compartments, as in the diagram. Assume that he always leaves any compartment he enters and that he is equally likely to take any tunnel entry.

   (a) If he starts in compartment 1, find the probability that he is in compartment 4 after 3 moves.

   (b) Find the compartment in which he spends most of his time if he is left for a long time.

9. If a stochastic matrix has a 1 on its main diagonal, show that it cannot be regular. Assume it is not 1×1.

10. If S_m is the stage-m state vector for a Markov chain, show that $S_{m+k} = P^k S_m$ holds for all $m \geq 1$ and $k \geq 1$ (where P is the transition matrix).

11. A stochastic matrix is **doubly stochastic** if all the row sums also equal 1. Find the steady-state vector for a doubly stochastic matrix.

12. Consider the 2×2 stochastic matrix
$$P = \begin{bmatrix} 1-p & q \\ p & 1-q \end{bmatrix}, \text{ where } 0 < p < 1 \text{ and } 0 < q < 1.$$

   (a) Show that $\dfrac{1}{p+q} \begin{bmatrix} q \\ p \end{bmatrix}$ is the steady-state vector for P.

   (b) Show that P^m converges to the matrix $\dfrac{1}{p+q} \begin{bmatrix} q & q \\ p & p \end{bmatrix}$ by first verifying inductively that
   $$P^m = \frac{1}{p+q} \begin{bmatrix} q & q \\ p & p \end{bmatrix} + \frac{(1-p-q)^m}{p+q} \begin{bmatrix} p & -q \\ -p & q \end{bmatrix}$$
   for $m = 1, 2, \dots$. (It can be shown that the sequence of powers $P, P^2, P^3, \dots$ of any regular transition matrix converges to the matrix each of whose columns equals the steady-state vector for P.)

Supplementary Exercises for Chapter 2

1. Solve for the matrix X if: (a) $PXQ = R$;
   (b) $XP = S$; where

   $$P = \begin{bmatrix} 1 & 0 \\ 2 & -1 \\ 0 & 3 \end{bmatrix}, \qquad Q = \begin{bmatrix} 1 & 1 & -1 \\ 2 & 0 & 3 \end{bmatrix},$$

   $$R = \begin{bmatrix} -1 & 1 & -4 \\ -4 & 0 & -6 \\ 6 & 6 & -6 \end{bmatrix}, \qquad S = \begin{bmatrix} 1 & 6 \\ 3 & 1 \end{bmatrix}$$

2. Consider $p(X) = X^3 - 5X^2 + 11X - 4I$.

   (a) If $p(A) = \begin{bmatrix} 1 & 3 \\ -1 & 0 \end{bmatrix}$, compute $p(A^T)$.

   ◆(b) If $p(U) = 0$ where U is $n \times n$, find U^{-1} in terms of U.

3. Show that, if a (possibly nonhomogeneous) system of equations is consistent and has more variables than equations, then it must have infinitely many solutions. [*Hint:* Use Theorem 2 §2.2 and Theorem 1 §1.3.]

4. Assume that a system $AX = B$ of linear equations has at least two distinct solutions Y and Z.

   (a) Show that $X_k = Y + k(Y - Z)$ is a solution for every k.

   ◆(b) Show that $X_k = X_m$ implies $k = m$. [*Hint:* See Example 7 §2.1.]

   (c) Deduce that $AX = B$ has infinitely many solutions.

5. (a) Let A be a 3×3 matrix with all entries on and below the main diagonal zero. Show that $A^3 = 0$.

   (b) Generalize to the $n \times n$ case and prove your answer.

6. Let I_{pq} denote the $n \times n$ matrix with (p, q)-entry equal to 1 and all other entries 0. Show that:

   (a) $I_n = I_{11} + I_{22} + \cdots + I_{nn}$.

   (b) $I_{pq}I_{rs} = \begin{cases} I_{ps} & \text{if } q = r \\ 0 & \text{if } q \neq r. \end{cases}$

   (c) If $A = [a_{ij}]$ is $n \times n$, then $A = \sum_{i=1}^{n}\sum_{j=1}^{n} a_{ij}I_{ij}$.

   ◆(d) If $A = [a_{ij}]$, then $I_{pq}AI_{rs} = a_{qr}I_{ps}$ for all $p, q, r,$ and s.

7. A matrix of the form aI_n, where a is a number, is called an $n \times n$ **scalar matrix**.

   (a) Show that each $n \times n$ scalar matrix commutes with every $n \times n$ matrix.

   (b) Show that A is a scalar matrix if it commutes with every $n \times n$ matrix. [*Hint:* See part (d) of Exercise 6.]

8. Let $M = \begin{bmatrix} A & B \\ C & D \end{bmatrix}$, where $A, B, C,$ and D are all $n \times n$ and each commutes with all the others. If $M^2 = 0$, show that $(A + D)^3 = 0$. [*Hint:* First show that $A^2 = -BC = D^2$ and that $B(A + D) = 0 = C(A + D)$.]

9. If A is 2×2, show that $A^{-1} = A^T$ if and only if

   $$A = \begin{bmatrix} \cos\theta & \sin\theta \\ -\sin\theta & \cos\theta \end{bmatrix} \text{ for some } \theta$$

   or $A = \begin{bmatrix} \cos\theta & \sin\theta \\ \sin\theta & -\cos\theta \end{bmatrix}$ for some θ.

   [*Hint:* If $a^2 + b^2 = 1$, then $a = \cos\theta$, $b = \sin\theta$ for some θ. Use $\cos(\theta - \varphi) = \cos\theta\cos\varphi + \sin\theta\sin\varphi$.]

10. (a) If $A = \begin{bmatrix} 0 & 1 \\ 1 & 0 \end{bmatrix}$, show that $A^2 = I$.

    (b) What is wrong with the following argument? If $A^2 = I$, then $A^2 - I = 0$, so $(A - I)(A + I) = 0$, whence $A = I$ or $A = -I$.

11. Let E and F be elementary matrices obtained from the identity matrix by adding multiples of row k to rows p and q. If $k \neq p$ and $k \neq q$, show that $EF = FE$.

12. If A is a 2×2 real matrix, $A^2 = A$ and $A^T = A$, show that either A is one of $\begin{bmatrix} 0 & 0 \\ 0 & 0 \end{bmatrix}$, $\begin{bmatrix} 1 & 0 \\ 0 & 0 \end{bmatrix}$, $\begin{bmatrix} 0 & 0 \\ 0 & 1 \end{bmatrix}$, $\begin{bmatrix} 1 & 0 \\ 0 & 1 \end{bmatrix}$, or $A = \begin{bmatrix} a & b \\ b & 1-a \end{bmatrix}$ where $a^2 + b^2 = a$, $-\frac{1}{2} \leq b \leq \frac{1}{2}$ and $b \neq 0$.

13. Show that the following are equivalent for matrices P, Q:

    (1) $P, Q,$ and $P + Q$ are all invertible and $(P + Q)^{-1} = P^{-1} + Q^{-1}$.

    (2) P is invertible and $Q = PG$ where $G^2 + G + I = 0$.

3 Determinants and Diagonalization

With each square matrix we can calculate a number, called the determinant of the matrix, which tells us whether or not the matrix is invertible. In fact, determinants can be used to give a formula for the inverse of a matrix, and they provide a method (called Cramer's rule) for solving certain systems of linear equations. They also arise in calculating certain numbers (called eigenvalues) associated with a matrix. These eigenvalues are essential to a technique called diagonalization that is used in many applications where it is desired to predict the future behaviour of a system. For example, we use it to predict whether a species of birds will become extinct.

Determinants are older than matrices: Cramer's rule was published in 1750 but Cayley did not introduce matrix algebra until 1878. Determinants were used extensively in the eighteenth and nineteenth centuries, primarily because of their significance in geometry. Although they are somewhat less important today, determinants still play a role in the theory and application of matrix algebra.

SECTION 3.1 The Cofactor Expansion

In Section 2.3 we defined the determinant of a 2×2 matrix $A = \begin{bmatrix} a & b \\ c & d \end{bmatrix}$ as follows:[1]

$$\det A = \begin{vmatrix} a & b \\ c & d \end{vmatrix} = ad - bc$$

and showed (in Example 4) that A has an inverse if and only if $\det A \neq 0$. One objective of this chapter is to do this for any square matrix A. There is no difficulty for 1×1 matrices: If $A = [a]$, we define $\det A = \det[a] = a$ and note that A is invertible if and only if $a \neq 0$.

If A is 3×3, we look for a condition that A is invertible by trying to carry it to the identity matrix by row operations. The first column is not zero (A is invertible); suppose the $(1, 1)$-entry a is not zero. Then row operations give

$$A = \begin{bmatrix} a & b & c \\ d & e & f \\ g & h & i \end{bmatrix} \rightarrow \begin{bmatrix} a & b & c \\ ad & ae & af \\ ag & ah & ai \end{bmatrix} \rightarrow \begin{bmatrix} a & b & c \\ 0 & ae - bd & af - cd \\ 0 & ah - bg & ai - cg \end{bmatrix} = \begin{bmatrix} a & b & c \\ 0 & u & af - cd \\ 0 & v & ai - cg \end{bmatrix}$$

---

1 Determinants are commonly written $|A| = \det A$ using vertical bars. We will use both notations.

where $u = ae - bd$ and $v = ah - bg$. Since A is invertible, one of u and v is nonzero (by Exercise 24(c) §2.3); suppose that $u \neq 0$. Then the reduction proceeds

$$A \to \begin{bmatrix} a & b & c \\ 0 & u & af - cd \\ 0 & v & ai - cg \end{bmatrix} \to \begin{bmatrix} a & b & c \\ 0 & u & af - cd \\ 0 & uv & u(ai - cg) \end{bmatrix} \to \begin{bmatrix} a & b & c \\ 0 & u & af - cd \\ 0 & 0 & w \end{bmatrix}$$

where $w = u(ai - cg) - v(af - cd) = a(aei + bfg + cdh - ceg - afh - bdi)$. We define

$$\det A = aei + bfg + cdh - ceg - afh - bdi \tag{$*$}$$

and observe that $\det A \neq 0$ because $a \det A = w \neq 0$ (is invertible).

To motivate the definition below, collect the terms in $(*)$ involving the entries a, b, and c in row 1 of A:

$$\det A = \begin{vmatrix} a & b & c \\ d & e & f \\ g & h & i \end{vmatrix} = aei + bfg + cdh - ceg - afh - bdi$$

$$= a(ei - fh) - b(di - fg) + c(dh - eg)$$

$$= a\begin{vmatrix} e & f \\ h & i \end{vmatrix} - b\begin{vmatrix} d & f \\ g & i \end{vmatrix} + c\begin{vmatrix} d & e \\ g & h \end{vmatrix}$$

This last expression can be described as follows: To compute the determinant of a 3×3 matrix A, multiply each entry in row 1 by a sign times the determinant of the 2×2 matrix obtained by deleting the row and column of that entry, and add the results. The signs alternate down row 1, starting with $+$. It is this observation that we generalize below.[2]

Example 1

$$\det\begin{bmatrix} 2 & 3 & 7 \\ -4 & 0 & 6 \\ 1 & 5 & 0 \end{bmatrix} = 2\begin{vmatrix} 0 & 6 \\ 5 & 0 \end{vmatrix} - 3\begin{vmatrix} -4 & 6 \\ 1 & 0 \end{vmatrix} + 7\begin{vmatrix} -4 & 0 \\ 1 & 5 \end{vmatrix}$$

$$= 2(-30) - 3(-6) + 7(-20)$$

$$= -182.$$

This suggests an inductive method of defining the determinant of any square matrix in terms of determinants of matrices one size smaller. Thus we define determinants of 3×3 matrices in terms of determinants of 2×2 matrices, then we do 4×4 matrices in terms of 3×3 matrices, and so on.

To describe this, we need some terminology. Assume that determinants of $(n-1) \times (n-1)$ matrices have been defined. Given the $n \times n$ matrix A, let

A_{ij} denote the $(n-1) \times (n-1)$ matrix obtained from A by deleting row i and column j.

Then the (i, j)-**cofactor** $c_{ij}(A)$ is the scalar defined by

$$c_{ij}(A) = (-1)^{i+j} \det(A_{ij}).$$

---

2 One way to remember $(*)$ is to adjoin columns 1 and 2 on the right of A to obtain $\begin{bmatrix} a & b & c & a & b \\ d & e & f & d & e \\ g & h & i & g & h \end{bmatrix}$. Then the positive terms

aei, bfg, and cdh are the products down and to the right starting at a, b, and c, while the terms $-ceg$, $-afh$, and $-bdi$ are the products down and to the left starting at c, a, and b. **Warning**: This rule does **not** extend to $n \times n$ matrices for $n > 3$.

Here $(-1)^{i+j}$ is called the **sign** of the (i, j)-position. It is clearly 1 or -1, and the following diagram is useful for remembering the sign of a position:

$$\begin{bmatrix} + & - & + & - & \cdots \\ - & + & - & + & \cdots \\ + & - & + & - & \cdots \\ - & + & - & + & \cdots \\ \vdots & \vdots & \vdots & \vdots & \end{bmatrix}$$

Note that the signs alternate along each row and column with + in the upper left corner.

Example 2

Find the cofactors of positions $(1, 2)$, $(3, 1)$, and $(2, 3)$ in the following matrix.

$$A = \begin{bmatrix} 3 & -1 & 6 \\ 5 & 2 & 7 \\ 8 & 9 & 4 \end{bmatrix}$$

Solution   Here A_{12} is the matrix $\begin{bmatrix} 5 & 7 \\ 8 & 4 \end{bmatrix}$ that remains when row 1 and column 2 are deleted. The sign of position $(1, 2)$ is $(-1)^{1+2} = -1$ (this is also the $(1, 2)$-entry in the sign diagram), so the $(1, 2)$-cofactor is

$$c_{12}(A) = (-1)^{1+2} \begin{vmatrix} 5 & 7 \\ 8 & 4 \end{vmatrix} = (-1)(5 \cdot 4 - 7 \cdot 8) = (-1)(-36) = 36$$

Turning to position $(3, 1)$, we find

$$c_{31}(A) = (-1)^{3+1} \det A_{31} = (-1)^{3+1} \begin{vmatrix} -1 & 6 \\ 2 & 7 \end{vmatrix} = (+1)(-7 - 12) = -19$$

Finally, the $(2, 3)$-cofactor is

$$c_{23}(A) = (-1)^{2+3} \det A_{23} = (-1)^{2+3} \begin{vmatrix} 3 & -1 \\ 8 & 9 \end{vmatrix} = (-1)(27 + 8) = -35$$

Clearly other cofactors can be found—there are nine in all, one for each position in the matrix.

With the notion of cofactor in hand, we can define the determinant of any $n \times n$ matrix $A = [a_{ij}]$. If determinants of $(n - 1) \times (n - 1)$ matrices have been defined, we define

$$\det A = a_{11}c_{11}(A) + a_{12}c_{12}(A) + \cdots + a_{1n}c_{1n}(A).$$

This is called the **cofactor expansion** of $\det A$ along row 1. It asserts that $\det A$ can be computed by multiplying the entries of row 1 by the corresponding cofactors, and adding the results. The astonishing thing is that $\det A$ can be computed by taking the cofactor expansion along *any row or column*: Simply multiply each entry of that row or column by the corresponding cofactor and add.[3]

---

3 The cofactor expansion is due to Pierre Simon de Laplace (1749–1827), who discovered it in 1772 as part of a study of linear differential equations. Laplace is primari y remembered for his work in astronomy and applied mathematics.

Theorem 1 Cofactor Expansion Theorem[4]

The determinant of an $n \times n$ matrix A can be computed by using the cofactor expansion along any row or column of A. More precisely, if $A = [a_{ij}]$ so that a_{ij} is the (i, j)-entry of A, then the expansion along row i is

$$\det A = a_{i1}c_{i1}(A) + a_{i2}c_{i2}(A) + a_{i3}c_{i3}(A) + \cdots + a_{in}c_{in}(A)$$

The expansion along column j is given by

$$\det A = a_{1j}c_{1j}(A) + a_{2j}c_{2j}(A) + a_{3j}c_{3j}(A) + \cdots + a_{nj}c_{nj}(A)$$

The proof will be given in Section 3.6.

Example 3

Compute the determinant of $A = \begin{bmatrix} 3 & 4 & 5 \\ 1 & 7 & 2 \\ 9 & 8 & -6 \end{bmatrix}$.

Solution   The cofactor expansion along the first row is as follows:

$$\det A = 3c_{11}(A) + 4c_{12}(A) + 5c_{13}(A)$$

$$= 3\begin{vmatrix} 7 & 2 \\ 8 & -6 \end{vmatrix} - 4\begin{vmatrix} 1 & 2 \\ 9 & -6 \end{vmatrix} + 5\begin{vmatrix} 1 & 7 \\ 9 & 8 \end{vmatrix}$$

$$= 3(-58) - 4(-24) + 5(-55)$$

$$= -353$$

Note that the signs alternate along the row (indeed along *any* row or column). Now we compute $\det A$ by expanding along the first column.

$$\det A = 3c_{11}(A) + 1c_{21}(A) + 9c_{31}(A)$$

$$= 3\begin{vmatrix} 7 & 2 \\ 8 & -6 \end{vmatrix} - \begin{vmatrix} 4 & 5 \\ 8 & -6 \end{vmatrix} + 9\begin{vmatrix} 4 & 5 \\ 7 & 2 \end{vmatrix}$$

$$= 3(-58) - (-64) + 9(-27)$$

$$= -353$$

The reader is invited to verify that $\det A$ can be computed by expanding along any other row or column.

The fact that the cofactor expansion along *any row or column* of a matrix A always gives the same result (the determinant of A) is remarkable, to say the least. The choice of a particular row or column can simplify the calculation.

Example 4

Compute $\det A$ where $A = \begin{bmatrix} 3 & 0 & 0 & 0 \\ 5 & 1 & 2 & 0 \\ 2 & 6 & 0 & -1 \\ -6 & 3 & 1 & 0 \end{bmatrix}$.

---

4 This is sometimes called the Laplace expansion theorem.

Solution   The first choice we must make is which row or column to use in the cofactor expansion. The expansion involves multiplying entries by cofactors, so the work is minimized when the row or column contains as many zero entries as possible. Row 1 is a best choice in this matrix (column 4 would do as well), and the expansion is

$$\det A = 3c_{11}(A) + 0c_{12}(A) + 0c_{13}(A) + 0c_{14}(A)$$

$$= 3\begin{vmatrix} 1 & 2 & 0 \\ 6 & 0 & -1 \\ 3 & 1 & 0 \end{vmatrix}$$

This is the first stage of the calculation, and we have succeeded in expressing the determinant of (the 4×4 matrix) A in terms of the determinant of a 3×3 matrix. The next stage involves this 3×3 matrix. Again, we can use any row or column for the cofactor expansion. The third column is preferred (with two zeros), so

$$\det A = 3\left(0\begin{vmatrix} 6 & 0 \\ 3 & 1 \end{vmatrix} - (-1)\begin{vmatrix} 1 & 2 \\ 3 & 1 \end{vmatrix} + 0\begin{vmatrix} 1 & 2 \\ 6 & 0 \end{vmatrix} \right)$$

$$= 3[0 + 1(-5) + 0]$$

$$= -15$$

This completes the calculation.

Computing the determinant of matrix A can be tedious. For example, if A is a 4×4 matrix, the cofactor expansion along any row or column involves calculating four cofactors, each of which involves the determinant of a 3×3 matrix. And if A is 5×5, the expansion involves five determinants of 4×4 matrices! There is a clear need for some techniques to cut down the work.

The motivation for the method is the observation (see Example 4) that calculating a determinant is simplified a great deal when a row or column consists mostly of zeros. (In fact, when a row or column consists *entirely* of zeros, the determinant is zero—simply expand along that row or column.)

Recall next that one method of *creating* zeros in a matrix is to apply elementary row operations to it. Hence, a natural question to ask is what effect such a row operation has on the determinant of the matrix. It turns out that the effect is easy to determine and that elementary *column* operations can be used in the same way. These observations lead to a technique for evaluating determinants that greatly reduces the labour involved. The necessary information is given in Theorem 2.

Theorem 2

Let A denote an $n \times n$ matrix.
1. If A has a row or column of zeros, $\det A = 0$.
2. If two distinct rows (or columns) of A are interchanged, the determinant of the resulting matrix is $-\det A$.
3. If a row (or column) of A is multiplied by a constant u, the determinant of the resulting matrix is $u(\det A)$.
4. If two distinct rows (or columns) of A are identical, $\det A = 0$.
5. If a multiple of one row of A is added to a different row (or if a multiple of a column is added to a different column), the determinant of the resulting matrix is $\det A$.

PROOF

We prove properties 2, 4, and 5 and leave the rest as exercises.

Property 2. If A is $n \times n$, this follows by induction on n. If $n = 2$, the verification is left to the reader. If $n > 2$ and two rows are interchanged, let B denote the resulting matrix. Expand $\det A$ and $\det B$ along a row *other than* the two that were interchanged. The entries in this row are the same for both A and B, but the cofactors in B are the negatives of those in A (by induction) because the corresponding $(n-1) \times (n-1)$ matrices have two rows interchanged. Hence, $\det B = -\det A$, as required. A similar argument works if two columns are interchanged.

Property 4. If two rows of A are equal, let B be the matrix obtained by interchanging them. Then $B = A$, so $\det B = \det A$. But $\det B = -\det A$ by property 2, so $\det A = \det B = 0$. Again, the same argument works for columns.

Property 5. Let B be obtained from $A = [a_{ij}]$ by adding u times row p to row q. Then row q of B is $(a_{q1} + ua_{p1}, a_{q2} + ua_{p2}, \dots, a_{qn} + ua_{pn})$. The cofactors of these elements in B are the same as in A (they do not involve row q): in symbols, $c_{qj}(B) = c_{qj}(A)$ for each j. Hence, expanding B along row q gives

$$\det B = \sum_{j=1}^{n}(a_{qj} + ua_{pj})c_{qj}(B)$$

$$= \sum_{j=1}^{n}a_{qj}c_{qj}(A) + u\sum_{j=1}^{n}a_{pj}c_{qj}(A)$$

$$= \det A + u \det c$$

where c is the matrix obtained from A by replacing row q by row p (and both expansions are along row q). Because rows p and q of c are equal, $\det c = 0$ by property 4. Hence, $\det B = \det A$, as required. As before, a similar proof holds for columns.

To illustrate Theorem 2, consider the following determinants.

$$\begin{vmatrix} 3 & -1 & 2 \\ 2 & 5 & 1 \\ 0 & 0 & 0 \end{vmatrix} = 0 \qquad \text{(because the last row consists of zeros)}$$

$$\begin{vmatrix} 3 & -1 & 5 \\ 2 & 8 & 7 \\ 1 & 2 & -1 \end{vmatrix} = -\begin{vmatrix} 5 & -1 & 3 \\ 7 & 8 & 2 \\ -1 & 2 & 1 \end{vmatrix} \qquad \text{(because two columns are interchanged)}$$

$$\begin{vmatrix} 8 & 1 & 2 \\ 3 & 0 & 9 \\ 1 & 2 & -1 \end{vmatrix} = 3\begin{vmatrix} 8 & 1 & 2 \\ 1 & 0 & 3 \\ 1 & 2 & -1 \end{vmatrix} \qquad \text{(because the second row of the matrix on the left is 3 times the second row of the matrix on the right)}$$

$$\begin{vmatrix} 2 & 1 & 2 \\ 4 & 0 & 4 \\ 1 & 3 & 1 \end{vmatrix} = 0 \qquad \text{(because two columns are identical)}$$

$$\begin{vmatrix} 2 & 5 & 2 \\ -1 & 2 & 9 \\ 3 & 1 & 1 \end{vmatrix} = \begin{vmatrix} 0 & 9 & 20 \\ -1 & 2 & 9 \\ 3 & 1 & 1 \end{vmatrix} \qquad \text{(because twice the second row of the matrix on the left was added to the first row)}$$

The following four examples illustrate how Theorem 2 is used to evaluate determinants.

Example 5

Evaluate det A when $A = \begin{bmatrix} 1 & -1 & 3 \\ 1 & 0 & -1 \\ 2 & 1 & 6 \end{bmatrix}$.

Solution   The matrix does have zero entries, so expansion along (say) the second row would involve somewhat less work. However, a column operation can be used to get a zero in position $(2, 3)$—namely, add column 1 to column 3. Because this does not change the value of the determinant, we obtain

$$\det A = \begin{vmatrix} 1 & -1 & 3 \\ 1 & 0 & -1 \\ 2 & 1 & 6 \end{vmatrix} = \begin{vmatrix} 1 & -1 & 4 \\ 1 & 0 & 0 \\ 2 & 1 & 8 \end{vmatrix} = - \begin{vmatrix} -1 & 4 \\ 1 & 8 \end{vmatrix} = 12$$

where we expanded the second 3×3 matrix along row 2.

Example 6

If $\det \begin{bmatrix} a & b & c \\ p & q & r \\ x & y & z \end{bmatrix} = 6$, evaluate $\det A$ where $A = \begin{bmatrix} a+x & b+y & c+z \\ 3x & 3y & 3z \\ -p & -q & -r \end{bmatrix}$.

Solution   First take common factors out of rows 2 and 3.

$$\det A = 3(-1) \det \begin{bmatrix} a+x & b+y & c+z \\ x & y & z \\ p & q & r \end{bmatrix}$$

Now subtract the second row from the first and interchange the last two rows.

$$\det A = -3 \det \begin{bmatrix} a & b & c \\ x & y & z \\ p & q & r \end{bmatrix} = 3 \det \begin{bmatrix} a & b & c \\ p & q & r \\ x & y & z \end{bmatrix} = 3 \cdot 6 = 18$$

The determinant of a matrix is a sum of products of its entries. In particular, if these entries are polynomials in x, then the determinant itself is a polynomial in x. It is often of interest to determine which values of x make the determinant zero, so it is very useful if the determinant is given in factored form. Theorem 2 can help.

Example 7

Find the values of x for which $\det A = 0$, where $A = \begin{bmatrix} 1 & x & x \\ x & 1 & x \\ x & x & 1 \end{bmatrix}$.

Solution   To evaluate $\det A$, first subtract x times row 1 from rows 2 and 3.

$$\det A = \begin{vmatrix} 1 & x & x \\ x & 1 & x \\ x & x & 1 \end{vmatrix} = \begin{vmatrix} 1 & x & x \\ 0 & 1-x^2 & x-x^2 \\ 0 & x-x^2 & 1-x^2 \end{vmatrix} = \begin{vmatrix} 1-x^2 & x-x^2 \\ x-x^2 & 1-x^2 \end{vmatrix}$$

At this stage we could simply evaluate the determinant (the result is $2x^3 - 3x^2 + 1$). Then we would have to factor this polynomial to find the values of x that make it zero. However, this factorization can be obtained directly by first factoring each entry in the determinant and taking a common factor of $(1 - x)$ from each row.

$$\det A = \begin{vmatrix} (1-x)(1+x) & x(1-x) \\ x(1-x) & (1-x)(1+x) \end{vmatrix} = (1-x)^2 \begin{vmatrix} 1+x & x \\ x & 1+x \end{vmatrix}$$

$$= (1-x)^2(2x+1)$$

Hence, $\det A = 0$ means $(1-x)^2(2x+1) = 0$, so $x = 1$ or $x = -\frac{1}{2}$.

Example 8

If a_1, a_2, and a_3 are given show that

$$\det \begin{bmatrix} 1 & a_1 & a_1^2 \\ 1 & a_2 & a_2^2 \\ 1 & a_3 & a_3^2 \end{bmatrix} = (a_3 - a_1)(a_3 - a_2)(a_2 - a_1)$$

Solution   Begin by subtracting row 1 from rows 2 and 3, and then expand along column 1:

$$\det \begin{bmatrix} 1 & a_1 & a_1^2 \\ 1 & a_2 & a_2^2 \\ 1 & a_3 & a_3^2 \end{bmatrix} = \det \begin{bmatrix} 1 & a_1 & a_1^2 \\ 0 & a_2 - a_1 & a_2^2 - a_1^2 \\ 0 & a_3 - a_1 & a_3^2 - a_1^2 \end{bmatrix} = \det \begin{bmatrix} a_2 - a_1 & a_2^2 - a_1^2 \\ a_3 - a_1 & a_3^2 - a_1^2 \end{bmatrix}$$

Now $(a_2 - a_1)$ and $(a_3 - a_1)$ are common factors in rows 1 and 2, respectively, so

$$\det \begin{bmatrix} 1 & a_1 & a_1^2 \\ 1 & a_2 & a_2^2 \\ 1 & a_3 & a_3^2 \end{bmatrix} = (a_2 - a_1)(a_3 - a_1) \det \begin{bmatrix} 1 & a_2 + a_1 \\ 1 & a_3 + a_1 \end{bmatrix}$$

$$= (a_2 - a_1)(a_3 - a_1)(a_3 - a_2)$$

The matrix in Example 8 is called a **Vandermonde matrix**, and the formula for its determinant can be generalized to the $n \times n$ case (see Theorem 7 §3.2).

If A is an $n \times n$ matrix, forming uA means multiplying *every* row of A by u. Applying property 3 of Theorem 2, we can take the common factor out of each row and so obtain the following useful result.

Theorem 3

If A is an $n \times n$ matrix, then $\det(uA) = u^n \det A$ for any number u.

The next example displays a type of matrix whose determinant is easy to compute.

Example 9

Evaluate $\det A$ if $A = \begin{bmatrix} a & 0 & 0 & 0 \\ u & b & 0 & 0 \\ v & w & c & 0 \\ x & y & z & d \end{bmatrix}$.

Solution   Expand along row 1 to get $\det A = a\begin{vmatrix} b & 0 & 0 \\ w & c & 0 \\ y & z & d \end{vmatrix}$. Now expand this along the top row to get $\det A = ab\begin{vmatrix} c & 0 \\ z & d \end{vmatrix} = abcd$, the product of the main diagonal entries.

A square matrix is called a **lower triangular matrix** if all entries above the main diagonal are zero (as in Example 9). Similarly, an **upper triangular matrix** is one for which all entries below the main diagonal are zero. A **triangular matrix** is one that is either upper or lower triangular. Theorem 4 gives an easy rule for calculating the determinant of any triangular matrix. The proof is like the solution to Example 9.

Theorem 4

If A is a square triangular matrix, then $\det A$ is the product of the entries on the main diagonal.

Theorem 4 is useful in computer calculations because it is a routine matter to carry a matrix to triangular form using row operations.

Block matrices such as those in the next theorem arise frequently in practice, and the theorem gives an easy method for computing their determinants.

Theorem 5

Consider matrices $\begin{bmatrix} A & X \\ 0 & B \end{bmatrix}$ and $\begin{bmatrix} A & 0 \\ Y & B \end{bmatrix}$ in block form, where A and B are square matrices. Then

$$\det\begin{bmatrix} A & X \\ 0 & B \end{bmatrix} = \det A \det B \quad \text{and} \quad \det\begin{bmatrix} A & 0 \\ Y & B \end{bmatrix} = \det A \det B$$

PROOF

Write $T = \begin{bmatrix} A & X \\ 0 & B \end{bmatrix}$ and proceed by induction on k where A is $k \times k$. If $k = 1$, it is the Laplace expansion along column 1. In general let $S_i(T)$ denote the matrix obtained from T by deleting row i and column 1. Then the cofactor expansion of $\det T$ along the first column is

$$\det T = a_{11}\det(S_1(T)) - a_{21}\det(S_2(T)) + \cdots \pm a_{k1}\det(S_k(T)) \qquad (*)$$

where $a_{11}, a_{21}, \ldots, a_{k1}$ are the entries in the first column of A. But

$$S_i(T) = \begin{bmatrix} S_i(A) & X_i \\ 0 & B \end{bmatrix}$$ for each $i = 1, 2, \ldots, k$, so $\det(S_i(T)) = \det(S_i(A))\cdot\det B$ by induction. Hence, equation $(*)$ becomes

$$\det T = \{a_{11}\det(S_1(A)) - a_{21}\det(S_2(A)) + \cdots \pm a_{k1}\det(S_k(A))\}\det B$$
$$= \{\det A\}\det B$$

as required. The lower triangular case is similar.

Example 10

$$\det\begin{bmatrix} 2 & 3 & 1 & 3 \\ 1 & -2 & -1 & 1 \\ 0 & 1 & 0 & 1 \\ 0 & 4 & 0 & 1 \end{bmatrix} = -\begin{vmatrix} 2 & 1 & 3 & 3 \\ 1 & -1 & -2 & 1 \\ 0 & 0 & 1 & 1 \\ 0 & 0 & 4 & 1 \end{vmatrix} = -\begin{vmatrix} 2 & 1 \\ 1 & -1 \end{vmatrix}\begin{vmatrix} 1 & 1 \\ 4 & 1 \end{vmatrix} = -(-3)(-3) = 9.$$

Exercises 3.1

1. Compute the determinants of the following matrices.

   (a) $\begin{bmatrix} 2 & -1 \\ 3 & 2 \end{bmatrix}$ ⊛(b) $\begin{bmatrix} 6 & 9 \\ 8 & 12 \end{bmatrix}$

   (c) $\begin{bmatrix} a^2 & ab \\ ab & b^2 \end{bmatrix}$ ⊛(d) $\begin{bmatrix} a+1 & a \\ a & a-1 \end{bmatrix}$

   (e) $\begin{bmatrix} \cos\theta & -\sin\theta \\ \sin\theta & \cos\theta \end{bmatrix}$ ⊛(f) $\begin{bmatrix} 2 & 0 & -3 \\ 1 & 2 & 5 \\ 0 & 3 & 0 \end{bmatrix}$

   (g) $\begin{bmatrix} 1 & 2 & 3 \\ 4 & 5 & 6 \\ 7 & 8 & 9 \end{bmatrix}$ ⊙◆(h) $\begin{bmatrix} 0 & a & 0 \\ b & c & d \\ 0 & e & 0 \end{bmatrix}$

   (i) $\begin{bmatrix} 1 & b & c \\ b & c & 1 \\ c & 1 & b \end{bmatrix}$ ◆(j) $\begin{bmatrix} 0 & a & b \\ a & 0 & c \\ b & c & 0 \end{bmatrix}$

   (k) $\begin{bmatrix} 0 & 1 & -1 & 0 \\ 3 & 0 & 0 & 2 \\ 0 & 1 & 2 & 1 \\ 5 & 0 & 0 & 7 \end{bmatrix}$ ◆(l) $\begin{bmatrix} 1 & 0 & 3 & 1 \\ 2 & 2 & 6 & 0 \\ -1 & 0 & -3 & 1 \\ 4 & 1 & 12 & 0 \end{bmatrix}$

   (m) $\begin{bmatrix} 3 & 1 & -5 & 2 \\ 1 & 3 & 0 & 1 \\ 1 & 0 & 5 & 2 \\ 1 & 1 & 2 & -1 \end{bmatrix}$ ◆(n) $\begin{bmatrix} 4 & -1 & 3 & -1 \\ 3 & 1 & 0 & 2 \\ 0 & 1 & 2 & 2 \\ 1 & 2 & -1 & 1 \end{bmatrix}$

   (o) $\begin{bmatrix} 1 & -1 & 5 & 5 \\ 3 & 1 & 2 & 4 \\ -1 & -3 & 8 & 0 \\ 1 & 1 & 2 & -1 \end{bmatrix}$ ⊛(p) $\begin{bmatrix} 0 & 0 & 0 & a \\ 0 & 0 & b & p \\ 0 & c & q & k \\ d & s & t & u \end{bmatrix}$

2. Show that $\det A = 0$ if A has a row or column consisting of zeros.

3. Show that the sign of the position in the last row and the last column of A is always +1.

4. Show that $\det I = 1$ for any identity matrix I.

5. Evaluate each determinant by reducing it to upper triangular form.

   (a) $\begin{bmatrix} 1 & -1 & 2 \\ 3 & 1 & 1 \\ 2 & -1 & 3 \end{bmatrix}$ ⊛(b) $\begin{bmatrix} -1 & 3 & 1 \\ 2 & 5 & 3 \\ 1 & -2 & 1 \end{bmatrix}$

   (c) $\begin{bmatrix} -1 & -1 & 1 & 0 \\ 2 & 1 & 1 & 3 \\ 0 & 1 & 1 & 2 \\ 1 & 3 & -1 & 2 \end{bmatrix}$ ◆(d) $\begin{bmatrix} 2 & 3 & 1 & 1 \\ 0 & 2 & -1 & 3 \\ 0 & 5 & 1 & 1 \\ 1 & 1 & 2 & 5 \end{bmatrix}$

6. Evaluate by inspection:

   (a) $\det\begin{bmatrix} a & b & c \\ a+1 & b+1 & c+1 \\ a-1 & b-1 & c-1 \end{bmatrix}$

   ⊙◆(b) $\det\begin{bmatrix} a & b & c \\ a+b & 2b & c+b \\ 2 & 2 & 2 \end{bmatrix}$

7. If $\det\begin{bmatrix} a & b & c \\ p & q & r \\ x & y & z \end{bmatrix} = -1$, compute:

   (a) $\det\begin{bmatrix} -x & -y & -z \\ 3p+a & 3q+b & 3r+c \\ 2p & 2q & 2r \end{bmatrix}$

   ◆(b) $\det\begin{bmatrix} -2a & -2b & -2c \\ 2p+x & 2q+y & 2r+z \\ 3x & 3y & 3z \end{bmatrix}$

8. Show that:

   (a) $\det\begin{bmatrix} p+x & q+y & r+z \\ a+x & b+y & c+z \\ a+p & b+q & c+r \end{bmatrix} = 2\det\begin{bmatrix} a & b & c \\ p & q & r \\ x & y & z \end{bmatrix}$

   (b) $\det\begin{bmatrix} 2a+p & 2b+q & 2c+r \\ 2p+x & 2q+y & 2r+z \\ 2x+a & 2y+b & 2z+c \end{bmatrix} = 9\det\begin{bmatrix} a & b & c \\ p & q & r \\ x & y & z \end{bmatrix}$

9. In each case either prove the statement or give an example showing that it is false:

   (a) $\det(A + B) = \det A + \det B$.

   ◆(b) If $\det A = 0$, then A has two equal rows.

   (c) If A is 2×2, then $\det(A^T) = \det A$.

   ◆(d) If R is the reduced row-echelon form of A, then $\det A = \det R$.

   (e) If A is 2×2, then $\det(7A) = 49\det A$.

   ◆(f) $\det(A^T) = -\det A$.

   (g) $\det(-A) = -\det A$.

   ◆(h) If $\det(A) = \det(B)$ where A and B are the same size, then $A = B$.

10. Compute the determinants of each matrix, using Theorem 5.

   (a) $\begin{bmatrix} 1 & -1 & 2 & 0 & -2 \\ 0 & 1 & 0 & 4 & 1 \\ 1 & 1 & 5 & 0 & 0 \\ 0 & 0 & 0 & 3 & -1 \\ 0 & 0 & 0 & 1 & 1 \end{bmatrix}$
   ◆(b) $\begin{bmatrix} 1 & 2 & 0 & 3 & 0 \\ -1 & 3 & 1 & 4 & 0 \\ 0 & 0 & 2 & 1 & 1 \\ 0 & 0 & -1 & 0 & 2 \\ 0 & 0 & 3 & 0 & 1 \end{bmatrix}$

11. If $\det A = 2$, $\det B = -1$, and $\det C = 3$, find:

   (a) $\det \begin{bmatrix} A & X & Y \\ 0 & B & Z \\ 0 & 0 & C \end{bmatrix}$
   ◆(b) $\det \begin{bmatrix} A & 0 & 0 \\ X & B & 0 \\ Y & Z & C \end{bmatrix}$

   (c) $\det \begin{bmatrix} A & X & Y \\ 0 & B & 0 \\ 0 & Z & C \end{bmatrix}$
   ◆(d) $\det \begin{bmatrix} A & X & 0 \\ 0 & B & 0 \\ Y & Z & C \end{bmatrix}$

12. If A has three columns with only the top two entries nonzero, show that $\det A = 0$.

13. (a) Find $\det A$ if A is 3×3 and $\det(2A) = 6$.

   (b) Under what conditions is $\det(-A) = \det A$?

14. Evaluate by first adding all other rows to the first row.

   (a) $\det \begin{bmatrix} x-1 & 2 & 3 \\ 2 & -3 & x-2 \\ -2 & x & -2 \end{bmatrix}$

   ◆(b) $\det \begin{bmatrix} x-1 & -3 & 1 \\ 2 & -1 & x-1 \\ -3 & x+2 & -2 \end{bmatrix}$

15. (a) Find b if $\det \begin{bmatrix} 5 & -1 & x \\ 2 & 6 & y \\ -5 & 4 & z \end{bmatrix} = ax + by + cz$.

   ◆(b) Find c if $\det \begin{bmatrix} 2 & x & -1 \\ 1 & y & 3 \\ -3 & z & 4 \end{bmatrix} = ax + by + cz$.

16. Find the real numbers x and y such that $\det A = 0$ if:

   (a) $A = \begin{bmatrix} 0 & x & y \\ y & 0 & x \\ x & y & 0 \end{bmatrix}$
   ◆(b) $A = \begin{bmatrix} 1 & x & x \\ -x & -2 & x \\ -x & -x & -3 \end{bmatrix}$

   (c) $A = \begin{bmatrix} 1 & x & x^2 & x^3 \\ x & x^2 & x^3 & 1 \\ x^2 & x^3 & 1 & x \\ x^3 & 1 & x & x^2 \end{bmatrix}$
   ◆(d) $A = \begin{bmatrix} x & y & 0 & 0 \\ 0 & x & y & 0 \\ 0 & 0 & x & y \\ y & 0 & 0 & x \end{bmatrix}$

17. Show that $\det \begin{bmatrix} 0 & 1 & 1 & 1 \\ 1 & 0 & x & x \\ 1 & x & 0 & x \\ 1 & x & x & 0 \end{bmatrix} = -3x^2$.

18. Show that $\det \begin{bmatrix} 1 & x & x^2 & x^3 \\ a & 1 & x & x^2 \\ p & b & 1 & x \\ q & r & c & 1 \end{bmatrix}$

   $= (1 - ax)(1 - bx)(1 - cx)$.

19. Given the polynomial $p(x) = a + bx + cx^2 + dx^3 + x^4$, the matrix

   $$C = \begin{bmatrix} 0 & 1 & 0 & 0 \\ 0 & 0 & 1 & 0 \\ 0 & 0 & 0 & 1 \\ -a & -b & -c & -d \end{bmatrix} \text{ is called}$$

   the **companion matrix** of $p(x)$. Show that $\det(xI - C) = p(x)$.

20. Show that $\det \begin{bmatrix} a+x & b+x & c+x \\ b+x & c+x & a+x \\ c+x & a+x & b+x \end{bmatrix}$

   $= (a + b + c + 3x)[(ab + ac + bc) - (a^2 + b^2 + c^2)]$.

21. Given $n \geq 2$, let $C_2, C_3, \ldots, C_n$ be $n - 1$ columns in $\mathbb{R}^n$. Define $T: \mathbb{R}^n \to \mathbb{R}$ by $T(X) = \det([X \; C_2 \; C_3 \; \cdots \; C_n])$ where $[X \; C_2 \; C_3 \; \cdots \; C_n]$ is the $n \times n$ matrix with columns $X, C_2, C_3, \ldots, C_n$. Show that T is a linear transformation; that is:

   (a) Show that $T(aX) = aT(X)$ for all X in $\mathbb{R}^n$ and all real numbers a.

♦(b) Show that $T(X + Y) = T(X) + T(Y)$ for all X and Y in $\mathbb{R}^n$. [*Hint*: Expand the determinants along column 1.]

22. Show that

$$\det\begin{bmatrix} 0 & 0 & \cdots & 0 & a_1 \\ 0 & 0 & \cdots & a_2 & * \\ \vdots & \vdots & & \vdots & \vdots \\ 0 & a_{n-1} & \cdots & * & * \\ a_n & * & \cdots & * & * \end{bmatrix} = (-1)^k a_1 a_2 \cdots a_n$$

where either $n = 2k$ or $n = 2k + 1$, and ∗-entries are arbitrary.

23. By expanding along the first column, show that:

$$\det\begin{bmatrix} 1 & 1 & 0 & 0 & \cdots & 0 & 0 \\ 0 & 1 & 1 & 0 & \cdots & 0 & 0 \\ 0 & 0 & 1 & 1 & \cdots & 0 & 0 \\ \vdots & \vdots & \vdots & \vdots & & \vdots & \vdots \\ 0 & 0 & 0 & 0 & \cdots & 1 & 1 \\ 1 & 0 & 0 & 0 & \cdots & 0 & 1 \end{bmatrix} = 1 + (-1)^{n+1}$$

if the matrix is $n \times n$, $n \geq 2$.

♦24. Form matrix B from a matrix A by writing the columns of A in reverse order. Express $\det B$ in terms of $\det A$.

25. Prove property 3 of Theorem 2 by expanding along the row (or column) in question.

26. Show that the line through two distinct points (x_1, y_1) and (x_2, y_2) in the plane has equation

$$\det\begin{bmatrix} x & y & 1 \\ x_1 & y_1 & 1 \\ x_2 & y_2 & 1 \end{bmatrix} = 0.$$

27. Let A be an $n \times n$ matrix. Given a polynomial $p(x) = a_0 + a_1 x + \cdots + a_m x^m$, we write

$$p(A) = a_0 I + a_1 A + \cdots + a_m A^m.$$

For example, if $p(x) = 2 - 3x + 5x^2$, then $p(A) = 2I - 3A + 5A^2$. The *characteristic polynomial* of A is defined to be $c_A(x) = \det[xI - A]$ and the Cayley–Hamilton theorem asserts that $c_A(A) = 0$ for any matrix A.

(a) Verify the theorem for (i) $A = \begin{bmatrix} 3 & 2 \\ 1 & -1 \end{bmatrix}$ and

(ii) $A = \begin{bmatrix} 1 & -1 & 1 \\ 0 & 1 & 0 \\ 8 & 2 & 2 \end{bmatrix}$.

(b) Prove the theorem for $A = \begin{bmatrix} a & b \\ c & d \end{bmatrix}$.

SECTION 3.2   Determinants and Matrix Inverses

In this section, several theorems about determinants are derived. One consequence of these theorems is that a square matrix A is invertible if and only if $\det A \neq 0$. Moreover, determinants are used to give a formula for A^{-1} that, in turn, yields a formula (called Cramer's rule) for the solution of any system of linear equations with an invertible coefficient matrix.

We begin with a remarkable theorem about the determinant of a product of matrices. The proof is given at the end of this section.

Theorem 1   Product Theorem

If A and B are $n \times n$ matrices, then $\det(AB) = \det A \det B$.

The complexity of matrix multiplication makes the product theorem quite unexpected. Here is an example where it reveals an important numerical identity.

Example 1

If $A = \begin{bmatrix} a & b \\ -b & a \end{bmatrix}$ and $B = \begin{bmatrix} c & d \\ -d & c \end{bmatrix}$, then $AB = \begin{bmatrix} ac - bd & ad + bc \\ -(ad + bc) & ac - bd \end{bmatrix}$.

Hence $\det A \det B = \det(AB)$ gives the identity

$$(a^2 + b^2)(c^2 + d^2) = (ac - bd)^2 + (ad + bc)^2$$

Theorem 1 extends easily to $\det(ABC) = \det A \det B \det C$. In fact, induction gives

$$\det(A_1 A_2 \cdots A_{k-1} A_k) = \det A_1 \det A_2 \cdots \det A_{k-1} \det A_k$$

for any square matrices $A_1, \ldots, A_k$ of the same size. In particular, if each $A_i = A$, we obtain

$$\det(A^k) = (\det A)^k \quad \text{for any } k \geq 1$$

We can now give the invertibility condition.

Theorem 2

An $n \times n$ matrix A is invertible if and only if $\det A \neq 0$. When this is the case,

$$\det(A^{-1}) = \frac{1}{\det A}.$$

PROOF

If A is invertible, then $AA^{-1} = I$; so, the product theorem gives

$$1 = \det I = \det(AA^{-1}) = \det A \det A^{-1}$$

Hence, $\det A \neq 0$ and $\det A^{-1} = \frac{1}{\det A}$.

Conversely, if $\det A \neq 0$, we show that A can be carried to I by elementary row operations (and invoke Theorem 5 §2.3). Certainly, A can be carried to its reduced row-echelon form R, so $R = E_k \cdots E_2 E_1 A$ where the E_i are elementary matrices (Theorem 1 §2.4). Hence the product theorem gives

$$\det R = \det E_k \cdots \det E_2 \det E_1 \det A$$

Since $\det E \neq 0$ for all elementary matrices E, this shows $\det R \neq 0$. In particular, R has no row of zeros, so $R = I$ because R is square and reduced row-echelon. This is what we wanted.

Example 2

For which values of c does $A = \begin{bmatrix} 1 & 0 & -c \\ -1 & 3 & 1 \\ 0 & 2c & -4 \end{bmatrix}$ have an inverse?

Solution

Compute $\det A$ by first adding c times column 1 to column 3 and then expanding along row 1.

$$\det A = \det \begin{bmatrix} 1 & 0 & -c \\ -1 & 3 & 1 \\ 0 & 2c & -4 \end{bmatrix} = \det \begin{bmatrix} 1 & 0 & 0 \\ -1 & 3 & 1 - c \\ 0 & 2c & -4 \end{bmatrix} = 2(c + 2)(c - 3).$$

Hence, $\det A = 0$ if $c = -2$ or $c = 3$, and A has an inverse if $c \neq -2$ and $c \neq 3$.

Example 3

If a product $A_1 A_2 \cdots A_k$ of square matrices is invertible, show that each A_i is invertible.

Solution   We have $(\det A_1)(\det A_2)\cdots(\det A_k) = \det(A_1 A_2 \cdots A_k)$ by the product theorem, and $\det(A_1 A_2 \cdots A_k) \neq 0$ by Theorem 2 because $A_1 A_2 \cdots A_k$ is invertible. Hence

$$(\det A_1)(\det A_2)\cdots(\det A_k) \neq 0,$$

so $\det A_i \neq 0$ for each i. This shows that each A_i is invertible, again by Theorem 2.

Theorem 3

If A is any square matrix, $\det A^T = \det A$.

PROOF

Consider first the case of an elementary matrix E. If E is of type I or II, then $E^T = E$; so certainly $\det E^T = \det E$. If E is of type III, then E^T is also of type III; so $\det E^T = 1 = \det E$ by Theorem 2 §3.1. Hence, $\det E^T = \det E$ for every elementary matrix E.

Now let A be any square matrix. If A is not invertible, then neither is A^T; so $\det A^T = 0 = \det A$ by Theorem 2. If A is invertible, then $A = E_k \cdots E_2 E_1$, where the E_i are elementary matrices (Theorem 2 §2.4). Hence, $A^T = E_1^T E_2^T \cdots E_k^T$, so the product theorem gives

$$\begin{aligned}
\det A^T &= \det E_1^T \det E_2^T \cdots \det E_k^T = \det E_1 \det E_2 \cdots \det E_k \\
&= \det E_k \cdots \det E_2 \det E_1 \\
&= \det A
\end{aligned}$$

This completes the proof.

Example 4

If $\det A = 2$ and $\det B = 5$, calculate $\det(A^3 B^{-1} A^T B^2)$.

Solution   We use several of the facts just derived.

$$\begin{aligned}
\det(A^3 B^{-1} A^T B^2) &= \det(A^3)\det(B^{-1})\det(A^T)\det(B^2) \\
&= (\det A)^3 \frac{1}{\det B} \det A \, (\det B)^2 \\
&= 2^3 \cdot \tfrac{1}{5} \cdot 2 \cdot 5^2 \\
&= 80
\end{aligned}$$

Example 5

A square matrix is called **orthogonal** if $A^{-1} = A^T$. What are the possible values of $\det A$ if A is orthogonal?

Solution   If A is orthogonal, we have $I = AA^T$. Take determinants to obtain
$1 = \det I = \det(AA^T) = \det A \det A^T = (\det A)^2$. Since $\det A$ is a number, this means $\det A = \pm 1$.

Hence Examples 5 and 10 of Section 2.5 imply that rotation about the origin and reflection about a line through the origin in $\mathbb{R}^2$ have orthogonal matrices with determinants 1 and −1 respectively. In fact they are the only such transformations of $\mathbb{R}^2$. We have more to say about this in Section 4.2.

Adjugates

In Section 2.3 we defined the adjugate of a 2×2 matrix $A = \begin{bmatrix} a & b \\ c & d \end{bmatrix}$ to be

$\text{adj}(A) = \begin{bmatrix} d & -b \\ -c & a \end{bmatrix}$. Then we verified that $A(\text{adj } A) = (\det A)I = (\text{adj } A)A$ and hence

that, if $\det A \neq 0$, $A^{-1} = \dfrac{1}{\det A} \text{adj } A$. We are now able to define the adjugate of an

arbitrary square matrix and to show that this formula for the inverse remains valid (when the inverse exists).

Recall that the (i, j)-cofactor $c_{ij}(A)$ of a square matrix A is a number defined for each position (i, j) in the matrix. If A is a square matrix, the **cofactor matrix of A** is defined to be the matrix $[c_{ij}(A)]$ whose (i, j)-entry is the (i, j)-cofactor of A. The **adjugate**[5] of A, denoted $\text{adj}(A)$, is the transpose of this cofactor matrix; in symbols,

$$\text{adj}(A) = [c_{ij}(A)]^T$$

This agrees with the earlier definition for a 2×2 matrix A.

Example 6

Compute the adjugate of $A = \begin{bmatrix} 1 & 3 & -2 \\ 0 & 1 & 5 \\ -2 & -6 & 7 \end{bmatrix}$ and calculate $A(\text{adj } A)$ and

$(\text{adj } A)A$.

Solution   We first find the cofactor matrix.

$$\begin{bmatrix} c_{11}(A) & c_{12}(A) & c_{13}(A) \\ c_{21}(A) & c_{22}(A) & c_{23}(A) \\ c_{31}(A) & c_{32}(A) & c_{33}(A) \end{bmatrix} = \begin{bmatrix} \begin{vmatrix} 1 & 5 \\ -6 & 7 \end{vmatrix} & -\begin{vmatrix} 0 & 5 \\ -2 & 7 \end{vmatrix} & \begin{vmatrix} 0 & 1 \\ -2 & -6 \end{vmatrix} \\ -\begin{vmatrix} 3 & -2 \\ -6 & 7 \end{vmatrix} & \begin{vmatrix} 1 & -2 \\ -2 & 7 \end{vmatrix} & -\begin{vmatrix} 1 & 3 \\ -2 & -6 \end{vmatrix} \\ \begin{vmatrix} 3 & -2 \\ 1 & 5 \end{vmatrix} & -\begin{vmatrix} 1 & -2 \\ 0 & 5 \end{vmatrix} & \begin{vmatrix} 1 & 3 \\ 0 & 1 \end{vmatrix} \end{bmatrix}$$

$$= \begin{bmatrix} 37 & -10 & 2 \\ -9 & 3 & 0 \\ 17 & -5 & 1 \end{bmatrix}$$

---

5 This is also called the adjoint of A, but this term has another meaning.

Then the adjugate of A is the transpose of this cofactor matrix.

$$\text{adj } A = \begin{bmatrix} 37 & -10 & 2 \\ -9 & 3 & 0 \\ 17 & -5 & 1 \end{bmatrix}^T = \begin{bmatrix} 37 & -9 & 17 \\ -10 & 3 & -5 \\ 2 & 0 & 1 \end{bmatrix}$$

The computation of $A(\text{adj } A)$ gives

$$A(\text{adj } A) = \begin{bmatrix} 1 & 3 & -2 \\ 0 & 1 & 5 \\ -2 & -6 & 7 \end{bmatrix} \begin{bmatrix} 37 & -9 & 17 \\ -10 & 3 & -5 \\ 2 & 0 & 1 \end{bmatrix} = \begin{bmatrix} 3 & 0 & 0 \\ 0 & 3 & 0 \\ 0 & 0 & 3 \end{bmatrix} = 3I$$

and the reader can verify that also $(\text{adj } A)A = 3I$. Hence, analogy with the 2×2 case would indicate that $\det A = 3$; this is, in fact, the case.

The relationship $A(\text{adj } A) = (\det A)I$ holds for any square matrix A. To see why this is so, consider the general 3×3 case. Writing $c_{ij}(A) = c_{ij}$ for short, we have

$$\text{adj } A = \begin{bmatrix} c_{11} & c_{12} & c_{13} \\ c_{21} & c_{22} & c_{23} \\ c_{31} & c_{32} & c_{33} \end{bmatrix}^T = \begin{bmatrix} c_{11} & c_{21} & c_{31} \\ c_{12} & c_{22} & c_{32} \\ c_{13} & c_{23} & c_{33} \end{bmatrix}$$

If $A = [a_{ij}]$ in the usual notation, we are to verify that $A(\text{adj } A) = (\det A)I$. That is,

$$A(\text{adj } A) = \begin{bmatrix} a_{11} & a_{12} & a_{13} \\ a_{21} & a_{22} & a_{23} \\ a_{31} & a_{32} & a_{33} \end{bmatrix} \begin{bmatrix} c_{11} & c_{21} & c_{31} \\ c_{12} & c_{22} & c_{32} \\ c_{13} & c_{23} & c_{33} \end{bmatrix} = \begin{bmatrix} \det A & 0 & 0 \\ 0 & \det A & 0 \\ 0 & 0 & \det A \end{bmatrix}$$

Consider the $(1, 1)$-entry in the product. It is given by $a_{11}c_{11} + a_{12}c_{12} + a_{13}c_{13}$, and this is just the cofactor expansion of $\det A$ along the first row of A. Similarly, the $(2, 2)$-entry and the $(3, 3)$-entry are the cofactor expansions of $\det A$ along rows 2 and 3, respectively.

So it remains to be seen why the off-diagonal elements in the matrix product $A(\text{adj } A)$ are all zero. Consider the $(1, 2)$-entry of the product. It is given by $a_{11}c_{21} + a_{12}c_{22} + a_{13}c_{23}$. This *looks* like the cofactor expansion of the determinant of *some* matrix. To see which, observe that c_{21}, c_{22}, and c_{23} are all computed by *deleting* row 2 of A (and one of the columns), so they remain the same if row 2 of A is changed. In particular, if row 2 of A is replaced by row 1, we obtain

$$a_{11}c_{21} + a_{12}c_{22} + a_{13}c_{23} = \det \begin{bmatrix} a_{11} & a_{12} & a_{13} \\ a_{11} & a_{12} & a_{13} \\ a_{31} & a_{32} & a_{33} \end{bmatrix} = 0$$

where the expansion is along row 2 and where the determinant is zero because two rows are identical. A similar argument shows that the other off-diagonal entries are zero.

This argument works in general and yields the first part of Theorem 4. The second assertion follows from the first by multiplying through by the scalar $\dfrac{1}{\det A}$.

Theorem 4 Adjugate Formula

If A is any square matrix, then

$$A(\operatorname{adj} A) = (\det A)I = (\operatorname{adj} A)A$$

In particular, if $\det A \neq 0$, the inverse of A is given by

$$A^{-1} = \frac{1}{\det A} \operatorname{adj} A$$

It is important to note that this theorem is *not* an efficient way to find the inverse of the matrix A. For example, if A were 10×10, the calculation of adj A would require computing $10^2 = 100$ determinants of 9×9 matrices! On the other hand, the matrix inversion algorithm would find A^{-1} with about the same effort as finding $\det A$. Clearly, Theorem 4 is not a *practical* result: its virtue is that it gives a formula for A^{-1} that is useful for *theoretical* purposes.

Example 7

Find the $(2, 3)$-entry of A^{-1} if $A = \begin{bmatrix} 2 & 1 & 3 \\ 5 & -7 & 1 \\ 3 & 0 & -6 \end{bmatrix}$.

Solution We have $\det A = \begin{vmatrix} 2 & 1 & 3 \\ 5 & -7 & 1 \\ 3 & 0 & -6 \end{vmatrix} = \begin{vmatrix} 2 & 1 & 7 \\ 5 & -7 & 11 \\ 3 & 0 & 0 \end{vmatrix} = 3 \begin{vmatrix} 1 & 7 \\ -7 & 11 \end{vmatrix} = 180$. Since

$A^{-1} = \dfrac{1}{\det A} \operatorname{adj} A = \tfrac{1}{180} [c_{ij}(A)]^T$, the $(2, 3)$-entry of A^{-1} is the $(3, 2)$-entry

of the transpose $\tfrac{1}{180}[c_{ij}(A)]$ of this matrix; that is, it equals

$$\tfrac{1}{180} c_{32}(A) = \tfrac{1}{180}\left(-\begin{vmatrix} 2 & 3 \\ 5 & 1 \end{vmatrix}\right) = \tfrac{13}{180}.$$

Example 8

If A is $n \times n$, $n \geq 2$, show that $\det(\operatorname{adj} A) = (\det A)^{n-1}$.

Solution Write $d = \det A$ so that $A(\operatorname{adj} A) = dI$ by Theorem 4. Taking determinants gives $d \det(\operatorname{adj} A) = d^n$, so we are done if $d \neq 0$. Assume $d = 0$; we must show that $\det(\operatorname{adj} A) = 0$, that is, adj A is not invertible. If $A \neq 0$, this follows from $A(\operatorname{adj} A) = dI = 0$; if $A = 0$, it follows because adj $A = 0$.

Cramer's Rule

Theorem 4 has a nice application to linear equations. Suppose

$$AX = B$$

is a system of n equations in n variables $x_1, x_2, \ldots, x_n$. Here A is the $n \times n$ coefficient matrix, and X and B are the columns

$$X = \begin{bmatrix} x_1 \\ x_2 \\ \vdots \\ x_n \end{bmatrix} \qquad B = \begin{bmatrix} b_1 \\ b_2 \\ \vdots \\ b_n \end{bmatrix}$$

of variables and constants, respectively. If $\det A \neq 0$, we left multiply by A^{-1} to obtain the solution $X = A^{-1}B$. When we use the adjugate formula, this becomes

$$\begin{bmatrix} x_1 \\ x_2 \\ \vdots \\ x_n \end{bmatrix} = \frac{1}{\det A}(\text{adj } A)B$$

$$= \frac{1}{\det A}\begin{bmatrix} c_{11}(A) & c_{21}(A) & \cdots & c_{n1}(A) \\ c_{12}(A) & c_{22}(A) & \cdots & c_{n2}(A) \\ \vdots & \vdots & & \vdots \\ c_{1n}(A) & c_{2n}(A) & \cdots & c_{nn}(A) \end{bmatrix}\begin{bmatrix} b_1 \\ b_2 \\ \vdots \\ b_n \end{bmatrix}$$

Hence, the variables $x_1, x_2, \ldots, x_n$ are given by

$$x_1 = \frac{1}{\det A}[b_1 c_{11}(A) + b_2 c_{21}(A) + \cdots + b_n c_{n1}(A)]$$

$$x_2 = \frac{1}{\det A}[b_1 c_{12}(A) + b_2 c_{22}(A) + \cdots + b_n c_{n2}(A)]$$

$$\vdots \qquad\qquad\qquad\qquad \vdots$$

$$x_n = \frac{1}{\det A}[b_1 c_{1n}(A) + b_2 c_{2n}(A) + \cdots + b_n c_{nn}(A)]$$

Now the quantity $b_1 c_{11}(A) + b_2 c_{21}(A) + \cdots + b_n c_{n1}(A)$ occurring in the formula for x_1 looks like the cofactor expansion of the determinant of a matrix. The cofactors involved are $c_{11}(A), c_{21}(A), \ldots, c_{n1}(A)$, corresponding to the first column of A. If A_1 is obtained from A by replacing the first column of A by B, then $c_{i1}(A_1) = c_{i1}(A)$ for each i. Hence, expanding $\det(A_1)$ by the first column gives

$$\det A_1 = b_1 c_{11}(A_1) + b_2 c_{21}(A_1) + \cdots + b_n c_{n1}(A_1)$$
$$= b_1 c_{11}(A) + b_2 c_{21}(A) + \cdots + b_n c_{n1}(A)$$
$$= (\det A)x_1$$

Hence, $x_1 = \dfrac{\det A_1}{\det A}$, and similar results hold for the other variables.

Theorem 5 Cramer's Rule[6]

If A is an invertible $n \times n$ matrix, the solution to the system

$$AX = B$$

of n equations in the variables $x_1, x_2, \ldots, x_n$ is given by

$$x_1 = \frac{\det A_1}{\det A}, \quad x_2 = \frac{\det A_2}{\det A}, \quad \ldots, \quad x_n = \frac{\det A_n}{\det A}$$

where, for each k, A_k is the matrix obtained from A by replacing column k by B.

6 Gabriel Cramer (1704–1752) was a Swiss mathematician who wrote an introductory work on algebraic curves. He popularized the rule that bears his name, but the idea was known earlier.

Example 9

Find x_1, given the following system of equations.

$$5x_1 + x_2 - x_3 = 4$$
$$9x_1 + x_2 - x_3 = 1$$
$$x_1 - x_2 + 5x_3 = 2$$

Solution   Compute the determinants of the coefficient matrix A and the matrix A_1 obtained from it by replacing the first column by the column of constants.

$$\det A = \det \begin{bmatrix} 5 & 1 & -1 \\ 9 & 1 & -1 \\ 1 & -1 & 5 \end{bmatrix} = -16$$

$$\det A_1 = \det \begin{bmatrix} 4 & 1 & -1 \\ 1 & 1 & -1 \\ 2 & -1 & 5 \end{bmatrix} = 12$$

Hence, $x_1 = (\det A_1)/\det A = -\frac{3}{4}$ by Cramer's rule.

Note that Cramer's rule enabled us to calculate x_1 here without computing x_2 or x_3. Although this might seem an advantage, the truth of the matter is that, for large systems of equations, the number of computations needed to find *all* the variables by the gaussian algorithm is comparable to the number required to find *one* of the determinants involved in Cramer's rule. Furthermore, the algorithm works when the matrix of the system is not invertible and even when the coefficient matrix is not square. Like the adjugate formula, then, Cramer's rule is *not* a practical numerical technique; its virtue is theoretical.

Polynomial Interpolation

Example 10

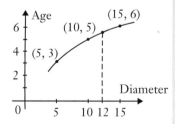

A forester wants to estimate the age (in years) of a tree by measuring the diameter of the trunk (in cm). She obtains the following data:

	Tree 1	Tree 2	Tree 3
Trunk Diameter	5	10	15
Age	3	5	6

Estimate the age of a tree with trunk diameter of 12 cm.

Solution   The forester decides to "fit" a quadratic polynomial $p(x) = r_0 + r_1 x + r_2 x^2$ to the data, that is choose the coefficients r_0, r_1, and r_2 so that $p(5) = 3$, $p(10) = 5$, and $p(15) = 6$, and then use $p(12)$ as the estimate. The conditions give three linear equations:

$$r_0 + 5r_1 + 25r_2 = 3$$
$$r_0 + 10r_1 + 100r_2 = 5$$
$$r_0 + 15r_1 + 225r_2 = 6$$

The (unique) solution is $r_0 = 0$, $r_1 = \frac{7}{10}$ and $r_2 = -\frac{1}{50}$, so

$p(x) = \frac{7}{10}x - \frac{1}{50}x^2 = \frac{1}{50}x(35 - x)$. Hence the estimate is $p(12) = 5.52$.

As in Example 10, it often happens that two variables x and y are related but the actual functional form $y = f(x)$ of the relationship is unknown. Suppose that for certain values $x_1, x_2, \ldots, x_n$ of x the corresponding values $y_1, y_2, \ldots, y_n$ are known (say from experimental measurements). One way to estimate the value of y corresponding to some other value a of x is to find a polynomial[7] $p(x)$

$$p(x) = r_0 + r_1 x + r_2 x + \cdots + r_{n-1} x^{n-1}$$

that "fits" the data, that is $p(x_i) = y_i$ holds for each $i = 1, 2, \ldots, n$. Then the estimate for y is $p(a)$. Such a polynomial always exists if the x_i are distinct.

The conditions that $p(x_i) = y_i$ are

$$r_0 + r_1 x_1 + r_2 x_1^2 + \cdots + r_{n-1} x_1^{n-1} = y_1$$
$$r_0 + r_1 x_2 + r_2 x_2^2 + \cdots + r_{n-1} x_2^{n-1} = y_2$$
$$\vdots \quad \vdots \quad \vdots \quad \quad \vdots \quad \quad \vdots$$
$$r_0 + r_1 x_n + r_2 x_n^2 + \cdots + r_{n-1} x_n^{n-1} = y_n$$

In matrix form, this is

$$\begin{bmatrix} 1 & x_1 & x_1^2 & \cdots & x_1^{n-1} \\ 1 & x_2 & x_2^2 & \cdots & x_2^{n-1} \\ \vdots & \vdots & \vdots & & \vdots \\ 1 & x_n & x_n^2 & \cdots & x_n^{n-1} \end{bmatrix} \begin{bmatrix} r_0 \\ r_1 \\ \vdots \\ r_{n-1} \end{bmatrix} = \begin{bmatrix} y_1 \\ y_2 \\ \vdots \\ y_n \end{bmatrix} \tag{$*$}$$

It can be shown (see Theorem 7) that the determinant of the coefficient matrix equals the product of all terms $(x_i - x_j)$ with $i > j$ and so is nonzero (because the x_i are distinct). Hence the equations have a unique solution $r_0, r_1, \ldots, r_{n-1}$. This proves

---

Theorem 6

Let n data pairs $(x_1, y_1), (x_2, y_2), \ldots, (x_n, y_n)$ be given, and assume that the x_i are distinct. Then there exists a unique polynomial

$$p(x) = r_0 + r_1 x + r_2 x^2 + \cdots + r_{n-1} x^{n-1}$$

such that $p(x_i) = y_i$ for each $i = 1, 2, \ldots, n$.

The polynomial in Theorem 6 is called the **interpolating polynomial** for the data.

We conclude by evaluating the determinant of the coefficient matrix in $(*)$. If $a_1, a_2, \ldots, a_n$ are numbers, the determinant

$$\det \begin{bmatrix} 1 & a_1 & a_1^2 & \cdots & a_1^{n-1} \\ 1 & a_2 & a_2^2 & \cdots & a_2^{n-1} \\ 1 & a_3 & a_3^2 & \cdots & a_3^{n-1} \\ \vdots & \vdots & \vdots & & \vdots \\ 1 & a_n & a_n^2 & \cdots & a_n^{n-1} \end{bmatrix}$$

is called a **Vandermonde determinant**.[8] There is a simple formula for this determinant. If $n = 2$, it equals $(a_2 - a_1)$; if $n = 3$, it is $(a_3 - a_2)(a_3 - a_1)(a_2 - a_1)$ by Example 8 §3.1. The general result is the product

$$\Pi_{1 \le j < i \le n}(a_i - a_j)$$

---

7 A **polynomial** is an expression of the form $a_0 + a_1 x + a_2 x^2 + \cdots + a_n x^n$ where the a_i are numbers and x is a variable. If $a_n \neq 0$, the integer n is called the **degree** of the polynomial, and a_n is called the **leading coefficient**.

8 Alexandre Théophile Vandermonde (1735–1796) was a French mathematician who made contributions to the theory of equations.

of all factors $(a_i - a_j)$ where $1 \le j < i \le n$. For example, if $n = 4$, it is

$$(a_4 - a_3)(a_4 - a_2)(a_4 - a_1)(a_3 - a_2)(a_3 - a_1)(a_2 - a_1).$$

Theorem 7

Let $a_1, a_2, \ldots, a_n$ be numbers where $n \ge 2$. Then the corresponding Vandermonde determinant is given by

$$\det \begin{bmatrix} 1 & a_1 & a_1^2 & \cdots & a_1^{n-1} \\ 1 & a_2 & a_2^2 & \cdots & a_2^{n-1} \\ 1 & a_3 & a_3^2 & \cdots & a_3^{n-1} \\ \vdots & \vdots & \vdots & & \vdots \\ 1 & a_n & a_n^2 & \cdots & a_n^{n-1} \end{bmatrix} = \Pi_{1 \le j < i \le n}(a_i - a_j).$$

PROOF

We may assume that the a_i are distinct; otherwise both sides are zero. We proceed by induction on $n \ge 2$; we have it for $n = 2, 3$. So assume it holds for $n - 1$. The trick is to replace a_n by a variable x, and consider the determinant

$$p(x) = \det \begin{bmatrix} 1 & a_1 & a_1^2 & \cdots & a_1^{n-1} \\ 1 & a_2 & a_2^2 & \cdots & a_2^{n-1} \\ \vdots & \vdots & \vdots & & \vdots \\ 1 & a_{n-1} & a_{n-1}^2 & \cdots\cdots & a_{n-1}^{n-1} \\ 1 & x & x^2 & \cdots & x^{n-1} \end{bmatrix}.$$

Then $p(x)$ is a polynomial of degree at most $n - 1$ (expand along the last row), and $p(a_i) = 0$ for $i = 1, 2, \ldots, n - 1$ because in each case there are two identical rows in the determinant. In particular, $p(a_1) = 0$, so we have $p(x) = (x - a_1)p_1(x)$ by the factor theorem. Since $a_2 \ne a_1$, we obtain $p_1(a_2) = 0$, and so $p_1(x) = (x - a_2)p_2(x)$. Thus $p(x) = (x - a_1)(x - a_2)p_2(x)$. As the a_i are distinct, this process continues to obtain

$$p(x) = (x - a_1)(x - a_2)\cdots(x - a_{n-1})d \qquad (**)$$

where d is the coefficient of x^{n-1} in $p(x)$. By the cofactor expansion of $p(x)$ along the last row we get

$$d = (-1)^{n+n} \det \begin{bmatrix} 1 & a_1 & a_1^2 & \cdots & a_1^{n-2} \\ 1 & a_2 & a_2^2 & \cdots & a_2^{n-2} \\ \vdots & \vdots & \vdots & & \vdots \\ 1 & a_{n-1} & a_{n-1}^2 & \cdots\cdots & a_{n-1}^{n-2} \end{bmatrix}.$$

Because $(-1)^{n+n} = 1$, the induction hypothesis shows that d is the product of all factors $(a_i - a_j)$ where $1 \le j < i \le n - 1$. The result now follows from $(**)$ by substituting a_n for x in $p(x)$.

PROOF OF THEOREM 1

If A and B are $n \times n$ matrices we must show that

$$\det(AB) = \det A \det B. \qquad (*)$$

Recall that if E is an elementary matrix obtained by doing one row operation to I_n, then doing that operation to a matrix C (Lemma 1 §2.4) results in EC. By looking at the three types of elementary matrices separately, Theorem 2 §3.1 shows that

$$\det(EC) = \det E \det C \quad \text{for any matrix } C. \qquad (**)$$

Thus if $E_1, E_2, \ldots, E_k$ are all elementary matrices, it follows by induction that

$$\det(E_k \cdots E_2 E_1 C) = \det E_k \cdots \det E_2 \det E_1 \det C \quad \text{for any matrix } C. \qquad (***)$$

Lemma. If A has no inverse, then $\det A = 0$.

Proof. Let $A \to R$ where R is reduced row-echelon, say $E_k \cdots E_2 E_1 A = R$. Then R has a row of zeros by Theorem 5(3) §2.3, and hence $\det R = 0$. But then $(***)$ gives $\det A = 0$ because $\det E \neq 0$ for any elementary matrix E. This proves the Lemma.

Now we can prove $(*)$ by considering two cases.
Case 1. A has no inverse. Then AB also has no inverse (otherwise $A[B(AB)^{-1}] = I$ so A is invertible by the Corollary to Theorem 5 §2.3). Hence the above Lemma (twice) gives

$$\det(AB) = 0 = 0 \det B = \det A \det B,$$

proving $(*)$ in this case.
Case 2. A has an inverse. Then A is a product of elementary matrices by Theorem 2 §2.4, say $A = E_1 E_2 \cdots E_k$. Then $(***)$ with $C = I$ gives

$$\det A = \det(E_1 E_2 \cdots E_k) = \det E_1 \det E_2 \cdots \det E_k.$$

But then $(***)$ with $C = B$ gives

$$\det(AB) = \det[(E_1 E_2 \cdots E_k)B] = \det E_1 \det E_2 \cdots \det E_1 \det B = \det A \det B,$$

and $(*)$ holds in this case too.

Exercises 3.2

1. Find the adjugate of each of the following matrices.

(a) $\begin{bmatrix} 5 & 1 & 3 \\ -1 & 2 & 3 \\ 1 & 4 & 8 \end{bmatrix}$

◆(b) $\begin{bmatrix} 1 & -1 & 2 \\ 3 & 1 & 0 \\ 0 & -1 & 1 \end{bmatrix}$

(c) $\begin{bmatrix} 1 & 0 & -1 \\ -1 & 1 & 0 \\ 0 & -1 & 1 \end{bmatrix}$

◆(d) $\frac{1}{3}\begin{bmatrix} -1 & 2 & 2 \\ 2 & -1 & 2 \\ 2 & 2 & -1 \end{bmatrix}$

2. Use determinants to find which real values of c make each of the following matrices invertible.

(a) $\begin{bmatrix} 1 & 0 & 3 \\ 3 & -4 & c \\ 2 & 5 & 8 \end{bmatrix}$ (b) $\begin{bmatrix} 0 & c & -c \\ -1 & 2 & -1 \\ c & -c & c \end{bmatrix}$

(c) $\begin{bmatrix} c & 1 & 0 \\ 0 & 2 & c \\ -1 & c & 5 \end{bmatrix}$ (d) $\begin{bmatrix} 4 & c & 3 \\ c & 2 & c \\ 5 & c & 4 \end{bmatrix}$

(e) $\begin{bmatrix} 1 & 2 & -1 \\ 0 & -1 & c \\ 2 & c & 1 \end{bmatrix}$ (f) $\begin{bmatrix} 1 & c & -1 \\ c & 1 & 1 \\ 0 & 1 & c \end{bmatrix}$

3. Let A, B, and C denote $n \times n$ matrices and assume that $\det A = -1$, $\det B = 2$, and $\det C = 3$. Evaluate:

  (a) $\det(A^3 BC^T B^{-1})$ (b) $\det(B^2 C^{-1}AB^{-1}C^T)$

4. Let A and B be invertible $n \times n$ matrices. Evaluate:

  (a) $\det(B^{-1}AB)$ (b) $\det(A^{-1}B^{-1}AB)$

5. If A is 3×3 and $\det(2A^{-1}) = -4 = \det(A^3(B^{-1})^T)$, find $\det A$ and $\det B$.

6. Let $A = \begin{bmatrix} a & b & c \\ p & q & r \\ u & v & w \end{bmatrix}$ and assume that $\det A = 3$.

  Compute:

  (a) $\det(2B^{-1})$ where $B = \begin{bmatrix} 4u & 2a & -p \\ 4v & 2b & -q \\ 4w & 2c & -r \end{bmatrix}$

  (b) $\det(2C^{-1})$ where $C = \begin{bmatrix} 2p & -a+u & 3u \\ 2q & -b+v & 3v \\ 2r & -c+w & 3w \end{bmatrix}$

7. If $\det \begin{bmatrix} a & b \\ c & d \end{bmatrix} = -2$, calculate:

  (a) $\det \begin{bmatrix} 2 & -2 & 0 \\ c+1 & -1 & 2a \\ d-2 & 2 & 2b \end{bmatrix}$

  (b) $\det \begin{bmatrix} 2b & 0 & 4d \\ 1 & 2 & -2 \\ a+1 & 2 & 2(c-1) \end{bmatrix}$

8. Solve each of the following by Cramer's rule:

  (a) $2x + y = 1$ (b) $3x + 4y = 9$
    $3x + 7y = -2$ $2x - y = -1$

  (c) $5x + y - z = -7$ (d) $4x - y + 3z = 1$
    $2x - y - 2z = 6$ $6x + 2y - z = 0$
    $3x + 2z = -7$ $3x + 3y + 2z = -1$

9. Use Theorem 4 to find the (2, 3)-entry of A^{-1} if:

  (a) $A = \begin{bmatrix} 3 & 2 & 1 \\ 1 & 1 & 2 \\ -1 & 2 & 1 \end{bmatrix}$ (b) $A = \begin{bmatrix} 1 & 2 & -1 \\ 3 & 1 & 1 \\ 0 & 4 & 7 \end{bmatrix}$

10. Explain what can be said about $\det A$ if:

  (a) $A^2 = A$

  (b) $A^2 = I$

  (c) $A^3 = A$

  (d) $PA = P$, P invertible

  (e) $A^2 = uA$, A is $n \times n$

  (f) $A = -A^T$, A is $n \times n$

  (g) $A^2 + I = 0$, A is $n \times n$

11. Let A be $n \times n$. Show that $uA = (uI)A$, and use this with Theorem 1 to deduce the result in Theorem 3 §3.1: $\det(uA) = u^n \det A$.

12. If A and B are $n \times n$ matrices, $AB = -BA$, and n is odd, show that either A or B has no inverse.

13. Show that $\det AB = \det BA$ holds for any two $n \times n$ matrices A and B.

14. If $A^k = 0$ for some $k \geq 1$, show that A is not invertible.

15. If $A^{-1} = A^T$, describe the cofactor matrix of A in terms of A.

16. Show that no 3×3 matrix A exists such that $A^2 + I = 0$. Find a 2×2 matrix A with this property.

17. Show that $\det(A + B^T) = \det(A^T + B)$ for any $n \times n$ matrices A and B.

18. Let A and B be invertible $n \times n$ matrices. Show that $\det A = \det B$ if and only if $A = UB$, where U is a matrix with $\det U = 1$.

19. For each of the matrices in Exercise 2, find the inverse for those values of c for which it exists.

20. In each case either prove the statement or give an example showing that it is false:

  (a) If adj A exists, then A is invertible.

  (b) If A is invertible and adj $A = A^{-1}$, then $\det A = 1$.

  (c) $\det(AB) = \det(B^T A)$.

  (d) If $\det A \neq 0$ and $AB = AC$, then $B = C$.

  (e) If $A^T = -A$, then $\det A = -1$.

  (f) If adj $A = 0$, then $A = 0$.

(g) If A is invertible, then adj A is invertible.

♦(h) If A has a row of zeros, so also does adj A.

(i) $\det(A^T A) > 0$.

♦(j) $\det(I + A) = 1 + \det A$.

21. If A is 2×2 and $\det A = 0$, show that one column of A is a scalar multiple of the other. [*Hint:* Theorem 4 §2.2 and Theorem 5(2) §2.3.]

22. Find a polynomial $p(x)$ of degree 2 such that:
    (a) $p(0) = 2$, $p(1) = 3$, $p(3) = 8$
    ♦(b) $p(0) = 5$, $p(1) = 3$, $p(2) = 5$

23. Find a polynomial $p(x)$ of degree 3 such that:
    (a) $p(0) = p(1) = 1$, $p(-1) = 4$, $p(2) = -5$
    ♦(b) $p(0) = p(1) = 1$, $p(-1) = 2$, $p(-2) = -3$

24. Given the following data pairs, find the interpolating polynomial of degree 3 and estimate the value of y corresponding to $x = 1.5$.
    (a) $(0, 1), (1, 2), (2, 5), (3, 10)$
    ♦(b) $(0, 1), (1, 1.49), (2, -0.42), (3, -11.33)$
    (c) $(0, 2), (1, 2.03), (2, -0.40), (-1, 0.89)$

25. If $A = \begin{bmatrix} 1 & a & b \\ -a & 1 & c \\ -b & -c & 1 \end{bmatrix}$, show that

    $\det A = 1 + a^2 + b^2 + c^2$.

    Hence, find A^{-1} for any a, b, and c.

26. (a) Show that $A = \begin{bmatrix} a & p & q \\ 0 & b & r \\ 0 & 0 & c \end{bmatrix}$ has an inverse if

    and only if $abc \neq 0$, and find A^{-1} in that case.

    ♦(b) Show that if an upper triangular matrix is invertible, the inverse is also upper triangular.

27. Let A be a matrix each of whose entries are integers. Show that each of the following conditions implies the other.
    (1) A is invertible and A^{-1} has integer entries.
    (2) $\det A = 1$ or -1.

♦28. If $A^{-1} = \begin{bmatrix} 3 & 0 & 1 \\ 0 & 2 & 3 \\ 3 & 1 & -1 \end{bmatrix}$, find adj A.

29. If A is 3×3 and $\det A = 2$, find $\det(A^{-1} + 4 \text{ adj } A)$.

30. If A and B are 2×2, show that $\det \begin{bmatrix} 0 & A \\ B & X \end{bmatrix}$

    $= \det A \det B$. What if A and B are 3×3?

    [*Hint:* Multiply by $\begin{bmatrix} 0 & I \\ I & 0 \end{bmatrix}$.]

31. Let A be $n \times n$, $n \geq 2$, and assume one column of A consists of zeros. Find the possible values of rank(adj A).

32. If A is 3×3 and invertible, compute $\det(-A^2(\text{adj } A)^{-1})$.

33. Show that $\text{adj}(uA) = u^{n-1} \text{ adj } A$ for all $n \times n$ matrices A.

34. Let A and B denote invertible $n \times n$ matrices. Show that:
    (a) $\text{adj}(\text{adj } A) = (\det A)^{n-2}A$ (here $n \geq 2$)
       [*Hint:* See Example 8.]
    ♦(b) $\text{adj}(A^{-1}) = (\text{adj } A)^{-1}$
    (c) $\text{adj}(A^T) = (\text{adj } A)^T$
    (d) $\text{adj}(AB) = (\text{adj } B)(\text{adj } A)$ [*Hint:* Show that $AB \text{ adj}(AB) = AB \text{ adj } B \text{ adj } A$.]

SECTION 3.3   Diagonalization and Eigenvalues

Describing systems that are changing with time often comes down to efficiently calculating powers A, A^2, A^3, ... of a square matrix A. In this section we outline a technique for doing this, called diagonalization, which is one of the most important methods in linear algebra. To motivate the idea, consider the following example.

Example 1

A nuclear reaction produces X-particles and Y-particles. If x_k and y_k, respectively, denote the number of X-particles and Y-particles present after k hours, it is known that the following relations hold:

$$x_{k+1} = x_k + 2y_k \qquad \text{for } k = 0, 1, \dots \qquad (*)$$
$$y_{k+1} = 3x_k + 2y_k$$

Suppose that the reaction begins with $x_0 = y_0 = 1$. Assuming the conditions for the reaction remain satisfied, how many X-particles and Y-particles will be present after 10 hours?

Solution   Of course we can put $x_0 = 1$ and $y_0 = 1$ into the equations (*) to obtain $x_1 = 3$, and $y_1 = 5$, and then substitute these to get x_2 and y_2, and so on. It turns out that a better idea is to calculate the column $V_k = \begin{bmatrix} x_k \\ y_k \end{bmatrix}$ for each k. Then $V_0 = \begin{bmatrix} 1 \\ 1 \end{bmatrix}$, and the equations (*) take the matrix form

$$V_{k+1} = AV_k \quad \text{for } k = 0, 1, \dots$$

where $A = \begin{bmatrix} 1 & 2 \\ 3 & 2 \end{bmatrix}$. Thus $V_1 = AV_0$ by (*) with $k = 0$, and then $V_2 = AV_1 = A^2V_0$ using (*) with $k = 1$. Continuing in this way we see that

$$V_k = A^kV_0 \quad \text{for } k = 0, 1, \dots$$

so the problem of computing V_k comes down to finding the powers A^k of the matrix A. We will complete the calculation in Example 10.

The columns V_k here are an example of a dynamical system, and we will have more to say about such systems later.

Direct computation of the powers A^k of a square matrix A can be time-consuming, so we adopt an indirect method that is commonly used. The idea is to first **diagonalize** the matrix A, that is, to find an invertible matrix P such that

$$P^{-1}AP = D \quad \text{is a diagonal matrix} \qquad (**)$$

This works because the powers D^k of the diagonal matrix D are easy to compute, and (**) enables us to compute powers A^k of the matrix A in terms of powers D^k of D. Indeed, we can solve (**) for A to get $A = PDP^{-1}$. Squaring this gives

$$A^2 = (PDP^{-1})(PDP^{-1}) = PD^2P^{-1}$$

Using this we can compute A^3 as follows:

$$A^3 = AA^2 = (PDP^{-1})(PD^2P^{-1}) = PD^3P^{-1}$$

Continuing in this way we obtain

Theorem 1

If $A = PDP^{-1}$ then $A^k = PD^kP^{-1}$ for each $k = 1, 2, \dots$.

Note that this works even if D is not diagonal.

Hence computing A^k comes down to finding an invertible matrix P as in equation (**). To do this it is necessary to first compute certain numbers (called eigenvalues) associated with the matrix A.

Eigenvalues and Eigenvectors

If A is an $n \times n$ matrix, a number λ is called an **eigenvalue** of A if

$$AX = \lambda X \quad \text{for some column } X \neq 0,$$

and such a nonzero column X is called an **eigenvector** of A corresponding to the eigenvalue λ, or a λ-**eigenvector** for short.

Example 2

If $A = \begin{bmatrix} 3 & 5 \\ 1 & -1 \end{bmatrix}$ and $X = \begin{bmatrix} 5 \\ 1 \end{bmatrix}$, then $AX = 4X$ so $\lambda = 4$ is an eigenvalue of A with corresponding eigenvector X.

The matrix A in Example 2 has another eigenvalue in addition to $\lambda = 4$. To find it, we develop a general procedure for *any* $n \times n$ matrix A.

By definition a number λ is an eigenvalue of the $n \times n$ matrix A if and only if $AX = \lambda X$ for some column $X \neq 0$. This is equivalent to asking that the homogeneous system

$$(\lambda I - A)X = 0$$

of linear equations has a nontrivial solution $X \neq 0$. By Theorem 5 §2.3 this happens if and only if the matrix $\lambda I - A$ is not invertible and this, in turn, holds if and only if the determinant of the coefficient matrix is zero:

$$\det(\lambda I - A) = 0$$

This last condition prompts the following definition: If A is an $n \times n$ matrix, the **characteristic polynomial** $c_A(x)$ of A is defined by

$$c_A(x) = \det(xI - A)$$

Note that $c_A(x)$ is indeed a polynomial in the variable x, and it has degree n where A is an $n \times n$ matrix (this is illustrated in the examples below). The above discussion shows that a number λ is an eigenvalue of A if and only if $c_A(\lambda) = 0$, that is if and only if λ is a **root** of the characteristic polynomial $c_A(x)$. We record these observations in

Theorem 2

Let A be an $n \times n$ matrix.
1. The eigenvalues λ of A are the roots of the characteristic polynomial $c_A(x)$ of A.
2. The λ-eigenvectors X are the nonzero solutions to the homogeneous system

$$(\lambda I - A)X = 0$$

of linear equations with $\lambda I - A$ as coefficient matrix.

In practice, solving the equations in part 2 of Theorem 2 is a routine application of gaussian elimination, but finding the eigenvalues can be difficult, often requiring computers (see Section 7.5). For now, the examples and exercises are constructed so that the roots of the characteristic polynomials are relatively easy to find (usually integers). However, the reader should not be misled by this into thinking that eigenvalues are so easily obtained for the matrices that occur in practical applications!

Example 3

Find the characteristic polynomial of the matrix $A = \begin{bmatrix} 3 & 5 \\ 1 & -1 \end{bmatrix}$ discussed in Example 2, and then find all the eigenvalues and their eigenvectors.

Solution Since $xI - A = \begin{bmatrix} x & 0 \\ 0 & x \end{bmatrix} - \begin{bmatrix} 3 & 5 \\ 1 & -1 \end{bmatrix} = \begin{bmatrix} x-3 & -5 \\ -1 & x+1 \end{bmatrix}$, we get

$$c_A(x) = \det \begin{bmatrix} x-3 & -5 \\ -1 & x+1 \end{bmatrix} = x^2 - 2x - 8 = (x-4)(x+2)$$

Hence, the roots of $c_A(x)$ are $\lambda_1 = 4$ and $\lambda_2 = -2$, so these are the eigenvalues of A. Note that $\lambda_1 = 4$ was the eigenvalue mentioned in Example 2, but we have found a new one: $\lambda_2 = -2$.

To find the eigenvectors corresponding to $\lambda_2 = -2$, observe that in this case

$$\lambda_2 I - A = \begin{bmatrix} \lambda_2 - 3 & -5 \\ -1 & \lambda_2 + 1 \end{bmatrix} = \begin{bmatrix} -5 & -5 \\ -1 & -1 \end{bmatrix}$$

so the general solution to $(\lambda_2 I - A)X = 0$ is $X = t \begin{bmatrix} -1 \\ 1 \end{bmatrix}$ where t is an arbitrary real number. Hence, the eigenvectors X corresponding to λ_2 are $X = t \begin{bmatrix} -1 \\ 1 \end{bmatrix}$ where $t \neq 0$ is arbitrary. Similarly, $\lambda_1 = 4$ gives rise to the eigenvectors $X = t \begin{bmatrix} 5 \\ 1 \end{bmatrix}$, $t \neq 0$, which includes the observation in Example 2.

Note that a square matrix A has *many* eigenvectors associated with any given eigenvalue λ. In fact *every* nonzero solution X of $(\lambda I - A)X = 0$ is an eigenvector. Recall that these solutions are all linear combinations of certain basic solutions determined by the gaussian algorithm (see Theorem 3 §2.2). Observe that any nonzero multiple of an eigenvector is again an eigenvector,[9] and such multiples are often more convenient.[10] Any set of nonzero multiples of the basic solutions of $(\lambda I - A)X = 0$ will be called a set of **basic eigenvectors** corresponding to λ.

Example 4

Find the characteristic polynomial, eigenvalues, and basic eigenvectors for

$$A = \begin{bmatrix} 2 & 0 & 0 \\ 1 & 2 & -1 \\ 1 & 3 & -2 \end{bmatrix}.$$

---

9 In fact, any nonzero linear combination of λ-eigenvectors is again a λ-eigenvector.

10 Allowing nonzero multiples helps eliminate round-off error when the eigenvectors involve fractions.

Solution   Here the characteristic polynomial is given by

$$c_A(x) = \det \begin{bmatrix} x-2 & 0 & 0 \\ -1 & x-2 & 1 \\ -1 & -3 & x+2 \end{bmatrix} = (x-2)(x-1)(x+1)$$

so the eigenvalues are $\lambda_1 = 2$, $\lambda_2 = 1$, and $\lambda_3 = -1$. To find all eigenvectors for $\lambda_1 = 2$, compute

$$\lambda_1 I - A = \begin{bmatrix} \lambda_1 - 2 & 0 & 0 \\ -1 & \lambda_1 - 2 & 1 \\ -1 & -3 & \lambda_1 + 2 \end{bmatrix} = \begin{bmatrix} 0 & 0 & 0 \\ -1 & 0 & 1 \\ -1 & -3 & 4 \end{bmatrix}$$

We want the (nonzero) solutions to $(\lambda_1 I - A)X = 0$. The augmented matrix becomes

$$\begin{bmatrix} 0 & 0 & 0 & | & 0 \\ -1 & 0 & 1 & | & 0 \\ -1 & -3 & 4 & | & 0 \end{bmatrix} \rightarrow \begin{bmatrix} 1 & 0 & -1 & | & 0 \\ 0 & 1 & -1 & | & 0 \\ 0 & 0 & 0 & | & 0 \end{bmatrix}$$

using row operations. Hence, the general solution X to $(\lambda_1 I - A)X = 0$ is $X = t \begin{bmatrix} 1 \\ 1 \\ 1 \end{bmatrix}$, where $t \neq 0$ is arbitrary, so we can use $X_1 = [1\ 1\ 1]^T$ as the basic eigenvector corresponding to $\lambda_1 = 2$. As the reader can verify, the gaussian algorithm gives basic eigenvectors $X_2 = [0\ 1\ 1]^T$ and $X_3 = [0\ \frac{1}{3}\ 1]^T$ corresponding to $\lambda_2 = 1$ and $\lambda_3 = -1$, respectively. Note that to eliminate fractions, we could instead use $3X_3 = [0\ 1\ 3]^T$ as the basic λ_3-eigenvector.

It is important to note that the eigenvalues of a real matrix need not be real numbers as above, and they need not be distinct. For example, the eigenvalues of $\begin{bmatrix} 0 & -1 \\ 1 & 0 \end{bmatrix}$ are the complex numbers $\lambda = i$ and $\lambda = -i$, and the characteristic polynomial of $\begin{bmatrix} 1 & 1 \\ 0 & 1 \end{bmatrix}$ is $(x-1)^2$ so the eigenvalue $\lambda = 1$ occurs twice. Furthermore, the eigenvalues of a square matrix A are usually not computed as the roots of the characteristic polynomial. There are iterative numerical methods (for example, the QR-algorithm in Section 8.5) that are much more efficient for large matrices. We return to this later.

Example 5

If A is a square matrix, show that A and A^T have the same characteristic polynomial, and hence the same eigenvalues.

Solution   We use the fact that $xI - A^T = (xI - A)^T$. Then

$$c_{A^T}(x) = \det(xI - A^T) = \det[(xI - A)^T] = \det(xI - A) = c_A(x)$$

by Theorem 3 §3.2. Hence $c_{A^T}(x)$ and $c_A(x)$ have the same roots, and so A^T and A have the same eigenvalues (by Theorem 2).

If A is a 2×2 matrix, we can characterize the real eigenvectors of A in terms of the action of A on lines through the origin in $\mathbb{R}^2$ (see Section 2.5).

Lemma

Let L be any line through the origin in $\mathbb{R}^2$. If $\vec{x}_0$ is any nonzero vector (point) in L, then the line L consists of all scalar multiples of $\vec{x}_0$, that is

$$L = \{t\vec{x}_0 \mid t \text{ a real number}\}$$

PROOF

Suppose first that L is not vertical, so it has equation $y = mx$ where m is the slope. Hence $\vec{x}_0 = \begin{bmatrix} x_0 \\ mx_0 \end{bmatrix}$ where $x_0 \neq 0$. It is easy to verify that $t\vec{x}_0$ is on the line for all real numbers t. Conversely, if $\vec{x} = \begin{bmatrix} x \\ mx \end{bmatrix}$ is *any* point on the line, then $\vec{x} = \dfrac{x}{x_0} \vec{x}_0$.

On the other hand, if L is vertical, then L is the Y axis and a similar argument goes through.

If A is any 2×2 matrix, consider A as a linear transformation $\mathbb{R}^2 \to \mathbb{R}^2$, where A carries a vector $\vec{x}$ in $\mathbb{R}^2$ to $A\vec{x}$ (see Section 2.5). We say that A **fixes** a line L through the origin if A carries L to L, that is if $A\vec{x}$ lies in L whenever $\vec{x}$ lies in L.

Lemma

If A is a 2×2 matrix, then

A has a real eigenvalue λ if and only if A fixes a line L through the origin in $\mathbb{R}^2$.

In this case every nonzero vector in L is an eigenvector of A corresponding to λ.

PROOF

If λ is a real eigenvector of A, let $\vec{x}_0$ be a corresponding eigenvector, and let $L = \{t\vec{x}_0 \mid t \text{ a real number}\}$ denote the line through the origin containing $\vec{x}_0$—by the preceding Lemma. If $\vec{x} = t\vec{x}_0$ is any vector in L, then $A\vec{x} = A(t\vec{x}_0) = tA\vec{x}_0 = t(\lambda\vec{x}_0) = (t\lambda)\vec{x}_0$ also lies in L. In other words, A fixes L. Conversely, if A fixes L, let $\vec{x}_0$ be any nonzero vector in L. Then $A\vec{x}_0$ is in L, so $A\vec{x}_0 = t\vec{x}_0$ for some real number t. Thus A has a real eigenvalue t. Moreover this shows that every nonzero vector in L is an eigenvector corresponding to t, and so completes the proof.

Thus, for example, if A is a reflection in a line L through the origin, then $A\vec{x} = \vec{x}$ for each vector $\vec{x}$ in L, so 1 is an eigenvector of A. On the other hand, if A is a rotation through $\pi/2$, then it is clear that A fixes *no* line through the origin, and so cannot have a real eigenvalue. This can be seen another way: By Example 5 §2.5 we have $A = \begin{bmatrix} 0 & -1 \\ 1 & 0 \end{bmatrix}$, and the characteristic polynomial $c_A(x) = \det(xI - A) = x^2 + 1$ has no real roots. In other words A has no real eigenvalues. Of course $x^2 + 1$ has complex roots i and $-i$, so there *are* eigenvalues, just not real ones.

Note that *every* polynomial has complex roots,[11] so every matrix has complex eigenvalues. This suggests that we really should be doing linear algebra over the complex numbers. Indeed, everything we have done (gaussian elimination, matrix algebra, determinants, etc.) works if all the scalars are complex.

Diagonalization

An $n \times n$ matrix D is called a **diagonal matrix** if all its entries off the main diagonal are zero, that is if D has the form

$$D = \begin{bmatrix} \lambda_1 & 0 & \cdots & 0 \\ 0 & \lambda_2 & \cdots & 0 \\ \vdots & \vdots & \ddots & \vdots \\ 0 & 0 & \cdots & \lambda_n \end{bmatrix} = \text{diag}(\lambda_1, \lambda_2, \dots, \lambda_n)$$

where $\lambda_1, \lambda_2, \dots, \lambda_n$ are numbers. Calculations with diagonal matrices are very easy. Indeed, if $D = \text{diag}(\lambda_1, \lambda_2, \dots, \lambda_n)$ and $E = \text{diag}(\mu_1, \mu_2, \dots, \mu_n)$ are two diagonal matrices, their product DE and sum $D + E$ are again diagonal, and are obtained by doing the same operations to corresponding diagonal elements:

$$DE = \text{diag}(\lambda_1\mu_1, \lambda_2\mu_2, \dots, \lambda_n\mu_n)$$
$$D + E = \text{diag}(\lambda_1 + \mu_1, \lambda_2 + \mu_2, \dots, \lambda_n + \mu_n)$$

Because of the simplicity of these formulas, and with an eye on Theorem 1 and the discussion preceding it, an $n \times n$ matrix A is called **diagonalizable** if

$$P^{-1}AP \quad \text{is diagonal for some invertible } n \times n \text{ matrix } P$$

Here the invertible matrix P is called a **diagonalizing matrix** for A.

To discover when such a matrix P exists, we let $X_1, X_2, \dots, X_n$ denote the columns of P and look for ways to determine when such X_i exist and how to compute them. To this end, write P in terms of its columns as follows:

$$P = [X_1 \quad X_2 \quad \cdots \quad X_n]$$

Observe that $P^{-1}AP = D$ for some diagonal matrix D holds if and only if

$$AP = PD$$

If we write $D = \text{diag}(\lambda_1, \lambda_2, \dots, \lambda_n)$, where the λ_i are numbers to be determined, the equation $AP = PD$ becomes

$$A[X_1 \quad X_2 \quad \cdots \quad X_n] = [X_1 \quad X_2 \quad \cdots \quad X_n]\begin{bmatrix} \lambda_1 & 0 & \cdots & 0 \\ 0 & \lambda_2 & \cdots & 0 \\ \vdots & \vdots & \ddots & \vdots \\ 0 & 0 & \cdots & \lambda_n \end{bmatrix}$$

By the definition of matrix multiplication, each side simplifies as follows

$$[AX_1 \quad AX_2 \quad \cdots \quad AX_n] = [\lambda_1 X_1 \quad \lambda_2 X_2 \quad \cdots \quad \lambda_n X_n]$$

Comparing columns shows that $AX_i = \lambda_i X_i$ for each i, so

$$P^{-1}AP = D \quad \text{if and only if} \quad AX_i = \lambda_i X_i \text{ for each } i.$$

---

11 This is called the *Fundamental Theorem of Algebra* and was first proved by Gauss in his doctoral dissertation.

In other words, $P^{-1}AP = D$ holds if and only if the diagonal entries of D are eigenvalues of A, and the columns of P are corresponding eigenvectors. This proves the following fundamental result.

Theorem 3

Let A be an $n \times n$ matrix.
1. A is diagonalizable if and only if it has eigenvectors $X_1, X_2, \ldots, X_n$ such that the matrix $P = [X_1 \ X_2 \ \cdots \ X_n]$ is invertible.
2. When this is the case, $P^{-1}AP = \text{diag}(\lambda_1, \lambda_2, \ldots, \lambda_n)$ where, for each i, λ_i is the eigenvalue of A corresponding to X_i.

Example 6

Diagonalize the matrix $A = \begin{bmatrix} 2 & 0 & 0 \\ 1 & 2 & -1 \\ 1 & 3 & -2 \end{bmatrix}$ in Example 4.

Solution By Example 4, the eigenvalues of A are $\lambda_1 = 2$, $\lambda_2 = 1$, and $\lambda_3 = -1$, with corresponding basic eigenvectors $X_1 = [1 \ 1 \ 1]^T$, $X_2 = [0 \ 1 \ 1]^T$, and

$X_3 = [0 \ 1 \ 3]^T$, respectively. Since the matrix $P = [X_1 \ X_2 \ X_3] = \begin{bmatrix} 1 & 0 & 0 \\ 1 & 1 & 1 \\ 1 & 1 & 3 \end{bmatrix}$ is

invertible, Theorem 3 guarantees that $P^{-1}AP = \begin{bmatrix} \lambda_1 & 0 & 0 \\ 0 & \lambda_2 & 0 \\ 0 & 0 & \lambda_3 \end{bmatrix} = \begin{bmatrix} 2 & 0 & 0 \\ 0 & 1 & 0 \\ 0 & 0 & -1 \end{bmatrix}$.

The reader can verify this directly.

In Example 6, suppose we let $Q = [X_2 \ X_1 \ X_3]$ be the matrix formed from the eigenvectors X_1, X_2, and X_3 of A, but in a *different order* than that used to form P. Then $Q^{-1}AQ = \text{diag}(\lambda_2, \lambda_1, \lambda_3)$ is diagonal by Theorem 3, but the eigenvalues are in the *new* order. Hence we can choose the diagonalizing matrix P so that the eigenvalues λ_i appear in any order we want along the main diagonal of D.

In every example above each eigenvalue has had only one basic eigenvector. Here is a diagonalizable matrix where this is not the case.

Example 7

Diagonalize the matrix $A = \begin{bmatrix} 0 & 1 & 1 \\ 1 & 0 & 1 \\ 1 & 1 & 0 \end{bmatrix}$.

Solution To compute the characteristic polynomial of A first add rows 2 and 3 of $xI - A$ to row 1:

$$c_A(x) = \det \begin{bmatrix} x & -1 & -1 \\ -1 & x & -1 \\ -1 & -1 & x \end{bmatrix} = \det \begin{bmatrix} x-2 & x-2 & x-2 \\ -1 & x & -1 \\ -1 & -1 & x \end{bmatrix}$$

$$= \det \begin{bmatrix} x-2 & 0 & 0 \\ -1 & x+1 & 0 \\ -1 & 0 & x+1 \end{bmatrix} = (x-2)(x+1)^2$$

Hence the eigenvalues are $\lambda_1 = 2$ and $\lambda_2 = -1$, with λ_2 repeated twice (we say that λ_2 has *multiplicity* two). However, A *is* diagonalizable. For $\lambda_1 = 2$, the equations $(\lambda_1 I - A)X = 0$ have general solution $X = t[1\ \ 1\ \ 1]^T$ as the reader can verify, so the basic λ_1-eigenvector is $X_1 = [1\ \ 1\ \ 1]^T$.

Turning to the repeated eigenvalue $\lambda_2 = -1$, we must solve $(\lambda_2 I - A)X = 0$. By gaussian elimination, the general solution is $X = s[-1\ \ 1\ \ 0]^T + t[-1\ \ 0\ \ 1]^T$ where s and t are arbitrary. Hence the gaussian algorithm produces *two* basic λ_2-eigenvectors $X_2 = [-1\ \ 1\ \ 0]^T$ and $Y_2 = [-1\ \ 0\ \ 1]^T$. If we take

$$P = [X_1\ \ X_2\ \ Y_2] = \begin{bmatrix} 1 & -1 & -1 \\ 1 & 1 & 0 \\ 1 & 0 & 1 \end{bmatrix}, \text{ we find that } P \text{ is invertible. Hence}$$

$P^{-1}AP = \mathrm{diag}(2, -1, -1)$ by Theorem 3.

Example 7 typifies every diagonalizable matrix. To describe the general case, we need some terminology. An eigenvalue λ of a square matrix A is said to have **multiplicity** m if it occurs m times as a root of the characteristic polynomial $c_A(x)$. Thus, for example, the eigenvalue $\lambda_2 = -1$ in Example 7 has multiplicity 2. In that example the gaussian algorithm yields two basic λ_2-eigenvectors, the same number as the multiplicity. This works in general.

Theorem 4

A square matrix A is diagonalizable if and only if every eigenvalue λ of multiplicity m yields exactly m basic eigenvectors; that is, if and only if the general solution of the system $(\lambda I - A)X = 0$ has exactly m parameters.

The proof requires more advanced techniques and is given in Chapter 5. The following procedure summarizes the method.

Diagonalization Algorithm

To diagonalize an $n \times n$ matrix A:

Step 1. Find the distinct eigenvalues λ of A.

Step 2. Compute the basic eigenvectors corresponding to each of these eigenvalues λ as basic solutions of the homogeneous system $(\lambda I - A)X = 0$.

Step 3. The matrix A is diagonalizable if and only if there are n basic eigenvectors in all.

Step 4. If A is diagonalizable, the $n \times n$ matrix P with these basic eigenvectors as its columns is a diagonalizing matrix for A, that is, P is invertible and $P^{-1}AP$ is diagonal.

The diagonalization algorithm is valid even if the eigenvalues are nonreal complex numbers. In this case the eigenvectors will also have complex entries, but we will not pursue this here.

Example 8

Show that $A = \begin{bmatrix} 1 & 1 \\ 0 & 1 \end{bmatrix}$ is not diagonalizable.

Solution 1 The characteristic polynomial is $c_A(x) = (x - 1)^2$, so A has only one eigenvalue $\lambda_1 = 1$ of multiplicity 2. But the system of equations $(\lambda_1 I - A)X = 0$ has general solution $t[1 \ 0]^T$, so there is only one basic eigenvector $[1 \ 0]^T$. Hence A is not diagonalizable.

Solution 2 We have $c_A(x) = (x - 1)^2$ so the only eigenvalue of A is $\lambda = 1$. Hence, if A were diagonalizable, Theorem 3 would give $P^{-1}AP = \begin{bmatrix} 1 & 0 \\ 0 & 1 \end{bmatrix} = I$ for some invertible matrix P. But then $A = PIP^{-1} = I$, which is not the case. So A cannot be diagonalizable.

Diagonalizable matrices share many properties of their eigenvalues. The following example illustrates why.

Example 9

If $\lambda^3 = 5\lambda$ for every eigenvalue of the diagonalizable matrix A, show that $A^3 = 5A$.

Solution Let $P^{-1}AP = D = \text{diag}(\lambda_1, \dots, \lambda_n)$. Because $\lambda_i^3 = 5\lambda_i$ for each i, we obtain

$$D^3 = \text{diag}(\lambda_1^3, \dots, \lambda_n^3) = \text{diag}(5\lambda_1, \dots, 5\lambda_n) = 5D$$

Hence $A^3 = (PDP^{-1})^3 = PD^3P^{-1} = P(5D)P^{-1} = 5(PDP^{-1}) = 5A$ using Theorem 1. This is what we wanted.

If $p(x)$ is any polynomial and $p(\lambda) = 0$ for every eigenvalue of the diagonalizable matrix A, an argument similar to that in Example 9 shows that $p(A) = 0$. Thus Example 9 deals with the case $p(x) = x^3 - 5x$. In general, $p(A)$ is called the *evaluation* of the polynomial $p(x)$ at the matrix A. For example, if $p(x) = 2x^3 - 3x + 5$, then $p(A) = 2A^3 - 3A + 5I$—note the identity matrix.

In particular, if $c_A(x)$ denotes the characteristic polynomial of A, we certainly have $c_A(\lambda) = 0$ for each eigenvalue λ of A (Theorem 2). Hence $c_A(A) = 0$ for every diagonalizable matrix A. This is, in fact, true for *any* square matrix, diagonalizable or not, and the general result is called the Cayley–Hamilton theorem. It is proved in Chapter 8.

Example 10

In Example 1 we asked for the numbers x_k and y_k where $x_0 = 1 = y_0$ and these numbers are related by the equations

$$\begin{aligned} x_{k+1} &= x_k + 2y_k \\ y_{k+1} &= 3x_k + 2y_k \end{aligned} \quad \text{for } k = 0, 1, \dots$$

Solution Taking $A = \begin{bmatrix} 1 & 2 \\ 3 & 2 \end{bmatrix}$ and $V_k = \begin{bmatrix} x_k \\ y_k \end{bmatrix}$, the above equations become the single matrix equation $V_{k+1} = AV_k$ for each k. The idea is to compute V_k in terms of k. We have $V_1 = AV_0$, then $V_2 = AV_1 = A^2V_0$, and continuing gives

$$V_k = A^kV_0 \quad \text{for } k = 0, 1, \dots, \text{ where } V_0 = \begin{bmatrix} 1 \\ 1 \end{bmatrix}$$

So we compute A^k for each k by diagonalizing A. The characteristic polynomial is $c_A(x) = (x - 4)(x + 1)$, so the eigenvalues are $\lambda_1 = 4$ and $\lambda_2 = -1$ with corresponding basic eigenvectors $X_1 = [2 \ 3]^T$ and $X_2 = [-1 \ 1]^T$. Since

$P = [X_1 \ \ X_2] = \begin{bmatrix} 2 & -1 \\ 3 & 1 \end{bmatrix}$ is invertible, it is a diagonalizing matrix for A. Thus Theorem 3 shows that

$$P^{-1}AP = \begin{bmatrix} \lambda_1 & 0 \\ 0 & \lambda_2 \end{bmatrix} = \begin{bmatrix} 4 & 0 \\ 0 & -1 \end{bmatrix} = D$$

is diagonal. Hence, $A = PDP^{-1}$ so $A^k = PD^kP^{-1}$ by Theorem 1, and a straightforward matrix calculation gives

$$\begin{bmatrix} x_k \\ y_k \end{bmatrix} = V_k = A^kV_0 = PD^kP^{-1}V_0 = \tfrac{1}{5}\begin{bmatrix} 4\cdot 4^k + (-1)^k \\ 6\cdot 4^k - (-1)^k \end{bmatrix} \quad \text{for } k = 0,\, 1,\, \ldots$$

Hence we obtain exact formulas for the numbers x_k and y_k:

$$x_k = \tfrac{1}{5}[4\cdot 4^k + (-1)^k] \quad \text{and} \quad y_k = \tfrac{1}{5}[6\cdot 4^k - (-1)^k]$$

for each $k \geq 0$. The reader can verify that $x_0 = 1 = y_0$ and that $x_k + 2y_k = x_{k+1}$ and $3x_k + 2y_k = y_{k+1}$ for each k, as required.

If $k = 10$ these formulas give $x_k = 838\,861$ and $y_k = 1\,258\,291$. Note that the formulas can be used to approximate x_k and y_k for large k (which is often all that is needed). Indeed, taking 4^k out of the formula for x_k gives $x_k = \tfrac{1}{5}4^k[4 + (-\tfrac{1}{4})^k]$. Since $(-\tfrac{1}{4})^k$ is very small for large values of k, we have the approximation $x_k \approx \tfrac{4}{5}4^k$ for large k. Similarly, $y_k \approx \tfrac{6}{5}4^k$ if k is large. Even if k is as small as 10, these give good approximations: $x_k \approx 838\,860.8$ and $y_k \approx 1\,258\,291.2$.

Linear Dynamical Systems

A sequence of columns $V_0,\ V_1,\ V_2, \ldots$ is called a **linear dynamical system** if the initial column V_0 is known, and the other columns are determined (as in Example 10) by the condition

$$V_{k+1} = AV_k \quad \text{for each } k \geq 0$$

where A is a square matrix. The condition $V_{k+1} = AV_k$ is called a **matrix recurrence** for the columns V_k. Linear dynamical systems arise in many parts of science and engineering, and also in other areas such as economics. For example, a Markov chain (Section 2.8) is a dynamical system where the matrix recurrence $S_{k+1} = PS_k$ relates the state vectors S_k, and P is the transition matrix.

The matrix recurrence $V_{k+1} = AV_k$ gives successively $V_1 = AV_0$, $V_2 = AV_1 = A(AV_0) = A^2V_0, \ldots$. As before we obtain (using induction)

$$V_k = A^kV_0 \quad \text{for each } k = 1,\, 2,\, \ldots \tag{$*$}$$

Hence, if V_0 is known, the columns V_k are determined by the powers A^k of the matrix A and, as we have seen, these powers can be efficiently computed if A is diagonalizable. In fact ($*$) can be used to give a nice "formula" for the columns V_k in this case.

Assume that A is diagonalizable with eigenvalues $\lambda_1, \lambda_2, \ldots, \lambda_n$ and corresponding basic eigenvectors $X_1, X_2, \ldots, X_n$. If $P = [X_1 \ X_2 \ \cdots \ X_n]$ is the diagonalizing matrix with the X_i as columns, then P is invertible and

$$P^{-1}AP = D = \text{diag}(\lambda_1, \lambda_2, \ldots, \lambda_n)$$

by Theorem 3. Hence $A = PDP^{-1}$ so (∗) and Theorem 1 give

$$V_k - A^k V_0 = (PDP^{-1})^k V_0 = (PD^k P^{-1})V_0 = PD^k(P^{-1}V_0)$$

for each $k = 1, 2, \ldots$. For convenience, we denote the column $P^{-1}V_0$ arising here as follows:

$$P^{-1}V_0 = \begin{bmatrix} b_1 \\ b_2 \\ \vdots \\ b_n \end{bmatrix} = [b_1 \quad b_2 \quad \cdots \quad b_n]^T$$

Then matrix multiplication gives (see Theorem 4 §2.2)

$$V_k = PD^k(P^{-1}V_0)$$

$$= [X_1 \quad X_2 \quad \cdots \quad X_n] \begin{bmatrix} \lambda_1^k & 0 & \cdots & 0 \\ 0 & \lambda_2^k & \cdots & 0 \\ \vdots & \vdots & \ddots & \vdots \\ 0 & 0 & \cdots & \lambda_n^k \end{bmatrix} \begin{bmatrix} b_1 \\ b_2 \\ \vdots \\ b_n \end{bmatrix}$$

$$= [X_1 \quad X_2 \quad \cdots \quad X_n] \begin{bmatrix} b_1 \lambda_1^k \\ b_2 \lambda_2^k \\ \vdots \\ b_n \lambda_n^k \end{bmatrix}$$

$$= b_1 \lambda_1^k X_1 + b_2 \lambda_2^k X_2 + \cdots + b_n \lambda_n^k X_n \qquad (**)$$

for each $k \geq 0$. This is a useful explicit formula for the columns V_k. Note that, in particular, $V_0 = b_1 X_1 + b_2 X_2 + \cdots + b_n X_n$.

However, such an exact formula for V_k is often not required in practice; all that is needed is to *estimate* V_k for large values of k (as was done in Example 10). This can be easily done if A has a largest eigenvalue. An eigenvalue λ of a matrix A is called a **dominant eigenvalue** of A if it has multiplicity 1 and

$$|\lambda| > |\mu| \quad \text{for all eigenvalues } \mu \neq \lambda$$

where $|\lambda|$ denotes the absolute value[12] of the number λ. For example, $\lambda_1 = 4$ is dominant in Example 10.

Returning to the above discussion, suppose that A *has* a dominant eigenvalue. By choosing the order in which the columns X_i are placed in P, we may assume that λ_1 is dominant among the eigenvalues $\lambda_1, \lambda_2, \ldots, \lambda_n$ of A (see the discussion following Example 6). Now recall the exact expression for V_k in (∗∗) above:

$$V_k = b_1 \lambda_1^k X_1 + b_2 \lambda_2^k X_2 + \cdots + b_n \lambda_n^k X_n$$

Take λ_1^k out as a common factor in this equation to get

$$V_k = \lambda_1^k \left[b_1 X_1 + b_2 \left(\frac{\lambda_2}{\lambda_1} \right)^k X_2 + \cdots + b_n \left(\frac{\lambda_n}{\lambda_1} \right)^k X_n \right]$$

---

12 The **absolute value** $|a|$ of a real number a is defined by $|a| = \begin{cases} a & \text{if } a \geq 0 \\ -a & \text{if } a < 0 \end{cases}$. Thus, for example, $|3| = 3$ and $|-2| = 2$.

It is useful to note that the absolute value is also given by the formula $|a| = \sqrt{a^2}$, where $\sqrt{a^2}$ denotes the positive square root of a^2.

for each $k \geq 0$. Since λ_1 is dominant, we have $|\lambda_i| < |\lambda_1|$ for each $i \geq 2$, so each of the numbers $(\lambda_i / \lambda_1)^k$ become small in absolute value as k increases. Hence V_k is approximately equal to the first term $\lambda_1^k b_1 X_1$, and we write this as $V_k \approx \lambda_1^k b_1 X_1$. These observations are summarized in the following theorem (together with the above exact formula for V_k).

Theorem 5

Consider the dynamical system with matrix recurrence

$$V_{k+1} = AV_k \quad \text{for } k \geq 0$$

where A and V_0 are given. Assume that A is a diagonalizable $n \times n$ matrix with eigenvalues $\lambda_1, \lambda_2, \ldots, \lambda_n$ and corresponding basic eigenvectors $X_1, X_2, \ldots, X_n$, and let $P = [X_1 \ X_2 \ \cdots \ X_n]$ be the diagonalizing matrix. Then an exact formula for V_k is

$$V_k = b_1 \lambda_1^k X_1 + b_2 \lambda_2^k X_2 + \cdots + b_n \lambda_n^k X_n \quad \text{for each } k \geq 0$$

where the coefficients b_i come from

$$P^{-1}V_0 = [b_1 \ b_2 \ \cdots \ b_n]^T.$$

Moreover, if A has dominant eigenvalue $\lambda_1,$[13] then V_k is approximated by

$$V_k \approx b_1 \lambda_1^k X_1 \quad \text{for sufficiently large } k.$$

Example 11

Returning to Example 10, we see that $\lambda_1 = 4$ is the dominant eigenvalue, with eigenvector $X_1 = [2 \ 3]^T$. Here $P = \begin{bmatrix} 2 & -1 \\ 3 & 1 \end{bmatrix}$ and $V_0 = \begin{bmatrix} 1 \\ 1 \end{bmatrix}$, so $P^{-1}V_0 = \frac{1}{5}\begin{bmatrix} 2 \\ -1 \end{bmatrix}$. Hence $b_1 = \frac{2}{5}$ in the notation of Theorem 5, so

$$\begin{bmatrix} x_k \\ y_k \end{bmatrix} = V_k \approx b_1 \lambda_1^k X_1 = \frac{2}{5} 4^k \begin{bmatrix} 2 \\ 3 \end{bmatrix} = \frac{1}{5} 4^k \begin{bmatrix} 4 \\ 6 \end{bmatrix}$$

where k is large. Hence $x_k \approx \frac{4}{5} 4^k$ and $y_k \approx \frac{6}{5} 4^k$, as in Example 10.

Google PageRank

Dominant eigenvalues are useful to the Google search engine for finding information on the Web. If an information query comes in from a client, Google has a sophisticated method of establishing the "relevance" of each site to that query. When the relevant sites have been determined, they are placed in order of importance using a ranking of *all* sites called the PageRank. The relevant sites with the highest PageRank are the ones presented to the client. It is the construction of the PageRank that is our interest here.

The Web contains many links from one site to another. Google interprets a link from site j to site i as a "vote" for the importance of site i. Hence if site i has more links to it than does site j, then i is regarded as more "important" and assigned a higher PageRank. One way to look at this is to view the sites as vertices in a huge directed graph (see Section 2.2). Then if site j links to site i there is an edge from j to i, and hence the (i, j)-entry is a 1 in the associated adjacency matrix (called the

13 Similar results can be found in other situations. If, for example, eigenvalues λ_1 and λ_2 (possibly equal) satisfy $|\lambda_1| = |\lambda_2| > |\lambda_i|$ for all $i > 2$, then we obtain $V_k \approx b_1 \lambda_1^k X_1 + b_2 \lambda_2^k X_2$ for large k.

connectivity matrix in this context). Thus a large number of 1's in row i of this matrix is a measure of the PageRank of site i.[14]

However this does not take into account the PageRank of the sites that link to i. Intuitively, the higher the rank of these sites, the higher the rank of site i. One approach is to compute a dominant eigenvector X for the connectivity matrix. In most cases the entries of X can be chosen to be positive with sum 1. Each site corresponds to an entry of X, so the sum of the entries of sites linking to a given site i is a measure of the rank of site i. In fact, Google chooses the PageRank of a site so that it is proportional to this sum.[15]

Exercises 3.3

1. In each case find the characteristic polynomial, eigenvalues, eigenvectors, and (if possible) an invertible matrix P such that $P^{-1}AP$ is diagonal.

(a) $A = \begin{bmatrix} 1 & 2 \\ 3 & 2 \end{bmatrix}$ (b) $A = \begin{bmatrix} 2 & -4 \\ -1 & -1 \end{bmatrix}$

(c) $A = \begin{bmatrix} 7 & 0 & -4 \\ 0 & 5 & 0 \\ 5 & 0 & -2 \end{bmatrix}$ (d) $A = \begin{bmatrix} 1 & 1 & -3 \\ 2 & 0 & 6 \\ 1 & -1 & 5 \end{bmatrix}$

(e) $A = \begin{bmatrix} 1 & -2 & 3 \\ 2 & 6 & -6 \\ 1 & 2 & -1 \end{bmatrix}$ (f) $A = \begin{bmatrix} 0 & 1 & 1 \\ 1 & 0 & 1 \\ 1 & 1 & 0 \end{bmatrix}$

(g) $A = \begin{bmatrix} 3 & 1 & 1 \\ -4 & -2 & -5 \\ 2 & 2 & 5 \end{bmatrix}$ (h) $A = \begin{bmatrix} 2 & 1 & 1 \\ 0 & 1 & 0 \\ 1 & -1 & 2 \end{bmatrix}$

(i) $A = \begin{bmatrix} \lambda & 0 & 0 \\ 0 & \lambda & 0 \\ 0 & 0 & \mu \end{bmatrix}$, $\lambda \neq \mu$

2. Consider a linear dynamical system $V_{k+1} = AV_k$ for $k \geq 0$. In each case approximate V_k using Theorem 5.

(a) $A = \begin{bmatrix} 2 & 1 \\ 4 & -1 \end{bmatrix}$, $V_0 = \begin{bmatrix} 1 \\ 2 \end{bmatrix}$

(b) $A = \begin{bmatrix} 3 & -2 \\ 2 & -2 \end{bmatrix}$, $V_0 = \begin{bmatrix} 3 \\ -1 \end{bmatrix}$

(c) $A = \begin{bmatrix} 1 & 0 & 0 \\ 1 & 2 & 3 \\ 1 & 4 & 1 \end{bmatrix}$, $V_0 = \begin{bmatrix} 1 \\ 1 \\ 1 \end{bmatrix}$

(d) $A = \begin{bmatrix} 1 & 3 & 2 \\ -1 & 2 & 1 \\ 4 & -1 & -1 \end{bmatrix}$, $V_0 = \begin{bmatrix} 2 \\ 0 \\ 1 \end{bmatrix}$

3. Show that A has $\lambda = 0$ as an eigenvalue if and only if A is not invertible.

4. Let A denote an $n \times n$ matrix and put $A_1 = A - \alpha I$, α in $\mathbb{R}$. Show that λ is an eigenvalue of A if and only if $\lambda - \alpha$ is an eigenvalue of A_1. How do the eigenvectors compare? (Hence, the eigenvalues of A_1 are just those of A "shifted" by α.)

5. Show that the eigenvalues of $\begin{bmatrix} \cos\theta & -\sin\theta \\ \sin\theta & \cos\theta \end{bmatrix}$ are $e^{i\theta}$ and $e^{-i\theta}$. (See Appendix A.)

6. Find the characteristic polynomial of the $n \times n$ identity matrix I. Show that I has exactly one eigenvalue and find the eigenvectors.

7. Given $A = \begin{bmatrix} a & b \\ c & d \end{bmatrix}$, show that:

(a) $c_A(x) = x^2 - \operatorname{tr} A\, x + \det A$, where $\operatorname{tr} A = a + d$ is called the **trace** of A.

(b) The eigenvalues are
$$\tfrac{1}{2}[(a+d) \pm \sqrt{(a-d)^2 + 4bc}]$$

8. In each case, find $P^{-1}AP$ and then compute A^n.

(a) $A = \begin{bmatrix} 6 & -5 \\ 2 & -1 \end{bmatrix}$, $P = \begin{bmatrix} 1 & 5 \\ 1 & 2 \end{bmatrix}$

(b) $A = \begin{bmatrix} -7 & -12 \\ 6 & 10 \end{bmatrix}$, $P = \begin{bmatrix} -3 & 4 \\ 2 & -3 \end{bmatrix}$

[Hint: $(PDP^{-1})^n = PD^nP^{-1}$ for each $n = 1, 2, \dots$.]

9. If $A = \begin{bmatrix} 1 & 3 \\ 0 & 2 \end{bmatrix}$ and $B = \begin{bmatrix} 2 & 0 \\ 0 & 1 \end{bmatrix}$, verify that A and B are diagonalizable, but AB is not.

14 For more on PageRank, visit http://www.google.com/technology/.

15 See the articles "Searching the web with eigenvectors" by Herbert S. Wilf, UMAP Journal 23(2), 2002, pages 101–103, and "The worlds largest matrix computation: Google's PageRank is an eigenvector of a matrix of order 2.7 billion" by Cleve Moler, Matlab News and Notes, October 2002, pages 12–13.

10. If A is an $n \times n$ matrix, show that A is diagonalizable if and only if A^T is diagonalizable.

11. If A is diagonalizable, show that each of the following is also diagonalizable.

    (a) A^n, $n \geq 1$ ◆(b) kA

    (c) $p(A)$, $p(x)$ any polynomial (Theorem 1)

    ◆(d) $U^{-1}AU$ for any invertible matrix U.

◆12. Give an example of two diagonalizable matrices A and B whose sum $A + B$ is not diagonalizable.

d13. If A is diagonalizable and 1 and -1 are the only eigenvalues, show that $A^{-1} = A$.

◉14. If A is diagonalizable and 0 and 1 are the only eigenvalues, show that $A^2 = A$.

◐15. If A is diagonalizable and $\lambda \geq 0$ for each eigenvalue of A, show that $A = B^2$ for some matrix B.

16. If $P^{-1}AP$ and $P^{-1}BP$ are both diagonal, show that $AB = BA$. [*Hint:* Diagonal matrices commute.]

17. A square matrix A is called **nilpotent** if $A^n = 0$ for some $n \geq 1$. Find all nilpotent diagonalizable matrices. [*Hint:* Theorem 1.]

18. Let A be any $n \times n$ matrix and $r \neq 0$ a real number.

    (a) Show that the eigenvalues of rA are precisely the numbers $r\lambda$, where λ is an eigenvalue of A.

    ◆(b) Show that $c_{rA}(x) = r^n c_A\left(\frac{x}{r}\right)$.

19. (a) If all rows of A have the same sum s, show that s is an eigenvalue.

    (b) If all columns of A have the same sum s, show that s is an eigenvalue.

20. Let A be an invertible $n \times n$ matrix.

    (a) Show that the eigenvalues of A are nonzero.

    ◆(b) Show that the eigenvalues of A^{-1} are precisely the numbers $1/\lambda$, where λ is an eigenvalue of A.

    (c) Show that $c_{A^{-1}}(x) = \frac{(-x)^n}{\det A} c_A\left(\frac{1}{x}\right)$.

21. Suppose λ is an eigenvalue of a square matrix A with eigenvector $X \neq 0$.

    (a) Show that λ^2 is an eigenvalue of A^2 (with the same X).

    ◆(b) Show that $\lambda^3 - 2\lambda + 3$ is an eigenvalue of $A^3 - 2A + 3I$.

    (c) Show that $p(\lambda)$ is an eigenvalue of $p(A)$ for any nonzero polynomial $p(x)$.

22. If A is an $n \times n$ matrix, show that $c_{A^2}(x^2) = (-1)^n c_A(x) c_A(-x)$.

23. An $n \times n$ matrix A is called nilpotent if $A^m = 0$ for some $m \geq 1$.

    (a) Show that every triangular matrix with zeros on the main diagonal is nilpotent.

    ◆(b) Show that $\lambda = 0$ is the only eigenvalue (even complex) of A, if A is nilpotent.

    (c) Deduce that $c_A(x) = x^n$, if A is $n \times n$ and nilpotent.

24. Given $p(x) = a_0 + a_1 x + a_2 x^2 + a_3 x^3 + x^4$, show

    that $A = \begin{bmatrix} 0 & 1 & 0 & 0 \\ 0 & 0 & 1 & 0 \\ 0 & 0 & 0 & 1 \\ -a_0 & -a_1 & -a_2 & -a_3 \end{bmatrix}$ is a matrix

    whose characteristic polynomial equals $p(x)$. [A is called the **companion matrix** for $p(x)$.]

25. (a) Show that the only diagonalizable matrix A that has only one eigenvalue λ is the scalar matrix $A = \lambda I$.

    (b) Is $\begin{bmatrix} 3 & -2 \\ 2 & -1 \end{bmatrix}$ diagonalizable?

26. Characterize the diagonalizable $n \times n$ matrices A such that $A^2 - 3A + 2I = 0$ in terms of their eigenvalues. [*Hint:* Theorem 1.]

27. Let $A = \begin{bmatrix} B & 0 \\ 0 & C \end{bmatrix}$ where B and C are square matrices.

    (a) If B and C are diagonalizable via Q and R (that is, $Q^{-1}BQ$ and $R^{-1}CR$ are diagonal), show that A is diagonalizable via $\begin{bmatrix} Q & 0 \\ 0 & R \end{bmatrix}$.

(b) Use (a) to diagonalize A if $B = \begin{bmatrix} 5 & 3 \\ 3 & 5 \end{bmatrix}$ and

$$C = \begin{bmatrix} 7 & -1 \\ -1 & 7 \end{bmatrix}.$$

28. Let $A = \begin{bmatrix} B & 0 \\ 0 & C \end{bmatrix}$, where B and C are square matrices.

(a) Show that $c_A(x) = c_B(x)c_C(x)$.

(b) If X and Y are eigenvectors of B and C, respectively, show that $\begin{bmatrix} X \\ 0 \end{bmatrix}$ and $\begin{bmatrix} 0 \\ Y \end{bmatrix}$ are eigenvectors of A, and show how every eigenvector of A arises from such eigenvectors.

SECTION 3.4 An Application to Linear Recurrences

It often happens that a problem can be solved by finding a sequence of numbers $x_0, x_1, x_2, \dots$ where the first few are known, and subsequent numbers are given in terms of earlier ones. Here is a combinatorial example where the object is to count the number of ways to do something.

Example 1

An urban planner wants to determine the number x_k of ways that a row of k parking spaces can be filled with cars and trucks if trucks take up two spaces each. Find the first few values of x_k.

Solution Clearly, $x_0 = 1$ and $x_1 = 1$, while $x_2 = 2$ since there can be two cars or one truck. We have $x_3 = 3$ (the 3 configurations are ccc, cT, and Tc) and $x_4 = 5$ ($cccc$, ccT, cTc, Tcc, and TT). The key to this method is to find a way to express each subsequent x_k in terms of earlier values. In this case we claim that

$$x_{k+2} = x_k + x_{k+1} \quad \text{for every } k \geq 0 \tag{$*$}$$

Indeed, every way to fill $k + 2$ spaces falls into one of two categories: Either a car is parked in the first space (and the remaining $k + 1$ spaces are filled in x_{k+1} ways), or a truck is parked in the first two spaces (with the other k spaces filled in x_k ways). Hence, there are $x_{k+1} + x_k$ ways to fill the $k + 2$ spaces. This is $(*)$.

The recurrence $(*)$ determines x_k for every $k \geq 2$ since x_0 and x_1 are given. In fact, the first few values are

$$\begin{aligned}
x_0 &= 1 \\
x_1 &= 1 \\
x_2 &= x_0 + x_1 = 2 \\
x_3 &= x_1 + x_2 = 3 \\
x_4 &= x_2 + x_3 = 5 \\
x_5 &= x_3 + x_4 = 8 \\
&\ \vdots \qquad \vdots \qquad \vdots
\end{aligned}$$

Clearly, we can find x_k for any value of k, but one wishes for a "formula" for x_k as a function of k. It turns out that such a formula can be found using diagonalization. We will return to this example later.

A sequence $x_0, x_1, x_2, \dots$ of numbers is said to be given **recursively** if each number in the sequence is completely determined by those that come before it.

Such sequences arise frequently in mathematics and computer science, and also occur in other parts of science. The formula $x_{k+2} = x_{k+1} + x_k$ in Example 1 is an example of a **linear recurrence relation** of length 2 because x_{k+2} is the sum of the two preceding terms x_{k+1} and x_k; in general, the **length** is m if x_{k+m} is a sum of multiples of $x_k, x_{k+1}, \dots, x_{k+m-1}$.

The simplest linear recursive sequences are of length 1, that is x_{k+1} is a fixed multiple of x_k for each k, say $x_{k+1} = ax_k$. If x_0 is specified, then $x_1 = ax_0$, $x_2 = ax_1 = a^2x_0$, and $x_3 = ax_2 = a^3x_0, \dots$. Continuing, we obtain $x_k = a^kx_0$ for each $k \geq 0$, which is an explicit formula for x_k as a function of k (when x_0 is given).

Such formulas are not always so easy to find for all choices of the initial values. Here is an example where diagonalization helps.

Example 2

Suppose the numbers $x_0, x_1, x_2, \dots$ are given by the linear recurrence relation

$$x_{k+2} = x_{k+1} + 6x_k \quad \text{for } k \geq 0$$

where x_0 and x_1 are specified. Find a formula for x_k when $x_0 = 1$ and $x_1 = 3$, and also when $x_0 = 1$ and $x_1 = 1$.

Solution If $x_0 = 1$ and $x_1 = 3$, then $x_2 = x_1 + 6x_0 = 9$, $x_3 = x_2 + 6x_1 = 27$, $x_4 = x_3 + 6x_2 = 81$, and it is apparent that

$$x_k = 3^k$$

for $k = 0, 1, 2, 3$, and 4. This formula holds for all k because it is true for $k = 0$ and $k = 1$, and it satisfies the recurrence $x_{k+2} = x_{k+1} + 6x_k$ for each k as is readily checked.

However, if we begin instead with $x_0 = 1$ and $x_1 = 1$, the sequence continues $x_2 = 7$, $x_3 = 13$, $x_4 = 55$, $x_5 = 133, \dots$. In this case, the sequence is uniquely determined but no formula is apparent. Nonetheless, a simple device transforms the recurrence into a matrix recurrence to which our diagonalization techniques apply.

The idea is to compute the sequence $V_0, V_1, V_2, \dots$ of columns instead of the numbers $x_0, x_1, x_2, \dots$, where

$$V_k = \begin{bmatrix} x_k \\ x_{k+1} \end{bmatrix} \quad \text{for each } k \geq 0$$

Then $V_0 = \begin{bmatrix} x_0 \\ x_1 \end{bmatrix} = \begin{bmatrix} 1 \\ 1 \end{bmatrix}$ is specified, and the numerical recurrence $x_{k+2} = x_{k+1} + 6x_k$ transforms into a matrix recurrence as follows:

$$V_{k+1} = \begin{bmatrix} x_{k+1} \\ x_{k+2} \end{bmatrix} = \begin{bmatrix} x_{k+1} \\ 6x_k + x_{k+1} \end{bmatrix} = \begin{bmatrix} 0 & 1 \\ 6 & 1 \end{bmatrix} \begin{bmatrix} x_k \\ x_{k+1} \end{bmatrix} = AV_k$$

where $A = \begin{bmatrix} 0 & 1 \\ 6 & 1 \end{bmatrix}$. Thus these columns V_k are a linear dynamical system, so Theorem 5 §3.3 applies provided the matrix A is diagonalizable.

We have $c_A(x) = (x - 3)(x + 2)$ so the eigenvalues are $\lambda_1 = 3$ and $\lambda_2 = -2$ with corresponding eigenvectors $X_1 = \begin{bmatrix} 1 \\ 3 \end{bmatrix}$ and $X_2 = \begin{bmatrix} -1 \\ 2 \end{bmatrix}$ as the reader can check.

Since $P = [X_1 \ X_2] = \begin{bmatrix} 1 & -1 \\ 3 & 2 \end{bmatrix}$ is invertible, it is a diagonalizing matrix for A.

The coefficients b_i in Theorem 5 §3.3 are given by $\begin{bmatrix} b_1 \\ b_2 \end{bmatrix} = P^{-1}V_0 = \begin{bmatrix} \frac{3}{5} \\ \frac{-2}{5} \end{bmatrix}$, so that the theorem gives

$$\begin{bmatrix} x_k \\ x_{k+1} \end{bmatrix} = V_k = b_1\lambda_1^k X_1 + b_2\lambda_2^k X_2 = \tfrac{3}{5}3^k\begin{bmatrix} 1 \\ 3 \end{bmatrix} + \tfrac{-2}{5}(-2)^k\begin{bmatrix} -1 \\ 2 \end{bmatrix}$$

Equating top entries yields

$$x_k = \tfrac{1}{5}\left[3^{k+1} - (-2)^{k+1} \right] \quad \text{for } k \geq 0$$

This gives $x_0 = 1 = x_1$, and it satisfies the recurrence $x_{k+2} = x_{k+1} + 6x_k$ as is easily verified. Hence, it is the desired formula for the x_k.

Returning to Example 1, these methods give an exact formula and a good approximation for the numbers x_k in that problem.

Example 3

In Example 1, an urban planner wants to determine x_k, the number of ways that a row of k parking spaces can be filled with cars and trucks if trucks take up two spaces each. Find a formula for x_k and estimate it for large k.

Solution   We saw that the numbers x_k satisfy a linear recurrence

$$x_{k+2} = x_k + x_{k+1} \quad \text{for every } k \geq 0$$

If we write $V_k = \begin{bmatrix} x_k \\ x_{k+1} \end{bmatrix}$ as before, this recurrence becomes a matrix recurrence for the V_k:

$$V_{k+1} = \begin{bmatrix} x_{k+1} \\ x_{k+2} \end{bmatrix} = \begin{bmatrix} x_{k+1} \\ x_k + x_{k+1} \end{bmatrix} = \begin{bmatrix} 0 & 1 \\ 1 & 1 \end{bmatrix}\begin{bmatrix} x_k \\ x_{k+1} \end{bmatrix} = AV_k$$

for all $k \geq 0$ where $A = \begin{bmatrix} 0 & 1 \\ 1 & 1 \end{bmatrix}$. Moreover, A is diagonalizable here. The characteristic polynomial is $c_A(x) = x^2 - x - 1$ with roots $\tfrac{1}{2}\left[1 \pm \sqrt{5} \right]$ by the quadratic formula, so A has eigenvalues

$$\lambda_1 = \tfrac{1}{2}\left[1 + \sqrt{5} \right] \quad \text{and} \quad \lambda_2 = \tfrac{1}{2}\left[1 - \sqrt{5} \right]$$

Corresponding eigenvectors are $X_1 = \begin{bmatrix} 1 \\ \lambda_1 \end{bmatrix}$ and $X_2 = \begin{bmatrix} 1 \\ \lambda_2 \end{bmatrix}$ respectively as the reader can verify. As the matrix $P = [X_1 \ X_2] = \begin{bmatrix} 1 & 1 \\ \lambda_1 & \lambda_2 \end{bmatrix}$ is invertible, it is a diagonalizing matrix for A. We compute the coefficients b_1 and b_2 (in Theorem 5 §3.3) as follows:

$$\begin{bmatrix} b_1 \\ b_2 \end{bmatrix} = P^{-1}V_0 = \tfrac{1}{-\sqrt{5}}\begin{bmatrix} \lambda_2 & -1 \\ -\lambda_1 & 1 \end{bmatrix}\begin{bmatrix} 1 \\ 1 \end{bmatrix} = \tfrac{1}{\sqrt{5}}\begin{bmatrix} \lambda_1 \\ -\lambda_2 \end{bmatrix}$$

where we used the fact that $\lambda_1 + \lambda_2 = 1$. Thus Theorem 5 §3.3 gives

$$\begin{bmatrix} x_k \\ x_{k+1} \end{bmatrix} = V_k = b_1\lambda_1^k X_1 + b_2\lambda_2^k X_2 = \tfrac{\lambda_1}{\sqrt{5}}\lambda_1^k\begin{bmatrix} 1 \\ \lambda_1 \end{bmatrix} - \tfrac{\lambda_2}{\sqrt{5}}\lambda_2^k\begin{bmatrix} 1 \\ \lambda_2 \end{bmatrix}$$

Comparing top entries gives an exact formula for the numbers x_k:

$$x_k = \tfrac{1}{\sqrt{5}}\left[\lambda_1^{k+1} - \lambda_2^{k+1}\right] \quad \text{for each } k \geq 0$$

Finally, observe that λ_1 is dominant here (in fact, $\lambda_1 = 1.618$ and $\lambda_2 = -0.618$ to three decimal places) so λ_2^{k+1} is negligible compared with λ_1^{k+1} if k is large. Thus,

$$x_k \approx \tfrac{1}{\sqrt{5}}\lambda_1^{k+1} \quad \text{for each } k \geq 0$$

This is a good approximation, even for as small a value as $k = 12$. Indeed, repeated use of the recurrence $x_{k+2} = x_k + x_{k+1}$ gives the exact value $x_{12} = 233$, while the approximation is $x_{12} \approx \frac{(1.618)^{13}}{\sqrt{5}} = 232.94$.

The sequence $x_0, x_1, x_2, \ldots$ in Example 3 was first discussed in 1202 by Leonardo Pisano of Pisa, also known as Fibonacci,[16] and is now called the **Fibonacci sequence**. It is completely determined by the conditions $x_0 = 1$, $x_1 = 1$ and the recurrence $x_{k+2} = x_k + x_{k+1}$ for each $k \geq 0$. These numbers have been studied for centuries and have many interesting properties (there is even a journal, the *Fibonacci Quarterly*, devoted exclusively to them). For example, biologists have discovered that the arrangement of leaves around the stems of some plants follow a Fibonacci pattern. The formula $x_k = \tfrac{1}{\sqrt{5}}\left[\lambda_1^{k+1} - \lambda_2^{k+1}\right]$ in Example 3 is called the **Binet formula**.

The Binet formula is remarkable in that the x_k are integers but λ_1 and λ_2 are not. This phenomenon can occur even if the eigenvalues λ_i are nonreal complex numbers.

We conclude with an example showing that *nonlinear* recurrences can be very complicated.

Example 4

Suppose a sequence $x_0, x_1, x_2, \ldots$ satisfies the following recurrence:

$$x_{k+1} = \begin{cases} \tfrac{1}{2}x_k & \text{if } x_k \text{ is even} \\ 3x_k + 1 & \text{if } x_k \text{ is odd} \end{cases}$$

If $x_0 = 1$, the sequence is $1, 4, 2, 1, 4, 2, 1, \ldots$ and so continues to cycle indefinitely. The same thing happens if $x_0 = 7$. Then the sequence is

$$7, 22, 11, 34, 17, 52, 26, 13, 40, 20, 10, 5, 16, 8, 4, 2, 1, \ldots$$

and it again cycles. However, it is not known whether every choice of x_0 will lead eventually to 1. It is quite possible that, for some x_0, the sequence will continue to produce different values indefinitely, or will repeat a value and cycle without reaching 1. No one knows for sure.

---

16 The problem Fibonacci discussed was: "How many pairs of rabbits will be produced in a year, beginning with a single pair, if in every month each pair brings forth a new pair that becomes productive from the second month on? Assume no pairs die." The number of pairs satisfies the Fibonacci recurrence.

Exercises 3.4

1. Solve the following linear recurrences.

   (a) $x_{k+2} = 3x_k + 2x_{k+1}$, where $x_0 = 1$ and $x_1 = 1$.

   ◆(b) $x_{k+2} = 2x_k - x_{k+1}$, where $x_0 = 1$ and $x_1 = 2$.

   (c) $x_{k+2} = 2x_k + x_{k+1}$, where $x_0 = 0$ and $x_1 = 1$.

   ◆(d) $x_{k+2} = 6x_k - x_{k+1}$, where $x_0 = 1$ and $x_1 = 1$.

2. Solve the following linear recurrences.

   (a) $x_{k+3} = 6x_{k+2} - 11x_{k+1} + 6x_k$, where $x_0 = 1$, $x_1 = 0$, and $x_2 = 1$.

   ◆(b) $x_{k+3} = -2x_{k+2} + x_{k+1} + 2x_k$, where $x_0 = 1$, $x_1 = 0$, and $x_2 = 1$.
   [*Hint:* Use $V_k = [x_k \ x_{k+1} \ x_{k+2}]^T$.]

3. In Example 1 suppose busses are also allowed to park, and let x_k denote the number of ways a row of k parking spaces can be filled with cars, trucks, and busses.

   (a) If trucks and busses take up 2 and 3 spaces respectively, show that $x_{k+3} = x_k + x_{k+1} + x_{k+2}$ for each k, and use this recurrence to compute x_{10}. [*Hint:* The eigenvalues are of little use.]

   ◆(b) If busses take up 4 spaces, find a recurrence for the x_k and compute x_{10}.

4. A man must climb a flight of k steps. He always takes one or two steps at a time. Thus he can climb 3 steps in the following ways: 111, 12, or 21. Find s_k, the number of ways he can climb the flight of k steps. [*Hint:* Fibonacci.]

◆5. How many "words" of k letters can be made from the letters $\{a, b\}$ if there are no adjacent a's?

6. How many sequences of k flips of a coin are there with no HH?

◆7. Find x_k, the number of ways to make a stack of k poker chips if only red, blue, and gold chips are used and no two gold chips are adjacent. [*Hint:* Show that $x_{k+2} = 2x_{k+1} + 2x_k$ by considering how many stacks have a red, blue, or gold chip on top.]

8. A nuclear reactor contains α- and β-particles. In every second each α-particle splits into three β-particles, and each β-particle splits into an α-particle and two β-particles. If there is a single α-particle in the reactor at time $t = 0$, how many α-particles are there at $t = 20$ seconds? [*Hint:* Let x_k and y_k denote the number of α- and β-particles at time $t = k$ seconds. Find x_{k+1} and y_{k+1} in terms of x_k and y_k.]

◆9. The annual yield of wheat in a certain country has been found to equal the average of the yield in the previous two years. If the yields in 1990 and 1991 were 10 and 12 million tons respectively, find a formula for the yield k years after 1990. What is the long-term average yield?

10. Find the general solution to the recurrence $x_{k+1} = rx_k + c$ where r and c are constants. [*Hint:* Consider the cases $r = 1$ and $r \neq 1$ separately. If $r \neq 1$, you will need the identity
$$1 + r + r^2 + \cdots + r^{n-1} = \frac{1 - r^n}{1 - r} \text{ for } n \geq 1.]$$

11. Consider the length 3 recurrence $x_{k+3} = ax_k + bx_{k+1} + cx_{k+2}$.

    (a) If $V_k = [x_k \ x_{k+1} \ x_{k+2}]^T$, show that
    $$V_{k+1} = AV_k \text{ where } A = \begin{bmatrix} 0 & 1 & 0 \\ 0 & 0 & 1 \\ a & b & c \end{bmatrix}.$$

    ◆(b) If λ is any eigenvalue of A, show that $X = [1 \ \lambda \ \lambda^2]^T$ is a λ-eigenvector. [*Hint:* Show directly that $AX = \lambda X$.]

    (c) Generalize (a) and (b) to a recurrence $x_{k+4} = ax_k + bx_{k+1} + cx_{k+2} + dx_{k+3}$ of length 4.

12. Consider the recurrence $x_{k+2} = ax_{k+1} + bx_k + c$ where c may not be zero.

    (a) If $a + b \neq 1$ show that p can be found such that, if we set $y_k = x_k + p$, then $y_{k+2} = ay_{k+1} + by_k$. [Hence, the sequence x_k can be found provided y_k can be found by the methods of this section (or otherwise).]

    ◆(b) Use (a) to solve the recurrence $x_{k+2} = x_{k+1} + 6x_k + 5$ where $x_0 = 1$ and $x_1 = 1$.

13. Consider the recurrence

$$x_{k+2} = ax_{k+1} + bx_k + c(k) \qquad (*)$$

where $c(k)$ is a function of k, and consider the related recurrence

$$x_{k+2} = ax_{k+1} + bx_k \qquad (**)$$

Suppose that $x_k = p_k$ is a particular solution of $(*)$.

(a) If q_k is any solution of $(**)$, show that $q_k + p_k$ is a solution of $(*)$.

(b) Show that every solution of $(*)$ arises as in (a) as the sum of a solution of $(**)$ plus the particular solution p_k of $(*)$.

SECTION 3.5   An Application to Population Growth

One of the most fertile applications of diagonalization is in modelling the growth of the population of an animal species. This has attracted more attention in recent years with the ever increasing awareness that many species are endangered. In this section we merely touch the surface of this subject. We look first at how the population of a single species evolves in time, and then examine the interaction of two species and the effect on the populations. In both cases the model we use is a linear dynamical system.

A Single Species Model

We begin by examining how the population of a species changes over time, making assumptions about survival and reproduction rates. The example we use involves birds, but the model can be modified to suit other species.

Example 1

Consider the evolution of the population of a species of birds. Because the number of males and females are nearly equal, we count only females. We assume that each female remains a juvenile for one year and then becomes an adult, and that only adults have offspring. We make three assumptions about reproduction and survival rates:

1. The number of juvenile females hatched in any year is on average twice the number of adult females alive the year before (we say the **reproduction rate** is 2).
2. Half of the adult females in any year survive to the next year (the **adult survival rate** is $\frac{1}{2}$).
3. One quarter of the juvenile females in any year survive into adulthood (the **juvenile survival rate** is $\frac{1}{4}$).

If there were 100 adult females and 40 juvenile females alive initially, compute the population of females k years later.

Solution   Let a_k and j_k denote, respectively, the number of adult and juvenile females after k years, so that the total female population is the sum $a_k + j_k$. Assumption 1 shows that $j_{k+1} = 2a_k$, while assumptions 2 and 3 show that $a_{k+1} = \frac{1}{2}a_k + \frac{1}{4}j_k$. Hence the numbers a_k and j_k in successive years are related by the following equations:

$$a_{k+1} = \tfrac{1}{2}a_k + \tfrac{1}{4}j_k$$
$$j_{k+1} = 2a_k$$

If we write $V_k = \begin{bmatrix} a_k \\ j_k \end{bmatrix}$ and $A = \begin{bmatrix} \frac{1}{2} & \frac{1}{4} \\ 2 & 0 \end{bmatrix}$, these equations take the matrix form of a linear dynamical system

$$V_{k+1} = AV_k \quad \text{for each } k = 0, 1, 2, \ldots$$

Here the column V_k is called the **population profile** for the species.

We have $c_A(x) = (x-1)(x + \frac{1}{2})$, so the eigenvalues are $\lambda_1 = 1$ and $\lambda_2 = -\frac{1}{2}$ with corresponding eigenvectors $X_1 = \begin{bmatrix} 1 \\ 2 \end{bmatrix}$ and $X_2 = \begin{bmatrix} -1 \\ 4 \end{bmatrix}$ respectively.

The matrix $P = [X_1 \ X_2] = \begin{bmatrix} 1 & -1 \\ 2 & 4 \end{bmatrix}$ is invertible and so is a diagonalizing matrix for A. The coefficients b_i in Theorem 5 §3.3 are given by

$$\begin{bmatrix} b_1 \\ b_2 \end{bmatrix} = P^{-1} V_0 = \frac{1}{6}\begin{bmatrix} 4 & 1 \\ -2 & 1 \end{bmatrix}\begin{bmatrix} 100 \\ 40 \end{bmatrix} = \frac{1}{3}\begin{bmatrix} 220 \\ -80 \end{bmatrix}$$

Hence Theorem 5 §3.3 gives

$$\begin{bmatrix} a_k \\ j_k \end{bmatrix} = V_k = b_1 \lambda_1^k X_1 + b_2 \lambda_2^k X_2 = \frac{220}{3} 1^k \begin{bmatrix} 1 \\ 2 \end{bmatrix} - \frac{80}{3}\left(\frac{-1}{2}\right)^k \begin{bmatrix} -1 \\ 4 \end{bmatrix}$$

Equating top and bottom entries we obtain exact formulas for a_k and j_k:

$$a_k = \frac{220}{3} + \frac{80}{3}\left(-\frac{1}{2}\right)^k \quad \text{and} \quad j_k = \frac{440}{3} - \frac{320}{3}\left(-\frac{1}{2}\right)^k \quad \text{for } k = 0, 1, 2, \ldots$$

Clearly, if k is large we obtain $a_k \approx \frac{220}{3} \approx 73$ and $j_k \approx \frac{440}{3} \approx 147$. This means that the population will stabilize eventually with approximately twice as many juveniles as adults.

The fact that the dominant eigenvalue λ_1 is 1 in Example 1 is the reason that the population stabilizes. With different survival and reproduction rates the model predicts that the population will become extinct (when $|\lambda_1| < 1$), or diverge and become very large (when $|\lambda_1| < 1$). Here is another example.

Example 2

If the juvenile survival rate is reduced to $\frac{1}{8}$ in Example 1, show that the population becomes extinct.

Solution   Now the equations are

$$a_{k+1} = \frac{1}{2} a_k + \frac{1}{8} j_k$$
$$j_{k+1} = 2a_k$$

so the matrix $A = \begin{bmatrix} \frac{1}{2} & \frac{1}{8} \\ 2 & 0 \end{bmatrix}$. Hence $c_A(x) = x^2 - \frac{1}{2}x - \frac{1}{4}$, so the eigenvalues are $\lambda_1 = \frac{1}{4}(1 + \sqrt{5}) = 0.809$ and $\lambda_2 = \frac{1}{4}(1 - \sqrt{5}) = -0.309$. Hence λ_1 is dominant and the population becomes extinct because $V_R \approx b_1 \lambda_1^k X_1 = b_1(0.089)^k X_1$ for large k, and this approaches zero.

A Predator–Prey Model

Ecosystems in nature involve several species interacting in various ways. We now look at a model that describes how this happens with two species X and Y. Let x_k and y_k, respectively, denote the populations after k years. We assume that x_0 and y_0 are known, and that the populations in successive years are related by linear equations

$$\begin{aligned} x_{k+1} &= ax_k + by_k \\ y_{k+1} &= cx_k + dy_k \end{aligned} \quad \text{for } k \geq 0$$

where a, b, c, and d are numbers reflecting the nature of the interaction between the species X and Y. Hence the population profiles $V_k = \begin{bmatrix} x_k \\ y_k \end{bmatrix}$ are a linear dynamical system with matrix $A = \begin{bmatrix} a & b \\ c & d \end{bmatrix}$. We give an example where one species preys upon the other.

Example 3

Let h_k and m_k denote the populations of hawks and mice respectively in a certain region in year k. Assume that they are related as follows:

$$h_{k+1} = \tfrac{1}{2} h_k + \tfrac{1}{100} m_k \qquad \text{for } k \geq 0$$
$$m_{k+1} = -\tfrac{50}{4} h_k + \tfrac{5}{4} m_k$$

Note that an increase in the hawk population (the predator) causes a sharp decrease in the mouse population (the prey), while more mice means, marginally, more hawks. If there are initially 50 hawks and 1600 mice in the region, determine the limiting populations.

Solution   If we write $V_k = \begin{bmatrix} h_k \\ m_k \end{bmatrix}$ and $A = \begin{bmatrix} \tfrac{1}{2} & \tfrac{1}{100} \\ -\tfrac{50}{4} & \tfrac{5}{4} \end{bmatrix}$, then $c_A(x) = (x-1)(x-\tfrac{3}{4})$, so the eigenvalues are $\lambda_1 = 1$ and $\lambda_2 = \tfrac{3}{4}$ with corresponding eigenvectors $X_1 = \begin{bmatrix} 1 \\ 50 \end{bmatrix}$ and $X_2 = \begin{bmatrix} 1 \\ 25 \end{bmatrix}$. Since λ_1 is dominant here, we can estimate the limiting value of V_k. Because $P = \begin{bmatrix} 1 & 1 \\ 50 & 25 \end{bmatrix}$ is invertible, it is a diagonalizing matrix, and we obtain $\begin{bmatrix} b_1 \\ b_2 \end{bmatrix} = P^{-1} V_0 = \begin{bmatrix} 14 \\ 36 \end{bmatrix}$ using the notation in Theorem 5 §3.3. Hence, for large k,

$$V_k \approx b_1 \lambda_1^k X_1 = 14 \begin{bmatrix} 1 \\ 50 \end{bmatrix}$$

so the populations stabilize at approximately 14 hawks and 700 mice.

Exercises 3.5

1. Referring to the model in Example 1, determine if the population stabilizes, becomes extinct, or becomes large in each case. Denote the adult and juvenile survival rates as A and J, and the reproduction rate as R.

	R	A	J
(a)	2	$\tfrac{1}{2}$	$\tfrac{1}{2}$
◆(b)	3	$\tfrac{1}{4}$	$\tfrac{1}{4}$
(c)	2	$\tfrac{1}{4}$	$\tfrac{1}{3}$
◆(d)	3	$\tfrac{3}{5}$	$\tfrac{1}{5}$

2. In the model of Example 1, does the final outcome depend on the initial population of adult and juvenile females? Support your answer.

3. In Example 1, keep the same reproduction rate of 2 and the same adult survival rate of $\tfrac{1}{2}$, but suppose that the juvenile survival rate is ρ. Determine which values of ρ cause the population to become extinct or to become large.

◆4. In Example 1, let the juvenile survival rate be $\tfrac{2}{5}$, and let the reproduction rate be 2. What values of the adult survival rate α will ensure that the population stabilizes?

5. In Example 3, change the entry $\frac{1}{100}$ in the matrix A to $\frac{1}{160}$. This means that an increased mouse population one year gives a decrease in the hawk population the next year. What is the effect on the populations in the long term?

SECTION 3.6 Proof of the Cofactor Expansion Theorem (Optional)

Recall that our definition of the term *determinant* is inductive: The determinant of any 1×1 matrix is defined first; then it is used to define the determinants of 2×2 matrices. Then that is used for the 3×3 case, and so on. The case of a 1×1 matrix $[a]$ poses no problem. We simply define

$$\det[a] = a$$

as in Section 3.1. Given an $n \times n$ matrix A, define A_{ij} to be the $(n-1) \times (n-1)$ matrix obtained from A by deleting row i and column j. Now assume that the determinant of any $(n-1) \times (n-1)$ matrix has been defined. Then the determinant of A is *defined* to be

$$\det A = a_{11} \det A_{11} - a_{21} \det A_{21} + \cdots + (-1)^{n+1} a_{n1} \det A_{n1}$$

$$= \sum_{i=1}^{n} (-1)^{i+1} a_{i1} \det A_{i1}$$

Observe that, in the terminology of Section 3.1, this is just the cofactor expansion of $\det A$ along the first column, and that $(-1)^{i+j} \det A_{ij}$ is the (i, j)-cofactor (previously denoted as $c_{ij}(A)$).[17] To illustrate the definition, consider the 2×2 matrix $A = \begin{bmatrix} a_{11} & a_{12} \\ a_{21} & a_{22} \end{bmatrix}$. Then the definition gives

$$\det \begin{bmatrix} a_{11} & a_{12} \\ a_{21} & a_{22} \end{bmatrix} = a_{11} \det[a_{22}] - a_{21} \det[a_{12}] = a_{11}a_{22} - a_{21}a_{12}$$

and this is the same as the definition in Section 3.1.

Of course, the task now is to use this definition to *prove* that the cofactor expansion along *any* row or column yields $\det A$ (this is Theorem 1 §3.1). The proof proceeds by first establishing the properties of determinants stated in Theorem 2 §3.1, but for *rows* only (see Lemma 2). This being done, the full proof of Theorem 1 §3.1 is not difficult. The proof of Lemma 2 requires the following preliminary result.

Lemma 1

Let A, B, and C be $n \times n$ matrices that are identical except that the pth row of A is the sum of the pth rows of B and C. Then

$$\det A = \det B + \det C$$

PROOF

We proceed by induction on n, the cases $n = 1$ and $n = 2$ being easily checked. Consider a_{i1} and A_{i1}:

Case 1: If $i \neq p$,

$$a_{i1} = b_{i1} = c_{i1} \quad \text{and} \quad \det A_{i1} = \det B_{i1} + \det C_{i1}$$

---

17 Note that we used the expansion along row 1 at the beginning of Section 3.1. The column 1 expansion definition is more convenient here.

by induction because A_{i1}, B_{i1}, C_{i1} are identical except that one row of A_{i1} is the sum of the corresponding rows of B_{i1} and C_{i1}.

Case 2: If $i = p$,

$$a_{p1} = b_{p1} + c_{p1} \quad \text{and} \quad A_{p1} = B_{p1} = C_{p1}$$

Now write out the defining sum for det A, splitting off the pth term for special attention.

$$\det A = \sum_{i \neq p} a_{i1}(-1)^{i+1} \det A_{i1} + a_{p1}(-1)^{p+1} \det A_{p1}$$

$$= \sum_{i \neq p} a_{i1}(-1)^{i+1}[\det B_{i1} + \det C_{i1}] + (b_{p1} + c_{p1})(-1)^{p+1} \det A_{p1}$$

where $\det A_{i1} = \det B_{i1} + \det C_{i1}$ by induction. But the terms here involving B_{i1} and b_{p1} add up to det B because $a_{i1} = b_{i1}$ if $i \neq p$ and $A_{p1} = B_{p1}$. Similarly, the terms involving C_{i1} and c_{p1} add up to det C. Hence $\det A = \det B + \det C$, as required.

Lemma 2

Let $A = [a_{ij}]$ denote an $n \times n$ matrix.
1. If $B = [b_{ij}]$ is formed from A by multiplying a row of A by a number u, then $\det B = u \det A$.
2. If A contains a row of zeros, then $\det A = 0$.
3. If $B = [b_{ij}]$ is formed by interchanging two rows of A, then $\det B = -\det A$.
4. If A contains two identical rows, then $\det A = 0$.
5. If $B = [b_{ij}]$ is formed by adding a multiple of one row of A to a different row, then $\det B = \det A$.

PROOF

For later reference the defining sums for det A and det B are as follows:

$$\det A = \sum_{i=1}^{n} a_{i1}(-1)^{i+1} \det A_{i1} \qquad (*)$$

$$\det B = \sum_{i=1}^{n} b_{i1}(-1)^{i+1} \det B_{i1} \qquad (**)$$

Property 1. The proof is by induction on n, the cases $n = 1$ and $n = 2$ being easily verified. Consider the ith term in the sum $(**)$ for det B where B is the result of multiplying row p of A by u.

  a. If $i \neq p$, then $b_{i1} = a_{i1}$ and $\det B_{i1} = u \det A_{i1}$ by induction because B_{i1} comes from A_{i1} by multiplying a row by u.
  b. If $i = p$, then $b_{p1} = ua_{p1}$ and $B_{p1} = A_{p1}$.

In either case, each term in equation $(**)$ is u times the corresponding term in equation $(*)$, so it is clear that $\det B = u \det A$.

Property 2. This is clear by property 1 because the row of zeros has a common factor $u = 0$.

Property 3. Observe first that it suffices to prove property 3 for interchanges of adjacent rows. (Rows p and q $(q > p)$ can be interchanged by carrying out $2(q - p) - 1$ adjacent changes, which results in an *odd* number of sign changes in the determinant.) So suppose that rows p and $p + 1$ of A are interchanged to obtain B. Again consider the ith term in $(**)$.

a. If $i \neq p$ and $i \neq p + 1$, then $b_{i1} = a_{i1}$ and $\det B_{i1} = -\det A_{i1}$ by induction because B_{i1} results from interchanging adjacent rows in A_{i1}. Hence the ith term in $(**)$ is the negative of the ith term in $(*)$, and so $\det B = -\det A$.

b. If $i = p$ or $i = p + 1$, then $b_{p1} = a_{p+1,1}$ and $B_{p1} = A_{p+1,1}$, whereas $b_{p+1,1} = a_{p1}$ and $B_{p+1,1} = A_{p1}$. Hence terms p and $p + 1$ in $(**)$ are

$$b_{p1}(-1)^{p+1} \det B_{p1} = -a_{p+1,1}(-1)^{(p+1)+1} \det(A_{p+1,1})$$

$$b_{p+1,1}(-1)^{(p+1)+1} \det(B_{p+1,1}) = -a_{p1}(-1)^{p+1} \det A_{p1}$$

This means that terms p and $p + 1$ in $(**)$ are the same as these terms in $(*)$, except that the order is reversed and the signs are changed. Thus the sum $(**)$ is the negative of the sum $(*)$; that is, $\det B = -\det A$.

Property 4. If rows p and q in A are identical, let B be obtained from A by interchanging these rows. Then $B = A$ so $\det A = \det B$. But $\det B = -\det A$ by property 3 so $\det A = -\det A$. This implies that $\det A = 0$.

Property 5. Suppose B results from adding u times row q of A to row p. Then Lemma 1 applies to B to show that $\det B = \det A + \det C$, where C is obtained from A by replacing row p by u times row q. It now follows from properties 1 and 4 that $\det C = 0$ so $\det B = \det A$, as asserted.

These facts are enough to enable us to prove Theorem 1 §3.1. For convenience, it is restated here in the notation of the foregoing lemmas. The only difference between the notations is that the (i, j)-cofactor of an $n \times n$ matrix A was denoted earlier by

$$c_{ij}(A) = (-1)^{i+j} \det A_{ij}$$

Theorem 1

If $A = [a_{ij}]$ is an $n \times n$ matrix, then

1. $\det A = \displaystyle\sum_{i=1}^{n} a_{ij}(-1)^{i+j} \det A_{ij}$ (cofactor expansion along column j).

2. $\det A = \displaystyle\sum_{j=1}^{n} a_{ij}(-1)^{i+j} \det A_{ij}$ (cofactor expansion along row i).

Here A_{ij} denotes the matrix obtained from A by deleting row i and column j.

PROOF

Lemma 2 establishes the truth of Theorem 2 §3.1 for *rows*. With this information, the arguments in Section 3.2 proceed exactly as written to establish that $\det A = \det A^T$ holds for any $n \times n$ matrix A. Now suppose B is obtained from A by interchanging two columns. Then B^T is obtained from A^T by interchanging two rows so, by property 3 of Lemma 2,

$$\det B = \det B^T = -\det A^T = -\det A$$

Hence property 3 of Lemma 2 holds for *columns* too.

This enables us to prove the cofactor expansion for columns.

Given an $n \times n$ matrix $A = [a_{ij}]$, let $B = [b_{ij}]$ be obtained by moving column j to the left side, using $j-1$ interchanges of adjacent columns. Then $\det B = (-1)^{j-1}\det A$ and, because $B_{i1} = A_{ij}$ and $b_{i1} = a_{ij}$ for all i, we obtain

$$\det A = (-1)^{j-1}\det B = (-1)^{j-1}\sum_{i=1}^{n} b_{i1}(-1)^{1+i}\det B_{i1}$$

$$= \sum_{i+1}^{n} a_{ij}(-1)^{i+j}\det A_{ij}$$

This is the cofactor expansion of $\det A$ along column j.

Finally, to prove the row expansion, write $B = A^T$. Then $B_{ij} = (A^T_{ji})$ and $b_{ij} = a_{ji}$ for all i and j. Expanding $\det B$ along column j gives

$$\det A = \det A^T = \det B = \sum_{i=1}^{n} b_{ij}(-1)^{i+j}\det B_{ij}$$

$$= \sum_{i=1}^{n} a_{ji}(-1)^{j+i}\det\left[(A^T_{ji})\right] = \sum_{i=1}^{n} a_{ji}(-1)^{j+i}\det A_{ji}$$

This is the required expansion of $\det A$ along row j.

Exercises 3.6

1. Prove Lemma 1 for columns.
♦2. Verify that interchanging rows p and q $(q > p)$ can be accomplished using $2(q-p)-1$ adjacent interchanges.

3. If u is a number and A is an $n \times n$ matrix, prove that $\det(uA) = u^n \det A$ by induction on n, using only the definition of $\det A$.

Supplementary Exercises for Chapter 3

1. Show that

$$\det\begin{bmatrix} a+px & b+qx & c+rx \\ p+ux & q+vx & r+wx \\ u+ax & v+bx & w+cx \end{bmatrix}$$

$$= (1+x^3)\det\begin{bmatrix} a & b & c \\ p & q & r \\ u & v & w \end{bmatrix}$$

2. (a) Show that $(A_{ij})^T = (A^T)_{ji}$ for all i, j and all square matrices A.

   (b) Use (a) to prove that $\det A^T = \det A$. [*Hint:* Induction on n where A is $n \times n$.]

3. Show that $\det\begin{bmatrix} 0 & I_n \\ I_m & 0 \end{bmatrix} = (-1)^{nm}$ for all $n \geq 1$ and $m \geq 1$.

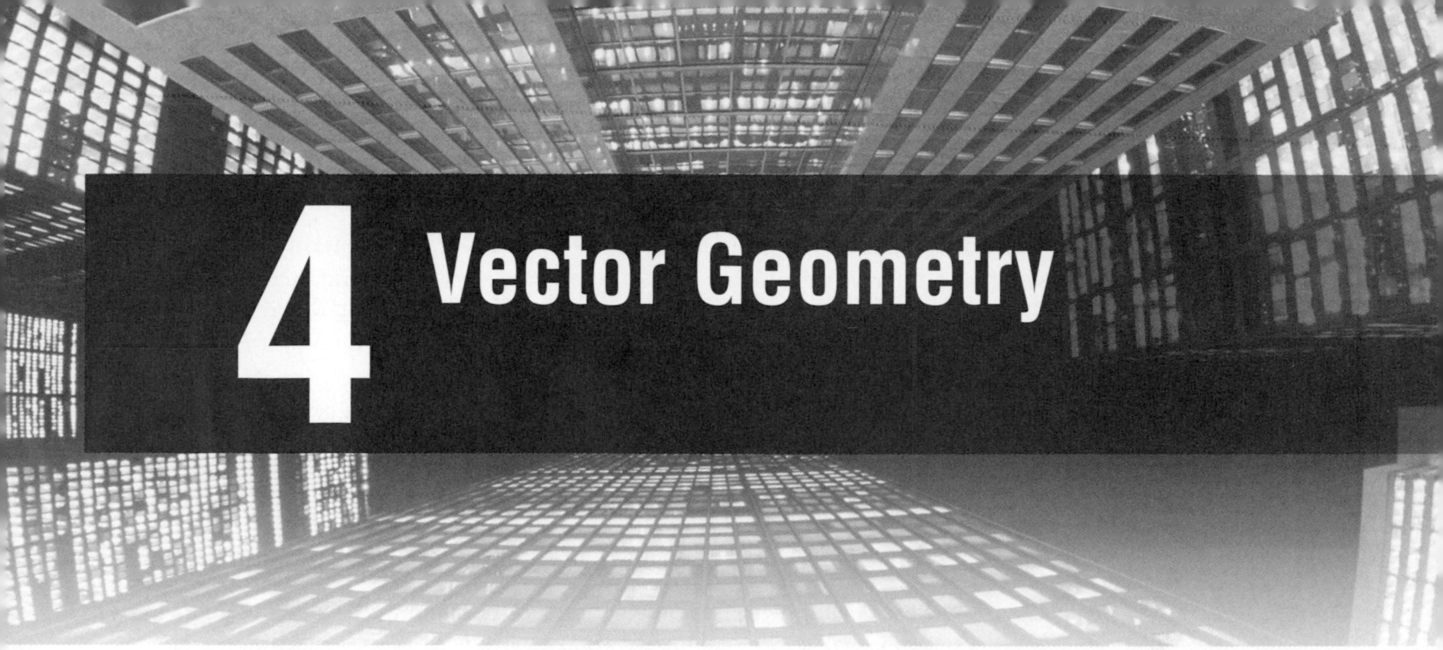

4 Vector Geometry

SECTION 4.1 Vectors and Lines

Many quantities in nature are completely specified by one number (called the magnitude of the quantity) and are referred to as **scalar** quantities. Examples include temperature, time, length, and pressure. However, for some quantities the magnitude is not enough. Consider displacement: To say that a boat sailed 10 kilometres does not specify where it went—it is necessary to give the direction too, say 10 km north-west. Quantities that require both a **magnitude** and a **direction** to specify them are called **vector** quantities. Hence:

> Two vector quantities are the same if and only if they have the same magnitude and the same direction.

For example, displacement, velocity, and force are fundamental vector quantities in physics.

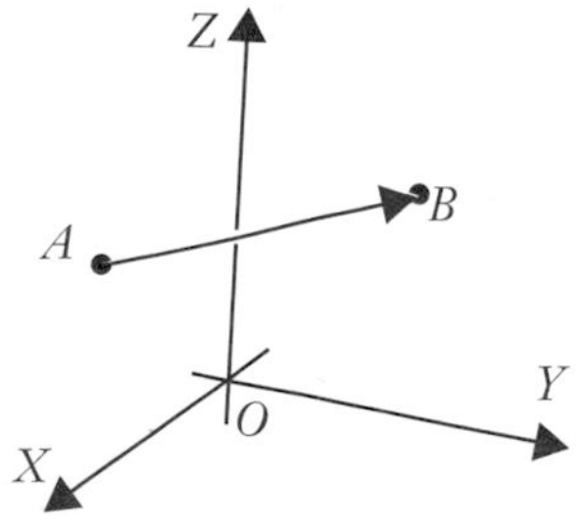

Geometric Vectors

In this chapter we are concerned only with geometric vectors. Suppose that A and B are any two points in $\mathbb{R}^3$ (or in $\mathbb{R}^2$). In Figure 4.1 the line segment from A to B is denoted $\overrightarrow{AB}$ and is called the **geometric vector** from A to B. As the name implies, $\overrightarrow{AB}$ is a vector quantity: It has magnitude (the distance from A to B) and direction (from A to B). The point A is called the **tail** of $\overrightarrow{AB}$, B is called the **tip** of $\overrightarrow{AB}$, and the magnitude of $\overrightarrow{AB}$ is called its **length** and denoted $\| \overrightarrow{AB} \|$.

Two geometric vectors can be the same even if the tips and tails are different. For example, $\overrightarrow{AB} = \overrightarrow{PQ}$ in Figure 4.2 because they have the same length and direction (they both go 1 unit left and 2 units up). Thus the same geometric vector can be positioned *anywhere* in the plane; what is important is that the length and direction are the same, and not where the tips and tails are located. One way to think of it is that $\overrightarrow{AB}$ and $\overrightarrow{PQ}$ are two *representations* of the same underlying vector (1 unit left and 2 units up).[1]

Now recall that in Section 2.5 the columns in $\mathbb{R}^n$ are *also* called vectors, and the above discussion shows why those in $\mathbb{R}^2$ and $\mathbb{R}^3$ are written as $\vec{v}$, $\vec{w}$, etc. In particular,

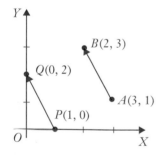

Figure 4.1

Figure 4.2

---

1 Fractions provide another example of quantities that can be the same but *look* different. For example $\frac{6}{9}$ and $\frac{14}{21}$ certainly look different, but they are equal fractions—both equal $\frac{2}{3}$.

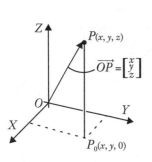

Figure 4.3

a vector $\vec{v} = \begin{bmatrix} x \\ y \\ z \end{bmatrix} = [x \ y \ z]^T$ in $\mathbb{R}^3$ is identified with the point $P(x, y, z)$ and

represented by the arrow from the origin O to P. In our present terminology this says that $\vec{v} = \overrightarrow{OP}$ is a geometric vector. Moreover, *every* geometric vector $\overrightarrow{AB}$ arises in this way. Indeed, if we move $\overrightarrow{AB}$ to **standard position** with its tail at the origin, then $\overrightarrow{AB} = \overrightarrow{OP}$ where P is the uniquely determined point at its tip. But if $P = P(x, y, z)$, then $\overrightarrow{OP} = [x \ y \ z]^T$ as in Figure 4.3. Hence $\overrightarrow{AB} = \overrightarrow{OP} = [x \ y \ z]^T$ is a vector in $\mathbb{R}^3$. In this way the (column) vectors in $\mathbb{R}^3$ are the same as the geometric vectors, and we call them all "vectors". The vector $\vec{v} = [x \ y \ z]^T$ is called the **position vector** of the point $P(x, y, z)$.

Of course, there is an $\mathbb{R}^2$-analog of all this: Each geometric vector in $\mathbb{R}^2$ has the form $\vec{v} = \begin{bmatrix} x \\ y \end{bmatrix} = \overrightarrow{OP}$ for a unique point $P(x, y)$, and $\vec{v}$ is called the position vector of P.

Clearly, we are dealing with $\mathbb{R}^2$ and $\mathbb{R}^3$ in much the same way. Since the geometry of $\mathbb{R}^2$ is easier to visualize, we will nearly always work with $\mathbb{R}^3$ in what follows, and leave the (natural) analog in $\mathbb{R}^2$ to the reader.

We now have two ways to view a vector in $\mathbb{R}^3$: As a quantity with magnitude and direction, or as a 3×1 column matrix. Part (1) of the following theorem shows that equality is the same from both points of view.

Theorem 1

Let $\vec{v} = \begin{bmatrix} x \\ y \\ z \end{bmatrix}$ and $\vec{w} = \begin{bmatrix} x_1 \\ y_1 \\ z_1 \end{bmatrix}$ be vectors. Then:

1. $\vec{v} = \vec{w}$ as vectors if and only if $x = x_1$, $y = y_1$ and $z = z_1$.
2. $\|\vec{v}\| = \sqrt{x^2 + y^2 + z^2}$.
3. $\vec{v} = \vec{0}$ if and only if $\|\vec{v}\| = 0$.
4. $\|a\vec{v}\| = |a|\|\vec{v}\|$ for any scalar a.

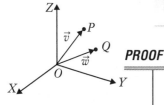

Figure 4.4

PROOF

1. If $\vec{v}$ and $\vec{w}$ are in standard position, their tips are the points $P(x, y, z)$ and $Q(x_1, y_1, z_1)$. By Figure 4.4, $\vec{v}$ and $\vec{w}$ have the same magnitude and direction if and only if P and Q are the same point; that is, if and only if $x = x_1$, $y = y_1$, and $z = z_1$. This proves (1).
2. In Figure 4.5, $\vec{v} = \overrightarrow{OP}$ where $P = P(x, y, z)$. Since OQP is a right triangle, Pythagoras' theorem[2] gives $\|\vec{v}\|^2 = b^2 + z^2$, and it gives $x^2 + y^2 = b^2$ when applied to the right triangle ORQ. Now (2) follows by eliminating b^2.
3. If $\|\vec{v}\| = 0$, then $\sqrt{x^2 + y^2 + z^2} = 0$ by (2), so $x^2 + y^2 + z^2 = 0$. Since squares of real numbers are nonnegative, it follows that $x = y = z = 0$; that is $\vec{v} = \vec{0}$. The converse is clear because $\|\vec{0}\| = 0$.
4. We have $a\vec{v} = [ax \ ay \ az]^T$, so (3) gives $\|a\vec{v}\|^2 = (ax)^2 + (ay)^2 + (az)^2 = a^2\|\vec{v}\|^2$. Hence $\|a\vec{v}\| = \sqrt{a^2}\|\vec{v}\|$, and we are done because $\sqrt{a^2} = |a|$.

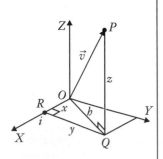

Figure 4.5

---

2 Pythagoras' theorem states that if a and b are the sides of a right triangle with hypotenuse h, then $a^2 + b^2 = h^2$. A proof is given at the end of this section.

Of course, the $\mathbb{R}^2$-version of Theorem 1 also holds.

Example 1

If $\vec{v} = [2 \;-1\; 3]^T$ and $\vec{w} = [3 \; 4]^T$ then $\|\vec{v}\| = \sqrt{4+1+9} = \sqrt{14}$, and
$\|\vec{w}\| = \sqrt{9+16} = 5$.

Many properties of vectors can be described entirely in terms of the length and direction of the vectors involved. Such descriptions are called **intrinsic** because they make no reference to a coordinate system, an important fact in applications. Thus (1) in Theorem 1 gives an intrinsic description of vector equality:

> Two vectors are equal as vectors (same length and direction) if and only if they are equal as matrices.

Similarly, (2) describes the zero vector intrinsically:

> The zero vector $\vec{0}$ is the only vector of length zero.

Finally, we give an intrinsic description of the negative of a vector.

> The negative $-\vec{v}$ of a vector $\vec{v}$ is the vector with the same length as $\vec{v}$ but opposite direction.

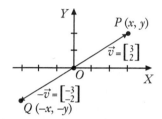

Figure 4.6

In fact, $\|-\vec{v}\| = \|(-1)\vec{v}\| = |-1|\|\vec{v}\| = \|\vec{v}\|$, so $\vec{v}$ and $-\vec{v}$ have the same length. Moreover, if $\vec{v} = \overrightarrow{OP}$ in standard position, then $-\vec{v} = \overrightarrow{PO}$ has the opposite direction (an illustration appears in Figure 4.6 where $\vec{v} = [3\; 2]^T$ and $-\vec{v} = [-3\; -2]^T = \overrightarrow{OQ}$).

The Parallelogram Law

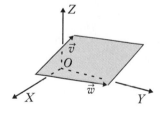

Figure 4.7

Two vectors $\vec{v}$ and $\vec{w}$ can be added (as column matrices), and we now turn to an intrinsic description of the vector $\vec{v} + \vec{w}$. If $\vec{v}$ and $\vec{w}$ are positioned with a common tail, they form a parallelogram[3] as in Figure 4.7, called the parallelogram **determined** by $\vec{v}$ and $\vec{w}$. Since this parallelogram is completely described by the lengths and directions of $\vec{v}$ and $\vec{w}$, the following "law" gives an intrinsic description of the sum $\vec{v} + \vec{w}$.

The Parallelogram Law

In the parallelogram determined by two vectors $\vec{v}$ and $\vec{w}$, the vector $\vec{v} + \vec{w}$ is the diagonal with the same tail as $\vec{v}$ and $\vec{w}$.

PROOF

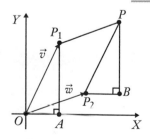

Figure 4.8

We may assume that $\vec{v}$ and $\vec{w}$ are in standard position since the parallelogram they determine is the same for any choice of a common tail. The situation in $\mathbb{R}^2$ is shown in Figure 4.8. If $\vec{v} = \begin{bmatrix} x_1 \\ y_1 \end{bmatrix}$ and $\vec{w} = \begin{bmatrix} x_2 \\ y_2 \end{bmatrix}$ then $\vec{v} + \vec{w} = \begin{bmatrix} x_1 + x_2 \\ y_1 + y_2 \end{bmatrix}$, so the tips of $\vec{v}$, $\vec{w}$, and $\vec{v} + \vec{w}$ are $P_1(x_1, y_1)$, $P_2(x_2, y_2)$, and $P(x_1 + x_2, y_1 + y_2)$ respectively in Figure 4.8. We must verify that OP_1PP_2 is a parallelogram (that the sides OP_1 and P_2P are parallel and have equal length). But this follows because the triangles OP_1A and P_2PB are identical.

---

3 **A parallelogram** is a four-sided figure with straight sides such that opposite sides are parallel and of the same length. For example, a rectangle is a parallelogram in which all interior angles are right angles.

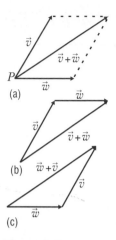

(a)

(b) $\vec{w}+\vec{v}$

(c)

Figure 4.9

It is worth noting that the parallelogram law holds for physical vectors like velocity and force, as can be verified by experiments. These are important facts in physics.

Recall that a vector can be positioned with its tail at any point. This leads to another way to view vector addition. In Figure 4.9(a) the sum $\vec{v}+\vec{w}$ of two vectors $\vec{v}$ and $\vec{w}$ is shown as given by the parallelogram law. If $\vec{w}$ is moved so its tail coincides with the tip of $\vec{v}$ (Figure 4.9(b)) then the sum $\vec{v}+\vec{w}$ is seen as "first $\vec{v}$ and then $\vec{w}$". Similarly, moving the tail of $\vec{v}$ to the tip of $\vec{w}$ shows in Figure 4.9(c) that $\vec{v}+\vec{w}$ is "first $\vec{w}$ and then $\vec{v}$". This will be referred to as the **tip-to-tail rule**, and it gives a graphic illustration of why $\vec{v}+\vec{w}=\vec{w}+\vec{v}$.

If $\overrightarrow{AB}$ denotes the vector from a point A to a point B, the tip-to-tail rule takes the easily remembered form $\overrightarrow{AB}+\overrightarrow{BC}=\overrightarrow{AC}$ for any points A, B, and C. The next example uses this to derive a theorem in geometry without using coordinates.

Example 2

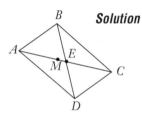

Solution

Show that the diagonals of a parallelogram bisect each other.

Let the parallelogram have vertices A, B, C, and D, as shown; let E denote the intersection of the two diagonals; and let M denote the midpoint of diagonal AC. We must show that $M=E$ and that this is the midpoint of diagonal BD. This is accomplished by showing that $\overrightarrow{BM}=\overrightarrow{MD}$. (Then the fact that $\overrightarrow{BM}$ and $\overrightarrow{MD}$ have the same direction means that $M=E$, and the fact that they have the same length means that $M=E$ is the midpoint of BD.) Now $\overrightarrow{AM}=\overrightarrow{MC}$ because M is the midpoint of AC, and $\overrightarrow{BA}=\overrightarrow{CD}$ because the figure is a parallelogram. Hence

$$\overrightarrow{BM}=\overrightarrow{BA}+\overrightarrow{AM}=\overrightarrow{CD}+\overrightarrow{MC}=\overrightarrow{MC}+\overrightarrow{CD}=\overrightarrow{MD}$$

where the first and last equalities use the tip-to-tail rule of vector addition.

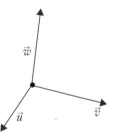

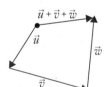

Figure 4.10

One reason for the importance of the tip-to-tail rule is that it means two or more vectors can be added by placing them tip-to-tail in sequence. This gives a useful "picture" of the sum of several vectors, and is illustrated for three vectors in Figure 4.10 where $\vec{u}+\vec{v}+\vec{w}$ is viewed as first $\vec{u}$, then $\vec{v}$, then $\vec{w}$.

There is a simple geometrical way to visualize the (matrix) **difference** $\vec{v}-\vec{w}$ of two vectors. If $\vec{v}$ and $\vec{w}$ are positioned so that they have a common tail A (see Figure 4.11), and if B and C are their respective tips, then the tip-to-tail rule gives $\vec{w}+\overrightarrow{CB}=\vec{v}$. Hence $\vec{v}-\vec{w}=\overrightarrow{CB}$ is the vector from the tip of $\vec{w}$ to the tip of $\vec{v}$. Thus both $\vec{v}-\vec{w}$ and $\vec{v}+\vec{w}$ appear as diagonals in the parallelogram determined by $\vec{v}$ and $\vec{w}$ (see Figure 4.11). We record this for reference.

Theorem 2

If $\vec{v}$ and $\vec{w}$ have a common tail, then $\vec{v}-\vec{w}$ is the vector from the tip of $\vec{w}$ to the tip of $\vec{v}$.

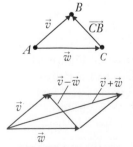

Figure 4.11

One of the most useful applications of vector subtraction is that it gives a simple formula for the vector from one point to another, and for the distance between the points.

Theorem 3

Let $P_1(x_1, y_1, z_1)$ and $P_2(x_2, y_2, z_2)$ be two points.

1. $\overrightarrow{P_1 P_2} = \begin{bmatrix} x_2 - x_1 \\ y_2 - y_1 \\ z_2 - z_1 \end{bmatrix}$.

2. The distance between P_1 and P_2 is $\sqrt{(x_2 - x_1)^2 + (y_2 - y_1)^2 + (z_2 - z_1)^2}$.

PROOF

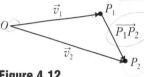

Figure 4.12

If O is the origin, write $\vec{v}_1 = \overrightarrow{OP_1} = [x_1 \ y_1 \ z_1]^T$ and $\vec{v}_2 = \overrightarrow{OP_2} = [x_2 \ y_2 \ z_2]^T$ as in Figure 4.12. Then Theorem 2 gives $\overrightarrow{P_1 P_2} = \vec{v}_2 - \vec{v}_1$, and (1) follows. But the distance between P_1 and P_2 is $\| \overrightarrow{P_1 P_2} \|^2$, so (2) follows from (1) and Theorem 1.

Of course the $\mathbb{R}^2$-version of Theorem 3 is also valid: If $P_1(x_1, y_1)$ and $P_2(x_2, y_2)$ are points in $\mathbb{R}^2$, then $\overrightarrow{P_1 P_2} = \begin{bmatrix} x_2 - x_1 \\ y_2 - y_1 \end{bmatrix}$, and the distance between P_1 and P_2 is $\sqrt{(x_2 - x_1)^2 + (y_2 - y_1)^2}$.

Example 3

The distance between $P_1(2, -1, 3)$ and $P_2(1, 1, 4)$ is $\sqrt{(-1)^2 + (2)^2 + (1)^2} = \sqrt{6}$, and the vector from P_1 to P_2 is $\overrightarrow{P_1 P_2} = [-1 \ 2 \ 1]^T$.

Scalar Multiplication

As for vector addition, there is a simple intrinsic description of the scalar product $a\vec{v}$ in terms of the real number a and the length and direction of the vector $\vec{v}$. To understand it, we require the following result:

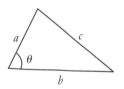

Law of Cosines

If a triangle has sides a, b, and c, and if θ is the angle opposite c (as in Figure 4.13), then

$$c^2 = a^2 + b^2 - 2ab \cos \theta.$$

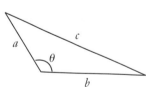

Figure 4.13

PROOF

We prove it if θ is acute, that is $0 \leq \theta \leq \frac{\pi}{2}$.[4] In Figure 4.14, we have $p = a \sin \theta$ and $q = b - a \cos \theta$, so Pythagoras' theorem (Theorem 5 below) gives
$$c^2 = p^2 + q^2 = a^2 (\sin^2 \theta + \cos^2 \theta) + b^2 - 2ab \cos \theta.$$
The law of cosines follows because $\sin^2\theta + \cos^2\theta = 1$.

Note that the law of cosines reduces to Pythagoras' theorem if $\theta = \frac{\pi}{2}$ (a right angle) because, in that case, $\cos \theta = \cos \frac{\pi}{2} = 0$.

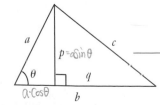

Figure 4.14

4 Recall that we use radian measure for angles. Thus $180° = \pi$ radians, $90° = \frac{\pi}{2}$, etc.

Intrinsic Description of Scalar Multiplication

If a is a real number and $\vec{v} \neq \vec{0}$ is a vector, then:

1. The length of $a\vec{v}$ is $\|a\vec{v}\| = |a|\|\vec{v}\|$.

2. If $a\vec{v} \neq \vec{0}^5$, the direction of $a\vec{v}$ is $\begin{cases} \text{the same as } \vec{v} \text{ if } a > 0, \\ \text{opposite to } \vec{v} \text{ if } a < 0. \end{cases}$

a > 0   same

a < 0   opposite

PROOF

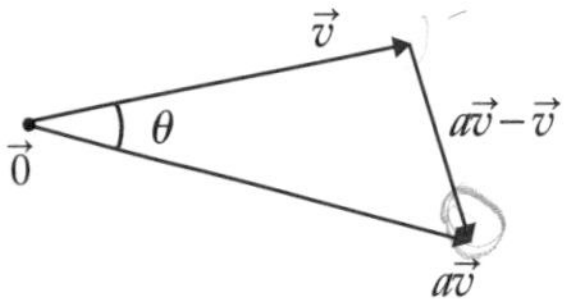

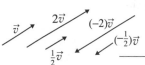

Figure 4.15

$\overrightarrow{a\vec{v}} - \vec{v}$

(1) is part of Theorem 1. Turning to (2), assume $a\vec{v} \neq \vec{0}$ and consider the angle θ between the nonzero vectors $\vec{v}$ and $a\vec{v}$ (see Figure 4.15). Then $a\vec{v}$ has the same or opposite direction as $\vec{v}$ exactly when $\theta = 0$ or $\theta = \pi$, respectively. Hence (2) translates to:

$$\theta = 0 \text{ if } a > 0, \quad \text{and} \quad \theta = \pi \text{ if } a < 0.$$

In the triangle with vertices $\vec{0}$, $\vec{v}$, and $a\vec{v}$ in Figure 4.15, the vector from $\vec{v}$ to $a\vec{v}$ is $a\vec{v} - \vec{v} = (a - 1)\vec{v}$ by Theorem 2. Applying the law of cosines (and using Theorem 1) gives

$$\|(1 - a)\vec{v}\|^2 = \|\vec{v}\|^2 + \|a\vec{v}\|^2 - 2\|\vec{v}\|\|a\vec{v}\|\cos\theta$$
$$= \|\vec{v}\|^2 + a^2\|\vec{v}\|^2 - 2\|\vec{v}\|^2|a|\cos\theta.$$

Since $\|\vec{v}\| \neq 0$ this gives $(1 - a)^2 = 1 + a^2 - 2|a|\cos\theta$, which reduces to $a = |a|\cos\theta$. If $a > 0$ this gives $\cos\theta = 1$ so $\theta = 0$; while $a < 0$ gives $\cos\theta = -1$ so $\theta = \pi$. This is what we wanted.

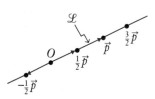

Figure 4.16

Figure 4.16 gives several examples of scalar multiples of a vector $\vec{v}$.

Consider a line $\mathcal{L}$ through the origin, let P be any point on $\mathcal{L}$ other than the origin O, and let $\vec{p} = \overrightarrow{OP}$ be the position vector of P. If $t \neq 0$, then $t\vec{p}$ is a point on $\mathcal{L}$ because it has direction the same or opposite as that of $\vec{p}$. Moreover $t > 0$ or $t < 0$ according as the point $t\vec{p}$ lies on the same or opposite side of the origin as P. This is illustrated in Figure 4.17.

A vector $\vec{u}$ is called a **unit vector** if $\|\vec{u}\| = 1$. For example $\vec{i} = [1\ 0\ 0]^T$, $\vec{j} = [0\ 1\ 0]^T$, and $\vec{k} = [0\ 0\ 1]^T$ are unit vectors, called the **coordinate** vectors. We discuss them in more detail in Section 4.2.

Figure 4.17

Example 4

If $\vec{v} \neq \vec{0}$ show that $\dfrac{1}{\|\vec{v}\|}\vec{v}$ is the unique unit vector in the same direction as $\vec{v}$.

Solution ▶ The vectors in the same direction as $\vec{v}$ are $a\vec{v}$ where $a > 0$. Because $\|a\vec{v}\| = |a|\|\vec{v}\| = a\|\vec{v}\|$ when $a > 0$, this is a unit vector when $a = \dfrac{1}{\|\vec{v}\|}$.

The **midpoint** of two points is the point on the line segment between them that is halfway from one to the other. The next example shows how to find the coordinates of the midpoint; the technique used is important and will be used again below.

---

5 Since the zero vector $\vec{0}$ has no direction, we deal only with the case $a\vec{v} \neq \vec{0}$.

Example 5

The midpoint $\vec{m}$ between the points $\vec{p}_1 = P_1(x_1, y_1, z_1)$ and $\vec{p}_2 = P_2(x_2, y_2, z_2)$ is $\vec{m} = \frac{1}{2}(\vec{p}_1 + \vec{p}_2) = M(\frac{1}{2}(x_1 + x_2), \frac{1}{2}(y_1 + y_2), \frac{1}{2}(z_1 + z_2))$.

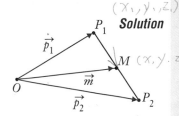

Figure 4.18

Solution   Let $\vec{m} = M(x, y, z)$. The vectors $\vec{p}_1$, $\vec{p}_2$, and $\vec{m}$ are shown in Figure 4.18. We have $\overrightarrow{P_1M} = \frac{1}{2}\overrightarrow{P_1P_2}$ because M is the midpoint of P_1P_2. By Theorem 2, $\overrightarrow{P_1P_2} = \vec{p}_2 - \vec{p}_1$, so tip-to-tail addition gives

$$\vec{m} = \vec{p}_1 + \overrightarrow{P_1M} = \vec{p}_1 + \frac{1}{2}(\vec{p}_2 - \vec{p}_1) = \frac{1}{2}(\vec{p}_1 + \vec{p}_2)$$

as required. Note that the midpoint $\vec{m}$ is the "average" of the points $\vec{p}_1$ and $\vec{p}_2$.

$$\overrightarrow{P_1M} = (x - x_1, \ y - y_1, \ z - z_1) = \frac{1}{2}\overrightarrow{P_1P_2} = \frac{1}{2}(\vec{P_2} - \vec{P_1})$$

Example 6

Show that the midpoints of the four sides of any quadrilateral are the vertices of a parallelogram. Here a quadrilateral is any two-dimensional figure with four vertices.

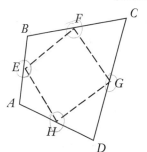

Solution   Suppose that the vertices of the quadrilateral are A, B, C, and D (in that order) and that E, F, G, and H are the midpoints of the sides as shown in the diagram. It suffices to show $\overrightarrow{EF} = \overrightarrow{HG}$ (because then these two sides are parallel and of equal length). Now the fact that E is the midpoint of AB means that $\overrightarrow{EB} = \frac{1}{2}\overrightarrow{AB}$. Similarly, $\overrightarrow{BF} = \frac{1}{2}\overrightarrow{BC}$, so

$$\overrightarrow{EF} = \overrightarrow{EB} + \overrightarrow{BF} = \frac{1}{2}\overrightarrow{AB} + \frac{1}{2}\overrightarrow{BC} = \frac{1}{2}(\overrightarrow{AB} + \overrightarrow{BC}) = \frac{1}{2}\overrightarrow{AC}$$

A similar argument shows that $\overrightarrow{HG} = \frac{1}{2}\overrightarrow{AC}$ too, so $\overrightarrow{EF} = \overrightarrow{HG}$ as required.

Many geometrical theorems involve the notion of being parallel, so the following theorem will be referred to repeatedly.

Theorem 4

Two nonzero vectors $\vec{v}$ and $\vec{w}$ are parallel if and only if one is a scalar multiple of the other.

PROOF

If one vector is a scalar multiple of the other vector, they are parallel by the intrinsic description of scalar multiplication. Conversely, let $\vec{v}$ and $\vec{w}$ be parallel. If they have the same direction, we show that $\vec{v} = \frac{\|\vec{v}\|}{\|\vec{w}\|}\vec{w}$ by showing that these vectors have the same direction and length. Indeed, $\frac{\|\vec{v}\|}{\|\vec{w}\|}\vec{w}$ has the same direction as $\vec{w}$, and hence the same as $\vec{v}$; and $\left\| \frac{\|\vec{v}\|}{\|\vec{w}\|}\vec{w} \right\| = \frac{\|\vec{v}\|}{\|\vec{w}\|}\|\vec{w}\| = \|\vec{v}\|$. On the other hand, if $\vec{v}$ and $\vec{w}$ have opposite directions we can similarly show that $\vec{v} = -\frac{\|\vec{v}\|}{\|\vec{w}\|}\vec{w}$.

Example 7

Given points $P(2, -1, 4)$, $Q(3, -1, 3)$, $A(0, 2, 1)$, and $B(1, 3, 0)$, determine if $\overrightarrow{PQ}$ and $\overrightarrow{AB}$ are parallel.

Solution   By Theorem 3, $\overrightarrow{PQ} = \begin{bmatrix} 1 \\ 0 \\ -1 \end{bmatrix}$ and $\overrightarrow{AB} = \begin{bmatrix} 1 \\ 1 \\ -1 \end{bmatrix}$. If $\overrightarrow{PQ} = t\overrightarrow{AB}$ then $\begin{bmatrix} 1 \\ 0 \\ -1 \end{bmatrix} = \begin{bmatrix} t \\ t \\ -t \end{bmatrix}$,

so $1 = t$ and $0 = t$, which is impossible. Hence $\overrightarrow{PQ}$ is *not* a scalar multiple of $\overrightarrow{AB}$, so these vectors are not parallel by Theorem 4.

Lines in Space

These vector techniques can be used to give a very simple way of describing straight lines in space. We use the fact that there is exactly one line through a particular point in space that is parallel to a given nonzero vector.

Consequently, given a straight line, any nonzero vector that is parallel to the line is called a **direction vector** for the line. Every line has many direction vectors. In fact, any nonzero multiple of a direction vector also serves as a direction vector.

Suppose $P_0 = P_0(x_0, y_0, z_0)$ is any point and $\vec{d} = [a\ b\ c]^T$ is any vector (assumed to be nonzero). There is a unique line through P_0 with direction vector $\vec{d}$, and we want to give a condition that a point $P = P(x, y, z)$ lies on that line. Let $\vec{p}_0 = [x_0\ y_0\ z_0]^T$ and $\vec{p} = [x\ y\ z]^T$ be the position vectors of P_0 and P, respectively (see Figure 4.19). Then

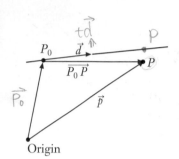

Origin

Figure 4.19

$$\vec{p} = \vec{p}_0 + \overrightarrow{P_0P}$$

Hence P lies on the line if and only if $\overrightarrow{P_0P}$ is parallel to $\vec{d}$—that is, if and only if $\overrightarrow{P_0P} = t\vec{d}$ for some scalar t by Theorem 2. Thus $\vec{p}$ is the position vector of a point on the line if and only if $\vec{p} = \vec{p}_0 + t\vec{d}$ for some scalar t.

Vector Equation of a Line

The line parallel to $\vec{d} \neq \vec{0}$ through the point with position vector $\vec{p}_0$ is given by

$$\vec{p} = \vec{p}_0 + t\vec{d} \quad \text{for some scalar } t$$

In other words, the point with position vector $\vec{p}$ is on this line if and only if a real number t exists such that $\vec{p} = \vec{p}_0 + t\vec{d}$.

In component form the vector equation becomes

$$\begin{bmatrix} x \\ y \\ z \end{bmatrix} = \begin{bmatrix} x_0 \\ y_0 \\ z_0 \end{bmatrix} + t\begin{bmatrix} a \\ b \\ c \end{bmatrix}$$

Equating components gives a different description of the line.

Parametric Equations of a Line

The line through $P_0(x_0, y_0, z_0)$ with direction vector $\vec{d} = [a \; b \; c]^T \neq \vec{0}$ is given by

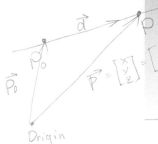

$$\begin{cases} x = x_0 + ta \\ y = y_0 + tb \quad t \text{ any scalar} \\ z = z_0 + tc \end{cases}$$

In other words, the point $P(x, y, z)$ is on this line if and only if a real number t exists such that $x = x_0 + ta$, $y = y_0 + tb$, and $z = z_0 + tc$.

Example 8

Find the equations of the line through the points $P_0(2, 0, 1)$ and $P_1(4, -1, 1)$.

Solution   Let $\vec{d} = \overrightarrow{P_0P_1} = [2 \; -1 \; 0]^T$ denote the vector from P_0 to P_1. Then $\vec{d}$ is parallel to the line (P_0 and P_1 are *on* the line), so $\vec{d}$ serves as a direction vector for the line. Using P_0 as the point on the line leads to the parametric equations

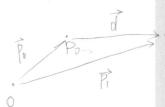

$$\begin{aligned} x &= 2 + 2t \\ y &= -t \qquad t \text{ a parameter} \\ z &= 1 \end{aligned}$$

Note that if P_1 is used (rather than P_0), the equations are

$$\begin{aligned} x &= 4 + 2s \\ y &= -1 - s \quad s \text{ a parameter} \\ z &= 1 \end{aligned}$$

These are different from the preceding equations, but this is merely the result of a change of parameter. In fact, $s = t - 1$.

Example 9

Find the equations of the line through $P_0(3, -1, 2)$ parallel to the line with equations

$$\begin{aligned} x &= -1 + 2t \\ y &= 1 + t \\ x &= -3 + 4t \end{aligned}$$

Solution   The coefficients of t give a direction vector $\vec{d} = [2 \; 1 \; 4]^T$ of the given line. Because the line we seek is parallel to this line, $\vec{d}$ also serves as a direction vector for the new line.
Thus the parametric equations are

$$\begin{aligned} x &= 3 + 2t \\ y &= -1 + t \\ z &= 2 + 4t \end{aligned}$$

Example 10

Handwritten notes in left margin:

$Q \to P$ $[6, 1, -1]$

$2, 10, ...$

P

R

$\vec{d} = [2, 10, ...$

$\vec{QR} = [x-0, y+5, z-1]$

$R = [x \ y \ z]^T$

$\begin{bmatrix} x \\ y \\ z \end{bmatrix} = \begin{matrix} 0 + 2t \\ -5 + 10t \\ 1 - 4t \end{matrix}$

$0 + (x-0) = 0 + 2t$
$-5 + (y+5)s = -5 + 10...$
$1 + (z-1)s = 1 - 4t$

Determine whether the following lines intersect and, if so, find the point of intersection.

$$\begin{array}{ll} x = 1 - 3t & x = -1 + s \\ y = 2 + 5t & y = 3 - 4s \\ z = 1 + t & z = 1 - s \end{array}$$

Solution Suppose $\vec{p} = P(x, y, z)$ lies on both lines. Then

$$\begin{bmatrix} 1 - 3t \\ 2 + 5t \\ 1 + t \end{bmatrix} = \begin{bmatrix} x \\ y \\ z \end{bmatrix} = \begin{bmatrix} -1 + s \\ 3 - 4s \\ 1 - s \end{bmatrix} \quad \text{for some } t \text{ and } s,$$

where the first (second) equation is because P lies on the first (second) line. This means that the three equations

$$\begin{array}{rcl} 1 - 3t & = & -1 + s \\ 2 + 5t & = & 3 - 4s \\ 1 + t & = & 1 - s \end{array}$$

must have a solution. In this case, $t = 1$ and $s = -1$ satisfy all three equations, so the lines *do* intersect and the point of intersection is

(handwritten: $x = 2$, $y = 5$)

$$\vec{p} = [1 - 3t \ \ 2 + 5t \ \ 1 + t]^T = [-2 \ 7 \ 2]^T$$

using $t = 1$. Of course, this point can also be found from

$$\vec{p} = [-1 + s \ \ 3 - 4s \ \ 1 - s]^T \text{ using } s = -1.$$

Example 11

Show that the line through $P_0(x_0, y_0)$ with slope m has direction vector $\vec{d} = [1 \ m]^T$ and equation $y - y_0 = m(x - x_0)$. This equation is called the point-slope formula.

Solution Let $P_1(x_1, y_1)$ be the point on the line one unit to the right of P_0—so that $x_1 = x_0 + 1$. Then $\vec{d} = \overrightarrow{P_0P_1}$ serves as direction vector of the line, and $\vec{d} = [x_1 - x_0 \ \ y_1 - y_0]^T = [1 \ \ y_1 - y_0]^T$. But the slope m can be computed as follows:

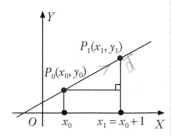

$$m = \frac{y_1 - y_0}{x_1 - x_0} = \frac{y_1 - y_0}{1} = y_1 - y_0$$

Hence $\vec{d} = [1 \ m]^T$ and the parametric equations are $x = x_0 + t$, $y = y_0 + mt$. Eliminating t gives $y - y_0 = mt = m(x - x_0)$, as asserted.

Note that the vertical line through $P_0(x_0, y_0)$ has a direction vector $\vec{d} = [0 \ 1]^T$ that is not of the form $[1 \ m]^T$ for any m. This result confirms that the notion of slope makes no sense in this case. However, the vector method gives parametric equations for the line:

$$\begin{array}{rcl} x & = & x_0 \\ y & = & y_0 + t \end{array}$$

Because y is arbitrary here (t is arbitrary), this is usually written simply as $x = x_0$.

Pythagoras' Theorem

The pythagorean theorem was known earlier, but Pythagoras (c. 550 B.C.) is credited with giving the first rigorous, logical, deductive proof of the result. The proof we give depends on the basic properties of similar triangles.

Theorem 5 Pythagoras' Theorem

Given a right-angled triangle with hypotenuse c and sides a and b, then $a^2 + b^2 = c^2$.

PROOF

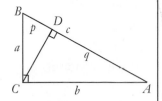

Let A, B, and C be the vertices of the triangle as in Figure 4.20. Draw a perpendicular from C to the point D on the hypotenuse, and let p and q be the lengths of BD and DA respectively. Then CBA and DBC are similar triangles so $\frac{p}{a} = \frac{a}{c}$. This means $a^2 = pc$. In the same way, the similarity of CBA and DCA gives $\frac{q}{b} = \frac{b}{c}$, whence $b^2 = qc$. But then

$$a^2 + b^2 = pc + qc = (p + q)c = c^2$$

because $p + q = c$. This proves Pythagoras' theorem.

Figure 4.20

Exercises 4.1

1. Compute $\|\vec{v}\|$ if $\vec{v}$ equals:
   (a) $[2 \ {-1} \ 2]^T$ ◆(b) $[1 \ {-1} \ 2]^T$
   (c) $[1 \ 0 \ {-1}]^T$ ◆(d) $[{-1} \ 0 \ 2]^T$
   (e) $2[1 \ {-1} \ 2]^T$ ◆(f) $-3[1 \ 1 \ 2]^T$

2. Find a unit vector in the direction of:
   (a) $[7 \ {-1} \ 5]^T$ ◆(b) $[{-2} \ {-1} \ 2]^T$

3. (a) Find a unit vector in the direction from $[3 \ {-1} \ 4]^T$ to $[1 \ 3 \ 5]^T$.
   (b) If $\vec{u} \neq \vec{0}$, for which value of a is $a\vec{u}$ a unit vector?

4. Find the distance between the following pairs of points.
   (a) $[3 \ {-1} \ 0]^T$ and $[2 \ {-1} \ 1]^T$
   ◆(b) $[2 \ {-1} \ 2]$ and $[2 \ 0 \ 1]^T$
   (c) $[{-3} \ 5 \ 2]$ and $[1 \ 3 \ 3]^T$
   ◆(d) $[4 \ 0 \ {-2}]$ and $[3 \ 2 \ 0]^T$

5. Use vectors to show that the line joining the midpoints of two sides of a triangle is parallel to the third side and half as long.

6. Let A, B, and C denote the three vertices of a triangle.
   (a) If E is the midpoint of side BC, show that
   $$\overrightarrow{AE} = \tfrac{1}{2}(\overrightarrow{AB} + \overrightarrow{AC}).$$
   ◆(b) If F is the midpoint of side AC, show that
   $$\overrightarrow{FE} = \tfrac{1}{2}\overrightarrow{AB}.$$

7. Determine whether $\vec{u}$ and $\vec{v}$ are parallel in each of the following cases.
   (a) $\vec{u} = [{-3} \ {-6} \ 3]^T$; $\vec{v} = [5 \ 10 \ {-5}]^T$
   ◆(b) $\vec{u} = [3 \ {-6} \ 3]^T$; $\vec{v} = [{-1} \ 2 \ {-1}]^T$
   (c) $\vec{u} = [1 \ 0 \ 1]^T$; $\vec{v} = [{-1} \ 0 \ 1]^T$
   ◆(d) $\vec{u} = [2 \ 0 \ {-1}]^T$; $\vec{v} = [{-8} \ 0 \ 4]^T$

8. Let $\vec{u}$ and $\vec{v}$ be the position vectors of points P and Q, respectively, and let R be the point whose position vector is $\vec{u} + \vec{v}$. Express the following in terms of $\vec{u}$ and $\vec{v}$.
   (a) $\overrightarrow{QP}$ ◆(b) $\overrightarrow{QR}$
   (c) $\overrightarrow{RP}$ ◆(d) $\overrightarrow{RO}$ where O is the origin

9. In each case, find $\overrightarrow{PQ}$ and $\|\overrightarrow{PQ}\|$.

  (a) $P(1, -1, 3)$, $Q(3, 1, 0)$

  ◆(b) $P(2, 0, 1)$, $Q(1, -1, 6)$

  (c) $P(1, 0, 1)$, $Q(1, 0, -3)$

  ◆(d) $P(1, -1, 2)$, $Q(1, -1, 2)$

  (e) $P(1, 0, -3)$, $Q(-1, 0, 3)$

  ◆(f) $P(3, -1, 6)$, $Q(1, 1, 4)$

10. In each case, find a point Q such that $\overrightarrow{PQ}$ has (i) the same direction as $\vec{v}$; (ii) the opposite direction to $\vec{v}$.

  (a) $P(-1, 2, 2)$, $\vec{v} = [1\ 3\ 1]^T$

  ◆(b) $P(3, 0, -1)$, $\vec{v} = [2\ -1\ 3]^T$

11. Let $\vec{u} = [3\ -1\ 0]^T$, $\vec{v} = [4\ 0\ 1]^T$, and $\vec{w} = [-1\ 1\ 5]^T$. In each case, find $\vec{x}$ such that:

  (a) $3(2\vec{u} + \vec{x}) + \vec{w} = 2\vec{x} - \vec{v}$

  ◆(b) $2(3\vec{v} - \vec{x}) = 5\vec{w} + \vec{u} - 3\vec{x}$

12. Let $\vec{u} = [1\ 1\ 2]^T$, $\vec{v} = [0\ 1\ 2]^T$, and $\vec{w} = [1\ 0\ -1]^T$. In each case, find numbers a, b, and c such that $\vec{x} = a\vec{u} + b\vec{v} + c\vec{w}$.

  (a) $\vec{x} = [2\ -1\ 6]^T$ ◆(b) $\vec{x} = [1\ 3\ 0]^T$

13. Let $\vec{u} = [3\ -1\ 0]^T$, $\vec{v} = [4\ 0\ 1]^T$, and $\vec{z} = [1\ 1\ 1]^T$. In each case, show that there are no numbers a, b, and c such that:

  (a) $a\vec{u} + b\vec{v} + c\vec{z} = [1\ 2\ 1]^T$

  (b) $a\vec{u} + b\vec{v} + c\vec{z} = [5\ 6\ -1]^T$

14. Let $P_1 = P_1(2, 1, -2)$ and $P_2 = P_2(1, -2, 0)$. Find the coordinates of the point P:

  (a) $\frac{1}{5}$ the way from P_1 to P_2

  ◆(b) $\frac{1}{4}$ the way from P_2 to P_1

15. Find the two points trisecting the segment between $P(2, 3, 5)$ and $Q(8, -6, 2)$.

16. Let $P_1 = P_1(x_1, y_1, z_1)$ and $P_2 = P_2(x_2, y_2, z_2)$ be two points with position vectors $\vec{p}_1$ and $\vec{p}_2$, respectively. If r and s are positive integers, show that the point P lying $\frac{r}{r+s}$ the way from P_1 to P_2 has position vector

$$\vec{p} = \left(\frac{s}{r+s}\right)\vec{p}_1 + \left(\frac{r}{r+s}\right)\vec{p}_2.$$

17. In each case, find the point Q:

  (a) $\overrightarrow{PQ} = [2\ 0\ -3]^T$ and $P = P(2, -3, 1)$

  ◆(b) $\overrightarrow{PQ} = [-1\ 4\ 7]^T$ and $P = P(1, 3, -4)$

18. Let $\vec{u} = [2\ 0\ -4]^T$ and $\vec{v} = [2\ 1\ -2]^T$. In each case find $\vec{x}$:

  (a) $2\vec{u} - \|\vec{v}\|\vec{v} = \frac{3}{2}(\vec{u} - 2\vec{x})$

  ◆(b) $3\vec{u} + 7\vec{v} = \|\vec{u}\|^2(2\vec{x} + \vec{v})$

19. Find all vectors $\vec{u}$ that are parallel to $\vec{v} = [3\ -2\ 1]^T$ and satisfy $\|\vec{u}\| = 3\|\vec{v}\|$.

20. Let P, Q, and R be the vertices of a parallelogram with adjacent sides PQ and PR. In each case, find the other vertex S.

  (a) $P(3, -1, -1)$, $Q(1, -2, 0)$, $R(1, -1, 2)$

  ◆(b) $P(2, 0, -1)$, $Q(-2, 4, 1)$, $R(3, -1, 0)$

21. In each case either prove the statement or give an example showing that it is false.

  (a) The zero vector $\vec{0}$ is the only vector of length 0.

  ◆(b) If $\|\vec{v} - \vec{w}\| = 0$, then $\vec{v} = \vec{w}$.

  (c) If $\vec{v} = -\vec{v}$, then $\vec{v} = \vec{0}$.

  ◆(d) If $\|\vec{v}\| = \|\vec{w}\|$, then $\vec{v} = \vec{w}$.

  (e) If $\|\vec{v}\| = \|\vec{w}\|$, then $\vec{v} = \pm\vec{w}$.

  ◆(f) If $\vec{v} = t\vec{w}$ for some scalar t, then $\vec{v}$ and $\vec{w}$ have the same direction.

  (g) If $\vec{v}$ and $\vec{v} + \vec{w}$ are parallel, then $\vec{v}$ and $\vec{w}$ are parallel.

  ◆(h) $\|-5\vec{v}\| = -5\|\vec{v}\|$ for all $\vec{v}$.

  (i) If $\|\vec{v}\| = \|2\vec{v}\|$, then $\vec{v} = \vec{0}$.

  ◆(j) $\|\vec{v} + \vec{w}\| = \|\vec{v}\| + \|\vec{w}\|$ for all $\vec{v}$ and $\vec{w}$.

22. Find the vector and parametric equations of the following lines.

  (a) The line parallel to $[2\ -1\ 0]^T$ and passing through $P(1, -1, 3)$.

  ◆(b) The line passing through $P(3, -1, 4)$ and $Q(1, 0, -1)$.

  (c) The line passing through $P(3, -1, 4)$ and $Q(3, -1, 5)$.

  ◆(d) The line parallel to $[1\ 1\ 1]^T$ and passing through $P(1, 1, 1)$.

  (e) The line passing through $P(1, 0, -3)$ and parallel to the line with parametric equations $x = -1 + 2t$, $y = 2 - t$, and $z = 3 + 3t$.

  ◆(f) The line passing through $P(2, -1, 1)$ and parallel to the line with parametric equations $x = 2 - t$, $y = 1$, and $z = t$.

(g) The lines through $P(1, 0, 1)$ that meet the line with vector equation
$$\vec{p} = [1\ 2\ 0]^T + t[2\ -1\ 2]^T$$ at points at distance 3 from $P_0(1, 2, 0)$.

23. In each case, verify that the points P and Q lie on the line.

    (a) $x = 3 - 4t$ $P(-1, 3, 0), Q(11, 0, 3)$
        $y = 2 + t$
        $z = 1 - t$

    ◆(b) $x = 4 - t$ $P(2, 3, -3), Q(-1, 3, -9)$
        $y = 3$
        $z = 1 - 2t$

24. Find the point of intersection (if any) of the following pairs of lines.

    (a) $x = 3 + t$ $x = 4 + 2s$
        $y = 1 - 2t$ $y = 6 + 3s$
        $z = 3 + 3t$ $z = 1 + s$

    ◆(b) $x = 1 - t$ $x = 2s$
        $y = 2 + 2t$ $y = 1 + s$
        $z = -1 + 3t$ $z = 3$

    (c) $[x\ y\ z]^T = [3\ -1\ 2]^T + t[1\ 1\ -1]^T$
        $[x\ y\ z]^T = [1\ 1\ -2]^T + s[2\ 0\ 3]^T$

    ◆(d) $[x\ y\ z]^T = [4\ -1\ 5]^T + t[1\ 0\ 1]^T$
        $[x\ y\ z]^T = [2\ -7\ 12]^T + s[0\ -2\ 3]^T$

25. Show that if a line passes through the origin, the position vectors of points on the line are all scalar multiples of some fixed nonzero vector.

26. Show that every line parallel to the Z axis has parametric equations $x = x_0, y = y_0, z = t$ for some fixed numbers x_0 and y_0.

27. Let $\vec{d} = [a\ b\ c]^T$ be a vector where a, b, and c are *all* nonzero. Show that the equations of the line through $P_0(x_0, y_0, z_0)$ with direction vector $\vec{d}$ can be written in the form
$$\frac{x - x_0}{a} = \frac{y - y_0}{b} = \frac{z - z_0}{c}$$

    This is called the **symmetric form** of the equations.

28. A parallelogram has sides AB, BC, CD, and DA. Given $A(1, -1, 2), C(2, 1, 0)$, and the midpoint $M(1, 0, -3)$ of AB, find $\overrightarrow{BD}$.

◆29. Find all points C on the line through $A(1, -1, 2)$ and $B = (2, 0, 1)$ such that $\|\overrightarrow{AC}\| = 2\|\overrightarrow{BC}\|$.

30. Let A, B, C, D, E, and F be the vertices of a regular hexagon, taken in order. Show that
$$\overrightarrow{AB} + \overrightarrow{AC} + \overrightarrow{AD} + \overrightarrow{AE} + \overrightarrow{AF} = 3\overrightarrow{AD}.$$

31. (a) Let P_1, P_2, P_3, P_4, P_5, and P_6 be six points equally spaced on a circle with centre C. Show that
$$\overrightarrow{CP_1} + \overrightarrow{CP_2} + \overrightarrow{CP_3} + \overrightarrow{CP_4} + \overrightarrow{CP_5} + \overrightarrow{CP_6} = \vec{0}.$$

    ◆(b) Show that the conclusion in part (a) holds for any *even* set of points evenly spaced on the circle.

    (c) Show that the conclusion in part (a) holds for *three* points.

    (d) Do you think it works for *any* finite set of points evenly spaced around the circle?

32. Consider a quadrilateral with vertices A, B, C, and D in order (as shown in the diagram).

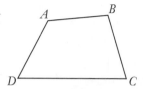

If the diagonals AC and BD bisect each other, show that the quadrilateral is a parallelogram. (This is the converse of Example 2.) [*Hint:* Let E be the intersection of the diagonals. Show that $\overrightarrow{AB} = \overrightarrow{DC}$ by writing $\overrightarrow{AB} = \overrightarrow{AE} + \overrightarrow{EB}$.]

◆33. Consider the parallelogram $ABCD$ (see diagram), and let E be the midpoint of side AD.

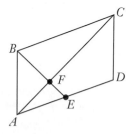

Show that BE and AC trisect each other; that is, show that the intersection point is one-third of the way from E to B and from A to C. [*Hint:* If F is one-third of the way from A to C, show that $\overrightarrow{FB} = 2\overrightarrow{EF}$ and argue as in Example 2.]

34. The line from a vertex of a triangle to the midpoint of the opposite side is called a **median** of the triangle. If the vertices of a triangle have position vectors $\vec{u}, \vec{v}$, and $\vec{w}$, show that the point

on each median that is $\frac{1}{3}$ the way from the midpoint to the vertex has position vector $\frac{1}{3}(\vec{u} + \vec{v} + \vec{w})$. Conclude that the point C with position vector $\frac{1}{3}(\vec{u} + \vec{v} + \vec{w})$ lies on all three medians. This point C is called the **centroid** of the triangle.

35. Given four noncoplanar points in space, the figure with these points as vertices is called a **tetrahedron**. The line from a vertex through the centroid (see previous exercise) of the triangle formed by the remaining vertices is called a **median** of the tetrahedron. If $\vec{u}, \vec{v}, \vec{w}$, and $\vec{x}$ are the position vectors of the four vertices, show that the point on a median one-fourth the way from the centroid to the vertex has position vector $\frac{1}{4}(\vec{u} + \vec{v} + \vec{w} + \vec{x})$. Conclude that the four medians are concurrent.

SECTION 4.2   Projections and Planes

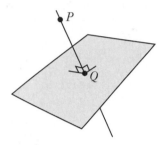

Figure 4.21

Any student of geometry soon realizes that the notion of perpendicular lines is fundamental. As an illustration, suppose a point P and a plane are given and it is desired to find the point Q that lies in the plane and is closest to P, as shown in Figure 4.21. Clearly, what is required is to find the line through P that is perpendicular to the plane and then to obtain Q as the point of intersection of this line with the plane. Finding the line *perpendicular* to the plane requires a way to determine when two vectors are perpendicular. Surprisingly, this can be done using the idea of the dot product of two vectors.

The Dot Product and Angles

Given vectors $\vec{v} = \begin{bmatrix} x_1 \\ y_1 \\ z_1 \end{bmatrix}$ and $\vec{w} = \begin{bmatrix} x_2 \\ y_2 \\ z_2 \end{bmatrix}$, their **dot product** $\vec{v} \cdot \vec{w}$ is a number defined by

$$\vec{v} \cdot \vec{w} = x_1 x_2 + y_1 y_2 + z_1 z_2 = \vec{v}^T \vec{w}$$

Because $\vec{v} \cdot \vec{w}$ is a number, it is sometimes called the **scalar product** of $\vec{v}$ and $\vec{w}$.[6]

Example 1

If $\vec{v} = [2\ -1\ 3]^T$ and $\vec{w} = [1\ 4\ -1]^T$, then $\vec{v} \cdot \vec{w} = 2 \cdot 1 + (-1) \cdot 4 + 3 \cdot (-1) = -5$.

The next theorem lists several basic properties of the dot product.

Theorem 1

Let $\vec{u}, \vec{v}$, and $\vec{w}$ denote vectors in $\mathbb{R}^3$ (or $\mathbb{R}^2$).
1. $\vec{v} \cdot \vec{w}$ is a real number.
2. $\vec{v} \cdot \vec{w} = \vec{w} \cdot \vec{v}$.
3. $\vec{v} \cdot \vec{0} = 0 = \vec{0} \cdot \vec{v}$.
4. $\vec{v} \cdot \vec{v} = \|\vec{v}\|^2$.
5. $(k\vec{v}) \cdot \vec{w} = k(\vec{v} \cdot \vec{w}) = \vec{v} \cdot (k\vec{w})$ for all scalars k.
6. $\vec{u} \cdot (\vec{v} \pm \vec{w}) = \vec{u} \cdot \vec{v} \pm \vec{u} \cdot \vec{w}$.

---

6 Similarly, if $\vec{v} = \begin{bmatrix} x_1 \\ y_1 \end{bmatrix}$ and $\vec{w} = \begin{bmatrix} x_2 \\ y_2 \end{bmatrix}$ in $\mathbb{R}^2$, then $\vec{v} \cdot \vec{w} = x_1 y_1 + x_2 y_2$.

PROOF

(1), (2), and (3) are easily verified, and (4) comes from Theorem 1 §4.1. The rest are properties of matrix arithmetic (because $\vec{v} \cdot \vec{w} = \vec{v}^T\vec{w}$), and are left to the reader.

The properties in Theorem 1 enable us to do calculations like

$$3\vec{u} \cdot (2\vec{v} - 3\vec{w} + 4\vec{z}) = 6\vec{u} \cdot \vec{v} - 9\vec{u} \cdot \vec{w} + 12\vec{u} \cdot \vec{z}$$

and such computations will be used without comment below. Here is an example.

Example 2

Verify that $\|\vec{v} - 3\vec{w}\|^2 = 1$ if $\|\vec{v}\| = 2$, $\|\vec{w}\| = 1$ and $\vec{v} \cdot \vec{w} = 2$.

Solution

We apply Theorem 1 several times:

$$\begin{aligned}
\|\vec{v} - 3\vec{w}\|^2 &= (\vec{v} - 3\vec{w}) \cdot (\vec{v} - 3\vec{w}) \\
&= \vec{v} \cdot \vec{v} - 3(\vec{v} \cdot \vec{w}) - 3(\vec{w} \cdot \vec{v}) + 9(\vec{w} \cdot \vec{w}) \\
&= \|\vec{v}\|^2 - 6(\vec{v} \cdot \vec{w}) + 9\|\vec{w}\|^2 \\
&= 4 - 12 + 9 = 1.
\end{aligned}$$

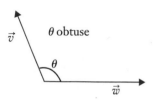

Now let $\vec{v}$ and $\vec{w}$ be nonzero vectors positioned with a common tail as in Figure 4.22. Then they determine a unique angle θ in the range

$$0 \leq \theta \leq \pi$$

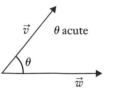

This angle θ will be called the **angle between** $\vec{v}$ and $\vec{w}$. Figure 4.22 illustrates when θ is acute (less than $\frac{\pi}{2}$) and obtuse (greater than $\frac{\pi}{2}$). Clearly $\vec{v}$ and $\vec{w}$ are parallel if θ is either 0 or π, and we say that they are **orthogonal** if $\theta = \frac{\pi}{2}$. Note that we do not define the angle between $\vec{v}$ and $\vec{w}$ if one of these vectors is $\vec{0}$.

The next result gives an easy way to compute the angle between two nonzero vectors using the dot product.

Figure 4.22

Theorem 2

Let $\vec{v}$ and $\vec{w}$ be nonzero vectors. If θ is the angle between $\vec{v}$ and $\vec{w}$, then

$$\vec{v} \cdot \vec{w} = \|\vec{v}\|\|\vec{w}\| \cos \theta.$$

PROOF

We calculate $\|\vec{v} - \vec{w}\|^2$ two ways. First apply the law of cosines (Section 4.1) to the triangle in Figure 4.23 to obtain:

$$\|\vec{v} - \vec{w}\|^2 = \|\vec{v}\|^2 + \|\vec{w}\|^2 - 2\|\vec{v}\|\|\vec{w}\| \cos \theta.$$

On the other hand, we use Theorem 1:

$$\begin{aligned}
\|\vec{v} - \vec{w}\|^2 &= (\vec{v} - \vec{w}) \cdot (\vec{v} - \vec{w}) \\
&= \vec{v} \cdot \vec{v} - \vec{v} \cdot \vec{w} - \vec{w} \cdot \vec{v} + \vec{w} \cdot \vec{w} \\
&= \|\vec{v}\|^2 - 2\vec{v} \cdot \vec{w} + \|\vec{w}\|^2
\end{aligned}$$

Comparing these we see that $-2\|\vec{v}\|\|\vec{w}\| \cos \theta = -2\vec{v} \cdot \vec{w}$, and the result follows.

Figure 4.23

If $\vec{u}$ and $\vec{v}$ are nonzero vectors, Theorem 2 gives an intrinsic description of $\vec{v} \cdot \vec{w}$ because $\|\vec{v}\|, \|\vec{w}\|$, and the angle θ between $\vec{v}$ and $\vec{w}$ do not depend on the choice of coordinate system. Moreover, since $\|\vec{v}\|$ and $\|\vec{w}\|$ are nonzero (since $\vec{v}$ and $\vec{w}$ are nonzero vectors), it gives a formula for the angle θ:

$$\cos \theta = \frac{\vec{v} \cdot \vec{w}}{\|\vec{v}\|\|\vec{w}\|} \tag{*}$$

This can be used to find θ. In this connection, it is worth noting that $\cos \theta$ has the same sign as $\vec{u} \cdot \vec{v}$, so

$$\vec{u} \cdot \vec{v} > 0 \quad \text{if and only if} \quad \theta \text{ is acute } (0 \le \theta < \tfrac{\pi}{2})$$
$$\vec{u} \cdot \vec{v} < 0 \quad \text{if and only if} \quad \theta \text{ is obtuse } (\tfrac{\pi}{2} < \theta \le \pi)$$
$$\vec{u} \cdot \vec{v} = 0 \quad \text{if and only if} \quad \theta = \tfrac{\pi}{2}$$

In this last case, the (nonzero) vectors are perpendicular. The following terminology is used in linear algebra: Two vectors $\vec{u}$ and $\vec{v}$ are said to be **orthogonal** if $\vec{u} = \vec{0}$ or $\vec{v} = \vec{0}$ or the angle between them is $\tfrac{\pi}{2}$. Since $\vec{u} \cdot \vec{v} = 0$ if either $\vec{u} = \vec{0}$ or $\vec{v} = \vec{0}$, we have the following theorem:

Theorem 3

Two vectors $\vec{u}$ and $\vec{v}$ are orthogonal if and only if $\vec{u} \cdot \vec{v} = 0$.

Example 3

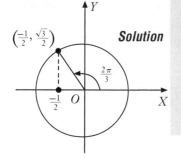

Compute the angle between $\vec{u} = [-1 \ 1 \ 2]^T$ and $\vec{v} = [2 \ 1 \ -1]^T$.

Solution Compute $\cos \theta = \frac{\vec{u} \cdot \vec{v}}{\|\vec{u}\|\|\vec{v}\|} = \frac{-2+1-2}{\sqrt{6}\sqrt{6}} = \frac{-1}{2}$. Now recall that $\cos \theta$ and $\sin \theta$ are defined so that $(\cos \theta, \sin \theta)$ is the point on the unit circle determined by the angle θ (drawn counterclockwise, starting from the positive X axis). In the present case, we know that $\cos \theta = -\tfrac{1}{2}$ and that $0 \le \theta \le \pi$. Because $\cos \tfrac{\pi}{3} = \tfrac{1}{2}$, it follows that $\theta = \tfrac{2\pi}{3}$ (see the diagram).

Example 4

Show that the points $P(3, -1, 1)$, $Q(4, 1, 4)$, and $R(6, 0, 4)$ are the vertices of a right triangle.

Solution The vectors along the sides of the triangle are

$$\overrightarrow{PQ} = [1 \ 2 \ 3]^T, \quad \overrightarrow{PR} = [3 \ 1 \ 3]^T, \quad \text{and} \quad \overrightarrow{QR} = [2 \ -1 \ 0]^T$$

Evidently $\overrightarrow{PQ} \cdot \overrightarrow{QR} = [1 \ 2 \ 3]^T \cdot [2 \ -1 \ 0]^T = 2 - 2 + 0 = 0$, so $\overrightarrow{PQ}$ and $\overrightarrow{QR}$ are orthogonal vectors. This means sides PQ and QR are perpendicular—that is, the angle at Q is a right angle.

Examples 5 and 6 demonstrate how the dot product can be used to verify geometrical theorems involving perpendicular lines.

Example 5

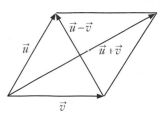

A parallelogram with sides of equal length is called a **rhombus**. Show that the diagonals of a rhombus are perpendicular.

Solution   Let $\vec{u}$ and $\vec{v}$ denote vectors along two adjacent sides of a rhombus, as shown in the diagram. Then the diagonals are $\vec{u} - \vec{v}$ and $\vec{u} + \vec{v}$, and we compute

$$(\vec{u} - \vec{v}) \cdot (\vec{u} + \vec{v}) = \vec{u} \cdot (\vec{u} + \vec{v}) - \vec{v} \cdot (\vec{u} + \vec{v})$$
$$= \vec{u} \cdot \vec{u} + \vec{u} \cdot \vec{v} - \vec{v} \cdot \vec{u} - \vec{v} \cdot \vec{v}$$
$$= \|\vec{u}\|^2 - \|\vec{v}\|^2$$
$$= 0$$

because $\|\vec{u}\| = \|\vec{v}\|$ (it is a rhombus). Hence $\vec{u} - \vec{v}$ and $\vec{u} + \vec{v}$ are orthogonal.

Example 6

The line through a vertex of a triangle, perpendicular to the opposite side, is called an **altitude** of the triangle. Show that the three altitudes of any triangle are concurrent.

Solution   Let the vertices be A, B, and C, and let P be the point of intersection of the altitudes through A and B, as in the diagram. We must show that $\overrightarrow{PC}$ is orthogonal to $\overrightarrow{AB}$; that is, $\overrightarrow{PC} \cdot \overrightarrow{AB} = 0$.
We have $\overrightarrow{AB} = \overrightarrow{AC} - \overrightarrow{BC}$, so

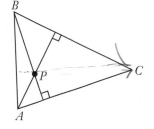

$$\overrightarrow{PC} \cdot \overrightarrow{AB} = \overrightarrow{PC} \cdot \overrightarrow{AC} - \overrightarrow{PC} \cdot \overrightarrow{BC}$$
$$= (\overrightarrow{PB} + \overrightarrow{BC}) \cdot \overrightarrow{AC} - (\overrightarrow{PA} + \overrightarrow{AC}) \cdot \overrightarrow{BC}$$
$$= \overrightarrow{PB} \cdot \overrightarrow{AC} + \overrightarrow{BC} \cdot \overrightarrow{AC} - \overrightarrow{PA} \cdot \overrightarrow{BC} - \overrightarrow{AC} \cdot \overrightarrow{BC}$$
$$= 0 + \overrightarrow{BC} \cdot \overrightarrow{AC} - 0 - \overrightarrow{AC} \cdot \overrightarrow{BC}$$
$$= 0$$

Projections

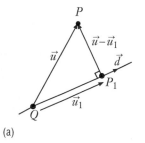

If a nonzero vector $\vec{d}$ is given, it is often useful to be able to write an arbitrary vector $\vec{u}$ as a sum of two vectors,

$$\vec{u} = \vec{u}_1 + \vec{u}_2$$

where $\vec{u}_1$ is parallel to $\vec{d}$ and $\vec{u}_2 = \vec{u} - \vec{u}_1$ is orthogonal to $\vec{d}$. Suppose that $\vec{u}$ and $\vec{d} \neq \vec{0}$ emanate from a common tail Q (see Figure 4.24). Let P be the tip of $\vec{u}$, and let P_1 denote the foot of the perpendicular from P to the line through Q parallel to $\vec{d}$. Then $\vec{u}_1 = \overrightarrow{QP_1}$ has the required properties:

1. $\vec{u}_1$ is parallel to $\vec{d}$. $\vec{u}_1 = t\vec{d}$
2. $\vec{u}_2 = \vec{u} - \vec{u}_1$ is orthogonal to $\vec{d}$. $(\vec{u} - \vec{u}_1) \cdot \vec{d} = 0$
3. $\vec{u} = \vec{u}_1 + \vec{u}_2$.

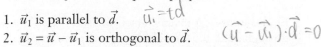

The vector $\vec{u}_1 = \overrightarrow{QP_1}$ is called **the projection of $\vec{u}$ on $\vec{d}$**. It is denoted

$$\vec{u}_1 = \mathrm{proj}_{\vec{d}}\, \vec{u}$$

(a)

(b)

Figure 4.24

In Figure 4.24(a) the vector $\vec{u}_1 = \text{proj}_{\vec{d}}\,\vec{u}$ has the same direction as $\vec{d}$; however, it has the opposite direction from $\vec{d}$ if the angle between $\vec{u}$ and $\vec{d}$ is greater than $\frac{\pi}{2}$ (Figure 4.24(b)). Note that the projection $\vec{u}_1 = \text{proj}_{\vec{d}}\,\vec{u}$ is zero if and only if $\vec{u}$ and $\vec{d}$ are orthogonal.

Calculating the projection of $\vec{u}$ on $\vec{d} \neq \vec{0}$ is remarkably easy.

Theorem 4

Let $\vec{u}$ and $\vec{d} \neq \vec{0}$ be vectors.

1. The projection $\vec{u}_1$ of $\vec{u}$ on $\vec{d}$ is given by $\text{proj}_{\vec{d}}\,\vec{u} = \dfrac{\vec{u} \bullet \vec{d}}{\|\vec{d}\|^2}\,\vec{d}$.

2. The vector $\vec{u} - \text{proj}_{\vec{d}}\,\vec{u}$ is orthogonal to $\vec{d}$.

PROOF

The vector $\vec{u}_1 = \text{proj}_{\vec{d}}\,\vec{u}$ is parallel to $\vec{d}$ and so has the form $\vec{u}_1 = t\vec{d}$ for some scalar t. The requirement that $\vec{u} - \vec{u}_1$ and $\vec{d}$ are orthogonal determines t. In fact, it means that $(\vec{u} - \vec{u}_1) \bullet \vec{d} = 0$ by Theorem 3; and if $\vec{u}_1 = t\vec{d}$ is substituted here, the condition is

$$0 = (\vec{u} - t\vec{d}) \bullet \vec{d} = \vec{u} \bullet \vec{d} - t(\vec{d} \bullet \vec{d}) = \vec{u} \bullet \vec{d} - t\|\vec{d}\|^2$$

It follows that $t = \dfrac{\vec{u} \bullet \vec{d}}{\|\vec{d}\|^2}$, where the assumption that $\vec{d} \neq \vec{0}$ guarantees that $\|\vec{d}\|^2 \neq 0$.

Example 7

Find the projection of $\vec{u} = [2\ -3\ 1]^T$ on $\vec{d} = [1\ -1\ 3]^T$ and express $\vec{u} = \vec{u}_1 + \vec{u}_2$ where $\vec{u}_1$ is parallel to $\vec{d}$ and $\vec{u}_2$ is orthogonal to $\vec{d}$.

Solution   The projection $\vec{u}_1$ of $\vec{u}$ on $\vec{d}$ is

$$\vec{u}_1 = \text{proj}_{\vec{d}}\,\vec{u} = \frac{\vec{u} \bullet \vec{d}}{\|\vec{d}\|^2}\,\vec{d} = \frac{2+3+3}{1^2 + (-1)^2 + 3^2}\,[1\ -1\ 3]^T = \tfrac{8}{11}[1\ -1\ 3]^T$$

The vector $\vec{u}_2 = \vec{u} - \vec{u}_1 = \frac{1}{11}[14\ -25\ -13]^T$ is orthogonal to $\vec{d}$ by Theorem 4 (alternatively, observe that $\vec{d} \bullet \vec{u}_2 = 0$), and $\vec{u} = \vec{u}_1 + \vec{u}_2$ as required.

Example 8

Find the shortest distance (see diagram) from the point $P(1, 3, -2)$ to the line through $P_0(2, 0, -1)$ with direction vector $\vec{d} = [1\ -1\ 0]^T$. Also find the point P_1 that lies on the line and is closest to P.

Solution   Let $\vec{u} = [1\ 3\ -2]^T - [2\ 0\ -1]^T = [-1\ 3\ -1]^T$ denote the vector from P_0 to P, and let $\vec{u}_1$ denote the projection of $\vec{u}$ on $\vec{d}$. Thus

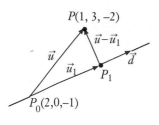

$$\vec{u}_1 = \frac{\vec{u} \cdot \vec{d}}{\|\vec{d}\|^2}\,\vec{d} = \frac{-1-3+0}{1^2+(-1)^2+0^2}\,\vec{d} = -2\vec{d} = [-2 \ \ 2 \ \ 0]^T$$

by Theorem 4. We see geometrically that the point P_1 on the line is closest to P, so the distance is

$$\|\overrightarrow{P_1P}\| = \|\vec{u} - \vec{u}_1\| = \|(1, \ 1, \ -1)\| = \sqrt{3}$$

To find the coordinates of P_1, let $\vec{p}_0$ and $\vec{p}_1$ denote the position vectors of P_0 and P_1, respectively. Then $\vec{p}_0 = [2 \ 0 \ -1]^T$ and $\vec{p}_1 = \vec{p}_0 + \vec{u}_1 = [0 \ 2 \ -1]^T$. Hence $P_1(0, 2, -1)$ is the required point. It can be checked that the distance from P_1 to P is $\sqrt{3}$, as expected.

Planes

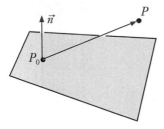

Figure 4.25

It is evident geometrically that among all planes that are perpendicular to a given straight line there is exactly one containing a given point. This fact can be used to give a very simple description of a plane. To do this, it is necessary to introduce the following notion: A nonzero vector $\vec{n}$ is called a **normal** to a plane if it is orthogonal to every vector in the plane. For example, the coordinate vector $\vec{k} = [0 \ 0 \ 1]^T$ is a normal to the X-Y plane.

Given a point $P_0 = P_0(x_0, y_0, z_0)$ and a nonzero vector $\vec{n}$, there is a unique plane through P_0 with normal $\vec{n}$. By Figure 4.25, a point $P = P(x, y, z)$ lies on this plane if and only if the vector $\overrightarrow{P_0P}$ is orthogonal to $\vec{n}$—that is, if and only if $\vec{n} \cdot \overrightarrow{P_0P} = 0$. Because $\overrightarrow{P_0P} = [x - x_0 \ \ y - y_0 \ \ z - z_0]^T$, this gives the following:

Scalar Equation of a Plane

The plane through $P_0(x_0, y_0, z_0)$ with normal $\vec{n} = [a \ b \ c]^T \neq \vec{0}$ is given by

$$a(x - x_0) + b(y - y_0) + c(z - z_0) = 0$$

In other words, the point $P(x, y, z)$ is on this plane if and only if x, y, and z satisfy this equation.

Example 9

Find an equation of the plane through $P_0(1, -1, 3)$ with normal $\vec{n} = [3 \ -1 \ 2]^T$.

Solution   Here the general scalar equation becomes

$$3(x - 1) - (y + 1) + 2(z - 3) = 0$$

This simplifies to $3x - y + 2z = 10$.

If we write $d = ax_0 + by_0 + cz_0$, the scalar equation shows that every plane with normal $\vec{n} = [a \ b \ c]^T$ has a linear equation of the form

$$ax + by + cz = d \qquad\qquad (*)$$

for some constant d. Conversely, the graph of this equation is a plane with $\vec{n} = [a \ b \ c]^T$ as a normal vector (assuming that a, b, and c are not all zero).

Example 10

Find an equation of the plane through $P_0(3, -1, 2)$ that is parallel to the plane with equation $2x - 3y = 6$.

normal vector

Solution   The plane with equation $2x - 3y = 6$ has normal $\vec{n} = [2\ \ -3\ \ 0]^T$. Because the two planes are parallel, $\vec{n}$ serves as a normal to the plane we seek, so the equation is $2x - 3y = d$ for some d by equation (∗). Insisting that $P_0(3, -1, 2)$ lies on the plane determines d: $d = 2 \cdot 3 - 3(-1) = 9$. Hence, the equation is $2x - 3y = 9$.

Consider points $P_0(x_0, y_0, z_0)$ and $P(x, y, z)$ with position vectors $\vec{p}_0 = [x_0\ \ y_0\ \ z_0]^T$ and $\vec{p} = [x\ \ y\ \ z]^T$. Given a nonzero vector $\vec{n}$, the scalar equation of the plane through $P_0(x_0, y_0, z_0)$ with normal $\vec{n} = [a\ \ b\ \ c]^T$ takes the vector form:

Vector Equation of a Plane

The plane with normal $\vec{n} \neq \vec{0}$ through the point with position vector $\vec{p}_0$ is given by

$$\vec{n} \cdot (\vec{p} - \vec{p}_0) = 0$$

In other words, the point with position vector $\vec{p}$ is on the plane if and only if $\vec{p}$ satisfies this condition.

Moreover, equation (∗) translates as follows: *Every plane with normal $\vec{n}$ has vector equation*

$$\vec{n} \cdot \vec{p} = d$$

for some number d. This is useful in the second solution of Example 11.

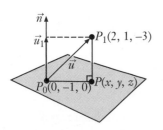

Example 11

Find the shortest distance from the point $P_1(2, 1, -3)$ to the plane with equation $3x - y + 4z = 1$. Also find the point on this plane closest to P_1.

Solution 1   The plane in question has normal $\vec{n} = [3\ \ -1\ \ 4]^T$. Choose any point P_0 on the plane—say $P_0(0, -1, 0)$—and let $P(x, y, z)$ be the point on the plane closest to P_1 (see the diagram). The vector from P_0 to P_1 is $\vec{u} = [2\ \ 2\ \ -3]^T$. Now erect $\vec{n}$ with its tail at P_0. Then $\overrightarrow{PP_1} = \vec{u}_1$ is the projection of $\vec{u}$ on $\vec{n}$:

$$\vec{u}_1 = \frac{\vec{n} \cdot \vec{u}}{\|\vec{n}\|^2}\, \vec{n} = \tfrac{-8}{26} [3\ \ -1\ \ 4]^T = \tfrac{-4}{13} [3\ \ -1\ \ 4]^T$$

Hence the distance is $\|\overrightarrow{PP_1}\| = \|\vec{u}_1\| = \frac{4\sqrt{26}}{13}$. To calculate the point P, let $\vec{p} = [x\ \ y\ \ z]^T$ and $\vec{p}_0 = [0\ \ -1\ \ 0]^T$ be the position vectors of P and P_0. Then

$$\vec{p} = \vec{p}_0 + \vec{u} - \vec{u}_1 = [0\ \ -1\ \ 0]^T + [2\ \ 2\ \ -3]^T + \tfrac{4}{13} [3\ \ -1\ \ 4]^T = \left[\tfrac{38}{13}\ \ \tfrac{9}{13}\ \ \tfrac{-23}{13}\right]^T$$

This gives the coordinates of P.

Solution 2   Let $\vec{p} = [x \ y \ z]^T$ and $\vec{p}_1 = [2 \ 1 \ -3]^T$ be the position vectors of P and P_1. Then P is on the line through P_1 with direction vector $\vec{n}$, so $\vec{p} = \vec{p}_1 + t\vec{n}$ for some value of t. In addition, P lies on the plane, so $\vec{n} \cdot \vec{p} = 3x - y + 4z = 1$. This determines t:

$$1 = \vec{n} \cdot \vec{p} = \vec{n} \cdot (\vec{p}_1 + t\vec{n}) = \vec{n} \cdot \vec{p}_1 + t\|\vec{n}\|^2 = -7 + t(26)$$

This gives $t = \frac{8}{26} = \frac{4}{13}$, so

$$[x \ y \ z]^T = \vec{p} = \vec{p}_1 + t\vec{n} = [2 \ 1 \ -3]^T + \tfrac{4}{13}[3 \ -1 \ 4]^T = \tfrac{1}{13}[38 \ 9 \ -23]^T$$

as before. This determines P (in the diagram), and the reader can verify that the required distance is $\|\overrightarrow{PP_1}\| = \frac{4}{13}\sqrt{26}$, as before.

The Cross Product

If P, Q, and R are three distinct points in $\mathbb{R}^3$ that are not all on some line, it is clear geometrically that there is a unique plane containing all three. The vectors $\overrightarrow{PQ}$ and $\overrightarrow{PR}$ both lie in this plane, so finding a normal amounts to finding a nonzero vector orthogonal to both $\overrightarrow{PQ}$ and $\overrightarrow{PR}$. The cross product provides a systematic way to do this.

Given vectors $\vec{v}_1 = \begin{bmatrix} x_1 \\ y_1 \\ z_1 \end{bmatrix}$ and $\vec{v}_2 = \begin{bmatrix} x_2 \\ y_2 \\ z_2 \end{bmatrix}$, define the **cross product** $\vec{v} \times \vec{w}$ by

$$\vec{v}_1 \times \vec{v}_2 = \begin{bmatrix} y_1 z_2 - z_1 y_2 \\ -(x_1 z_2 - z_1 x_2) \\ x_1 y_2 - y_1 x_2 \end{bmatrix}.$$

There is an easy way to remember this definition using the **coordinate vectors**:

$$\vec{i} = \begin{bmatrix} 1 \\ 0 \\ 0 \end{bmatrix}, \quad \vec{j} = \begin{bmatrix} 0 \\ 1 \\ 0 \end{bmatrix}, \quad \text{and} \quad \vec{k} = \begin{bmatrix} 0 \\ 0 \\ 1 \end{bmatrix}.$$

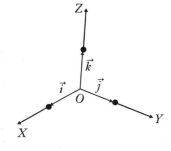

They are vectors of length 1 pointing along the positive X, Y, and Z axes, respectively, as in the diagram. The reason for the name is that any vector $\vec{v} = [x \ y \ z]^T$ can be written as

$$\vec{v} = x\vec{i} + y\vec{j} + z\vec{k}.$$

With this, the cross product can be described as follows:

Determinant Form of the Cross Product

If $\vec{v}_1 = \begin{bmatrix} x_1 \\ y_1 \\ z_1 \end{bmatrix}$ and $\vec{v}_2 = \begin{bmatrix} x_2 \\ y_2 \\ z_2 \end{bmatrix}$ are two vectors, then

$$\vec{v}_1 \times \vec{v}_2 = \det\begin{bmatrix} \vec{i} & x_1 & x_2 \\ \vec{j} & y_1 & y_2 \\ \vec{k} & z_1 & z_2 \end{bmatrix} = \begin{vmatrix} y_1 & y_2 \\ z_1 & z_2 \end{vmatrix}\vec{i} - \begin{vmatrix} x_1 & x_2 \\ z_1 & z_2 \end{vmatrix}\vec{j} + \begin{vmatrix} x_1 & x_2 \\ y_1 & y_2 \end{vmatrix}\vec{k}$$

where the determinant is expanded along the first column.

Example 12

If $\vec{v} = \begin{bmatrix} 2 \\ -1 \\ 4 \end{bmatrix}$ and $\vec{w} = \begin{bmatrix} 1 \\ 3 \\ 7 \end{bmatrix}$, then

$$\vec{v} \times \vec{w} = \det \begin{bmatrix} \vec{i} & 2 & 1 \\ \vec{j} & -1 & 3 \\ \vec{k} & 4 & 7 \end{bmatrix} = \begin{vmatrix} -1 & 3 \\ 4 & 7 \end{vmatrix} \vec{i} - \begin{vmatrix} 2 & 1 \\ 4 & 7 \end{vmatrix} \vec{j} + \begin{vmatrix} 2 & 1 \\ -1 & 3 \end{vmatrix} \vec{k}$$

$$= -19\vec{i} - 10\vec{j} + 7\vec{k}$$

$$= \begin{bmatrix} -19 \\ -10 \\ 7 \end{bmatrix}$$

Observe that $\vec{v} \times \vec{w}$ is orthogonal to both $\vec{v}$ and $\vec{w}$ in Example 12. This holds in general as can be verified directly by computing $\vec{v} \cdot (\vec{v} \times \vec{w})$ and $\vec{w} \cdot (\vec{v} \times \vec{w})$, and is recorded as the first part of the following theorem. It will follow from a more general result which, together with the second part, will be proved in Section 4.3 where a more detailed study of the cross product will be undertaken.

Theorem 5

Let $\vec{v}$ and $\vec{w}$ be vectors in $\mathbb{R}^3$.
1. $\vec{v} \times \vec{w}$ is a vector orthogonal to both $\vec{v}$ and $\vec{w}$.
2. If $\vec{v}$ and $\vec{w}$ are nonzero, then $\vec{v} \times \vec{w} = \vec{0}$ if and only if $\vec{v}$ and $\vec{w}$ are parallel.

It is interesting to contrast Theorem 5(2) with the assertion (in Theorem 3) that

$$\vec{v} \cdot \vec{w} = 0 \quad \text{if and only if} \quad \vec{v} \text{ and } \vec{w} \text{ are orthogonal.}$$

Example 13

Find the equation of the plane through $P(1, 3, -2)$, $Q(1, 1, 5)$, and $R(2, -2, 3)$.

Solution The vectors $\vec{PQ} = [0\ -2\ 7]^T$ and $\vec{PR} = [1\ -5\ 5]^T$ lie in the plane, so

$$\vec{PQ} \times \vec{PR} = \det \begin{bmatrix} \vec{i} & 0 & 1 \\ \vec{j} & -2 & -5 \\ \vec{k} & 7 & 5 \end{bmatrix} = 25\vec{i} + 7\vec{j} + 2\vec{k} = \begin{bmatrix} 25 \\ 7 \\ 2 \end{bmatrix}$$

is a normal to the plane (being orthogonal to both $\vec{PQ}$ and $\vec{PR}$). Hence the plane has equation

$$25x + 7y + 2z = d \text{ for some number } d.$$

Since $P(1, 3, -2)$ lies in the plane we have $25 \cdot 1 + 7 \cdot 3 + 2(-2) = d$. Hence $d = 42$ and the equation is $25x + 7y + 2z = 42$. Incidentally, the reader can verify that the same equation is obtained if either Q or R is used as the point lying in the plane.

Example 14

Find the shortest distance between the nonparallel lines

$$[x \ y \ z]^T = [1 \ 0 \ -1]^T + t[2 \ 0 \ 1]^T$$
$$[x \ y \ z]^T = [3 \ 1 \ 0]^T + s[1 \ 1 \ -1]^T$$

Then find the points A and B on the lines that are closest together.

Solution Direction vectors for the two lines are $\vec{d}_1 = [2 \ 0 \ 1]^T$ and $\vec{d}_2 = [1 \ 1 \ -1]^T$, so

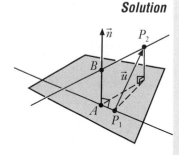

$$\vec{n} = \vec{d}_1 \times \vec{d}_2 = \det \begin{bmatrix} \vec{i} & 2 & 1 \\ \vec{j} & 0 & 1 \\ \vec{k} & 1 & -1 \end{bmatrix} = [-1 \ 3 \ 2]^T$$

is perpendicular to both lines. Consider the plane containing the first line with $\vec{n}$ as normal. This plane contains $P_1(1, 0, -1)$ and is parallel to the second line. Because $P_2(3, 1, 0)$ is on the second line, the distance in question is just the shortest distance between $P_2(3, 1, 0)$ and this plane. The vector $\vec{u}$ from P_1 to P_2 is $\vec{u} = \overrightarrow{P_1P_2} = [2 \ 1 \ 1]^T$ and so, as in Example 11, the distance is the length of the projection of $\vec{u}$ on $\vec{n}$.

$$\text{distance} = \left\| \frac{\vec{u} \cdot \vec{n}}{\|\vec{n}\|^2} \vec{n} \right\| = \frac{|\vec{u} \cdot \vec{n}|}{\|\vec{n}\|} = \frac{3}{\sqrt{14}} = \frac{3\sqrt{14}}{14}$$

Note that it is necessary that $\vec{n} = \vec{d}_1 \times \vec{d}_2$ not be zero for this calculation to be possible. As is shown later (Theorem 4 §4.3), this is guaranteed by the fact that $\vec{d}_1$ and $\vec{d}_2$ are *not* parallel.

The points A and B have coordinates $A(1 + 2t, 0, t - 1)$ and $B(3 + s, 1 + s, -s)$ for some s and t, so $\overrightarrow{AB} = [2 + s - 2t \ 1 + s \ 1 - s - t]^T$. This vector is orthogonal to $\vec{d}_1$ and $\vec{d}_2$, and the conditions $\overrightarrow{AB} \cdot \vec{d}_1 = 0$ and $\overrightarrow{AB} \cdot \vec{d}_2 = 0$ give equations $5t - s = 5$ and $t - 3s = 2$. The solution is $s = \frac{-5}{14}$ and $t = \frac{13}{14}$, so the points are $A(\frac{40}{14}, 0, \frac{-1}{14})$ and $B(\frac{37}{14}, \frac{9}{14}, \frac{5}{14})$. We have $\|\overrightarrow{AB}\| = \frac{3\sqrt{14}}{14}$, as before.

Exercises 4.2

1. Compute $\vec{u} \cdot \vec{v}$ where:

   (a) $\vec{u} = [2 \ -1 \ 3]^T$, $\vec{v} = [-1 \ 1 \ 1]^T$

   ◆(b) $\vec{u} = [1 \ 2 \ -1]^T$, $\vec{v} = \vec{u}$

   (c) $\vec{u} = [1 \ 1 \ -3]^T$, $\vec{v} = 2[2 \ -1 \ 1]^T$

   ◆(d) $\vec{u} = [3 \ -1 \ 5]^T$, $\vec{v} = [6 \ -7 \ -5]^T$

   (e) $\vec{u} = [x \ y \ z]^T$, $\vec{v} = [a \ b \ c]^T$

   ◆(f) $\vec{u} = [a \ b \ c]^T$, $\vec{v} = \vec{0}$

2. Find the angle between the following pairs of vectors.

   (a) $\vec{u} = [1 \ 0 \ 3]^T$, $\vec{v} = [2 \ 0 \ 1]^T$

   ◆(b) $\vec{u} = [3 \ -1 \ 0]^T$, $\vec{v} = [-6 \ 2 \ 0]^T$

   (c) $\vec{u} = [7 \ -1 \ 3]^T$, $\vec{v} = [1 \ 4 \ -1]^T$

   ◆(d) $\vec{u} = [2 \ 1 \ -1]^T$, $\vec{v} = [3 \ 6 \ 3]^T$

   (e) $\vec{u} = [1 \ -1 \ 0]^T$, $\vec{v} = [0 \ 1 \ 1]^T$

   ◆(f) $\vec{u} = [0 \ 3 \ 4]^T$, $\vec{v} = [5\sqrt{2} \ -7 \ -1]^T$

3. Find all real numbers x such that:

   (a) $[2 \ -1 \ 3]^T$ and $[x \ -2 \ 1]^T$ are orthogonal.

   ◆(b) $[2 \ -1 \ 1]^T$ and $[1 \ x \ 2]^T$ are at an angle of $\frac{\pi}{3}$.

4. Find all vectors $\vec{v} = [x \ y \ z]^T$ orthogonal to both:

   (a) $\vec{u}_1 = [-1 \ -3 \ 2]^T$ and $\vec{u}_2 = [0 \ 1 \ 1]^T$

   ◆(b) $\vec{u}_1 = [3 \ -1 \ 2]^T$ and $\vec{u}_2 = [2 \ 0 \ 1]^T$

(c) $\vec{u}_1 = [2 \; 0 \; -1]^T$ and $\vec{u}_2 = [-4 \; 0 \; 2]^T$

◆(d) $\vec{u}_1 = [2 \; -1 \; 3]^T$ and $\vec{u}_2 = [0 \; 0 \; 0]^T$

5. Find two orthogonal vectors $\vec{x}$ and $\vec{y}$ that are both orthogonal to $\vec{v} = [1 \; 2 \; 0]^T$.

6. Consider the triangle with vertices $P(2, 0, -3)$, $Q(5, -2, 1)$, and $R(7, 5, 3)$.

   (a) Show that it is a right-angled triangle.

   ◆(b) Find the lengths of the three sides and verify the pythagorean theorem.

7. Show that the triangle with vertices $A(4, -7, 9)$, $B(6, 4, 4)$, and $C(7, 10, -6)$ is not a right-angled triangle.

8. Find the three internal angles of the triangle with vertices:

   (a) $A(3, 1, -2)$, $B(3, 0, -1)$, and $C(5, 2, -1)$

   ◆(b) $A(3, 1, -2)$, $B(5, 2, -1)$, and $C(4, 3, -3)$

9. Show that the line through $P_0(3, 1, 4)$ and $P_1(2, 1, 3)$ is perpendicular to the line through $P_2(1, -1, 2)$ and $P_3(0, 5, 3)$.

10. In each case, compute the projection of $\vec{u}$ on $\vec{v}$.

    (a) $\vec{u} = [5 \; 7 \; 1]^T$, $\vec{v} = [2 \; -1 \; 3]^T$

    ◆(b) $\vec{u} = [3 \; -2 \; 1]^T$, $\vec{v} = [4 \; 1 \; 1]^T$

    (c) $\vec{u} = [1 \; -1 \; 2]^T$, $\vec{v} = [3 \; -1 \; 1]^T$

    ◆(d) $\vec{u} = [3 \; -2 \; -1]^T$, $\vec{v} = [-6 \; 4 \; 2]^T$

11. In each case, write $\vec{u} = \vec{u}_1 + \vec{u}_2$, where $\vec{u}_1$ is parallel to $\vec{v}$ and $\vec{u}_2$ is orthogonal to $\vec{v}$.

    (a) $\vec{u} = [2 \; -1 \; 1]^T$, $\vec{v} = [1 \; -1 \; 3]^T$

    ◆(b) $\vec{u} = [3 \; 1 \; 0]^T$, $\vec{v} = [-2 \; 1 \; 4]^T$

    (c) $\vec{u} = [2 \; -1 \; 0]^T$, $\vec{v} = [3 \; 1 \; -1]^T$

    ◆(d) $\vec{u} = [3 \; -2 \; 1]^T$, $\vec{v} = [-6 \; 4 \; -1]^T$

12. Calculate the distance from the point P to the line in each case and find the point Q on the line closest to P.

    (a) $P(3, 2, -1)$
        line: $[x \; y \; z]^T = [2 \; 1 \; 3]^T + t[3 \; -1 \; -2]^T$

    ◆(b) $P(1, -1, 3)$
        line: $[x \; y \; z]^T = [1 \; 0 \; -1]^T + t[3 \; 1 \; 4]^T$

13. Compute $\vec{u} \times \vec{v}$ where:

    (a) $\vec{u} = [1 \; 2 \; 3]^T$, $\vec{v} = [1 \; 1 \; 2]^T$

    ◆(b) $\vec{u} = [3 \; -1 \; 0]^T$, $\vec{v} = [-6 \; 2 \; 0]^T$

    (c) $\vec{u} = [3 \; -2 \; 1]^T$, $\vec{v} = [1 \; 1 \; -1]^T$

    ◆(d) $\vec{u} = [2 \; 0 \; -1]^T$, $\vec{v} = [1 \; 4 \; 7]^T$

14. Find the equation of each of the following planes.

    (a) Passing through $A(2, 1, 3)$, $B(3, -1, 5)$, and $C(1, 2, -3)$.

    ◆(b) Passing through $A(1, -1, 6)$, $B(0, 0, 1)$, and $C(4, 7, -11)$.

    (c) Passing through $P(2, -3, 5)$ and parallel to the plane with equation $3x - 2y - z = 0$.

    ◆(d) Passing through $P(3, 0, -1)$ and parallel to the plane with equation $2x - y + z = 3$.

    (e) Containing $P(3, 0, -1)$ and the line $[x \; y \; z]^T = [0 \; 0 \; 2]^T + t[1 \; 0 \; 1]^T$.

    ◆(f) Containing $P(2, 1, 0)$ and the line $[x \; y \; z]^T = [3 \; -1 \; 2]^T + t[1 \; 0 \; -1]^T$.

    (g) Containing the lines $[x \; y \; z]^T = [1 \; -1 \; 2]^T + t[1 \; 1 \; 1]^T$ and $[x \; y \; z]^T = [0 \; 0 \; 2]^T + t[1 \; -1 \; 0]^T$.

    ◆(h) Containing the lines $[x \; y \; z]^T = [3 \; 1 \; 0]^T + t[1 \; -1 \; 3]^T$ and $[x \; y \; z]^T = [0 \; -2 \; 5]^T + t[2 \; 1 \; -1]^T$.

    (i) Each point of which is equidistant from $P(2, -1, 3)$ and $Q(1, 1, -1)$.

    ◆(j) Each point of which is equidistant from $P(0, 1, -1)$ and $Q(2, -1, -3)$.

15. In each case, find the equation of the line.

    (a) Passing through $P(3, -1, 4)$ and perpendicular to the plane $3x - 2y - z = 0$.

    ◆(b) Passing through $P(2, -1, 3)$ and perpendicular to the plane $2x + y = 1$.

    (c) Passing through $P(0, 0, 0)$ and perpendicular to the lines $[x \; y \; z]^T = [1 \; 1 \; 0]^T + t[2 \; 0 \; -1]^T$ and $[x \; y \; z]^T = [2 \; 1 \; -3]^T + t[1 \; -1 \; 5]^T$.

    ◆(d) Passing through $P(1, 1, -1)$, and perpendicular to the lines $[x \; y \; z]^T = [2 \; 0 \; 1]^T + t[1 \; 1 \; -2]^T$ and $[x \; y \; z]^T = [5 \; 5 \; -2]^T + t[1 \; 2 \; -3]^T$.

(e) Passing through $P(2, 1, -1)$, intersecting the line $[x \ y \ z]^T = [1 \ 2 \ -1]^T + t[3 \ 0 \ 1]^T$, and perpendicular to that line.

♦(f) Passing through $P(1, 1, 2)$, intersecting the line $[x \ y \ z]^T = [2 \ 1 \ 0]^T + t[1 \ 1 \ 1]^T$, and perpendicular to that line.

16. In each case, find the shortest distance from the point P to the plane and find the point Q on the plane closest to P.

    (a) $P(2, 3, 0)$; plane with equation $5x + y + z = 1$.

  ♦(b) $P(3, 1, -1)$; plane with equation $2x + y - z = 6$.

17. (a) Does the line through $P(1, 2, -3)$ with direction vector $\vec{d} = [1 \ 2 \ -3]^T$ lie in the plane $2x - y - z = 3$? Explain.

  ♦(b) Does the plane through $P(4, 0, 5)$, $Q(2, 2, 1)$, and $R(1, -1, 2)$ pass through the origin? Explain.

18. Show that every plane containing $P(1, 2, -1)$ and $Q(2, 0, 1)$ must also contain $R(-1, 6, -5)$.

19. Find the equations of the line of intersection of the following planes.

    (a) $2x - 3y + 2z = 5$ and $x + 2y - z = 4$.

  ♦(b) $3x + y - 2z = 1$ and $x + y + z = 5$.

20. In each case, find all points of intersection of the given plane and the line $[x \ y \ z]^T = [1 \ -2 \ 3]^T + t[2 \ 5 \ -1]^T$.

    (a) $x - 3y + 2z = 4$

  ♦(b) $2x - y - z = 5$

    (c) $3x - y + z = 8$

  ♦(d) $-x - 4y - 3z = 6$

21. Find the equation of *all* planes:

    (a) Perpendicular to the line $[x \ y \ z]^T = [2 \ -1 \ 3]^T + t[2 \ 1 \ 3]^T$.

  ♦(b) Perpendicular to the line $[x \ y \ z]^T = [1 \ 0 \ -1]^T + t[3 \ 0 \ 2]^T$.

    (c) Containing the origin.

  ♦(d) Containing $P(3, 2, -4)$.

    (e) Containing $P(1, 1, -1)$ and $Q(0, 1, 1)$.

  ♦(f) Containing $P(2, -1, 1)$ and $Q(1, 0, 0)$.

    (g) Containing the line $[x \ y \ z]^T = [2 \ 1 \ 0]^T + t[1 \ -1 \ 0]^T$.

♦(h) Containing the line $[x \ y \ z]^T = [3 \ 0 \ 2]^T + t[1 \ -2 \ -1]^T$.

22. If a plane contains two distinct points P_1 and P_2, show that it contains every point on the line through P_1 and P_2.

23. Find the shortest distance between the following pairs of parallel lines.

    (a) $[x \ y \ z]^T = [2 \ -1 \ 3]^T + t[1 \ -1 \ 4]^T$
       $[x \ y \ z]^T = [1 \ 0 \ 1]^T + t[1 \ -1 \ 4]^T$

  ♦(b) $[x \ y \ z]^T = [3 \ 0 \ 2]^T + t[3 \ 1 \ 0]^T$
       $[x \ y \ z]^T = [-1 \ 2 \ 2]^T + t[3 \ 1 \ 0]^T$

24. Find the shortest distance between the following pairs of nonparallel lines and find the points on the lines that are closest together.

    (a) $[x \ y \ z]^T = [3 \ 0 \ 1]^T + s[2 \ 1 \ -3]^T$
       $[x \ y \ z]^T = [1 \ 1 \ -1]^T + t[1 \ 0 \ 1]^T$

  ♦(b) $[x \ y \ z]^T = [1 \ -1 \ 0]^T + s[1 \ 1 \ 1]^T$
       $[x \ y \ z]^T = [2 \ -1 \ 3]^T + t[3 \ 1 \ 0]^T$

    (c) $[x \ y \ z]^T = [3 \ 1 \ -1]^T + s[1 \ 1 \ -1]^T$
       $[x \ y \ z]^T = [1 \ 2 \ 0]^T + t[1 \ 0 \ 2]^T$

  ♦(d) $[x \ y \ z]^T = [1 \ 2 \ 3]^T + s[2 \ 0 \ -1]^T$
       $[x \ y \ z]^T = [3 \ -1 \ 0]^T + t[1 \ 1 \ 0]^T$

25. Show that two lines in the plane with slopes m_1 and m_2 are perpendicular if and only if $m_1 m_2 = -1$. [*Hint*: Example 11 §4.1.]

26. (a) Show that, of the four diagonals of a cube, no pair is perpendicular.

    (b) Show that each diagonal is perpendicular to the face diagonals it does not meet.

27. Given a rectangular solid with sides of lengths 1, 1, and $\sqrt{2}$, find the angle between a diagonal and one of the longest sides.

♦28. Consider a rectangular solid with sides of lengths a, b, and c. Show that it has two orthogonal diagonals if and only if the sum of two of a^2, b^2, and c^2 equals the third.

29. Let A, B, and $C(2, -1, 1)$ be the vertices of a triangle where $\overrightarrow{AB}$ is parallel to $[1 \ -1 \ 1]^T$, $\overrightarrow{AC}$ is parallel to $[2 \ 0 \ -1]^T$, and angle $C = 90°$. Find the equation of the line through B and C.

30. If the diagonals of a parallelogram have equal length, show that the parallelogram is a rectangle.

31. Given $\vec{v} = [x\ y\ z]^T$ in component form, show that the projections of $\vec{v}$ on $\vec{i}, \vec{j},$ and $\vec{k}$ are $x\vec{i}, y\vec{j},$ and $z\vec{k}$, respectively.

32. Can $\vec{u} \cdot \vec{v} = -7$ if $\|\vec{u}\| = 3$ and $\|\vec{v}\| = 2$? Defend your answer.

33. Show that $(\vec{u} + \vec{v}) \cdot (\vec{u} - \vec{v}) = \|\vec{u}\|^2 - \|\vec{v}\|^2$ for any vectors $\vec{u}$ and $\vec{v}$.

34. (a) Show that $\|\vec{u} + \vec{v}\|^2 + \|\vec{u} - \vec{v}\|^2 = 2(\|\vec{u}\|^2 + \|\vec{v}\|^2)$ for any vectors $\vec{u}$ and $\vec{v}$.

    ◆ (b) What does this say about parallelograms?

35. Show that if the diagonals of a parallelogram are perpendicular, it is necessarily a rhombus. [*Hint:* Example 5.]

36. Let A and B be the end points of a diameter of a circle (see the diagram). If C is any point on the circle, show that AC and BC are perpendicular. [*Hint:* Express $\overrightarrow{AC}$ and $\overrightarrow{BC}$ in terms of $\vec{u} = \overrightarrow{OA}$ and $\vec{v} = \overrightarrow{OC}$, where O is the centre.]

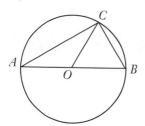

37. If $\vec{u}$ and $\vec{v}$ are orthogonal, show that $\|\vec{u} + \vec{v}\|^2 = \|\vec{u}\|^2 + \|\vec{v}\|^2$.

38. Let $\vec{u}, \vec{v},$ and $\vec{w}$ be pairwise orthogonal vectors.

    (a) Show that $\|\vec{u} + \vec{v} + \vec{w}\|^2 = \|\vec{u}\|^2 + \|\vec{v}\|^2 + \|\vec{w}\|^2$.

    ◆ (b) If $\vec{u}, \vec{v},$ and $\vec{w}$ are all the same length, show that they all make the same angle with $\vec{u} + \vec{v} + \vec{w}$.

39. (a) Show that $\vec{n} = [a\ b]^T$ is perpendicular to the line $ax + by + c = 0$.

    (b) Show that the shortest distance from $P_0(x_0, y_0)$ to the line is $\dfrac{|ax_0 + by_0 + c|}{\sqrt{a^2 + b^2}}$.

    [*Hint:* If P_1 is on the line, project $\vec{u} = \overrightarrow{P_1 P_0}$ on $\vec{n}$.]

40. Assume $\vec{u}$ and $\vec{v}$ are nonzero vectors that are not parallel. Show that $\vec{w} = \|\vec{u}\|\vec{v} + \|\vec{v}\|\vec{u}$ is a nonzero vector that bisects the angle between $\vec{u}$ and $\vec{v}$.

41. Let $\alpha, \beta,$ and γ be the angles a vector $\vec{v} \neq \vec{0}$ makes with the positive $X, Y,$ and Z axes, respectively. Then $\cos\alpha, \cos\beta,$ and $\cos\gamma$ are called the **direction cosines** of the vector $\vec{v}$.

    (a) If $\vec{v} = [a\ b\ c]^T$, show that
    $$\cos\alpha = \frac{a}{\|\vec{v}\|}, \quad \cos\beta = \frac{b}{\|\vec{v}\|}, \quad \text{and} \cos\gamma = \frac{c}{\|\vec{v}\|}.$$

    ◆ (b) Show that $\cos^2\alpha + \cos^2\beta + \cos^2\gamma = 1$.

42. Let $\vec{v} \neq \vec{0}$ be any nonzero vector and suppose that a vector $\vec{u}$ can be written as $\vec{u} = \vec{p} + \vec{q}$, where $\vec{p}$ is parallel to $\vec{v}$ and $\vec{q}$ is orthogonal to $\vec{v}$. Show that $\vec{p}$ is necessarily the projection of $\vec{u}$ on $\vec{v}$. [*Hint:* Argue as in the proof of Theorem 4.]

43. Let $\vec{v} \neq \vec{0}$ be a nonzero vector and let $a \neq 0$ be a scalar. If $\vec{u}$ is any vector, show that the projection of $\vec{u}$ on $\vec{v}$ equals the projection of $\vec{u}$ on $a\vec{v}$.

44. (a) Show that the **Cauchy–Schwarz inequality** $|\vec{u} \cdot \vec{v}| \leq \|\vec{u}\|\|\vec{v}\|$ holds for all vectors $\vec{u}$ and $\vec{v}$. [*Hint:* $|\cos\theta| \leq 1$ for all angles θ.]

    (b) Show that $|\vec{u} \cdot \vec{v}| = \|\vec{u}\|\|\vec{v}\|$ if and only if $\vec{u}$ and $\vec{v}$ are parallel. [*Hint:* When is $\cos\theta = \pm 1$?]

    (c) Show that
    $$|x_1 x_2 + y_1 y_2 + z_1 z_2| \leq \sqrt{x_1^2 + y_1^2 + z_1^2}\,\sqrt{x_2^2 + y_2^2 + z_2^2}$$
    holds for all numbers $x_1, x_2, y_1, y_2, z_1,$ and z_2.

    ◆ (d) Show that $|xy + yz + zx| \leq x^2 + y^2 + z^2$ for all $x, y,$ and z.

    (e) Show that $(x + y + z)^2 \leq 3(x^2 + y^2 + z^2)$ holds for all $x, y,$ and z.

45. Prove that the **triangle inequality** $\|\vec{u} + \vec{v}\| \leq \|\vec{u}\| + \|\vec{v}\|$ holds for all vectors $\vec{u}$ and $\vec{v}$. [*Hint:* Consider the triangle with $\vec{u}$ and $\vec{v}$ as two sides.]

SECTION 4.3   The Cross Product[7]

In Section 4.2 we defined the cross product $\vec{v} \times \vec{w}$ of two $\mathbb{R}^3$-vectors, $\vec{v} = \begin{bmatrix} x_1 \\ y_1 \\ z_1 \end{bmatrix}$ and

$\vec{w} = \begin{bmatrix} x_2 \\ y_2 \\ z_2 \end{bmatrix}$, and observed that it can be best remembered using a determinant:

$$\vec{v} \times \vec{w} = \det \begin{bmatrix} \vec{i} & x_1 & x_2 \\ \vec{j} & y_1 & y_2 \\ \vec{k} & z_1 & z_2 \end{bmatrix} = \begin{vmatrix} y_1 & y_2 \\ z_1 & z_2 \end{vmatrix} \vec{i} - \begin{vmatrix} x_1 & x_2 \\ z_1 & z_2 \end{vmatrix} \vec{j} + \begin{vmatrix} x_1 & x_2 \\ y_1 & y_2 \end{vmatrix} \vec{k} \qquad (*)$$

where $\vec{i} = [1\ 0\ 0]^T$, $\vec{j} = [0\ 1\ 0]^T$, and $\vec{k} = [0\ 0\ 1]^T$ are the coordinate vectors, and the determinant is expanded along the first column. We observed (but did not prove) in Theorem 5 §4.2 that $\vec{v} \times \vec{w}$ is orthogonal to both $\vec{v}$ and $\vec{w}$. This follows easily from the next result.

Theorem 1

If $\vec{u} = \begin{bmatrix} x_0 \\ y_0 \\ z_0 \end{bmatrix}$, $\vec{v} = \begin{bmatrix} x_1 \\ y_1 \\ z_1 \end{bmatrix}$, and $\vec{w} = \begin{bmatrix} x_2 \\ y_2 \\ z_2 \end{bmatrix}$, then $\vec{u} \cdot (\vec{v} \times \vec{w}) = \det \begin{bmatrix} x_0 & x_1 & x_2 \\ y_0 & y_1 & y_2 \\ z_0 & z_1 & z_2 \end{bmatrix}$.

PROOF

Recall that $\vec{u} \cdot (\vec{v} \times \vec{w})$ is computed by multiplying corresponding components of $\vec{u}$ and $\vec{v} \times \vec{w}$ and then adding. Using $(*)$, the result is:

$$\vec{u} \cdot (\vec{v} \times \vec{w}) = x_0 \left(\begin{vmatrix} y_1 & y_2 \\ z_1 & z_2 \end{vmatrix} \right) + y_0 \left(- \begin{vmatrix} x_1 & x_2 \\ z_1 & z_2 \end{vmatrix} \right) + z_0 \left(\begin{vmatrix} x_1 & x_2 \\ y_1 & y_2 \end{vmatrix} \right) = \det \begin{bmatrix} x_0 & x_1 & x_2 \\ y_0 & y_1 & y_2 \\ z_0 & z_1 & z_2 \end{bmatrix}$$

where the last determinant is expanded along column 1.

The result in Theorem 1 can be succinctly stated as follows: If $\vec{u}$, $\vec{v}$, and $\vec{w}$ are three vectors in $\mathbb{R}^3$, then

$$\vec{u} \cdot (\vec{v} \times \vec{w}) = \det[\vec{u}\ \vec{v}\ \vec{w}]$$

where $[\vec{u}\ \vec{v}\ \vec{w}]$ is the matrix with $\vec{u}$, $\vec{v}$, and $\vec{w}$ as its columns. Now it is clear that $\vec{v} \times \vec{w}$ is orthogonal to both $\vec{v}$ and $\vec{w}$ because the determinant of a matrix is zero if two columns are identical.

Because of $(*)$ and Theorem 1, several of the following properties of the cross product follow from properties of determinants (they can also be verified directly).

---

7 Apart from the following section, this material is used very little in the rest of the book.

Theorem 2

Let $\vec{u}, \vec{v}$, and $\vec{w}$ denote arbitrary vectors in $\mathbb{R}^3$.
1. $\vec{u} \times \vec{v}$ is a vector.
2. $\vec{u} \times \vec{v}$ is orthogonal to both $\vec{u}$ and $\vec{v}$.
3. $\vec{u} \times \vec{0} = \vec{0} = \vec{0} \times \vec{u}$.
4. $\vec{u} \times \vec{u} = \vec{0}$.
5. $\vec{u} \times \vec{v} = -(\vec{v} \times \vec{u})$.
6. $(k\vec{u}) \times \vec{v} = k(\vec{u} \times \vec{v}) = \vec{u} \times (k\vec{v})$ for any scalar k.
7. $\vec{u} \times (\vec{v} + \vec{w}) = (\vec{u} \times \vec{v}) + (\vec{u} \times \vec{w})$.
8. $(\vec{v} \times \vec{w}) \times \vec{u} = (\vec{v} \times \vec{u}) + (\vec{w} \times \vec{u})$.

PROOF

(1) is clear; (2) follows from Theorem 1; and (3) and (4) follow because the determinant of a matrix is zero if one column is zero or if two columns are identical. If two columns are interchanged, the determinant changes sign, and this proves (5). The proofs of (6), (7), and (8) are left as Exercise 15.

Joseph Louis Lagrange.
Photo © Corbis.

Theorem 3   Lagrange Identity[8]

If $\vec{u}$ and $\vec{v}$ are any two vectors in $\mathbb{R}^3$, then

$$\|\vec{u} \times \vec{v}\|^2 = \|\vec{u}\|^2 \|\vec{v}\|^2 - (\vec{u} \cdot \vec{v})^2$$

PROOF

Given $\vec{u}$ and $\vec{v}$, introduce a coordinate system and write $\vec{u} = [x_1 \ y_1 \ z_1]^T$ and $\vec{v} = [x_2 \ y_2 \ z_2]^T$ in component form. Then all the terms in the identity can be computed in terms of the components. The detailed proof is left as Exercise 14.

An expression for the magnitude of the vector $\vec{u} \times \vec{v}$ can be easily obtained from the Lagrange identity. If θ is the angle between $\vec{u}$ and $\vec{v}$, substituting $\vec{u} \cdot \vec{v} = \|\vec{u}\| \ \|\vec{v}\| \cos \theta$ into the Lagrange identity gives

$$\|\vec{u} \times \vec{v}\|^2 = \|\vec{u}\|^2 \|\vec{v}\|^2 - \|\vec{u}\|^2 \|\vec{v}\|^2 \cos^2 \theta = \|\vec{u}\|^2 \|\vec{v}\|^2 \sin^2 \theta$$

using the fact that $1 - \cos^2 \theta = \sin^2 \theta$. But $\sin \theta$ is nonnegative on the range $0 \le \theta \le \pi$, so taking the positive square root of both sides gives

$$\|\vec{u} \times \vec{v}\| = \|\vec{u}\| \ \|\vec{v}\| \sin \theta$$

This expression for $\|\vec{u} \times \vec{v}\|$ makes no reference to a coordinate system and, moreover, it has a nice geometrical interpretation. The parallelogram determined

---

8 Joseph Louis Lagrange (1736–1813) was born in Italy and spent his early years in Turin. At the age of 19 he solved a famous problem by inventing an entirely new method, known today as the calculus of variations, and went on to become one of the greatest mathematicians of all time. His work brought a new level of rigour to analysis and his *Mécanique Analytique* is a masterpiece in which he introduced methods still in use. In 1766 he was appointed to the Berlin Academy by Frederik the Great who asserted that the "greatest mathematician in Europe" should be at the court of the "greatest king in Europe." After the death of Frederick, Lagrange went to Paris at the invitation of Louis XVI. He remained there throughout the revolution and was made a count by Napoleon, who called him the "lofty pyramid of the mathematical sciences."

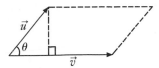

Figure 4.26

by the vectors $\vec{u}$ and $\vec{v}$ has base length $\|\vec{v}\|$ and altitude $\|\vec{u}\|\sin\theta$ (see Figure 4.26). Hence the area of the parallelogram formed by $\vec{u}$ and $\vec{v}$ is

$$(\|\vec{u}\|\sin\theta)\|\vec{v}\| = \|\vec{u}\times\vec{v}\|$$

This is also valid if $\vec{u}$ and $\vec{v}$ are parallel because then $\vec{u}\times\vec{v} = \vec{0}$ by Theorem 2 and $\sin\theta = 0$ (as $\theta = 0$ or π). This proves the first part of Theorem 4.

Theorem 4

If $\vec{u}$ and $\vec{v}$ are two nonzero vectors and θ is the angle between $\vec{u}$ and $\vec{v}$, then
1. $\|\vec{u}\times\vec{v}\| = \|\vec{u}\|\,\|\vec{v}\|\sin\theta$ = area of the parallelogram determined by $\vec{u}$ and $\vec{v}$.
2. $\vec{u}$ and $\vec{v}$ are parallel if and only if $\vec{u}\times\vec{v} = \vec{0}$.

PROOF OF (2)

By (1), $\vec{u}\times\vec{v} = \vec{0}$ if and only if the area of the parallelogram is zero. But the area vanishes if and only if $\vec{u}$ and $\vec{v}$ have the same or opposite direction—that is, if and only if they are parallel.

Note that (2) in Theorem 4 holds if $\vec{v} = \vec{0}$ or $\vec{w} = \vec{0}$. Of course θ makes no sense in this case.

Example 1

Find the area of the triangle with vertices $P(2, 1, 0)$, $Q(3, -1, 1)$, and $R(1, 0, 1)$.

Solution   We have $\overrightarrow{RP} = [1\ 1\ -1]^{T}$ and $\overrightarrow{RQ} = [2\ -1\ 0]^{T}$. The area of the triangle is half the area of the parallelogram (see the diagram), and so equals $\frac{1}{2}\|\overrightarrow{RP}\times\overrightarrow{RQ}\|$. We have

$$\overrightarrow{RP}\times\overrightarrow{RQ} = \det\begin{bmatrix} \vec{i} & 1 & 2 \\ \vec{j} & 1 & -1 \\ \vec{k} & -1 & 0 \end{bmatrix} = \begin{bmatrix} -1 \\ -2 \\ -3 \end{bmatrix},$$

so the area of the triangle is $\frac{1}{2}\|\overrightarrow{RP}\times\overrightarrow{RQ}\| = \frac{1}{2}\sqrt{1+4+9} = \frac{1}{2}\sqrt{14}$.

If three vectors $\vec{u}$, $\vec{v}$, and $\vec{w}$ are given, they determine a "squashed" rectangular solid called a **parallelepiped** (Figure 4.27), and it is often useful to be able to find the volume of such a solid. The base of the solid is the parallelogram determined by $\vec{u}$ and $\vec{v}$, so it has area $A = \|\vec{u}\times\vec{v}\|$ by Theorem 4. The height of the solid is the length h of the projection of $\vec{w}$ on $\vec{u}\times\vec{v}$. Hence

$$h = \left|\frac{\vec{w}\cdot(\vec{u}\times\vec{v})}{\|\vec{u}\times\vec{v}\|^{2}}\right|\|\vec{u}\times\vec{v}\| = \frac{|\vec{w}\cdot(\vec{u}\times\vec{v})|}{\|\vec{u}\times\vec{v}\|} = \frac{|\vec{w}\cdot(\vec{u}\times\vec{v})|}{A}$$

Thus the volume of the parallelepiped is $hA = |\vec{w}\cdot(\vec{u}\times\vec{v})|$.

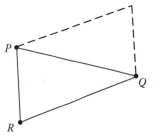

Figure 4.27

Theorem 5

The volume of the parallelepiped determined by three vectors $\vec{w}$, $\vec{u}$, and $\vec{v}$ (Figure 4.27) is given by $|\vec{w}\cdot(\vec{u}\times\vec{v})|$.

Example 2

Find the volume of the parallelepiped determined by the vectors $\vec{w} = [1\ 2\ -1]^T$, $\vec{u} = [1\ 1\ 0]^T$, and $\vec{v} = [-2\ 0\ 1]^T$.

Solution

We use Theorem 1.

$$\vec{w} \cdot (\vec{u} \times \vec{v}) = \det \begin{bmatrix} 1 & 1 & -2 \\ 2 & 1 & 0 \\ -1 & 0 & 1 \end{bmatrix} = -3$$

Hence the volume is $|\vec{w} \cdot (\vec{u} \times \vec{v})| = |-3| = 3$ by Theorem 5.

We can now give a coordinate-free description of the cross product $\vec{u} \times \vec{v}$. Its magnitude is given by $\|\vec{u} \times \vec{v}\| = \|\vec{u}\|\,\|\vec{v}\| \sin \theta$, and if $\vec{u} \times \vec{v} \neq \vec{0}$, its direction is very nearly determined by the fact that it is orthogonal to both $\vec{u}$ and $\vec{v}$ and so points along the line normal to the plane determined by $\vec{u}$ and $\vec{v}$. It remains only to decide which of the two possible directions is correct.

Before this can be done, the basic issue of how coordinates are assigned must be clarified. When coordinate axes are chosen in space, the procedure is as follows: An origin is selected, two perpendicular lines (the X and Y axes) are chosen through the origin, and a positive direction on each of these axes is selected quite arbitrarily. Then the line through the origin normal to this X-Y plane is called the Z axis, but there is a choice of which direction on this axis is the positive one. The two possibilities are shown in Figure 4.28, and it is a standard convention that Cartesian coordinates are always **right-hand coordinate systems**. The reason for this terminology is that, in such a system, if the Z axis is grasped in the right hand with the thumb pointing in the positive Z direction, then the fingers curl around from the positive X axis to the positive Y axis (through a right angle).

Suppose now that $\vec{u}$ and $\vec{v}$ are given and that θ is the angle between them (so $0 \leq \theta \leq \pi$). Then the direction of $\|\vec{u} \times \vec{v}\|$ is given by the right-hand rule.

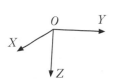

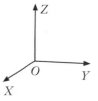

Left-hand system

Right-hand system

Figure 4.28

Right-hand Rule

If the vector $\vec{u} \times \vec{v}$ is grasped in the right hand and the fingers curl around from $\vec{u}$ to $\vec{v}$ through the angle θ, the thumb points in the direction for $\vec{u} \times \vec{v}$.

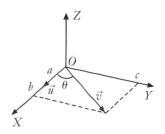

Figure 4.29

To indicate why this is true, introduce coordinates in $\mathbb{R}^3$ as follows: Let $\vec{u}$ and $\vec{v}$ have a common tail O, choose the origin at O, choose the X axis so that $\vec{u}$ points in the positive X direction, and then choose the Y axis so that $\vec{v}$ is in the X-Y plane and the positive Y axis is on the same side of the X axis as $\vec{v}$. Then, in this system, $\vec{u}$ and $\vec{v}$ have component form $\vec{u} = [a\ 0\ 0]^T$ and $\vec{v} = [b\ c\ 0]^T$ where $a > 0$ and $c > 0$. The situation is depicted in Figure 4.29. The right-hand rule asserts that $\vec{u} \times \vec{v}$ should point in the positive Z direction. But our definition of $\vec{u} \times \vec{v}$ gives

$$\vec{u} \times \vec{v} = \det \begin{bmatrix} \vec{i} & a & b \\ \vec{j} & 0 & c \\ \vec{k} & 0 & 0 \end{bmatrix} = \begin{bmatrix} 0 \\ 0 \\ ac \end{bmatrix} = (ac)\vec{k}$$

and $(ac)\,\vec{k}$ has the positive Z direction because $ac > 0$.

Exercises 4.3

1. If $\vec{i}, \vec{j}$, and $\vec{k}$ are the coordinate vectors, verify that $\vec{i} \times \vec{j} = \vec{k}, \vec{j} \times \vec{k} = \vec{i}$, and $\vec{k} \times \vec{i} = \vec{j}$.

2. Show that $\vec{u} \times (\vec{v} \times \vec{w})$ need not equal $(\vec{u} \times \vec{v}) \times \vec{w}$ by calculating both when $\vec{u} = [1\ 1\ 1]^T$, $\vec{v} = [1\ 1\ 0]^T$, and $\vec{w} = [0\ 0\ 1]^T$.

3. Find two unit vectors orthogonal to both $\vec{u}$ and $\vec{v}$ if:

    (a) $\vec{u} = [1\ 2\ 2]^T$, $\vec{v} = [2\ -1\ 2]^T$

    ◆(b) $\vec{u} = [1\ 2\ -1]^T$, $\vec{v} = [3\ 1\ 2]^T$

4. Find the area of the triangle with the following vertices.

    (a) $A(3, -1, 2)$, $B(1, 1, 0)$, and $C(1, 2, -1)$

    ◆(b) $A(3, 0, 1)$, $B(5, 1, 0)$, and $C(7, 2, -1)$

    (c) $A(1, 1, -1)$, $B(2, 0, 1)$, and $C(1, -1, 3)$

    ◆(d) $A(3, -1, 1)$, $B(4, 1, 0)$, and $C(2, -3, 0)$

5. Find the volume of the parallelepiped determined by $\vec{w}, \vec{u}$, and $\vec{v}$ when:

    (a) $\vec{w} = [2\ 1\ 1]^T$, $\vec{v} = [1\ 0\ 2]^T$, and $\vec{u} = [2\ 1\ -1]^T$

    ◆(b) $\vec{w} = [1\ 0\ 3]^T$, $\vec{v} = [2\ 1\ -3]^T$, and $\vec{u} = [1\ 1\ 1]^T$

6. Let P_0 be a point with position vector $\vec{p}_0$, and let $ax + by + cz = d$ be the equation of a plane with normal $\vec{n} = [a\ b\ c]^T$.

    (a) Show that the point on the plane closest to P_0 has position vector $\vec{p}$ given by

    $$\vec{p} = \vec{p}_0 + \frac{d - (\vec{p}_0 \bullet \vec{n})}{\|\vec{n}\|^2}\, \vec{n}.$$

    [*Hint:* $\vec{p} = \vec{p}_0 + t\vec{n}$ for some t, and $\vec{p} \bullet \vec{n} = d$.]

    ◆(b) Show that the shortest distance from P_0 to the plane is $\dfrac{|d - (\vec{p}_0 \bullet \vec{n})|}{\|\vec{n}\|}$.

    (c) Let P_0' denote the reflection of P_0 in the plane —that is, the point on the opposite side of the plane such that the line through P_0 and P_0' is perpendicular to the plane. Show that $\vec{p}_0 + 2\dfrac{d - (\vec{p}_0 \bullet \vec{n})}{\|\vec{n}\|^2}\, \vec{n}$ is the position vector of P_0'.

7. Simplify $(a\vec{u} + b\vec{v}) \times (c\vec{u} + d\vec{v})$.

8. Show that the shortest distance from a point P to the line through P_0 with direction vector $\vec{d}$ is $\dfrac{\|\overrightarrow{P_0P} \times \vec{d}\|}{\|\vec{d}\|}$.

9. Let $\vec{u}$ and $\vec{v}$ be nonzero, nonorthogonal vectors. If θ is the angle between them, show that $\tan \theta = \dfrac{\|\vec{u} \times \vec{v}\|}{\vec{u} \bullet \vec{v}}$.

◆10. Show that points A, B, and C are all on one line if and only if $\overrightarrow{AB} \times \overrightarrow{AC} = \vec{0}$.

11. Show that points A, B, C, and D are all on one plane if and only if $\overrightarrow{AB} \bullet (\overrightarrow{AC} \times \overrightarrow{AD}) = 0$.

◆12. Use Theorem 5 to confirm that, if $\vec{u}, \vec{v}$, and $\vec{w}$ are mutually perpendicular, the (rectangular) parallelepiped they determine has volume $\|\vec{u}\|\|\vec{v}\|\|\vec{w}\|$.

13. Show that the volume of the parallelepiped determined by $\vec{u}, \vec{v}$, and $\vec{u} \times \vec{v}$ is $\|\vec{u} \times \vec{v}\|^2$.

14. Complete the proof of Theorem 3.

15. Prove the following properties in Theorem 2.

    (a) Property 6

    ◆(b) Property 7

    (c) Property 8

16. (a) Show that $\vec{w} \bullet (\vec{u} \times \vec{v}) = \vec{u} \bullet (\vec{v} \times \vec{w}) = \vec{v} \bullet (\vec{w} \times \vec{u})$ holds for all vectors $\vec{w}, \vec{u}$, and $\vec{v}$.

    ◆(b) Show that $\vec{v} - \vec{w}$ and $(\vec{u} \times \vec{v}) + (\vec{v} \times \vec{w}) + (\vec{w} \times \vec{u})$ are orthogonal.

17. Show that $\vec{u} \times (\vec{v} \times \vec{w}) = (\vec{u} \bullet \vec{w})\vec{v} - (\vec{u} \bullet \vec{v})\vec{w}$.
    [*Hint:* First do it for $\vec{u} = \vec{i}, \vec{j}$, and $\vec{k}$; then write $\vec{u} = x\vec{i} + y\vec{j} + z\vec{k}$ and use Theorem 2.]

18. Prove the **Jacobi identity**:
    $\vec{u} \times (\vec{v} \times \vec{w}) + \vec{v} \times (\vec{w} \times \vec{u}) + \vec{w} \times (\vec{u} \times \vec{v}) = \vec{0}$.
    [*Hint:* The preceding exercise.]

19. Show that

    $$(\vec{u} \times \vec{v}) \bullet (\vec{w} \times \vec{z}) = \det\begin{bmatrix} \vec{u} \bullet \vec{w} & \vec{u} \bullet \vec{z} \\ \vec{v} \bullet \vec{w} & \vec{v} \bullet \vec{z} \end{bmatrix}.$$

    [*Hint:* Exercises 16 and 17.]

20. Let P, Q, R, and S be four points, not all on one plane, as in the diagram on the left at the top of page 186. Show that the volume of the pyramid

they determine is $\frac{1}{6}\left|\overrightarrow{PQ} \cdot (\overrightarrow{PR} \times \overrightarrow{PS})\right|$. [*Hint:* The volume of a cone with base area A and height h as in the diagram below right is $\frac{1}{3}Ah$.]

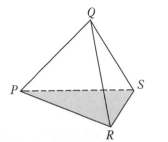
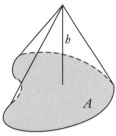

21. Consider a triangle with vertices A, B, and C, as in the diagram below. Let α, β, and γ denote the angles at A, B, and C, respectively, and let a, b, and c denote the lengths of the sides opposite A, B, and C, respectively. Write $\vec{u} = \overrightarrow{AB}$, $\vec{v} = \overrightarrow{BC}$, and $\vec{w} = \overrightarrow{CA}$.

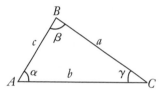

(a) Deduce that $\vec{u} + \vec{v} + \vec{w} = \vec{0}$.

(b) Show that $\vec{u} \times \vec{v} = \vec{w} \times \vec{u} = \vec{v} \times \vec{w}$.
[*Hint:* Compute $\vec{u} \times (\vec{u} + \vec{v} + \vec{w})$ and $\vec{v} \times (\vec{u} + \vec{v} + \vec{w})$.]

(c) Deduce the **law of sines**:
$$\frac{\sin \alpha}{a} = \frac{\sin \beta}{b} = \frac{\sin \gamma}{c}$$

♦22. Show that the (shortest) distance between two planes $\vec{n} \cdot \vec{p} = d_1$ and $\vec{n} \cdot \vec{p} = d_2$ with $\vec{n}$ as normal is $\dfrac{|d_2 - d_1|}{\|\vec{n}\|}$.

23. Let A and B be points other than the origin, and let $\vec{a}$ and $\vec{b}$ be their position vectors. If $\vec{a}$ and $\vec{b}$ are not parallel, show that the plane through A, B, and the origin is given by
$$\{P(x, y, z) \mid [x \;\; y \;\; z]^T = s\vec{a} + t\vec{b} \text{ for some } s \text{ and } t\}.$$

24. Let A be a 2×3 matrix of rank 2 with rows $\vec{r}_1$ and $\vec{r}_2$. Show that $P = \{XA \mid X = [x \;\; y]; x, y \text{ arbitrary}\}$ is the plane through the origin with normal $\vec{r}_1 \times \vec{r}_2$.

25. Given the cube with vertices $P(x, y, z)$, where each of x, y, and z is either 0 or 2, consider the plane perpendicular to the diagonal through $P(0, 0, 0)$ and $P(2, 2, 2)$ and bisecting it.

(a) Show that the plane meets six of the edges of the cube and bisects them.

(b) Show that the six points in (a) are the vertices of a regular hexagon.

SECTION 4.4   Matrix Transformations II

Recall that an $n \times n$ matrix A induces a linear transformation

$$T: \mathbb{R}^n \to \mathbb{R}^n \quad \text{where } T(X) = AX \text{ for all columns } X \text{ in } \mathbb{R}^n.$$

Such transformations were introduced in Section 2.5, with most examples in the case $n = 2$. In this section we use the vector techniques we have been developing to compute the matrices for many important transformations, and to gain more insight into the geometric effect of these transformations, particularly when $n = 3$.

Projections, Reflections, and Rotations

A transformation $T: \mathbb{R}^n \to \mathbb{R}^m$ is called **linear** if

$$T(X + Y) = T(X) + T(Y) \quad \text{and} \quad T(aX) = aT(X)$$

hold for all columns X and Y in $\mathbb{R}^n$, and for all real numbers a. In this case Theorem 2 §2.5 shows that there exists a unique $m \times n$ matrix A such that

$$T(X) = AX \text{ for all columns } X \text{ in } \mathbb{R}^n,$$

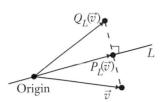

Figure 4.30

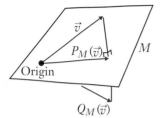

Figure 4.31

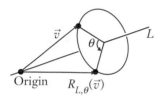

Figure 4.32

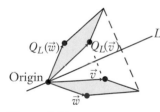

Figure 4.33

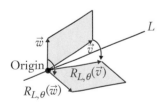

Figure 4.34

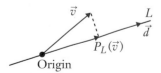

Figure 4.35

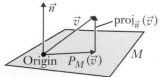

Figure 4.36

and we say that T is the **matrix transformation** induced by A. A linear transformation $T : \mathbb{R}^n \to \mathbb{R}^n$ is called a **linear operator** on $\mathbb{R}^n$, and we showed in Section 2.5 that rotations about the origin in $\mathbb{R}^2$, and reflections in a line through the origin in $\mathbb{R}^2$, are linear operators. Our method in that case was to find the matrices of these operators explicitly; we now extend this to $\mathbb{R}^3$.

Given a line L through the origin in $\mathbb{R}^3$ and a vector $\vec{v}$ in $\mathbb{R}^3$, the **projection** $P_L(\vec{v})$ of $\vec{v}$ on L, and the **reflection** $Q_L(\vec{v})$ of $\vec{v}$ in L are defined as in Figure 4.30. Similarly, if M is a plane through the origin, Figure 4.31 defines the **projection** $P_M(\vec{v})$ of $\vec{v}$ on M, and the **reflection** $Q_M(\vec{v})$ of $\vec{v}$ in M. Finally, the **rotation** $R_{L,\theta}(\vec{v})$ of $\vec{v}$ about the line L through an angle θ is shown in Figure 4.32. Hence

$$P_L : \mathbb{R}^3 \to \mathbb{R}^3, \quad Q_L : \mathbb{R}^3 \to \mathbb{R}^3 \quad \text{and} \quad R_{L,\theta} : \mathbb{R}^3 \to \mathbb{R}^3$$

are all operators on $\mathbb{R}^3$, and we claim that they are all linear. A direct geometric argument is available for reflections and rotations.

Let $T : \mathbb{R}^3 \to \mathbb{R}^3$ be either a reflection (in a line or plane), or a rotation (about a line); we claim that T is linear. In each case one sees geometrically that T preserves lengths and angles; that is, if $\vec{v}$ and $\vec{w}$ are vectors, then:

$\|T(\vec{v})\| = \|\vec{v}\|$,

$\|T(\vec{w})\| = \|\vec{w}\|$, and

The angle between $\vec{v}$ and $\vec{w}$ is the same as the angle between $T(\vec{v})$ and $T(\vec{w})$.

Hence the effect of the operator T is to carry the *entire parallelogram* determined by $\vec{v}$ and $\vec{w}$ to the parallelogram determined by $T(\vec{v})$ and $T(\vec{w})$. These parallelograms are identical—see Figures 4.33 and 4.34 in the case of Q_L and $R_{L,\theta}$. In particular, the diagonal $\vec{v} + \vec{w}$ of the first parallelogram is carried to the diagonal $T(\vec{v}) + T(\vec{w})$ of the second. But T carries $\vec{v} + \vec{w}$ to $T(\vec{v} + \vec{w})$, and it follows that $T(\vec{v} + \vec{w}) = T(\vec{v}) + T(\vec{w})$. A similar argument shows that $T(a\vec{v}) = aT(\vec{v})$ for all scalars a, so T is linear. Moreover, this argument works in $\mathbb{R}^2$ for rotations about the origin and reflections in a line through the origin.

However, this method breaks down for projections because they preserve neither lengths nor angles, so we must give an algebraic argument to prove linearity in this case. First consider the projection Q_L on the line L through the origin in $\mathbb{R}^3$ with (nonzero) direction vector $\vec{d} = [a\ b\ c]^T$. Given a vector $\vec{v}$, Figure 4.35 and Theorem 4 §4.2 give

$$P_L(\vec{v}) = \text{proj}_{\vec{d}}(\vec{v}) = \left(\frac{\vec{v} \cdot \vec{d}}{\|\vec{d}\|^2} \right) \vec{d} \quad \text{for all vectors } \vec{v}$$

Now the linearity of P_L follows because, for all $\vec{v}$ and $\vec{w}$,

$$(\vec{v} + \vec{w}) \cdot \vec{d} = \vec{v} \cdot \vec{d} + \vec{w} \cdot \vec{d} \quad \text{and} \quad (a\vec{v}) \cdot \vec{d} = a(\vec{v} \cdot \vec{d})$$

as the reader can verify.

Finally, consider the projection P_M on the plane M through the origin in $\mathbb{R}^3$ with (nonzero) normal $\vec{n} = [a\ b\ c]^T$. Then Figure 4.36 shows that

$$P_M(\vec{v}) = \vec{v} - \text{proj}_{\vec{n}}(\vec{v}) = \vec{v} - \left(\frac{\vec{v} \cdot \vec{n}}{\|\vec{n}\|^2} \right) \vec{n} \quad \text{for all } \vec{v}$$

Hence, if $\vec{v} = [x\ y\ z]^T$, some matrix arithmetic gives

$$P_M(\vec{v}) = \begin{bmatrix} x \\ y \\ z \end{bmatrix} - \frac{ax + by + cz}{a^2 + b^2 + c^2} \begin{bmatrix} a \\ b \\ c \end{bmatrix} = \frac{1}{a^2 + b^2 + c^2} \begin{bmatrix} b^2 + c^2 & -ab & -ac \\ -ab & a^2 + c^2 & -bc \\ -ac & -bc & a^2 + b^2 \end{bmatrix} \begin{bmatrix} x \\ y \\ z \end{bmatrix}$$

Thus, not only is P_M linear but it has matrix $A = \dfrac{1}{a^2 + b^2 + c^2} \begin{bmatrix} b^2 + c^2 & -ab & -ac \\ -ab & a^2 + c^2 & -bc \\ -ac & -bc & a^2 + b^2 \end{bmatrix}$.

This proves

Theorem 1

The following transformations are all linear:
1. Projections on a line or plane through the origin in $\mathbb{R}^3$.
2. Reflections in a line or plane through the origin in $\mathbb{R}^3$.
3. Rotations about a line through the origin in $\mathbb{R}^3$.
4. Rotations about the origin in $\mathbb{R}^2$, and reflections in a line through the origin in $\mathbb{R}^2$.

Computations like the one for projections prior to Theorem 1 give the following explicit formulas for the matrices of some of these transformations (the details are left to the reader):

- If L is the line through the origin with direction vector $\vec{d} = [a \ b \ c]^T$, then

$$P_L \text{ has matrix } \frac{1}{a^2 + b^2 + c^2} \begin{bmatrix} a^2 & ab & ac \\ ab & b^2 & bc \\ ac & bc & c^2 \end{bmatrix}$$

$$Q_L \text{ has matrix } \frac{1}{a^2 + b^2 + c^2} \begin{bmatrix} a^2 - b^2 - c^2 & 2ab & 2ac \\ 2ab & b^2 - a^2 - c^2 & 2bc \\ 2ac & 2bc & c^2 - a^2 - b^2 \end{bmatrix}$$

- If M is the plane through the origin with normal $\vec{n} = [a \ b \ c]^T$, then

$$P_M \text{ has matrix } \frac{1}{a^2 + b^2 + c^2} \begin{bmatrix} b^2 + c^2 & -ab & -ac \\ -ab & a^2 + c^2 & -bc \\ -ac & -bc & a^2 + b^2 \end{bmatrix}$$

$$Q_M \text{ has matrix } \frac{1}{a^2 + b^2 + c^2} \begin{bmatrix} b^2 + c^2 - a^2 & -2ab & -2ac \\ -2ab & a^2 + c^2 - b^2 & -2bc \\ -2ac & -2bc & a^2 + b^2 - c^2 \end{bmatrix}$$

A similar argument shows that, if L is the line through the origin in $\mathbb{R}^2$ with direction vector $\vec{d} = \begin{bmatrix} a \\ b \end{bmatrix} \neq \vec{0}$, then

$$P_L \text{ has matrix } \frac{1}{a^2 + b^2} \begin{bmatrix} a^2 & ab \\ ab & b^2 \end{bmatrix} \quad \text{and} \quad Q_L \text{ has matrix } \frac{1}{a^2 + b^2} \begin{bmatrix} a^2 - b^2 & 2ab \\ 2ab & b^2 - a^2 \end{bmatrix}.$$

In particular, if the line L has slope m, then (as in Example 11 §4.1) $\vec{d} = \begin{bmatrix} 1 \\ m \end{bmatrix}$ is a direction vector for L, and so

$$P_L \text{ has matrix } \frac{1}{1 + m^2} \begin{bmatrix} 1 & m \\ m & m^2 \end{bmatrix} \quad \text{and} \quad Q_L \text{ has matrix } \frac{1}{1 + m^2} \begin{bmatrix} 1 - m^2 & 2m \\ 2m & m^2 - 1 \end{bmatrix}.$$

Example 1

If L denotes the Z axis, find the matrix A of the rotation $R_{L,\theta}$ of $\mathbb{R}^3$ about the Z axis through the angle θ clockwise in the X-Y plane.

Solution

In the notation of Section 2.5, the standard basis E_1, E_2, E_3 of $\mathbb{R}^3$ consists of the coordinate vectors: $E_1 = \vec{i}$, $E_2 = \vec{j}$, and $E_3 = \vec{k}$. Hence Theorem 2 §2.5 shows that

$$A = [R_{L,\theta}(E_1)\ R_{L,\theta}(E_2)\ R_{L,\theta}(E_3)] = [R_{L,\theta}(\vec{i})\ R_{L,\theta}(\vec{j})\ R_{L,\theta}(\vec{k})].$$

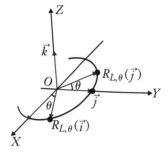

We have $R_{L,\theta}(\vec{k}) = \vec{k}$ because $R_{L,\theta}$ fixes the Z axis. The effect of $R_{L,\theta}$ on the X-Y plane is to rotate it counterclockwise through the angle θ as in the diagram. This transformation of $\mathbb{R}^2$ is denoted R_θ in Section 2.5, and has matrix $\begin{bmatrix} \cos\theta & -\sin\theta \\ \sin\theta & \cos\theta \end{bmatrix}$ by Example 5 §2.5. In particular,

$$R_{L,\theta}\begin{bmatrix} 1 \\ 0 \end{bmatrix} = \begin{bmatrix} \cos\theta & -\sin\theta \\ \sin\theta & \cos\theta \end{bmatrix}\begin{bmatrix} 1 \\ 0 \end{bmatrix} = \begin{bmatrix} \cos\theta \\ \sin\theta \end{bmatrix}, \quad \text{and}$$

$$R_{L,\theta}\begin{bmatrix} 0 \\ 1 \end{bmatrix} = \begin{bmatrix} \cos\theta & -\sin\theta \\ \sin\theta & \cos\theta \end{bmatrix}\begin{bmatrix} 0 \\ 1 \end{bmatrix} = \begin{bmatrix} -\sin\theta \\ \cos\theta \end{bmatrix}.$$

Since $R_{L,\theta}(\vec{i})$ and $R_{L,\theta}(\vec{j})$ both lie in the X-Y plane, this gives

$$R_{L,\theta}(\vec{i}) = \begin{bmatrix} \cos\theta \\ \sin\theta \\ 0 \end{bmatrix} \quad \text{and} \quad R_{L,\theta}(\vec{j}) = \begin{bmatrix} -\sin\theta \\ \cos\theta \\ 0 \end{bmatrix}.$$

Hence the matrix of $R_{L,\theta}$ is $A = \begin{bmatrix} \cos\theta & -\sin\theta & 0 \\ \sin\theta & \cos\theta & 0 \\ 0 & 0 & 1 \end{bmatrix}$.

Transformations of Areas and Volumes

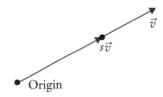

Let $\vec{v}$ be a nonzero vector in $\mathbb{R}^3$. Each vector in the same direction as $\vec{v}$ whose length is a fraction s of the length of $\vec{v}$ has the form $s\vec{v}$ (see Figure 4.37). With this, scrutiny of Figure 4.38 shows that a vector $\vec{u}$ is in the parallelogram determined by $\vec{v}$ and $\vec{w}$ if and only if it has the form $\vec{u} = s\vec{v} + t\vec{w}$ where $0 \le s \le 1$ and $0 \le t \le 1$. But then, if $T : \mathbb{R}^3 \to \mathbb{R}^3$ is a linear transformation, we have

$$T(s\vec{v} + t\vec{w}) = T(s\vec{v}) + T(t\vec{w}) = sT(\vec{v}) + tT(\vec{w}).$$

Figure 4.37

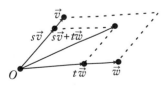

Hence $T(s\vec{v} + t\vec{w})$ is in the parallelogram determined by $T(\vec{v})$ and $T(\vec{w})$ and, conversely, every vector in this parallelogram has the form $T(s\vec{v} + t\vec{w})$ where $s\vec{v} + t\vec{w}$ is in the parallelogram determined by $\vec{v}$ and $\vec{w}$. For this reason, the parallelogram determined by $T(\vec{v})$ and $T(\vec{w})$ is called the **image** of the parallelogram determined by $\vec{v}$ and $\vec{w}$. We record this discussion as:

Figure 4.38

Theorem 2

If $T : \mathbb{R}^3 \to \mathbb{R}^3$ (or $\mathbb{R}^2 \to \mathbb{R}^2$) is a linear operator, the image of the parallelogram determined by vectors $\vec{v}$ and $\vec{w}$ is the parallelogram determined by $T(\vec{v})$ and $T(\vec{w})$.

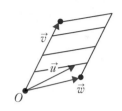

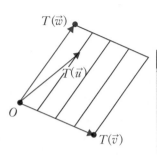

Figure 4.39

This result is illustrated in Figure 4.39, and was used in Examples 3 and 4 §2.5 to reveal the effect of expansion and shear transformations.

Now we are interested in the effect of a linear transformation $T : \mathbb{R}^3 \to \mathbb{R}^3$ on the parallelepiped determined by three vectors $\vec{u}$, $\vec{v}$, and $\vec{w}$ in $\mathbb{R}^3$ (see the discussion preceding Theorem 5 §4.3). By Theorem 2 this parallelepiped is carried to the parallelepiped determined by $T(\vec{u}) = A(\vec{u})$, $T(\vec{v}) = A(\vec{v})$, and $T(\vec{w}) = A(\vec{w})$ where A is the matrix of T. In particular, we want to discover how the volume changes, and it turns out to be closely related to the matrix A.

Theorem 3

Let $\operatorname{vol}(\vec{u}, \vec{v}, \vec{w})$ denote the volume of the parallelepiped determined by three vectors $\vec{u}$, $\vec{v}$, and $\vec{w}$ in $\mathbb{R}^3$, and let $\operatorname{area}(\vec{p}, \vec{q})$ denote the area of the parallelogram determined by two vectors $\vec{p}$ and $\vec{q}$ in $\mathbb{R}^2$. Then:
1. If A is a 3×3 matrix, then $\operatorname{vol}(A\vec{u}, A\vec{v}, A\vec{w}) = |\det(A)| \cdot \operatorname{vol}(\vec{u}, \vec{v}, \vec{w})$.
2. If A is a 2×2 matrix, then $\operatorname{area}(A\vec{p}, A\vec{q}) = |\det(A)| \cdot \operatorname{area}(\vec{p}, \vec{q})$.

PROOF

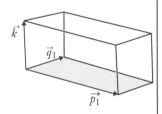

1. Let $[\vec{u}\ \vec{v}\ \vec{w}]$ denote the 3×3 matrix with columns $\vec{u}$, $\vec{v}$, and $\vec{w}$. Then

   $$\operatorname{vol}(A\vec{u}, A\vec{v}, A\vec{w}) = |A\vec{u} \cdot (A\vec{v} \times A\vec{w})|$$

   by Theorem 5 §4.3. Now apply Theorem 1 §4.3 twice to get

   $$\begin{aligned} A\vec{u} \cdot (A\vec{v} \times A\vec{w}) = \det[A\vec{u}\ A\vec{v}\ A\vec{w}] &= \det\{A[\vec{u}\ \vec{v}\ \vec{w}]\} \\ &= \det(A)\det[\vec{u}\ \vec{v}\ \vec{w}] \\ &= \det(A)(\vec{u} \cdot (\vec{v} \times \vec{w})) \end{aligned}$$

   where we used block multiplication and the product theorem for determinants. Finally (1) follows by taking absolute values.

2. Given $\vec{p} = \begin{bmatrix} x \\ y \end{bmatrix}$ in $\mathbb{R}^2$, write $\vec{p}_1 = \begin{bmatrix} x \\ y \\ 0 \end{bmatrix}$ in $\mathbb{R}^3$.

   By the diagram, $\operatorname{area}(\vec{p}, \vec{q}) = \operatorname{vol}(\vec{p}_1, \vec{q}_1, \vec{k})$ where $\vec{k}$ is the (length 1) coordinate vector along the Z axis. If A is a 2×2 matrix, write $A_1 = \begin{bmatrix} A & 0 \\ 0 & 1 \end{bmatrix}$ in block form, and observe that $(A\vec{v})_1 = (A_1\vec{v}_1)$ and $A_1\vec{k} = \vec{k}$. Hence part (1) if this theorem shows

   $$\begin{aligned} \operatorname{area}(A\vec{p}, A\vec{q}) = \operatorname{vol}((A\vec{p})_1, (A\vec{q})_1, \vec{k}) &= \operatorname{vol}(A_1\vec{p}_1, A_1\vec{q}_1, A_1\vec{k}) \\ &= |\det(A_1)| \operatorname{vol}(\vec{p}_1, \vec{q}_1, \vec{k}) \\ &= |\det(A)| \operatorname{area}(\vec{p}, \vec{q}) \end{aligned}$$

   as required.

Define the **unit square** and **unit cube** to be the square and cube corresponding to the coordinate vectors in $\mathbb{R}^2$ and $\mathbb{R}^3$, respectively. Hence Theorem 3 gives a geometrical meaning to the determinant:

- If A is a 2×2 matrix then $|\det(A)|$ is the area of the image of the unit square under multiplication by A;
- If A is a 3×3 matrix then $|\det(A)|$ is the volume of the image of the unit cube under multiplication by A.

These results, together with the importance of areas and volumes in geometry, were among the reasons for the initial development of determinants.

Exercises 4.4

1. In each case show that that T is either projection on a line, reflection in a line, or rotation through an angle, and find the line or angle.

   (a) $T[x \ y]^T = \frac{1}{5}[x + 2y \ \ 2x + 4y]^T$

   ◆(b) $T[x \ y]^T = \frac{1}{2}[x - y \ \ y - x]^T$

   (c) $T[x \ y]^T = \frac{1}{\sqrt{2}}[-x - y \ \ x - y]^T$

   ◆(d) $T[x \ y]^T = \frac{1}{5}[-3x + 4y \ \ 4x + 3y]^T$

   (e) $T[x \ y]^T = [-y \ -x]^T$

   ◆(f) $T[x \ y]^T = \frac{1}{2}[x - \sqrt{3}y \ \ \sqrt{3}x + y]^T$

2. Determine the effect of the following transformations.

   (a) Rotation through $\frac{\pi}{2}$, followed by projection on the Y axis, followed by reflection in the line $y = x$.

   ◆(b) Projection on the line $y = x$ followed by projection on the line $y = -x$.

   (c) Projection on the X axis followed by reflection in the line $y = x$.

3. In each case solve the problem by finding the matrix of the operator.

   (a) Find the projection of $\vec{v} = [1 \ -2 \ 3]^T$ on the plane with equation $3x - 5y + 2z = 0$.

   ◆(b) Find the projection of $\vec{v} = [0 \ 1 \ -3]^T$ on the plane with equation $2x - y + 4z = 0$.

   (c) Find the reflection of $\vec{v} = [1 \ -2 \ 3]^T$ in the plane with equation $x - y + 3z = 0$.

   ◆(d) Find the reflection of $\vec{v} = [0 \ 1 \ -3]^T$ in the plane with equation $2x + y - 5z = 0$.

   (e) Find the reflection of $\vec{v} = [2 \ 5 \ -1]^T$ in the line with equation $\begin{bmatrix} x \\ y \\ z \end{bmatrix} = t \begin{bmatrix} 1 \\ 1 \\ -2 \end{bmatrix}$.

◆(f) Find the projection of $\vec{v} = [1 \ -1 \ 7]^T$ on the line with equation $\begin{bmatrix} x \\ y \\ z \end{bmatrix} = t \begin{bmatrix} 3 \\ 0 \\ 4 \end{bmatrix}$.

(g) Find the projection of $\vec{v} = [1 \ 1 \ -3]^T$ on the line with equation $\begin{bmatrix} x \\ y \\ z \end{bmatrix} = t \begin{bmatrix} 2 \\ 0 \\ -3 \end{bmatrix}$.

◆(h) Find the reflection of $\vec{v} = [2 \ -5 \ 0]^T$ in the line with equation $\begin{bmatrix} x \\ y \\ z \end{bmatrix} = t \begin{bmatrix} 1 \\ 1 \\ -3 \end{bmatrix}$.

4. (a) Find the rotation of $\vec{v} = [2 \ 3 \ -1]^T$ about the Z axis through $\theta = \frac{\pi}{4}$.

   ◆(b) Find the rotation of $\vec{v} = [1 \ 0 \ 3]^T$ about the Z axis through $\theta = \frac{\pi}{6}$.

5. Find the matrix of the rotation about the X axis through the angle θ (from the positive Y axis to the positive Z axis).

◆6. Find the matrix of the rotation about the Y axis through the angle θ (from the positive X axis to the positive Z axis).

7. If A is 3×3, show that the image of the line in $\mathbb{R}^3$ through $\vec{p}_0$ with direction vector $\vec{d}$ is the line through $A\vec{p}_0$ with direction vector $A\vec{d}$, assuming that $A\vec{d} \neq \vec{0}$. What happens if $A\vec{d} = \vec{0}$?

8. If A is 3×3 and invertible, show that the image of the plane through the origin with normal $\vec{n}$ is the plane through the origin with normal $\vec{n}_1 = B\vec{n}$ where $B = (A^{-1})^T$. [*Hint*: Use the fact that $\vec{v} \cdot \vec{w} = \vec{v}^T\vec{w}$ to show that $\vec{n}_1 \cdot (A\vec{p}) = \vec{n} \cdot \vec{p}$ for each $\vec{p}$ in $\mathbb{R}^3$.]

9. Let P_L and Q_L denote, respectively, projection on and reflection in the line L through the origin with direction vector $\vec{d} = [a \ b \ c]^T \neq \vec{0}$.

   (a) Show that P_L has matrix

   $$\frac{1}{a^2+b^2+c^2}\begin{bmatrix} a^2 & ab & ac \\ ab & b^2 & bc \\ ac & bc & c^2 \end{bmatrix}.$$

   ◆(b) Show that Q_L has matrix

   $$\frac{1}{a^2+b^2+c^2}\begin{bmatrix} a^2-b^2-c^2 & 2ab & 2ac \\ 2ab & b^2-a^2-c^2 & 2bc \\ 2ac & 2bc & c^2-a^2-b^2 \end{bmatrix}.$$

10. Let P_M and Q_M denote, respectively, projection on and reflection in the plane M through the origin with normal $\vec{n} = [a \ b \ c]^T \neq \vec{0}$.

    (a) Show that P_M has matrix

    $$\frac{1}{a^2+b^2+c^2}\begin{bmatrix} b^2+c^2 & -ab & -ac \\ -ab & a^2+c^2 & -bc \\ -ac & -bc & a^2+b^2 \end{bmatrix}.$$

    ◆(b) Show that Q_M has matrix

    $$\frac{1}{a^2+b^2+c^2}\begin{bmatrix} b^2+c^2-a^2 & -2ab & -2ac \\ -2ab & a^2+c^2-b^2 & -2bc \\ -2ac & -2bc & a^2+b^2-c^2 \end{bmatrix}.$$

SECTION 4.5   An Application to Computer Graphics

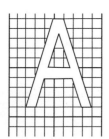

Figure 4.40

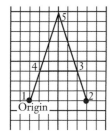

Figure 4.41

Computer graphics deals with images displayed on a computer screen, and so arises in a variety of applications, ranging from word processors, to Star Wars animations, to video games, to wire-frame images of an airplane. These images consist of a number of points on the screen, together with instructions on how to fill in areas bounded by lines and curves. Often curves are approximated by a set of short straight-line segments, so that the curve is specified by a series of points on the screen at the end of these segments. Matrix transformations are important here because matrix images of straight line segments are again line segments.[9] Note that a colour image requires that three images are sent, one to each of the red, green, and blue phosphorus dots on the screen, in varying intensities.

Consider displaying the letter A. In reality, it is depicted on the screen, as in Figure 4.40, by specifying the coordinates of the 11 corners and filling in the interior. For simplicity, we will disregard the thickness of the letter, so we require only five coordinates as in Figure 4.41. This simplified letter can then be stored as a data matrix

$$\begin{array}{c} \textit{Vertex} \quad 1 \ 2 \ 3 \ 4 \ 5 \\ D = \begin{bmatrix} 0 & 6 & 5 & 1 & 3 \\ 0 & 0 & 3 & 3 & 9 \end{bmatrix} \end{array}$$

where the columns are the coordinates of the vertices in order (and the letter is 9 units high and 6 units wide). Then if we want to transform the letter by a 2×2 matrix A, we left-multiply this data matrix by A (the effect is to multiply each column by A and so transform each vertex).

For example, we can slant the letter to the right by multiplying by an X-Shear matrix $A = \begin{bmatrix} 1 & 0.2 \\ 0 & 1 \end{bmatrix}$—see Section 2.5. The result is the letter with data matrix

---

9 If $\vec{v}_0$ and $\vec{v}_1$ are vectors, the vector from $\vec{v}_0$ to $\vec{v}_1$ is $\vec{d} = \vec{v}_1 - \vec{v}_0$. So a vector $\vec{v}$ lies on the line segment between $\vec{v}_0$ and $\vec{v}_1$ if and only if $\vec{v} = \vec{v}_0 + t\vec{d}$ for some number t in the range $0 \le t \le 1$. Thus the image of this segment is the set of vectors $A\vec{v} = A\vec{v}_0 + tA\vec{d}$ with $0 \le t \le 1$, that is the image is the segment between $A\vec{v}_0$ and $A\vec{v}_1$.

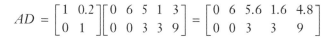

$$AD = \begin{bmatrix} 1 & 0.2 \\ 0 & 1 \end{bmatrix} \begin{bmatrix} 0 & 6 & 5 & 1 & 3 \\ 0 & 0 & 3 & 3 & 9 \end{bmatrix} = \begin{bmatrix} 0 & 6 & 5.6 & 1.6 & 4.8 \\ 0 & 0 & 3 & 3 & 9 \end{bmatrix}$$

which is shown in Figure 4.42. If we want to make this slanted matrix narrower, we can now apply an *X*-scale matrix $B = \begin{bmatrix} 0.8 & 0 \\ 0 & 1 \end{bmatrix}$ that shrinks the *X*-coordinate by 0.8. The result is the composite transformation

$$BAD = \begin{bmatrix} 0.8 & 0 \\ 0 & 1 \end{bmatrix} \begin{bmatrix} 1 & 0.2 \\ 0 & 1 \end{bmatrix} \begin{bmatrix} 0 & 6 & 5 & 1 & 3 \\ 0 & 0 & 3 & 3 & 9 \end{bmatrix} = \begin{bmatrix} 0 & 4.8 & 4.48 & 1.28 & 3.84 \\ 0 & 0 & 3 & 3 & 9 \end{bmatrix}$$

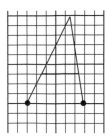

Figure 4.42

which is drawn in Figure 4.43.

On the other hand, we can rotate the letter about the origin through $\frac{\pi}{6}$ (or 30°) by multiplying by the matrix $R_{\frac{\pi}{6}} = \begin{bmatrix} \cos(\frac{\pi}{6}) & -\sin(\frac{\pi}{6}) \\ \sin(\frac{\pi}{6}) & \cos(\frac{\pi}{6}) \end{bmatrix} = \begin{bmatrix} 0.866 & -0.5 \\ 0.5 & 0.866 \end{bmatrix}$. This gives

$$R_{\frac{\pi}{6}}D = \begin{bmatrix} 0.866 & -0.5 \\ 0.5 & 0.866 \end{bmatrix} \begin{bmatrix} 0 & 6 & 5 & 1 & 3 \\ 0 & 0 & 3 & 3 & 9 \end{bmatrix} = \begin{bmatrix} 0 & 5.196 & 2.83 & -0.634 & -1.902 \\ 0 & 3 & 5.098 & 3.098 & 9.294 \end{bmatrix}$$

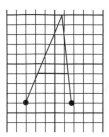

Figure 4.43

and is plotted in Figure 4.44.

This poses a problem: How do we rotate at a point other than the origin? It turns out that we can do this when we have solved another more basic problem. It is clearly important to be able to translate a screen image by a fixed vector $\vec{w}$, that is apply the transformation $T_{\vec{w}} : \mathbb{R}^2 \to \mathbb{R}^2$ given by $T_{\vec{w}}(\vec{v}) = \vec{v} + \vec{w}$ for all $\vec{v}$ in $\mathbb{R}^2$. The problem is that these translations are not matrix transformations $\mathbb{R}^2 \to \mathbb{R}^2$ because they do not carry $\vec{0}$ to $\vec{0}$ (unless $\vec{w} = \vec{0}$). However, there is a clever way around this.

The idea is to represent a point $\vec{v} = \begin{bmatrix} x \\ y \end{bmatrix}$ as a 3×1 column $\begin{bmatrix} x \\ y \\ 1 \end{bmatrix}$, called the

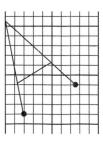

Figure 4.44

homogeneous coordinates of $\vec{v}$. Then translation by $\vec{w} = \begin{bmatrix} p \\ q \end{bmatrix}$ can be achieved by multiplying by a 3×3 matrix:

$$\begin{bmatrix} 1 & 0 & p \\ 0 & 1 & q \\ 0 & 0 & 1 \end{bmatrix} \begin{bmatrix} x \\ y \\ 1 \end{bmatrix} = \begin{bmatrix} x + p \\ y + q \\ 1 \end{bmatrix} = \begin{bmatrix} T_{\vec{w}}(\vec{v}) \\ 1 \end{bmatrix}$$

Thus, by using homogeneous coordinates we can implement the translation $T_{\vec{w}}$ in the top two coordinates. On the other hand, the matrix transformation induced by $A = \begin{bmatrix} a & b \\ c & d \end{bmatrix}$ is also given by a 3×3 matrix:

$$\begin{bmatrix} a & b & 0 \\ c & d & 0 \\ 0 & 0 & 1 \end{bmatrix} \begin{bmatrix} x \\ y \\ 1 \end{bmatrix} = \begin{bmatrix} ax + by \\ cx + dy \\ 1 \end{bmatrix} = \begin{bmatrix} A\vec{v} \\ 1 \end{bmatrix}$$

So everything can be accomplished at the expense of using 3×3 matrices and homogeneous coordinates.

Example 1

Rotate the letter A in Figure 4.41 through $\frac{\pi}{6}$ about the point $\begin{bmatrix} 4 \\ 5 \end{bmatrix}$.

Solution   Using homogenous coordinates for the vertices of the letter results in a data matrix with three rows:

$$K_d = \begin{bmatrix} 0 & 6 & 5 & 1 & 3 \\ 0 & 0 & 3 & 3 & 9 \\ 1 & 1 & 1 & 1 & 1 \end{bmatrix}$$

If we write $\vec{w} = \begin{bmatrix} 4 \\ 5 \end{bmatrix}$, the idea is to use a composite of transformations: First translate the letter by $-\vec{w}$ so that the point $\vec{w}$ moves to the origin, then rotate this translated letter, and then translate it by $\vec{w}$ back to its original position. The matrix arithmetic is as follows (remember the order of composition!):

$$\begin{bmatrix} 1 & 0 & 4 \\ 0 & 1 & 5 \\ 0 & 0 & 1 \end{bmatrix} \begin{bmatrix} 0.866 & -0.5 & 0 \\ 0.5 & 0.866 & 0 \\ 0 & 0 & 1 \end{bmatrix} \begin{bmatrix} 1 & 0 & -4 \\ 0 & 1 & -5 \\ 0 & 0 & 1 \end{bmatrix} \begin{bmatrix} 0 & 6 & 5 & 1 & 3 \\ 0 & 0 & 3 & 3 & 9 \\ 1 & 1 & 1 & 1 & 1 \end{bmatrix}$$

$$= \begin{bmatrix} 3.036 & 8.232 & 5.866 & 2.402 & 1.134 \\ -1.33 & 1.67 & 3.768 & 1.768 & 7.964 \\ 1 & 1 & 1 & 1 & 1 \end{bmatrix}$$

This is plotted in Figure 4.45.

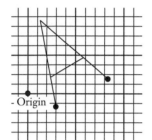

- Origin

Figure 4.45

This discussion merely touches the surface of computer graphics, and the reader is referred to specialized books on the subject. Realistic graphic rendering requires an enormous number of matrix calculations. In fact, matrix multiplication algorithms are now embedded in microchip circuits, and can perform over 100 million matrix multiplications per second. This is particularly important in the field of three-dimensional graphics where the homogeneous coordinates have four components and 4×4 matrices are required.

Exercises 4.5

1. Consider the letter A described in Figure 4.41. Find the data matrix for the letter obtained by:

   (a) Rotating the letter through $\frac{\pi}{4}$ about the origin.

   ◆(b) Rotating the letter through $\frac{\pi}{4}$ about the point $\begin{bmatrix} 1 \\ 2 \end{bmatrix}$.

2. Find the matrix for turning the letter A in Figure 4.41 upside-down in place.

3. Find the 3×3 matrix for reflecting in the line $y = mx + b$. Use $\begin{bmatrix} 1 \\ m \end{bmatrix}$ as direction vector for the line.

4. Find the 3×3 matrix for rotating through the angle θ about the point $P(a, b)$.

5. Find the reflection of the point P in the line $y = 1 + 2x$ in $\mathbb{R}^2$ if:

   (a) $P = P(1, 1)$

   ◆(b) $P = P(1, 4)$

   (c) What about $P = P(1, 3)$? Explain. [*Hint:* Example 1 and Section 4.4.]

Supplementary Exercises for Chapter 4

1. Suppose that $\vec{u}$ and $\vec{v}$ are nonzero vectors. If $\vec{u}$ and $\vec{v}$ are not parallel, and $a\vec{u} + b\vec{v} = a_1\vec{u} + b_1\vec{v}$, show that $a = a_1$ and $b = b_1$.

2. Consider a triangle with vertices A, B, and C. Let E and F be the midpoints of sides AB and AC, respectively, and let the medians EC and FB meet at O. Write $\overrightarrow{EO} = s\overrightarrow{EC}$ and $\overrightarrow{FO} = t\overrightarrow{FB}$, where s and t are scalars. Show that $s = t = \frac{1}{3}$ by expressing $\overrightarrow{AO}$ two ways in the form $a\overrightarrow{AB} + b\overrightarrow{AC}$, and applying Exercise 1. Conclude that the medians of a triangle meet at the point on each that is one-third of the way from the midpoint to the vertex (and so are concurrent).

3. A river flows at 1 km/h and a swimmer moves at 2 km/h (relative to the water). At what angle must he swim to go straight across? What is his resulting speed?

♦4. A wind is blowing from the south at 75 knots, and an airplane flies heading east at 100 knots. Find the resulting velocity of the airplane.

5. An airplane pilot flies at 300 km/h in a direction $30°$ south of east. The wind is blowing from the south at 150 km/h.

   (a) Find the resulting direction and speed of the airplane.

   (b) Find the speed of the airplane if the wind is from the west (at 150 km/h).

♦6. A rescue boat has a top speed of 13 knots. The captain wants to go due east as fast as possible in water with a current of 5 knots due south. Find the velocity vector $\vec{v} = (x, y)$ that she must achieve, assuming the X and Y axes point east and north, respectively, and find her resulting speed.

7. A boat goes 12 knots heading north. The current is 5 knots from the west. In what direction does the boat actually move and at what speed?

5 The Vector Space $\mathbb{R}^n$

SECTION 5.1   Subspaces and Spanning

In Section 2.5 we introduced the set $\mathbb{R}^n$ of all $n \times 1$ columns, investigated the linear transformations $\mathbb{R}^n \to \mathbb{R}^m$, and showed that they are all given by left multiplication by an $m \times n$ matrix. Particular attention was paid to the euclidean plane $\mathbb{R}^2$ and to euclidean space $\mathbb{R}^3$, where geometric transformations like rotations and reflections were shown to be matrix transformations. We returned to this in Section 4.4 where projections in $\mathbb{R}^2$ or $\mathbb{R}^3$ were also shown to be matrix transformations, and where determinants were related to areas and volumes.

In this chapter we investigate $\mathbb{R}^n$ in full generality, and introduce some of the most important concepts and methods in linear algebra. While the n-tuples in $\mathbb{R}^n$ can be written as rows or as columns, we will primarily denote them as column matrices X, Y, etc. The main exception is that the geometric vectors in $\mathbb{R}^2$ and $\mathbb{R}^3$ will be written as $\vec{v}$, $\vec{w}$, etc.

Subspaces of $\mathbb{R}^n$

A set[1] U of vectors in $\mathbb{R}^n$ is called a **subspace** of $\mathbb{R}^n$ if it satisfies the following properties:

  S1. The zero vector 0 is in U.

  S2. If X and Y are in U, then $X + Y$ is also in U.

  S3. If X is in U, then aX is in U for every real number a.

We say that the subset U is **closed under addition** if S2 holds, and that U is **closed under scalar multiplication** if S3 holds.

Clearly $\mathbb{R}^n$ is a subspace of itself. The set $U = \{0\}$, consisting of only the zero vector, is also a subspace because $0 + 0 = 0$ and $a0 = 0$ for each a in $\mathbb{R}$; it is called the **zero subspace**. Any subspace of $\mathbb{R}^n$ other than $\{0\}$ or $\mathbb{R}^n$ is called a **proper** subspace.

---

1 We use the language of sets. Informally, a **set** X is a collection of objects, called the **elements** of the set. Two sets X and Y are called equal (written $X = Y$) if they have the same elements. If every element of X is in the set Y, we say that X is a **subset** of Y, and write $X \subseteq Y$. Hence both $X \subseteq Y$ and $Y \subseteq X$ if and only if $X = Y$.

We saw in Section 4.2 that every plane M through the origin in $\mathbb{R}^3$ has equation $ax + by + cz = 0$ where a, b, and c are not all zero. Here $\vec{n} = [a \ \ b \ \ c]^T$ is a normal to the plane and

$$M = \{\vec{v} \text{ in } \mathbb{R}^3 \mid \vec{n} \cdot \vec{v} = 0\}$$

where $\vec{v} = [x \ \ y \ \ z]^T$ and $\vec{n} \cdot \vec{v}$ denotes the dot product introduced in Section 4.2.[2] Then M is a subspace of $\mathbb{R}^3$. Indeed we show that M satisfies S1, S2, and S3 as follows:

S1. $\vec{0}$ is in M because $\vec{n} \cdot \vec{0} = 0$;

S2. If $\vec{v}$ and $\vec{v}_1$ are in M, then $\vec{n} \cdot (\vec{v} + \vec{v}_1) = \vec{n} \cdot \vec{v} + \vec{n} \cdot \vec{v}_1 = 0 + 0 = 0$, so $\vec{v} + \vec{v}_1$ is in M;

S3. If $\vec{v}$ is in M, then $\vec{n} \cdot (a\vec{v}) = a(\vec{n} \cdot \vec{v}) = a(0) = 0$, so $a\vec{v}$ is in M.

This proves the first part of

Example 1

Planes and lines through the origin in $\mathbb{R}^3$ are all subspaces of $\mathbb{R}^3$.

Solution   We dealt with planes above. If L is a line through the origin with direction vector $\vec{d}$, then $L = \{t\vec{d} \mid t \text{ in } \mathbb{R}\}$. We leave it as an exercise to verify that L satisfies S1, S2, and S3.

Example 1 shows that lines through the origin in $\mathbb{R}^2$ are subspaces; in fact, they are the *only* proper subspaces of $\mathbb{R}^2$ (Exercise 24). Indeed, we shall see in Example 11 §5.2 that lines and planes through the origin in $\mathbb{R}^3$ are the only proper subspaces of $\mathbb{R}^3$. Thus the geometry of lines and planes through the origin is captured by the subspace concept. (Note that *every* line or plane is just a translation of one of these.)

Subspaces can also be used to describe important features of an $m \times n$ matrix A. The **null space** of A, denoted null A, and the **image space** of A, denoted im A, are defined by

$$\text{null } A = \{X \text{ in } \mathbb{R}^n \mid AX = 0\} \quad \text{and} \quad \text{im } A = \{AX \mid X \text{ in } \mathbb{R}^n\}.$$

In the language of Chapter 2, null A consists of all solutions X in $\mathbb{R}^n$ of the homogeneous system $AX = 0$, and im A is the set of all vectors Y in $\mathbb{R}^m$ such that $AX = Y$ *has* a solution for some X.

Note that X is in null A if it satisfies the *condition* $AX = 0$, while im A consists of vectors of the *form* AX for some X in $\mathbb{R}^n$. These two ways to describe subsets occur frequently.

Example 2

If A is an $m \times n$ matrix, then:
1. null A is a subspace of $\mathbb{R}^n$.
2. im A is a subspace of $\mathbb{R}^m$.

Solution   1. The zero vector 0 in $\mathbb{R}^n$ lies in null A because $A0 = 0$.[3] If X and X_1 are in null A, then $X + X_1$ and aX are in null A because they satisfy the required condition:

---

2 We are using set notation here. In general $\{q \mid p\}$ means the set of all objects q with property p.

3 We are using 0 to represent the zero vector in both $\mathbb{R}^m$ and $\mathbb{R}^n$. This abuse of notation is common, and causes no confusion once everyone knows that it is going on.

$$A(X + X_1) = AX + AX_1 = 0 + 0 = 0 \quad \text{and} \quad A(aX) = a(AX) = a0 = 0.$$

Hence null A satisfies S1, S2, and S3, and so is a subspace of $\mathbb{R}^n$.

2. The zero vector 0 in $\mathbb{R}^m$ lies in im A because $0 = A0$. Suppose that Y and Y_1 are in im A, say $Y = AX$ and $Y_1 = AX_1$ where X and X_1 are in $\mathbb{R}^n$. Then

$$Y + Y_1 = AX + AX_1 = A(X + X_1) \quad \text{and} \quad aY = a(AX) = A(aX)$$

show that $Y + Y_1$ and aY are both in im A (they have the required form). Hence im A is a subspace of $\mathbb{R}^m$.

There are other important subspaces associated with a matrix A that clarify basic properties of A. If A is an $m \times n$ matrix and λ is any number, let

$$E_\lambda(A) = \{X \text{ in } \mathbb{R}^n \mid AX = \lambda X\}.$$

A vector X is in $E_\lambda(A)$ if and only if $(\lambda I - A)X = \vec{0}$, so Example 2 gives:

Example 3

$E_\lambda(A) = \text{null}(\lambda I - A)$ is a subspace of $\mathbb{R}^n$ for each $n \times n$ matrix A and number λ.

$E_\lambda(A)$ is called the **eigenspace** of A corresponding to λ. The reason for the name is that, in the terminology of Section 3.3, λ is an **eigenvalue** of A if $E_\lambda(A) \neq \{0\}$. In this case the nonzero vectors in $F_\lambda(A)$ are called the **eigenvectors** of A corresponding to λ.

The reader should not get the impression that *every* subset of $\mathbb{R}^n$ is a subspace. For example:

$$U_1 = \left\{ \begin{bmatrix} x \\ y \end{bmatrix} \middle| x \geq 0 \right\} \text{ satisfies S1 and S2, but not S3;}$$

$$U_2 = \left\{ \begin{bmatrix} x \\ y \end{bmatrix} \middle| x^2 = y^2 \right\} \text{ satisfies S1 and S3, but not S2.}$$

Hence neither U_1 nor U_2 is a subspace of $\mathbb{R}^2$. (However, see Exercise 20.)

Spanning Sets

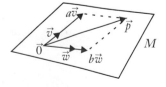

Let $\vec{v}$ and $\vec{w}$ be two nonzero, nonparallel vectors in $\mathbb{R}^3$ with their tails at the origin. The plane M through the origin containing these vectors is described in Section 4.2 by saying that $\vec{n} = \vec{v} \times \vec{w}$ is a *normal* for M, and that M consists of all vectors $\vec{p}$ such that $\vec{n} \cdot \vec{p} = 0.$[4] While this is a very useful way to look at planes, there is another approach that is at least as useful in $\mathbb{R}^3$ and, more importantly, works for *all* subspaces of $\mathbb{R}^n$.

The idea is as follows: Observe that, by the diagram, a vector $\vec{p}$ is in M if and only if it has the form

$$\vec{p} = a\vec{v} + b\vec{w}$$

for certain real numbers a and b (we say that $\vec{p}$ is a *linear combination* of $\vec{v}$ and $\vec{w}$).

---

4 The vector $\vec{n} = \vec{v} \times \vec{w}$ is nonzero because $\vec{v}$ and $\vec{w}$ are *not* parallel.

Hence we can describe M as

$$M = \{a\vec{v} + b\vec{w} \mid a, b \text{ in } \mathbb{R}\}^{\,5}$$

and we say that $\{\vec{v}, \vec{w}\}$ is a *spanning set* for M. It is this notion of a spanning set that provides a way to describe all subspaces of $\mathbb{R}^n$.

Given vectors $X_1, X_2, \ldots, X_k$ in $\mathbb{R}^n$, a vector of the form

$$t_1 X_1 + t_2 X_2 + \cdots + t_k X_k \text{ where the } t_i \text{ are scalars}$$

is called a **linear combination** of the X_i, and t_i is called the **coefficient** of X_i in the linear combination. The set of *all* such linear combinations is called the **span** of the X_i and is denoted

$$\text{span}\{X_1, X_2, \ldots, X_k\} = \{t_1 X_1 + t_2 X_2 + \cdots + t_k X_k \mid t_i \text{ in } \mathbb{R}\}.$$

Thus $\text{span}\{X, Y\} = \{sX + tY \mid s, t \text{ in } \mathbb{R}\}$, and $\text{span}\{X\} = \{tX \mid t \text{ in } \mathbb{R}\}$.

In particular, the above discussion shows that, if $\vec{v}$ and $\vec{w}$ are two nonzero, nonparallel vectors in $\mathbb{R}^3$, then

$$M = \text{span}\{\vec{v}, \vec{w}\}$$

is the plane in $\mathbb{R}^3$ containing $\vec{v}$ and $\vec{w}$. Moreover, if $\vec{d}$ is any nonzero vector in $\mathbb{R}^3$ (or $\mathbb{R}^2$), then

$$L = \text{span}\{\vec{v}\} = \{t\vec{d} \mid t \text{ in } \mathbb{R}\}$$

is the line with direction vector $\vec{d}$. Hence lines and planes can both be described in terms of spanning sets.

Example 4

Let $X = [2 \ {-1} \ 2 \ 1]^T$ and $Y = [3 \ 4 \ {-1} \ 1]^T$ in $\mathbb{R}^4$. Determine whether $P = [0 \ {-11} \ 8 \ 1]^T$ or $Q = [2 \ 3 \ 1 \ 2]^T$ is in $U = \text{span}\{X, Y\}$.

Solution   The vector P is in U if and only if $P = sX + tY$ for scalars s and t. Equating components gives equations

$$2s + 3t = 0, \quad -s + 4t = -11, \quad 2s - t = 8, \quad \text{and} \quad s + t = 1.$$

This linear system has solution $s = 3$ and $t = -2$, so P is in U. On the other hand, $Q = sX + tY$ leads to equations

$$2s + 3t = 2, \quad -s + 4t = 3, \quad 2s - t = 1, \quad \text{and} \quad s + t = 2$$

and this system has *no* solution. So Q does *not* lie in U.

Theorem 1

Let $U = \text{span}\{X_1, X_2, \ldots, X_k\}$ in $\mathbb{R}^n$. Then:
1. U is a subspace of $\mathbb{R}^n$ containing each X_i.
2. If W is a subspace of $\mathbb{R}^n$ and each X_i is in W, then $U \subseteq W$.

---

5 In particular, this implies that any vector $\vec{p}$ orthogonal to $\vec{v} \times \vec{w}$ must be a linear combination $\vec{p} = a\vec{v} + b\vec{w}$ of $\vec{v}$ and $\vec{w}$ for some a and b. Can you prove this directly?

PROOF

Write $U = \text{span}\{X_1, X_2, \ldots, X_k\}$ for convenience.
1. The zero vector 0 is in U because $0 = 0X_1 + 0X_2 + \cdots + 0X_k$ is a linear combination of the X_i. If $X = t_1X_1 + t_2X_2 + \cdots + t_kX_k$ and $Y = s_1X_1 + s_2X_2 + \cdots + s_kX_k$ are in U, then $X + Y$ and aX are in U because

$$X + Y = (s_1 + t_1)X_1 + (s_2 + t_2)X_2 + \cdots + (s_k + t_k)X_k, \quad \text{and}$$
$$aX = (at_1)X_1 + (at_2)X_2 + \cdots + (at_k)X_k.$$

Hence S1, S2, and S3 are satisfied for U, proving (1).
2. Let $X = t_1X_1 + t_2X_2 + \cdots + t_kX_k$ where the t_i are scalars and each X_i is in W. Then each t_iX_i is in W because W satisfies S3. But then X is in W because W satisfies S2 (verify). This proves (2).

Condition (2) in Theorem 1 can be expressed by saying that $\text{span}\{X_1, X_2, \ldots, X_k\}$ is the *smallest* subspace of $\mathbb{R}^n$ that contains each X_i. Here is an example of how it is used.
   If $U = \text{span}\{X_1, X_2, \ldots, X_k\}$ we say that the vectors $X_1, X_2, \ldots, X_k$ **span** the subspace U.

Example 5

If X and Y are in $\mathbb{R}^n$, show that $\text{span}\{X, Y\} = \text{span}\{X + Y, X - Y\}$.

Solution   Since both $X + Y$ and $X - Y$ are in $\text{span}\{X, Y\}$, Theorem 1 gives

$$\text{span}\{X + Y, X - Y\} \subseteq \text{span}\{X, Y\}.$$

But $X = \frac{1}{2}(X + Y) + \frac{1}{2}(X - Y)$ and $Y = \frac{1}{2}(X + Y) - \frac{1}{2}(X - Y)$ are both in $\text{span}\{X + Y, X - Y\}$, so

$$\text{span}\{X, Y\} \subseteq \text{span}\{X + Y, X - Y\}$$

again by Theorem 1. Thus $\text{span}\{X, Y\} = \text{span}\{X + Y, X - Y\}$.

It turns out that many important subspaces are best described by giving a spanning set. Here are three examples, beginning with an important spanning set for $\mathbb{R}^n$ itself. Column j of the $n \times n$ identity matrix I_n is denoted E_j and called the jth **coordinate vector** in $\mathbb{R}^n$, and the set $\{E_1, E_2, \ldots, E_n\}$ is called the **standard basis** of $\mathbb{R}^n$. If $X = [x_1 \quad x_2 \quad \cdots \quad x_n]^T$ is any vector in $\mathbb{R}^n$, then

$$X = x_1E_1 + x_2E_2 + \cdots + x_nE_n$$

as the reader can verify. This proves:

Example 6

$\mathbb{R}^n = \text{span}\{E_1, E_2, \ldots, E_k\}$.

If A is an $m \times n$ matrix A, the next two examples show that it is a routine matter to find spanning sets for null A and im A.

Example 7

Given an $m \times n$ matrix A, let $X_1, X_2, \ldots, X_k$ denote the basic solutions to the system $AX = 0$ given by the gaussian algorithm. Then

$$\text{null } A = \text{span}\{X_1, X_2, \ldots, X_k\}.$$

Solution   If X is in null A, then $AX = 0$ so Theorem 3 §2.2 shows that X is a linear combination of the basic solutions; that is, null $A \subseteq \text{span}\{X_1, X_2, \ldots, X_k\}$. On the other hand, if X is in $\text{span}\{X_1, X_2, \ldots, X_k\}$, then $X = t_1 X_1 + t_2 X_2 + \cdots + t_k X_k$ for scalars t_i, so

$$AX = t_1 A X_1 + t_2 A X_2 + \cdots + t_k A X_k = t_1 0 + t_2 0 + \cdots + t_k 0 = 0.$$

This shows that X is in null A, and hence that $\text{span}\{X_1, X_2, \ldots, X_k\} \subseteq \text{null } A$. Thus we have equality.

Example 8

Let $C_1, C_2, \ldots, C_n$ denote the columns of the $m \times n$ matrix A. Then

$$\text{im } A = \text{span}\{C_1, C_2, \ldots, C_n\}.$$

Solution   Observe first that $AE_j = C_j$ for each j where E_j is the jth coordinate vector in $\mathbb{R}^n$. Hence each C_j is in im A, and so $\text{span}\{C_1, C_2, \ldots, C_n\} \subseteq \text{im } A$ by Theorem 1. Conversely, let Y be a vector in im A, say $Y = AX$ for some X in $\mathbb{R}^n$. If $X = [x_1 \; x_2 \; \cdots \; x_n]^T$, then Theorem 4 §2.2 gives

$$Y = AX = [C_1 \; C_2 \; \cdots \; C_n] \begin{bmatrix} x_1 \\ x_2 \\ \vdots \\ x_n \end{bmatrix} = x_1 C_1 + x_2 C_2 + \cdots + x_n C_n$$

so Y is in $\text{span}\{C_1, C_2, \ldots, C_n\}$. Hence im $A \subseteq \text{span}\{C_1, C_2, \ldots, C_n\}$, and the result is proved.

Exercises 5.1

1. In each case determine whether U is a subspace of $\mathbb{R}^3$. Support your answer.
   (a) $U = \{[1 \; s \; t]^T \mid s \text{ and } t \text{ in } \mathbb{R}\}$.
   ◆(b) $U = \{[0 \; s \; t]^T \mid s \text{ and } t \text{ in } \mathbb{R}\}$.
   (c) $U = \{[r \; s \; t]^T \mid r, s, \text{ and } t \text{ in } \mathbb{R}, -r + 3s + 2t = 0\}$.
   ◆(d) $U = \{[r \; 3s \; r - 2]^T \mid r \text{ and } s \text{ in } \mathbb{R}\}$.
   (e) $U = \{[r \; 0 \; s]^T \mid r^2 + s^2 = 0, r \text{ and } s \text{ in } \mathbb{R}\}$.
   ◆(f) $U = \{[2r \; -s^2 \; t]^T \mid r, s, \text{ and } t \text{ in } \mathbb{R}\}$.

2. In each case determine if X lies in $U = \text{span}\{Y, Z\}$. If X is in U, write it as a linear combination of Y and Z; if X is not in U, show why not.
   (a) $X = [2 \; -1 \; 0 \; 1]^T$, $Y = [1 \; 0 \; 0 \; 1]^T$, and $Z = [0 \; 1 \; 0 \; 1]^T$.
   ◆(b) $X = [1 \; 2 \; 15 \; 11]^T$, $Y = [2 \; -1 \; 0 \; 2]^T$, and $Z = [1 \; -1 \; -3 \; 1]^T$.
   (c) $X = [8 \; 3 \; -13 \; 20]^T$, $Y = [2 \; 1 \; -3 \; 5]^T$, and $Z = [-1 \; 0 \; 2 \; -3]^T$.
   ◆(d) $X = [2 \; 5 \; 8 \; 3]^T$, $Y = [2 \; -1 \; 0 \; 5]^T$, and $Z = [-1 \; 2 \; 2 \; -3]^T$.

3. In each case determine if the given vectors span $\mathbb{R}^4$. Support your answer.
   (a) $\{[1 \; 1 \; 1 \; 1]^T, [0 \; 1 \; 1 \; 1]^T, [0 \; 0 \; 1 \; 1]^T, [0 \; 0 \; 0 \; 1]^T\}$.
   ◆(b) $\{[1 \; 3 \; -5 \; 0]^T, [-2 \; 1 \; 0 \; 0]^T, [0 \; 2 \; 1 \; -1]^T, [1 \; -4 \; 5 \; 0]^T\}$.

4. Is it possible that $\{[1 \; 2 \; 0]^T, [2 \; 0 \; 3]^T\}$ can span the subspace $U = \{[r \; s \; 0]^T \mid r \text{ and } s \text{ in } \mathbb{R}\}$? Defend your answer.

5. Give a spanning set for the zero subspace $\{0\}$ of $\mathbb{R}^n$.

6. Is $\mathbb{R}^2$ a subspace of $\mathbb{R}^3$? Defend your answer.

7. If $U = \text{span}\{X, Y, Z\}$ in $\mathbb{R}^n$, show that $U = \text{span}\{X + tZ, Y, Z\}$ for every t in $\mathbb{R}$.

8. If $U = \text{span}\{X, Y, Z\}$ in $\mathbb{R}^n$, show that $U = \text{span}\{X + Y, Y + Z, Z + X\}$.

9. If $a \neq 0$ is a scalar, show that $\text{span}\{aX\} = \text{span}\{X\}$ for every vector X in $\mathbb{R}^n$.

♦10. If $a_1, a_2, \ldots, a_k$ are nonzero scalars, show that
$\text{span}\{a_1X_1, a_2X_2, \ldots, a_kX_k\} = \text{span}\{X_1, X_2, \ldots, X_k\}$
for any vectors X_i in $\mathbb{R}^n$.

11. If $X \neq 0$ in $\mathbb{R}^n$, determine all subspaces of $\text{span}\{X\}$.

12. Suppose that $U = \text{span}\{X_1, X_2, \ldots, X_k\}$ where each X_i is in $\mathbb{R}^n$. If A is an $m \times n$ matrix and $AX_i = 0$ for each i, show that $AY = 0$ for every vector Y in U.

13. If A is an $m \times n$ matrix, show that, for each invertible $m \times m$ matrix U, $\text{null}(A) = \text{null}(UA)$.

14. If A is an $m \times n$ matrix, show that, for each invertible $n \times n$ matrix V, $\text{im}(A) = \text{im}(AV)$.

15. Let U be a subspace of $\mathbb{R}^n$, and let X be a vector in $\mathbb{R}^n$.

   (a) If aX is in U where $a \neq 0$ is a number, show that X is in U.

   ♦(b) If Y and $X + Y$ are in U where Y is a vector in $\mathbb{R}^n$, show that X is in U.

16. In each case either show that the statement is true or give an example showing that it is false.

   (a) If $U \neq \mathbb{R}^n$ is a subspace of $\mathbb{R}^n$ and $X + Y$ is in U, then X and Y are both in U.

   ♦(b) If U is a subspace of $\mathbb{R}^n$ and rX is in U for all r in $\mathbb{R}$, then X is in U.

   (c) If U is a subspace of $\mathbb{R}^n$ and X is in U, then $-X$ is also in U.

   ♦(d) If X is in U and $U = \text{span}\{Y, Z\}$, then $U = \text{span}\{X, Y, Z\}$.

   (e) The empty set of vectors in $\mathbb{R}^n$ is a subspace of $\mathbb{R}^n$.

17. (a) If A and B are $m \times n$ matrices, show that $U = \{X \text{ in } \mathbb{R}^n \mid AX = BX\}$ is a subspace of $\mathbb{R}^n$.

   (b) What if A is $m \times n$, B is $k \times n$, and $m \neq k$?

18. Suppose that $X_1, X_2, \ldots, X_k$ are vectors in $\mathbb{R}^n$. If $Y = a_1X_1 + a_2X_2 + \cdots + a_kX_k$ where $a_1 \neq 0$, show that $\text{span}\{X_1, X_2, \ldots, X_k\} = \text{span}\{Y, X_2, \ldots, X_k\}$.

19. If $U \neq \{0\}$ is a subspace of $\mathbb{R}$, show that $U = \mathbb{R}$.

♦20. Let U be a nonempty subset of $\mathbb{R}^n$. Show that U is a subspace if and only if S2 and S3 hold.

21. If S and T are nonempty sets of vectors in $\mathbb{R}^n$, and if $S \subseteq T$, show that $\text{span}\{S\} \subseteq \text{span}\{T\}$.

22. Let U and W be subspaces of $\mathbb{R}^n$. Define their **intersection** $U \cap W$ and their **sum** $U + W$ as follows:

$$U \cap W = \{X \text{ in } \mathbb{R}^n \mid X \text{ belongs to both } U \text{ and } W\}.$$
$$U + W = \{X \text{ in } \mathbb{R}^n \mid X \text{ is a sum of a vector in } U \text{ and a vector in } W\}.$$

   (a) Show that $U \cap W$ is a subspace of $\mathbb{R}^n$.

   ♦(b) Show that $U + W$ is a subspace of $\mathbb{R}^n$.

23. Let P denote an invertible $n \times n$ matrix. If λ is a number, show that
$$E_\lambda(PAP^{-1}) = \{PX \mid X \text{ is in } E_\lambda(A)\}$$
for each $n \times n$ matrix A. [Here $E_\lambda(A)$ is the set of eigenvectors of A.]

24. Show that every proper subspace U of $\mathbb{R}^2$ is a line through the origin. [*Hint:* If $\vec{d}$ is a nonzero vector in U, let $L = \mathbb{R}\vec{d} = \{r\vec{d} \mid r \text{ in } \mathbb{R}\}$ denote the line with direction vector $\vec{d}$. If $\vec{u}$ is in U but not in L, argue geometrically that every vector $\vec{v}$ in $\mathbb{R}^2$ is a linear combination of $\vec{u}$ and $\vec{d}$.]

SECTION 5.2   Independence and Dimension

Some spanning sets are better than others. If $U = \text{span}\{X_1, X_2, \ldots, X_k\}$ is a subspace of $\mathbb{R}^n$, then every vector in U can be written as a linear combination of the X_i in at least one way. Our interest here is in spanning sets where each vector in U has a *exactly one* representation as a linear combination of these vectors.

Linear Independence

Suppose that two linear combinations are equal in $\mathbb{R}^n$:

$$r_1X_1 + r_2X_2 + \cdots + r_kX_k = s_1X_1 + s_2X_2 + \cdots + s_kX_k.$$

We are looking for a condition on the set $\{X_1, X_2, \ldots, X_k\}$ of vectors that guarantees that this representation is *unique*; that is, $r_i = s_i$ for each i. Taking all terms to the left side gives

$$(r_1 - s_1)X_1 + (r_2 - s_2)X_2 + \cdots + (r_k - s_k)X_k = 0.$$

so the required condition is that this equation forces all the coefficients $r_i - s_i$ to be zero.

With this in mind, we call a set $\{X_1, X_2, \ldots, X_k\}$ of vectors **linearly independent** (or simply **independent**) if it satisfies the following condition:

$$\text{If } \quad t_1 X_1 + t_2 X_2 + \cdots + t_k X_k = 0 \quad \text{then} \quad t_1 = t_2 = \cdots = t_k = 0.$$

We record the result of the above discussion for reference.

Theorem 1

If $\{X_1, X_2, \ldots, X_k\}$ is an independent set of vectors in $\mathbb{R}^n$, then every vector in span$\{X_1, X_2, \ldots, X_k\}$ has a *unique* representation as a linear combination of the X_i.

It is useful to state the definition of independence in different language. Let us say that a linear combination **vanishes** if it equals the zero vector, and call a linear combination **trivial** if every coefficient is zero. Then the definition of independence can be compactly stated as follows:

A set of vectors is independent if and only if the only linear combination that vanishes is the trivial one.

Hence the procedure for checking that a set of vectors is independent is:

Independence Test

To verify that a set $\{X_1, X_2, \ldots, X_k\}$ of vectors in $\mathbb{R}^n$ is independent, proceed as follows:
1. Set a linear combination equal to zero: $t_1 X_1 + t_2 X_2 + \cdots + t_k X_k = 0$.
2. Show that $t_i = 0$ for each i (that is, the linear combination is trivial).
Of course, if some nontrivial linear combination vanishes, the vectors are *not* independent.

Example 1

Determine whether $\{[1 \ 0 \ -2 \ 5]^T, [2 \ 1 \ 0 \ -1]^T, [1 \ 1 \ 2 \ 1]^T\}$ is independent in $\mathbb{R}^4$.

Solution   Suppose a linear combination vanishes:

$$r[1 \ 0 \ -2 \ 5]^T + s[2 \ 1 \ 0 \ -1]^T + t[1 \ 1 \ 2 \ 1]^T = [0 \ 0 \ 0 \ 0]^T.$$

Equating corresponding entries gives a system of four equations:

$$r + 2s + t = 0, \quad s + t = 0, \quad -2r + 2t = 0, \quad \text{and} \quad 5r - s + t = 0.$$

The only solution is the trivial one $r = s = t = 0$ (verify), so these vectors are independent by the independence test.

Example 2

Show that the standard basis $\{E_1, E_2, \ldots, E_k\}$ of $\mathbb{R}^n$ is independent.

Solution   We have $t_1E_1 + t_2E_2 + \cdots + t_nE_n = [t_1 \ \ t_2 \ \ \cdots \ \ t_n]^T$ for all scalars t_i, so the linear combination vanishes if and only if each $t_i = 0$. Hence the independence test applies.

Example 3

If $\{X, Y\}$ is independent, show that $\{2X + 3Y, X - 5Y\}$ is also independent.

Solution   If $s(2X + 3Y) + t(X - 5Y) = 0$, collect terms to get $(2s + t)X + (3s - 5t)Y = 0$. Since $\{X, Y\}$ is independent this combination must be trivial; that is, $2s + t = 0$ and $3s - 5t = 0$. These equations have only the trivial solution $s = t = 0$, as required.

Example 4

Show that the zero vector in $\mathbb{R}^n$ does not belong to any independent set.

Solution   Given a set $\{0, X_1, X_2, \ldots, X_k\}$ of vectors containing 0, we have a vanishing, nontrivial linear combination $1 \cdot 0 + 0X_1 + 0X_2 + \cdots + 0X_k = 0$. Hence the set is not independent.

Example 5

Given X in $\mathbb{R}^n$, show that $\{X\}$ is independent if and only if $X \neq 0$.

Solution   A vanishing linear combination from $\{X\}$ takes the form $tX = 0$, t in $\mathbb{R}$. This implies that $t = 0$ because $X \neq 0$.

A set of vectors in $\mathbb{R}^n$ is called **linearly dependent** (or simply **dependent**) if it is *not* linearly independent, equivalently if some nontrivial linear combination vanishes.

Example 6

If $\vec{v}$ and $\vec{w}$ are nonzero vectors in $\mathbb{R}^3$, show that $\{\vec{v}, \vec{w}\}$ is dependent if and only if $\vec{v}$ and $\vec{w}$ are parallel.

Solution   If $\vec{v}$ and $\vec{w}$ are parallel, then one is a scalar multiple of the other (Theorem 4 §4.1), say $\vec{v} = a\vec{w}$ for some scalar a. Then the nontrivial linear combination $\vec{v} - a\vec{w} = \vec{0}$ vanishes, so $\{\vec{v}, \vec{w}\}$ is dependent. Conversely, if $\{\vec{v}, \vec{w}\}$ is dependent, let $s\vec{v} + t\vec{w} = \vec{0}$ be nontrivial, say $s \neq 0$. Then $\vec{v} = -\frac{t}{s}\vec{w}$, so X and $\vec{w}$ are parallel (by Theorem 4 §4.1). A similar argument works if $t \neq 0$.

By Theorem 5 §2.3, the following conditions are equivalent for an $n \times n$ matrix A:

1. A is invertible.
2. If $AX = 0$ where X is in $\mathbb{R}^n$, then $X = 0$.
3. $AX = B$ has a solution X for every vector B in $\mathbb{R}^n$.

While (1) makes no sense if A is not square, conditions (2) and (3) are meaningful for any matrix A and, in fact, are related to independence and spanning. To see how, let $C_1, C_2, \dots, C_n$ denote the columns of an $m \times n$ matrix A. If $X = [x_1\, x_2 \cdots x_n]^T$ is a column in $\mathbb{R}^n$, then

$$AX = x_1C_1 + x_2C_2 + \cdots x_nC_n \tag{*}$$

by Theorem 4 §2.2. With this, we get the following theorem:

Theorem 2

If A is an $m \times n$ matrix, let $\{C_1, C_2, \dots, C_n\}$ denote the columns of A.
  1. $\{C_1, C_2, \dots, C_n\}$ is independent in $\mathbb{R}^m$ if and only if $AX = 0$, X in $\mathbb{R}^n$, implies $X = 0$.
  2. $\mathbb{R}^m = \text{span}\{C_1, C_2, \dots, C_n\}$ if and only if $AX = B$ has a solution X for every vector B in $\mathbb{R}^n$.

PROOF

Write $X = [x_1 \quad x_2 \quad \cdots \quad x_n]^T$. Then $AX = 0$ means $x_1C_1 + x_2C_2 + \cdots x_nC_n = 0$ by (*), so (1) follows from the definition of independence. Similarly, (*) shows that a vector B in $\mathbb{R}^n$ satisfies $AX = B$ if and only if X is a linear combination of the columns C_j, so (2) follows from the definition of a spanning set.

For a *square* matrix A, Theorem 2 characterizes the invertibility of A in terms of the spanning and independence of its columns (see the discussion preceding Theorem 2). It is important to be able to discuss these notions for *rows*. If $X_1, X_2, \dots, X_k$ are $1 \times n$ rows, we define $\text{span}\{X_1, X_2, \dots, X_k\}$ to be the set of all linear combinations of the X_i (as matrices), and we say that $\{X_1, X_2, \dots, X_k\}$ is linearly independent if the only vanishing linear combination is the trivial one (that is, if $\{X_1^T, X_2^T, \dots, X_k^T\}$ is independent in $\mathbb{R}^n$, as the reader can verify).[6]

Theorem 3

The following are equivalent for an $n \times n$ matrix A:
  1. A is invertible.
  2. The columns of A are linearly independent.
  3. The columns of A span $\mathbb{R}^n$.
  4. The rows of A are linearly independent.
  5. The rows of A span the set of all $1 \times n$ rows.

PROOF

Let $C_1, C_2, \dots, C_n$ denote the columns of A.
    (1) $\Leftrightarrow$ (2). By Theorem 5 §2.3, A is invertible if and only if $AX = 0$ implies $X = 0$; this holds if and only if $\{C_1, C_2, \dots, C_n\}$ is independent by Theorem 2.
    (1) $\Leftrightarrow$ (3). Again by Theorem 5 §2.3, A is invertible if and only if $AX = B$ has a solution for every column B in $\mathbb{R}^n$; this holds if and only if $\text{span}\{C_1, C_2, \dots, C_n\} = \mathbb{R}^n$ by Theorem 2.

---

6 It is best to view columns and rows as just two different *notations* for ordered n-tuples. This discussion will become redundant in Chapter 6 where we define the general notion of a vector space.

(1) ⇔ (4). The matrix A is invertible if and only if A^T is invertible (by the Corollary to Theorem 4 §2.3); this in turn holds if and only if A^T has independent columns (by (1) ⇔ (2)); finally, this last statement holds if and only if A has independent rows (because the rows of A are the transposes of the columns of A^T).

(1) ⇔ (5). The proof is similar to (1) ⇔ (4).

Dimension

It is common geometrical language to say that $\mathbb{R}^3$ is 3-dimensional, that planes are 2-dimensional and that lines are 1-dimensional. The next theorem is a basic tool for clarifying this idea of "dimension". Its importance is difficult to exaggerate.

Theorem 4 Fundamental Theorem

Let U be a subspace of $\mathbb{R}^n$. If U is spanned by m vectors, and if U contains k linearly independent vectors, then $k \leq m$.

We give a proof at the end of this section.

The main use of the fundamental theorem depends on the following concept. If U is a subspace of $\mathbb{R}^n$, a set $\{X_1, X_2, \dots, X_m\}$ of vectors in U is called a **basis** of U if it satisfies the following two conditions:

1. $\{X_1, X_2, \dots, X_m\}$ is linearly independent.
2. $U = \text{span}\{X_1, X_2, \dots, X_m\}$.

The most remarkable result about bases[7] is:

Theorem 5 Invariance Theorem

If $\{X_1, X_2, \dots, X_m\}$ and $\{Y_1, Y_2, \dots, Y_k\}$ are bases of a subspace U of $\mathbb{R}^n$, then $m = k$.

PROOF

We have $k \leq m$ by the fundamental theorem because $\{X_1, X_2, \dots, X_m\}$ spans U, and $\{Y_1, Y_2, \dots, Y_k\}$ is independent. Similarly $m \leq k$, so $m = k$.

The invariance theorem guarantees that there is no ambiguity in the following definition: If U is a subspace of $\mathbb{R}^n$ and $\{X_1, X_2, \dots, X_m\}$ is any basis of U, the number m of vectors in the basis is called the **dimension** of U, and is denoted

$$\dim U = m.$$

The importance of the invariance theorem is that the dimension of U can be determined by counting the number of vectors in *any* basis.[8] This is very useful as we shall see.

---

7 The plural of "basis" is "bases".

8 We will show in Theorem 6 that every subspace of $\mathbb{R}^n$ does indeed have a basis.

Let $\{E_1, E_2, \ldots, E_n\}$ denote the standard basis of $\mathbb{R}^n$, that is the set of columns of the identity matrix. Then $\mathbb{R}^n = \text{span}\{E_1, E_2, \ldots, E_n\}$ by Example 6 §5.1, and $\{E_1, E_2, \ldots, E_n\}$ is independent by Example 2. Hence it is indeed a basis of $\mathbb{R}^n$ in the present terminology, and we have

---

Example 7

$\dim(\mathbb{R}^n) = n$ and $\{E_1, E_2, \ldots, E_n\}$ is a basis.

This agrees with our sense that $\mathbb{R}^2$ is two-dimensional and $\mathbb{R}^3$ is three-dimensional. It also says that $\mathbb{R}^1 = \mathbb{R}$ is one-dimensional, and $\{1\}$ is a basis. Returning to subspaces of $\mathbb{R}^n$, we define

$$\dim\{0\} = 0.$$

This amounts to saying $\{0\}$ has a basis containing *no* vectors. This makes sense because 0 cannot belong to *any* independent set (Example 4).

Example 8

Let $U = \{[r \ \ s \ \ t \ \ s]^T \mid r, s, \text{ and } t \text{ in } \mathbb{R}\}$. Show that U is a subspace of $\mathbb{R}^4$, find a basis of U, and calculate $\dim U$.

Solution   Clearly, $[r \ \ s \ \ t \ \ s]^T = rX_1 + sX_2 + tX_3$ where $X_1 = [1 \ \ 0 \ \ 0 \ \ 0]^T$, $X_2 = [0 \ \ 1 \ \ 0 \ \ 1]^T$, and $X_3 = [0 \ \ 0 \ \ 1 \ \ 0]^T$. It follows that $U = \text{span}\{X_1, X_2, X_3\}$, and hence that U is a subspace of $\mathbb{R}^4$. Moreover, if a linear combination $rX_1 + sX_2 + tX_3 = [r \ \ s \ \ t \ \ s]^T$ vanishes, it is clear that $r = s = t = 0$, so $\{X_1, X_2, X_3\}$ is independent. Hence $\{X_1, X_2, X_3\}$ is a basis of U and so $\dim U = 3$.

Example 9

Let $B = \{X_1, X_2, \ldots, X_n\}$ be a basis of $\mathbb{R}^n$. If A is an invertible $n \times n$ matrix, then $D = \{AX_1, AX_2, \ldots, AX_n\}$ is also a basis of $\mathbb{R}^n$.

Solution   Let X be a vector in $\mathbb{R}^n$. Then $A^{-1}X$ is in $\mathbb{R}^n$ so, since B is a basis, we have $A^{-1}X = t_1 X_1 + t_2 X_2 + \cdots + t_n X_n$ for t_i in $\mathbb{R}$. Left multiplication by A gives $X = t_1(AX_1) + t_2(AX_2) + \cdots + t_n(AX_n)$, and it follows that D spans $\mathbb{R}^n$. To show independence, let $s_1(AX_1) + s_2(AX_2) + \cdots + s_n(AX_n) = 0$, where the s_i are in $\mathbb{R}$. Then $A(s_1 X_1 + s_2 X_2 + \cdots + s_n X_n) = 0$, so left multiplication by A^{-1} gives $s_1 X_1 + s_2 X_2 + \cdots + s_n X_n = 0$. Now the independence of B shows that each $s_i = 0$, and so proves the independence of D. Hence D is a basis of $\mathbb{R}^n$.

While we have found bases in many subspaces of $\mathbb{R}^n$, we have not yet shown that *every* subspace *has* a basis. This is part of the next theorem, the proof of which is deferred to Section 6.4 where it will be proved in more generality.

Theorem 6

Let $U \neq \{0\}$ be a subspace of $\mathbb{R}^n$. Then:
1. U has a basis and $\dim U \leq n$.
2. Any independent set in U can be enlarged (by adding vectors) to a basis of U.
3. If B spans U, then B can be cut down (by deleting vectors) to a basis of U.

Theorem 6 has a number of useful consequences. Here is the first.

Theorem 7

Let U be a subspace of $\mathbb{R}^n$, and let $B = \{X_1, X_2, \ldots, X_m\}$ be a set of m vectors in U where $m = \dim U$. Then

$$B \text{ is independent} \quad \text{if and only if} \quad B \text{ spans } U.$$

PROOF

Suppose B is independent. If B does not span U then, by Theorem 6, B can be enlarged to a basis of U containing more than m vectors. This contradicts the invariance theorem because $\dim U = m$, so B spans U. Conversely, if B spans U but is not independent, then B can be cut down to a basis of U containing fewer than m vectors, again a contradiction. So B is independent, as required.

Theorem 7 is a "labour-saving" result. It asserts that, given a subspace U of dimension m and a set B of exactly m vectors in U, to prove that B is a basis of U it suffices to show either that B spans U or that B is independent. It is not necessary to verify both properties.

Example 10

Find a basis of $\mathbb{R}^4$ containing $B = \{X_1, X_2, X_3\}$ where $X_1 = [1 \ \ 2 \ \ {-1} \ \ 0]^T$, $X_2 = [0 \ \ 1 \ \ 1 \ \ 3]^T$, and $X_3 = [1 \ \ 1 \ \ 1 \ \ 1]^T$.

Solution   If $E_1 = [1 \ \ 0 \ \ 0 \ \ 0]^T$, then it is routine to verify that $\{E_1, X_1, X_2, X_3\}$ is linearly independent. Since $\mathbb{R}^4$ has dimension 4 it follows by Theorem 7 that $\{E_1, X_1, X_2, X_3\}$ is a basis.[9]

Theorem 8

Let $U \subseteq W$ be subspaces of $\mathbb{R}^n$. Then:
1. $\dim U \leq \dim W$.
2. If $\dim U = \dim W$, then $U = W$.

PROOF

Write $\dim W = k$, and let B be a basis of U.
1. If $\dim U > k$, then B is an independent set in W containing more than k vectors, contradicting the fundamental theorem. So $\dim U \leq \dim W$, proving (1).
2. If $\dim U = k$, then B is an independent set in W containing $k = \dim W$ vectors, so B spans W by Theorem 7. Hence $W = \operatorname{span} B = U$, proving (2).

---

9 In fact, an independent subset of $\mathbb{R}^n$ can always be enlarged to a basis by adding vectors from the standard basis of $\mathbb{R}^n$. (See Example 7 §6.4.)

It follows from Theorem 8 that if U is a subspace of $\mathbb{R}^n$, then dim $U = 0, 1, 2, \ldots, n$, and that

$$\text{dim } U = 0 \quad \text{if and only if} \quad U = \{0\}, \quad \text{and}$$
$$\text{dim } U = n \quad \text{if and only if} \quad U = \mathbb{R}^n$$

The other subspaces are called **proper**. The following example uses Theorem 8 to show that the proper subspaces of $\mathbb{R}^2$ are the lines through the origin, while the proper subspaces of $\mathbb{R}^3$ are the lines and planes through the origin.

Example 11

1. If U is a subspace of $\mathbb{R}^2$ or $\mathbb{R}^3$, then dim $U = 1$ if and only if U is a line through the origin.
2. If U is a subspace of $\mathbb{R}^3$, then dim $U = 2$ if and only if U is a plane through the origin.

PROOF

1. Since dim $U = 1$, let $\{\vec{u}\}$ be a basis of U. Then $U = \{t\vec{u} \mid t \text{ in } \mathbb{R}\}$, so U is the line through the origin with direction vector $\vec{u}$. Conversely each line L with direction vector $\vec{d}$ has the form $L = \{t\vec{d} \mid t \text{ in } \mathbb{R}\}$. Hence $\{\vec{d}\}$ is a basis of U, so U has dimension 1.
2. If $U \subseteq \mathbb{R}^3$ has dimension 2, let $\{\vec{v}, \vec{w}\}$ be a basis of U. Then $\vec{v}$ and $\vec{w}$ are not parallel (by Example 6) so $\vec{n} = \vec{v} \times \vec{w} \neq \vec{0}$. Let $P = \{\vec{x} \text{ in } \mathbb{R}^3 \mid \vec{n} \cdot \vec{x} = 0\}$ denote the plane through the origin with normal $\vec{n}$. Then P is a subspace of $\mathbb{R}^3$ (Example 1 §5.1) and both $\vec{v}$ and $\vec{w}$ lie in P (they are orthogonal to $\vec{n}$), so $U = \text{span}\{\vec{v}, \vec{w}\} \subseteq P$ by Theorem 1 §5.1. Hence

$$U \subseteq P \subseteq \mathbb{R}^3.$$

Since dim $U = 2$ and dim$(\mathbb{R}^3) = 3$, it follows from Theorem 8 that dim $P = 2$ or 3, whence $P = U$ or $\mathbb{R}^3$. But $P \neq \mathbb{R}^3$ (for example, $\vec{n}$ is not in P) and so $U = P$ is a plane through the origin.

Conversely, if U is a plane through the origin, then dim $U = 0, 1, 2,$ or 3 by Theorem 8. But dim $U \neq 0$ or 3 because $U \neq \{\vec{0}\}$ and $U \neq \mathbb{R}^3$, and dim $U \neq 1$ by (1). So dim $U = 2$.

Note that this proof shows that if $\vec{v}$ and $\vec{w}$ are nonzero, nonparallel vectors in $\mathbb{R}^3$, then $\text{span}\{\vec{v}, \vec{w}\}$ is the plane with normal $\vec{n} = \vec{v} \times \vec{w}$. We gave a geometrical verification of this fact in Section 5.1.

Proof of the Fundamental Theorem

Fundamental Theorem (Theorem 4)

Let U be a subspace of $\mathbb{R}^n$. If $U = \text{span}\{X_1, X_2, \ldots, X_m\}$ and if $\{Y_1, Y_2, \ldots, Y_k\}$ is an independent set in U, then $k \leq m$.

PROOF

We assume that $k > m$ and show that this leads to a contradiction. Each Y_j is in $U = \text{span}\{X_1, X_2, \ldots, X_m\}$, so write

$$Y_j = a_{1j}X_1 + a_{2j}X_2 + \cdots + a_{mj}X_m = \sum_{i=1}^{m} a_{ij}X_i, \qquad j = 1, 2, \ldots, k,$$

where the coefficients a_{ij} are real numbers. These coefficients form column j of a matrix $A = [a_{ij}]$ of size $m \times k$. Since $k > m$ by assumption, the homogeneous system $AX = 0$ has a nontrivial solution $X = [x_1 \ x_2 \ \cdots \ x_k]^T \neq 0$. Consider the linear combination of the Y_j with these x_j as coefficients:

$$\sum_{j=1}^{k} x_j Y_j = \sum_{j=1}^{k} x_j \left(\sum_{i=1}^{m} a_{ij}X_i \right) = \sum_{i=1}^{m} \left(\sum_{j=1}^{k} a_{ij}x_j \right) X_i = \sum_{i=1}^{m} (0)X_i = 0,$$

where $\sum_{j=1}^{k} a_{ij}x_j = 0$ for each i because it is the ith entry of $AX = 0$. Since the Y_j are independent, the equation $\sum_{j=1}^{k} x_j Y_j = 0$ implies that each $x_j = 0$, a contradiction because $X \neq 0$.

Exercises 5.2

1. Which of the following subsets are independent? Support your answer.
   (a) $\{[1 \ -1 \ 0]^T, [3 \ 2 \ -1]^T, [3 \ 5 \ -2]^T\}$ in $\mathbb{R}^3$.
   ◆(b) $\{[1 \ 1 \ 1]^T, [1 \ -1 \ 1]^T, [0 \ 0 \ 1]^T\}$ in $\mathbb{R}^3$.
   (c) $\{[1 \ -1 \ 1 \ -1]^T, [2 \ 0 \ 1 \ 0]^T, [0 \ -2 \ 1 \ -2]^T\}$ in $\mathbb{R}^4$.
   ◆(d) $\{[1 \ 1 \ 0 \ 0]^T, [1 \ 0 \ 1 \ 0]^T, [0 \ 0 \ 1 \ 1]^T, [0 \ 1 \ 0 \ 1]^T\}$ in $\mathbb{R}^4$.

2. Let $\{X, Y, Z, W\}$ be an independent set in $\mathbb{R}^n$. Which of the following sets is independent? Support your answer.
   (a) $\{X - Y, Y - Z, Z - X\}$
   ◆(b) $\{X + Y, Y + Z, Z + X\}$
   (c) $\{X - Y, Y - Z, Z - W, W - X\}$
   ◆(d) $\{X + Y, Y + Z, Z + W, W + X\}$

3. Find a basis and calculate the dimension of the following subspaces of $\mathbb{R}^4$.
   (a) $\text{span}\{[1 \ -1 \ 2 \ 0]^T, [2 \ 3 \ 0 \ 3]^T, [1 \ 9 \ -6 \ 6]^T\}$.
   ◆(b) $\text{span}\{[2 \ 1 \ 0 \ -1]^T, [-1 \ 1 \ 1 \ 2]^T, [2 \ 7 \ 4 \ 5]^T\}$.
   (c) $\text{span}\{[-1 \ 2 \ 1 \ 0]^T, [2 \ 0 \ 3 \ -1]^T, [4 \ 4 \ 11 \ -3]^T, [3 \ -2 \ 2 \ -1]^T\}$.
   ◆(d) $\text{span}\{[-2 \ 0 \ 3 \ 1]^T, [1 \ 2 \ -1 \ 0]^T, [-2 \ 8 \ 5 \ 3]^T, [-1 \ 2 \ 2 \ 1]^T\}$.

4. Find a basis and calculate the dimension of the following subspaces of $\mathbb{R}^4$.
   (a) $U = \{[a \ a+b \ a-b \ b]^T \mid a \text{ and } b \text{ in } \mathbb{R}\}$.

◆(b) $U = \{[a+b \ a-b \ b \ a]^T \mid a \text{ and } b \text{ in } \mathbb{R}\}$.
(c) $U = \{[a \ b \ c+a \ c]^T \mid a, b, \text{ and } c \text{ in } \mathbb{R}\}$.
◆(d) $U = \{[a-b \ b+c \ a \ b+c]^T \mid a, b, \text{ and } c \text{ in } \mathbb{R}\}$.
(e) $U = \{[a \ b \ c \ d]^T \mid a+b-c+d=0 \text{ in } \mathbb{R}\}$.
◆(f) $U = \{[a \ b \ c \ d]^T \mid a+b=c+d \text{ in } \mathbb{R}\}$.

5. Suppose that $\{X, Y, Z, W\}$ is a basis of $\mathbb{R}^4$. Show that:
   (a) $\{X + aW, Y, Z, W\}$ is also a basis of $\mathbb{R}^4$ for any choice of the scalar a.
   ◆(b) $\{X + W, Y + W, Z + W, W\}$ is also a basis of $\mathbb{R}^4$.
   (c) $\{X, X+Y, X+Y+Z, X+Y+Z+W\}$ is also a basis of $\mathbb{R}^4$.

6. Use Theorem 3 to determine if the following sets of vectors are a basis of the indicated space.
   (a) $\{[3 \ -1]^T, [2 \ 2]^T\}$ in $\mathbb{R}^2$.
   ◆(b) $\{[1 \ 1 \ -1]^T, [1 \ -1 \ 1]^T, [0 \ 0 \ 1]^T\}$ in $\mathbb{R}^3$.
   (c) $\{[-1 \ 1 \ -1]^T, [1 \ -1 \ 2]^T, [0 \ 0 \ 1]^T\}$ in $\mathbb{R}^3$.
   ◆(d) $\{[5 \ 2 \ -1]^T, [1 \ 0 \ 1]^T, [3 \ -1 \ 0]^T\}$ in $\mathbb{R}^3$.
   (e) $\{[2 \ 1 \ -1 \ 3]^T, [1 \ 1 \ 0 \ 2]^T, [0 \ 1 \ 0 \ -3]^T, [-1 \ 2 \ 3 \ 1]^T\}$ in $\mathbb{R}^4$.
   ◆(f) $\{[1 \ 0 \ -2 \ 5]^T, [4 \ 4 \ -3 \ 2]^T, [0 \ 1 \ 0 \ -3]^T, [1 \ 3 \ 3 \ -10]^T\}$ in $\mathbb{R}^4$.

7. In each case show that the statement is true or give an example showing that it is false.

   (a) If $\{X, Y\}$ is independent, then $\{X, Y, X + Y\}$ is independent.

   ♦(b) If $\{X, Y, Z\}$ is independent, then $\{Y, Z\}$ is independent.

   (c) If $\{Y, Z\}$ is dependent, then $\{X, Y, Z\}$ is dependent.

   ♦(d) If all of $X_1, X_2, \ldots, X_k$ are nonzero, then $\{X_1, X_2, \ldots, X_k\}$ is independent.

   (e) If one of $X_1, X_2, \ldots, X_k$ is zero, then $\{X_1, X_2, \ldots, X_k\}$ is dependent.

   ♦(f) If $aX + bY + cZ = 0$, then $\{X, Y, Z\}$ is independent.

   (g) If $\{X, Y, Z\}$ is independent, then $aX + bY + cZ = 0$ for some a, b, and c in $\mathbb{R}$.

   ♦(h) If $\{X_1, X_2, \ldots, X_k\}$ is dependent, then $t_1X_1 + t_2X_2 + \cdots + t_kX_k = 0$ for some numbers t_i in $\mathbb{R}$ not all zero.

   (i) If $\{X_1, X_2, \ldots, X_k\}$ is independent, then $t_1X_1 + t_2X_2 + \cdots + t_kX_k = 0$ for some t_i in $\mathbb{R}$.

8. If A is an $n \times n$ matrix, show that $\det A = 0$ if and only if some column of A is a linear combination of the other columns.

9. Let $\{X, Y, Z\}$ be a linearly independent set in $\mathbb{R}^4$. Show that $\{X, Y, Z, E_k\}$ is a basis of $\mathbb{R}^4$ for some E_k in the standard basis $\{E_1, E_2, E_3, E_4\}$.

♦10. If $\{X_1, X_2, X_3, X_4, X_5, X_6\}$ is an independent set of vectors, show that the subset $\{X_2, X_3, X_5\}$ is also independent.

11. Let A be any $m \times n$ matrix, and let $B_1, B_2, B_3, \ldots, B_k$ be columns in $\mathbb{R}^m$ such that the system $AX = B_i$ has a solution X_i for each i. If $\{B_1, B_2, B_3, \ldots, B_k\}$ is independent in $\mathbb{R}^m$, show that $\{X_1, X_2, X_3, \ldots, X_k\}$ is independent in $\mathbb{R}^n$.

♦12. If $\{X_1, X_2, X_3, \ldots, X_k\}$ is independent, show that $\{X_1, X_1 + X_2, X_1 + X_2 + X_3, \ldots, X_1 + X_2 + \cdots + X_k\}$ is also independent.

13. If $\{Y, X_1, X_2, X_3, \ldots, X_k\}$ is independent, show that $\{Y + X_1, Y + X_2, Y + X_3, \ldots, Y + X_k\}$ is also independent.

14. Suppose that $\{X, Y\}$ is a basis of $\mathbb{R}^2$, and let $A = \begin{bmatrix} a & b \\ c & d \end{bmatrix}$.

   (a) If A is invertible, show that $\{aX + bY, cX + dY\}$ is a basis of $\mathbb{R}^2$.

   ♦(b) If $\{aX + bY, cX + dY\}$ is a basis of $\mathbb{R}^2$, show that A is invertible.

15. Let A denote an $m \times n$ matrix.

   (a) Show that $\operatorname{null} A = \operatorname{null}(UA)$ for every invertible $m \times m$ matrix U.

   ♦(b) Show that $\dim(\operatorname{null} A) = \dim(\operatorname{null}(AV))$ for every invertible $n \times n$ matrix V. [*Hint:* If $\{X_1, X_2, \ldots, X_k\}$ is a basis of $\operatorname{null} A$, show that $\{V^{-1}X_1, V^{-1}X_2, \ldots, V^{-1}X_k\}$ is a basis of $\operatorname{null}(AV)$.]

16. Let A denote an $m \times n$ matrix.

   (a) Show that $\operatorname{im} A = \operatorname{im}(AV)$ for every invertible $n \times n$ matrix V.

   (b) Show that $\dim(\operatorname{im} A) = \dim(\operatorname{im}(UA))$ for every invertible $m \times m$ matrix U. [*Hint:* If $\{Y_1, Y_2, \ldots, Y_k\}$ is a basis of $\operatorname{im}(UA)$, show that $\{U^{-1}Y_1, U^{-1}Y_2, \ldots, U^{-1}Y_k\}$ is a basis of $\operatorname{im} A$.]

17. Let U and W denote subspaces of $\mathbb{R}^n$, and assume that $U \subseteq W$. If $\dim U = n - 1$, show that either $W = U$ or $W = \mathbb{R}^n$.

♦18. Let U and W denote subspaces of $\mathbb{R}^n$, and assume that $U \subseteq W$. If $\dim W = 1$, show that either $U = \{0\}$ or $U = W$.

SECTION 5.3 Orthogonality

Length and orthogonality are basic concepts in geometry and, in $\mathbb{R}^2$ and $\mathbb{R}^3$, they can both can be defined using the dot product. In this section we extend these concepts to $\mathbb{R}^n$, introduce the idea of an orthogonal basis—one of the most useful concepts in linear algebra, and begin exploring some of its applications.

Dot Product, Length, and Distance

Let $X = [x_1 \ x_2 \ \cdots \ x_n]^T$ and $Y = [y_1 \ y_2 \ \cdots \ y_n]^T$ be two vectors in $\mathbb{R}^n$. The **dot product** $X \cdot Y$ of X and Y is defined by

$$X \cdot Y = X^T Y = x_1 y_2 + x_2 y_2 + \cdots + x_n y_n.$$

Note that technically $X^T Y$ is a 1×1 matrix, which we take to be a number. The **length** $\|X\|$ of the vector X is defined by

$$\|X\| = \sqrt{X \cdot X} = \sqrt{x_1^2 + x_2^2 + \cdots + x_n^2}$$

where $\sqrt{\ }$ indicates the positive square root. A vector of length 1 is called a **unit vector**.

Example 1

> If $X = [1 \ 0 \ -3 \ 2 \ 1]^T$ and $Y = [2 \ -1 \ 5 \ 1 \ -1]^T$ in $\mathbb{R}^5$, then
> $X \cdot Y = 2 + 0 - 15 + 2 - 1 = -12$, and $\|X\| = \sqrt{1 + 0 + 9 + 4 + 1} = \sqrt{15}$.

These definitions agree with those in $\mathbb{R}^2$ and $\mathbb{R}^3$, and many properties carry over to $\mathbb{R}^n$:

Theorem 1

Let X, Y, and Z denote vectors in $\mathbb{R}^n$. Then:
  1. $X \cdot Y = Y \cdot X$.
  2. $X \cdot (Y + Z) = X \cdot Y + X \cdot Z$.
  3. $(aX) \cdot Y = a(X \cdot Y) = X \cdot (aY)$ for all scalars a.
  4. $\|X\|^2 = X \cdot X$.
  5. $\|X\| \geq 0$, and $\|X\| = 0$ if and only if $X = 0$.
  6. $\|aX\| = |a| \|X\|$ for all scalars a.

PROOF

(1), (2), and (3) follow from matrix arithmetic because $X \cdot Y = X^T Y$; (4) is clear from the definition; and (6) is a routine verification since $|a| = \sqrt{a^2}$. If $X = [x_1 \ x_2 \ \cdots \ x_n]^T$, then $\|X\| = \sqrt{x_1^2 + x_2^2 + \cdots + x_n^2}$, so $\|X\| = 0$ if and only if $x_1^2 + x_2^2 + \cdots + x_n^2 = 0$. Since each x_i is a real number this happens if and only if $x_i = 0$ for each i; that is, if and only if $X = 0$. This proves (5).

Because of Theorem 1, computations with dot products in $\mathbb{R}^n$ are similar to those in $\mathbb{R}^3$. In particular, the dot product

$$(X_1 + X_2 + \cdots + X_m) \cdot (Y_1 + Y_2 + \cdots + Y_n)$$

equals the sum of mn terms, $X_i \cdot Y_j$, one for each choice of i and j. For example:

$$(3X - 4Y) \cdot (7X + 2Y) = 21 X \cdot X + 6 X \cdot Y - 28 Y \cdot X - 8 Y \cdot Y$$
$$= 21\|X\|^2 - 22 X \cdot Y - 8\|Y\|^2$$

holds for all vectors X and Y.

Example 2

Show that $\|X + Y\|^2 = \|X\|^2 + 2X \cdot Y + \|Y\|^2$ for any X and Y in $\mathbb{R}^n$.

Solution   Using Theorem 1 several times:

$$\|X + Y\|^2 = (X + Y) \cdot (X + Y) = X \cdot X + X \cdot Y + Y \cdot X + Y \cdot Y$$
$$= \|X\|^2 + 2X \cdot Y + \|Y\|^2.$$

Example 3

Suppose that $\mathbb{R}^n = \text{span}\{F_1, F_2, \ldots, F_k\}$ for some vectors F_i. If $X \cdot F_i = 0$ for each i where X is in $\mathbb{R}^n$, show that $X = 0$.

Solution   We show $X = 0$ by showing that $\|X\| = 0$ and using (5) of Theorem 1. Since the F_i span $\mathbb{R}^n$, write $X = t_1F_1 + t_2F_2 + \cdots + t_kF_k$ where the t_i are in $\mathbb{R}$. Then

$$\|X\|^2 = X \cdot X = X \cdot (t_1F_1 + t_2F_2 + \cdots + t_kF_k)$$
$$= t_1(X \cdot F_1) + t_2(X \cdot F_2) + \cdots + t_k(X \cdot F_k)$$
$$= t_1(0) + t_2(0) + \cdots + t_k(0) = 0.$$

We saw in Section 4.2 that if $\vec{u}$ and $\vec{v}$ are nonzero vectors in $\mathbb{R}^3$, then
$\frac{\vec{u} \cdot \vec{v}}{\|\vec{u}\|\|\vec{u}\|} = \cos\theta$ where θ is the angle between $\vec{u}$ and $\vec{v}$. Since $|\cos\theta| \leq 1$ for any angle θ, this shows that $|\vec{u} \cdot \vec{v}| \leq \|\vec{u}\|\|\vec{v}\|$. In this form the result holds in $\mathbb{R}^n$.

Theorem 2 Cauchy Inequality[10].

If X and Y are vectors in $\mathbb{R}^n$, then

$$|X \cdot Y| \leq \|X\|\|Y\|.$$

Moreover $|X \cdot Y| = \|X\|\|Y\|$ if and only if one of X and Y is a multiple of the other.

Augustin Louis Cauchy.
Photo © Corbis.

PROOF

The inequality holds if $X = 0$ or $Y = 0$ (in fact it is equality). Otherwise, write $\|X\| = a > 0$ and $\|Y\| = b > 0$ for convenience. A computation like that in Example 2 gives

$$\|bX - aY\|^2 = 2ab(ab - X \cdot Y) \quad \text{and} \quad \|bX + aY\|^2 = 2ab(ab + X \cdot Y). \quad (*)$$

It follows that $ab - X \cdot Y \geq 0$ and $ab + X \cdot Y \geq 0$, and so that $-ab \leq X \cdot Y \leq ab$. Hence $|X \cdot Y| \leq ab = \|X\|\|Y\|$, proving the Cauchy inequality.

---

10 Augustin Louis Cauchy (1789–1857) was born in Paris and became a professor at the École Polytechnique at the age of 26. He was one of the great mathematicians, producing more than 700 papers, and is best remembered for his work in analysis in which he established new standards of rigour and founded the theory of functions of a complex variable. He was a devout Catholic with a long-term interest in charitable work, and he was a royalist, following King Charles X into exile in Prague after he was deposed in 1830.

If equality holds, then $|X \cdot Y| = ab$, so $X \cdot Y = ab$ or $X \cdot Y = -ab$. Hence (∗) shows that $bX - aY = 0$ or $bX + aY = 0$, so one of X and Y is a multiple of the other (even if $a = 0$ or $b = 0$).

The Cauchy inequality is equivalent to $(X \cdot Y)^2 \leq \|X\|^2 \|X\|^2$; for example, in $\mathbb{R}^5$ this becomes

$$(x_1 y_1 + x_2 y_2 + x_3 y_3 + x_4 y_4 + x_5 y_5)^2$$
$$\leq (x_1^2 + x_2^2 + x_3^2 + x_4^2 + x_5^2)^2 (y_1^2 + y_2^2 + y_3^2 + y_4^2 + y_5^2)^2$$

for all x_i and y_i in $\mathbb{R}$.

There is an important consequence of the Cauchy inequality. Given X and Y in $\mathbb{R}^n$, use Example 2 and the fact that $X \cdot Y \leq \|X\| \|Y\|$ to compute

$$\|X + Y\|^2 = \|X\|^2 + 2X \cdot Y + \|Y\|^2 \leq \|X\|^2 + 2\|X\|\|Y\| + \|Y\|^2 = (\|X\| + \|Y\|)^2.$$

Taking positive square roots gives:

Corollary Triangle Inequality

If X and Y are vectors in $\mathbb{R}^n$, then $\|X + Y\| \leq \|X\| + \|Y\|$.

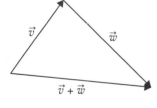

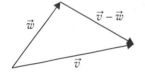

The reason for the name comes from the observation that in $\mathbb{R}^2$ the inequality asserts that the sum of the lengths of two sides of a triangle is not less than the length of the third side. This is illustrated in the first diagram.

If X and Y are two vectors in $\mathbb{R}^n$, we define the **distance** $d(X, Y)$ between X and Y by

$$d(X, Y) = \|X - Y\|$$

The motivation again comes from $\mathbb{R}^2$ as is clear in the second diagram. This distance function has all the intuitive properties of distance in $\mathbb{R}^2$, including another version of the triangle inequality.

Theorem 3

If X, Y, and Z are three vectors in $\mathbb{R}^n$ we have:
1. $d(X, Y) \geq 0$ for all X and Y.
2. $d(X, Y) = 0$ if and only if $X = Y$.
3. $d(X, Y) = d(Y, X)$.
4. **Triangle inequality.** $d(X, Z) \leq d(X, Y) + d(Y, Z)$.

PROOF

(1) and (2) restate part (5) of Theorem 1 because $d(X, Y) = \|X - Y\|$, and (3) follows because $\|U\| = \|-U\|$ for every vector U in $\mathbb{R}^n$. To prove (4) use the Corollary to Theorem 2:

$$d(X, Z) = \|X - Z\| = \|(X - Y) + (Y - Z)\|$$
$$\leq \|(X - Y)\| + \|(Y - Z)\| = d(X, Y) + d(Y, Z).$$

Orthogonal Sets and the Expansion Theorem

Two nonzero vectors $\vec{v}$ and $\vec{w}$ in $\mathbb{R}^3$ are orthogonal if and only if $\vec{v} \cdot \vec{w} = 0$ (Theorem 3 §4.2). More generally, a set $\{X_1, X_2, \ldots, X_k\}$ of vectors in $\mathbb{R}^n$ is called an **orthogonal set** if

$$X_i \cdot X_j = 0 \text{ for all } i \neq j \quad \text{and} \quad X_i \neq 0 \text{ for all } i \text{ }^{11}.$$

Note that $\{X\}$ is an orthogonal set if $X \neq 0$. A set $\{X_1, X_2, \ldots, X_k\}$ of vectors in $\mathbb{R}^n$ is called **orthonormal** if it is orthogonal and, in addition, each X_i is a unit vector:

$$\|X_i\| = 1 \quad \text{for each } i.$$

Example 4

The standard basis $\{E_1, E_2, \ldots, E_n\}$ is an orthonormal set in $\mathbb{R}^n$.

The routine verification is left to the reader, as is the proof of:

Example 5

If $\{X_1, X_2, \ldots, X_k\}$ is orthogonal, so also is $\{a_1 X_1, a_2 X_2, \ldots, a_k X_k\}$ for any nonzero scalars a_i.

If $X \neq 0$, it follows from item (6) of Theorem 1 that $\dfrac{1}{\|X\|} X$ is a unit vector, that is it has length 1. Hence if $\{X_1, X_2, \ldots, X_k\}$ is an orthogonal set, then $\left\{\dfrac{1}{\|X_1\|} X_1, \dfrac{1}{\|X_2\|} X_2, \ldots, \dfrac{1}{\|X_k\|} X_k\right\}$ is an orthonormal set, and we say that it is the result of **normalizing** the orthogonal set $\{X_1, X_2, \ldots, X_k\}$.

Example 6

If $F_1 = [1 \ \ 1 \ \ 1 \ \ -1]^T$, $F_2 = [1 \ \ 0 \ \ 1 \ \ 2]^T$, $F_3 = [-1 \ \ 0 \ \ 1 \ \ 0]^T$, and $F_4 = [-1 \ \ 3 \ \ -1 \ \ 1]^T$, then $\{F_1, F_2, F_3, F_4\}$ is an orthogonal set in $\mathbb{R}^4$ as is easily verified. After normalizing, the corresponding orthonormal set is $\{\frac{1}{2} F_1, \frac{1}{\sqrt{6}} F_2, \frac{1}{\sqrt{2}} F_3, \frac{1}{2\sqrt{3}} F_4\}$.

The most important result about orthogonality is Pythagoras' theorem. Given orthogonal vectors $\vec{v}$ and $\vec{w}$ in $\mathbb{R}^3$, it asserts that $\|\vec{v} + \vec{w}\|^2 = \|\vec{v}\|^2 + \|\vec{w}\|^2$ as in the diagram. In this form the result holds for any orthogonal set in $\mathbb{R}^n$.

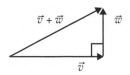

Theorem 4 Pythagoras' Theorem

If $\{X_1, X_2, \ldots, X_k\}$ is a orthogonal set in $\mathbb{R}^n$, then

$$\|X_1 + X_2 + \cdots + X_k\|^2 = \|X_1\|^2 + \|X_2\|^2 + \cdots + \|X_k\|^2.$$

PROOF

The fact that $X_i \cdot X_j = 0$ whenever $i \neq j$ gives

$$
\begin{aligned}
\|X_1 + X_2 + \cdots + X_k\|^2 &= (X_1 + X_2 + \cdots + X_k) \cdot (X_1 + X_2 + \cdots + X_k) \\
&= (X_1 \cdot X_1 + X_2 \cdot X_2 + \cdots + X_k \cdot X_k) \\
&\quad + (X_1 \cdot X_2 + X_1 \cdot X_3 + X_2 \cdot X_3 + \cdots) \\
&= (\|X_1\|^2 + \|X_2\|^2 + \cdots + \|X_k\|^2) + (0 + 0 + 0 + \cdots).
\end{aligned}
$$

This is what we wanted.

---

11 The reason for insisting that orthogonal sets consist of *nonzero* vectors is that we will be primarily concerned with orthogonal *bases*.

If $\vec{v}$ and $\vec{w}$ are orthogonal, nonzero vectors in $\mathbb{R}^3$, then they are certainly not parallel, and so are linearly independent by Example 6 §5.2. The next theorem gives a far-reaching extension of this observation.

Theorem 5

Every orthogonal set in $\mathbb{R}^n$ is linearly independent.

PROOF

Let $\{X_1, X_2, \ldots, X_k\}$ be an orthogonal set in $\mathbb{R}^n$ and suppose a linear combination vanishes: $t_1 X_1 + t_2 X_2 + \cdots + t_k X_k = 0$. Then

$$
\begin{aligned}
0 = X_1 \bullet 0 &= X_1 \bullet (t_1 X_1 + t_2 X_2 + \cdots + t_k X_k) \\
&= t_1(X_1 \bullet X_1) + t_2(X_1 \bullet X_2) + \cdots + t_k(X_1 \bullet X_k) \\
&= t_1 \|X_1\|^2 + t_2(0) + \cdots + t_k(0) \\
&= t_1 \|X_1\|^2.
\end{aligned}
$$

Since $\|X_1\|^2 \neq 0$, this implies that $t_1 = 0$. Similarly $t_i = 0$ for each i.

Theorem 5 suggests considering orthogonal bases for $\mathbb{R}^n$, that is orthogonal sets that span $\mathbb{R}^n$. These turn out to be the best bases. One reason is that, when expanding a vector as a linear combination of the basis vectors, there are explicit formulas for the coefficients.

Theorem 6 Expansion Theorem

Let $\{F_1, F_2, \ldots, F_m\}$ be an orthogonal basis of a subspace U of $\mathbb{R}^n$. If X is any vector in U, we have

$$
X = \left(\frac{X \bullet F_1}{\|F_1\|^2} \right) F_1 + \left(\frac{X \bullet F_2}{\|F_2\|^2} \right) F_2 + \cdots + \left(\frac{X \bullet F_m}{\|F_m\|^2} \right) F_m.
$$

PROOF

Since $\{F_1, F_2, \ldots, F_m\}$ spans U, we have $X = t_1 F_1 + t_2 F_2 + \cdots + t_m F_m$ where the t_i are scalars. To find t_1 we take the dot product of both sides with F_1:

$$
\begin{aligned}
X \bullet F_1 &= (t_1 F_1 + t_2 F_2 + \cdots + t_m F_m) \bullet F_1 \\
&= t_1(F_1 \bullet F_1) + t_2(F_2 \bullet F_1) + \cdots + t_m(F_m \bullet F_1) \\
&= t_1 \|F_1\|^2 + t_2(0) + \cdots + t_m(0) \\
&= t_1 \|F_1\|^2.
\end{aligned}
$$

Since $F_1 \neq 0$, this gives $t_1 = \dfrac{X \bullet F_1}{\|F_1\|^2}$. Similarly, $t_i = \dfrac{X \bullet F_i}{\|F_i\|^2}$ for each i.

The expansion of X as a linear combination of the orthogonal basis $\{F_1, F_2, \ldots, F_m\}$ is called the **Fourier expansion** of X, and the coefficients $t_i = \dfrac{X \bullet F_i}{\|F_i\|^2}$ are called the **Fourier coefficients**. Note that if $\{F_1, F_2, \ldots, F_m\}$ is actually orthonormal, then $t_i = X \bullet F_i$ for each i. We will have a great deal more to say about this in Section 10.5.

Example 7

Expand $X = [a \; b \; c \; d]^T$ as a linear combination of the orthogonal basis $\{F_1, F_2, F_3, F_4\}$ of $\mathbb{R}^4$ given in Example 6.

Solution   We have $F_1 = [1 \; 1 \; 1 \; -1]^T$, $F_2 = [1 \; 0 \; 1 \; 2]^T$, $F_3 = [-1 \; 0 \; 1 \; 0]^T$, and $F_4 = [-1 \; 3 \; -1 \; 1]^T$, so the Fourier coefficients are

$$t_1 = \frac{X \bullet F_1}{\|F_1\|^2} = \tfrac{1}{4}(a + b + c - d) \qquad t_3 = \frac{X \bullet F_3}{\|F_3\|^2} = \tfrac{1}{2}(-a + c)$$

$$t_2 = \frac{X \bullet F_2}{\|F_2\|^2} = \tfrac{1}{6}(a + c + 2d) \qquad t_4 = \frac{X \bullet F_4}{\|F_4\|^2} = \tfrac{1}{12}(-a + 3b - c + d)$$

The reader can verify that indeed $X = t_1 F_1 + t_2 F_2 + t_3 F_3 + t_4 F_4$.

A natural question arises here: Does every subspace U of $\mathbb{R}^n$ *have* an orthogonal basis? The answer is "yes"; in fact, there is a systematic procedure, called the Gram–Schmidt algorithm, for turning any basis of U into an orthogonal one. This leads to a definition of the projection onto a subspace U that generalizes the projection along a vector used in $\mathbb{R}^2$ and $\mathbb{R}^3$. All this is discussed in Section 8.1.

Exercises 5.3

1. Obtain an orthonormal basis of $\mathbb{R}^3$ by normalizing the following.

   (a) $\{[1 \; -1 \; 2]^T, [0 \; 2 \; 1]^T, [5 \; 1 \; -2]^T\}$

   ◆(b) $\{[1 \; 1 \; 1]^T, [4 \; 1 \; -5]^T, [2 \; -3 \; 1]^T\}$

2. In each case, show that the set of vectors is orthogonal in $\mathbb{R}^4$.

   (a) $\{[1 \; -1 \; 2 \; 5]^T, [4 \; 1 \; 1 \; -1]^T, [-7 \; 28 \; 5 \; 5]^T\}$

   (b) $\{[2 \; -1 \; 4 \; 5]^T, [0 \; -1 \; 1 \; -1]^T, [0 \; 3 \; 2 \; -1]^T\}$

3. In each case, show that B is an orthogonal basis of $\mathbb{R}^3$ and use Theorem 6 to expand $X = [a \; b \; c]^T$ as a linear combination of the basis vectors.

   (a) $B = \{[1 \; -1 \; 3]^T, [-2 \; 1 \; 1]^T, [4 \; 7 \; 1]^T\}$

   ◆(b) $B = \{[1 \; 0 \; -1]^T, [1 \; 4 \; 1]^T, [2 \; -1 \; 2]^T\}$

   (c) $B = \{[1 \; 2 \; 3]^T, [-1 \; -1 \; 1]^T, [5 \; -4 \; 1]^T\}$

   ◆(d) $B = \{[1 \; 1 \; 1]^T, [1 \; -1 \; 0]^T, [1 \; 1 \; -2]^T\}$

4. In each case, write X as a linear combination of the orthogonal basis of the subspace U.

   (a) $X = [13 \; -20 \; 15]^T$; $U = \text{span}\{[1 \; -2 \; 3]^T, [-1 \; 1 \; 1]^T\}$

   ◆(b) $X = [14 \; 1 \; -8 \; 5]^T$; $U = \text{span}\{[2 \; -1 \; 0 \; 3]^T, [2 \; 1 \; -2 \; -1]^T\}$

5. In each case, find all $[a \; b \; c \; d]^T$ in $\mathbb{R}^4$ such that the given set is orthogonal.

   (a) $\{[1 \; 2 \; 1 \; 0]^T, [1 \; -1 \; 1 \; 3]^T, [2 \; -1 \; 0 \; -1]^T, [a \; b \; c \; d]^T\}$

   ◆(b) $\{[1 \; 0 \; -1 \; 1]^T, [2 \; 1 \; 1 \; -1]^T, [1 \; -3 \; 1 \; 0]^T, [a \; b \; c \; d]^T\}$

6. If $\|X\| = 3$, $\|Y\| = 1$, and $X \bullet Y = -2$, compute:

   (a) $\|3X - 5Y\|$

   ◆(b) $\|2X + 7Y\|$

   (c) $(3X - Y) \bullet (2Y - X)$

   ◆(d) $(X - 2Y) \bullet (3X + 5Y)$

7. In each case either show that the statement is true or give an example showing that it is false.

   (a) Every independent set in $\mathbb{R}^n$ is orthogonal.

   ◆(b) If $\{X, Y\}$ is an orthogonal set in $\mathbb{R}^n$, then $\{X, X + Y\}$ is also orthogonal.

   (c) If $\{X, Y\}$ and $\{Z, W\}$ are both orthogonal in $\mathbb{R}^n$, then $\{X, Y, Z, W\}$ is also orthogonal.

   ◆(d) If $\{X_1, X_2\}$ and $\{Y_1, Y_2, Y_3\}$ are both orthogonal and $X_i \bullet Y_j = 0$ for all i and j, then $\{X_1, X_2, Y_1, Y_2, Y_3\}$ is orthogonal.

   (e) If $\{X_1, X_2, \dots, X_n\}$ is orthogonal in $\mathbb{R}^n$, then $\mathbb{R}^n = \text{span}\{X_1, X_2, \dots, X_n\}$.

   ◆(f) If $X \neq 0$ in $\mathbb{R}^n$, then $\{X\}$ is an orthogonal set.

8. If A is an $m \times n$ matrix with orthonormal columns, show that $A^T A = I_n$. [*Hint:* If $C_1, C_2, \ldots, C_n$ are the columns of A, show that column j of $A^T A$ is $[C_1 \bullet C_j \quad C_2 \bullet C_j \quad \cdots \quad C_n \bullet C_j]^T$.]

9. Use the Cauchy inequality to show that $\sqrt{xy} \le \frac{1}{2}(x + y)$ for all $x \ge 0$ and $y \ge 0$. Here $\sqrt{xy}$ and $\frac{1}{2}(x + y)$ are called, respectively, the *geometric mean* and *arithmetic mean* of x and y.

   [*Hint:* Use $X = \begin{bmatrix} \sqrt{x} \\ \sqrt{y} \end{bmatrix}$ and $Y = \begin{bmatrix} \sqrt{y} \\ \sqrt{x} \end{bmatrix}$.]

10. Use the Cauchy inequality to prove that:

    (a) $(r_1 + r_2 + \cdots + r_n)^2 \le n(r_1^2 + r_2^2 + \cdots + r_n^2)$ for all r_i in $\mathbb{R}$ and all $n \ge 1$.

    ◆(b) $r_1 r_2 + r_1 r_3 + r_2 r_3 \le r_1^2 + r_2^2 + r_3^2$ for all r_1, r_2, and r_3 in $\mathbb{R}$. [*Hint:* See part (a).]

11. (a) Show that X and Y are orthogonal in $\mathbb{R}^n$ if and only if $\|X + Y\| = \|X - Y\|$.

    ◆(b) Show that $X + Y$ and $X - Y$ are orthogonal

in $\mathbb{R}^n$ if and only if $\|X\| = \|Y\|$.

12. Show that $\|X + Y\|^2 = \|X\|^2 + \|Y\|^2$ if and only if X is orthogonal to Y.

13. (a) Show that $X \bullet Y = \frac{1}{4}[\|X + Y\|^2 - \|X - Y\|^2]$ for all X, Y in $\mathbb{R}^n$.

    (b) Show that $\|X\|^2 + \|Y\|^2 = \frac{1}{2}[\|X + Y\|^2 + \|X - Y\|^2]$ for all X, Y in $\mathbb{R}^n$.

14. If A is $n \times n$, show that every eigenvalue of $A^T A$ is nonnegative. [*Hint:* Compute $\|AX\|^2$ where X is an eigenvector.]

15. If $\mathbb{R}^n = \text{span}\{X_1, \ldots, X_m\}$ and $X \bullet X_i = 0$ for all i, show that $X = 0$. [*Hint:* Show $\|X\| = 0$.]

16. If $\mathbb{R}^n = \text{span}\{X_1, \ldots, X_m\}$ and $X \bullet X_i = Y \bullet X_i$ for all i, show that $X = Y$.

17. Let $\{E_1, \ldots, E_n\}$ be an orthogonal basis of $\mathbb{R}^n$. Given X and Y in $\mathbb{R}^n$, show that

$$X \bullet Y = \frac{(X \bullet E_1)(Y \bullet E_1)}{\|E_1\|^2} + \cdots + \frac{(X \bullet E_n)(Y \bullet E_n)}{\|E_n\|^2}.$$

SECTION 5.4 Rank of a Matrix

In this section we use independence and spanning to properly define the rank of a matrix and to study its properties. This requires that we deal with rows and columns in the same way. While it has been our custom to write the n-tuples in $\mathbb{R}^n$ as columns, in this section we will frequently write them as rows. Subspaces, independence, spanning, and dimension are defined for rows using matrix operations, just as for columns. If A is an $m \times n$ matrix, we define:

> The **column space**, col A, of A is the subspace of $\mathbb{R}^m$ spanned by the columns of A.
>
> The **row space**, row A, of A is the subspace of $\mathbb{R}^n$ spanned by the rows of A.

Much of what we do in this section involves these subspaces. Recall from Theorem 4 §2.2 that if $C_1, C_2, \ldots, C_n$ are the columns of an $m \times n$ matrix A, and if $X = [x_1 \quad x_2 \quad \cdots \quad x_n]^T$ is any column in $\mathbb{R}^n$, then

$$AX = [C_1 \quad C_2 \quad \cdots \quad C_n] \begin{bmatrix} x_1 \\ x_2 \\ \vdots \\ x_n \end{bmatrix} = x_1 C_1 + x_2 C_2 + \cdots + x_n C_n. \qquad (*)$$

With this we can prove:

Theorem 1

Let A, U, and V be matrices of sizes $m \times n$, $p \times m$, and $n \times q$ respectively. Then:
1. $\text{col}(AV) \subseteq \text{col } A$, with equality if V is (square and) invertible.
2. $\text{row}(UA) \subseteq \text{row } A$, with equality if U is (square and) invertible.

PROOF

Let $C_1, C_2, \ldots, C_n$ denote the columns of A, and let $X_j = [x_1 \quad x_2 \quad \cdots \quad x_n]^T$ denote column j of V. Then column j of AV is AX_j, and by (∗) this is $AX_j = x_1 C_1 + x_2 C_2 + \cdots + x_n C_n$. Since this is in col A for each j, it follows that $\text{col}(AV) \subseteq \text{col } A$. If V is invertible, we obtain col $A = \text{col}[(AV)V^{-1}] \subseteq \text{col}(AV)$ in the same way, and (1) follows.

As to (2), we have $\text{col}[(UA)^T] = \text{col}(A^T U^T) \subseteq \text{col}(A^T)$ by (1), from which it follows that row$(UA) \subseteq$ row A. If U is invertible, we obtain row$(UA) =$ row A as in the proof of (1).

Now suppose that a matrix A is carried to some row-echelon matrix R by row operations. Then $R = UA$ for some invertible matrix U by Theorem 1 §2.4, so Theorem 1 shows that row $R =$ row A. Moreover, the next lemma shows that $\dim(\text{row } R)$ is the rank of A defined in Section 1.2, and hence shows that rank A is independent of the particular row-echelon matrix to which A can be carried. This fact was not proved in Section 1.2.

Lemma 1

Let R denote an $m \times n$ row-echelon matrix.
  1. The rows of R are a basis of row R.
  2. The columns of R containing the leading ones are a basis of col R.

PROOF

1. If $R_1, R_2, \ldots, R_r$ denote the nonzero rows of R, we have row $R = \text{span}\{R_1, R_2, \ldots, R_r\}$ by definition. Suppose $a_1 R_1 + a_2 R_2 + \cdots + a_r R_r = 0$ where each a_i is in $\mathbb{R}$. Then $a_1 = 0$ because the leading 1 in R_1 is to the left of any nonzero entry in any other R_i. But then $a_2 R_2 + \cdots + a_r R_r = 0$ and so $a_2 = 0$ in the same way (because the matrix R with R_1 deleted is also row-echelon). This continues to show that each $a_i = 0$.

2. The r columns containing leading ones are independent because the leading ones are in different rows (and have zeros below them). It is clear that col R is contained in the subspace of all columns in $\mathbb{R}^m$ with the last $m - r$ entries zero. This space has dimension r, so the r independent columns containing leading ones are a basis by Theorem 7 §5.2.

Somewhat surprisingly, Lemma 1 is instrumental in showing that $\dim(\text{col } A) = \dim(\text{row } A)$ for any matrix A. This is the main result in the following fundamental theorem.

Theorem 2 Rank Theorem

Let A denote any $m \times n$ matrix. Then

$$\dim(\text{row } A) = \dim(\text{col } A).$$

Moreover, suppose A can be carried to a matrix R in row-echelon form by a series of elementary row operations. If r denotes the number of nonzero rows in R, then

1. The r nonzero rows of R are a basis of row A.
2. If the leading 1's lie in columns $j_1, j_2, \ldots, j_r$ of R, then the corresponding columns $j_1, j_2, \ldots, j_r$ of A are a basis of col A.

PROOF

We have $R = UA$ for some invertible matrix U. Hence row $A =$ row R by Theorem 1, and (1) follows from Lemma 1.

To prove (2), let $C_1, C_2, \ldots, C_n$ denote the columns of A. Then $A = [C_1 \ C_2 \ \cdots \ C_n]$ in block form, and

$$R = UA = U[C_1 \ C_2 \ \cdots \ C_n] = [UC_1 \ UC_2 \ \cdots \ UC_n].$$

Hence, in the notation of (2), the set $B = \{UC_{j_1}, UC_{j_2}, \ldots, UC_{j_r}\}$ consists of the columns of R that contain a leading 1, so B is a basis of col R by Lemma 1. But then the fact that U is invertible implies that $\{C_{j_1}, C_{j_2}, \ldots, C_{j_r}\}$ is linearly independent. Furthermore, if C_j is any column of A, then UC_j is a linear combination of the columns in the set B. Again, the invertibility of U implies that C_j is a linear combination of $C_{j_1}, C_{j_2}, \ldots, C_{j_r}$. This proves (2).

Finally, dim(row A) $= r =$ dim(col A) by (1) and (2).

The common dimension of the row and column spaces of an $m \times n$ matrix A is called the **rank** of A and is denoted rank A. By (1) of Theorem 2, this agrees with the definition in Section 1.2 and we record the result for reference.

Corollary 1

Suppose a matrix A can be carried to a matrix R in row-echelon form by a series of elementary row operations. Then the rank of A is equal to the number of nonzero rows of R.

Example 1

Compute the rank of matrix $A = \begin{bmatrix} 1 & 2 & 2 & -1 \\ 3 & 6 & 5 & 0 \\ 1 & 2 & 1 & 2 \end{bmatrix}$ and find bases for the row space and the column space of A.

Solution   The reduction of A to row-echelon form is as follows:

$$\begin{bmatrix} 1 & 2 & 2 & -1 \\ 3 & 6 & 5 & 0 \\ 1 & 2 & 1 & 2 \end{bmatrix} \rightarrow \begin{bmatrix} 1 & 2 & 2 & -1 \\ 0 & 0 & -1 & 3 \\ 0 & 0 & -1 & 3 \end{bmatrix} \rightarrow \begin{bmatrix} 1 & 2 & 2 & -1 \\ 0 & 0 & 1 & -3 \\ 0 & 0 & 0 & 0 \end{bmatrix}$$

Hence rank $A = 2$, and $\{[\,1 \ \ 2 \ \ 2 \ \ -1\,], [\,0 \ \ 0 \ \ 1 \ \ -3\,]\}$ is a basis of the row space of A. Moreover, the leading 1's are in columns 1 and 3 of the row-echelon matrix, so Theorem 2 shows that columns 1 and 3 of A are

a basis $\left\{ \begin{bmatrix} 1 \\ 3 \\ 1 \end{bmatrix}, \begin{bmatrix} 2 \\ 5 \\ 1 \end{bmatrix} \right\}$ of col A.

The rank theorem has several other important consequences. Corollary 2 follows because the rows of A are independent (respectively, span row A) if and only if their transposes are independent (respectively, span $\text{col}(A^T)$).

Corollary 2

If A is any matrix, then rank $A = \text{rank}(A^T)$.

Corollary 3

If A is an $m \times n$ matrix, then rank $A \leq m$ and rank $A \leq n$.

PROOF

If A is carried to the row-echelon matrix R by row operations, then Corollary 1 shows that rank $A = r$ where r is the number of nonzero rows of R. Since R is $m \times n$ too, it follows that $r \leq m$. Applying this to A^T gives $\text{rank}(A^T) \leq n$ because A^T is $n \times m$. Hence we are done by Corollary 2.

Theorem 1 immediately yields

Corollary 4

rank $A = \text{rank}(UA) = \text{rank}(AV)$ whenever U and V are invertible.

Corollary 5

An $n \times n$ matrix A is invertible if and only if rank $A = n$.

PROOF

If A is invertible, then $A \to I_n$ by row operations (by Theorem 5 §2.3), so rank $A = n$ by Corollary 1. Conversely, let $A \to R$ by row operations where R is an $n \times n$ reduced row-echelon matrix. If rank $A = n$, then R has n leading ones by Corollary 1, and so $R = I_n$. Hence $A \to I_n$ so A is invertible, again by Theorem 5 §2.3.

The rank theorem can be used to find bases of subspaces of the space of all $n \times 1$ rows. Here is an example where $n = 4$.

Example 2

Find a basis for the following subspace of $\mathbb{R}^4$, (written as rows).

$$U = \text{span}\{[1 \ 1 \ 2 \ 3], [2 \ 4 \ 1 \ 0], [1 \ 5 \ -4 \ -9]\}$$

Solution U is just the row space of $\begin{bmatrix} 1 & 1 & 2 & 3 \\ 2 & 4 & 1 & 0 \\ 1 & 5 & -4 & -9 \end{bmatrix}$, so we reduce this to row-echelon

form: $\begin{bmatrix} 1 & 1 & 2 & 3 \\ 2 & 4 & 1 & 0 \\ 1 & 5 & -4 & -9 \end{bmatrix} \rightarrow \begin{bmatrix} 1 & 1 & 2 & 3 \\ 0 & 2 & -3 & -6 \\ 0 & 4 & -6 & -12 \end{bmatrix} \rightarrow \begin{bmatrix} 1 & 1 & 2 & 3 \\ 0 & 1 & -\frac{3}{2} & -3 \\ 0 & 0 & 0 & 0 \end{bmatrix}$

The required basis is $\{[1\ 1\ 2\ 3], [0\ 1\ -\frac{3}{2}\ -3]\}$. Thus $\{[1\ 1\ 2\ 3], [0\ 2\ -3\ -6]\}$ is
also a basis and avoids fractions.

In Section 5.1 we discussed two other subspaces associated with an $m \times n$ matrix A,
the null space null $A = \{X \mid X$ in $\mathbb{R}^n$ and $AX = 0\}$ and the image im $A = \{AX \mid X$ in $\mathbb{R}^n\}$.
Using the rank, there are simple ways to find bases for these spaces.
    We already know (Theorem 3 §2.2) that null A is spanned by the basic solutions
to the system $AX = 0$. The following example is instructive in showing that these
basic solutions are, in fact, independent (and so are a basis of null A).

Example 3

If $A = \begin{bmatrix} 1 & -2 & 1 & 1 \\ -1 & 2 & 0 & 1 \\ 2 & -4 & 1 & 0 \end{bmatrix}$ find a basis of null A and so find its dimension.

Solution   If X is in null A, then $AX = 0$, so X is given by solving the system $AX = 0$.
The reduction of the augmented matrix to reduced form is

$$\begin{bmatrix} 1 & -2 & 1 & 1 & | & 0 \\ -1 & 2 & 0 & 1 & | & 0 \\ 2 & -4 & 1 & 0 & | & 0 \end{bmatrix} \rightarrow \begin{bmatrix} 1 & -2 & 0 & -1 & | & 0 \\ 0 & 0 & 1 & 2 & | & 0 \\ 0 & 0 & 0 & 0 & | & 0 \end{bmatrix}$$

Hence, writing $X = [x_1\ x_2\ x_3\ x_4]^T$, the leading variables are x_1 and x_3, and the
nonleading variables x_2 and x_4 become parameters: $x_2 = s$ and $x_4 = t$. Then the
equations corresponding to the reduced matrix determine the leading variables
in terms of the parameters:

$$x_1 = 2s + t \quad \text{and} \quad x_3 = -2t.$$

This means that the general solution is

$$X = [2s + t\ \ s\ \ -2t\ \ t]^T = s[2\ 1\ 0\ 0]^T + t[1\ 0\ -2\ 1]^T. \tag{$*$}$$

Hence X is in span$\{X_1, X_2\}$ where $X_1 = [2\ 1\ 0\ 0]^T$ and $X_2 = [1\ 0\ -2\ 1]^T$ are
the basic solutions, and we have shown that null$(A) \subseteq$ span$\{X_1, X_2\}$. But X_1 and
X_2 are in null A (they are solutions of $AX = 0$), so

$$\text{null } A = \text{span}\{X_1, X_2\}$$

by Theorem 1 §5.1. We claim further that $\{X_1, X_2\}$ is linearly independent.
To see this, let $sX_1 + tX_2 = 0$ be a linear combination that vanishes. Then $(*)$
shows that $[2s + t\ \ s\ \ -2t\ \ t]^T = 0$, whence $s = t = 0$. Thus $\{X_1, X_2\}$ is a basis of
null(A), and so dim(null A) = 2.

The calculation in Example 3 is typical of what happens in general. If A is an $m \times n$ matrix with rank $A = r$, there are exactly r leading variables, and hence exactly $n - r$ nonleading variables. These lead to exactly $n - r$ basic solutions $X_1, X_2, \ldots, X_{n-r}$, and the reduced equations give the leading variables as linear combinations of the X_i. Hence

$$\text{null } A = \text{span}\{X_1, X_2, \ldots, X_{n-r}\}.$$

(This is Theorem 3 §2.2.). We now claim that these basic solutions X_i are independent. The general solution is a linear combination $X = t_1 X_1 + t_2 X_2 + \cdots + t_{n-r} X_{n-r}$ where each coefficient t_i is a parameter equal to a nonleading variable. Thus, if this linear combination vanishes, then each $t_i = 0$ (as for s and t in Example 3, each t_i is a coefficient when X is expressed as a linear combination of the standard basis of $\mathbb{R}^n$). This proves that $\{X_1, X_2, \ldots, X_{n-r}\}$ is linearly independent, and so is a basis of null A. This proves the first part of the following theorem.

Theorem 3

Let A denote an $m \times n$ matrix of rank r.
1. If $X_1, X_2, \ldots, X_{n-r}$ are the basic solutions of the homogeneous system $AX = 0$ that are produced by the gaussian algorithm, then $\{X_1, X_2, \ldots, X_{n-r}\}$ is a basis of null A. In particular

$$\dim(\text{null } A) = n - r.$$

2. We have im A = col A so the rank theorem provides a basis of im A. In particular,

$$\dim(\text{im } A) = r.$$

PROOF

It remains to prove (2). But im A = col A by Example 8 §5.1, so $\dim(\text{im } A) = \dim(\text{col } A) = r$. The rest follows from Theorem 2.

Let A be an $m \times n$ matrix. Corollary 3 of the rank theorem asserts that rank $A \leq m$ and rank $A \leq n$, and it is natural to ask when these extreme cases arise. If $C_1, C_2, \ldots, C_n$ are the columns of A, Theorem 2 §5.2 shows that $\{C_1, C_2, \ldots, C_n\}$ spans $\mathbb{R}^m$ if and only if the system $AX = B$ is consistent for every B in $\mathbb{R}^m$, and that $\{C_1, C_2, \ldots, C_n\}$ is independent if and only if $AX = 0$, X in $\mathbb{R}^n$, implies $X = 0$. The next two theorems improve on both these results, and relate them to when the rank of A is n or m.

Theorem 4

The following are equivalent for an $m \times n$ matrix A:
1. rank $A = n$.
2. The rows of A span $\mathbb{R}^n$.
3. The columns of A are linearly independent in $\mathbb{R}^m$.

4. The $n \times n$ matrix $A^T A$ is invertible.
5. $CA = I_n$ for some $n \times m$ matrix C.
6. If $AX = 0$, X in $\mathbb{R}^n$, then $X = 0$.

PROOF

(1) $\Rightarrow$ (2). We have row $A \subseteq \mathbb{R}^n$, and dim(row A) $= n$ by (1), so row $A = \mathbb{R}^n$ by Theorem 8 §5.2. This is (2).

(2) $\Rightarrow$ (3). By (2), row $A = \mathbb{R}^n$, so rank $A = n$. This means dim(col A) $= n$. Since the n columns of A span col A, they are independent by Theorem 7 §5.2.

(3) $\Rightarrow$ (4). If $(A^T A)X = 0$, X in $\mathbb{R}^n$, we show that $X = 0$ (Theorem 5 §2.3). We have

$$\|AX\|^2 \ = \ (AX)^T AX \ = \ X^T A^T AX \ = \ X^T 0 \ = \ 0.$$

Hence $AX = 0$, so $X = 0$ by (3) and Theorem 2 §5.2.

(4) $\Rightarrow$ (5). Given (4), take $C = (A^T A)^{-1} A^T$.

(5) $\Rightarrow$ (6). If $AX = 0$, then left multiplication by C (from (5)) gives $X = 0$.

(6) $\Rightarrow$ (1). Given (6), the columns of A are independent by Theorem 2 §5.2. Hence dim(col A) $= n$, and (1) follows.

Theorem 5

The following are equivalent for an $m \times n$ matrix A:
1. rank $A = m$.
2. The columns of A span $\mathbb{R}^m$.
3. The rows of A are independent in $\mathbb{R}^n$.
4. The $m \times m$ matrix AA^T is invertible.
5. $AC = I_m$ for some $n \times m$ matrix C.
6. The system $AX = B$ is consistent for every B in $\mathbb{R}^m$.

PROOF

(1) $\Rightarrow$ (2). By (1), dim(col A) $= m$, so col $A = \mathbb{R}^m$ by Theorem 8 §5.2.

(2) $\Rightarrow$ (3). By (2), col $A = \mathbb{R}^m$, so rank $A = m$. This means dim(row A) $= m$. Since the m rows of A span row A, they are independent by Theorem 7 §5.2.

(3) $\Rightarrow$ (4). We have rank $A = m$ by (3), so the $n \times m$ matrix A^T has rank m. Hence applying Theorem 4 to A^T in place of A shows that $(A^T)^T A^T$ is invertible, proving (4).

(4) $\Rightarrow$ (5). Given (4), take $C = A^T (AA^T)^{-1}$ in (5).

(5) $\Rightarrow$ (6). Comparing columns in $AC = I_m$ gives $AC_j = E_j$ for each j, where C_j and E_j denote column j of C and I_m respectively. Given B in $\mathbb{R}^m$, write $B = \sum_{j=1}^{m} r_j E_j$, r_j in $\mathbb{R}$. Then (6) holds with $X = \sum_{j=1}^{m} r_j C_j$ as the reader can verify.

(6) $\Rightarrow$ (1). Given (6), the columns of A span $\mathbb{R}^m$ by Theorem 2 §5.2. Thus col $A = \mathbb{R}^m$ and (1) follows.

Example 4

Show that $\begin{bmatrix} 3 & x + y + z \\ x + y + z & x^2 + y^2 + z^2 \end{bmatrix}$ is invertible if x, y, and z are all distinct.

Solution   The given matrix has the form $A^T A$ where $A = \begin{bmatrix} 1 & x \\ 1 & y \\ 1 & z \end{bmatrix}$ has independent

columns (verify). Hence Theorem 4 applies.

Theorems 4 and 5 relate several important properties of an $m \times n$ matrix A to the invertibility of the square, symmetric matrices $A^T A$ and $A A^T$. In fact, even if the columns of A are not independent or do not span $\mathbb{R}^m$, the matrices $A^T A$ and $A A^T$ are both symmetric and, as such, have real eigenvalues as we shall see. We return to this in Chapter 7.

Exercises 5.4

1. In each case find bases for the row and column spaces of A and determine the rank of A.

   (a) $A = \begin{bmatrix} 2 & -4 & 6 & 8 \\ 2 & -1 & 3 & 2 \\ 4 & -5 & 9 & 10 \\ 0 & -1 & 1 & 2 \end{bmatrix}$ ◆(b) $A = \begin{bmatrix} 2 & -1 & 1 \\ -2 & 1 & 1 \\ 4 & -2 & 3 \\ -6 & 3 & 0 \end{bmatrix}$

   (c) $A = \begin{bmatrix} 1 & -1 & 5 & -2 & 2 \\ 2 & -2 & -2 & 5 & 1 \\ 0 & 0 & -12 & 9 & -3 \\ -1 & 1 & 7 & -7 & 1 \end{bmatrix}$

   ◆(d) $A = \begin{bmatrix} 1 & 2 & -1 & 3 \\ -3 & -6 & 3 & -2 \end{bmatrix}$

2. In each case find a basis of the subspace U.

   (a) $U = \text{span}\{[1 \ -1 \ 0 \ 3], [2 \ 1 \ 5 \ 1], [4 \ -2 \ 5 \ 7]\}$

   ◆(b) $U = \text{span}\{[1 \ -1 \ 2 \ 5 \ 1], [3 \ 1 \ 4 \ 2 \ 7],$
   $[1 \ 1 \ 0 \ 0 \ 0], [5 \ 1 \ 6 \ 7 \ 8]\}$

   (c) $U = \text{span}\left\{ \begin{bmatrix} 1 \\ 1 \\ 0 \\ 0 \end{bmatrix}, \begin{bmatrix} 0 \\ 0 \\ 1 \\ 1 \end{bmatrix}, \begin{bmatrix} 1 \\ 0 \\ 1 \\ 0 \end{bmatrix}, \begin{bmatrix} 0 \\ 1 \\ 0 \\ 1 \end{bmatrix} \right\}$

   ◆(d) $U = \text{span}\left\{ \begin{bmatrix} 1 \\ 5 \\ -6 \end{bmatrix}, \begin{bmatrix} 2 \\ 6 \\ -8 \end{bmatrix}, \begin{bmatrix} 3 \\ 7 \\ -10 \end{bmatrix}, \begin{bmatrix} 4 \\ 8 \\ 12 \end{bmatrix} \right\}$

3. (a) Can a 3×4 matrix have independent columns? Independent rows? Explain.

   ◆(b) If A is 4×3 and rank $A = 2$, can A have independent columns? Independent rows? Explain.

   (c) If A is an $m \times n$ matrix and rank $A = m$, show that $m \le n$.

◆(d) Can a nonsquare matrix have its rows independent and its columns independent? Explain.

(e) Can the null space of a 3×6 matrix have dimension 2? Explain.

◆(f) If A is not square, show that either the rows of A or the columns of A are not linearly independent.

4. (a) Show that rank $UA \le$ rank A, with equality if U is invertible.

   (b) Show that rank $AV \le$ rank A, with equality if V is invertible.

5. Show that rank $(AB) \le$ rank A and that rank $(AB) \le$ rank B.

6. Show that the rank does not change when an elementary row or column operation is performed on a matrix.

7. In each case find a basis of the null space of A. Then compute rank A and verify (1) of Theorem 3.

   (a) $A = \begin{bmatrix} 3 & 1 & 1 \\ 2 & 0 & 1 \\ 4 & 2 & 1 \\ 1 & -1 & 1 \end{bmatrix}$

   ◆(b) $A = \begin{bmatrix} 3 & 5 & 5 & 2 & 0 \\ 1 & 0 & 2 & 2 & 1 \\ 1 & 1 & 1 & -2 & -2 \\ -2 & 0 & -4 & -4 & -2 \end{bmatrix}$

8. Let $A = CR$ where $C \ne 0$ is a column in $\mathbb{R}^m$ and $R \ne 0$ is a row in $\mathbb{R}^n$.

   (a) Show that col $A = \text{span}\{C\}$ and row $A = \text{span}\{R\}$.

   ◆(b) Find dim(null A).

   (c) Show that null A = null R.

9. Show that null $A = 0$ if and only if the columns of A are independent.

10. Let A be an $n \times n$ matrix.

    (a) Show that $A^2 = 0$ if and only if col $A \subseteq$ null A.

    ◆(b) Conclude that if $A^2 = 0$, then rank $A \leq \frac{n}{2}$.

    (c) Find a matrix A for which col $A =$ null A.

11. If A is $m \times n$ and B is $n \times m$, show that $AB = 0$ if and only if col $B \subseteq$ null A.

◆12. If A is an $m \times n$ matrix, show that col $A = \{AX \mid X$ in $\mathbb{R}^n\}$.

13. Let A be an $m \times n$ matrix with columns $C_1, C_2, \ldots, C_n$. If rank $A = n$, show that $\{A^T C_1, A^T C_2, \ldots, A^T C_n\}$ is a basis of $\mathbb{R}^n$.

14. If A is $m \times n$ and B is $m \times 1$, show that B lies in the column space of A if and only if rank$[A \ B] =$ rank A.

15. (a) Show that $AX = B$ has a solution if and only if rank $A =$ rank$[A \ B]$. [*Hint:* Exercises 12 and 14.]

    ◆(b) If $AX = B$ has no solution, show that rank$[A \ B] = 1 +$ rank A.

16. Let X be a $k \times m$ matrix. If I is the $m \times m$ identity matrix, show that $I + X^T X$ is invertible.

[*Hint:* $I + X^T X = A^T A$ where $A = \begin{bmatrix} I \\ X \end{bmatrix}$ in block form.]

17. If A is $m \times n$ of rank r, show that A can be factored as $A = PQ$ where P is $m \times r$ with r independent columns, and Q is $r \times n$ with r independent rows. [*Hint:* Let $UAV = \begin{bmatrix} I_r & 0 \\ 0 & 0 \end{bmatrix}$ by Theorem 3, §2.4, and write $U^{-1} = \begin{bmatrix} U_1 & U_2 \\ U_3 & U_4 \end{bmatrix}$ and $V^{-1} = \begin{bmatrix} V_1 & V_2 \\ V_3 & V_4 \end{bmatrix}$ in block form, where U_1 and V_1 are $r \times r$.]

18. (a) Show that if A and B have independent columns, so does AB.

    (b) Show that if A and B have independent rows, so does AB.

19. A matrix obtained from A by deleting rows and columns is called a **submatrix** of A. If A has an invertible $k \times k$ submatrix, show that rank $A \geq k$. [*Hint:* Show that row and column operations carry $A \to \begin{bmatrix} I_k & P \\ 0 & Q \end{bmatrix}$ in block form.] *Remark:* It can be shown that rank A is the largest integer r such that A has an invertible $r \times r$ submatrix.

SECTION 5.5   Similarity and Diagonalization

In Section 3.3 we studied diagonalization of a square matrix A, and found important applications (for example to linear dynamical systems). We can now utilize the concepts of subspace, basis, and dimension to clarify the diagonalization process, reveal some new results, and prove some theorems which could not be demonstrated in Section 3.3.

Before proceeding, we introduce a notion that simplifies the discussion of diagonalization, and is used throughout the book.

Similar Matrices

If A and B are $n \times n$ matrices, we say that A and B are **similar**, and write $A \sim B$, if $B = P^{-1}AP$ for some invertible matrix P, equivalently (writing $Q = P^{-1}$) if $B = QAQ^{-1}$ for an invertible matrix Q. The language of similarity is used throughout linear algebra. For example, a matrix A is diagonalizable if and only if it is similar to a diagonal matrix.

If $A \sim B$, then necessarily $B \sim A$. To see why, suppose that $B = P^{-1}AP$. Then $A = PBP^{-1} = Q^{-1}BQ$ where $Q = P^{-1}$ is invertible. This proves the second of the following properties of similarity (the others are left as an exercise):

1. $A \sim A$ for all square matrices A.
2. If $A \sim B$, then $B \sim A$.   ($*$)
3. If $A \sim B$ and $B \sim C$, then $A \sim C$.

These properties are often expressed by saying that the similarity relation ~ is an **equivalence relation** on the set of $n \times n$ matrices. Here is an example showing how these properties are used.

Example 1

If A is similar to B and either A or B is diagonalizable, show that the other is also diagonalizable.

Solution   We have $A \sim B$. Suppose that A is diagonalizable, say $A \sim D$ where D is diagonal. Since $B \sim A$ by (2) of (∗), we have $B \sim A$ and $A \sim D$. Hence $B \sim D$ by (3) of (∗), so B is diagonalizable too. An analogous argument works if we assume instead that B is diagonalizable.

Similarity is compatible with inverses, transposes, and powers in the following sense: If $A \sim B$, then $A^{-1} \sim B^{-1}$, $A^T \sim B^T$, and $A^k \sim B^k$ for all integers $k \geq 1$ (the proofs are routine matrix computations using Theorem 1 §3.3). Thus, for example, if A is diagonalizable, so also is A^T, A^{-1} (if it exists), and A^k (for each $k \geq 1$). Indeed, if $A \sim D$ where D is a diagonal matrix, we obtain $A^T \sim D^T$, $A^{-1} \sim D^{-1}$, and $A^k \sim D^k$, and each of the matrices D^T, D^{-1}, and D^k is diagonal.

We pause to introduce a simple matrix function that will be referred to later. The **trace** tr A of an $n \times n$ matrix A is defined to be the sum of the main diagonal elements of A. In other words:

$$\text{If } A = [a_{ij}], \text{ then } \text{tr}\,A = a_{11} + a_{22} + \cdots + a_{nn}.$$

It is evident that $\text{tr}(A + B) = \text{tr}\,A + \text{tr}\,B$ and that $\text{tr}(cA) = c\,\text{tr}\,A$ holds for all $n \times n$ matrices A and B and all scalars c. The following fact is more surprising.

Lemma 1

Let A and B be $n \times n$ matrices. Then $\text{tr}(AB) = \text{tr}(BA)$.

PROOF

Write $A = [a_{ij}]$ and $B = [b_{ij}]$. For each i, the (i, i)-entry of the matrix AB is $d_i = a_{i1}b_{1i} + a_{i2}b_{2i} + \cdots + a_{in}b_{ni} = \sum_j a_{ij}b_{ji}$. Hence

$$\text{tr}(AB) = d_1 + d_2 + \cdots + d_n = \sum_i d_i = \sum_i \left(\sum_j a_{ij}b_{ji} \right).$$

Similarly we have $\text{tr}(BA) = \sum_i (\sum_j b_{ij}a_{ji})$. Since these two double sums are the same, Lemma 1 is proved.

As the name indicates, similar matrices share many properties, some of which are collected in the next theorem for reference.

Theorem 1

If A and B are similar $n \times n$ matrices, then A and B have the same determinant, rank, trace, characteristic polynomial, and eigenvalues.

PROOF

Let $B = P^{-1}AP$ for some invertible matrix P. Then we have
$\det B = \det(P^{-1}) \det A \det P = \det A$ because $\det(P^{-1}) = 1/\det P$. Similarly,
$\operatorname{rank} B = \operatorname{rank}(P^{-1}AP) = \operatorname{rank} A$ by Corollary 4 of Theorem 2 §5.4. Next Lemma 1
gives

$$\operatorname{tr}(P^{-1}AP) = \operatorname{tr}[P^{-1}(AP)] = \operatorname{tr}[(AP)P^{-1}] = \operatorname{tr} A.$$

As to the characteristic polynomial,

$$\begin{aligned}
c_B(x) = \det\{xI - B\} &= \det\{x(P^{-1}IP) - P^{-1}AP\} \\
&= \det\{P^{-1}(xI - A)P\} \\
&= \det(xI - A) \\
&= c_A(x).
\end{aligned}$$

Finally, this shows that A and B have the same eigenvalues because the eigenvalues
of a matrix are the roots of its characteristic polynomial.

Example 2

The matrices $A = \begin{bmatrix} 1 & 1 \\ 0 & 1 \end{bmatrix}$ and $I = \begin{bmatrix} 1 & 0 \\ 0 & 1 \end{bmatrix}$ have the same determinant, rank,
trace, characteristic polynomial, and eigenvalues, but they are not similar
because $P^{-1}IP = I$ for any invertible matrix P. Hence sharing the five properties
in Theorem 1 does not guarantee that two matrices are similar.

Diagonalization Revisited

Recall that a square matrix A is **diagonalizable** if there exists an invertible matrix P
such that $P^{-1}AP = D$ is a diagonal matrix, that is if A is similar to a diagonal matrix D.
Unfortunately, not all matrices are diagonalizable, for example the matrix $\begin{bmatrix} 1 & 1 \\ 0 & 1 \end{bmatrix}$
(see Example 8 §3.3). Determining whether A is diagonalizable is closely related to
the eigenvalues and eigenvectors of A. Recall that a number λ is called an
eigenvalue of A if $AX = \lambda X$ for some nonzero column X in $\mathbb{R}^n$, and any such
nonzero vector X is called an **eigenvector** of A corresponding to λ (or simply
a λ-eigenvector of A). The eigenvalues and eigenvectors of A are closely related to
the **characteristic polynomial** $c_A(x)$ of A, defined by

$$c_A(x) = \det(xI - A).$$

If A is $n \times n$ this is a polynomial of degree n, and its relationship to the eigenvalues
is given in the following theorem (a repeat of Theorem 2 §3.3).

Theorem 2

Let A be an $n \times n$ matrix.
1. The eigenvalues λ of A are the roots of the characteristic polynomial $c_A(x)$ of A.
2. The λ-eigenvectors X are the nonzero solutions to the homogeneous system

$$(\lambda I - A)X = 0$$

of linear equations with $\lambda I - A$ as coefficient matrix.

Example 3

Show that the eigenvalues of a triangular matrix are the main diagonal entries.

Solution   Assume that A is triangular. Then the matrix $xI - A$ is also triangular and has diagonal entries $(x - a_{11})$, $(x - a_{22})$, ..., $(x - a_{nn})$ where $A = [a_{ij}]$. Hence Theorem 4 §3.1 gives

$$c_A(x) = (x - a_{11})(x - a_{22}) \cdots (x - a_{nn})$$

and the result follows because the eigenvalues are the roots of $c_A(x)$.

Theorem 3 §3.3 asserts (in part) that an $n \times n$ matrix A is diagonalizable if and only if it has n eigenvectors $X_1, \ldots, X_n$ such that the matrix $P = [X_1 \cdots X_n]$ with the X_i as columns is invertible. This is equivalent to requiring that $\{X_1, \ldots, X_n\}$ is a basis of $\mathbb{R}^n$ consisting of eigenvectors of A. Hence we can restate Theorem 3 §3.3 as follows:

Theorem 3

Let A be an $n \times n$ matrix.
1. A is diagonalizable if and only if $\mathbb{R}^n$ has a basis $\{X_1, X_2, \ldots, X_n\}$ consisting of eigenvectors of A.
2. When this is the case, the matrix $P = [X_1 \cdots X_n]$ is invertible and $P^{-1}AP = \text{diag}(\lambda_1, \lambda_2, \ldots, \lambda_n)$ where, for each i, λ_i is the eigenvalue of A corresponding to X_i.

The next result is a basic tool for determining when a matrix is diagonalizable. It reveals an important connection between eigenvalues and linear independence: Eigenvectors corresponding to distinct eigenvalues are necessarily linearly independent.

Theorem 4

Let $X_1, X_2, \ldots, X_k$ be eigenvectors corresponding to distinct eigenvalues $\lambda_1, \lambda_2, \ldots, \lambda_k$ of an $n \times n$ matrix A. Then $\{X_1, X_2, \ldots, X_k\}$ is a linearly independent set.

PROOF

We use induction on k. If $k = 1$, then $\{X_1\}$ is independent because $X_1 \neq 0$. In general, suppose the theorem is true for some $k \geq 1$. Given eigenvectors $\{X_1, X_2, \ldots, X_{k+1}\}$, suppose a linear combination vanishes:
$$t_1X_1 + t_2X_2 + \cdots + t_{k+1}X_{k+1} = 0. \tag{$*$}$$
We must show that each $t_i = 0$. Left multiply $(*)$ by A and use the fact that $AX_i = \lambda_i X_i$ to get
$$t_1\lambda_1 X_1 + t_2\lambda_2 X_2 + \cdots + t_{k+1}\lambda_{k+1}X_{k+1} = 0. \tag{$**$}$$
If we multiply $(*)$ by λ_1 and subtract the result from $(**)$, the first terms cancel and we obtain
$$t_2(\lambda_2 - \lambda_1)X_2 + t_3(\lambda_3 - \lambda_1)X_3 + \cdots + t_{k+1}(\lambda_{k+1} - \lambda_1)X_{k+1} = 0.$$

Since $X_2, X_3, \ldots, X_{k+1}$ correspond to distinct eigenvalues $\lambda_2, \lambda_3, \ldots, \lambda_{k+1}$, the set $\{X_2, X_3, \ldots, X_{k+1}\}$ is independent by the induction hypothesis. Hence,

$$t_2(\lambda_2 - \lambda_1) = 0, \quad t_3(\lambda_3 - \lambda_1) = 0, \quad \ldots, \quad t_{k+1}(\lambda_{k+1} - \lambda_1) = 0$$

and so $t_2 = t_3 = \cdots = t_{k+1} = 0$ because the λ_i are distinct. Hence (∗) becomes $t_1 X_1 = 0$, which implies that $t_1 = 0$ because $X_1 \neq 0$. This is what we wanted.

Theorem 4 will be applied several times; we begin by using it to give a useful test for when a matrix is diagonalizable.

Theorem 5

If A is an $n \times n$ matrix with n distinct eigenvalues, then A is diagonalizable.

PROOF

Choose one eigenvector for each of the n distinct eigenvalues. Then these eigenvectors are independent by Theorem 4, and so are a basis of $\mathbb{R}^n$ by Theorem 7 §5.2. Now use Theorem 3.

Example 4

Show that $A = \begin{bmatrix} 1 & 0 & 0 \\ 1 & 2 & 3 \\ -1 & 1 & 0 \end{bmatrix}$ is diagonalizable.

Solution   A routine computation shows that $c_A(x) = (x - 1)(x - 3)(x + 1)$ and so has distinct eigenvalues 1, 3, and −1. Hence Theorem 5 applies.

However, a matrix can have multiple eigenvalues as we saw in Section 3.3. To deal with this situation, we prove an important lemma which formalizes a technique that is basic to diagonalization, and which will be used three times below.

Lemma 2

Let $\{X_1, X_2, \ldots, X_k\}$ be a linearly independent set of eigenvectors of an $n \times n$ matrix A, extend it to a basis $\{X_1, X_2, \ldots, X_k, \ldots, X_n\}$ of $\mathbb{R}^n$ (by Theorem 6 §5.2), and let

$$P = [X_1 \ \ X_2 \ \cdots \ X_n]$$

be the (invertible) matrix with the X_i as its columns. If $\lambda_1, \lambda_2, \ldots, \lambda_k$ are the (not necessarily distinct) eigenvalues of A corresponding to $X_1, X_2, \ldots, X_k$ respectively, then $P^{-1}AP$ has block form

$$P^{-1}AP = \begin{bmatrix} \text{diag}(\lambda_1, \lambda_2, \ldots, \lambda_k) & B \\ 0 & A_1 \end{bmatrix}$$

where B and A_1 are matrices of size $k \times (n - k)$ and $(n - k) \times (n - k)$ respectively.

PROOF

If $\{E_1, E_2, \ldots, E_n\}$ is the standard basis of $\mathbb{R}^n$, then

$$[E_1 \quad E_2 \quad \cdots \quad E_n] = I_n = P^{-1}P = P^{-1}[X_1 \quad X_2 \quad \cdots \quad X_n]$$
$$= [P^{-1}X_1 \quad P^{-1}X_2 \quad \cdots \quad P^{-1}X_n].$$

Comparing columns, we have $P^{-1}X_i = E_i$ for each $1 \le i \le n$. On the other hand, observe that

$$P^{-1}AP = P^{-1}A[X_1 \quad X_2 \quad \cdots \quad X_n] = [P^{-1}AX_1 \quad P^{-1}AX_2 \quad \cdots \quad P^{-1}AX_n].$$

Hence, if $1 \le i \le k$, column i of $P^{-1}AP$ is

$$(P^{-1}A)X_i = P^{-1}(\lambda_i X_i) = \lambda_i(P^{-1}X_i) = \lambda_i E_i.$$

This describes the first k columns of $P^{-1}AP$, and Lemma 2 follows.

Note that Lemma 2 (with $k = n$) shows that an $n \times n$ matrix A is diagonalizable if $\mathbb{R}^n$ has a basis of eigenvectors of A, as in (1) of Theorem 3.

If λ is an eigenvalue of an $n \times n$ matrix A, write

$$E_\lambda(A) = \{X \text{ in } \mathbb{R}^n \mid AX = \lambda X\}.$$

This is a subspace of $\mathbb{R}^n$ called the eigenspace of A corresponding to λ (see Example 3 §5.1) and the eigenvectors corresponding to λ are just the nonzero vectors in $E_\lambda(A)$. In fact $E_\lambda(A)$ is the null space of the matrix $(\lambda I - A)$:

$$E_\lambda(A) = \{X \mid (\lambda I - A)X = 0\} = \text{null}(\lambda I - A).$$

Hence, by Example 7 §5.1, the basic solutions of the homogeneous system $(\lambda I - A)X = 0$ given by the gaussian algorithm form a basis for $E_\lambda(A)$. In particular

$$\dim[E_\lambda(A)] \text{ is the number of basic solutions to the system } (\lambda I - A)X = 0. \quad (***)$$

Now recall that the **multiplicity** of an eigenvalue λ of A is the number of times λ occurs as a root of the characteristic polynomial. In other words, the multiplicity of λ is the largest integer $m \ge 1$ such that

$$c_A(x) = (x - \lambda)^m g(x)$$

for some polynomial $g(x)$. Because of $(***)$, the assertion (without proof) in Theorem 4 §3.3 can be stated as follows: A square matrix is diagonalizable if and only if the multiplicity of each eigenvalue λ equals $\dim[E_\lambda(A)]$. We are going to prove this, and the proof requires the following result which is valid for *any* square matrix, diagonalizable or not.

Lemma 3

Let λ be an eigenvalue of multiplicity m of a square matrix A. Then $\dim[E_\lambda(A)] \le m$.

PROOF

Write $\dim[E_\lambda(A)] = d$. It suffices to show that $c_A(x) = (x - \lambda)^d g(x)$ for some polynomial $g(x)$, because m is the highest power of $(x - \lambda)$ that divides $c_A(x)$. To this end, let $\{X_1, X_2, \ldots, X_d\}$ be a basis of $E_\lambda(A)$. Then Lemma 2 shows that an invertible $n \times n$ matrix P exists such that

$$P^{-1}AP = \begin{bmatrix} \lambda I_d & B \\ 0 & A_1 \end{bmatrix}$$

in block form, where I_d denotes the $d \times d$ identity matrix. Now write $A' = P^{-1}AP$ and observe that $c_{A'}(x) = c_A(x)$ by Theorem 1. But Theorem 5 §3.1 gives

$$c_{A'}(x) = \det(xI_n - A') = \det\begin{bmatrix} (x - \lambda)I_d & -B \\ 0 & xI_{n-d} - A_1 \end{bmatrix}$$

$$= \det[(x - \lambda)I_d] \, \det[xI_{n-d} - A_1]$$

$$= (x - \lambda)^d c_{A_1}(x).$$

Hence $c_A(x) = c_{A'}(x) = (x - \lambda)^d g(x)$ where $g(x) = c_{A_1}(x)$. This is what we wanted.

It is impossible to ignore the question when equality holds in Lemma 3 for each eigenvalue λ. It turns out that this characterizes the diagonalizable matrices. This was stated without proof in Theorem 4 §3.3.

Theorem 6

The following are equivalent for a square matrix A:
1. A is diagonalizable.
2. $\dim[E_\lambda(A)]$ equals the multiplicity of λ for every eigenvalue λ of the matrix A.

PROOF

Let A be $n \times n$ and let $\lambda_1, \lambda_2, \ldots, \lambda_k$ be the distinct eigenvalues of A. For each i, let m_i denote the multiplicity of λ_i and write $d_i = \dim[E_{\lambda_i}(A)]$. Then $c_A(x) = (x - \lambda_1)^{m_1}(x - \lambda_2)^{m_2} \cdots (x - \lambda_k)^{m_k}$ so $m_1 + \cdots + m_k = n$ because $c_A(x)$ has degree n. Moreover, $d_i \le m_i$ for each i by Lemma 3.

(1) $\Rightarrow$ (2). By (1), $\mathbb{R}^n$ has a basis of n eigenvectors of A, so let t_i of them lie in $E_{\lambda_i}(A)$ for each i. Since the subspace spanned by these t_i eigenvectors has dimension t_i, we have $t_i \le d_i$ for each i by Theorem 4 §5.2. Hence

$$n = t_1 + \cdots + t_k \le d_1 + \cdots + d_k \le m_1 + \cdots + m_k = n.$$

It follows that $d_1 + \cdots + d_k = m_1 + \cdots + m_k$, so, since $d_i \le m_i$ for each i, we must have $d_i = m_i$. This is (2).

(2) $\Rightarrow$ (1). Let B_i denote a basis of $E_{\lambda_i}(A)$ for each i, and let $B = B_1 \cup \cdots \cup B_k$. Since each B_i contains m_i vectors by (2), and since the B_i are pairwise disjoint (the λ_i are distinct), it follows that B contains n vectors. So it suffices to show that B is linearly independent (then B is a basis of $\mathbb{R}^n$). Suppose a linear combination of the vectors in B vanishes, and let Y_i denote the sum of all terms that come from B_i.

Then Y_i lies in $E_{\lambda_i}(A)$ for each i, so the nonzero Y_i are independent by Theorem 4 (as the λ_i are distinct). Since the sum of the Y_i is zero, it follows that $Y_i = 0$ for each i. Hence all coefficients of terms in Y_i are zero (because B_i is independent). This shows that B is independent.

Example 5

If $A = \begin{bmatrix} 5 & 8 & 16 \\ 4 & 1 & 8 \\ -4 & -4 & -11 \end{bmatrix}$ and $B = \begin{bmatrix} 2 & 1 & 1 \\ 2 & 1 & -2 \\ -1 & 0 & -2 \end{bmatrix}$, show that A is diagonalizable but B is not.

Solution We have $c_A(x) = (x + 3)^2(x - 1)$ so the eigenvalues are $\lambda_1 = -3$ and $\lambda_2 = 1$. The corresponding eigenspaces are $E_{\lambda_1}(A) = \text{span}\{X_1, X_2\}$ and $E_{\lambda_2}(A) = \text{span}\{X_3\}$ where

$$X_1 = [-1 \ 1 \ 0]^T, \quad X_2 = [-2 \ 0 \ 1]^T, \quad X_3 = [2 \ 1 \ -1]^T$$

as the reader can verify. Since $\{X_1, X_2\}$ is independent, we have $\dim(E_{\lambda_1}(A)) = 2$ which is the multiplicity of λ_1. Similarly, $\dim(E_{\lambda_2}(A)) = 1$ equals the multiplicity of λ_2. Hence A is diagonalizable by Theorem 6, and a diagonalizing matrix is $P = [X_1 \ X_2 \ X_3]$.

Turning to B, $c_B(x) = (x + 1)^2(x - 3)$ so the eigenvalues are $\lambda_1 = -1$ and $\lambda_2 = 3$. The corresponding eigenspaces are $E_{\lambda_1}(B) = \text{span}\{Y_1\}$ and $E_{\lambda_2}(B) = \text{span}\{Y_2\}$ where

$$Y_1 = [-1 \ 2 \ 1]^T, \quad Y_2 = [5 \ 6 \ -1]^T.$$

Here $\dim(E_{\lambda_1}(B)) = 1$ is *smaller* than the multiplicity of λ_1, so the matrix B is *not* diagonalizable, again by Theorem 6. The fact that $\dim(E_{\lambda_1}(B)) = 1$ means that there is no possibility of finding *three* linearly independent eigenvectors.

Complex Eigenvalues

All the matrices we have considered have had real eigenvalues. But this need not be the case: The matrix $A = \begin{bmatrix} 0 & -1 \\ 1 & 0 \end{bmatrix}$ has characteristic polynomial $c_A(x) = x^2 + 1$ which has no real roots. Nonetheless, this matrix is diagonalizable; the only difference is that we must use a larger set of scalars, the complex numbers. The basic properties of these numbers are outlined in Appendix A.

Indeed, nearly everything we have done for real matrices can be done for complex matrices. The methods are the same; the only difference is that the arithmetic is carried out with complex numbers rather than real ones. For example, the gaussian algorithm works in exactly the same way to solve systems of linear equations with complex coefficients, matrix multiplication is defined the same way, and the matrix inversion algorithm works in the same way.

But the complex numbers are better than the real numbers in one respect: While there are polynomials like $x^2 + 1$ with real coefficients that have no real root, this problem does not arise with the complex numbers: *Every* nonconstant polynomial with complex coefficients has a complex root, and hence factors completely as a product of linear factors. This fact is known as the Fundamental Theorem of Algebra, and was first proved by Gauss.[12]

12 This was a famous open problem in 1799 when Gauss solved it at the age of 22 in his Ph.D. dissertation.

Example 6

Diagonalize the matrix $A = \begin{bmatrix} 0 & -1 \\ 1 & 0 \end{bmatrix}$.

Solution   The characteristic polynomial of A is

$$c_A(x) = \det(xI - A) = x^2 + 1 = (x - i)(x + i)$$

where $i^2 = -1$. Hence the eigenvalues are $\lambda_1 = i$ and $\lambda_2 = -i$, with corresponding eigenvectors $X_1 = \begin{bmatrix} 1 \\ -i \end{bmatrix}$ and $X_2 = \begin{bmatrix} 1 \\ i \end{bmatrix}$. Hence A is diagonalizable by the complex version of Theorem 5, and the complex version of Theorem 3 shows that $P = [X_1 \ X_2] = \begin{bmatrix} 1 & 1 \\ -i & i \end{bmatrix}$ is invertible and $P^{-1}AP = \begin{bmatrix} \lambda_1 & 0 \\ 0 & \lambda_2 \end{bmatrix} = \begin{bmatrix} i & 0 \\ 0 & -i \end{bmatrix}$. Of course, this can be checked directly.

We shall return to complex linear algebra in Section 8.6.

Symmetric Matrices[13]

On the other hand, many of the applications of linear algebra involve a real matrix A and, while A will have complex eigenvalues by the Fundamental Theorem of Algebra, it is always of interest to know when the eigenvalues are, in fact, real. While this can happen in a variety of ways, it turns out to hold whenever A is symmetric. This important theorem will be used extensively later. Surprisingly, the theory of *complex* eigenvalues can be used to prove this useful result about *real* eigenvalues.

If Z is a complex matrix, the **conjugate matrix** $\bar{Z}$ is defined to be the matrix obtained from Z by conjugating every entry. Thus, if $Z = [z_{ij}]$, then $\bar{Z} = [\bar{z}_{ij}]$. For example,

$$\text{If } Z = \begin{bmatrix} -i + 2 & 5 \\ i & 3 + 4i \end{bmatrix} \quad \text{then} \quad \bar{Z} = \begin{bmatrix} i + 2 & 5 \\ -i & 3 - 4i \end{bmatrix}$$

Recall that $\overline{z + w} = \bar{z} + \bar{w}$ and $\overline{zw} = \bar{z}\bar{w}$ holds for all complex numbers z and w. It follows that if Z and W are two complex matrices, then

$$\overline{Z + W} = \bar{Z} + \bar{W}, \qquad \overline{ZW} = \bar{Z}\bar{W} \quad \text{and} \quad \overline{(\lambda Z)} = \bar{\lambda}\bar{Z}$$

hold for all complex scalars λ. These facts are used in the proof of the following theorem.

Theorem 7

Let A be a symmetric real matrix. If λ is any eigenvalue of A, then λ is real.[14]

PROOF

Observe that $\bar{A} = A$ because A is real. If λ is an eigenvalue of A, we show that λ is real by showing that $\bar{\lambda} = \lambda$. Let X be a (possibly complex) eigenvector corresponding to λ, so that $X \neq 0$ and $AX = \lambda X$. Define $c = X^T\bar{X}$.

---

13 This discussion uses complex conjugation and absolute value. These topics are discussed in Appendix A.

14 This theorem was first proved in 1829 by the great French mathematician Augustin Louis Cauchy (1789–1857) who is most remembered for his work in analysis.

If we write $X = [z_1 \; z_2 \; \cdots \; z_n]^T$ where the z_i are complex numbers, we have

$$c = X^T \bar{X} = [z_1 \; z_2 \; \cdots \; z_n] \begin{bmatrix} \bar{z_1} \\ \bar{z_2} \\ \vdots \\ \bar{z_n} \end{bmatrix} = z_1 \bar{z_1} + z_2 \bar{z_2} + \cdots + z_n \bar{z_n}$$

$$= |z_1|^2 + |z_2|^2 + \cdots + |z_n|^2.$$

Thus c is a real number, and $c > 0$ because at least one of the $z_i \neq 0$ (as $X \neq 0$). We show that $\bar{\lambda} = \lambda$ by verifying that $\lambda c = \bar{\lambda} c$. We have

$$\lambda c = \lambda(X^T \bar{X}) = (\lambda X)^T \bar{X} = (AX)^T \bar{X} = X^T A^T \bar{X}.$$

At this point we use the hypothesis that A is symmetric and real. This means $A^T = A = \bar{A}$, so we continue

$$\lambda c = X^T A^T \bar{X} = X^T(\bar{A}\bar{X}) = X^T(\overline{AX}) = X^T(\overline{\lambda X})$$
$$= X^T(\bar{\lambda}\bar{X})$$
$$= \bar{\lambda} X^T \bar{X}$$
$$= \bar{\lambda} c$$

as required.

The technique in the proof of Theorem 7 will be used again when we return to complex linear algebra in Section 8.6.

Example 7

Verify Theorem 7 for every real, symmetric 2×2 matrix A.

Solution If $A = \begin{bmatrix} a & b \\ b & c \end{bmatrix}$ we have $c_A(x) = x^2 - (a+c)x + (ac - b^2)$, so the eigenvalues are given by $\lambda = \frac{1}{2}\left[(a+c) \pm \sqrt{(a+c)^2 - 4(ac - b^2)}\right]$. But the discriminant

$$(a+c)^2 - 4(ac - b^2) = (a-c)^2 + 4b^2 \geq 0$$

for any choice of a, b, and c. Hence, the eigenvalues are real numbers.

Exercises 5.5

1. By computing the trace, determinant, and rank, show that A and B are *not* similar in each case.

(a) $A = \begin{bmatrix} 1 & 2 \\ 2 & 1 \end{bmatrix}$, $B = \begin{bmatrix} 1 & 1 \\ -1 & 1 \end{bmatrix}$

♦(b) $A = \begin{bmatrix} 3 & 1 \\ 2 & -1 \end{bmatrix}$, $B = \begin{bmatrix} 1 & 1 \\ 2 & 1 \end{bmatrix}$

(c) $A = \begin{bmatrix} 2 & 1 \\ 1 & -1 \end{bmatrix}$, $B = \begin{bmatrix} 3 & 0 \\ 1 & -1 \end{bmatrix}$

♦(d) $A = \begin{bmatrix} 3 & 1 \\ -1 & 2 \end{bmatrix}$, $B = \begin{bmatrix} 2 & -1 \\ 3 & 2 \end{bmatrix}$

(e) $A = \begin{bmatrix} 2 & 1 & 1 \\ 1 & 0 & 1 \\ 1 & 1 & 0 \end{bmatrix}$, $B = \begin{bmatrix} 1 & -2 & 1 \\ -2 & 4 & -2 \\ -3 & 6 & -3 \end{bmatrix}$

♦(f) $A = \begin{bmatrix} 1 & 2 & -3 \\ 1 & -1 & 2 \\ 0 & 3 & -5 \end{bmatrix}$, $B = \begin{bmatrix} -2 & 1 & 3 \\ 6 & -3 & -9 \\ 0 & 0 & 0 \end{bmatrix}$

2. Show that $\begin{bmatrix} 1 & 2 & -1 & 0 \\ 2 & 0 & 1 & 1 \\ 1 & 1 & 0 & -1 \\ 4 & 3 & 0 & 0 \end{bmatrix}$ and $\begin{bmatrix} 1 & -1 & 3 & 0 \\ -1 & 0 & 1 & 1 \\ 0 & -1 & 4 & 1 \\ 5 & -1 & -1 & -4 \end{bmatrix}$

   are *not* similar.

3. If $A \sim B$, show that:

   (a) $A^T \sim B^T$ ◆(b) $A^{-1} \sim B^{-1}$

   (c) $rA \sim rB$ for r in $\mathbb{R}$ (d) $A^n \sim B^n$ for $n \geq 1$

4. In each case, decide whether the matrix A is diagonalizable. If so, find P such that $P^{-1}AP$ is diagonal.

   (a) $\begin{bmatrix} 1 & 0 & 0 \\ 1 & 2 & 1 \\ 0 & 0 & 1 \end{bmatrix}$ ◆(b) $\begin{bmatrix} 3 & 0 & 6 \\ 0 & -3 & 0 \\ 5 & 0 & 2 \end{bmatrix}$

   (c) $\begin{bmatrix} 3 & 1 & 6 \\ 2 & 1 & 0 \\ -1 & 0 & -3 \end{bmatrix}$ ◆(d) $\begin{bmatrix} 4 & 0 & 0 \\ 0 & 2 & 2 \\ 2 & 3 & 1 \end{bmatrix}$

5. If A is invertible, show that AB is similar to BA for all B.

6. Show that the only matrix similar to a scalar matrix $A = rI$, r in $\mathbb{R}$, is A itself.

7. Let λ be an eigenvalue of A with corresponding eigenvector X. If $B = P^{-1}AP$ is similar to A, show that $P^{-1}X$ is an eigenvector of B corresponding to λ.

8. If $A \sim B$ and A has any of the following properties, show that B has the same property.

   (a) Idempotent, that is $A^2 = A$.

   ◆(b) Nilpotent, that is $A^k = 0$ for some $k \geq 1$.

   (c) Invertible.

9. Let A denote an $n \times n$ upper triangular matrix.

   (a) If all the main diagonal entries of A are distinct, show that A is diagonalizable.

   ◆(b) If all the main diagonal entries of A are equal, show that A is diagonalizable only if it is *already* diagonal.

   (c) Show that $\begin{bmatrix} 1 & 0 & 1 \\ 0 & 1 & 0 \\ 0 & 0 & 2 \end{bmatrix}$ is diagonalizable but

   that $\begin{bmatrix} 1 & 1 & 0 \\ 0 & 1 & 0 \\ 0 & 0 & 2 \end{bmatrix}$ is not.

10. Let A be a diagonalizable $n \times n$ matrix with eigenvalues $\lambda_1, \lambda_2, \ldots, \lambda_n$ (including multiplicities). Show that:

    (a) $\det A = \lambda_1 \lambda_2 \cdots \lambda_n$

    ◆(b) $\text{tr} A = \lambda_1 + \lambda_2 + \cdots + \lambda_n$

11. Given a polynomial $p(x) = r_0 + r_1 x + \cdots + r_n x^n$ and a square matrix A, the matrix $p(A) = r_0 I + r_1 A + \cdots + r_n A^n$ is called the **evaluation** of $p(x)$ at A. Let $B = P^{-1}AP$. Show that $p(B) = P^{-1}p(A)P$ for all polynomials $p(x)$.

12. Let P be an invertible $n \times n$ matrix. If A is any $n \times n$ matrix, write $T_P(A) = P^{-1}AP$. Verify that:

    (a) $T_P(I) = I$

    ◆(b) $T_P(AB) = T_P(A) T_P(B)$

    (c) $T_P(A + B) = T_P(A) + T_P(B)$

    (d) $T_P(rA) = r T_P(A)$

    (e) $T_P(A^k) = [T_P(A)]^k$ for $k \geq 1$

    (f) If A is invertible, $T_P(A^{-1}) = [T_P(A)]^{-1}$.

    (g) If Q is invertible, $T_Q[T_P(A)] = T_{PQ}(A)$.

13. (a) Show that two diagonalizable matrices are similar if and only if they have the same eigenvalues with the same multiplicities.

    ◆(b) If A is diagonalizable, show that $A \sim A^T$.

14. If A is 2×2 and diagonalizable, show that $C(A) = \{X \mid XA = AX\}$ has dimension 2 or 4. [*Hint:* If $P^{-1}AP = D$, show that X is in $C(A)$ if and only if $P^{-1}XP$ is in $C(D)$.]

15. If A is diagonalizable and $p(x)$ is a polynomial such that $p(\lambda) = 0$ for all eigenvalues λ of A, show that $p(A) = 0$ (see Example 9 §3.3). In particular, show $c_A(A) = 0$. [*Remark:* $c_A(A) = 0$ for *all* square matrices A—this is the Cayley–Hamilton theorem (see Theorem 2 §9.4).]

16. Let A be $n \times n$ with n distinct real eigenvalues. If $AC = CA$, show that C is diagonalizable.

17. Let $A = \begin{bmatrix} 0 & a & b \\ a & 0 & c \\ b & c & 0 \end{bmatrix}$ and $B = \begin{bmatrix} c & a & b \\ a & b & c \\ b & c & a \end{bmatrix}$.

    (a) Show that $x^3 - (a^2 + b^2 + c^2)x - 2abc$ has real roots by considering A.

    ◆(b) Show that $a^2 + b^2 + c^2 \geq ab + ac + bc$ by considering B.

18. Assume the 2×2 matrix A is similar to an upper triangular matrix. If $\operatorname{tr} A = 0 = \operatorname{tr} A^2$, show that $A^2 = 0$.

19. Show that A is similar to A^T for all 2×2 matrices A. [*Hint:* Let $A = \begin{bmatrix} a & b \\ c & d \end{bmatrix}$. If $c = 0$, treat the cases $b = 0$ and $b \neq 0$ separately. If $c \neq 0$, reduce to the case $c = 1$ using Exercise 12(d).]

20. Refer to Section 3.4 on linear recurrences. Assume that the sequence $x_0, x_1, x_2, \ldots$ satisfies

    $$x_{n+k} = r_0 x_n + r_1 x_{n+1} + \cdots + r_{k-1} x_{n+k-1}$$

    for all $n \geq 0$. Define

$$A = \begin{bmatrix} 0 & 1 & 0 & \cdots & 0 \\ 0 & 0 & 1 & \cdots & 0 \\ \vdots & \vdots & \vdots & & \vdots \\ 0 & 0 & 0 & \cdots & 1 \\ r_0 & r_1 & r_2 & \cdots & r_{k-1} \end{bmatrix}, \quad V_n = \begin{bmatrix} x_n \\ x_{n+1} \\ \vdots \\ x_{n+k-1} \end{bmatrix}.$$

Then show that:

(a) $V_n = A^n V_0$ for all n.

(b) $c_A(x) = x^k - r_{k-1} x^{k-1} - \cdots - r_1 x - r_0$.

(c) If λ is an eigenvalue of A, the eigenspace E_λ has dimension 1, and $X = (1, \lambda, \lambda^2, \ldots, \lambda^{k-1})^T$ is an eigenvector. [*Hint:* Use $c_A(\lambda) = 0$ to show that $E_\lambda = \mathbb{R}X$.]

(d) A is diagonalizable if and only if the eigenvalues of A are distinct. [*Hint:* See part (c) and Theorem 4.]

(e) If $\lambda_1, \lambda_2, \ldots, \lambda_k$ are distinct real eigenvalues, there exist constants $t_1, t_2, \ldots, t_k$ such that $x_n = t_1 \lambda_1^n + \cdots + t_k \lambda_k^n$ holds for all n. [*Hint:* If D is diagonal with $\lambda_1, \lambda_2, \ldots, \lambda_k$ as the main diagonal entries, show that $A^n = PD^nP^{-1}$ has entries that are linear combinations of $\lambda_1^n, \lambda_2^n, \ldots, \lambda_k^n$.]

SECTION 5.6   An Application to Correlation and Variance

Suppose the heights $h_1, h_2, \ldots, h_n$ of n men are measured. Such a data set is called a **sample** of the heights of all the men in the population under study, and various questions are often asked about such a sample: What is the average height in the sample? How much variation is there in the sample heights, and how can it be measured? What can be inferred from the sample about the heights of all men in the population? How do these heights compare to heights of men in neighbouring countries? Does the prevalence of smoking affect the height of a man?

The analysis of samples, and of inferences that can be drawn from them, is a subject called *mathematical statistics*, and an extensive body of information has been developed to answer many such questions. In this section we will describe a few ways that linear algebra can be used.

It is convenient to represent a sample $\{x_1, x_2, \ldots, x_n\}$ as a **sample vector** $X = [x_1 \quad x_2 \quad \cdots \quad x_n]^T$ in $\mathbb{R}^n$. This being done, the dot product in $\mathbb{R}^n$ provides a convenient tool to study the sample and describe some of the statistical concepts related to it. The most widely known statistic for describing a data set is the **sample mean** $\bar{x}$ defined by[15]

$$\bar{x} = \frac{1}{n}(x_1 + x_2 + \cdots + x_n) = \frac{1}{n}\sum_{i=1}^{n} x_i.$$

The mean $\bar{x}$ is "typical" of the sample values x_i, but may not itself be one of them. The number $x_i - \bar{x}$ is called the **deviation** of x_i from the mean $\bar{x}$. The deviation is

---

15 The mean is often called the "average" of the sample values x_i, but statisticians use the term "mean".

positive if $x_i > \bar{x}$ and it is negative if $x_i < \bar{x}$. Moreover, the sum of these deviations is zero:

$$\sum_{i=1}^{n}(x_i - \bar{x}) = \left(\sum_{i=1}^{n} x_i\right) - n\bar{x} = n\bar{x} - n\bar{x} = 0. \qquad (*)$$

This is described by saying that the sample mean $\bar{x}$ is *central* to the sample values x_i.

If the mean $\bar{x}$ is subtracted from each data value x_i, the resulting data $x_i - \bar{x}$ are said to be **centred**. The corresponding data vector is

$$X_c = [x_1 - \bar{x} \quad x_2 - \bar{x} \cdots x_n - \bar{x}]^T$$

and ($*$) shows that the mean $\bar{x}_c = 0$. For example, the sample $X = [-1 \ 0 \ 1 \ 4 \ 6]^T$ is plotted in the diagram. The mean is $\bar{x} = 2$, and the centred sample $X_c = [-3 \ -2 \ -1 \ 2 \ 4]^T$ is also plotted. Thus, the effect of centring is to shift the data by an amount $\bar{x}$ (to the left if $\bar{x}$ is positive) so that the mean moves to 0.

Another question that arises about samples is how much variability there is in the sample $X = [x_1 \ x_2 \ \cdots \ x_n]^T$; that is, how widely are the data "spread out" around the sample mean $\bar{x}$. A natural measure of variability would be the sum of the deviations of the x_i about the mean, but this sum is zero by ($*$); these deviations cancel out. To avoid this cancellation, statisticians use the *squares* $(x_i - \bar{x})^2$ of the deviations as a measure of variability. More precisely, they compute a statistic called the **sample variance** s_x^2, defined[16] as follows:

$$s_x^2 = \tfrac{1}{n-1}\left[(x_1 - \bar{x})^2 + (x_2 - \bar{x})^2 + \cdots + (x_n - \bar{x})^2\right] = \tfrac{1}{n-1}\sum_{i=1}^{n}(x_i - \bar{x})^2.$$

The sample variance will be large if there are many x_i at a large distance from the mean $\bar{x}$, and it will be small if all the x_i are tightly clustered about the mean. The variance is clearly nonnegative (hence the notation s_x^2), and the square root s_x of the variance is called the **sample standard deviation**.

The sample mean and variance can be conveniently described using the dot product. Let

$$\mathbf{1} = [1 \ 1 \ \cdots \ 1]^T$$

denote the column with every entry equal to 1. If $X = [x_1 \ x_2 \ \cdots \ x_n]^T$, then $X \bullet \mathbf{1} = x_1 + x_2 + \cdots + x_n$, so the sample mean is given by the formula

$$\bar{x} = \frac{1}{n}(X \bullet \mathbf{1}).$$

Moreover, remembering that $\bar{x}$ is a scalar, we have $\bar{x}\mathbf{1} = [\bar{x} \ \bar{x} \ \cdots \ \bar{x}]^T$, so the centred sample vector X_c is given by

$$X_c = X - \bar{x}\mathbf{1} = [x_1 - \bar{x} \quad x_2 - \bar{x} \quad \cdots \quad x_n - \bar{x}]^T.$$

Thus we obtain a formula for the sample variance:

$$s_x^2 = \tfrac{1}{n-1}\|X_c\|^2 = \tfrac{1}{n-1}\|X - \bar{x}\mathbf{1}\|^2.$$

Linear algebra is also useful for comparing two different samples. To illustrate how, consider two examples.

Sample X

$-1 \quad 0 \ 1 \ 2 \qquad 4 \qquad 6$

$\bar{x}$

Centred Sample X_c

$-3 \ -2 \ -1 \quad 0 \qquad 2 \qquad 4$

$\bar{x}_c$

---

16 Since there are n sample values, it seems more natural to divide by n here, rather than by $n-1$. The reason for using $n-1$ is that then the sample variance s_x^2 provides a better estimate of the variance of the entire population from which the sample was drawn.

The following table represents the number of sick days at work per year and the yearly number of visits to a physician for 10 individuals.

Individual	1	2	3	4	5	6	7	8	9	10
Doctor visits	2	6	8	1	5	10	3	9	7	4
Sick days	2	4	8	3	5	9	4	7	7	2

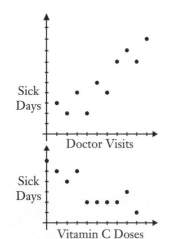

Sick Days

Doctor Visits

Sick Days

Vitamin C Doses

The data are plotted in the **scatter diagram** where it is evident that, roughly speaking, the more visits to the doctor the more sick days. This is an example of a *positive correlation* between sick days and doctor visits.

Now consider the following table representing the daily doses of vitamin C and the number of sick days.

Individual	1	2	3	4	5	6	7	8	9	10
Vitamin C	1	5	7	0	4	9	2	8	6	3
Sick days	5	2	2	6	2	1	4	3	2	5

The scatter diagram is plotted as shown and it appears that the more vitamin C taken, the fewer sick days. In this case there is a *negative correlation* between daily vitamin C and sick days.

In both these situations, we have **paired samples**, that is observations of two variables are made for ten individuals: doctor visits and sick days in the first case; daily vitamin C and sick days in the second case. The scatter diagrams point to a relationship between these variables, and there is a way to use the sample to compute a number, called the correlation coefficient, that measures the degree to which the variables are associated.

To motivate the definition of the correlation coefficient, suppose two paired samples $X = [x_1 \quad x_2 \quad \cdots \quad x_n]^T$ and $Y = [y_1 \quad y_2 \quad \cdots \quad y_n]^T$ are given and consider the centred samples

$$X_c = [x_1 - \bar{x} \quad x_2 - \bar{x} \quad \cdots \quad x_n - \bar{x}]^T \quad \text{and} \quad Y_c = [y_1 - \bar{y} \quad y_2 - \bar{y} \quad \cdots \quad y_n - \bar{y}]^T.$$

If x_k is large among the x_i's, then the deviation $x_k - \bar{x}$ will be positive; and $x_k - \bar{x}$ will be negative if x_k is small among the x_i's. The situation is similar for Y, and the following table displays the sign of the quantity $(x_i - \bar{x})(y_i - \bar{y})$ in all four cases:

Sign of $(x_i - \bar{x})(y_i - \bar{y})$:

	x_i large	x_i small
y_i large	positive	negative
y_i small	negative	positive

Intuitively, if X and Y are positively correlated, then two things happen:

1. Large values of the x_i tend to be associated with large values of the y_i, and
2. Small values of the x_i tend to be associated with small values of the y_i.

It follows from the table that, if X and Y are positively correlated, then the dot product

$$X_c \bullet Y_c = \sum_{i=1}^{n} (x_i - \bar{x})(y_i - \bar{y})$$

is positive. Similarly $X_c \bullet Y_c$ is negative if X and Y are negatively correlated. With this in mind, the **sample correlation coefficient**[17] r is defined by

---

[17] The idea of using a single number to measure the degree of relationship between different variables was pioneered by Francis Galton (1822–1911). He was studying the degree to which characteristics of an offspring relate to those of its parents. The idea was refined by Karl Pearson (1857–1936) and r is often referred to as the Pearson correlation coefficient.

$$r = r(X,Y) = \frac{X_c \cdot Y_c}{\|X_c\| \|Y_c\|}.$$

Bearing the situation in $\mathbb{R}^3$ in mind, r is the cosine of the "angle" between the vectors X_c and Y_c, and so we would expect it to lie between -1 and 1. Moreover, we would expect r to be near 1 (or -1) if these vectors were pointing in the same (opposite) direction, that is the "angle" is near zero (or π).

This is confirmed by Theorem 1 below, and it is also borne out in the examples above. If we compute the correlation between sick days and visits to the physician (in the first scatter diagram above) the result is $r = 0.90$ as expected. On the other hand, the correlation between daily vitamin C doses and sick days (second scatter diagram) is $r = -0.84$.

However, a word of caution is in order here. We *cannot* conclude from the second example that taking more vitamin C will reduce the number of sick days at work. The (negative) correlation may arise because of some third factor that is related to both variables. In this case it may be that less healthy people are inclined to take more vitamin C. Correlation does *not* imply causation. Similarly, the correlation between sick days and visits to the doctor does not mean that having many sick days *causes* more visits to the doctor. A correlation between two variables may point to the existence of other underlying factors, but it does not necessarily mean that there is a causality relationship between the variables.

Our discussion of the dot product in $\mathbb{R}^n$ provides the basic properties of the correlation coefficient:

Theorem 1

Let $X = [x_1 \ \ x_2 \ \ \cdots \ \ x_n]^T$ and $Y = [y_1 \ \ y_2 \ \ \cdots \ \ y_n]^T$ be (nonzero) paired samples, and let $r = r(X, Y)$ denote the correlation coefficient. Then:
1. $-1 \le r \le 1$.
2. $r = 1$ if and only if there exist a and $b > 0$ such that $y_i = a + bx_i$ for each i.
3. $r = -1$ if and only if there exist a and $b < 0$ such that $y_i = a + bx_i$ for each i.

PROOF

The Cauchy inequality (Theorem 2 §5.3) proves (1), and also shows that $r = \pm 1$ if and only if one of X_c and Y_c is a scalar multiple of the other. This in turn holds if and only if $Y_c = bX_c$ for some $b \ne 0$, and it is easy to verify that $r = 1$ when $b > 0$ and $r = -1$ when $b < 0$.

Finally, $Y_c = bX_c$ means $y_i - \bar{y} = b(x_i - \bar{x})$ for each i; that is, $y_i = a + bx_i$ where $a = \bar{y} - b\bar{x}$. Conversely, if $y_i = a + bx_i$, then $\bar{y} = a + b\bar{x}$ (verify), so $y_i - \bar{y} = (a + bx_i) - (a + b\bar{x}) = b(x_i - \bar{x})$ for each i. In other words, $Y_c = bX_c$. This completes the proof.

Properties (2) and (3) in Theorem 1 show that $r(X, Y) = 1$ means that there is a linear relation with *positive* slope between the paired data (so large x values are paired with large y values). Similarly, $r(X, Y) = -1$ means that there is a linear relation with *negative* slope between the paired data (so small x values are paired with small y values). This is borne out in the two scatter diagrams above.

We conclude by using the dot product to derive some useful formulas for computing variances and correlation coefficients. Given samples $X = [x_1 \ \ x_2 \ \ \cdots \ \ x_n]^T$ and $Y = [y_1 \ \ y_2 \ \ \cdots \ \ y_n]^T$, the key observation is the following formula:

$$X_c \cdot Y_c = X \cdot Y - n\bar{x}\,\bar{y}. \qquad (**)$$

Indeed, remembering that $\bar{x}$ and $\bar{y}$ are scalars:

$$
\begin{aligned}
X_c \bullet Y_c &= (X - \bar{x}\mathbf{1}) \bullet (Y - \bar{y}\mathbf{1}) \\
&= X \bullet Y - \bar{y}(X \bullet \mathbf{1}) - \bar{x}(\mathbf{1} \bullet Y) + \bar{x}\bar{y}(\mathbf{1} \bullet \mathbf{1}) \\
&= X \bullet Y - \bar{y}(n\bar{x}) - \bar{x}(n\bar{y}) + \bar{x}\bar{y}(n) \\
&= X \bullet Y - n\bar{x}\bar{y}.
\end{aligned}
$$

Taking $Y = X$ in (**) gives a formula for the variance $s_x^2 = \frac{1}{n-1}\|X_c\|^2$ of X.

Variance Formula

If x is a sample vector, then $s_x^2 = \frac{1}{n-1}\left(\|X\|^2 - n\bar{x}^2\right)$.

We also get a convenient formula for the correlation coefficient, $r = r(X, Y) = \dfrac{X_c \bullet Y_c}{\|X_c\|\|Y_c\|}$.

Moreover, (**) and the fact that $s_x^2 = \frac{1}{n-1}\|X_c\|^2$ give:

Correlation Formula

If X and Y are sample vectors, then

$$
r = r(X, Y) = \frac{X \bullet Y - n\bar{x}\,\bar{y}}{(n-1)s_x\,s_y}.
$$

Finally, we give a method that simplifies the computations of variances and correlations.

Data Scaling

Let $X = [x_1 \;\; x_2 \;\; \cdots \;\; x_n]^T$ and $Y = [y_1 \;\; y_2 \;\; \cdots \;\; y_n]^T$ be sample vectors. Given constants a, b, c, and d, consider new samples $Z = [z_1 \;\; z_2 \;\; \cdots \;\; z_n]^T$ and $W = [w_1 \;\; w_2 \;\; \cdots \;\; w_n]^T$ where $z_i = a + bx_i$ for each i and $w_i = c + dy_i$ for each i.
Then:
  (a) $\bar{z} = a + b\bar{x}$.
  (b) $s_z^2 = b^2 s_x^2$, so $s_z = |b|\,s_x$.
  (c) If b and d have the same sign, then $r(X, Y) = r(Z, W)$.

The verification is left as an exercise.
   For example, if $X = [101 \;\; 98 \;\; 103 \;\; 99 \;\; 100 \;\; 97]^T$, subtracting 100 yields $Z = [1 \;\; -2 \;\; 3 \;\; -1 \;\; 0 \;\; -3]^T$. A routine calculation shows that $\bar{z} = -\frac{1}{3}$ and $s_z^2 = \frac{14}{3}$, so $\bar{x} = 100 - \frac{1}{3} = 99.67$, and $s_x^2 = \frac{14}{3} = 4.67$.

Exercises 5.6

1. The following table gives IQ scores for 10 fathers and their eldest sons. Calculate the means, the variances, and the correlation coefficient r. (The data scaling formula is useful.)

	1	2	3	4	5	6	7	8	9	10
Father's IQ	140	131	120	115	110	106	100	95	91	86
Son's IQ	130	138	110	99	109	120	105	99	100	94

♦2. The following table gives the number of years of education and the annual income (in thousands) of 10 individuals. Find the means, the variances, and the correlation coefficient. (Again the data scaling formula is useful.)

Individual	1	2	3	4	5	6	7	8	9	10
Years of education	12	16	13	18	19	12	18	19	12	14
Yearly income (1000's)	31	48	35	28	55	40	39	60	32	35

3. If X is a sample vector, and X_c is the centred sample, show that $\bar{x}_c = 0$ and the standard deviation of X_c is s_x.

4. Prove the data scaling formulas:
   (a)   ♦(b)   (c)
   found on page 242.

SECTION 5.7   An Application to Least Squares Approximation

In many scientific investigations, data are collected that relate two variables. For example, if x is the number of dollars spent on advertising by a manufacturer and y is the value of sales in the region in question, the manufacturer could generate data by spending $x_1, x_2, \ldots, x_n$ dollars at different times and measuring the corresponding sales values $y_1, y_2, \ldots, y_n$.

Suppose it is known that a linear relationship exists between the variables x and y—in other words, that $y = a + bx$ for some constants a and b. If the data are plotted, the points $(x_1, y_1), (x_2, y_2), \ldots, (x_n, y_n)$ may appear to lie on a straight line and estimating a and b requires finding the "best-fitting" line through these data points. For example, if five data points occur as shown in Figure 5.1, line 1 is clearly a better fit than line 2. In general, the problem is to find the values of the constants a and b such that the line $y = a + bx$ best approximates the data in question. Note that an exact fit would be obtained if a and b were such that $y_i = a + bx_i$ were true for each data point (x_i, y_i). But this is too much to expect. Experimental errors in measurement are bound to occur, so the choice of a and b should be made in such a way that the errors between the observed values y_i and the corresponding fitted values $a + bx_i$ are in some sense minimized.

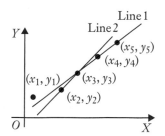

Figure 5.1

The first thing we must do is explain exactly what we mean by the *best fit* of a line $y = a + bx$ to an observed set of data points $(x_1, y_1), (x_2, y_2), \ldots, (x_n, y_n)$. For convenience, write the linear function $r + sx$ as

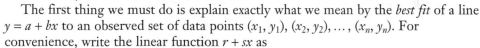
$$f(x) = r + sx$$

so that the fitted points (on the line) have coordinates $(x_1, f(x_1)), \ldots, (x_n, f(x_n))$. Figure 5.2 is a sketch of what the line $y = f(x)$ might look like. For each i the observed data point (x_i, y_i) and the fitted point $(x_i, f(x_i))$ need not be the same, and the distance d_i between them measures how far the line misses the observed point. For this reason d_i is often called the **error** at x_i, and a natural measure of how close the line $y = f(x)$ is to the observed data points is the sum $d_1 + d_2 + \cdots + d_n$ of all these errors. However, it turns out to be better to use the sum of squares

$$S = d_1^2 + d_2^2 + \cdots + d_n^2$$

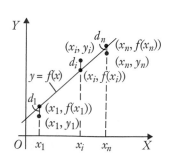

Figure 5.2

as the measure of error, and the line $y = f(x)$ is to be chosen so as to make this sum as small as possible. This line is said to be the **least squares approximating line** for the data points $(x_1, y_1), (x_2, y_2), \ldots, (x_n, y_n)$.

The square of the error d_i is given by $d_i^2 = [y_i - f(x_i)]^2$ for each i, so the quantity S to be minimized is the sum:

$$S = [y_1 - f(x_1)]^2 + [y_2 - f(x_2)]^2 + \cdots + [y_n - f(x_n)]^2.$$

Note that all the numbers x_i and y_i are *given* here: what is required is that the *function f* be chosen in such a way as to minimize S. Because $f(x) = r + sx$, this amounts to choosing r and s so as to minimize S, and the problem can be solved using vector techniques. The following notations simplify the discussion.

$$Y = \begin{bmatrix} y_1 \\ y_2 \\ \vdots \\ y_n \end{bmatrix} \qquad M = \begin{bmatrix} 1 & x_1 \\ 1 & x_2 \\ \vdots & \vdots \\ 1 & x_n \end{bmatrix} \qquad Z = \begin{bmatrix} r \\ s \end{bmatrix}$$

Observe that

$$Y - MZ = \begin{bmatrix} y_1 - (r + sx_1) \\ y_2 - (r + sx_2) \\ \vdots \\ y_n - (r + sx_n) \end{bmatrix} = \begin{bmatrix} y_1 - f(x_1) \\ y_2 - f(x_2) \\ \vdots \\ y_n - f(x_n) \end{bmatrix}$$

so the quantity S to be minimized is

$$S = \|Y - MZ\|^2.$$

Here Y and M are given and we are asked to find Z such that the length of the vector $Y - MZ$ is as small as possible. To this end, consider the set U of all vectors MZ where Z varies:

$$U = \left\{ MZ \mid Z = \begin{bmatrix} r \\ s \end{bmatrix} \text{arbitrary} \right\} = \left\{ \begin{bmatrix} r + sx_1 \\ r + sx_2 \\ \vdots \\ r + sx_n \end{bmatrix} \middle| r \text{ and } s \text{ arbitrary} \right\}.$$

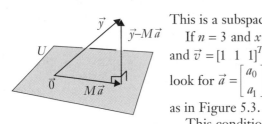

Figure 5.3

This is a subspace of $\mathbb{R}^n$, and the task is to choose MZ in U as close as possible to Y. If $n = 3$ and x_1, x_2 and x_3 are distinct, U is the plane containing $\vec{x} = [x_1 \ x_2 \ x_3]^T$ and $\vec{v} = [1 \ 1 \ 1]^T$ with normal $\vec{x} \times \vec{v} = [x_2 - x_3 \ x_3 - x_1 \ x_1 - x_2]^T$. In this case, we look for $\vec{a} = \begin{bmatrix} a_0 \\ a_1 \end{bmatrix}$ so that $\vec{y} - M\vec{a}$ is orthogonal to every vector $M\vec{z}$ in the plane U, as in Figure 5.3.

This condition[18] makes sense in $\mathbb{R}^n$ so we look for $A = \begin{bmatrix} a_0 \\ a_1 \end{bmatrix}$ such that $(MZ) \cdot (Y - MA) = 0$ for all Z in $\mathbb{R}^2$. This dot product is in $\mathbb{R}^n$, and it can be written as a dot product in $\mathbb{R}^2$:

$$0 = (MZ)^T(Y - MA) = Z^T[M^T(Y - MA)] = Z \cdot (M^TY - M^TMA)$$

for all Z in $\mathbb{R}^2$. This means that $M^TY - M^TMA$ is orthogonal to *every* vector in $\mathbb{R}^2$. In particular, it is orthogonal to itself, and so must be zero, and we obtain

$$(M^TM)A = M^TY.$$

These are called the **normal equations** for A, and can be solved using gaussian elimination. Moreover, if at least two of the x_i are distinct, the matrix M^TM can be shown to be invertible, so the solution A is unique. If the solution is $A = \begin{bmatrix} a_0 \\ a_1 \end{bmatrix}$, the best fitting line is $y = a_0 + a_1x$. This proves the following useful theorem.

---

18 We will revisit this in Chapter 8 where a more rigorous argument will be given.

Theorem 1

Suppose that n data points $(x_1, y_1), (x_2, y_2), \ldots, (x_n, y_n)$ are given, where at least two of $x_1, x_2, \ldots, x_n$ are distinct. Put

$$Y = \begin{bmatrix} y_1 \\ y_2 \\ \vdots \\ y_n \end{bmatrix} \qquad M = \begin{bmatrix} 1 & x_1 \\ 1 & x_2 \\ \vdots & \vdots \\ 1 & x_n \end{bmatrix}$$

Then the least squares approximating line for these data points has the equation

$$y = a_0 + a_1 x$$

where $A = \begin{bmatrix} a_0 \\ a_1 \end{bmatrix}$ is found by gaussian elimination from the normal equations

$$(M^T M)A = M^T Y.$$

The condition that at least two of $x_1, x_2, \ldots, x_n$ are distinct ensures that $M^T M$ is an invertible matrix, so A is unique:

$$A = (M^T M)^{-1} M^T Y.$$

Example 1

Let data points $(x_1, y_1), (x_2, y_2), \ldots, (x_5, y_5)$ be given as in the accompanying table. Find the least squares approximating line for these data.

Solution   In this case we have

$$M^T M = \begin{bmatrix} 1 & 1 & \cdots & 1 \\ x_1 & x_2 & \cdots & x_5 \end{bmatrix} \begin{bmatrix} 1 & x_1 \\ 1 & x_2 \\ \vdots & \vdots \\ 1 & x_5 \end{bmatrix}$$

$$= \begin{bmatrix} 5 & x_1 + \cdots + x_5 \\ x_1 + \cdots + x_5 & x_1^2 + \cdots + x_5^2 \end{bmatrix} = \begin{bmatrix} 5 & 21 \\ 21 & 111 \end{bmatrix}$$

x	y
1	1
3	2
4	3
6	4
7	5

$$M^T Y = \begin{bmatrix} 1 & 1 & \cdots & 1 \\ x_1 & x_2 & \cdots & x_5 \end{bmatrix} \begin{bmatrix} y_1 \\ y_2 \\ \vdots \\ y_5 \end{bmatrix}$$

$$= \begin{bmatrix} y_1 + y_2 + \cdots + y_5 \\ x_1 y_1 + x_2 y_2 + \cdots + x_5 y_5 \end{bmatrix} = \begin{bmatrix} 15 \\ 78 \end{bmatrix}$$

so the normal equations $(M^T M)A = M^T Y$ for $A = \begin{bmatrix} a_0 \\ a_1 \end{bmatrix}$ become

$$\begin{bmatrix} 5 & 21 \\ 21 & 111 \end{bmatrix} \begin{bmatrix} a_0 \\ a_1 \end{bmatrix} = \begin{bmatrix} 15 \\ 78 \end{bmatrix}$$

The solution (using gaussian elimination) is $\begin{bmatrix} a_0 \\ a_1 \end{bmatrix} = \begin{bmatrix} 0.24 \\ 0.66 \end{bmatrix}$ to two decimal places, so the least squares approximating line for these data is $y = 0.24 + 0.66x$. Note that $M^T M$ is indeed invertible here (the determinant is 114), and the exact solution is

$$A = (M^TM)^{-1}M^TY = \frac{1}{114}\begin{bmatrix} 111 & -21 \\ -21 & 5 \end{bmatrix}\begin{bmatrix} 15 \\ 78 \end{bmatrix} = \frac{1}{114}\begin{bmatrix} 27 \\ 75 \end{bmatrix} = \frac{1}{38}\begin{bmatrix} 9 \\ 25 \end{bmatrix}$$

Suppose now that, rather than a straight line, we want to find the parabola $y = a_0 + a_1x + a_2x^2$ that is the least squares approximation to the data points $(x_1, y_1), \ldots, (x_n, y_n)$. In the function $f(x) = a_0 + a_1x + a_2x^2$, the *three* constants a_0, a_1, and a_2 must be chosen to minimize the sum of squares of the errors:

$$S = [y_1 - f(x_1)]^2 + [y_2 - f(x_2)]^2 + \cdots + [y_n - f(x_n)]^2.$$

Choosing a_0, a_1, and a_2 amounts to choosing the (parabolic) function f that minimizes S.

In general, there is a relationship $y = f(x)$ between the variables, and the range of candidate functions is limited—say, to all lines or to all parabolas. The task is to find, among the suitable candidates, the function that makes the quantity S as small as possible. The function that does so is called the least squares approximating function (of that type) for the data points.

As might be imagined, this is not always an easy task. However, if the functions $f(x)$ are restricted to polynomials of degree m,

$$f(x) = a_0 + a_1x + \cdots + a_mx^m$$

the analysis proceeds much as before (where $m = 1$). The problem is to choose the numbers $a_0, a_1, \ldots, a_m$ so as to minimize the sum

$$S = [y_1 - f(x_1)]^2 + [y_2 - f(x_2)]^2 + \cdots + [y_n - f(x_n)]^2.$$

The resulting function $y = f(x) = a_0 + a_1x + \cdots + a_mx^m$ is called the **least squares approximating polynomial of degree m** for the data $(x_1, y_1), \ldots, (x_n, y_n)$. By analogy with the preceding analysis, define

$$Y = \begin{bmatrix} y_1 \\ y_2 \\ \vdots \\ y_n \end{bmatrix} \quad M = \begin{bmatrix} 1 & x_1 & x_1^2 & \cdots & x_1^m \\ 1 & x_2 & x_2^2 & \cdots & x_2^m \\ \vdots & \vdots & \vdots & & \vdots \\ 1 & x_n & x_n^2 & \cdots & x_n^m \end{bmatrix} \quad A = \begin{bmatrix} a_0 \\ a_1 \\ \vdots \\ a_m \end{bmatrix}$$

Then

$$Y - MA = \begin{bmatrix} y_1 - (a_0 + a_1x_1 + \cdots + a_mx_1^m) \\ y_2 - (a_0 + a_1x_2 + \cdots + a_mx_2^m) \\ \vdots \\ y_n - (a_0 + a_1x_n + \cdots + a_mx_n^m) \end{bmatrix} = \begin{bmatrix} y_1 - f(x_1) \\ y_2 - f(x_2) \\ \vdots \\ y_n - f(x_n) \end{bmatrix}$$

so S is the sum of the squares of the entries of $Y - MA$. An analysis similar to that for Theorem 1 can be used to prove Theorem 2.

Theorem 2

Suppose n data points $(x_1, y_1), (x_2, y_2), \ldots, (x_n, y_n)$ are given, where at least $m + 1$ of $x_1, x_2, \ldots, x_n$ are distinct (in particular $n \geq m + 1$). Put

$$Y = \begin{bmatrix} y_1 \\ y_2 \\ \vdots \\ y_n \end{bmatrix} \qquad M = \begin{bmatrix} 1 & x_1 & x_1^2 & \cdots & x_1^m \\ 1 & x_2 & x_2^2 & \cdots & x_2^m \\ \vdots & \vdots & \vdots & & \vdots \\ 1 & x_n & x_n^2 & \cdots & x_n^m \end{bmatrix}$$

Then the least squares approximating polynomial of degree m for the data points has the equation

$$y = a_0 + a_1 x + \cdots + a_m x^m$$

where $A = \begin{bmatrix} a_0 \\ a_1 \\ \vdots \\ a_m \end{bmatrix}$ is found by gaussian elimination from the normal equations

$$(M^T M)A = M^T Y .$$

The condition that at least $m + 1$ of $x_1, x_2, \ldots, x_n$ be distinct ensures that the matrix MM^T is invertible, so A is unique:

$$A = (M^T M)^{-1} M^T Y.$$

A proof of this theorem is given in Section 8.7 (Theorem 2).

Example 2

Find the least squares approximating quadratic $y = a_0 + a_1 x + a_2 x^2$ for the following data points.

$$(-3, 3), \ (-1, 1), \ (0, 1), \ (1, 2), \ (3, 4)$$

Solution   This is an instance of Theorem 2 with $m = 2$. Here

$$Y = \begin{bmatrix} 3 \\ 1 \\ 1 \\ 2 \\ 4 \end{bmatrix} \qquad M = \begin{bmatrix} 1 & -3 & 9 \\ 1 & -1 & 1 \\ 1 & 0 & 0 \\ 1 & 1 & 1 \\ 1 & 3 & 9 \end{bmatrix}$$

Hence,

$$M^T M = \begin{bmatrix} 1 & 1 & 1 & 1 & 1 \\ -3 & -1 & 0 & 1 & 3 \\ 9 & 1 & 0 & 1 & 9 \end{bmatrix} \begin{bmatrix} 1 & -3 & 9 \\ 1 & -1 & 1 \\ 1 & 0 & 0 \\ 1 & 1 & 1 \\ 1 & 3 & 9 \end{bmatrix} = \begin{bmatrix} 5 & 0 & 20 \\ 0 & 20 & 0 \\ 20 & 0 & 164 \end{bmatrix}$$

$$M^T Y = \begin{bmatrix} 1 & 1 & 1 & 1 & 1 \\ -3 & -1 & 0 & 1 & 3 \\ 9 & 1 & 0 & 1 & 9 \end{bmatrix} \begin{bmatrix} 3 \\ 1 \\ 1 \\ 2 \\ 4 \end{bmatrix} = \begin{bmatrix} 11 \\ 4 \\ 66 \end{bmatrix}$$

The normal equations for A are

$$\begin{bmatrix} 5 & 0 & 20 \\ 0 & 20 & 0 \\ 20 & 0 & 164 \end{bmatrix} A = \begin{bmatrix} 11 \\ 4 \\ 66 \end{bmatrix} \quad \text{whence } A = \begin{bmatrix} 1.15 \\ 0.20 \\ 0.26 \end{bmatrix}$$

This means that the least squares approximating quadratic for these data is $y = 1.15 + 0.20x + 0.26x^2$.

Least squares approximation can be used to estimate physical constants, as is illustrated by the next example.

Example 3

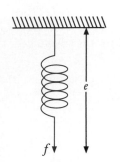

Hooke's law in mechanics asserts that the magnitude of the force f required to hold a spring is a linear function of the extension e of the spring (see the accompanying diagram). That is,

$$f = ke + e_0$$

where k and e_0 are constants depending only on the spring. The following data were collected for a particular spring.

e	9	11	12	16	19
f	33	38	43	54	61

Find the least squares approximating line $f = a_0 + a_1 e$ for these data, and use it to estimate k.

Solution   Here f and e play the role of y and x in the general theory. We have

$$Y = \begin{bmatrix} 33 \\ 38 \\ 43 \\ 54 \\ 61 \end{bmatrix} \qquad M = \begin{bmatrix} 1 & 9 \\ 1 & 11 \\ 1 & 12 \\ 1 & 16 \\ 1 & 19 \end{bmatrix}$$

as in Theorem 1, so

$$M^T M = \begin{bmatrix} 5 & 67 \\ 67 & 963 \end{bmatrix} \quad \text{and} \quad M^T Y = \begin{bmatrix} 229 \\ 3254 \end{bmatrix}$$

Hence the normal equations for A are

$$\begin{bmatrix} 5 & 67 \\ 67 & 963 \end{bmatrix} A = \begin{bmatrix} 229 \\ 3254 \end{bmatrix} \quad \text{whence} \quad A = \begin{bmatrix} 7.70 \\ 2.84 \end{bmatrix}$$

The least squares approximating line is $f = 7.70 + 2.84e$, so the estimate for k is $k = 2.84$.

Exercises 5.7

1. Find the least squares approximating line $y = a_0 + a_1 x$ for each of the following sets of data points.

   (a) (1, 1), (3, 2), (4, 3), (6, 4)

   ♦(b) (2, 4), (4, 3), (7, 2), (8, 1)

   (c) (−1, −1), (0, 1), (1, 2), (2, 4), (3, 6)

   ♦(d) (−2, 3), (−1, 1), (0, 0), (1, −2), (2, −4)

2. Find the least squares approximating quadratic $y = a_0 + a_1 x + a_2 x^2$ for each of the following sets of data points.

   (a) (0, 1), (2, 2), (3, 3), (4, 5)

   ◆(b) (−2, 1), (0, 0), (3, 2), (4, 3)

3. If M is a square invertible matrix, show that $A = M^{-1}Y$ (in the notation of Theorem 2).

◆4. Newton's laws of motion imply that an object dropped from rest at a height of 100 metres will be at a height $s = 100 - \frac{1}{2}gt^2$ metres t seconds later, where g is a constant called the acceleration due to gravity. The values of s and t given in the table are observed. Write $x = t^2$, find the least squares approximating line $s = a + bx$ for these data, and use b to estimate g.

   Then find the least squares approximating quadratic $s = a_0 + a_1 t + a_2 t^2$ and use the value of a_2 to estimate g.

t	1	2	3
s	95	80	56

5. A naturalist measured the heights y_i (in metres) of several spruce trees with trunk diameters x_i (in centimetres). The data are as given in the table. Find the least squares approximating line for these data and use it to estimate the height of a spruce tree with a trunk of diameter 10 cm.

x_i	5	7	8	12	13	16
y_i	2	3.3	4	7.3	7.9	10.1

6. (a) Use $m = 0$ in Theorem 2 to show that the best-fitting horizontal line $y = a_0$ through the data points $(x_1, y_1), \ldots, (x_n, y_n)$ is $y = \frac{1}{n}(y_1 + y_2 + \cdots + y_n)$, the average of the y coordinates.

   ◆(b) Deduce the conclusion in (a) without using Theorem 2.

7. Assume $n = m + 1$ in Theorem 2 (so M is square). If the x_i are distinct, use Theorem 6 §3.2 to show that M is invertible. Deduce that $A = M^{-1}Y$ and that the least squares polynomial is the interpolating polynomial (Theorem 6 §3.2) and actually passes through all the data points.

Supplementary Exercises for Chapter 5

1. In each case either show that the statement is true or give an example showing that it is false. Throughout, $X, Y, Z, X_1, X_2, \ldots, X_n$ denote vectors in $\mathbb{R}^n$.

   (a) If U is a subspace of $\mathbb{R}^n$ and $X + Y$ is in U, then X and Y are both in U.

   ◆(b) If U is a subspace of $\mathbb{R}^n$ and rX is in U, then X is in U.

   (c) If U is a nonempty set and $sX + tY$ is in U for any s and t whenever X and Y are in U, then U is a subspace.

   ◆(d) If U is a subspace of $\mathbb{R}^n$ and X is in U, then $-X$ is in U.

   (e) If $\{X, Y\}$ is independent, then $\{X, Y, X + Y\}$ is independent.

   ◆(f) If $\{X, Y, Z\}$ is independent, then $\{X, Y\}$ is independent.

   (g) If $\{X, Y\}$ is not independent, then $\{X, Y, Z\}$ is not independent.

   ◆(h) If all of $X_1, X_2, \ldots, X_n$ are nonzero, then $\{X_1, X_2, \ldots, X_n\}$ is independent.

   (i) If one of $X_1, X_2, \ldots, X_n$ is zero, then $\{X_1, X_2, \ldots, X_n\}$ is not independent.

   ◆(j) If $aX + bY + cZ = 0$ where a, b, and c are in $\mathbb{R}$, then $\{X, Y, Z\}$ is independent.

   (k) If $\{X, Y, Z\}$ is independent, then $aX + bY + cZ = 0$ for some a, b, and c in $\mathbb{R}$.

   ◆(l) If $\{X_1, X_2, \ldots, X_n\}$ is not independent, then $t_1 X_1 + t_2 X_2 + \cdots + t_n X_n = 0$ for t_i in $\mathbb{R}$ not all zero.

   (m) If $\{X_1, X_2, \ldots, X_n\}$ is independent, then $t_1 X_1 + t_2 X_2 + \cdots + t_n X_n = 0$ for some t_i in $\mathbb{R}$.

   ◆(n) Every set of four nonzero vectors in $\mathbb{R}^4$ is a basis.

   (o) No basis of $\mathbb{R}^3$ can contain a vector with a component 0.

   ◆(p) $\mathbb{R}^3$ has a basis of the form $\{X, X + Y, Y\}$ where X and Y are vectors.

   (q) Every basis of $\mathbb{R}^5$ contains one column of I_5.

   ◆(r) Every nonempty subset of a basis of $\mathbb{R}^3$ is again a basis of $\mathbb{R}^3$.

   (s) If $\{X_1, X_2, X_3, X_4\}$ and $\{Y_1, Y_2, Y_3, Y_4\}$ are bases of $\mathbb{R}^4$, then $\{X_1 + Y_1, X_2 + Y_2, X_3 + Y_3, X_4 + Y_4\}$ is also a basis of $\mathbb{R}^4$.

6 Vector Spaces

In this chapter we introduce vector spaces in full generality. The reader will notice some similarity with the discussion of the space $\mathbb{R}^n$ in Chapter 5. In fact much of the present material has been developed there, and there is some repetition. However, Chapter 6 deals with the notion of an *abstract* vector space, a concept that will be new to most readers. It turns out that there are many systems in which addition and scalar multiplication are defined and satisfy the usual rules familiar from $\mathbb{R}^n$. The study of abstract vector spaces is a way to deal with all these examples simultaneously. The new aspect is that we are dealing with an abstract system in which *all we know* about the vectors is that they are objects that can be added and multiplied by a scalar and satisfy rules familiar from $\mathbb{R}^n$. The novel thing is the *abstraction*. Getting used to this new conceptual level is facilitated by the work done in Chapter 5: First, the vector manipulations are familiar, giving the reader more time to become accustomed to the abstract setting; and, second, the mental images developed in the concrete setting of $\mathbb{R}^n$ serve as an aid to doing many of the exercises in Chapter 6.

The concept of a vector space was first introduced in 1844 by the German mathematician Hermann Grassmann (1809–1877), but his work did not receive the attention it deserved. It was not until 1888 that the Italian mathematician Guiseppe Peano (1858–1932) clarified Grassmann's work in his book *Calcolo Geometrico* and gave the vector space axioms in their present form. Vector spaces became established with the work of the Polish mathematician Stephan Banach (1892–1945), and the idea was finally accepted in 1918 when Hermann Weyl (1885–1955) used it in his widely read book *Raum-Zeit-Materie* ("Space-Time-Matter"), an introduction to the general theory of relativity.

SECTION 6.1   Examples and Basic Properties

Many mathematical entities have the property that they can be added and multiplied by a number. Numbers themselves have this property, as do $m \times n$ matrices : The sum of two such matrices is again $m \times n$ as is any scalar multiple of such a matrix. Polynomials are another familiar example, as are the geometric vectors in Chapter 4. It turns out that there are many other types of mathematical objects that can be added and multiplied by a scalar, and the general study of such systems is introduced in this chapter. Remarkably, much of what we could say in Chapter 5 about the dimension of subspaces in $\mathbb{R}^n$ can be formulated in this generality.

A **vector space** consists of a nonempty set V of objects (called **vectors**) that can be added, that can be multiplied by a real number (called a **scalar** in this context), and for which certain axioms hold.[1] If **v** and **w** are two vectors in V, their sum is expressed as $\mathbf{v} + \mathbf{w}$, and the scalar product of **v** by a real number a is denoted as $a\mathbf{v}$. These operations are called **vector addition** and **scalar multiplication**, respectively, and the following axioms are assumed to hold.

Axioms for vector addition

A1. If **u** and **v** are in V, then $\mathbf{u} + \mathbf{v}$ is in V.

A2. $\mathbf{u} + \mathbf{v} = \mathbf{v} + \mathbf{u}$ for all **u** and **v** in V.

A3. $\mathbf{u} + (\mathbf{v} + \mathbf{w}) = (\mathbf{u} + \mathbf{v}) + \mathbf{w}$ for all **u**, **v**, and **w** in V.

A4. An element **0** in V exists such that $\mathbf{v} + \mathbf{0} = \mathbf{v} = \mathbf{0} + \mathbf{v}$ for every **v** in V.

A5. For each **v** in V, an element $-\mathbf{v}$ in V exists such that $-\mathbf{v} + \mathbf{v} = \mathbf{0}$ and $\mathbf{v} + (-\mathbf{v}) = \mathbf{0}$.

Axioms for scalar multiplication

S1. If **v** is in V, then $a\mathbf{v}$ is in V for all a in $\mathbb{R}$.

S2. $a(\mathbf{v} + \mathbf{w}) = a\mathbf{v} + a\mathbf{w}$ for all **v** and **w** in V and all a in $\mathbb{R}$.

S3. $(a + b)\mathbf{v} = a\mathbf{v} + b\mathbf{v}$ for all **v** in V and all a and b in $\mathbb{R}$.

S4. $a(b\mathbf{v}) = (ab)\mathbf{v}$ for all **v** in V and all a and b in $\mathbb{R}$.

S5. $1\mathbf{v} = \mathbf{v}$ for all **v** in V.

The content of axioms A1 and S1 is described by saying that V is **closed** under vector addition and scalar multiplication. The element **0** in axiom A4 is called the **zero vector**, and the vector $-\mathbf{v}$ in axiom A5 is called the **negative** of **v**.

The rules of matrix arithmetic, when applied to $\mathbb{R}^n$ give

Example 1

$\mathbb{R}^n$ is a vector space using matrix addition and scalar multiplication.[2]

It is important to realize that, in a general vector space, the vectors need not be n-tuples as in $\mathbb{R}^n$. They can be any kind of object at all as long as the addition and scalar multiplication are defined and the axioms are satisfied. The following examples illustrate the diversity of the concept.

The space $\mathbb{R}^n$ consists of special types of matrices. More generally, let $\mathbf{M}_{mn}$ denote the set of all $m \times n$ matrices with real entries. Then Theorem 1 §2.1 gives the following information.

Example 2

The set $\mathbf{M}_{mn}$ of all $m \times n$ matrices is a vector space using matrix addition and scalar multiplication. The zero element in this vector space is the zero matrix of size $m \times n$, and the vector space negative of a matrix (required by axiom A5) is the usual matrix negative discussed in Section 2.1.

---

1 The scalars will usually be real numbers, but they could be complex numbers, or elements of an algebraic system called a field. Another example is the field $\mathbb{Q}$ of rational numbers. We will look briefly at finite fields in Section 8.8.

2 For the most part we will continue to write the n-tuples in $\mathbb{R}^n$ as columns. However, if it is convenient, we will sometimes denote them as rows or simply as n-tuples.

In Chapter 5 we identified many important subspaces of $\mathbb{R}^n$ such as im A and null A for a matrix A. These are all vector spaces.

Example 3

Show that every subspace of $\mathbb{R}^n$ is a vector space in its own right using the addition and scalar multiplication of $\mathbb{R}^n$.

Solution Axioms A1 and S1 are two of the defining conditions for a subspace U of $\mathbb{R}^n$ (see Section 5.1). The other eight axioms for a vector space are inherited from $\mathbb{R}^n$. For example, if X and Y are in U and a is a scalar, then $a(X + Y) = aX + aY$ because X and Y are in $\mathbb{R}^n$. This shows that axiom S2 holds for U; similarly, the other axioms also hold for U.

Example 4

Let V denote the set of all ordered pairs (x, y) and define addition in V as in $\mathbb{R}^2$. However, define a new scalar multiplication in V by

$$a(x, y) = (ay, ax)$$

Determine if V is a vector space with these operations.

Solution Axioms A1 to A5 are valid for V because they hold for matrices. Also $a(x, y) = (ay, ax)$ is again in V, so axiom S1 holds. To verify axiom S2, let $\mathbf{v} = (x, y)$ and $\mathbf{w} = (x_1, y_1)$ be typical elements in V and compute

$$a(\mathbf{v} + \mathbf{w}) = a(x + x_1, y + y_1) = (a(y + y_1), a(x + x_1))$$
$$a\mathbf{v} + a\mathbf{w} = (ay, ax) + (ay_1, ax_1) = (ay + ay_1, ax + ax_1)$$

Because these are equal, axiom S2 holds. Similarly, the reader can verify that axiom S3 holds. However, axiom S4 fails because

$$a(b(x, y)) = a(by, bx) = (abx, aby)$$

need not equal $ab(x, y) = (aby, abx)$. Hence, V is *not* a vector space. (In fact, axiom S5 also fails.)

Sets of polynomials provide another important source of examples of vector spaces, so we review some basic facts. A **polynomial** in an indeterminate x is an expression

$$p(x) = a_0 + a_1 x + a_2 x^2 + \cdots + a_n x^n$$

where $a_0, a_1, a_2, \ldots, a_n$ are real numbers called the **coefficients** of the polynomial. If all the coefficients are zero, the polynomial is called the **zero polynomial** and is denoted simply as 0. If $p(x) \neq 0$, the highest power of x with a nonzero coefficient is called the **degree** of $p(x)$ and is denoted as $\deg p(x)$. Hence $\deg(3 + 5x) = 1$, $\deg(1 + x + x^2) = 2$, and $\deg(4) = 0$. (The degree of the zero polynomial is not defined.)

Let **P** denote the set of all polynomials and suppose that

$$p(x) = a_0 + a_1 x + a_2 x^2 + \cdots$$
$$q(x) = b_0 + b_1 x + b_2 x^2 + \cdots$$

are two polynomials in **P** (possibly of different degrees). Then $p(x)$ and $q(x)$ are called **equal** [written $p(x) = q(x)$] if and only if all the corresponding coefficients are equal—that is, $a_0 = b_0$, $a_1 = b_1$, $a_2 = b_2$, and so on. In particular, $a_0 + a_1x + a_2x^2 + \cdots = 0$ means that $a_0 = 0$, $a_1 = 0$, $a_2 = 0, \ldots$, and this is the reason for calling x an **indeterminate**. The set $\mathbf{P}$ has an addition and scalar multiplication defined on it as follows: if $p(x)$ and $q(x)$ are as before and a is a real number,

$$p(x) + q(x) = (a_0 + b_0) + (a_1 + b_1)x + (a_2 + b_2)x^2 + \cdots$$

$$ap(x) = aa_0 + (aa_1)x + (aa_2)x^2 + \cdots$$

Evidently, these are again polynomials, so $\mathbf{P}$ is closed under these operations. The other vector space axioms are easily verified, and we have

Example 5

The set $\mathbf{P}$ of all polynomials is a vector space with the foregoing addition and scalar multiplication. The zero vector is the zero polynomial, and the negative of a polynomial $p(x) = a_0 + a_1x + a_2x^2 + \cdots$ is the polynomial $-p(x) = -a_0 - a_1x - a_2x^2 - \cdots$ obtained by negating all the coefficients.

If a and b are real numbers and $a < b$, the **interval** $[a, b]$ is defined to be the set of all real numbers x such that $a \leq x \leq b$. A (real-valued) **function** f on $[a, b]$ is a rule that associates to every number x in $[a, b]$ a real number denoted $f(x)$. The rule is frequently specified by giving a formula for $f(x)$ in terms of x. For example, $f(x) = 2^x$, $f(x) = \sin x$, and $f(x) = x^2 + 1$ are familiar functions. In fact, every polynomial $p(x)$ can be regarded as the formula for a function p. The set of all functions on $[a, b]$ is denoted $\mathbf{F}[a, b]$. Two functions f and g in $\mathbf{F}[a, b]$ are **equal** if $f(x) = g(x)$ for every x in $[a, b]$, and we describe this by saying that f and g have the **same action**. Note that two polynomials are equal in $\mathbf{P}$ (defined prior to Example 5) if and only if they are equal as functions.

If f and g are two functions in $\mathbf{F}[a, b]$, and if r is a real number, define the sum $f + g$ and the scalar product rf by

$$(f + g)(x) = f(x) + g(x) \quad \text{for each } x \text{ in } [a, b]$$

$$(rf)(x) = rf(x) \qquad\quad \text{for each } x \text{ in } [a, b]$$

In other words, the action of $f + g$ upon x is to associate x with the number $f(x) + g(x)$, and rf associates x with $rf(x)$. These operations on $\mathbf{F}[a, b]$ are called **pointwise** addition and scalar multiplication of functions and they are the usual operations familiar from elementary algebra and calculus.

Example 6

The set $\mathbf{F}[a, b]$ of all functions on the interval $[a, b]$ is a vector space using pointwise addition and scalar multiplication. The zero function (in axiom A4) is denoted as 0 and has action defined by

$$0(x) = 0 \quad \text{for all } x \text{ in } [a, b].$$

The negative of a function f is denoted $-f$ and has action defined by

$$(-f)(x) = -f(x) \quad \text{for all } x \text{ in } [a, b].$$

Axioms A1 and S1 are clearly satisfied because, if f and g are functions on $[a, b]$, then $f + g$ and rf are again such functions. The verification of the remaining axioms is left as Exercise 14.

Other examples of vector spaces will appear later, but these are sufficiently varied to indicate the scope of the concept and to illustrate the properties of vector spaces to be discussed. With such a variety of examples, it may come as a surprise that a well-developed *theory* of vector spaces exists. That is, many properties can be shown to hold for *all* vector spaces and hence hold in every example. Such properties are called *theorems* and can be deduced from the axioms. Here is an important example.

Theorem 1 Cancellation

Let **u**, **v**, and **w** be vectors in a vector space V. If $\mathbf{v} + \mathbf{u} = \mathbf{v} + \mathbf{w}$, then $\mathbf{u} = \mathbf{w}$.

PROOF

We are given $\mathbf{v} + \mathbf{u} = \mathbf{v} + \mathbf{w}$. If these were numbers instead of vectors, we would simply subtract **v** from both sides of the equation to obtain $\mathbf{u} = \mathbf{w}$. This can be accomplished with vectors by adding $-\mathbf{v}$ to both sides of the equation. The steps (using only the axioms) are as follows.

$$\mathbf{v} + \mathbf{u} = \mathbf{v} + \mathbf{w}$$
$$-\mathbf{v} + (\mathbf{v} + \mathbf{u}) = -\mathbf{v} + (\mathbf{v} + \mathbf{w}) \quad \text{(axiom A5)}$$
$$(-\mathbf{v} + \mathbf{v}) + \mathbf{u} = (-\mathbf{v} + \mathbf{v}) + \mathbf{w} \quad \text{(axiom A3)}$$
$$\mathbf{0} + \mathbf{u} = \mathbf{0} + \mathbf{w} \quad \text{(axiom A5)}$$
$$\mathbf{u} = \mathbf{w} \quad \text{(axiom A4)}$$

This is the desired conclusion.[3]

As with many good mathematical theorems, the technique of the proof of Theorem 1 is at least as important as the theorem itself. The idea was to mimic the well-known process of numerical subtraction in a vector space V as follows: To subtract a vector **v** from both sides of a vector equation, we added $-\mathbf{v}$ to both sides. With this in mind, we define **difference** $\mathbf{u} - \mathbf{v}$ of two vectors in V as

$$\mathbf{u} - \mathbf{v} = \mathbf{u} + (-\mathbf{v}).$$

We shall say that this vector is the result of having **subtracted v** from **u** and, as in arithmetic, this operation has the property given in Theorem 2.

Theorem 2

If **u** and **v** are vectors in a vector space V, the equation

$$\mathbf{x} + \mathbf{v} = \mathbf{u}$$

has one and only one solution **x** in V given by

$$\mathbf{x} = \mathbf{u} - \mathbf{v}.$$

PROOF

The difference $\mathbf{x} = \mathbf{u} - \mathbf{v}$ is a solution to the equation because (using several axioms)
$$\mathbf{x} + \mathbf{v} = (\mathbf{u} - \mathbf{v}) + \mathbf{v} = [\mathbf{u} + (-\mathbf{v})] + \mathbf{v} = \mathbf{u} + (-\mathbf{v} + \mathbf{v}) = \mathbf{u} + \mathbf{0} = \mathbf{u}.$$

---

3 Observe that none of the scalar multiplication axioms are needed here.

To see that this is the only solution, suppose x_1 is another solution so that $x_1 + v = u$. Then $x + v = x_1 + v$ (they both equal u), so $x = x_1$ by cancellation.

Similarly, cancellation shows that there is only one zero vector in any vector space and only one negative of each vector. (Exercises 10 and 11). Hence we speak of *the* zero vector and *the* negative of a vector.

The next theorem introduces some basic properties that hold in every vector space, and will be used extensively later.

Theorem 3

Let v denote a vector in a vector space V and let a denote a real number.
1. $0v = 0$.
2. $a0 = 0$.
3. If $av = 0$, then either $a = 0$ or $v = 0$.
4. $(-1)v = -v$.
5. $(-a)v = -(av) = a(-v)$.

PROOF

The proofs of (2) and (5) are left as Exercise 12.
1. Observe that $0v + 0v = (0 + 0)v = 0v = 0v + 0$ where the first equality is by axiom S3. It follows that $0v = 0$ by cancellation.
3. Assume $av = 0$; it suffices to show that if $a \neq 0$, then necessarily $v = 0$. But $a \neq 0$ means we can scalar-multiply the given equation $av = 0$ by $1/a$ to obtain

$$v = 1v = (\tfrac{1}{a}a)v = \tfrac{1}{a}(av) = \tfrac{1}{a}0 = 0$$

   using (2) and axioms S4 and S5.
4. We have $-v + v = 0$ by axiom A5. On the other hand,

$$(-1)v + v = (-1)v + 1v = (-1 + 1)v = 0v = 0$$

   using (1) and axioms S5 and S3. Hence $(-1)v + v = -v + v$ (because both are equal to 0), so $(-1)v = -v$ by cancellation.

The properties in Theorem 3 are familiar for matrices; the point here is that they hold in *every* vector space.

Axioms S2 and S3 extend. For example $a(u + v + w) = au + av + aw$ and $(a + b + c)v = av + bv + cv$ hold for all values of the scalars and vectors involved. More generally,[4]

$$a(v_1 + v_2 + \cdots + v_n) = av_1 + av_2 + \cdots + av_n$$
$$(a_1 + a_2 + \cdots + a_n)v = a_1v + a_2v + \cdots + a_nv$$

hold for all $n \geq 1$, all numbers $a, a_1, \ldots, a_n$, and all vectors, $v, v_1, \ldots, v_n$. The verifications are by induction and are left to the reader (Exercise 13). These facts—together with the axioms, Theorem 3, and the definition of subtraction—enable us to simplify expressions involving sums of scalar multiples of vectors by collecting

4 It is a consequence of axiom A3 that we can omit parentheses when writing a sum $v_1 + v_2 + \cdots + v_n$ of vectors.

like terms, expanding, and taking out common factors. This has been discussed for the vector space of matrices in Section 2.1 (and for geometric vectors in Section 4.1); the manipulations in an arbitrary vector space are carried out in the same way. Here is an illustration.

Example 7

If **u**, **v**, and **w** are vectors in a vector space V, simplify

$$2(\mathbf{u} + 3\mathbf{w}) - 3(2\mathbf{w} - \mathbf{v}) - 3[2(2\mathbf{u} + \mathbf{v} - 4\mathbf{w}) - 4(\mathbf{u} - 2\mathbf{w})].$$

Solution   The reduction proceeds as though **u**, **v**, and **w** were matrices or variables.

$$2(\mathbf{u} + 3\mathbf{w}) - 3(2\mathbf{w} - \mathbf{v}) - 3[2(2\mathbf{u} + \mathbf{v} - 4\mathbf{w}) - 4(\mathbf{u} - 2\mathbf{w})]$$
$$= 2\mathbf{u} + 6\mathbf{w} - 6\mathbf{w} + 3\mathbf{v} - 3[4\mathbf{u} + 2\mathbf{v} - 8\mathbf{w} - 4\mathbf{u} + 8\mathbf{w}]$$
$$= 2\mathbf{u} + 3\mathbf{v} - 3[2\mathbf{v}]$$
$$= 2\mathbf{u} + 3\mathbf{v} - 6\mathbf{v}$$
$$= 2\mathbf{u} - 3\mathbf{v}.$$

Condition (2) in Theorem 3 points to another example of a vector space.

Example 8

A set $\{\mathbf{0}\}$ with one element becomes a vector space if we define

$$\mathbf{0} + \mathbf{0} = \mathbf{0} \quad \text{and} \quad a\mathbf{0} = \mathbf{0} \text{ for all scalars } a.$$

The resulting space is called the **zero vector space** and is denoted $\{\mathbf{0}\}$.

The vector space axioms are easily verified for $\{\mathbf{0}\}$. In any vector space V, Theorem 3(2) shows that the zero subspace (consisting of the zero vector of V alone) is a copy of the zero vector space.

Exercises 6.1

1. Let V denote the set of ordered triples (x, y, z) and define addition in V as in $\mathbb{R}^3$. For each of the following definitions of scalar multiplication, decide whether V is a vector space.

   (a) $a(x, y, z) = (ax, y, az)$

   ◆(b) $a(x, y, z) = (ax, 0, az)$

   (c) $a(x, y, z) = (0, 0, 0)$

   ◆(d) $a(x, y, z) = (2ax, 2ay, 2az)$

2. Are the following sets vector spaces with the indicated operations? If not, why not?

   (a) The set V of nonnegative real numbers; ordinary addition and scalar multiplication.

   ◆(b) The set V of all polynomials of degree ≥ 3, together with 0; operations of **P**.

   (c) The set of all polynomials of degree ≤ 3; operations of **P**.

   ◆(d) The set $\{1, x, x^2, \dots\}$; operations of **P**.

   (e) The set V of all 2×2 matrices of the form $\begin{bmatrix} a & b \\ 0 & c \end{bmatrix}$; operations of $\mathbf{M}_{22}$.

   ◆(f) The set V of 2×2 matrices with equal column sums; operations of $\mathbf{M}_{22}$.

   (g) The set V of 2×2 matrices with zero determinant; usual matrix operations.

   ◆(h) The set V of real numbers; usual operations.

   (i) The set V of complex numbers; usual addition and multiplication by a real number.

   ◆(j) The set V of all ordered pairs (x, y) with the addition of $\mathbb{R}^2$, but scalar multiplication $a(x, y) = (ax, -ay)$.

(k) The set V of all ordered pairs (x, y) with the addition of $\mathbb{R}^2$, but scalar multiplication $a(x, y) = (x, y)$ for all a in $\mathbb{R}$.

♦(l) The set V of all functions $f: \mathbb{R} \to \mathbb{R}$ with pointwise addition, and scalar multiplication defined by $(af)(x) = f(ax)$.

(m) The set V of all 2×2 matrices whose entries sum to 0; operations of $\mathbf{M}_{22}$.

♦(n) The set V of all 2×2 matrices with the addition of $\mathbf{M}_{22}$ but scalar multiplication $*$ defined by $a * X = aX^T$.

3. Let V be the set of positive real numbers with vector addition being ordinary multiplication, and scalar multiplication being $a \cdot v = v^a$. Show that V is a vector space.

♦4. If V is the set of ordered pairs (x, y) of real numbers, show that it is a vector space if $(x, y) + (x_1, y_1) = (x + x_1, y + y_1 + 1)$ and $a(x, y) = (ax, ay + a - 1)$. What is the zero vector in V?

5. Find $\mathbf{x}$ and $\mathbf{y}$ (in terms of $\mathbf{u}$ and $\mathbf{v}$) such that:

(a) $2\mathbf{x} + \mathbf{y} = \mathbf{u}$ ♦(b) $3\mathbf{x} - 2\mathbf{y} = \mathbf{u}$
    $5\mathbf{x} + 3\mathbf{y} = \mathbf{v}$ $4\mathbf{x} - 5\mathbf{y} = \mathbf{v}$

6. In each case show that the condition $a\mathbf{u} + b\mathbf{v} + c\mathbf{w} = \mathbf{0}$ in V implies that $a = b = c = 0$.

(a) $V = \mathbb{R}^4$; $\mathbf{u} = (2, 1, 0, 2)$, $\mathbf{v} = (1, 1, -1, 0)$, $\mathbf{w} = (0, 1, 2, 1)$

♦(b) $V = \mathbf{M}_{22}$; $\mathbf{u} = \begin{bmatrix} 1 & 0 \\ 0 & 1 \end{bmatrix}$, $\mathbf{v} = \begin{bmatrix} 0 & 1 \\ 1 & 0 \end{bmatrix}$, $\mathbf{w} = \begin{bmatrix} 1 & 1 \\ 1 & -1 \end{bmatrix}$

(c) $V = \mathbf{P}$; $\mathbf{u} = x^3 + x$, $\mathbf{v} = x^2 + 1$, $\mathbf{w} = x^3 - x^2 + x + 1$

♦(d) $V = \mathbf{F}[0, \pi]$; $\mathbf{u} = \sin x$, $\mathbf{v} = \cos x$, $\mathbf{w} = 1$

7. Simplify each of the following.

(a) $3[2(\mathbf{u} - 2\mathbf{v} - \mathbf{w}) + 3(\mathbf{w} - \mathbf{v})] - 7(\mathbf{u} - 3\mathbf{v} - \mathbf{w})$

♦(b) $4(3\mathbf{u} - \mathbf{v} + \mathbf{w}) - 2[(3\mathbf{u} - 2\mathbf{v}) - 3(\mathbf{v} - \mathbf{w})] + 6(\mathbf{w} - \mathbf{u} - \mathbf{v})$

8. Show that $\mathbf{x} = \mathbf{v}$ is the only solution to the equation $\mathbf{x} + \mathbf{x} = 2\mathbf{v}$ in a vector space V. Cite all axioms used.

9. Show that $-\mathbf{0} = \mathbf{0}$ in any vector space. Cite all axioms used.

10. Show that the zero vector $\mathbf{0}$ is uniquely determined by the property in axiom A4.

♦11. Given a vector $\mathbf{v}$, show that its negative $-\mathbf{v}$ is uniquely determined by the property in axiom A5.

12. (a) Prove (2) of Theorem 3. [*Hint*: Axiom S2.]

♦(b) Prove that $(-a)\mathbf{v} = -(a\mathbf{v})$ in Theorem 3 by first computing $(-a)\mathbf{v} + a\mathbf{v}$. Then do it using (4) of Theorem 3 and axiom S4.

(c) Prove that $a(-\mathbf{v}) = -(a\mathbf{v})$ in Theorem 3 in two ways, as in part (b).

13. Let $\mathbf{v}, \mathbf{v}_1, \ldots, \mathbf{v}_n$ denote vectors in a vector space V and let $a, a_1, \ldots, a_n$ denote numbers. Use induction on n to prove each of the following.

(a) $a(\mathbf{v}_1 + \mathbf{v}_2 + \cdots + \mathbf{v}_n) = a\mathbf{v}_1 + a\mathbf{v}_2 + \cdots + a\mathbf{v}_n$

(b) $(a_1 + a_2 + \cdots + a_n)\mathbf{v} = a_1\mathbf{v} + a_2\mathbf{v} + \cdots + a_n\mathbf{v}$

14. Verify axioms A2–A5 and S2–S5 for the space $\mathbf{F}[a, b]$ of functions on $[a, b]$ (Example 6).

15. Prove each of the following for vectors $\mathbf{u}$ and $\mathbf{v}$ and scalars a and b.

(a) If $a\mathbf{v} = b\mathbf{v}$ and $\mathbf{v} \neq \mathbf{0}$, then $a = b$.

♦(b) If $a\mathbf{v} = a\mathbf{w}$ and $a \neq 0$, then $\mathbf{v} = \mathbf{w}$.

16. By calculating $(1 + 1)(\mathbf{v} + \mathbf{w})$ in two ways (using axioms S2 and S3), show that axiom A2 follows from the other axioms.

17. Let V be a vector space, and define V^n to be the set of all n-tuples $(\mathbf{v}_1, \mathbf{v}_2, \ldots, \mathbf{v}_n)$ of n vectors $\mathbf{v}_i$, each belonging to V. Define addition and scalar multiplication in V^n as follows:

$(\mathbf{u}_1, \mathbf{u}_2, \ldots, \mathbf{u}_n) + (\mathbf{v}_1, \mathbf{v}_2, \ldots, \mathbf{v}_n)$
$\quad = (\mathbf{u}_1 + \mathbf{v}_1, \mathbf{u}_2 + \mathbf{v}_2, \ldots, \mathbf{u}_n + \mathbf{v}_n)$
$a(\mathbf{v}_1, \mathbf{v}_2, \ldots, \mathbf{v}_n) = (a\mathbf{v}_1, a\mathbf{v}_2, \ldots, a\mathbf{v}_n)$

Show that V^n is a vector space.

18. Let V^n be the vector space of n-tuples from the preceding exercise, written as columns. If A is an $m \times n$ matrix, and X is in V^n, define AX in V^m by matrix multiplication. More precisely, if

$A = [a_{ij}]$ and $X = \begin{bmatrix} \mathbf{v}_1 \\ \vdots \\ \mathbf{v}_n \end{bmatrix}$, let $AX = \begin{bmatrix} \mathbf{u}_1 \\ \vdots \\ \mathbf{u}_n \end{bmatrix}$, where

$\mathbf{u}_i = a_{i1}\mathbf{v}_1 + a_{i2}\mathbf{v}_2 + \cdots + a_{in}\mathbf{v}_n$ for each i. Prove that:

(a) $B(AX) = (BA)X$

(b) $(A + A_1)X = AX + A_1X$

(c) $A(X + X_1) = AX + AX_1$

(d) $(kA)X = k(AX) = A(kX)$ if k is any number

(e) $IX = X$ if I is the $n \times n$ identity matrix

(f) Let E be an elementary matrix obtained by performing a row operation on (the rows of) I_n (see Section 2.4). Show that

EX is the column resulting from performing that same row operation on the vectors (call them rows) of X. [*Hint:* Theorem 1 §2.4.]

SECTION 6.2 Subspaces and Spanning Sets

If V is a vector space, a nonempty subset $U \subseteq V$ is called a **subspace** of V if U is itself a vector space using the addition and scalar multiplication of V. Subspaces of $\mathbb{R}^n$ (as defined in Section 5.1) are subspaces in the present sense by Example 3 §6.1. Moreover, the defining properties for a subspace of $\mathbb{R}^n$ actually *characterize* subspaces in general.

Theorem 1 Subspace Test

A subset U of a vector space is a subspace of V if and only if it satisfies the following three conditions:[5]
   1. $\mathbf{0}$ lies in U where $\mathbf{0}$ is the zero vector of V.
   2. If $\mathbf{u}_1$ and $\mathbf{u}_2$ are in U, then $\mathbf{u}_1 + \mathbf{u}_2$ is also in U.
   3. If $\mathbf{u}$ is in U, then $a\mathbf{u}$ is also in U for each scalar a.

PROOF

If U is a subspace of V, then (2) and (3) hold by axioms A1 and S1 respectively, applied to the vector space U. Since U is nonempty, choose $\mathbf{u}$ in U. Then (1) holds because $\mathbf{0} = 0\mathbf{u}$ is in U by (3) and Theorem 3 §6.1.

Conversely, if (1), (2), and (3) hold, then axioms A1 and S1 hold because of (2) and (3), and axioms A2, A3, S2, S3, S4, and S5 hold in U because they hold in V. Axiom A4 holds because the zero vector $\mathbf{0}$ of V is actually in U by (1), and so serves as the zero of U. Finally, given $\mathbf{u}$ in U, then its negative $-\mathbf{u}$ in V is again in U by (3) because $-\mathbf{u} = (-1)\mathbf{u}$ (again using Theorem 3 §6.1), and so serves as the negative of $\mathbf{u}$ in U.

Note that the proof of Theorem 1 shows that if U is a subspace of V, then U and V share the same zero vector, and that the negative of a vector in U is the same as its negative in V.

Example 1

If V is any vector space, show that $\{\mathbf{0}\}$ and V are subspaces of V.

Solution $U = V$ clearly satisfies the conditions of the test. As to $U = \{\mathbf{0}\}$, it satisfies the conditions because $\mathbf{0} + \mathbf{0} = \mathbf{0}$ and $a\mathbf{0} = \mathbf{0}$ for all a in $\mathbb{R}$.

The vector space $\{\mathbf{0}\}$ is called the **zero subspace** of V.

---

5 Condition (1) can be replaced by the requirement that U is nonempty, but we prefer (1) because it is usually easy to check and, if it fails, U *cannot* be a subspace.

Example 2

Let $\mathbf{v}$ be a vector in a vector space V. Show that the set

$$\mathbb{R}\mathbf{v} = \{a\mathbf{v}\,|\,a \text{ in } \mathbb{R}\}$$

of all scalar multiples of $\mathbf{v}$ is a subspace of V.

Solution   Because $\mathbf{0} = 0\mathbf{v}$, it is clear that $\mathbf{0}$ lies in $\mathbb{R}\mathbf{v}$. Given two vectors $a\mathbf{v}$ and $a_1\mathbf{v}$ in $\mathbb{R}\mathbf{v}$, their sum $a\mathbf{v} + a_1\mathbf{v} = (a + a_1)\mathbf{v}$ is also a scalar multiple of $\mathbf{v}$ and so lies in $\mathbb{R}\mathbf{v}$. Hence $\mathbb{R}\mathbf{v}$ is closed under addition. Finally, given $a\mathbf{v}$, $r(a\mathbf{v}) = (ra)\mathbf{v}$ lies in $\mathbb{R}\mathbf{v}$, so $\mathbb{R}\mathbf{v}$ is closed under scalar multiplication. Hence the subspace test applies.

The set $\mathbb{R}\mathbf{v}$ in Example 2 is described by giving the *form* of each vector in $\mathbb{R}\mathbf{v}$. The next example describes a subset U of the space $\mathbf{M}_{nn}$ by giving a *condition* that each matrix of U must satisfy.

Example 3

Let A be a fixed matrix in $\mathbf{M}_{nn}$. Show that $U = \{X \text{ in } \mathbf{M}_{nn} \mid AX = XA\}$ is a subspace of $\mathbf{M}_{nn}$.

Solution   If 0 is the $n \times n$ zero matrix, then $A0 = 0A$, so 0 satisfies the condition for membership in U. Next suppose that X and X_1 lie in U so that $AX = XA$ and $AX_1 = X_1A$. Then

$$A(X + X_1) = AX + AX_1 = XA + X_1A = (X + X_1)A$$
$$A(aX) = a(AX) = a(XA) = (aX)A$$

for all a in $\mathbb{R}$, so both $X + X_1$ and aX lie in U. Hence U is a subspace of $\mathbf{M}_{nn}$.

Suppose $p(x)$ is a polynomial and a is a number. Then the number $p(a)$ obtained by replacing x by a in the expression for $p(x)$ is called the **evaluation** of $p(x)$ at a. For example, if $p(x) = 5 - 6x + 2x^2$, then the evaluation of $p(x)$ at $a = 2$ is $p(2) = 1$. If $p(a) = 0$, the number a is called a **root** of $p(x)$.

Example 4

Consider the set U of all polynomials in $\mathbf{P}$ that have 3 as a root:

$$U = \{p(x) \text{ in } \mathbf{P} \mid p(3) = 0\}.$$

Show that U is a subspace of $\mathbf{P}$.

Solution   Clearly, the zero polynomial lies in U. Now let $p(x)$ and $q(x)$ lie in U so $p(3) = 0$ and $q(3) = 0$. Then $(p + q)(x) = p(x) + q(x)$ for all x, so $(p + q)(3) = p(3) + q(3) = 0 + 0 = 0$, and U is closed under addition. The verification that U is closed under scalar multiplication is similar.

There are other important examples of vector spaces consisting of polynomials. Let $\mathbf{P}_n$ denote the set of all polynomials of degree at most n, together with the zero polynomial. In other words, $\mathbf{P}_n$ consists of all polynomials of the form

$$a_0 + a_1x + a_2x^2 + \cdots + a_nx^n$$

where $a_0, a_1, a_2, \ldots, a_n$ are real numbers, and so is closed under the addition and scalar multiplication in $\mathbf{P}$. Moreover, the zero polynomial is included in $\mathbf{P}_n$. Thus the subspace test gives Example 5.

Example 5

For each $n \geq 0$, $\mathbf{P}_n$ is a subspace of $\mathbf{P}$.

The next example involves the notion of the derivative f' of a function f. (If the reader is not familiar with calculus, this example may be omitted.) A function f defined on the interval $[a, b]$ is called **differentiable** if the derivative $f'(r)$ exists at every r in $[a, b]$.

Example 6

Show that the subset $\mathbf{D}[a, b]$ of all **differentiable functions** on $[a, b]$ is a subspace of the vector space $\mathbf{F}[a, b]$ of all functions on $[a, b]$.

Solution The derivative of any constant function is the constant function 0; in particular, 0 itself is differentiable and so lies in $\mathbf{D}[a, b]$. If f and g both lie in $\mathbf{D}[a, b]$ (so that f' and g' exist), then it is a theorem of calculus that $f + g$ and af are both differentiable [in fact, $(f + g)' = f' + g'$ and $(af)' = af'$], so both lie in $\mathbf{D}[a, b]$. This shows that $\mathbf{D}[a, b]$ is a subspace of $\mathbf{F}[a, b]$.

Linear Combinations and Spanning Sets

Let $\{\mathbf{v}_1, \mathbf{v}_2, \ldots, \mathbf{v}_n\}$ be a set of vectors in a vector space V. As in $\mathbb{R}^n$, a vector $\mathbf{v}$ is called a **linear combination** of the vectors $\mathbf{v}_1, \mathbf{v}_2, \ldots, \mathbf{v}_n$ if it can be expressed in the form

$$\mathbf{v} = a_1\mathbf{v}_1 + a_2\mathbf{v}_2 + \cdots + a_n\mathbf{v}_n$$

where $a_1, a_2, \ldots, a_n$ are scalars, called the **coefficients** of $\mathbf{v}_1, \mathbf{v}_2, \ldots, \mathbf{v}_n$. The set of *all* linear combinations of these vectors is called their **span**, and is denoted by

$$\text{span}\{\mathbf{v}_1, \mathbf{v}_2, \ldots, \mathbf{v}_n\}.$$

If it happens that $V = \text{span}\{\mathbf{v}_1, \mathbf{v}_2, \ldots, \mathbf{v}_n\}$, these vectors are called a **spanning set** for V. For example, the span of two vectors $\mathbf{v}$ and $\mathbf{w}$ is the set

$$\text{span}\{\mathbf{v}, \mathbf{w}\} = \{s\mathbf{v} + t\mathbf{w} \mid s \text{ and } t \text{ in } \mathbb{R}\}$$

of all sums of scalar multiples of the vectors.

Example 7

Consider the vectors $p_1 = 1 + x + 4x^2$ and $p_2 = 1 + 5x + x^2$ in $\mathbf{P}_2$. Determine whether p_1 and p_2 lie in $\text{span}\{1 + 2x - x^2, 3 + 5x + 2x^2\}$.

Solution For p_1, we want to determine if s and t exist such that

$$p_1 = s(1 + 2x - x^2) + t(3 + 5x + 2x^2).$$

Equating coefficients of powers of x (where $x^0 = 1$) gives

$$1 = s + 3t, \quad 1 = 2s + 5t, \quad \text{and} \quad 4 = -s + 2t.$$

These equations have the solutions $s = -2$ and $t = 1$, so p_1 is indeed in $\text{span}\{1 + 2x - x^2, 3 + 5x + 2x^2\}$. Turning to $p_2 = 1 + 5x + x^2$, we are looking for s and t such that $p_2 = s(1 + 2x - x^2) + t(3 + 5x + 2x^2)$. Again equating coefficients of powers of x gives equations $1 = s + 3t$, $5 = 2s + 5t$, and $1 = -s + 2t$. But in this case there is no solution, so p_2 is *not* in $\text{span}\{1 + 2x - x^2, 3 + 5x + 2x^2\}$.

We saw in Example 6 §5.1 that $\mathbb{R}^m = \text{span}\{E_1, E_2, \dots, E_m\}$ where the vectors $E_1, E_2, \dots, E_m$ are the columns of the $m \times m$ identity matrix. Of course $\mathbb{R}^m = \mathbf{M}_{m1}$ is the set of all $m \times 1$ matrices, and there is an analogous spanning set for each space $\mathbf{M}_{mn}$. For example, each 2×2 matrix has the form

$$\begin{bmatrix} a & b \\ c & d \end{bmatrix} = a\begin{bmatrix} 1 & 0 \\ 0 & 0 \end{bmatrix} + b\begin{bmatrix} 0 & 1 \\ 0 & 0 \end{bmatrix} + c\begin{bmatrix} 0 & 0 \\ 1 & 0 \end{bmatrix} + d\begin{bmatrix} 0 & 0 \\ 0 & 1 \end{bmatrix} \text{ so}$$

$$\mathbf{M}_{22} = \text{span}\left\{ \begin{bmatrix} 1 & 0 \\ 0 & 0 \end{bmatrix}, \begin{bmatrix} 0 & 1 \\ 0 & 0 \end{bmatrix}, \begin{bmatrix} 0 & 0 \\ 1 & 0 \end{bmatrix}, \begin{bmatrix} 0 & 0 \\ 0 & 1 \end{bmatrix} \right\}.$$

Similarly, we obtain

Example 8

$\mathbf{M}_{mn}$ is the span of the set of all $m \times n$ matrices with exactly one entry equal to 1, and all other entries zero.

The fact that every polynomial in $\mathbf{P}_n$ has the form $a_0 + a_1 x + a_2 x^2 + \dots + a_n x^n$ where each a_i is in $\mathbb{R}$ shows that

Example 9

$\mathbf{P}_n = \text{span}\{1, x, x^2, \dots, x^n\}.$

In Example 2 we saw that $\text{span}\{\mathbf{v}\} = \{a\mathbf{v} \mid a \text{ in } \mathbb{R}\} = \mathbb{R}\mathbf{v}$ is a subspace for any vector $\mathbf{v}$ in a vector space V. In fact the span of *any* set of vectors is a subspace.

Theorem 2

Let $U = \text{span}\{\mathbf{v}_1, \mathbf{v}_2, \dots, \mathbf{v}_n\}$ in a vector space V. Then:
1. U is a subspace of V containing each of $\mathbf{v}_1, \mathbf{v}_2, \dots, \mathbf{v}_n$.
2. U is the "smallest" subspace containing these vectors in the sense that any subspace that contains each of $\mathbf{v}_1, \mathbf{v}_2, \dots, \mathbf{v}_n$ must contain U.

PROOF

The proof of Theorem 1 §5.1 goes through verbatim in the general context.

Theorem 2 is used frequently to determine spanning sets, as the following examples show.

Example 10

Show that $\mathbf{P}_3 = \text{span}\{x^2 + x^3, x, 2x^2 + 1, 3\}$.

Solution Write $U = \text{span}\{x^2 + x^3, x, 2x^2 + 1, 3\}$. Then $U \subseteq \mathbf{P}_3$, and we use the fact that $\mathbf{P}_3 = \text{span}\{1, x, x^2, x^3\}$ to show that $\mathbf{P}_3 \subseteq U$. In fact, x and $1 = \frac{1}{3} \cdot 3$ clearly lie in U. But then successively,

$$x^2 = \tfrac{1}{2}[(2x^2 + 1) - 1] \quad \text{and} \quad x^3 = (x^2 + x^3) - x^2$$

also lie in U. Hence $\mathbf{P}_3 \subseteq U$ by Theorem 2.

Example 11

Let **u** and **v** be two vectors in a vector space V. Show that

$$\text{span}\{\mathbf{u}, \mathbf{v}\} = \text{span}\{\mathbf{u} + 2\mathbf{v}, \mathbf{u} - \mathbf{v}\}.$$

Solution We have span$\{\mathbf{u} + 2\mathbf{v}, \mathbf{u} - \mathbf{v}\} \subseteq$ span$\{\mathbf{u}, \mathbf{v}\}$ by Theorem 2 because both $\mathbf{u} + 2\mathbf{v}$ and $\mathbf{u} - \mathbf{v}$ lie in span$\{\mathbf{u}, \mathbf{v}\}$. On the other hand,

$$\mathbf{u} = \tfrac{1}{3}(\mathbf{u} + 2\mathbf{v}) + \tfrac{2}{3}(\mathbf{u} - \mathbf{v})$$
$$\mathbf{v} = \tfrac{1}{3}(\mathbf{u} + 2\mathbf{v}) - \tfrac{1}{3}(\mathbf{u} - \mathbf{v})$$

so span$\{\mathbf{u}, \mathbf{v}\} \subseteq$ span$\{\mathbf{u} + 2\mathbf{v}, \mathbf{u} - \mathbf{v}\}$, again by Theorem 2.

Exercises 6.2

1. Which of the following are subspaces of $\mathbf{P}_3$? Support your answer.
   (a) $U = \{f(x) \mid f(x) \text{ in } \mathbf{P}_3, f(2) = 1\}$
   ◆(b) $U = \{xg(x) \mid g(x) \text{ in } \mathbf{P}_2\}$
   (c) $U = \{xg(x) \mid g(x) \text{ in } \mathbf{P}_3\}$
   ◆(d) $U = \{xg(x) + (1 - x)h(x) \mid g(x) \text{ and } h(x) \text{ in } \mathbf{P}_2\}$
   (e) $U = $ The set of all polynomials in $\mathbf{P}_3$ with constant term 0
   ◆(f) $U = \{f(x) \mid f(x) \text{ in } \mathbf{P}_3, \deg f(x) = 3\}$

2. Which of the following are subspaces of $\mathbf{M}_{22}$? Support your answer.
   (a) $U = \left\{ \begin{bmatrix} a & b \\ 0 & c \end{bmatrix} \middle| a, b, \text{ and } c \text{ in } \mathbb{R} \right\}$
   ◆(b) $U = \left\{ \begin{bmatrix} a & b \\ c & d \end{bmatrix} \middle| a + b = c + d; a, b, c, \text{ and } d \text{ in } \mathbb{R} \right\}$
   (c) $U = \{A \mid A \text{ in } \mathbf{M}_{22}, A = A^T\}$
   ◆(d) $U = \{A \mid A \text{ in } \mathbf{M}_{22}, AB = 0\}$, B a fixed 2×2 matrix
   (e) $U = \{A \mid A \text{ in } \mathbf{M}_{22}, A^2 = A\}$
   ◆(f) $U = \{A \mid A \text{ in } \mathbf{M}_{22}, A \text{ is not invertible}\}$
   (g) $U = \{A \mid A \text{ in } \mathbf{M}_{22}, BAC = CAB\}$, B and C fixed 2×2 matrices

3. Which of the following are subspaces of $\mathbf{F}[0, 1]$? Support your answer.
   (a) $U = \{f \mid f(0) = 0\}$
   ◆(b) $U = \{f \mid f(0) = 1\}$
   (c) $U = \{f \mid f(0) = f(1)\}$
   ◆(d) $U = \{f \mid f(x) \geq 0 \text{ for all } x \text{ in } [0, 1]\}$
   (e) $U = \{f \mid f(x) = f(y) \text{ for all } x \text{ and } y \text{ in } [0, 1]\}$
   ◆(f) $U = \{f \mid f(x + y) = f(x) + f(y) \text{ for all } x \text{ and } y \text{ in } [0, 1]\}$
   (g) $U = \{f \mid \int_0^1 f(x)dx = 0\}$

4. Let A be an $m \times n$ matrix. For which columns B in $\mathbb{R}^m$ is $U = \{X \mid X \text{ in } \mathbb{R}^n, AX = B\}$ a subspace of $\mathbb{R}^n$? Support your answer.

5. Let X be a vector in $\mathbb{R}^n$ (written as a column), and define $U = \{AX \mid A \text{ in } \mathbf{M}_{mn}\}$.
   (a) Show that U is a subspace of $\mathbb{R}^m$.
   ◆(b) Show that $U = \mathbb{R}^m$ if $X \neq 0$.

6. Write each of the following as a linear combination of $x + 1$, $x^2 + x$, and $x^2 + 2$.
   (a) $x^2 + 3x + 2$ ◆(b) $2x^2 - 3x + 1$
   (c) $x^2 + 1$ ◆(d) x

7. Determine whether **v** lies in span$\{\mathbf{u}, \mathbf{w}\}$ in each case.
   (a) $\mathbf{v} = 3x^2 - 2x - 1; \mathbf{u} = x^2 + 1, \mathbf{w} = x + 2$
   ◆(b) $\mathbf{v} = x; \mathbf{u} = x^2 + 1, \mathbf{w} = x + 2$
   (c) $\mathbf{v} = \begin{bmatrix} 1 & 3 \\ -1 & 1 \end{bmatrix}; \mathbf{u} = \begin{bmatrix} 1 & -1 \\ 2 & 1 \end{bmatrix}, \mathbf{w} = \begin{bmatrix} 2 & 1 \\ 1 & 0 \end{bmatrix}$
   ◆(d) $\mathbf{v} = \begin{bmatrix} 1 & -4 \\ 5 & 3 \end{bmatrix}; \mathbf{u} = \begin{bmatrix} 1 & -1 \\ 2 & 1 \end{bmatrix}, \mathbf{w} = \begin{bmatrix} 2 & 1 \\ 1 & 0 \end{bmatrix}$

8. Which of the following functions lie in span$\{\cos^2 x, \sin^2 x\}$? (Work in $\mathbf{F}[0, \pi]$.)
   (a) $\cos 2x$ ◆(b) 1
   (c) x^2 ◆(d) $1 + x^2$

9. (a) Show that $\mathbb{R}^3$ is spanned by
$\{[1\ 0\ 1]^T, [1\ 1\ 0]^T, [0\ 1\ 1]^T\}$.

♦(b) Show that $\mathbf{P}_2$ is spanned by
$\{1 + 2x^2, 3x, 1 + x\}$.

(c) Show that $\mathbf{M}_{22}$ is spanned by
$$\left\{\begin{bmatrix} 1 & 0 \\ 0 & 0 \end{bmatrix}, \begin{bmatrix} 1 & 0 \\ 0 & 1 \end{bmatrix}, \begin{bmatrix} 0 & 1 \\ 1 & 0 \end{bmatrix}, \begin{bmatrix} 1 & 1 \\ 0 & 1 \end{bmatrix}\right\}.$$

10. If X and Y are two sets of vectors in a vector space V, and if $X \subseteq Y$, show that span $X \subseteq$ span Y.

11. Let $\mathbf{u}$, $\mathbf{v}$, and $\mathbf{w}$ denote vectors in a vector space V. Show that:

(a) span$\{\mathbf{u}, \mathbf{v}, \mathbf{w}\}$ = span$\{\mathbf{u} + \mathbf{v}, \mathbf{u} + \mathbf{w}, \mathbf{v} + \mathbf{w}\}$

♦(b) span$\{\mathbf{u}, \mathbf{v}, \mathbf{w}\}$ = span$\{\mathbf{u} - \mathbf{v}, \mathbf{u} + \mathbf{w}, \mathbf{w}\}$

12. Show that
span$\{\mathbf{v}_1, \mathbf{v}_2, \ldots, \mathbf{v}_n, \mathbf{0}\}$ = span$\{\mathbf{v}_1, \mathbf{v}_2, \ldots, \mathbf{v}_n\}$ holds for any set of vectors $\{\mathbf{v}_1, \mathbf{v}_2, \ldots, \mathbf{v}_n\}$.

13. If X and Y are nonempty subsets of a vector space V such that span X = span $Y = V$, must there be a vector common to both X and Y? Justify your answer.

♦14. Is it possible that $\{(1, 2, 0), (1, 1, 1)\}$ can span the subspace $U = \{(a, b, 0) \mid a$ and b in $\mathbb{R}\}$?

15. Describe span$\{\mathbf{0}\}$.

16. Let $\mathbf{v}$ denote any vector in a vector space V. Show that span$\{\mathbf{v}\}$ = span$\{a\mathbf{v}\}$ for any $a \neq 0$.

17. Determine all subspaces of $\mathbb{R}\mathbf{v}$ where $\mathbf{v} \neq \mathbf{0}$ in some vector space V.

♦18. Suppose $V = $ span$\{\mathbf{v}_1, \mathbf{v}_2, \ldots, \mathbf{v}_n\}$.
If $\mathbf{u} = a_1\mathbf{v}_1 + a_2\mathbf{v}_2 + \cdots + a_n\mathbf{v}_n$ where the a_i are in $\mathbb{R}$ and $a_1 \neq 0$, show that $V = $ span$\{\mathbf{u}, \mathbf{v}_2, \ldots, \mathbf{v}_n\}$.

19. If $\mathbf{M}_{nn} = $ span$\{A_1, A_2, \ldots, A_k\}$, show that
$\mathbf{M}_{nn} = $ span$\{A_1^T, A_2^T, \ldots, A_k^T\}$.

20. If $\mathbf{P}_n = $ span$\{p_1(x), p_2(x), \ldots, p_k(x)\}$ and a is in $\mathbb{R}$, show that $p_i(a) \neq 0$ for some i.

21. Let U be a subspace of a vector space V.

(a) If $a\mathbf{u}$ is in U where $a \neq 0$, show that $\mathbf{u}$ is in U.

♦(b) If $\mathbf{u}$ and $\mathbf{u} + \mathbf{v}$ are in U, show that $\mathbf{v}$ is in U.

22. Let U be a nonempty subset of a vector space V. Show that U is a subspace of V if and only if $\mathbf{u}_1 + a\mathbf{u}_2$ lies in U for all $\mathbf{u}_1$ and $\mathbf{u}_2$ in U and all a in $\mathbb{R}$.

23. Let U be the set in Example 4:
$U = \{p(x)$ in $\mathbf{P} \mid p(3) = 0\}$. Use the factor theorem (see §6.5) to show that U consists of multiples of $x - 3$; that is, show that $U = \{(x - 3)q(x) \mid q(x)$ in $\mathbf{P}\}$. Use this to show that U is a subspace of $\mathbf{P}$.

24. Let $A_1, A_2, \ldots, A_m$ denote $n \times n$ matrices. If Y is a nonzero column in $\mathbb{R}^n$ and $A_1Y = A_2Y = \cdots = A_mY = 0$, show that $\{A_1, A_2, \ldots, A_m\}$ cannot span $\mathbf{M}_{nn}$.

25. Let $\{\mathbf{v}_1, \mathbf{v}_2, \ldots, \mathbf{v}_n\}$ and $\{\mathbf{u}_1, \mathbf{u}_2, \ldots, \mathbf{u}_n\}$ be sets of vectors in a vector space, and let
$$X = \begin{bmatrix} \mathbf{v}_1 \\ \vdots \\ \mathbf{v}_n \end{bmatrix} \qquad Y = \begin{bmatrix} \mathbf{u}_1 \\ \vdots \\ \mathbf{u}_n \end{bmatrix}$$
as in Exercise 18 §6.1.

(a) Show that span$\{\mathbf{v}_1, \ldots, \mathbf{v}_n\} \subseteq$ span$\{\mathbf{u}_1, \ldots, \mathbf{u}_n\}$ if and only if $AY = X$ for some $n \times n$ matrix A.

(b) If $X = AY$ where A is invertible, show that span$\{\mathbf{v}_1, \ldots, \mathbf{v}_n\}$ = span$\{\mathbf{u}_1, \ldots, \mathbf{u}_n\}$.

26. If U and W are subspaces of a vector space V, let $U \cup W = \{\mathbf{v} \mid \mathbf{v}$ is in U or $\mathbf{v}$ is in $W\}$. Show that $U \cup W$ is a subspace if and only if $U \subseteq W$ or $W \subseteq U$.

27. Show that $\mathbf{P}$ cannot be spanned by a finite set of polynomials.

SECTION 6.3   Linear Independence and Dimension

As in $\mathbb{R}^n$, a set of vectors $\{\mathbf{v}_1, \mathbf{v}_2, \ldots, \mathbf{v}_n\}$ is called **linearly independent** (or simply **independent**) if it satisfies the following condition:

$$\text{If } s_1\mathbf{v}_1 + s_2\mathbf{v}_2 + \cdots + s_n\mathbf{v}_n = \mathbf{0}, \quad \text{then } s_1 = s_2 = \cdots = s_n = 0.$$

A set of vectors that is not linearly independent is said to be **linearly dependent** (or simply **dependent**). The **trivial linear combination** of the vectors $\mathbf{v}_1, \mathbf{v}_2, \ldots, \mathbf{v}_n$ is the one with every coefficient zero:

$$0\mathbf{v}_1 + 0\mathbf{v}_2 + \cdots + 0\mathbf{v}_n.$$

This is obviously one way of expressing **0** as a linear combination of these vectors, and they are linearly independent when it is the *only* way.

Example 1

Show that $\{1 + x,\ 3x + x^2,\ 2 + x - x^2\}$ is independent in $\mathbf{P}_2$.

Solution   Suppose a linear combination of these polynomials vanishes.

$$s_1(1 + x) + s_2(3x + x^2) + s_3(2 + x - x^2) = 0.$$

Equating the coefficients of 1, x, and x^2 gives a set of linear equations.

$$
\begin{aligned}
s_1 & & & + 2s_3 &= 0 \\
s_1 & + 3s_2 & + s_3 & &= 0 \\
& s_2 & - s_3 & &= 0
\end{aligned}
$$

The only solution is $s_1 = s_2 = s_3 = 0$.

Example 2

Show that $\{\sin x,\ \cos x\}$ is independent in the vector space $\mathbf{F}[0, 2\pi]$ of functions defined on the interval $[0, 2\pi]$.

Solution   Suppose that a linear combination of these functions vanishes.

$$s_1(\sin x) + s_2(\cos x) = 0.$$

This must hold for *all* values of x (by the definition of equality in $\mathbf{F}[0, 2\pi]$). Taking $x = 0$ yields $s_2 = 0$ (because $\sin 0 = 0$ and $\cos 0 = 1$). Similarly, $s_1 = 0$ follows from taking $x = \frac{\pi}{2}$ (because $\sin \frac{\pi}{2} = 1$ and $\cos \frac{\pi}{2} = 0$).

Example 3

Suppose that $\{\mathbf{u}, \mathbf{v}\}$ is an independent set in a vector space V. Show that $\{\mathbf{u} + 2\mathbf{v},\ \mathbf{u} - 3\mathbf{v}\}$ is also independent.

Solution   Suppose a linear combination of $\mathbf{u} + 2\mathbf{v}$ and $\mathbf{u} - 3\mathbf{v}$ vanishes.

$$s(\mathbf{u} + 2\mathbf{v}) + t(\mathbf{u} - 3\mathbf{v}) = \mathbf{0}.$$

We must deduce that $s = t = 0$. Collecting coefficients of $\mathbf{u}$ and $\mathbf{v}$ gives

$$(s + t)\mathbf{u} + (2s - 3t)\mathbf{v} = \mathbf{0}.$$

Now this is a linear combination of $\mathbf{u}$ and $\mathbf{v}$ that vanishes so, because $\{\mathbf{u}, \mathbf{v}\}$ is linearly independent, all the coefficients must be zero. This yields linear equations $s + t = 0$ and $2s - 3t = 0$, and the only solution is $s = t = 0$.

Example 4

Solution

Show that any set of polynomials of distinct degrees is independent.

Let $p_1, p_2, \ldots, p_m$ be polynomials where $\deg(p_i) = d_i$. By relabelling if necessary, we may assume that $d_1 > d_2 > \cdots > d_m$. Suppose that a linear combination vanishes:

$$t_1 p_1 + t_2 p_2 + \cdots + t_m p_m = 0$$

where each t_i is in $\mathbb{R}$. As $\deg(p_1) = d_1$, let ax^{d_1} be the term in p_1 of highest degree, where $a \neq 0$. Since $d_1 > d_2 > \cdots > d_m$, it follows that $t_1 ax^{d_1}$ is the only term of degree d_1 in the linear combination $t_1 p_1 + t_2 p_2 + \cdots + t_m p_m = 0$. This means that $t_1 ax^{d_1} = 0$, whence $t_1 a = 0$, hence $t_1 = 0$ (because $a \neq 0$). But then $t_2 p_2 + \cdots + t_m p_m = 0$ so we can repeat the argument to show that $t_2 = 0$. Continuing, we obtain $t_i = 0$ for each i, as desired.

Example 5

Solution

Suppose that A is an $n \times n$ matrix such that $A^k = 0$ but $A^{k-1} \neq 0$. Show that $B = \{I, A, A^2, \ldots, A^{k-1}\}$ is independent in $\mathbf{M}_{nn}$.

Suppose that $r_0 I + r_1 A + r_2 A^2 + \cdots + r_{k-1} A^{k-1} = 0$. Multiplying by A^{k-1} gives $r_0 A^{k-1} = 0$ (because $A^k = A^{k+1} = \cdots = A^{2k-1} = 0$). Since $A^{k-1} \neq 0$, this shows that $r_0 = 0$, and we have $r_1 A + r_2 A^2 + \cdots + r_{k-1} A^{k-1} = 0$. Now multiplication by A^{k-2} shows in the same way that $r_1 = 0$, and we continue to get $r_i = 0$ for each i. Hence B is independent.

The next example collects several useful properties of independence for reference.

Example 6

Solution

Let V denote a vector space.
1. If $\mathbf{v} \neq \mathbf{0}$ in V, then $\{\mathbf{v}\}$ is an independent set.
2. No independent set of vectors in V can contain the zero vector.
3. If $\{\mathbf{v}_1, \mathbf{v}_2, \ldots, \mathbf{v}_k\}$ is independent in V, so is $\{a_1 \mathbf{v}_1, a_2 \mathbf{v}_2, \ldots, a_k \mathbf{v}_k\}$ if each $a_i \neq 0$.

1. Let $t\mathbf{v} = \mathbf{0}$, t in $\mathbb{R}$. If $t \neq 0$, then $\mathbf{v} = 1\mathbf{v} = \frac{1}{t}(t\mathbf{v}) = \frac{1}{t}\mathbf{0} = \mathbf{0}$, contrary to assumption. So $t = 0$.
2. If $\{\mathbf{v}_1, \mathbf{v}_2, \ldots, \mathbf{v}_k\}$ is independent and (say) $\mathbf{v}_2 = \mathbf{0}$, then $0\mathbf{v}_1 + 1\mathbf{v}_2 + \cdots + 0\mathbf{v}_k = \mathbf{0}$ is a nontrivial linear combination that vanishes, contrary to the independence of $\{\mathbf{v}_1, \mathbf{v}_2, \ldots, \mathbf{v}_k\}$.
3. Suppose $t_1(a_1\mathbf{v}_1) + t_2(a_2\mathbf{v}_2) + \cdots + t_k(a_k\mathbf{v}_k) = \mathbf{0}$ where the t_i are scalars. Then $(t_1 a_1)\mathbf{v}_1 + (t_2 a_2)\mathbf{v}_2 + \cdots + (t_k a_k)\mathbf{v}_k = \mathbf{0}$, so each $t_i a_i = 0$ by the independence of $\{\mathbf{v}_1, \mathbf{v}_2, \ldots, \mathbf{v}_k\}$. Since each $a_i \neq 0$, this implies that each $t_i = 0$, as required.

A set of vectors is independent if $\mathbf{0}$ is a linear combination in a unique way. The following theorem shows that all linear combinations have uniquely determined coefficients, and so extends Theorem 1 §5.2.

Theorem 1

Let $\{\mathbf{v}_1, \mathbf{v}_2, \ldots, \mathbf{v}_n\}$ be a linearly independent set of vectors in a vector space V. If a vector $\mathbf{v}$ has two (ostensibly different) representations

$$\mathbf{v} = s_1\mathbf{v}_1 + s_2\mathbf{v}_2 + \cdots + s_n\mathbf{v}_n$$
$$\mathbf{v} = t_1\mathbf{v}_1 + t_2\mathbf{v}_2 + \cdots + t_n\mathbf{v}_n$$

as linear combinations of these vectors, then $s_1 = t_1, s_2 = t_2, \ldots, s_n = t_n$.

PROOF

Subtracting the equations given in the theorem gives

$$(s_1 - t_1)\mathbf{v}_1 + (s_2 - t_2)\mathbf{v}_2 + \cdots + (s_n - t_n)\mathbf{v}_n = \mathbf{0}$$

The independence of $\{\mathbf{v}_1, \mathbf{v}_2, \ldots, \mathbf{v}_n\}$ gives $s_i - t_i = 0$ for each i, as required.

The following theorem extends Theorem 4 §5.2, and the proof can be adapted. However, because of the importance of the result, we give a different proof.

Theorem 2 Fundamental Theorem

Suppose a vector space V can be spanned by n vectors. If any set of m vectors in V is linearly independent, then $m \leq n$.

PROOF

Let $V = \text{span}\{\mathbf{v}_1, \mathbf{v}_2, \ldots, \mathbf{v}_n\}$, and suppose that $\{\mathbf{u}_1, \mathbf{u}_2, \ldots, \mathbf{u}_m\}$ is an independent set in V. Then $\mathbf{u}_1 = a_1\mathbf{v}_1 + a_2\mathbf{v}_2 + \cdots + a_n\mathbf{v}_n$ where each a_i is in $\mathbb{R}$. As $\mathbf{u}_1 \neq \mathbf{0}$ (Example 6), not all of the a_i are zero, say $a_1 \neq 0$ (after relabelling the $\mathbf{v}_i$). Then $V = \text{span}\{\mathbf{u}_1, \mathbf{v}_2, \mathbf{v}_3, \ldots, \mathbf{v}_n\}$ as the reader can verify. Hence, write $\mathbf{u}_2 = b_1\mathbf{u}_1 + c_2\mathbf{v}_2 + c_3\mathbf{v}_3 + \cdots + c_n\mathbf{v}_n$. Then some $c_i \neq 0$ because $\{\mathbf{u}_1, \mathbf{u}_2\}$ is independent; so, as before, $V = \text{span}\{\mathbf{u}_1, \mathbf{u}_2, \mathbf{v}_3, \ldots, \mathbf{v}_n\}$, again after possible relabelling of the $\mathbf{v}_i$. If $m > n$, this procedure continues until all the vectors $\mathbf{v}_i$ are replaced by the vectors $\mathbf{u}_1, \mathbf{u}_2, \ldots, \mathbf{u}_n$. In particular, $V = \text{span}\{\mathbf{u}_1, \mathbf{u}_2, \ldots, \mathbf{u}_n\}$. But then $\mathbf{u}_{n+1}$ is a linear combination of $\mathbf{u}_1, \mathbf{u}_2, \ldots, \mathbf{u}_n$ contrary to the independence of the $\mathbf{u}_i$. Hence, the assumption $m > n$ cannot be valid, so $m \leq n$ and the theorem is proved.

If $V = \text{span}\{\mathbf{v}_1, \mathbf{v}_2, \ldots, \mathbf{v}_n\}$, and if $\{\mathbf{u}_1, \mathbf{u}_2, \ldots, \mathbf{u}_m\}$ is an independent set in V, the above proof shows not only that $m \leq n$ but also that m of the (spanning) vectors $\mathbf{v}_1, \mathbf{v}_2, \ldots, \mathbf{v}_n$ can be replaced by the (independent) vectors $\mathbf{u}_1, \mathbf{u}_2, \ldots, \mathbf{u}_m$ and the resulting set will still span V. In this form the result is called the **Steinitz Exchange Lemma**.

As in $\mathbb{R}^n$, a set $\{\mathbf{e}_1, \mathbf{e}_2, \ldots, \mathbf{e}_n\}$ of vectors in a vector space V is called a **basis** of V if it satisfies the following two conditions:

1. $\{\mathbf{e}_1, \mathbf{e}_2, \ldots, \mathbf{e}_n\}$ is linearly independent
2. $V = \text{span}\{\mathbf{e}_1, \mathbf{e}_2, \ldots, \mathbf{e}_n\}$

Thus if a set of vectors $\{\mathbf{e}_1, \mathbf{e}_2, \ldots, \mathbf{e}_n\}$ is a basis, then *every* vector in V can be written as a linear combination of these vectors in a *unique* way (Theorem 1). But even more is true: Any two (finite) bases of V contain the same number of vectors.

Theorem 3 Invariance Theorem

Let $\{e_1, e_2, \ldots, e_n\}$ and $\{f_1, f_2, \ldots, f_m\}$ be two bases of a vector space V. Then $n = m$.

PROOF

Because $V = \text{span}\{e_1, e_2, \ldots, e_n\}$, it follows from Theorem 2 that $m \leq n$. Similarly $n \leq m$, so $n = m$, as asserted.

Theorem 3 guarantees that no matter which basis of V is chosen it contains the same number of vectors as any other basis. Hence there is no ambiguity about the following definition. If $\{e_1, e_2, \ldots, e_n\}$ is a basis of the nonzero vector space V, the number n of vectors in the basis is called the **dimension** of V, and we write

$$\dim V = n.$$

The zero vector space $\{0\}$ is defined to have dimension 0:

$$\dim\{0\} = 0.$$

In our discussion to this point we have always assumed that a basis is nonempty and hence that the dimension of the space is at least 1. However, the zero space $\{0\}$ has *no* basis (by Example 6) so our insistence that $\dim\{0\} = 0$ amounts to saying that the *empty* set of vectors is a basis of $\{0\}$. Thus the statement that "the dimension of a vector space is the number of vectors in any basis" holds even for the zero space.

We saw in Section 5.2, Example 7 that $\dim(\mathbb{R}^n) = n$ and, if E_j denotes column j of I_n, that $\{E_1, E_2, \ldots, E_n\}$ is a basis (called the standard basis). In Example 7 below, similar considerations apply to the space $\mathbf{M}_{mn}$ of all $m \times n$ matrices; the verifications are left to the reader.

Example 7

The space $\mathbf{M}_{mn}$ has dimension mn, and one basis consists of all $m \times n$ matrices with exactly one entry equal to 1 and all other entries equal to 0. We call this the **standard basis** of $\mathbf{M}_{mn}$.

Example 8

Show that $\dim \mathbf{P}_n = n + 1$ and that $\{1, x, x^2, \ldots, x^n\}$ is a basis, called the **standard basis** of $\mathbf{P}_n$.

Solution

Each polynomial $p(x) = a_0 + a_1 x + \cdots + a_n x^n$ in $\mathbf{P}_n$ is clearly a linear combination of $1, x, \ldots, x^n$, so $\mathbf{P}_n = \text{span}\{1, x, \ldots, x^n\}$. However, if a linear combination of these vectors vanishes, $a_0 1 + a_1 x + \cdots + a_n x^n = 0$, then $a_0 = a_1 = \cdots = a_n = 0$ because x is an indeterminate. So $\{1, x, \ldots, x^n\}$ is linearly independent and hence is a basis containing $n + 1$ vectors. Thus, $\dim(\mathbf{P}_n) = n + 1$.

Example 9

If $v \neq 0$ is any nonzero vector in a vector space V, show that $\text{span}\{v\} = \mathbb{R}v$ has dimension 1.

Solution $\{\mathbf{v}\}$ clearly spans $\mathbb{R}\mathbf{v}$, and it is linearly independent by Example 6. Hence $\{\mathbf{v}\}$ is a basis of $\mathbb{R}\mathbf{v}$, and so $\dim \mathbb{R}\mathbf{v} = 1$.

Example 10

Let $A = \begin{bmatrix} 1 & 1 \\ 0 & 0 \end{bmatrix}$ and consider the subspace

$$U = \{X \text{ in } \mathbf{M}_{22} \mid AX = XA\}$$

of $\mathbf{M}_{22}$. Show that $\dim U = 2$ and find a basis of U.

Solution   It was shown in Example 3 §6.2 that U is a subspace for any choice of the matrix A. In the present case, if $X = \begin{bmatrix} x & y \\ z & w \end{bmatrix}$ is in U, the condition $AX = XA$ gives $z = 0$ and $x = y + w$. Hence each matrix X in U can be written

$$X = \begin{bmatrix} y + w & y \\ 0 & w \end{bmatrix} = y\begin{bmatrix} 1 & 1 \\ 0 & 0 \end{bmatrix} + w\begin{bmatrix} 1 & 0 \\ 0 & 1 \end{bmatrix}$$

so $U = \mathrm{span}\, B$ where $B = \left\{ \begin{bmatrix} 1 & 1 \\ 0 & 0 \end{bmatrix}, \begin{bmatrix} 1 & 0 \\ 0 & 1 \end{bmatrix} \right\}$. Moreover, the set B is linearly independent (verify this), so it is a basis of U and $\dim U = 2$.

Example 11

Show that the set V of all symmetric 2×2 matrices is a vector space, and find the dimension of V.

Solution   A matrix A is symmetric if $A^T = A$. If A and B lie in V, then

$$(A + B)^T = A^T + B^T = A + B$$
$$(kA)^T = kA^T = kA$$

using Theorem 2 §2.1. Hence $A + B$ and kA are also symmetric. As the 2×2 zero matrix is also in V, this shows that V is a vector space (being a subspace of $\mathbf{M}_{22}$). Now a matrix A is symmetric when entries directly across the main diagonal are equal, so each 2×2 symmetric matrix has the form

$$\begin{bmatrix} a & c \\ c & b \end{bmatrix} = a\begin{bmatrix} 1 & 0 \\ 0 & 0 \end{bmatrix} + b\begin{bmatrix} 0 & 0 \\ 0 & 1 \end{bmatrix} + c\begin{bmatrix} 0 & 1 \\ 1 & 0 \end{bmatrix}$$

Hence the set $B = \left\{ \begin{bmatrix} 1 & 0 \\ 0 & 0 \end{bmatrix}, \begin{bmatrix} 0 & 0 \\ 0 & 1 \end{bmatrix}, \begin{bmatrix} 0 & 1 \\ 1 & 0 \end{bmatrix} \right\}$ spans V, and the reader can verify that B is linearly independent. Thus B is a basis of V, so $\dim V = 3$.

It is frequently convenient to alter a basis by multiplying some of the basis vectors by nonzero scalars. This always produces another basis as the following example shows; the verification is left as an exercise for the reader.

Example 12

If $B = \{\mathbf{v}_1, \mathbf{v}_2, \ldots, \mathbf{v}_k\}$ is a basis of a subspace U of $\mathbb{R}^n$, then $D = \{a_1\mathbf{v}_1, a_2\mathbf{v}_2, \ldots, a_k\mathbf{v}_k\}$ is also a basis of U provided that each scalar $a_i \neq 0$.

Exercises 6.3

1. Show that each of the following sets of vectors is independent.

   (a) $\{(1 + x, 1 - x, x + x^2\}$ in $\mathbf{P}_2$

   ◆(b) $\{x^2, x + 1, 1 - x - x^2\}$ in $\mathbf{P}_2$

   (c) $\left\{ \begin{bmatrix} 1 & 1 \\ 0 & 0 \end{bmatrix}, \begin{bmatrix} 1 & 0 \\ 1 & 0 \end{bmatrix}, \begin{bmatrix} 0 & 0 \\ 1 & -1 \end{bmatrix}, \begin{bmatrix} 0 & 1 \\ 0 & 1 \end{bmatrix} \right\}$ in $\mathbf{M}_{22}$

   ◆(d) $\left\{ \begin{bmatrix} 1 & 1 \\ 1 & 0 \end{bmatrix}, \begin{bmatrix} 0 & 1 \\ 1 & 1 \end{bmatrix}, \begin{bmatrix} 1 & 0 \\ 1 & 1 \end{bmatrix}, \begin{bmatrix} 1 & 1 \\ 0 & 1 \end{bmatrix} \right\}$ in $\mathbf{M}_{22}$

2. Which of the following subsets of V are independent?

   (a) $V = \mathbf{P}_2$; $\{x^2 + 1, x + 1, x\}$

   ◆(b) $V = \mathbf{P}_2$; $\{x^2 - x + 3, 2x^2 + x + 5, x^2 + 5x + 1\}$

   (c) $V = \mathbf{M}_{22}$; $\left\{ \begin{bmatrix} 1 & 1 \\ 0 & 1 \end{bmatrix}, \begin{bmatrix} 1 & 0 \\ 1 & 1 \end{bmatrix}, \begin{bmatrix} 1 & 0 \\ 0 & 1 \end{bmatrix} \right\}$

   ◆(d) $V = \mathbf{M}_{22}$;
   $$\left\{ \begin{bmatrix} -1 & 0 \\ 0 & -1 \end{bmatrix}, \begin{bmatrix} 1 & -1 \\ -1 & 1 \end{bmatrix}, \begin{bmatrix} 1 & 1 \\ 1 & 1 \end{bmatrix}, \begin{bmatrix} 0 & -1 \\ -1 & 0 \end{bmatrix} \right\}$$

   (e) $V = \mathbf{F}[1, 2]$; $\left\{ \dfrac{1}{x}, \dfrac{1}{x^2}, \dfrac{1}{x^3} \right\}$

   ◆(f) $V = \mathbf{F}[0, 1]$;
   $$\left\{ \dfrac{1}{x^2 + x - 6}, \dfrac{1}{x^2 - 5x + 6}, \dfrac{1}{x^2 - 9} \right\}$$

3. Which of the following are independent in $\mathbf{F}[0, 2\pi]$?

   (a) $\{\sin^2 x, \cos^2 x\}$

   ◆(b) $\{1, \sin^2 x, \cos^2 x\}$

   (c) $\{x, \sin^2 x, \cos^2 x\}$

4. Find all values of x such that the following are independent in $\mathbb{R}^3$.

   (a) $\{(1, -1, 0), (x, 1, 0), (0, 2, 3)\}$

   ◆(b) $\{(2, x, 1), (1, 0, 1), (0, 1, 3)\}$

5. Show that the following are bases of the space V indicated.

   (a) $\{(1, 1, 0), (1, 0, 1), (0, 1, 1)\}$; $V = \mathbb{R}^3$

   ◆(b) $\{(-1, 1, 1), (1, -1, 1), (1, 1, -1)\}$; $V = \mathbb{R}^3$

   (c) $\left\{ \begin{bmatrix} 1 & 0 \\ 0 & 1 \end{bmatrix}, \begin{bmatrix} 0 & 1 \\ 1 & 0 \end{bmatrix}, \begin{bmatrix} 1 & 1 \\ 0 & 1 \end{bmatrix}, \begin{bmatrix} 1 & 0 \\ 0 & 0 \end{bmatrix} \right\}$; $V = \mathbf{M}_{22}$

   ◆(d) $\{1 + x, x + x^2, x^2 + x^3, x^3\}$; $V = \mathbf{P}_3$

6. Exhibit a basis and calculate the dimension of each of the following subspaces of $\mathbf{P}_2$.

   (a) $\{a(1 + x) + b(x + x^2) \mid a \text{ and } b \text{ in } \mathbb{R}\}$

   ◆(b) $\{a + b(x + x^2) \mid a \text{ and } b \text{ in } \mathbb{R}\}$

   (c) $\{p(x) \mid p(1) = 0\}$

   ◆(d) $\{p(x) \mid p(x) = p(-x)\}$

7. Exhibit a basis and calculate the dimension of each of the following subspaces of $\mathbf{M}_{22}$.

   (a) $\{A \mid A^T = -A\}$

   ◆(b) $\left\{ A \,\middle|\, A \begin{bmatrix} 1 & 1 \\ -1 & 0 \end{bmatrix} = \begin{bmatrix} 1 & 1 \\ -1 & 0 \end{bmatrix} A \right\}$

   (c) $\left\{ A \,\middle|\, A \begin{bmatrix} 1 & 0 \\ -1 & 0 \end{bmatrix} = \begin{bmatrix} 0 & 0 \\ 0 & 0 \end{bmatrix} \right\}$

   ◆(d) $\left\{ A \,\middle|\, A \begin{bmatrix} 1 & 1 \\ -1 & 0 \end{bmatrix} = \begin{bmatrix} 0 & 1 \\ -1 & 1 \end{bmatrix} A \right\}$

8. Let $A = \begin{bmatrix} 1 & 1 \\ 0 & 0 \end{bmatrix}$ and define

   $U = \{X \mid X \text{ is in } \mathbf{M}_{22} \text{ and } AX = X\}$.

   (a) Find a basis of U containing A.

   ◆(b) Find a basis of U not containing A.

9. Show that the set $\mathbb{C}$ of all complex numbers is a vector space with the usual operations, and find its dimension.

10. (a) Let V denote the set of all 2×2 matrices with equal column sums. Show that V is a subspace of $\mathbf{M}_{22}$, and compute $\dim V$.

    ◆(b) Repeat part (a) for 3×3 matrices.

    ◆(c) Repeat part (a) for $n \times n$ matrices.

11. (a) Let $V = \{(x^2 + x + 1)p(x) \mid p(x) \text{ in } \mathbf{P}_2\}$. Show that V is a subspace of $\mathbf{P}_4$ and find $\dim V$. [*Hint:* If $f(x)g(x) = 0$ in $\mathbf{P}$, then $f(x) = 0$ or $g(x) = 0$.]

    ◆(b) Repeat with $V = \{(x^2 - x)p(x) \mid p(x) \text{ in } \mathbf{P}_3\}$, a subset of $\mathbf{P}_5$.

    (c) Generalize.

12. In each case, either prove the assertion or give an example showing that it is false.

    (a) Every set of four nonzero polynomials in $\mathbf{P}_3$ is a basis.

    ◆(b) $\mathbf{P}_2$ has a basis of polynomials $f(x)$ such that $f(0) = 0$.

    (c) $\mathbf{P}_2$ has a basis of polynomials $f(x)$ such that $f(0) = 1$.

◆(d) Every basis of M_{22} contains a noninvertible matrix.

(e) No independent subset of M_{22} contains a matrix A with $A^2 = 0$.

◆(f) If $\{u, v, w\}$ is independent then, $au + bv + cw = 0$ for some a, b, c.

(g) $\{u, v, w\}$ is independent if $au + bv + cw = 0$ for some a, b, c.

◆(h) If $\{u, v\}$ is independent, so is $\{u, u + v\}$.

(i) If $\{u, v\}$ is independent, so is $\{u, v, u + v\}$.

◆(j) If $\{u, v, w\}$ is independent, so is $\{u, v\}$.

(k) If $\{u, v, w\}$ is independent, so is $\{u + w, v + w\}$.

13. Let $A \neq 0$ and $B \neq 0$ be $n \times n$ matrices, and assume that A is symmetric and B is skew-symmetric (that is, $B^T = -B$). Show that $\{A, B\}$ is independent.

14. Show that every set of vectors containing a dependent set is again dependent.

◆15. Show that every nonempty subset of an independent set of vectors is again independent.

16. Let f and g be functions on $[a, b]$, and assume that $f(a) = 1 = g(b)$ and $f(b) = 0 = g(a)$. Show that $\{f, g\}$ is independent in $F[a, b]$.

17. Let $\{A_1, A_2, \ldots, A_k\}$ be independent in M_{mn}, and suppose that U and V are invertible matrices of size $m \times m$ and $n \times n$, respectively. Show that $\{UA_1V, UA_2V, \ldots, UA_kV\}$ is independent.

18. Show that $\{v, w\}$ is independent if and only if neither v nor w is a scalar multiple of the other.

◆19. Assume that $\{u, v\}$ is independent in a vector space V. Write $u' = au + bv$ and $v' = cu + dv$, where a, b, c, and d are numbers. Show that $\{u', v'\}$ is independent if and only if the matrix $\begin{bmatrix} a & c \\ b & d \end{bmatrix}$ is invertible. [*Hint:* Theorem 5 §2.3.]

20. If $\{v_1, v_2, \ldots, v_k\}$ is independent and w is not in span $\{v_1, v_2, \ldots, v_k\}$, show that:

(a) $\{w, v_1, v_2, \ldots, v_k\}$ is independent.

(b) $\{v_1 + w, v_2 + w, \ldots, v_k + w\}$ is independent.

21. If $\{v_1, v_2, \ldots, v_k\}$ is independent, show that $\{v_1, v_1 + v_2, \ldots, v_1 + v_2 + \cdots + v_k\}$ is also independent.

22. Let $\{u, v, w, z\}$ be independent. Which of the following are dependent?

(a) $\{u - v, v - w, w - u\}$

◆(b) $\{u + v, v + w, w + u\}$

(c) $\{u - v, v - w, w - z, z - u\}$

◆(d) $\{u + v, v + w, w + z, z + u\}$

23. Let U and W be subspaces of V with bases $\{u_1, u_2, u_3\}$ and $\{w_1, w_2\}$ respectively. If U and W have only the zero vector in common, show that $\{u_1, u_2, u_3, w_1, w_2\}$ is independent.

24. Let $\{p, q\}$ be independent polynomials. Show that $\{p, q, pq\}$ is independent if and only if $\deg p \geq 1$ and $\deg q \geq 1$.

◆25. If z is a complex number, show that $\{z, z^2\}$ is independent if and only if z is not real.

26. If $V = F[a, b]$ as in Example 6 §6.1, show that the set of constant functions is a subspace of dimension 1 (f is **constant** if there is a number c such that $f(x) = c$ for all x).

27. (a) If U is an invertible $n \times n$ matrix and $\{A_1, A_2, \ldots, A_{mn}\}$ is a basis of M_{mn}, show that $\{A_1U, A_2U, \ldots, A_{mn}U\}$ is also a basis.

◆(b) Show that part (a) fails if U is not invertible. [*Hint:* Theorem 5 §2.3.]

28. Show that $\{(a, b), (a_1, b_1)\}$ is a basis of $\mathbb{R}^2$ if and only if $\{a + bx, a_1 + b_1x\}$ is a basis of P_1.

29. Find the dimension of the subspace $\text{span}\{1, \sin^2 \theta, \cos 2\theta\}$ of $F[0, 2\pi]$.

30. Show that $F[0, 1]$ is not finite dimensional.

31. If U and W are subspaces of V, define their intersection $U \cap W$ as follows:

$$U \cap W = \{v \mid v \text{ is in both } U \text{ and } W\}$$

(a) Show that $U \cap W$ is a subspace contained in U and W.

(b) Show that $U \cap W = \{0\}$ if and only if $\{u, w\}$ is independent for any nonzero vectors u in U and w in W.

(c) If B and D are bases of U and W, and if $U \cap W = \{0\}$, show that $B \cup D = \{v \mid v \text{ is in } B \text{ or } D\}$ is independent.

32. If U and W are vector spaces, let $V = \{(u, w) \mid u \text{ in } U \text{ and } w \text{ in } W\}$.

(a) Show that V is a vector space if $(u, w) + (u_1, w_1) = (u + u_1, w + w_1)$ and $a(u, w) = (au, aw)$.

(b) If $\dim U = m$ and $\dim W = n$, show that $\dim V = m + n$.

(c) If $V_1, \ldots, V_m$ are vector spaces, let
$V = V_1 \times \cdots \times V_m = \{(\mathbf{v}_1, \ldots, \mathbf{v}_m) \mid \mathbf{v}_i \text{ in } V_i \text{ for }$
each $i\}$ denote the space of n-tuples from
the V_i with componentwise operations (see
Exercise 17 §6.1). If $\dim V_i = n_i$ for each i,
show that $\dim V = n_1 + \cdots + n_m$.

33. Let $\mathbf{D}_n$ denote the set of all functions f from the
set $\{1, 2, \ldots, n\}$ to $\mathbb{R}$.

(a) Show that $\mathbf{D}_n$ is a vector space with pointwise addition and scalar multiplication.

(b) Show that $\{S_1, S_2, \ldots, S_n\}$ is a basis of $\mathbf{D}_n$
where, for each $k = 1, 2, \ldots, n$, the function
S_k is defined by $S_k(k) = 1$, whereas $S_k(j) = 0$
if $j \neq k$.

34. A polynomial $p(x)$ is **even** if $p(-x) = p(x)$ and
odd if $p(-x) = -p(x)$. Let E_n and O_n denote the
sets of even and odd polynomials in $\mathbf{P}_n$.

(a) Show that E_n is a subspace of $\mathbf{P}_n$ and find
$\dim E_n$.

♦(b) Show that O_n is a subspace of $\mathbf{P}_n$ and find
$\dim O_n$.

35. Suppose that $\{\mathbf{v}_1, \mathbf{v}_2, \ldots, \mathbf{v}_n\}$ is a maximal
independent set in a vector space V. That is,
$\{\mathbf{v}_1, \ldots, \mathbf{v}_n\}$ is independent and no set of more
than n vectors in V is independent. Show that
$\{\mathbf{v}_1, \ldots, \mathbf{v}_n\}$ is a basis of V.

36. Suppose that $\{\mathbf{v}_1, \mathbf{v}_2, \ldots, \mathbf{v}_n\}$ is a minimal
spanning set for a vector space V. That is,
$V = \mathrm{span}\{\mathbf{v}_1, \ldots, \mathbf{v}_n\}$ and V cannot be spanned by
fewer than n vectors. Show that $\{\mathbf{v}_1, \ldots, \mathbf{v}_n\}$ is
a basis of V.

37. Let $\{\mathbf{v}_1, \ldots, \mathbf{v}_n\}$ be independent in a vector
space V, and let A be an $n \times n$ matrix.
Define $\mathbf{u}_1, \ldots, \mathbf{u}_n$ by

$$\begin{bmatrix} \mathbf{u}_1 \\ \vdots \\ \mathbf{u}_n \end{bmatrix} = A \begin{bmatrix} \mathbf{v}_1 \\ \vdots \\ \mathbf{v}_n \end{bmatrix}$$

(See Exercise 18 §6.1.) Show that $\{\mathbf{u}_1, \ldots, \mathbf{u}_n\}$
is independent if and only if A is invertible.

SECTION 6.4   Finite Dimensional Spaces

Up to this point, we have had no guarantee that an arbitrary vector space *has* a basis
—and hence no guarantee that one can speak *at all* of the dimension of V. However,
Theorem 1 will show that any space that is spanned by a finite set of vectors has a
(finite) basis: The proof requires a basic lemma, of interest in itself, that gives a way
to enlarge a given independent set of vectors.

Lemma 1 Independent Lemma

Let $\{\mathbf{v}_1, \mathbf{v}_2, \ldots, \mathbf{v}_k\}$ be an independent set of vectors in a vector space V. If $\mathbf{u}$ is in V
but $\mathbf{u}$ is not in $\mathrm{span}\{\mathbf{v}_1, \mathbf{v}_2, \ldots, \mathbf{v}_k\}$, then $\{\mathbf{u}, \mathbf{v}_1, \mathbf{v}_2, \ldots, \mathbf{v}_k\}$ is also independent.

PROOF

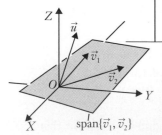

Let $t\mathbf{u} + t_1\mathbf{v}_1 + t_2\mathbf{v}_2 + \cdots + t_k\mathbf{v}_k = \mathbf{0}$; we must show that all the coefficients are zero.
First, $t = 0$ because, otherwise, $\mathbf{u} = -\frac{t_1}{t}\mathbf{v}_1 - \frac{t_2}{t}\mathbf{v}_2 - \cdots - \frac{t_k}{t}\mathbf{v}_k$ is in $\mathrm{span}\{\mathbf{v}_1, \mathbf{v}_2, \ldots, \mathbf{v}_k\}$,
contrary to our assumption. Hence $t = 0$. But then $t_1\mathbf{v}_1 + t_2\mathbf{v}_2 + \cdots + t_k\mathbf{v}_k = \mathbf{0}$ so the
rest of the t_i are zero by the independence of $\{\mathbf{v}_1, \mathbf{v}_2, \ldots, \mathbf{v}_k\}$. This is what we wanted.

Note that the converse of Lemma 1 is also true: if $\{\mathbf{u}, \mathbf{v}_1, \mathbf{v}_2, \ldots, \mathbf{v}_k\}$ is independent,
then $\mathbf{u}$ is not in $\mathrm{span}\{\mathbf{v}_1, \mathbf{v}_2, \ldots, \mathbf{v}_k\}$.

As an illustration, suppose that $\{\vec{v}_1, \vec{v}_2\}$ is independent in $\mathbb{R}^3$. Then $\vec{v}_1$ and $\vec{v}_2$ are
not parallel, so $\mathrm{span}\{\vec{v}_1, \vec{v}_2\}$ is a plane (shaded in the diagram) through the origin.
Hence $\vec{u}$ is not in the plane $\mathrm{span}\{\vec{v}_1, \vec{v}_2\}$ if and only if $\{\vec{u}, \vec{v}_1, \vec{v}_2\}$ is independent.

A vector space V is called **finite dimensional** if it is spanned by a finite set of vectors. Otherwise, V is called **infinite dimensional**. We regard the zero vector space $\{\mathbf{0}\}$ as finite dimensional (it has an *empty* basis).

Theorem 1

Let V be a finite dimensional vector space spanned by m vectors.
1. V has a finite basis and dim $V \leq m$.
2. Every independent set in V can be enlarged (by adding vectors) to a basis of V.
3. If U is a subspace of V, then:
   (a) U is finite dimensional and dim $U \leq$ dim V.
   (b) Every basis of U is part of a basis of V.

PROOF

Let $D = \{\mathbf{u}_1, \mathbf{u}_2, \ldots, \mathbf{u}_k\}$ be an independent subset of the subspace U.

Claim. D can be enlarged to a basis of U containing at most m vectors.

Proof. If $U = \text{span } D$, we are done. Otherwise, choose $\mathbf{v}_1$ in U but not in span D. Then the set $\{\mathbf{v}_1, \mathbf{u}_1, \mathbf{u}_2, \ldots, \mathbf{u}_k\}$ is independent by Lemma 1, and we are done if $U = \text{span}\{\mathbf{v}_1, \mathbf{u}_1, \mathbf{u}_2, \ldots, \mathbf{u}_k\}$. If not, there exists $\mathbf{v}_2$ in U but not in span$\{\mathbf{v}_1, \mathbf{u}_1, \mathbf{u}_2, \ldots, \mathbf{u}_k\}$. Hence $\{\mathbf{v}_2, \mathbf{v}_1, \mathbf{u}_1, \mathbf{u}_2, \ldots, \mathbf{u}_k\}$ is independent and again we are done if $U = \text{span}\{\mathbf{v}_2, \mathbf{v}_1, \mathbf{u}_1, \mathbf{u}_2, \ldots, \mathbf{u}_k\}$. Continue in this way. Either we are done at some stage, or the process ultimately creates an independent subset of U containing more than m vectors. Such a set is independent in V and so contradicts the fundamental theorem (Theorem 2 §6.3) because V is spanned by m vectors. This proves the Claim.

1. If $V = \{\mathbf{0}\}$, then V has an empty basis and dim $V = 0 \leq m$. Otherwise, let $\mathbf{v} \neq \mathbf{0}$ be a vector in V. Then $\{\mathbf{v}\}$ is independent and (1) follows from the Claim with $U = V$.
2. This follows from the Claim with $U = V$.
3. (a) This is clear if $U = \{\mathbf{0}\}$. Otherwise, if $\mathbf{u} \neq \mathbf{0}$ in U, then $\{\mathbf{u}\}$ can be enlarged to a finite basis B of U by the Claim, so U is finite dimensional. Moreover, B is independent in V, so dim $U \leq$ dim V by the fundamental theorem.
   (b) This is clear if $U = \{\mathbf{0}\}$; otherwise, it follows from (2).

Theorem 1 shows that a vector space V is finite dimensional if and only if it has a finite basis (possibly empty), and that every subspace of a finite dimensional space is again finite dimensional.

Example 1

Show that the space $\mathbf{P}$ of all polynomials is infinite dimensional.

Solution For each $n \geq 1$, $\mathbf{P}$ has a subspace $\mathbf{P}_n$ of dimension $n + 1$. Suppose $\mathbf{P}$ is finite dimensional, say dim $\mathbf{P} = m$. Then dim $\mathbf{P}_n \leq$ dim $\mathbf{P}$ by Theorem 1, that is $n + 1 \leq m$. This is impossible since n is arbitrary, so $\mathbf{P}$ must be infinite dimensional.

The next example illustrates how (2) of Theorem 1 can be used.

Example 2

If $C_1, C_2, \ldots, C_k$ are independent columns in $\mathbb{R}^n$, show that they are the first k columns in some invertible $n \times n$ matrix.

Solution   By Theorem 1, expand $\{C_1, C_2, \ldots, C_k\}$ to a basis $\{C_1, C_2, \ldots, C_k, C_{k+1}, \ldots, C_n\}$ of $\mathbb{R}^n$. Then the matrix $A = [C_1 \ C_2 \ \cdots \ C_k \ C_{k+1} \ \cdots \ C_n]$ with this basis as its columns is an $n \times n$ matrix and it is invertible by Theorem 3 §5.2.

Theorem 2

Let U and W be subspaces of the finite dimensional space V.
1. If $U \subseteq W$, then $\dim U \leq \dim W$.
2. If $U \subseteq W$ and $\dim U = \dim W$, then $U = W$.

PROOF

Since W is finite dimensional, (1) follows by taking $V = W$ in part (3) of Theorem 1. Now assume $\dim U = \dim W = n$, and let B be a basis of U. Then B is an independent set in W. If $U \neq W$, then $\text{span } B \neq W$, so B can be extended to an independent set of $n + 1$ vectors in W by Lemma 1. This contradicts the fundamental theorem (Theorem 2 §6.3) because W is spanned by $\dim W = n$ vectors. Hence $U = W$, proving (2).

Theorem 2 is very useful. This was illustrated in Example 11 §5.2 for $\mathbb{R}^2$ and $\mathbb{R}^3$; here is another example.

Example 3

If a is a number, let W denote the subspace of all polynomials in $\mathbf{P}_n$ with a as a root:
$$W = \{p(x) \mid p(x) \text{ is in } \mathbf{P}_n \text{ and } p(a) = 0\}.$$
Show that $\{(x - a), (x - a)^2, \ldots, (x - a)^n\}$ is a basis of W.

Solution   Observe first that $(x - a), (x - a)^2, \ldots, (x - a)^n$ are members of W, and that they are independent because they have distinct degrees (Example 4 §6.3). Write
$$U = \text{span}\{(x - a), (x - a)^2, \ldots, (x - a)^n\}.$$
Then we have $U \subseteq W \subseteq \mathbf{P}_n$, $\dim U = n$, and $\dim \mathbf{P}_n = n + 1$. Hence $n \leq \dim W \leq n + 1$ by Theorem 2, so $\dim W = n$ or $\dim W = n + 1$. But then $W = U$ or $W = \mathbf{P}_n$, again by Theorem 2. Because $W \neq \mathbf{P}_n$, it follows that $W = U$, as required.

A set of vectors is called **dependent** if it is *not* independent, that is if some nontrivial linear combination vanishes. The next result is a convenient test for dependence.

Lemma 2 Dependent Lemma

A set $D = \{\mathbf{v}_1, \mathbf{v}_2, \ldots, \mathbf{v}_k\}$ of vectors in a vector space V is dependent if and only if some vector in D is a linear combination of the others.

PROOF

Let $\mathbf{v}_2$ (say) be a linear combination of the rest: $\mathbf{v}_2 = s_1\mathbf{v}_1 + s_3\mathbf{v}_3 + \cdots + s_k\mathbf{v}_k$. Then $s_1\mathbf{v}_1 + (-1)\mathbf{v}_2 + s_3\mathbf{v}_3 + \cdots + s_k\mathbf{v}_k = \mathbf{0}$ is a nontrivial linear combination that vanishes, so D is dependent. Conversely, if D is dependent, let $t_1\mathbf{v}_1 + t_2\mathbf{v}_2 + \cdots + t_k\mathbf{v}_k = \mathbf{0}$ where some coefficient is nonzero. If (say) $t_2 \neq 0$, then $\mathbf{v}_2 = -\frac{t_1}{t_2}\mathbf{v}_1 - \frac{t_3}{t_2}\mathbf{v}_3 - \cdots - \frac{t_k}{t_2}\mathbf{v}_k$ is a linear combination of the others.

Lemma 1 gives a way to enlarge independent sets to a basis; by contrast, Lemma 2 shows that spanning sets can be cut down to a basis.

Theorem 3

Let V be a finite dimensional vector space. Any spanning set for V can be cut down (by deleting vectors) to a basis of V.

PROOF

Let S be a spanning set for V. We may assume that $V \neq \{\mathbf{0}\}$ since otherwise V has an empty basis (contained in S). Assume that V is spanned by n vectors. Each of these is a finite linear combination of vectors in S, so the vectors in S that arise in these linear combinations will span V (by Theorem 2 §6.2). Hence we may assume that S is finite, say $S = \{\mathbf{u}_1, \mathbf{u}_2, \ldots, \mathbf{u}_m\}$. If S is independent, it is already a basis of V. If not, some vector in S is a linear combination of the rest by Lemma 2. Relabel the vectors of S so that $\mathbf{u}_1$ is in $\mathrm{span}\{\mathbf{u}_2, \ldots, \mathbf{u}_m\}$. It follows that $V = \mathrm{span}\{\mathbf{u}_2, \ldots, \mathbf{u}_m\}$, again by Theorem 2 §6.2, so we are done if $\{\mathbf{u}_2, \ldots, \mathbf{u}_m\}$ is independent. If not, a similar argument shows (after relabelling if necessary) that $V = \mathrm{span}\{\mathbf{u}_3, \ldots, \mathbf{u}_m\}$. This process continues until either a basis is reached or, after relabelling, $V = \mathrm{span}\{\mathbf{u}_m\}$. Since $\mathbf{u}_m \neq \mathbf{0}$ (because $V \neq \{\mathbf{0}\}$) the set $\{\mathbf{u}_m\}$ is independent and so is a basis of V in this case too.

With Theorem 1, Theorem 3 completes the promised proof of Theorem 6 §5.2.

Example 4

Find a basis of $\mathbf{P}_3$ in the spanning set $S = \{1, x + x^2, 2x - 3x^2, 1 + 3x - 2x^2, x^3\}$.

Solution   Since dim $\mathbf{P}_3 = 4$, we must eliminate one polynomial from S. It cannot be x^3 because the span of the rest of S is contained in $\mathbf{P}_2$. But eliminating $1 + 3x - 2x^2$ does leave a basis (verify). Note that $1 + 3x - 2x^2$ is the sum of the first three polynomials in S.

Theorems 1 and 3 have other useful consequences.

Theorem 4

Let V be a vector space of dimension n. Then:
1. No set of more than n vectors in V can be independent.
2. No set of fewer than n vectors in V can span V.
3. If B is a set of n vectors in V, then B is independent if and only if B spans V.

PROOF

V can be spanned by n vectors (any basis) so (1) follows by the fundamental theorem. But the basis vectors are also independent, so the fundamental theorem gives (2) as well.

Turning to (3), suppose that B is independent. If B does not span V then, by Theorem 1, B can be enlarged to a basis of V containing more than n vectors, contradicting (1). So B spans V. Conversely, if B spans V but is not independent, then B can be cut down to a basis of V containing fewer than n vectors by Theorem 3, contradicting (2). This proves (3).

If V is a vector space of dimension n, part (3) of Theorem 4 asserts that to show that a set B of $n = \dim V$ vectors in V is a basis it is enough to show either that B is independent or that $B = \text{span } V$. Often one is easier than the other.

Example 5

Let V denote the space of all symmetric 2×2 matrices. Find a basis of V consisting of invertible matrices.

Solution   We know that $\dim V = 3$ (Example 11 §6.3), so what is needed is a set of three invertible, symmetric matrices that (using Theorem 4) is either independent or spans V. The set $\left\{ \begin{bmatrix} 1 & 0 \\ 0 & 1 \end{bmatrix}, \begin{bmatrix} 1 & 0 \\ 0 & -1 \end{bmatrix}, \begin{bmatrix} 0 & 1 \\ 1 & 0 \end{bmatrix} \right\}$ is independent (verify) and so is a basis of the required type.

Example 6

Let A be any $n \times n$ matrix. Show that there exist $n^2 + 1$ scalars $a_0, a_1, a_2, \ldots, a_{n^2}$, not all zero, such that

$$a_0 I + a_1 A + a_2 A^2 + \cdots + a_{n^2} A^{n^2} = 0$$

where I denotes the $n \times n$ identity matrix.

Solution   The space M_{nn} of all $n \times n$ matrices has dimension n^2 by Example 7 §6.3. Hence the $n^2 + 1$ matrices $I, A, A^2, \ldots, A^{n^2}$ cannot be independent by Theorem 4, so a nontrivial linear combination vanishes. This is the desired conclusion.

Note that the result in Example 6 can be written as $f(A) = 0$ where $f(x) = a_0 + a_1 x + a_2 x^2 + \cdots + a_{n^2} x^{n^2}$. In other words, A satisfies a nonzero polynomial $f(x)$ of degree at most n^2. In fact we know that A satisfies a nonzero polynomial of degree n (this is the Cayley–Hamilton theorem—see Theorem 10 §8.6 or Theorem 2 §9.4), but the brevity of the solution in Example 6 is an indication of the power of these methods.

If a finite basis of V is known, there is a systematic way to enlarge an independent set to a basis. It is based on a refinement of the proof of the claim in Theorem 1.

Example 7

Let $B = \{\mathbf{v}_1, \mathbf{v}_2, \ldots, \mathbf{v}_n\}$ be a fixed basis of a finite dimensional vector space V. If $D = \{\mathbf{u}_1, \mathbf{u}_2, \ldots, \mathbf{u}_m\}$ is any independent subset of V, show that D can be enlarged to a basis of V by adding vectors from B.

Solution   If $\text{span } D = V$, then D is already a basis. If $\text{span } D \neq V$, then V is not contained in $\text{span } D$, that is $\text{span}\{\mathbf{v}_1, \mathbf{v}_2, \ldots, \mathbf{v}_n\} \not\subseteq \text{span } D$. By Theorem 2 §6.2, some vector $\mathbf{v}_k$

is not in $\text{span } D$. Hence $\{\mathbf{v}_k, \mathbf{u}_1, \mathbf{u}_2, \dots, \mathbf{u}_m\}$ is independent by Lemma 1. If this spans V we are done; otherwise, a similar argument shows that $\{\mathbf{v}_l, \mathbf{v}_k, \mathbf{u}_1, \mathbf{u}_2, \dots, \mathbf{u}_m\}$ is independent for some $\mathbf{v}_l$ in B. This process cannot continue indefinitely by the fundamental theorem (since B spans V), so a basis is reached at some stage.

Example 8

Enlarge the independent set $D = \left\{ \begin{bmatrix} 1 & 1 \\ 1 & 0 \end{bmatrix}, \begin{bmatrix} 0 & 1 \\ 1 & 1 \end{bmatrix}, \begin{bmatrix} 1 & 0 \\ 1 & 1 \end{bmatrix} \right\}$ to a basis of $\mathbf{M}_{22}$.

Solution The standard basis of $\mathbf{M}_{22}$ is $\left\{ \begin{bmatrix} 1 & 0 \\ 0 & 0 \end{bmatrix}, \begin{bmatrix} 0 & 1 \\ 0 & 0 \end{bmatrix}, \begin{bmatrix} 0 & 0 \\ 1 & 0 \end{bmatrix}, \begin{bmatrix} 0 & 0 \\ 0 & 1 \end{bmatrix} \right\}$, so including one of these in D will produce a basis by Example 7. In fact including *any* of these matrices in D produces an independent set (verify), and hence a basis by Theorem 4. Of course these vectors are not the only possibilities, for example, including $\begin{bmatrix} 1 & 1 \\ 0 & 1 \end{bmatrix}$ works as well.

Example 9

Find a basis of $\mathbf{P}_3$ containing the independent set $\{1 + x, 1 + x^2\}$.

Solution The standard basis of $\mathbf{P}_3$ is $\{1, x, x^2, x^3\}$, so including two of these vectors will do. If we use 1 and x^3, the result is $\{1, 1 + x, 1 + x^2, x^3\}$. This is independent because the polynomials have distinct degrees (Example 4 §6.3), and so is a basis by Theorem 4. Of course, including $\{1, x\}$ or $\{1, x^2\}$ would not work!

If U and W are subspaces of a vector space V, there are two related subspaces that are of interest, their **sum** $U + W$ and their **intersection** $U \cap W$, defined by

$$U + W = \{\mathbf{u} + \mathbf{w} \mid \mathbf{u} \text{ in } U, \text{ and } \mathbf{w} \text{ in } W\}$$
$$U \cap W = \{\mathbf{v} \text{ in } V \mid \mathbf{v} \text{ is in both } U \text{ and } W\}$$

It is routine to verify that these are indeed subspaces of V, that $U \cap W$ is contained in both U and W, and that $U + W$ contains both U and W. We conclude this section with a useful fact about the dimensions of these spaces.

Theorem 5

Suppose that U and W are finite dimensional subspaces of a vector space V. Then $U + W$ is finite dimensional and

$$\dim(U + W) = \dim U + \dim W - \dim(U \cap W).$$

PROOF

Let $\{\mathbf{x}_1, \dots, \mathbf{x}_d\}$ be a basis of $U \cap W$, and extend it to a basis $\{\mathbf{x}_1, \dots, \mathbf{x}_d, \mathbf{u}_1, \dots, \mathbf{u}_m\}$ of U by Theorem 1. Similarly extend $\{\mathbf{x}_1, \dots, \mathbf{x}_d\}$ to a basis $\{\mathbf{x}_1, \dots, \mathbf{x}_d, \mathbf{w}_1, \dots, \mathbf{w}_k\}$ of W. Then

$$U + W = \text{span}\{\mathbf{x}_1, \dots, \mathbf{x}_d, \mathbf{u}_1, \dots, \mathbf{u}_m, \mathbf{w}_1, \dots, \mathbf{w}_k\}$$

as the reader can verify, so $U + W$ is finite dimensional. For the rest, it suffices to show that $\{\mathbf{x}_1, \ldots, \mathbf{x}_d, \mathbf{u}_1, \ldots, \mathbf{u}_m, \mathbf{w}_1, \ldots, \mathbf{w}_k\}$ is independent (verify). Suppose that

$$r_1\mathbf{x}_1 + \cdots + r_d\mathbf{x}_d + s_1\mathbf{u}_1 + \cdots + s_m\mathbf{u}_m + t_1\mathbf{w}_1 + \cdots + t_k\mathbf{w}_k = \mathbf{0} \qquad (*)$$

where the r_i, s_j, and t_k are scalars. Then

$$r_1\mathbf{x}_1 + \cdots + r_d\mathbf{x}_d + s_1\mathbf{u}_1 + \cdots + s_m\mathbf{u}_m = -(t_1\mathbf{w}_1 + \cdots + t_k\mathbf{w}_k)$$

is in U (left side) and also in W (right side), and so is in $U \cap W$. Hence $(t_1\mathbf{w}_1 + \cdots + t_k\mathbf{w}_k)$ is a linear combination of $\{\mathbf{x}_1, \ldots, \mathbf{x}_d\}$, so $t_1 = \cdots = t_k = 0$ because $\{\mathbf{x}_1, \ldots, \mathbf{x}_d, \mathbf{w}_1, \ldots, \mathbf{w}_k\}$ is independent. Similarly, $s_1 = \cdots = s_m = 0$, so $(*)$ becomes $r_1\mathbf{x}_1 + \cdots + r_d\mathbf{x}_d = \mathbf{0}$. It follows that $r_1 = \cdots = r_d = 0$, as required.

Theorem 5 is particularly interesting if $U \cap W = \{\mathbf{0}\}$. Then there are *no* vectors $\mathbf{x}_i$ in the above proof, and the argument shows that if $\{\mathbf{u}_1, \ldots, \mathbf{u}_m\}$ and $\{\mathbf{w}_1, \ldots, \mathbf{w}_k\}$ are bases of U and W respectively, then $\{\mathbf{u}_1, \ldots, \mathbf{u}_m, \mathbf{w}_1, \ldots, \mathbf{w}_k\}$ is a basis of $U + W$. In this case $U + W$ is said to be a **direct sum** (written $U \oplus W$); we return to this in Chapter 9.

Exercises 6.4

1. In each case, find a basis for V that includes the vector $\mathbf{v}$.

   (a) $V = \mathbb{R}^3$, $\mathbf{v} = (1, -1, 1)$

   ◆(b) $V = \mathbb{R}^3$, $\mathbf{v} = (0, 1, 1)$

   (c) $V = \mathbf{M}_{22}$, $\mathbf{v} = \begin{bmatrix} 1 & 1 \\ 1 & 1 \end{bmatrix}$

   ◆(d) $V = \mathbf{P}_2$, $\mathbf{v} = x^2 - x + 1$

2. In each case, find a basis for V among the given vectors.

   (a) $V = \mathbb{R}^3$, $\{(1, 1, -1), (2, 0, 1), (-1, 1, -2), (1, 2, 1)\}$

   ◆(b) $V = \mathbf{P}_2$, $\{x^2 + 3, x + 2, x^2 - 2x - 1, x^2 + x\}$

3. In each case, find a basis of V containing $\mathbf{v}$ and $\mathbf{w}$.

   (a) $V = \mathbb{R}^4$, $\mathbf{v} = (1, -1, 1, -1)$, $\mathbf{w} = (0, 1, 0, 1)$

   ◆(b) $V = \mathbb{R}^4$, $\mathbf{v} = (0, 0, 1, 1)$, $\mathbf{w} = (1, 1, 1, 1)$

   (c) $V = \mathbf{M}_{22}$, $\mathbf{v} = \begin{bmatrix} 1 & 0 \\ 0 & 1 \end{bmatrix}$, $\mathbf{w} = \begin{bmatrix} 0 & 1 \\ 1 & 0 \end{bmatrix}$

   ◆(d) $V = \mathbf{P}_3$, $\mathbf{v} = x^2 + 1$, $\mathbf{w} = x^2 + x$

4. (a) If z is not a real number, show that $\{z, z^2\}$ is a basis of the real vector space $\mathbb{C}$ of all complex numbers.

   ◆(b) If z is neither real nor pure imaginary, show that $\{z, \bar{z}\}$ is a basis of $\mathbb{C}$.

5. Find a basis of $\mathbf{M}_{22}$ consisting of matrices with the property that $A^2 = A$.

6. Find a basis of $\mathbf{P}_3$ consisting of polynomials whose coefficients sum to 4. What if they sum to 0?

7. If $\{\mathbf{u}, \mathbf{v}, \mathbf{w}\}$ is a basis of V, determine which of the following are bases.

   (a) $\{\mathbf{u} + \mathbf{v}, \mathbf{u} + \mathbf{w}, \mathbf{v} + \mathbf{w}\}$

   ◆(b) $\{2\mathbf{u} + \mathbf{v} + 3\mathbf{w}, 3\mathbf{u} + \mathbf{v} - \mathbf{w}, \mathbf{u} - 4\mathbf{w}\}$

   (c) $\{\mathbf{u}, \mathbf{u} + \mathbf{v} + \mathbf{w}\}$

   ◆(d) $\{\mathbf{u}, \mathbf{u} + \mathbf{w}, \mathbf{u} - \mathbf{w}, \mathbf{v} + \mathbf{w}\}$

8. (a) Can two vectors span $\mathbb{R}^3$? Can they be linearly independent? Explain.

   ◆(b) Can four vectors span $\mathbb{R}^3$? Can they be linearly independent? Explain.

9. Show that any nonzero vector in a finite dimensional vector space is part of a basis.

◆10. If A is a square matrix, show that $\det A = 0$ if and only if some row is a linear combination of the others.

11. Let D, I, and X denote finite, nonempty sets of vectors in a vector space V. Assume that D is dependent and I is independent. In each case answer yes or no, and defend your answer.

    (a) If $X \supseteq D$, must X be dependent?

    ◆(b) If $X \subseteq D$, must X be dependent?

    (c) If $X \supseteq I$, must X be independent?

    ◆(d) If $X \subseteq I$, must X be independent?

12. If U and W are subspaces of V and dim $U = 2$, show that either $U \subseteq W$ or dim$(U \cap W) \leq 1$. [See Exercise 31 §6.3.]

13. Let A be a nonzero 2×2 matrix and write $U = \{X \text{ in } \mathbf{M}_{22} \mid XA = AX\}$. Show that dim $U \geq 2$. [*Hint: I* and *A* are in *U*.]

14. If $U \subseteq \mathbb{R}^2$ is a subspace, show that $U = \{\vec{0}\}$, $U = \mathbb{R}^2$, or U is a line through the origin.

15. Given $\mathbf{v}_1, \mathbf{v}_2, \mathbf{v}_3, \ldots, \mathbf{v}_k$, and $\mathbf{v}$, let $U = \text{span}\{\mathbf{v}_1, \mathbf{v}_2, \ldots, \mathbf{v}_k\}$ and $W = \text{span}\{\mathbf{v}_1, \ldots, \mathbf{v}_k, \mathbf{v}\}$. Show that either dim $W = $ dim U or dim $W = 1 + $ dim U.

16. Suppose U is a subspace of $\mathbf{P}_1$ and $U \neq \{0\}$, $U \neq \mathbf{P}_1$. Show that either $U = \mathbb{R}$ or $U = \mathbb{R}(a + x)$ for some a in $\mathbb{R}$.

17. Let U be a subspace of V and assume dim $V = 4$ and dim $U = 2$. Does every basis of V result from adding (two) vectors to some basis of U (as in Theorem 4)? Defend your answer.

18. Let U and W be subspaces of a vector space V.

    (a) If dim $V = 3$, dim $U = $ dim $W = 2$, and $U \neq W$, show that dim$(U \cap W) = 1$. [*Hint: Exercise 31 §6.3.*]

    ◆(b) Interpret (a) geometrically if $V = \mathbb{R}^3$.

19. Let $U \subseteq W$ be subspaces of V with dim $U = k$ and dim $W = m$, where $k < m$. If $k < l < m$, show that a subspace X exists where $U \subseteq X \subseteq W$ and dim $X = l$.

20. (a) Let $p(x)$ and $q(x)$ lie in $\mathbf{P}_1$ and suppose that $p(1) \neq 0$, $q(2) \neq 0$, and $p(2) = 0 = q(1)$. Show that $\{p(x), q(x)\}$ is a basis of $\mathbf{P}_1$. [*Hint:* If $rp(x) + sq(x) = 0$, evaluate at $x = 1$, $x = 2$.]

    (b) Let $B = \{p_0(x), p_1(x), \ldots, p_n(x)\}$ be a set of polynomials in $\mathbf{P}_n$. Assume that there exist numbers $a_0, a_1, \ldots, a_n$ such that $p_i(a_i) \neq 0$ for each i but $p_i(a_j) = 0$ if i is different from j. Show that B is a basis of $\mathbf{P}_n$.

21. Let V be the set of all infinite sequences $(a_0, a_1, a_2, \ldots)$ of real numbers. Define addition and scalar multiplication by
$$(a_0, a_1, \ldots) + (b_0, b_1, \ldots) = (a_0 + b_0, a_1 + b_1, \ldots)$$
and $r(a_0, a_1, \ldots) = (ra_0, ra_1, \ldots)$.

    (a) Show that V is a vector space.

    ◆(b) Show that V is not finite dimensional.

    (c) [For those with some calculus.] Show that the set of convergent sequences (that is, $\lim_{n \to \infty} a_n$ exists) is a subspace, also of infinite dimension.

22. Let A be an $n \times n$ matrix of rank r. If $U = \{X \text{ in } \mathbf{M}_{nn} \mid AX = 0\}$, show that dim $U = n(n - r)$. [*Hint: Exercise 32 §6.3.*]

23. Let U and W be subspaces of V.

    (a) Show that $U + W$ is a subspace of V containing U and W.

    ◆(b) Show that span$\{\mathbf{u}, \mathbf{w}\} = \mathbb{R}\mathbf{u} + \mathbb{R}\mathbf{w}$ for any vectors $\mathbf{u}$ and $\mathbf{w}$.

    (c) Show that span$\{\mathbf{u}_1, \ldots, \mathbf{u}_m, \mathbf{w}_1, \ldots, \mathbf{w}_n\}$ $= \text{span}\{\mathbf{u}_1, \ldots, \mathbf{u}_m\} + \text{span}\{\mathbf{w}_1, \ldots, \mathbf{w}_n\}$ for any vectors $\mathbf{u}_i$ in U and $\mathbf{w}_j$ in W.

24. If A and B are $m \times n$ matrices, show that rank$(A + B) \leq$ rank $A +$ rank B. [*Hint:* If U and V are the column spaces of A and B, respectively, show that the column space of $A + B$ is contained in $U + V$ and that dim$(U + V) \leq$ dim $U +$ dim V. (See Theorem 5.)

SECTION 6.5   An Application to Polynomials[6]

The vector space of all polynomials of degree at most n is denoted $\mathbf{P}_n$, and it was established in Section 6.3 that $\mathbf{P}_n$ has dimension $n + 1$; in fact, $\{1, x, x^2, \ldots, x^n\}$ is a basis. More generally, *any* $n + 1$ polynomials of distinct degrees form a basis, by Theorem 4 §6.4 (they are independent by Example 4 §6.3). This proves

Theorem 1

Let $p_0(x)$, $p_1(x)$, $p_2(x)$, ..., $p_n(x)$ be polynomials in $\mathbf{P}_n$ of degrees $0, 1, 2, \ldots, n$, respectively. Then $\{p_0(x), \ldots, p_n(x)\}$ is a basis of $\mathbf{P}_n$.

An immediate consequence is that $\{1, (x - a), (x - a)^2, \ldots, (x - a)^n\}$ is a basis of $\mathbf{P}_n$ for any number a. Hence we have the following:

Corollary 1

If a is any number, every polynomial $f(x)$ of degree at most n has an expansion in powers of $(x - a)$:

$$f(x) = a_0 + a_1(x - a) + a_2(x - a)^2 + \cdots + a_n(x - a)^n. \tag{$*$}$$

If $f(x)$ is evaluated at $x = a$, then equation ($*$) becomes

$$f(a) = a_0 + a_1(a - a) + \cdots + a_n(a - a)^n = a_0.$$

Hence $a_0 = f(a)$, and equation ($*$) can be written $f(x) = f(a) + (x - a)g(x)$, where $g(x)$ is a polynomial of degree $n - 1$ (this assumes that $n \geq 1$). If it happens that $f(a) = 0$, then it is clear that $f(x)$ has the form $f(x) = (x - a)g(x)$. Conversely, every such polynomial certainly satisfies $f(a) = 0$, and we obtain:

Corollary 2

Let $f(x)$ be a polynomial of degree $n \geq 1$ and let a be any number. Then

Remainder Theorem
1. $f(x) = f(a) + (x - a)g(x)$ for some polynomial $g(x)$ of degree $n - 1$.

Factor Theorem
2. $f(a) = 0$ if and only if $f(x) = (x - a)g(x)$ for some polynomial $g(x)$.

The polynomial $g(x)$ can be computed easily by using "long division" to divide $f(x)$ by $(x - a)$.

All the coefficients in the expansion ($*$) of $f(x)$ in powers of $(x - a)$ can be determined in terms of the derivatives of $f(x)$.[7] These will be familiar to students of calculus. Let $f^{(n)}(x)$ denote the nth derivative of the polynomial $f(x)$, and write $f^{(0)}(x) = f(x)$. Then, if

$$f(x) = a_0 + a_1(x - a) + a_2(x - a)^2 + \cdots + a_n(x - a)^n$$

it is clear that $a_0 = f(a) = f^{(0)}(a)$. Differentiation gives

$$f^{(1)}(x) = a_1 + 2a_2(x - a) + 3a_3(x - a)^2 + \cdots + na_n(x - a)^{n-1}$$

---

6   The two applications in this chapter are independent and may be taken in any order.

7   The discussion of Taylor's theorem can be omitted with no loss of continuity.

and substituting $x = a$ yields $a_1 = f^{(1)}(a)$. This process continues to give

$$a_2 = \frac{f^{(2)}(a)}{2!}, \quad a_3 = \frac{f^{(3)}(a)}{3!}, \quad \dots, = \frac{f^{(k)}(a)}{k!} \text{ where } k! \text{ is defined as } k! = k(k-1) \cdots 2 \cdot 1.$$

Hence we obtain the following:

Corollary 3 Taylor's Theorem

If $f(x)$ is a polynomial of degree n, then

$$f(x) = f(a) + \frac{f^{(1)}(a)}{1!}(x-a) + \frac{f^{(2)}(a)}{2!}(x-a)^2 + \cdots + \frac{f^{(n)}(a)}{n!}(x-a)^n.$$

Example 1

Expand $f(x) = 5x^3 + 10x + 2$ as a polynomial in powers of $x - 1$.

Solution The derivatives are $f^{(1)}(x) = 15x^2 + 10$, $f^{(2)}(x) = 30x$, and $f^{(3)}(x) = 30$. Hence the Taylor expansion is

$$f(x) = f(1) + \frac{f^{(1)}(1)}{1!}(x-1) + \frac{f^{(2)}(1)}{2!}(x-1)^2 + \frac{f^{(3)}(1)}{3!}(x-1)^3$$

$$= 17 + 25(x-1) + 15(x-1)^2 + 5(x-1)^3.$$

Taylor's theorem is useful in that it provides a formula for the coefficients in the expansion. It is dealt with in calculus texts and will not be pursued here.

Theorem 1 produces bases of $\mathbf{P}_n$ consisting of polynomials of distinct degrees. A different criterion is involved in the next theorem.

Theorem 2

Let $f_0(x), f_1(x), \dots, f_n(x)$ be polynomials in $\mathbf{P}_n$. Assume that numbers $a_0, a_1, \dots, a_n$ exist such that

$$f_i(a_i) \neq 0 \quad \text{for each } i$$
$$f_i(a_j) = 0 \quad \text{if } i \neq j$$

Then

1. $\{f_0(x), \dots, f_n(x)\}$ is a basis of $\mathbf{P}_n$.
2. If $f(x)$ is any polynomial in $\mathbf{P}_n$, its expansion as a linear combination of these basis vectors is

$$f(x) = \frac{f(a_0)}{f_0(a_0)} f_0(x) + \frac{f(a_1)}{f_1(a_1)} f_1(x) + \cdots + \frac{f(a_n)}{f_n(a_n)} f_n(x).$$

PROOF

1. It suffices (by Theorem 4 §6.4) to show that $\{f_0(x), \dots, f_n(x)\}$ is linearly independent (because $\dim \mathbf{P}_n = n + 1$). Suppose that

$$r_0 f_0(x) + r_1 f_1(x) + \cdots + r_n f_n(x) = 0, \quad r_i \text{ in } \mathbb{R}.$$

Because $f_i(a_0) = 0$ for all $i > 0$, taking $x = a_0$ gives $r_0 f_0(a_0) = 0$. But then the fact that $f_0(a_0) \neq 0$ shows that $r_0 = 0$. The proof that $r_i = 0$ for $i > 0$ is analogous.

2. By (1), $f(x) = r_0 f_0(x) + \cdots + r_n f_n(x)$ for *some* numbers r_i. Again, evaluating at a_0 gives $f(a_0) = r_0 f_0(a_0)$, so $r_0 = f(a_0)/f_0(a_0)$. Similarly, $r_i = f(a_i)/f_i(a_i)$ for each i.

Example 2

Show that $\{x^2 - x, x^2 - 2x, x^2 - 3x + 2\}$ is a basis of $\mathbf{P}_2$.

Solution   Write $f_0(x) = x^2 - x = x(x - 1)$, $f_1(x) = x^2 - 2x = x(x - 2)$, and $f_2(x) = x^2 - 3x + 2 = (x - 1)(x - 2)$. Then the conditions of Theorem 2 are satisfied with $a_0 = 2$, $a_1 = 1$, and $a_2 = 0$.

We investigate one natural choice of the polynomials $f_i(x)$ in Theorem 2. To illustrate, let a_0, a_1, and a_2 be distinct numbers and write

$$f_0(x) = \frac{(x - a_1)(x - a_2)}{(a_0 - a_1)(a_0 - a_2)} \qquad f_1(x) = \frac{(x - a_0)(x - a_2)}{(a_1 - a_0)(a_1 - a_2)}$$

$$f_2(x) = \frac{(x - a_0)(x - a_1)}{(a_2 - a_0)(a_2 - a_1)}$$

Then $f_0(a_0) = f_1(a_1) = f_2(a_2) = 1$, and $f_i(a_j) = 0$ for $i \neq j$. Hence Theorem 2 applies, and because $f_i(a_i) = 1$ for each i, the formula for expanding any polynomial is simplified.

In fact, this can be generalized with no extra effort. If $a_0, a_1, \ldots, a_n$ are distinct numbers, define the **Lagrange polynomials** $\delta_0(x), \delta_1(x), \ldots, \delta_n(x)$ relative to these numbers as follows:

$$\delta_k(x) = \frac{\prod_{i \neq k}(x - a_i)}{\prod_{i \neq k}(a_k - a_i)} \qquad k = 0, 1, 2, \ldots, n$$

Here the numerator is the product of all the terms $(x - a_0), (x - a_1), \ldots, (x - a_n)$ with $(x - a_k)$ omitted, and a similar remark applies to the denominator. If $n = 2$, these are just the polynomials in the preceding paragraph. For example, if $n = 3$, the polynomial $\delta_1(x)$ takes the form

$$\delta_1(x) = \frac{(x - a_0)(x - a_2)(x - a_3)}{(a_1 - a_0)(a_1 - a_2)(a_1 - a_3)}$$

In the general case, it is clear that $\delta_i(a_i) = 1$ for each i and that $\delta_i(a_j) = 0$ if $i \neq j$. Hence Theorem 2 specializes as Theorem 3.

Theorem 3 Lagrange Interpolation Expansion

Let $a_0, a_1, \ldots, a_n$ be distinct numbers. The corresponding set

$$\{\delta_0(x), \delta_1(x), \ldots, \delta_n(x)\}$$

of Lagrange polynomials is a basis of $\mathbf{P}_n$, and any polynomial $f(x)$ in $\mathbf{P}_n$ has the following unique expansion as a linear combination of these polynomials.

$$f(x) = f(a_0)\delta_0(x) + f(a_1)\delta_1(x) + \cdots + f(a_n)\delta_n(x)$$

Example 3

Find the Lagrange interpolation expansion for $f(x) = x^2 - 2x + 1$ relative to $a_0 = -1$, $a_1 = 0$, and $a_2 = 1$.

Solution   The Lagrange polynomials are

$$\delta_0(x) = \frac{(x-0)(x-1)}{(-1-0)(-1-1)} = \tfrac{1}{2}(x^2 - x)$$

$$\delta_1(x) = \frac{(x+1)(x-1)}{(0+1)(0-1)} = -(x^2 - 1)$$

$$\delta_2(x) = \frac{(x+1)(x-0)}{(1+1)(1-0)} = \tfrac{1}{2}(x^2 + x)$$

Because $f(-1) = 4$, $f(0) = 1$, and $f(1) = 0$, the expansion is

$$f(x) = 2(x^2 - x) - (x^2 - 1).$$

The Lagrange interpolation expansion gives an easy proof of the following important fact.

Theorem 4

Let $f(x)$ be a polynomial in $\mathbf{P}_n$, and let $a_0, a_1, \ldots, a_n$ denote distinct numbers. If $f(a_i) = 0$ for all i, then $f(x)$ is the zero polynomial (that is, all coefficients are zero).

PROOF

All the coefficients in the Lagrange expansion of $f(x)$ are zero.

Exercises 6.5

1. If polynomials $f(x)$ and $g(x)$ satisfy $f(a) = g(a)$, show that $f(x) - g(x) = (x - a)h(x)$ for some polynomial $h(x)$.

Exercises 2, 3, 4, and 5 require polynomial differentiation.

2. Expand each of the following as a polynomial in powers of $x - 1$.

   (a) $f(x) = x^3 - 2x^2 + x - 1$

   ◆(b) $f(x) = x^3 + x + 1$

   (c) $f(x) = x^4$

   ◆(d) $f(x) = x^3 - 3x^2 + 3x$

3. Prove Taylor's theorem for polynomials.

4. Use Taylor's theorem to derive the **binomial theorem**:

$$(1 + x)^n = \binom{n}{0} + \binom{n}{1}x + \binom{n}{2}x^2 + \cdots + \binom{n}{n}x^n$$

Here the **binomial coefficients** $\binom{n}{r}$ are

defined by $\binom{n}{r} = \dfrac{n!}{r!(n-r!)}$ where

$n! = n(n-1)\cdots 2\cdot 1$ if $n \geq 1$ and $0! = 1$.

5. Let $f(x)$ be a polynomial of degree n. Show that, given any polynomial $g(x)$ in $\mathbf{P}_n$, there exist numbers $b_0, b_1, \ldots, b_n$ such that

$$g(x) = b_0 f(x) + b_1 f^{(1)}(x) + \cdots + b_n f^{(n)}(x)$$

where $f^{(k)}(x)$ denotes the kth derivative of $f(x)$.

6. Use Theorem 2 to show that the following are bases of $\mathbf{P}_2$.

   (a) $\{x^2 - 2x, \ x^2 + 2x, \ x^2 - 4\}$

   ◆(b) $\{x^2 - 3x + 2, \ x^2 - 4x + 3, \ x^2 - 5x + 6\}$

7. Find the Lagrange interpolation expansion of $f(x)$ relative to $a_0 = 1$, $a_1 = 2$, and $a_2 = 3$.

   (a) $f(x) = x^2 + 1$ ◆(b) $f(x) = x^2 + x + 1$

8. Let $a_0, a_1, \dots, a_n$ be distinct numbers. If $f(x)$ and $g(x)$ in $\mathbf{P}_n$ satisfy $f(a_i) = g(a_i)$ for all i, show that $f(x) = g(x)$. [*Hint:* See Theorem 4.]

9. Let $a_0, a_1, \dots, a_n$ be distinct numbers. If $f(x)$ in $\mathbf{P}_{n+1}$ satisfies $f(a_i) = 0$ for each $i = 0, 1, \dots, n$, show that $f(x) = r(x - a_0)(x - a_1) \cdots (x - a_n)$ for some r in $\mathbb{R}$. [*Hint:* r is the coefficient of x^{n+1} in $f(x)$. Consider $f(x) - r(x - a_0) \cdots (x - a_n)$ and use Theorem 4.]

10. Let a and b denote distinct numbers.

    (a) Show that $\{(x - a), (x - b)\}$ is a basis of $\mathbf{P}_1$.

    ◆(b) Show that $\{(x - a)^2, (x - a)(x - b), (x - b)^2\}$ is a basis of $\mathbf{P}_2$.

    (c) Show that $\{(x - a)^n, (x - a)^{n-1}(x - b), \dots,$

$(x - a)(x - b)^{n-1}, (x - b)^n\}$ is a basis of $\mathbf{P}_n$. [*Hint:* If a linear combination vanishes, evaluate at $x = a$ and $x = b$. Then reduce to the case $n - 2$ by using the fact that if $p(x)q(x) = 0$ in $\mathbf{P}$, then either $p(x) = 0$ or $q(x) = 0$.]

11. Let a and b be two distinct numbers. Assume that $n \geq 2$ and let

$$U_n = \{f(x) \text{ in } \mathbf{P}_n \mid f(a) = 0 = f(b)\}.$$

    (a) Show that $U_n = \{(x - a)(x - b)p(x) \mid p(x) \text{ in } \mathbf{P}_{n-2}\}$.

    ◆(b) Show that $\dim U_n = n - 1$. [*Hint:* If $p(x)q(x) = 0$ in $\mathbf{P}$, then either $p(x) = 0$, or $q(x) = 0$.]

    (c) Show that $\{(x - a)^{n-1}(x - b), (x - a)^{n-2}(x - b)^2, \dots, (x - a)^2(x - b)^{n-2}, (x - a)(x - b)^{n-1}\}$ is a basis of U_n. [*Hint:* Exercise 10.]

SECTION 6.6   An Application to Differential Equations[8]

Let f be a function of a real variable x, and let f' and f'' denote the first and second derivatives of f. Equations of the form

$$f' + 3f = 0$$
$$f'' + 2f' + f = 0$$

are called **differential equations**. Solving many practical problems comes down to finding functions f satisfying such an equation. The study of differential equations is a very large undertaking and the present book gives a short introduction to how linear algebra aids in the solution of these equations (we return to this subject in Section 8.10). Of course, an acquaintance with calculus is required.

The simplest example is the **first-order** equation

$$f' + af = 0$$

where a is a number. It is easily verified that $f(x) = e^{-ax}$ is one solution, and this equation is simple enough for us to find *all* solutions. In fact, suppose f is *any* solution so that $f'(x) + af(x) = 0$. Then consider the new function given by $g(x) = f(x)e^{ax}$. The product rule of differentiation gives

$$\begin{aligned} g'(x) &= f(x)[ae^{ax}] + f'(x)e^{ax} \\ &= af(x)e^{ax} - af(x)e^{ax} \\ &= 0 \end{aligned}$$

Hence the function $g(x)$ has zero derivative and so must be a constant—say, $g(x) = c$. But then $f(x)e^{ax} = c$, so

$$f(x) = ce^{-ax}$$

In other words, every solution $f(x)$ is just a multiple of the "basic" solution e^{-ax}.

---

8 Section 8.10 could also be discussed at this stage.

At this point we can see where linear algebra comes into play. The aim is to describe *all* solutions of the equation $f' + af = 0$—that is, to describe the set

$$U = \{f \mid f' \text{ exists and } f' + af = 0\}.$$

But this set U is a vector space. In fact, if f and f_1 both lie in U (so $f' + af = 0$ and $f_1' + af_1 = 0$), then given a number c the basic theory of differentiation shows that $(f + f_1)' = f' + f_1'$ and $(cf)' = cf'$ both exist, and that $f + f_1$ and cf lie in U:

$$(f + f_1)' + a(f + f_1) = (f' + af) + (f_1' + af_1) = 0$$

and

$$(cf)' + a(cf) = c(f' + af) = 0.$$

Hence U is a vector space (in fact, it is a subspace of the space of all real-valued functions), and the previous paragraph shows that e^{-ax} lies in U and *every* member of U is a scalar multiple of e^{-ax}. This can be expressed as in Theorem 1.

Theorem 1

The set of solutions of the first-order differential equation

$$f' + af = 0$$

is a one-dimensional vector space, and $\{e^{-ax}\}$ is a basis.

Example 1

Assume that the number $n(t)$ of bacteria in a culture at time t has the property that the rate of change of n is proportional to n itself. If there are n_0 bacteria present when $t = 0$, find the number at time t.

Solution   Let k denote the proportionality constant. The rate of change of $n(t)$ is its time-derivative $n'(t)$, so the given relationship is $n'(t) = kn(t)$. Thus, $n' - kn = 0$, and Theorem 1 shows that all solutions n are given by $n(t) = ce^{kt}$, where c is a constant. In this case, the constant c is determined by the requirement that there be n_0 bacteria present when $t = 0$. Hence $n_0 = n(0) = ce^{k0} = c$, so

$$n(t) = n_0 e^{kt}$$

gives the number at time t. Of course the constant k depends on the strain of bacteria.

The condition that $n(0) = n_0$ in Example 1 is called an **initial condition** or a **boundary condition** and serves to select one solution from the available solutions. Only one initial condition is needed here because the space of solutions is one-dimensional.

Now consider **second-order** differential equations of the form

$$f'' + af' + bf = 0$$

where a and b are constants. Again the set

$$U = \{f \mid f'' + af' + bf = 0\}$$

is a vector space, and here dim $U = 2$ (we omit the proof). To find a basis for U, it is necessary to introduce the **characteristic polynomial**

$$x^2 + ax + b$$

of the differential equation. Suppose that λ is a real root of this polynomial—that is, $\lambda^2 + a\lambda + b = 0$. Then the function

$$g(x) = e^{\lambda x}$$

is a solution to the differential equation. Indeed,

$$g''(x) + ag'(x) + bg(x) = \lambda^2 e^{\lambda x} + a\lambda e^{\lambda x} + be^{\lambda x}$$
$$= (\lambda^2 + a\lambda + b)e^{\lambda x}$$
$$= 0.$$

Hence if λ and μ are two distinct real roots of the characteristic polynomial, then $e^{\lambda x}$ and $e^{\mu x}$ are solutions to the differential equation and so lie in U. Moreover, they are linearly independent because, if $re^{\lambda x} + se^{\mu x} = 0$ for numbers r and s, and if $r \neq 0$, then $e^{(\lambda - \mu)x} = \frac{-s}{r}$, so $e^{(\lambda - \mu)x}$ is constant. This is not the case if $\lambda \neq \mu$, so the assumption that $r \neq 0$ is invalid. Thus $r = 0$, and similarly $s = 0$. Hence $\{e^{\lambda x}, e^{\mu x}\}$ is a linearly independent set in U and so, because dim $U = 2$, it is a basis. This establishes the first part of Theorem 2.

Theorem 2

Let U denote the space of solutions of the second-order differential equation

$$f'' + af' + bf = 0.$$

Assume that λ and μ are real roots of the characteristic polynomial $x^2 + ax + b$. Then
1. If $\lambda \neq \mu$, then $\{e^{\lambda x}, e^{\mu x}\}$ is a basis of U.
2. If $\lambda = \mu$, then $\{e^{\lambda x}, xe^{\lambda x}\}$ is a basis of U.

PROOF

It is known that dim $U = 2$ (we omit the proof), so (1) was just proved. If $\lambda = \mu$, the verification that $xe^{\lambda x}$ is a solution and that $\{e^{\lambda x}, xe^{\lambda x}\}$ is linearly independent is left as Exercise 4. Then (2) follows, again because dim $U = 2$.

Example 2

Find all solutions f of $f'' - f' - 6f = 0$.

Solution   The characteristic polynomial is $x^2 - x - 6 = (x - 3)(x + 2)$. The roots are 3 and -2, so $\{e^{3x}, e^{-2x}\}$ is a basis for the space of solutions. Hence every solution has the form

$$f(x) = ce^{3x} + de^{-2x}$$

where c and d are constants.

The function $f(x) = ce^{3x} + de^{-2x}$ in Example 2 is sometimes referred to as the **general solution** of the differential equation. The constants c and d are determined by two boundary conditions.

Example 3

Find the solution of $f'' + 4f' + 4f = 0$ that satisfies the boundary conditions $f(0) = 1$, $f(1) = -1$.

Solution The characteristic polynomial is $x^2 + 4x + 4 = (x + 2)^2$, so -2 is a double root. Hence $\{e^{-2x}, xe^{-2x}\}$ is a basis for the space of solutions, and the general solution takes the form $f(x) = ce^{-2x} + dxe^{-2x}$. Applying the boundary conditions gives $1 = f(0) = c$ and $-1 = f(1) = (c + d)e^{-2}$. Hence $c = 1$ and $d = -(1 + e^2)$, so the required solution is

$$f(x) = e^{-2x} - (1 + e^2)xe^{-2x}.$$

One other question remains: What happens if the roots of the characteristic polynomial are not real? To answer this, we must first state precisely what $e^{\lambda x}$ means when λ is not real. If q is a real number, define

$$e^{iq} = \cos q + i \sin q$$

where $i^2 = -1$. Then the relationship $e^{iq}e^{iq_1} = e^{i(q+q_1)}$ holds for all real q and q_1, as is easily verified. If $\lambda = p + iq$, where p and q are real numbers, we define

$$e^\lambda = e^p e^{iq} = e^p(\cos q + i \sin q).$$

Then it is a routine exercise to show that

1. $e^\lambda e^\mu = e^{\lambda + \mu}$
2. $e^\lambda = 1$ if and only if $\lambda = 0$
3. $(e^{\lambda x})' = \lambda e^{\lambda x}$

These easily imply that $f(x) = e^{\lambda x}$ is a solution to $f'' + af' + bf = 0$ if λ is a (possibly complex) root of the characteristic polynomial $x^2 + ax + b$. Now write $\lambda = p + iq$ so that

$$f(x) = e^{\lambda x} = e^{px}\cos(qx) + ie^{px}\sin(qx).$$

For convenience, denote the real and imaginary parts of $f(x)$ as $u(x) = e^{px}\cos(qx)$ and $v(x) = e^{px}\sin(qx)$. Then the fact that $f(x)$ satisfies the differential equation gives

$$0 = f'' + af' + bf = (u'' + au' + bu) + i(v'' + av' + bv).$$

Equating real and imaginary parts shows that $u(x)$ and $v(x)$ are both solutions to the differential equation. This proves part of Theorem 3.

Theorem 3

Let U denote the space of solutions of the second-order differential equation

$$f'' + af' + bf = 0$$

where a and b are real. Suppose λ is a nonreal root of the characteristic polynomial $x^2 + ax + b$. If $\lambda = p + iq$, where p and q are real, then

$$\{e^{px}\cos(qx),\ e^{px}\sin(qx)\}$$

is a basis of U.

PROOF

The foregoing discussion shows that these functions lie in U. Because $\dim U = 2$ (a fact that we have not proved), it suffices to show that they are linearly independent. But if

$$re^{px} \cos(qx) + se^{px} \sin(qx) = 0$$

for all x, then $r\cos(qx) + s\sin(qx) = 0$ for all x (because $e^{px} \neq 0$). Taking $x = 0$ gives $r = 0$, and taking $x = \frac{\pi}{2q}$ gives $s = 0$ ($q \neq 0$ because λ is not real).
This is what we wanted.

Example 4

Find the solution $f(x)$ to $f'' - 2f' + 2f = 0$ that satisfies $f(0) = 2$ and $f(\frac{\pi}{2}) = 0$.

Solution   The characteristic polynomial $x^2 - 2x + 2$ has roots $1 + i$ and $1 - i$. Taking $\lambda = 1 + i$ (quite arbitrarily) gives $p = q = 1$ in the notation of Theorem 3, so $\{e^x \cos x, e^x \sin x\}$ is a basis for the space of solutions. The general solution is thus $f(x) = e^x(r\cos x + s\sin x)$. The boundary conditions yield $2 = f(0) = r$ and $0 = f(\frac{\pi}{2}) = e^{\pi/2}s$. Thus $r = 2$ and $s = 0$, and the required solution is

$f(x) = 2e^x \cos x$.

The following theorem is an important special case of Theorem 3.

Theorem 4

If $q \neq 0$ is a real number, the space of solutions to the differential equation $f'' + q^2 f = 0$ has basis $\{\cos(qx), \sin(qx)\}$.

PROOF

The characteristic polynomial $x^2 + q^2$ has roots qi and $-qi$, so Theorem 3 applies with $p = 0$.

In many situations, the displacement $s(t)$ of some object at time t turns out to have an oscillating form $s(t) = c\sin(at) + d\cos(at)$. These are called **simple harmonic motions**. An example follows.

Example 5

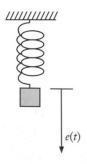

A weight is attached to an extension spring (see diagram). If it is pulled from the equilibrium position and released, it is observed to oscillate up and down. Let $e(t)$ denote the distance of the weight below the equilibrium position t seconds later. It is known (**Hooke's law**) that the acceleration $e''(t)$ of the weight is proportional to the displacement $e(t)$ and in the opposite direction. That is,

$$e''(t) = -ke(t)$$

where $k > 0$ is called the **spring constant**. Find $e(t)$ if the maximum extension is 10 cm below the equilibrium position and find the **period** of the oscillation (time taken for the weight to make a full oscillation).

Solution   It follows from Theorem 4 (with $q^2 = k$) that

$$e(t) = c\,\sin(\sqrt{k}\,t) + d\,\cos(\sqrt{k}\,t)$$

where c and d are constants. The condition $e(0) = 0$ gives $d = 0$, so $e(t) = c\,\sin(\sqrt{k}\,t)$. Now the maximum value of the function $\sin x$ is 1 (when $x = \frac{\pi}{2}$), so $c = 10$ (when $t = \frac{\pi}{2\sqrt{k}}$). Hence

$$e(t) = 10\,\sin(\sqrt{k}\,t).$$

Finally, the weight goes through a full oscillation as $\sqrt{k}\,t$ increases from 0 to 2π. The time taken is $t = \frac{2\pi}{\sqrt{k}}$, the period of the oscillation.

Exercises 6.6

1. Find a solution f to each of the following differential equations satisfying the given boundary conditions.

   (a) $f' - 3f = 0;\ f(1) = 2$

   ◆(b) $f' + f = 0;\ f(1) = 1$

   (c) $f'' + 2f' - 15f = 0;\ f(1) = f(0) = 0$

   ◆(d) $f'' + f' - 6f = 0;\ f(0) = 0,\ f(1) = 1$

   (e) $f'' - 2f' + f = 0;\ f(1) = f(0) = 1$

   ◆(f) $f'' - 4f' + 4f = 0;\ f(0) = 2,\ f(-1) = 0$

   (g) $f'' - 3af' + 2a^2 f = 0,\ a \neq 0;\ f(0) = 0,$
   $f(1) = 1 - e^a$

   ◆(h) $f'' - a^2 f = 0,\ a \neq 0;\ f(0) = 1,\ f(1) = 0$

   (i) $f'' - 2f' + 5f = 0;\ f(0) = 1,\ f(\frac{\pi}{4}) = 0$

   ◆(j) $f'' + 4f' + 5f = 0;\ f(0) = 0,\ f(\frac{\pi}{2}) = 1$

2. Show that the solution to $f' + af = 0$ satisfying $f(x_0) = k$ is $f(x) = k e^{a(x_0 - x)}$.

3. If the characteristic polynomial of $f'' + af' + bf = 0$ has real roots, show that $f = 0$ is the only solution satisfying $f(0) = 0 = f(1)$.

4. Complete the proof of Theorem 2. [*Hint:* If λ is a double root of $x^2 + ax + b$, show that $a = -2\lambda$ and $b = \lambda^2$.]

5. (a) Given the equation $f' + af = b,\ (a \neq 0)$, make the substitution $f(x) = g(x) + b/a$ and obtain a differential equation for g. Then derive the general solution for $f' + af = b$.

   ◆(b) Find the general solution to $f' + f = 2$.

6. Consider the differential equation $f'' + af' + bf = g$, where g is some fixed function. Assume that f_0 is one solution of this equation.

   (a) Show that the general solution is $cf_1 + df_2 + f_0$, where c and d are constants and $\{f_1, f_2\}$ is any basis for the solutions to $f'' + af' + bf = 0$.

   ◆(b) Find a solution to $f'' + f' - 6f = 2x^3 - x^2 - 2x$.
   [*Hint:* Try $f(x) = \frac{-1}{3}x^3.$]

7. A radioactive element decays at a rate proportional to the amount present. Suppose an initial mass of 10 grams decays to 8 grams in 3 hours.

   (a) Find the mass t hours later.

   ◆(b) Find the *half-life* of the element—the time it takes to decay to half its mass.

8. The population $N(t)$ of a region at time t increases at a rate proportional to the population. If the population doubles in 5 years and is 3 million initially, find $N(t)$.

◆9. Consider a spring, as in Example 5. If the period of the oscillation is 30 seconds, find the spring constant k.

10. As a pendulum swings (see the diagram), let t measure the time since it was vertical. The angle $\theta = \theta(t)$ from the vertical can be shown to satisfy the equation $\theta'' + k\theta = 0$, provided that θ is small. If the maximal angle is $\theta = 0.05$ radians, find $\theta(t)$ in terms of k. If the period is 0.5 seconds, find k. [Assume that $\theta = 0$ when $t = 0$.]

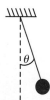

Supplementary Exercises for Chapter 6

1. (Requires calculus) Let V denote the space of all functions $f: \mathbb{R} \to \mathbb{R}$ for which the derivatives f' and f'' exist. Show that f_1, f_2, and f_3 in V are linearly independent provided that their **wronskian** $w(x)$ is nonzero for some x, where

$$w(x) = \det \begin{bmatrix} f_1(x) & f_2(x) & f_3(x) \\ f_1'(x) & f_2'(x) & f_3'(x) \\ f_1''(x) & f_2''(x) & f_3''(x) \end{bmatrix}$$

2. Let $\{\mathbf{v}_1, \mathbf{v}_2, \ldots, \mathbf{v}_n\}$ be a basis of $\mathbb{R}^n$ (written as columns), and let A be an $n \times n$ matrix.

    (a) If A is invertible, show that $\{A\mathbf{v}_1, A\mathbf{v}_2, \ldots, A\mathbf{v}_n\}$ is a basis of $\mathbb{R}^n$.

    ◆(b) If $\{A\mathbf{v}_1, A\mathbf{v}_2, \ldots, A\mathbf{v}_n\}$ is a basis of $\mathbb{R}^n$, show that A is invertible.

3. If A is an $m \times n$ matrix, show that A has rank m if and only if col A contains every column of I_m.

◆4. Show that null $A = \text{null}(A^T A)$ for any real matrix A.

5. Let A be an $m \times n$ matrix of rank r. Show that $\dim(\text{null } A) = n - r$ (Theorem 3 §5.4) as follows. Choose a basis $\{X_1, \ldots, X_k\}$ of null A and extend it to a basis $\{X_1, \ldots, X_k, Z_1, \ldots, Z_m\}$ of $\mathbb{R}^n$. Show that $\{AZ_1, \ldots, AZ_m\}$ is a basis of col A.

7 Linear Transformations[1]

SECTION 7.1   Examples and Elementary Properties

Much of mathematics is concerned with the study of functions. Polynomial functions such as $p(x) = 3x^2 - 5x + 1$ come up in a wide variety of situations. Functions such as the exponential function e^x, the logarithm function $\ln x$, and the trigonometric functions $\sin x$ and $\cos x$ play a fundamental role in calculus as well as in other areas of mathematics. If X and Y are sets, a **function** f from X to Y (written $f : X \to Y$ or $X \xrightarrow{f} Y$) is a rule that associates to every element x of X a uniquely determined element $f(x)$ of Y. In all these examples $X = \mathbb{R}$ and $Y = \mathbb{R}$, but we shall be considering functions where X and Y are both vector spaces.

If V and W are two vector spaces, a function $T : V \to W$ is called a **linear transformation** if it satisfies the following axioms.

$$\text{T1. } T(\mathbf{v} + \mathbf{v}_1) = T(\mathbf{v}) + T(\mathbf{v}_1) \quad \text{for all } \mathbf{v} \text{ and } \mathbf{v}_1 \text{ in } V$$

$$\text{T2. } T(r\mathbf{v}) = rT(\mathbf{v}) \qquad\qquad \text{for all } \mathbf{v} \text{ in } V \text{ and all } r \text{ in } \mathbb{R}$$

A linear transformation $T : V \to V$ is called a **linear operator** on V. The situation can be visualized as in the diagram.

Axiom T1 is just the requirement that T preserves vector addition. It asserts that the result $T(\mathbf{v} + \mathbf{v}_1)$ of adding $\mathbf{v}$ and $\mathbf{v}_1$ first and then applying T is the same as applying T first to get $T(\mathbf{v})$ and $T(\mathbf{v}_1)$ and then adding. Similarly, axiom T2 means that T preserves scalar multiplication. Note that, even though the additions in axiom T1 are both denoted by the same symbol +, the addition on the left forming $\mathbf{v} + \mathbf{v}_1$ is carried out in V, whereas the addition $T(\mathbf{v}) + T(\mathbf{v}_1)$ is done in W. Similarly, the scalar multiplications $r\mathbf{v}$ and $rT(\mathbf{v})$ in axiom T2 refer to the spaces V and W, respectively.

We have already seen many examples of linear transformations $T : \mathbb{R}^n \to \mathbb{R}^m$. In fact Theorem 2 §2.5 shows that, for each such T, there is an $m \times n$ matrix A such that $T(X) = AX$ for every X in $\mathbb{R}^n$. Moreover, the matrix A is given by $A = [T(E_1) \quad T(E_2) \quad \cdots \quad T(E_n)]$ where $\{E_1, E_2, \dots, E_n\}$ is the standard basis of $\mathbb{R}^n$. We will denote this transformation by T_A:

$$T_A : \mathbb{R}^n \to \mathbb{R}^m \quad \text{is defined by} \quad T_A(X) = AX \quad \text{for all } X \text{ in } \mathbb{R}^n.$$

The example that follows lists three important linear transformations that will be referred to later. The verification of axioms T1 and T2 is left to the reader.

---

1 Chapters 7 and 8 can be taken in either order.

Example 1

If V and W are vector spaces, the following are linear transformations:

Identity operator $V \to V$ $1_V : V \to V$ where $1_V(\mathbf{v}) = \mathbf{v}$ for all $\mathbf{v}$ in V
Zero transformation $V \to W$ $0 : V \to W$ where $0(\mathbf{v}) = \mathbf{0}$ for all $\mathbf{v}$ in V
Scalar operator $V \to V$ $a : V \to V$ where $a(\mathbf{v}) = a\mathbf{v}$ for all $\mathbf{v}$ in V
(Here a is any real number.)

The symbol 0 will be used to denote the zero transformation from V to W for *any* spaces V and W. It was also used earlier to denote the zero function $[a, b] \to \mathbb{R}$.

The next example gives two important transformations of matrices. Recall that the trace tr A of an $n \times n$ matrix A is the sum of the entries on the main diagonal.

Example 2

Show that the transposition and trace are linear transformations. More precisely,

$$T : \mathbf{M}_{mn} \to \mathbf{M}_{nm} \quad \text{where } T(A) = A^T \text{ for all } A \text{ in } \mathbf{M}_{mn}$$
$$T : \mathbf{M}_{nn} \to \mathbb{R} \quad \text{where } T(A) = \text{tr } A \text{ for all } A \text{ in } \mathbf{M}_{nn}$$

are both linear transformations.

Solution Axioms T1 and T2 for transposition are $(A + B)^T = A^T + B^T$ and $(rA)^T = r(A^T)$, respectively (using Theorem 2 §2.1). The verifications for the trace are left to the reader.

Example 3

If a is a scalar, define $E_a : \mathbf{P}_n \to \mathbb{R}$ by $E_a(p) = p(a)$ for each polynomial p in $\mathbf{P}_n$. Show that E_a is a linear transformation (called **evaluation** at a).

Solution If p and q are polynomials and r is in $\mathbb{R}$, we use the fact that the sum $p + q$ and scalar product rp are defined as for functions:

$$(p + q)(x) = p(x) + q(x) \quad \text{and} \quad (rp)(x) = rp(x)$$

for all x. Hence, for all p and q in $\mathbf{P}_n$ and all r in $\mathbb{R}$:

$$E_a(p + q) = (p + q)(a) = p(a) + q(a) = E_a(p) + E_a(q), \quad \text{and}$$
$$E_a(rp) = (rp)(a) = rp(a) = rE_a(p).$$

Hence E_a is a linear transformation.

The next example involves some calculus.

Example 4

Show that the differentiation and integration operations on $\mathbf{P}_n$ are linear transformations. More precisely,

$$D : \mathbf{P}_n \to \mathbf{P}_{n-1} \quad \text{where } D[p(x)] = p'(x) \text{ for all } p(x) \text{ in } \mathbf{P}_n$$
$$I : \mathbf{P}_n \to \mathbf{P}_{n+1} \quad \text{where } I[p(x)] = \int_0^x p(t)\,dt \text{ for all } p(x) \text{ in } \mathbf{P}_n$$

are linear transformations.

Solution   These restate the following fundamental properties of differentiation and integration.

$$[p(x) + q(x)]' = p'(x) + q'(x) \quad \text{and} \quad [rp(x)]' = rp'(x)$$

$$\int_0^x [p(t) + q(t)]\,dt = \int_0^x p(t)\,dt + \int_0^x q(t)\,dt \quad \text{and} \quad \int_0^x rp(t)\,dt = r\int_0^x p(t)\,dt$$

The next theorem collects three useful properties of *all* linear transformations. They can be described by saying that, in addition to preserving addition and scalar multiplication (these are the axioms), linear transformations preserve the zero vector, negatives, and linear combinations.

Theorem 1

Let $T : V \to W$ be a linear transformation.
1. $T(\mathbf{0}) = \mathbf{0}$.
2. $T(-\mathbf{v}) = -T(\mathbf{v})$ for all $\mathbf{v}$ in V.
3. $T(r_1\mathbf{v}_1 + r_2\mathbf{v}_2 + \cdots + r_k\mathbf{v}_k) = r_1T(\mathbf{v}_1) + r_2T(\mathbf{v}_2) + \cdots + r_kT(\mathbf{v}_k)$ for all $\mathbf{v}_i$ in V and all r_i in $\mathbb{R}$.

PROOF

1. $T(\mathbf{0}) = T(0\mathbf{v}) = 0T(\mathbf{v}) = \mathbf{0}$ for any $\mathbf{v}$ in V.
2. $T(-\mathbf{v}) = T[(-1)\mathbf{v}] = (-1)T(\mathbf{v}) = -T(\mathbf{v})$ for any $\mathbf{v}$ in V.
3. The proof of Theorem 1 §2.5 goes through.

The ability to use the last part of Theorem 1 effectively is vital to achieving any facility with linear transformations. The next two examples provide illustrations.

Example 5

Let $T : V \to W$ be a linear transformation. If $T(\mathbf{v} - 3\mathbf{v}_1) = \mathbf{w}$ and $T(2\mathbf{v} - \mathbf{v}_1) = \mathbf{w}_1$, find $T(\mathbf{v})$ and $T(\mathbf{v}_1)$ in terms of $\mathbf{w}$ and $\mathbf{w}_1$.

Solution   The given relations imply that

$$T(\mathbf{v}) - 3T(\mathbf{v}_1) = \mathbf{w}$$
$$2T(\mathbf{v}) - T(\mathbf{v}_1) = \mathbf{w}_1$$

by Theorem 1. Subtracting twice the first from the second gives $T(\mathbf{v}_1) = \frac{1}{5}(\mathbf{w}_1 - 2\mathbf{w})$. Then substitution gives $T(\mathbf{v}) = \frac{1}{5}(3\mathbf{w}_1 - \mathbf{w})$.

The full effect of property (3) in Theorem 1 is this: If $T : V \to W$ is a linear transformation and $T(\mathbf{v}_1), T(\mathbf{v}_2), \dots, T(\mathbf{v}_n)$ are known, then $T(\mathbf{v})$ can be computed for *every* vector $\mathbf{v}$ in span$\{\mathbf{v}_1, \mathbf{v}_2, \dots, \mathbf{v}_n\}$. In particular, if $\{\mathbf{v}_1, \mathbf{v}_2, \dots, \mathbf{v}_n\}$ spans V, then $T(\mathbf{v})$ is determined for all $\mathbf{v}$ in V by the choice of $T(\mathbf{v}_1), T(\mathbf{v}_2), \dots, T(\mathbf{v}_n)$. The next theorem states this somewhat differently. As for functions in general, two linear transformations $T : V \to W$ and $S : V \to W$ are called **equal** (written $T = S$) if they have the same **action**; that is, if $T(\mathbf{v}) = S(\mathbf{v})$ for all $\mathbf{v}$ in V.

Theorem 2

Let $T : V \to W$ and $S : V \to W$ be two linear transformations. Suppose that $V = \text{span}\{\mathbf{v}_1, \mathbf{v}_2, \dots, \mathbf{v}_n\}$. If $T(\mathbf{v}_i) = S(\mathbf{v}_i)$ for each i, then $T = S$.

PROOF

If $\mathbf{v}$ is any vector in $V = \text{span}\{\mathbf{v}_1, \mathbf{v}_2, \dots, \mathbf{v}_n\}$, write $\mathbf{v} = a_1\mathbf{v}_1 + a_2\mathbf{v}_2 + \dots + a_n\mathbf{v}_n$ where each a_i is in $\mathbb{R}$. Since $T(\mathbf{v}_i) = S(\mathbf{v}_i)$ for each i, Theorem 1 gives

$$
\begin{aligned}
T(\mathbf{v}) &= T(a_1\mathbf{v}_1 + a_2\mathbf{v}_2 + \dots + a_n\mathbf{v}_n) \\
&= a_1 T(\mathbf{v}_1) + a_2 T(\mathbf{v}_2) + \dots + a_n T(\mathbf{v}_n) \\
&= a_1 S(\mathbf{v}_1) + a_2 S(\mathbf{v}_2) + \dots + a_n S(\mathbf{v}_n) \\
&= S(a_1\mathbf{v}_1 + a_2\mathbf{v}_2 + \dots + a_n\mathbf{v}_n) \\
&= S(\mathbf{v}).
\end{aligned}
$$

Since $\mathbf{v}$ was arbitrary in V, this shows that $T = S$.

Example 6

Let $V = \text{span}\{\mathbf{v}_1, \dots, \mathbf{v}_n\}$. If $T : V \to W$ is a linear transformation and $T(\mathbf{v}_1) = \dots = T(\mathbf{v}_n) = \mathbf{0}$, show that $T = 0$, the zero transformation from V to W.

Solution   The zero transformation $0 : V \to W$ is defined by $0(\mathbf{v}) = \mathbf{0}$ for all $\mathbf{v}$ in V (Example 1), so $T(\mathbf{v}_i) = 0(\mathbf{v}_i)$ holds for each i. Hence $T = 0$ by Theorem 2.

Theorem 2 can be expressed as follows: If we know what a linear transformation $T : V \to W$ does to every vector in a spanning set for V, then we know what T does to *every* vector in V. If the spanning set is a basis, we can say more.

Theorem 3

Let V and W be vector spaces and let $\{\mathbf{e}_1, \mathbf{e}_2, \dots, \mathbf{e}_n\}$ be a basis of V. Given any vectors $\mathbf{w}_1, \mathbf{w}_2, \dots, \mathbf{w}_n$ in W (they need not be distinct), there exists a unique linear transformation $T : V \to W$ satisfying $T(\mathbf{e}_i) = \mathbf{w}_i$ for each $i = 1, 2, \dots, n$. In fact, the action of T is as follows:
Given $\mathbf{v} = v_1\mathbf{e}_1 + v_2\mathbf{e}_2 + \dots + v_n\mathbf{e}_n$ in V, then

$$
T(\mathbf{v}) = T(v_1\mathbf{e}_1 + v_2\mathbf{e}_2 + \dots + v_n\mathbf{e}_n) = v_1\mathbf{w}_1 + v_2\mathbf{w}_2 + \dots + v_n\mathbf{w}_n.
$$

PROOF

If such a transformation T *does* exist, and if S is any other such transformation, then $T(\mathbf{e}_i) = \mathbf{w}_i = S(\mathbf{e}_i)$ holds for each i, so $S = T$ by Theorem 2. Hence T is unique if it exists, and it remains to show that there really is such a linear transformation. Given $\mathbf{v}$ in V, we must specify $T(\mathbf{v})$ in W. Because $\{\mathbf{e}_1, \dots, \mathbf{e}_n\}$ is a basis of V, we have $\mathbf{v} = v_1\mathbf{e}_1 + \dots + v_n\mathbf{e}_n$, where $v_1, \dots, v_n$ are *uniquely* determined by $\mathbf{v}$ (this is Theorem 1 §6.3). Hence we can define $T : V \to W$ by

$$
T(\mathbf{v}) = T(v_1\mathbf{e}_1 + \dots + v_n\mathbf{e}_n) = v_1\mathbf{w}_1 + v_2\mathbf{w}_2 + \dots + v_n\mathbf{w}_n
$$

for all $\mathbf{v} = v_1\mathbf{e}_1 + \dots + v_n\mathbf{e}_n$ in V. This satisfies $T(\mathbf{e}_i) = \mathbf{w}_i$ for each i; the verification that T is linear is left to the reader.

This theorem shows that linear transformations can be defined almost at will: Simply specify where the basis vectors go, and the rest of the action is dictated by the linearity. Moreover, Theorem 2 shows that deciding whether two linear transformations are equal comes down to determining whether they have the same effect on the basis vectors. So, given a basis $\{\mathbf{e}_1, \ldots, \mathbf{e}_n\}$ of a vector space V, there is a different linear transformation $V \to W$ for every ordered selection $\mathbf{w}_1, \mathbf{w}_2, \ldots, \mathbf{w}_n$ of vectors in W (not necessarily distinct).

Example 7

Find a linear transformation $T : \mathbf{P}_2 \to \mathbf{M}_{22}$ such that

$$T(1 + x) = \begin{bmatrix} 1 & 0 \\ 0 & 0 \end{bmatrix}, \quad T(x + x^2) = \begin{bmatrix} 0 & 1 \\ 1 & 0 \end{bmatrix} \quad \text{and} \quad T(1 + x^2) = \begin{bmatrix} 0 & 0 \\ 0 & 1 \end{bmatrix}.$$

Solution The set $\{1 + x, x + x^2, 1 + x^2\}$ is a basis of $\mathbf{P}_2$, so every vector $p = a + bx + cx^2$ in $\mathbf{P}_2$ is a linear combination of these vectors. In fact

$$p = \tfrac{1}{2}(a + b - c)(1 + x) + \tfrac{1}{2}(-a + b + c)(x + x^2) + \tfrac{1}{2}(a - b + c)(1 + x^2)$$

Hence Theorem 3 gives

$$T(p) = \tfrac{1}{2}(a + b - c)\begin{bmatrix} 1 & 0 \\ 0 & 0 \end{bmatrix} + \tfrac{1}{2}(-a + b + c)\begin{bmatrix} 0 & 1 \\ 1 & 0 \end{bmatrix} + \tfrac{1}{2}(a - b + c)\begin{bmatrix} 0 & 0 \\ 0 & 1 \end{bmatrix}$$

$$= \tfrac{1}{2}\begin{bmatrix} a + b - c & -a + b + c \\ -a + b + c & a - b + c \end{bmatrix}$$

Exercises 7.1

1. Show that each of the following functions is a linear transformation.

   (a) $T : \mathbb{R}^2 \to \mathbb{R}^2$; $T(x, y) = (x, -y)$ (reflection in the X axis)

   ◆(b) $T : \mathbb{R}^3 \to \mathbb{R}^3$; $T(x, y, z) = (x, y, -z)$ (reflection in the X-Y plane)

   (c) $T : \mathbb{C} \to \mathbb{C}$; $T(z) = \bar{z}$ (conjugation)

   ◆(d) $T : \mathbf{M}_{mn} \to \mathbf{M}_{kl}$; $T(A) = PAQ$, P a $k \times m$ matrix, Q an $n \times l$ matrix, both fixed

   (e) $T : \mathbf{M}_{nn} \to \mathbf{M}_{nn}$; $T(A) = A^T + A$

   ◆(f) $T : \mathbf{P}_n \to \mathbb{R}$; $T[p(x)] = p(0)$

   (g) $T : \mathbf{P}_n \to \mathbb{R}$; $T(r_0 + r_1x + \cdots + r_nx^n) = r_n$

   ◆(h) $T : \mathbb{R}^n \to \mathbb{R}$; $T(X) = X \bullet Z$, Z a fixed vector in $\mathbb{R}^n$

   (i) $T : \mathbf{P}_n \to \mathbf{P}_n$; $T[p(x)] = p(x + 1)$

   ◆(j) $T : V \to \mathbb{R}$; $T(r_1\mathbf{e}_1 + \cdots + r_n\mathbf{e}_n) = r_1$, where $\{\mathbf{e}_1, \ldots, \mathbf{e}_n\}$ is a fixed basis of V

2. In each case, show that T is *not* a linear transformation.

   (a) $T : \mathbf{M}_{nn} \to \mathbb{R}$; $T(A) = \det A$

   ◆(b) $T : \mathbf{M}_{nm} \to \mathbb{R}$; $T(A) = \text{rank } A$

   (c) $T : \mathbb{R} \to \mathbb{R}$; $T(x) = x^2$

   ◆(d) $T : V \to V$; $T(\mathbf{v}) = \mathbf{v} + \mathbf{u}$ where $\mathbf{u} \neq \mathbf{0}$ is a fixed vector in V (T is called the **translation** by $\mathbf{u}$)

3. In each case, assume that T is a linear transformation.

   (a) If $T : V \to \mathbb{R}$ and $T(\mathbf{v}_1) = 1$, $T(\mathbf{v}_2) = -1$, find $T(3\mathbf{v}_1 - 5\mathbf{v}_2)$.

   ◆(b) If $T : V \to \mathbb{R}$ and $T(\mathbf{v}_1) = 2$, $T(\mathbf{v}_2) = -3$, find $T(3\mathbf{v}_1 + 2\mathbf{v}_2)$.

   (c) If $T : \mathbb{R}^2 \to \mathbb{R}^2$ and $T\begin{bmatrix} 1 \\ 3 \end{bmatrix} = \begin{bmatrix} 1 \\ 1 \end{bmatrix}$, $T\begin{bmatrix} 1 \\ 1 \end{bmatrix} = \begin{bmatrix} 0 \\ 1 \end{bmatrix}$, find $T\begin{bmatrix} -1 \\ 3 \end{bmatrix}$.

   ◆(d) If $T : \mathbb{R}^2 \to \mathbb{R}^2$ and $T\begin{bmatrix} 1 \\ -1 \end{bmatrix} = \begin{bmatrix} 0 \\ 1 \end{bmatrix}$, $T\begin{bmatrix} 1 \\ 1 \end{bmatrix} = \begin{bmatrix} 1 \\ 0 \end{bmatrix}$, find $T\begin{bmatrix} 1 \\ -7 \end{bmatrix}$.

   (e) If $T : \mathbf{P}_2 \to \mathbf{P}_2$ and $T(x + 1) = x$, $T(x - 1) = 1$, $T(x^2) = 0$, find $T(2 + 3x - x^2)$.

♦ (f) If $T : \mathbf{P}_2 \to \mathbb{R}$ and $T(x + 2) = 1$, $T(1) = 5$, $T(x^2 + x) = 0$, find $T(2 - x + 3x^2)$.

4. In each case, find a linear transformation with the given properties and compute $T(\mathbf{v})$.

   (a) $T : \mathbb{R}^2 \to \mathbb{R}^3$; $T(1, 2) = (1, 0, 1)$, $T(-1, 0) = (0, 1, 1)$; $\mathbf{v} = (2, 1)$

   ♦ (b) $T : \mathbb{R}^2 \to \mathbb{R}^3$; $T(2, -1) = (1, -1, 1)$, $T(1, 1) = (0, 1, 0)$; $\mathbf{v} = (-1, 2)$

   (c) $T : \mathbf{P}_2 \to \mathbf{P}_3$; $T(x^2) = x^3$, $T(x + 1) = 0$, $T(x - 1) = x$; $\mathbf{v} = x^2 + x + 1$

   ♦ (d) $T : \mathbf{M}_{22} \to \mathbb{R}$; $T \begin{bmatrix} 1 & 0 \\ 0 & 0 \end{bmatrix} = 3$, $T \begin{bmatrix} 0 & 1 \\ 1 & 0 \end{bmatrix} = -1$,

   $T \begin{bmatrix} 1 & 0 \\ 1 & 0 \end{bmatrix} = 0 = T \begin{bmatrix} 0 & 0 \\ 0 & 1 \end{bmatrix}$; $\mathbf{v} = \begin{bmatrix} a & b \\ c & d \end{bmatrix}$

5. If $T : V \to V$ is a linear transformation, find $T(\mathbf{v})$ and $T(\mathbf{w})$ if:

   (a) $T(\mathbf{v} + \mathbf{w}) = \mathbf{v} - 2\mathbf{w}$ and $T(2\mathbf{v} - \mathbf{w}) = 2\mathbf{v}$

   ♦ (b) $T(\mathbf{v} + 2\mathbf{w}) = 3\mathbf{v} - \mathbf{w}$ and $T(\mathbf{v} - \mathbf{w}) = 2\mathbf{v} - 4\mathbf{w}$

6. If $T : V \to W$ is a linear transformation, show that $T(\mathbf{v} - \mathbf{v}_1) = T(\mathbf{v}) - T(\mathbf{v}_1)$ for all $\mathbf{v}$ and $\mathbf{v}_1$ in V.

7. Let $\{\vec{e}_1, \vec{e}_2\}$ be the standard basis of $\mathbb{R}^2$. Is it possible to have a linear transformation T such that $T(\vec{e}_1)$ lies in $\mathbb{R}$ while $T(\vec{e}_2)$ lies in $\mathbb{R}^2$? Explain your answer.

8. Let $\{\mathbf{v}_1, \dots, \mathbf{v}_n\}$ be a basis of V and let $T : V \to V$ be a linear transformation.

   (a) If $T(\mathbf{v}_i) = \mathbf{v}_i$ for each i, show that $T = 1_V$.

   ♦ (b) If $T(\mathbf{v}_i) = -\mathbf{v}_i$ for each i, show that $T = -1$ is the scalar operator (see Example 1).

9. If A is an $m \times n$ matrix, let $C_k(A)$ denote column k of A. Show that $C_k : \mathbf{M}_{mn} \to \mathbb{R}^m$ is a linear transformation for each $k = 1, \dots, n$.

10. Let $\{E_1, \dots, E_n\}$ be a basis of $\mathbb{R}^n$. Given k, $1 \le k \le n$, define $P_k : \mathbb{R}^n \to \mathbb{R}^n$ by $P_k(r_1E_1 + \dots + r_nE_n) = r_kE_k$. Show that P_k a linear transformation for each k.

11. Let $S : V \to W$ and $T : V \to W$ be linear transformations. Given a in $\mathbb{R}$, define functions $(S + T) : V \to W$ and $(aT) : V \to W$ by $(S + T)(\mathbf{v}) = S(\mathbf{v}) + T(\mathbf{v})$ and $(aT)(\mathbf{v}) = aT(\mathbf{v})$ for all $\mathbf{v}$ in V. Show that $S + T$ and aT are linear transformations.

♦ 12. Describe all linear transformations $T : \mathbb{R} \to V$.

13. Let V and W be vector spaces, let V be finite dimensional, and let $\mathbf{v} \ne \mathbf{0}$ in V. Given any $\mathbf{w}$ in W, show that there exists a linear transformation $T : V \to W$ with $T(\mathbf{v}) = \mathbf{w}$. [*Hint:* Theorem 1 §6.4 and Theorem 3.]

14. Given Y in $\mathbb{R}^n$, define $T_Y : \mathbb{R}^n \to \mathbb{R}$ by $T_Y(X) = X \bullet Y$ for all X in $\mathbb{R}^n$ (where $\bullet$ is the dot product introduced in Section 5.3).

   (a) Show that $T_Y : \mathbb{R}^n \to \mathbb{R}$ is a linear transformation for any Y in $\mathbb{R}^n$.

   (b) Show that every linear transformation $T : \mathbb{R}^n \to \mathbb{R}$ arises in this way; that is, $T = T_Y$ for some Y in $\mathbb{R}^n$. [*Hint:* If $\{E_1, \dots, E_n\}$ is the standard basis of $\mathbb{R}^n$, write $T(E_i) = y_i$ for each i. Use Theorem 1.]

15. Let $T : V \to W$ be a linear transformation.

   (a) If U is a subspace of V, show that $T(U) = \{T(\mathbf{u}) \mid \mathbf{u} \text{ in } U\}$ is a subspace of W (called the **image** of U under T).

   ♦ (b) If P is a subspace of W, show that $T^{-1}(P) = \{\mathbf{v} \text{ in } V \mid T(\mathbf{v}) \text{ in } P\}$ is a subspace of V (called the **preimage** of P under T).

16. Show that differentiation is the only linear transformation $\mathbf{P}_n \to \mathbf{P}_n$ that satisfies $T(x^k) = kx^{k-1}$ for each $k = 0, 1, 2, \dots, n$.

17. Let $T : V \to W$ be a linear transformation and let $\mathbf{v}_1, \dots, \mathbf{v}_n$ denote vectors in V.

   (a) If $\{T(\mathbf{v}_1), \dots, T(\mathbf{v}_n)\}$ is linearly independent, show that $\{\mathbf{v}_1, \dots, \mathbf{v}_n\}$ is also independent.

   (b) Find $T : \mathbb{R}^2 \to \mathbb{R}^2$ for which the converse of part (a) is false.

♦ 18. Suppose $T : V \to V$ is a linear operator with the property that $T[T(\mathbf{v})] = \mathbf{v}$ for all $\mathbf{v}$ in V. (For example, transposition in $\mathbf{M}_{mn}$ or conjugation in $\mathbb{C}$.) If $\mathbf{v} \ne \mathbf{0}$ in V, show that $\{\mathbf{v}, T(\mathbf{v})\}$ is linearly independent if and only if $T(\mathbf{v}) \ne \mathbf{v}$ and $T(\mathbf{v}) \ne -\mathbf{v}$.

19. If a and b are real numbers, define $T_{a,b} : \mathbb{C} \to \mathbb{C}$ by $T_{a,b}(r + si) = ra + sbi$.

   (a) Show that $T_{a,b}$ is linear and $T_{a,b}(\bar{z}) = \overline{T_{a,b}(z)}$ for all z in $\mathbb{C}$. (Here $\bar{z}$ denotes the conjugate of z.)

   (b) If $T : \mathbb{C} \to \mathbb{C}$ is linear and $T(\bar{z}) = \overline{T(z)}$ for all z in $\mathbb{C}$, show that $T = T_{a,b}$ for some real a and b.

20. Show that the following conditions are equivalent for a linear transformation $T : \mathbf{M}_{22} \to \mathbf{M}_{22}$.

   (1) $\text{tr}[T(A)] = \text{tr } A$ for all A in $\mathbf{M}_{22}$.

   (2) $T \begin{bmatrix} r_{11} & r_{12} \\ r_{21} & r_{22} \end{bmatrix} = r_{11}B_{11} + r_{12}B_{12} + r_{21}B_{21} + r_{22}B_{22}$

for matrices B_{ij} such that $\operatorname{tr} B_{11} = 1 = \operatorname{tr} B_{22}$ and $\operatorname{tr} B_{12} = 0 = \operatorname{tr} B_{21}$.

21. Given a in $\mathbb{R}$, consider the **evaluation** map $E_a : \mathbf{P}_n \to \mathbb{R}$ defined in Example 3.

    (a) Show that E_a is a linear transformation satisfying the additional condition that $E_a(x^k) = [E_a(x)]^k$ holds for all $k = 0, 1, 2, \ldots$. [*Note:* $x^0 = 1$.]

    ◆(b) If $T : \mathbf{P}_n \to \mathbb{R}$ is a linear transformation satisfying $T(x^k) = [T(x)]^k$ for all $k = 0, 1, 2, \ldots$, show that $T = E_a$ for some a in $\mathbb{R}$.

22. If $T : \mathbf{M}_{nn} \to \mathbb{R}$ is any linear transformation satisfying $T(AB) = T(BA)$ for all A and B in $\mathbf{M}_{nn}$, show that there exists a number k such that $T(A) = k \operatorname{tr} A$ for all A. (See Lemma 1 §5.5.)

[*Hint:* Let E_{ij} denote the $n \times n$ matrix with 1 in the (i, j) position and zeros elsewhere. Show that $E_{ik} E_{lj} = \begin{cases} 0 & \text{if } k \neq l \\ E_{ij} & \text{if } k = l \end{cases}$. Use this to show that $T(E_{ij}) = 0$ if $i \neq j$ and $T(E_{11}) = T(E_{22}) = \cdots = T(E_{nn})$. Put $k = T(E_{11})$ and use the fact that $\{E_{ij} \mid 1 \leq i, j \leq n\}$ is a basis of $\mathbf{M}_{nn}$.]

23. Let $T : \mathbb{C} \to \mathbb{C}$ be a linear transformation of the real vector space $\mathbb{C}$, and assume that $T(a) = a$ for every real number a. Show that the following are equivalent:

    (a) $T(zw) = T(z)T(w)$ for all z and w in $\mathbb{C}$.

    (b) Either $T = 1_{\mathbb{C}}$ or $T(z) = \bar{z}$ for each z in $\mathbb{C}$ (where $\bar{z}$ denotes the conjugate).

SECTION 7.2   Kernel and Image of a Linear Transformation

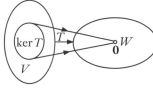

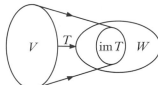

This section is devoted to two important subspaces associated with a linear transformation $T : V \to W$. The **kernel** of T (denoted $\ker T$) and the **image** of T (denoted $\operatorname{im} T$ or $T(V)$) are defined by

$$\ker T = \{\mathbf{v} \text{ in } V \mid T(\mathbf{v}) = \mathbf{0}\}$$

$$\operatorname{im} T = \{T(\mathbf{v}) \mid \mathbf{v} \text{ in } V\}$$

The kernel of T is often called the **nullspace** of T. It consists of all vectors $\mathbf{v}$ in V satisfying the *condition* that $T(\mathbf{v}) = \mathbf{0}$. The image of T is often called the **range** of T and consists of all vectors $\mathbf{w}$ in W of the *form* $\mathbf{w} = T(\mathbf{v})$ for some $\mathbf{v}$ in V. These subspaces are depicted in the diagrams.

Example 1

Let $T_A : \mathbb{R}^n \to \mathbb{R}^m$ be the linear transformation induced by the $m \times n$ matrix A, that is $T_A(X) = AX$ for all X in $\mathbb{R}^n$. Then

$$\ker T_A = \{X \mid AX = \mathbf{0}\} = \operatorname{null} A$$

$$\operatorname{im} T_A = \{AX \mid X \text{ in } \mathbb{R}^n\} = \operatorname{im} A$$

Hence the following theorem extends Example 2 §5.1.

Theorem 1

If $T : V \to W$ is a linear transformation, $\ker T$ is a subspace of V, and $\operatorname{im} T$ is a subspace of W.

PROOF

The fact that $T(\mathbf{0}) = \mathbf{0}$ shows that both $\ker T$ and $\operatorname{im} T$ contain the zero vector. If $\mathbf{v}$ and $\mathbf{v}_1$ lie in $\ker T$, then $T(\mathbf{v}) = \mathbf{0} = T(\mathbf{v}_1)$, so

$$T(\mathbf{v} + \mathbf{v}_1) = T(\mathbf{v}) + T(\mathbf{v}_1) = \mathbf{0} + \mathbf{0} = \mathbf{0}$$

$$T(r\mathbf{v}) = rT(\mathbf{v}) = r\mathbf{0} = \mathbf{0} \quad \text{for all } r \text{ in } \mathbb{R}$$

Hence $\mathbf{v} + \mathbf{v}_1$ and $r\mathbf{v}$ lie in ker T (they satisfy the required condition), so ker T is a subspace of V by Theorem 1 §6.2. If $\mathbf{w}$ and $\mathbf{w}_1$ lie in im T, write $\mathbf{w} = T(\mathbf{v})$ and $\mathbf{w}_1 = T(\mathbf{v}_1)$ where $\mathbf{v}$ and $\mathbf{v}_1$ lie in V. Then

$$\mathbf{w} + \mathbf{w}_1 = T(\mathbf{v}) + T(\mathbf{v}_1) = T(\mathbf{v} + \mathbf{v}_1)$$
$$r\mathbf{w} = rT(\mathbf{v}) = T(r\mathbf{v}) \quad \text{for all } r \text{ in } \mathbb{R}$$

Hence $\mathbf{w} + \mathbf{w}_1$ and $r\mathbf{w}$ both lie in im T (they have the required form), so im T is a subspace of W.

Given a linear transformation $T : V \to W$:

dim(ker T) is called the **nullity** of T and denoted as nullity(T)

dim(im T) is called the **rank** of T and denoted as rank(T)

The rank of a matrix A was defined earlier to be the dimension of col A, the column space of A. The two usages of the word *rank* are consistent in the following sense. Recall the definition of T_A in Example 1.

Example 2

Given an $m \times n$ matrix A, show that im $T_A = $ col A, so rank $T_A = $ rank A.

Solution Write $A = [C_1 \cdots C_n]$ in terms of its columns. Then

$$\text{im } T_A = \{AX \mid X \text{ in } \mathbb{R}^n\} = \{[C_1 \cdots C_n][x_1 \cdots x_n]^T \mid x_i \text{ in } \mathbb{R}\}$$
$$= \{x_1 C_1 + \cdots + x_n C_n \mid x_i \text{ in } \mathbb{R}\}$$

using Theorem 4 §2.2. Hence im T_A is the column space of A.

Often, a useful way to study a subspace of a vector space is to exhibit it as the kernel or image of a linear transformation. Here is an example.

Example 3

Define a transformation $T : \mathbf{M}_{nn} \to \mathbf{M}_{nn}$ by $T(A) = A - A^T$ for all A in $\mathbf{M}_{nn}$. Show that T is linear and that:

(a) ker T consists of all symmetric matrices.

(b) im T consists of all skew-symmetric matrices.

Solution The verification that T is linear is left to the reader. To prove part (a), note that a matrix A lies in ker T just when $0 = T(A) = A - A^T$, and this occurs if and only if $A = A^T$—that is, A is symmetric. Turning to part (b), the space im T consists of all matrices $T(A)$, A in $\mathbf{M}_{nn}$. Every such matrix is skew-symmetric because

$$T(A)^T = (A - A^T)^T = A^T - A = -T(A)$$

On the other hand, if S is skew-symmetric (that is, $S^T = -S$), then S lies in im T. In fact,

$$T[\tfrac{1}{2}S] = \tfrac{1}{2}S - [\tfrac{1}{2}S]^T = \tfrac{1}{2}(S + S) = S.$$

One-to-One and Onto Transformations

Let $T : V \to W$ be a linear transformation.

1. T is said to be **onto** if im $T = W$.

2. T is said to be **one-to-one** if $T(\mathbf{v}) = T(\mathbf{v}_1)$ implies $\mathbf{v} = \mathbf{v}_1$.

Thus $T : V \rightarrow W$ is onto if *every* vector $\mathbf{w}$ in W has the form $\mathbf{w} = T(\mathbf{v})$ for some (not necessarily unique) vector $\mathbf{v}$ in V, whereas T is one-to-one if *no* two distinct vectors $\mathbf{v} \neq \mathbf{v}_1$ in V are carried to the same image $T(\mathbf{v}) = T(\mathbf{v}_1)$ in W. The onto transformations T are those for which im T is as large a subspace of W as possible. By contrast, Theorem 2 shows that the one-to-one transformations T are the ones with ker T as *small* as possible.

Theorem 2

If $T : V \rightarrow W$ is a linear transformation, then T is one-to-one if and only if ker $T = \{\mathbf{0}\}$.

PROOF

If T is one-to-one, let $\mathbf{v}$ be any vector in ker T. Then $T(\mathbf{v}) = \mathbf{0}$, so $T(\mathbf{v}) = T(\mathbf{0})$. Hence $\mathbf{v} = \mathbf{0}$ because T is one-to-one. Conversely, assume that ker $T = \{\mathbf{0}\}$ and let $T(\mathbf{v}) = T(\mathbf{v}_1)$ with $\mathbf{v}$ and $\mathbf{v}_1$ in V. Then $T(\mathbf{v} - \mathbf{v}_1) = T(\mathbf{v}) - T(\mathbf{v}_1) = \mathbf{0}$, so $\mathbf{v} - \mathbf{v}_1$ lies in ker $T = \{\mathbf{0}\}$. This means that $\mathbf{v} - \mathbf{v}_1 = \mathbf{0}$, so $\mathbf{v} = \mathbf{v}_1$. This proves that T is one-to-one.

Example 4

The identity transformation $1_V : V \rightarrow V$ is both one-to-one and onto for any vector space V.

Example 5

Consider the linear transformations

$$S : \mathbb{R}^3 \rightarrow \mathbb{R}^2 \quad \text{given by } S(x, y, z) = (x + y, x - y)$$
$$T : \mathbb{R}^2 \rightarrow \mathbb{R}^3 \quad \text{given by } T(x, y) = (x + y, x - y, x)$$

Show that T is one-to-one but not onto, whereas S is onto but not one-to-one.

Solution

The verification that they are linear is omitted. T is one-to-one because

$$\text{ker } T = \{(x, y) \mid x + y = x - y = x = 0\} = \{(0, 0)\}$$

However, it is not onto. For example $(0, 0, 1)$ does not lie in im T because if $(0, 0, 1) = (x + y, x - y, x)$ for some x and y, then $x + y = 0 = x - y$ and $x = 1$, an impossibility. Turning to S, it is not one-to-one by Theorem 2 because $(0, 0, 1)$ lies in ker S. But every element (s, t) in $\mathbb{R}^2$ lies in im S because $(s, t) = (x + y, x - y) = S(x, y, z)$ for some x, y, and z (in fact, $x = \frac{1}{2}(s + t)$ and $y = \frac{1}{2}(s - t)$). Hence S is onto.

Example 6

Let U be an invertible $m \times m$ matrix and define

$$T : \mathbf{M}_{mn} \rightarrow \mathbf{M}_{mn} \quad \text{by} \quad T(X) = UX \text{ for all } X \text{ in } \mathbf{M}_{mn}$$

Show that T is a linear transformation that is both one-to-one and onto.

Solution

The verification that T is linear is left to the reader. To see that T is one-to-one, let $T(X) = 0$. Then $UX = 0$, so left-multiplication by U^{-1} gives

$X = 0$. Hence ker $T = \{\mathbf{0}\}$, so T is one-to-one. Finally, if Y is any member of $\mathbf{M}_{mn}$, then $U^{-1}Y$ lies in $\mathbf{M}_{mn}$ too, and $T(U^{-1}Y) = U(U^{-1}Y) = Y$. This shows that T is onto.

The linear transformations $\mathbb{R}^n \to \mathbb{R}^m$ all have the form T_A for some $m \times n$ matrix A (Theorem 2 §2.5). The next theorem gives conditions under which they are onto or one-to-one.

Theorem 3

Let A be an $m \times n$ matrix, and let $T_A : \mathbb{R}^n \to \mathbb{R}^m$ be the linear transformation induced by A, that is $T_A(X) = AX$ for all X in $\mathbb{R}^n$.
1. T_A is onto if and only if rank $A = m$.
2. T_A is one-to-one if and only if rank $A = n$.

PROOF

1. We have that im T_A is the column space of A (see Example 2), so T_A is onto if and only if the column space of A is $\mathbb{R}^m$. Because the rank of A is the dimension of the column space, this holds if and only if rank $A = m$.
2. ker $T_A = \{X$ in $\mathbb{R}^n \mid AX = 0\}$, so (using Theorem 2) T_A is one-to-one if and only if $AX = 0$ implies $X = 0$. This is equivalent to rank $A = n$ by Theorem 4 §5.4.

The Dimension Theorem

The following theorem is the main result of this section.

Theorem 4 Dimension Theorem

Let $T : V \to W$ be any linear transformation and assume that ker T and im T are both finite dimensional. Then V is also finite dimensional and
$$\dim V = \dim(\ker T) + \dim(\operatorname{im} T)$$
In other words, dim $V = \operatorname{nullity}(T) + \operatorname{rank}(T)$.

PROOF

Every vector in im $T = T(V)$ has the form $T(\mathbf{v})$ for some $\mathbf{v}$ in V. Hence let $\{T(\mathbf{e}_1), T(\mathbf{e}_2), \ldots, T(\mathbf{e}_r)\}$ be a basis of im T, where the $\mathbf{e}_i$ lie in V. Let $\{\mathbf{f}_1, \mathbf{f}_2, \ldots, \mathbf{f}_k\}$ be any basis of ker T. Then dim(im $T) = r$ and dim(ker $T) = k$, so it suffices to show that $B = \{\mathbf{e}_1, \ldots, \mathbf{e}_r, \mathbf{f}_1, \ldots, \mathbf{f}_k\}$ is a basis of V.

1. *B spans V.* If $\mathbf{v}$ lies in V, then $T(\mathbf{v})$ lies in im V, so
$$T(\mathbf{v}) = t_1 T(\mathbf{e}_1) + t_2 T(\mathbf{e}_2) + \cdots + t_r T(\mathbf{e}_r) \quad t_i \text{ in } \mathbb{R}$$
This implies that $\mathbf{v} - t_1\mathbf{e}_1 - t_2\mathbf{e}_2 - \cdots - t_r\mathbf{e}_r$ lies in ker T and so is a linear combination of $\mathbf{f}_1, \ldots, \mathbf{f}_k$. Hence $\mathbf{v}$ is a linear combination of the vectors in B.

2. *B is linearly independent.* Suppose that t_i and s_j in $\mathbb{R}$ satisfy
$$t_1\mathbf{e}_1 + \cdots + t_r\mathbf{e}_r + s_1\mathbf{f}_1 + \cdots + s_k\mathbf{f}_k = \mathbf{0} \qquad (*)$$

Applying T gives $t_1 T(\mathbf{e}_1) + \cdots + t_r T(\mathbf{e}_r) = \mathbf{0}$ (because $T(\mathbf{f}_i) = \mathbf{0}$ for each i), so the independence of $\{T(\mathbf{e}_1), \ldots, T(\mathbf{e}_r)\}$ yields $t_1 = \cdots = t_r = 0$. Hence (∗) becomes

$$s_1 \mathbf{f}_1 + \cdots + s_k \mathbf{f}_k = \mathbf{0}$$

so $s_1 = \cdots = s_k = 0$ by the independence of $\{\mathbf{f}_1, \ldots, \mathbf{f}_k\}$. This proves that B is linearly independent.

Note that $r + k = n$ in the proof so, after relabelling, we end up with a basis
$$B = \{\mathbf{e}_1, \mathbf{e}_2, \ldots, \mathbf{e}_r, \mathbf{e}_{r+1}, \ldots, \mathbf{e}_n\}$$
of V with the property that $\{\mathbf{e}_{r+1}, \ldots, \mathbf{e}_n\}$ is a basis of ker T and $\{T(\mathbf{e}_1), \ldots, T(\mathbf{e}_r)\}$ is a basis of im T. In fact, if V is known in advance to be finite dimensional, then *any* basis $\{\mathbf{e}_{r+1}, \ldots, \mathbf{e}_n\}$ of ker T can be extended to a basis $\{\mathbf{e}_1, \mathbf{e}_2, \ldots, \mathbf{e}_r, \mathbf{e}_{r+1}, \ldots, \mathbf{e}_n\}$ of V by Theorem 1 §6.4. Moreover, it turns out that, no matter how this is done, the vectors $\{T(\mathbf{e}_1), \ldots, T(\mathbf{e}_r)\}$ will be a basis of im T. This result is useful, and we record it for reference. The proof is much like that of Theorem 4 and is left as Exercise 26.

Theorem 5

Let $T : V \to W$ be a linear transformation, and let $\{\mathbf{e}_1, \ldots, \mathbf{e}_r, \mathbf{e}_{r+1}, \ldots, \mathbf{e}_n\}$ be a basis of V such that $\{\mathbf{e}_{r+1}, \ldots, \mathbf{e}_n\}$ is a basis of ker T. Then $\{T(\mathbf{e}_1), \ldots, T(\mathbf{e}_r)\}$ is a basis of im T, and hence $r =$ rank T.

The dimension theorem is one of the most useful results in all of linear algebra. It shows that if either dim(ker T) or dim(im T) can be found, then the other is automatically known. In many cases it is easier to compute one than the other, so the theorem is a real asset. The rest of this section is devoted to illustrations of this fact. The next example uses the dimension theorem to give a different proof of the first part of Theorem 3 §5.4.

Example 7

Let A be an $m \times n$ matrix of rank r. Show that the space null A of all solutions of the system $AX = 0$ of m homogeneous equations in n variables has dimension $n - r$.

Solution   The space in question is just ker T_A, where $T_A : \mathbb{R}^n \to \mathbb{R}^m$ is defined by $T_A(X) = AX$ for all columns X in $\mathbb{R}^n$. But dim(im T_A) = rank T_A = rank $A = r$ by Example 2, so dim(ker T_A) $= n - r$ by the dimension theorem.

Example 8

If $T : V \to W$ is a linear transformation where V is finite dimensional, then
$$\dim(\ker T) \le \dim V \quad \text{and} \quad \dim(\operatorname{im} T) \le \dim V$$
because dim $V =$ dim(ker T) + dim(im T). Of course, the first inequality also follows because ker T is a subspace of V.

Example 9

Let $D : \mathbf{P}_n \to \mathbf{P}_{n-1}$ be the differentiation map defined by $D[p(x)] = p'(x)$. Compute ker D and hence conclude that D is onto.

Solution   Because $p'(x) = 0$ means $p(x)$ is constant, we have $\dim(\ker D) = 1$. Because $\dim \mathbf{P}_n = n + 1$, the dimension theorem gives

$$\dim(\operatorname{im} D) = (n + 1) - \dim(\ker D) = n = \dim(\mathbf{P}_{n-1})$$

This implies that $\operatorname{im} D = \mathbf{P}_{n-1}$, so D is onto.

Of course it is not difficult to verify directly that each polynomial $q(x)$ in $\mathbf{P}_{n-1}$ is the derivative of some polynomial in $\mathbf{P}_n$ (simply integrate $q(x)$!), so the dimension theorem is not needed in this case. However, in some situations it is difficult to see directly that a linear transformation is onto, and the method used in Example 9 may be by far the easiest way to proceed. Here is another illustration.

Example 10

Given a in $\mathbb{R}$, the evaluation map $E_a : \mathbf{P}_n \to \mathbb{R}$ is given by $E_a[p(x)] = p(a)$. Show that E_a is linear and onto, and hence conclude that $\{(x - a), (x - a)^2, \ldots, (x - a)^n\}$ is a basis of $\ker E_a$, the subspace of all polynomials $p(x)$ for which $p(a) = 0$.

Solution E_a is linear by Example 3 §7.1; the verification that it is onto is left to the reader. Hence $\dim(\operatorname{im} E_a) = \dim(\mathbb{R}) = 1$, so $\dim(\ker E_a) = (n + 1) - 1 = n$ by the dimension theorem. Now each of the n polynomials $(x - a), (x - a)^2, \ldots, (x - a)^n$ clearly lies in $\ker E_a$, so they are a basis because they are linearly independent (they have distinct degrees).

We conclude by applying the dimension theorem to the rank of a matrix.

Example 11

If A is any $m \times n$ matrix, show that $\operatorname{rank} A = \operatorname{rank} A^T A = \operatorname{rank} AA^T$.

Solution   It suffices to show that $\operatorname{rank} A = \operatorname{rank} A^T A$ (the rest follows by replacing A with A^T). Write $B = A^T A$, and consider the associated matrix transformations

$$T_A : \mathbb{R}^n \to \mathbb{R}^m \quad \text{and} \quad T_B : \mathbb{R}^n \to \mathbb{R}^n$$

The dimension theorem and Example 2 give

$$\operatorname{rank} A = \operatorname{rank} T_A = \dim(\operatorname{im} T_A) = n - \dim(\ker T_A)$$
$$\operatorname{rank} B = \operatorname{rank} T_B = \dim(\operatorname{im} T_B) = n - \dim(\ker T_B)$$

so it suffices to show that $\ker T_A = \ker T_B$. Now $AX = 0$ implies that $BX = A^T AX = 0$, so $\ker T_A$ is contained in $\ker T_B$. On the other hand, if $BX = 0$, then $A^T AX = 0$, so

$$\|AX\|^2 = (AX)^T(AX) = X^T A^T AX = X^T 0 = 0$$

This implies that $AX = 0$, so $\ker T_B$ is contained in $\ker T_A$.

Exercises 7.2

1. For each matrix A, find a basis for the kernel and image of T_A, and find the rank and nullity of T_A.

(a) $\begin{bmatrix} 1 & 2 & -1 & 1 \\ 3 & 1 & 0 & 2 \\ 1 & -3 & 2 & 0 \end{bmatrix}$ ◆(b) $\begin{bmatrix} 2 & 1 & -1 & 3 \\ 1 & 0 & 3 & 1 \\ 1 & 1 & -4 & 2 \end{bmatrix}$

(c) $\begin{bmatrix} 1 & 2 & -1 \\ 3 & 1 & 2 \\ 4 & -1 & 5 \\ 0 & 2 & -2 \end{bmatrix}$ ◆(d) $\begin{bmatrix} 2 & 1 & 0 \\ 1 & -1 & 3 \\ 1 & 2 & -3 \\ 0 & 3 & -6 \end{bmatrix}$

2. In each case, (i) find a basis of ker T, and (ii) find a basis of im T.

   (a) $T : \mathbf{P}_2 \to \mathbb{R}^2$; $T(a + bx + cx^2) = (a, b)$

   ◆(b) $T : \mathbf{P}_2 \to \mathbb{R}^2$; $T(p(x)) = (p(0), p(1))$

   (c) $T : \mathbb{R}^3 \to \mathbb{R}^3$; $T(x, y, z) = (x + y, x + y, 0)$

   ◆(d) $T : \mathbb{R}^3 \to \mathbb{R}^4$; $T(x, y, z) = (x, x, y, y)$

   (e) $T : \mathbf{M}_{22} \to \mathbf{M}_{22}$; $T\begin{bmatrix} a & b \\ c & d \end{bmatrix} = \begin{bmatrix} a+b & b+c \\ c+d & d+a \end{bmatrix}$

   ◆(f) $T : \mathbf{M}_{22} \to \mathbb{R}$; $T\begin{bmatrix} a & b \\ c & d \end{bmatrix} = a + d$

   (g) $T : \mathbf{P}_n \to \mathbb{R}$; $T(r_0 + r_1 x + \cdots + r_n x^n) = r_n$

   ◆(h) $T : \mathbb{R}^n \to \mathbb{R}$; $T(r_1, r_2, \ldots, r_n) = r_1 + r_2 + \cdots + r_n$

   (i) $T : \mathbf{M}_{22} \to \mathbf{M}_{22}$; $T(X) = XA - AX$,
   where $A = \begin{bmatrix} 0 & 1 \\ 1 & 0 \end{bmatrix}$

   ◆(j) $T : \mathbf{M}_{22} \to \mathbf{M}_{22}$; $T(X) = XA$,
   where $A = \begin{bmatrix} 1 & 1 \\ 0 & 0 \end{bmatrix}$

3. Let $P : V \to \mathbb{R}$ and $Q : V \to \mathbb{R}$ be linear transformations, where V is a vector space. Define $T : V \to \mathbb{R}^2$ by $T(\mathbf{v}) = (P(\mathbf{v}), Q(\mathbf{v}))$.

   (a) Show that T is a linear transformation.

   ◆(b) Show that ker $T = $ ker $P \cap$ ker Q, the set of vectors in both ker P and ker Q.

4. In each case, find a basis $B = \{\mathbf{e}_1, \ldots, \mathbf{e}_r, \mathbf{e}_{r+1}, \ldots, \mathbf{e}_n\}$ of V such that $\{\mathbf{e}_{r+1}, \ldots, \mathbf{e}_n\}$ is a basis of ker T, and verify Theorem 5.

   (a) $T : \mathbb{R}^3 \to \mathbb{R}^4$; $T(x, y, z) = $
   $(x - y + 2z, x + y - z, 2x + z, 2y - 3z)$

   ◆(b) $T : \mathbb{R}^3 \to \mathbb{R}^4$; $T(x, y, z) = $
   $(x + y + z, 2x - y + 3z, z - 3y, 3x + 4z)$

5. Show that every matrix X in $\mathbf{M}_{nn}$ has the form $X = A^T - 2A$ for some matrix A in $\mathbf{M}_{nn}$. [*Hint*: The dimension theorem.]

6. In each case either prove the statement or give an example in which it is false. Throughout, let $T : V \to W$ be a linear transformation where V and W are finite dimensional.

   (a) If $V = W$, then ker $T \subseteq$ im T.

   ◆(b) If dim $V = 5$, dim $W = 3$, and dim(ker T) $= 2$, then T is onto.

   (c) If dim $V = 5$ and dim $W = 4$, then ker $T \neq \{\mathbf{0}\}$.

   ◆(d) If ker $T = V$, then $W = \{\mathbf{0}\}$.

   (e) If $W = \{\mathbf{0}\}$, then ker $T = V$.

   ◆(f) If $W = V$, and im $T \subseteq$ ker T, then $T = 0$.

   (g) If $\{\mathbf{e}_1, \mathbf{e}_2, \mathbf{e}_3\}$ is a basis of V and $T(\mathbf{e}_1) = \mathbf{0} = T(\mathbf{e}_2)$, then dim(im T) ≤ 1.

   ◆(h) If dim(ker T) $\leq$ dim W, then dim $W \geq \frac{1}{2}$dim V.

   (i) If T is one-to-one, then dim $V \leq$ dim W.

   ◆(j) If dim $V \leq$ dim W, then T is one-to-one.

   (k) If T is onto, then dim $V \geq$ dim W.

   ◆(l) If dim $V \geq$ dim W, then T is onto.

7. Show that linear independence is preserved by one-to-one transformations and that spanning sets are preserved by onto transformations. More precisely, if $T : V \to W$ is a linear transformation, show that:

   (a) If T is one-to-one and $\{\mathbf{v}_1, \ldots, \mathbf{v}_n\}$ is independent in V, then $\{T(\mathbf{v}_1), \ldots, T(\mathbf{v}_n)\}$ is independent in W.

   ◆(b) If T is onto and $V = $ span$\{\mathbf{v}_1, \ldots, \mathbf{v}_n\}$, then $W = $ span$\{T(\mathbf{v}_1), \ldots, T(\mathbf{v}_n)\}$.

8. Given $\{\mathbf{v}_1, \ldots, \mathbf{v}_n\}$ in a vector space V, define $T : \mathbb{R}^n \to V$ by $T(r_1, \ldots, r_n) = r_1\mathbf{v}_1 + \cdots + r_n\mathbf{v}_n$. Show that T is linear, and that:

   (a) T is one-to-one if and only if $\{\mathbf{v}_1, \ldots, \mathbf{v}_n\}$ is independent.

   ◆(b) T is onto if and only if $V = $ span$\{\mathbf{v}_1, \ldots, \mathbf{v}_n\}$.

9. Let $T : V \to V$ be a linear transformation where V is finite dimensional. Show that exactly one of (i) and (ii) holds: (i) $T(\mathbf{v}) = \mathbf{0}$ for some $\mathbf{v} \neq \mathbf{0}$ in V; (ii) $T(\mathbf{x}) = \mathbf{v}$ has a solution $\mathbf{x}$ in V for every $\mathbf{v}$ in V.

◆10. Let $T : \mathbf{M}_{nn} \to \mathbb{R}$ denote the trace map: $T(A) = $ tr A for all A in $\mathbf{M}_{nn}$. Show that dim(ker T) $= n^2 - 1$.

11. Show that the following are equivalent for a linear transformation $T : V \to W$.

   (a) ker $T = V$ (b) im $T = \{\mathbf{0}\}$ (c) $T = 0$

12. Let A and B be $m \times n$ and $k \times n$ matrices, respectively. Assume that $AX = 0$ implies $BX = 0$ for every n-column X. Show that rank $A \geq$ rank B.

13. Let A be an $m \times n$ matrix of rank r. Thinking of $\mathbb{R}^n$ as rows, define $V = \{X$ in $\mathbb{R}^m \mid XA = 0\}$. Show that dim $V = m - r$.

14. Consider $V = \left\{ \begin{bmatrix} a & b \\ c & d \end{bmatrix} \middle| a + c = b + d \right\}$.

(a) Consider $S : \mathbf{M}_{22} \to \mathbb{R}$ with

$$S\begin{bmatrix} a & b \\ c & d \end{bmatrix} = a + c - b - d.$$ Show that S is linear and onto and that V is a subspace of $\mathbf{M}_{22}$, and compute $\dim V$.

(b) Consider $T : V \to \mathbb{R}$ with $T\begin{bmatrix} a & b \\ c & d \end{bmatrix} = a + c.$
Show that T is linear and onto, and use this information to compute $\dim(\ker T)$.

15. Define $T : \mathbf{P}_n \to \mathbb{R}$ by $T[p(x)] =$ the sum of all the coefficients of $p(x)$.

   (a) Use the dimension theorem to show that $\dim(\ker T) = n$.

   ◆(b) Conclude that $\{x - 1, x^2 - 1, \dots, x^n - 1\}$ is a basis of $\ker T$.

16. Use the dimension theorem to prove Theorem 1 §1.3: If A is an $m \times n$ matrix with $m < n$, the system $AX = 0$ of m homogeneous equations in n variables always has a nontrivial solution.

17. Let B be an $n \times n$ matrix, and consider the subspaces $U = \{A \mid A \text{ in } \mathbf{M}_{mn}, AB = 0\}$ and $V = \{AB \mid A \text{ in } \mathbf{M}_{mn}\}$. Show that $\dim U + \dim V = mn$.

18. Let U and V denote, respectively, the spaces of even and odd polynomials in $\mathbf{P}_n$. Show that $\dim U + \dim V = n + 1$. [*Hint:* Consider $T : \mathbf{P}_n \to \mathbf{P}_n$ where $T[p(x)] = p(x) - p(-x)$.]

19. Show that every polynomial $f(x)$ in $\mathbf{P}_{n-1}$ can be written as $f(x) = p(x + 1) - p(x)$ for some polynomial $p(x)$ in $\mathbf{P}_n$. [*Hint:* Define $T : \mathbf{P}_n \to \mathbf{P}_{n-1}$ by $T[p(x)] = p(x + 1) - p(x)$.]

◆20. Let U and V denote the spaces of symmetric and skew-symmetric $n \times n$ matrices. Show that $\dim U + \dim V = n^2$.

21. Assume that B in $\mathbf{M}_{nn}$ satisfies $B^k = 0$ for some $k \geq 1$. Show that every matrix in $\mathbf{M}_{nn}$ has the form $BA - A$ for some A in $\mathbf{M}_{nn}$. [*Hint:* Show that $T : \mathbf{M}_{nn} \to \mathbf{M}_{nn}$ is linear and one-to-one where $T(A) = BA - A$ for each A.]

◆22. Fix a column $Y \neq 0$ in $\mathbb{R}^n$ and let $U = \{A \text{ in } \mathbf{M}_{nn} \mid AY = 0\}$. Show that $\dim U = n(n - 1)$.

23. If B in $\mathbf{M}_{mn}$ has rank r, let $U = \{A \text{ in } \mathbf{M}_{nm} \mid BA = 0\}$ and $W = \{BA \mid A \text{ in } \mathbf{M}_{nm}\}$. Show that $\dim U = n(n - r)$ and $\dim W = nr$. [*Hint:* Show that U consists of all matrices A whose columns are in the null space of B. Use Example 7.]

24. Let $T : V \to V$ be a linear transformation where $\dim V = n$. If $\mathbf{0}$ is the only vector in both $\ker T$ and $\operatorname{im} T$, show that every vector $\mathbf{v}$ in V can be written $\mathbf{v} = \mathbf{u} + \mathbf{w}$ for some $\mathbf{u}$ in $\ker T$ and $\mathbf{w}$ in $\operatorname{im} T$. [*Hint:* Exercise 31 §6.3.]

25. Let $T : \mathbb{R}^n \to \mathbb{R}^n$ be a linear transformation of rank 1, where $\mathbb{R}^n$ is written as rows. Show that there exist numbers $a_1, a_2, \dots, a_n$ and $b_1, b_2, \dots, b_n$ such that $T(X) = XA$ for all rows X in $\mathbb{R}^n$, where

$$A = \begin{bmatrix} a_1b_1 & a_1b_2 & \cdots & a_1b_n \\ a_2b_1 & a_2b_2 & \cdots & a_2b_n \\ \vdots & \vdots & & \vdots \\ a_nb_1 & a_nb_2 & \cdots & a_nb_n \end{bmatrix}$$

[*Hint:* $\operatorname{im} T = \mathbb{R}\mathbf{w}$ for some $\mathbf{w} = (b_1, \dots, b_n)$ in $\mathbb{R}^n$.]

26. Prove Theorem 5.

27. Let $T : V \to \mathbb{R}$ be a nonzero linear transformation, where $\dim V = n$. Show that there is a basis $\{\mathbf{e}_1, \dots, \mathbf{e}_n\}$ of V such that $T(r_1\mathbf{e}_1 + r_2\mathbf{e}_2 + \cdots + r_n\mathbf{e}_n) = r_1$.

28. Let U be a subspace of a finite dimensional vector space V.

   (a) Show that $U = \ker T$ for some linear transformation $T : V \to V$.

   ◆(b) Show that $U = \operatorname{im} S$ for some linear transformation $S : V \to V$. [*Hint:* Theorems 1 §6.4 and 3 §7.1.]

29. Let V and W be finite dimensional vector spaces.

   (a) Show that $\dim W \leq \dim V$ if and only if there exists an onto linear transformation $T : V \to W$. [*Hint:* Theorems 1 §6.4 and 3 §7.1.]

   (b) Show that $\dim W \geq \dim V$ if and only if there exists a one-to-one linear transformation $T : V \to W$. [*Hint:* Theorems 1 §6.4 and 3 §7.1.]

SECTION 7.3   Isomorphisms and Composition

Often two vector spaces can look quite different but, on closer examination, turn out to be the same vector space displayed in different symbols. The notion of isomorphism clarifies this.

A linear transformation $T: V \to W$ is called an **isomorphism** if it is both onto and one-to-one. The vector spaces V and W are called **isomorphic** if there exists an isomorphism $T: V \to W$.

Example 1

The identity transformation $1_V : V \to V$ is an isomorphism for any vector space V.

Example 2

If $T: \mathbf{M}_{mn} \to \mathbf{M}_{nm}$ is defined by $T(A) = A^T$ for all A in $\mathbf{M}_{mn}$, then T is an isomorphism.

Example 3

If U is any invertible $m \times m$ matrix, the map $T: \mathbf{M}_{mn} \to \mathbf{M}_{mn}$ given by $T(X) = UX$ is an isomorphism by Example 6 §7.2.

The word *isomorphism* comes from two Greek roots: *iso*, meaning "same," and *morphos*, meaning "form." An isomorphism $T: V \to W$ induces a pairing

$$\mathbf{v} \leftrightarrow T(\mathbf{v})$$

between vectors $\mathbf{v}$ in V and vectors $T(\mathbf{v})$ in W that preserves vector addition and scalar multiplication. Hence, *as far as their vector space properties are concerned*, the spaces V and W are identical except for notation. Because addition and scalar multiplication in either space are completely determined by the same operations in the other space, all *vector space* properties of either space are completely determined by those of the other.

One of the most important examples of isomorphic spaces was considered in Chapter 4. There the space $\mathbb{R}^3$ was identified with the space of geometric vectors by pairing each column matrix $[x\ y\ z]^T$ with the "arrow" $\vec{v}$ from the origin to the point $P(x, y, z)$. Matrix addition and scalar multiplication in $\mathbb{R}^3$ were shown in Section 4.1 to correspond to the parallelogram law of addition and the intrinsic scalar multiplication, respectively, of these "arrows". These operations make the set of all "arrows" into a vector space, so the function T from $\mathbb{R}^3$ to this space given by $T([x\ y\ z]^T) = \vec{v}$ is a linear transformation. Moreover, T is one-to-one by the first part of Theorem 1 §4.1, and T is onto because each "arrow" $\vec{v}$ has the form $\vec{v} = T([x\ y\ z]^T)$ where $P(x, y, z)$ is the tip of $\vec{v}$ when its tail is positioned at the origin. Hence T is an isomorphism, and this justifies the *identification*

$$\vec{v} = [x\ y\ z]^T$$

in Chapter 4 of the geometric arrows with the algebraic matrices. This identification is very useful. The arrows give a "picture" of the matrices and so bring geometric intuition into $\mathbb{R}^3$; the matrices are useful for doing detailed calculations and so bring analytic precision into geometry. This is one of the best examples of the power of an isomorphism to shed light on *both* spaces being considered.

The following theorem gives a very useful characterization of isomorphisms: They are the linear transformations that preserve bases.

Theorem 1

If V and W are finite dimensional spaces, the following conditions are equivalent for a linear transformation $T : V \to W$.

1. T is an isomorphism.
2. If $\{\mathbf{e}_1, \mathbf{e}_2, \dots, \mathbf{e}_n\}$ is any basis of V, then $\{T(\mathbf{e}_1), T(\mathbf{e}_2), \dots, T(\mathbf{e}_n)\}$ is a basis of W.
3. There exists a basis $\{\mathbf{e}_1, \mathbf{e}_2, \dots, \mathbf{e}_n\}$ of V such that $\{T(\mathbf{e}_1), T(\mathbf{e}_2), \dots, T(\mathbf{e}_n)\}$ is a basis of W.

PROOF

(1) $\Rightarrow$ (2). Let $\{\mathbf{e}_1, \dots, \mathbf{e}_n\}$ be a basis of V. If $t_1 T(\mathbf{e}_1) + \dots + t_n T(\mathbf{e}_n) = \mathbf{0}$ with t_i in $\mathbb{R}$, then $T(t_1\mathbf{e}_1 + \dots + t_n\mathbf{e}_n) = \mathbf{0}$, so $t_1\mathbf{e}_1 + \dots + t_n\mathbf{e}_n = \mathbf{0}$ (because ker $T = \{\mathbf{0}\}$). But then each $t_i = 0$ by the independence of the $\mathbf{e}_i$, so $\{T(\mathbf{e}_1), \dots, T(\mathbf{e}_n)\}$ is independent. To show that it spans W, choose $\mathbf{w}$ in W. Because T is onto, $\mathbf{w} = T(\mathbf{v})$ for some $\mathbf{v}$ in V, so write $\mathbf{v} = t_1\mathbf{e}_1 + \dots + t_n\mathbf{e}_n$. Then $\mathbf{w} = T(\mathbf{v}) = t_1 T(\mathbf{e}_1) + \dots + t_n T(\mathbf{e}_n)$, so $\{T(\mathbf{e}_1), \dots, T(\mathbf{e}_n)\}$ spans W.

(2) $\Rightarrow$ (3). This is because V *has* a basis.

(3) $\Rightarrow$ (1). If $T(\mathbf{v}) = \mathbf{0}$, write $\mathbf{v} = v_1\mathbf{e}_1 + \dots + v_n\mathbf{e}_n$ where each v_i is in $\mathbb{R}$. Then $\mathbf{0} = T(\mathbf{v}) = v_1 T(\mathbf{e}_1) + \dots + v_n T(\mathbf{e}_n)$, so $v_1 = \dots = v_n = 0$ by (3). Hence $\mathbf{v} = \mathbf{0}$, so ker $T = \{\mathbf{0}\}$ and T is one-to-one. To show that T is onto, let $\mathbf{w}$ be any vector in W. By (3) there exist $w_1, \dots, w_n$ in $\mathbb{R}$ such that $\mathbf{w} = w_1 T(\mathbf{e}_1) + \dots + w_n T(\mathbf{e}_n) = T(w_1\mathbf{e}_1 + \dots + w_n\mathbf{e}_n)$. Thus T is onto.

Theorem 1 dovetails nicely with Theorem 3 §7.1 as follows. Let V and W be vector spaces of dimension n, and suppose that $\{\mathbf{e}_1, \mathbf{e}_2, \dots, \mathbf{e}_n\}$ and $\{\mathbf{f}_1, \mathbf{f}_2, \dots, \mathbf{f}_n\}$ are bases of V and W, respectively. Theorem 3 §7.1 asserts that there exists a linear transformation $T : V \to W$ such that

$$T(\mathbf{e}_i) = \mathbf{f}_i \quad \text{for each } i = 1, 2, \dots, n$$

Then $\{T(\mathbf{e}_1), \dots, T(\mathbf{e}_n)\}$ is evidently a basis of W, so T is an isomorphism by Theorem 1. Furthermore, the action of T is prescribed by

$$T(r_1\mathbf{e}_1 + \dots + r_n\mathbf{e}_n) = r_1\mathbf{f}_1 + \dots + r_n\mathbf{f}_n$$

so isomorphisms between spaces of equal dimension can be easily defined as soon as bases are known. In particular, we have proved half of the following theorem.

Theorem 2

Two finite dimensional vector spaces V and W are isomorphic if and only if dim V = dim W.

PROOF

It remains to show that if $T : V \to W$ is an isomorphism, then dim V = dim W. But if $\{\mathbf{e}_1, \dots, \mathbf{e}_n\}$ is a basis of V, then $\{T(\mathbf{e}_1), \dots, T(\mathbf{e}_n)\}$ is a basis of W by Theorem 1, so dim $W = n =$ dim V.

In particular, every vector space of dimension n is isomorphic to $\mathbb{R}^n$.

Example 4

Solution

Let V denote the space of all 2×2 symmetric matrices. Find an isomorphism $T : \mathbf{P}_2 \to V$ such that $T(1) = I$.

$\{1, x, x^2\}$ is a basis of $\mathbf{P}_2$, and we want a basis of V containing I. The set
$$\left\{ \begin{bmatrix} 1 & 0 \\ 0 & 1 \end{bmatrix}, \begin{bmatrix} 0 & 1 \\ 1 & 0 \end{bmatrix}, \begin{bmatrix} 0 & 0 \\ 0 & 1 \end{bmatrix} \right\} \text{ is independent in } V, \text{ so it is a basis because}$$
dim $V = 3$ (by Example 11 §6.3). Hence define $T : \mathbf{P}_2 \to V$ by taking
$$T(1) = \begin{bmatrix} 1 & 0 \\ 0 & 1 \end{bmatrix}, \ T(x) = \begin{bmatrix} 0 & 1 \\ 1 & 0 \end{bmatrix}, \text{ and } T(x^2) = \begin{bmatrix} 0 & 0 \\ 0 & 1 \end{bmatrix} \text{ and extending linearly as in}$$
Theorem 3 §7.1. Then T is an isomorphism by Theorem 1, and its action is
given by $T(a + bx + cx^2) = aT(1) + bT(x) + cT(x^2) = \begin{bmatrix} a & b \\ b & a+c \end{bmatrix}.$

The dimension theorem (Theorem 4 §7.2) gives the following useful fact about isomorphisms.

Theorem 3

If dim $V =$ dim $W = n$, a linear transformation $T : V \to W$ is an isomorphism if it is either one-to-one or onto.

PROOF

The dimension theorem asserts that dim(ker T) + dim(im T) = n, so dim(ker T) = 0 if and only if dim(im T) = n. Thus T is one-to-one if and only if T is onto, and the result follows.

Composition

Suppose that $T : V \to W$ and $S : W \to U$ are linear transformations. They link together as in the diagram, so it is possible to define a new function $V \to U$ by first applying T and then S. Given linear transformations $V \xrightarrow{T} W \xrightarrow{S} U$, the **composite** $ST : V \to U$ of T and S is defined by

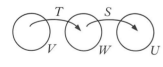

$$ST(\mathbf{v}) = S[T(\mathbf{v})] \quad \text{for all } \mathbf{v} \text{ in } V.^{2}$$

The operation of forming the new function ST is called **composition**. The action of ST can be described compactly as follows: ST means first T then S.

Not all pairs of linear transformations can be composed. For example, if $T : V \to W$ and $S : W \to U$ are linear transformations then $ST : V \to U$ is defined, but TS *cannot* be formed unless $U = V$ because $S : W \to U$ and $T : V \to W$ do not "link" in that order.

Moreover, even if ST and TS can both be formed, they may not be equal. In fact, if $S : \mathbb{R}^m \to \mathbb{R}^n$ and $T : \mathbb{R}^n \to \mathbb{R}^m$ are induced by matrices A and B respectively, then ST and TS can both be formed (they are induced by AB and BA respectively), but the matrix products AB and BA may not be equal (they may not be the same *size*). Here is another example.

---

2 In Section 2.5 we denoted the composite as $S \circ T$. However, it is more convenient to use the simpler notation ST.

Example 5

Define: $S : \mathbf{M}_{22} \to \mathbf{M}_{22}$ and $T : \mathbf{M}_{22} \to \mathbf{M}_{22}$ by $S\begin{bmatrix} a & b \\ c & d \end{bmatrix} = \begin{bmatrix} c & d \\ a & b \end{bmatrix}$ and $T(A) = A^T$.
Describe the action of ST and TS, and show that $ST \neq TS$.

Solution $ST\begin{bmatrix} a & b \\ c & d \end{bmatrix} = S\begin{bmatrix} a & c \\ b & d \end{bmatrix} = \begin{bmatrix} b & d \\ a & c \end{bmatrix}$, whereas $TS\begin{bmatrix} a & b \\ c & d \end{bmatrix} = T\begin{bmatrix} c & d \\ a & b \end{bmatrix} = \begin{bmatrix} c & a \\ d & b \end{bmatrix}$. It is

clear that $TS\begin{bmatrix} a & b \\ c & d \end{bmatrix}$ need not equal $ST\begin{bmatrix} a & b \\ c & d \end{bmatrix}$, so $TS \neq ST$.

The next theorem collects some basic properties[3] of the composition operation.

Theorem 4

Let $V \xrightarrow{T} W \xrightarrow{S} U \xrightarrow{R} Z$ be linear transformations.
1. The composite ST is again a linear transformation.
2. $T1_V = T$ and $1_W T = T$.
3. $(RS)T = R(ST)$.

PROOF

To prove (3), observe that, for all $\mathbf{v}$ in V:
$$\{(RS)T\}(\mathbf{v}) = (RS)[T(\mathbf{v})] = R\{S[T(\mathbf{v})]\} = R\{(ST)(\mathbf{v})\} = \{R(ST)\}(\mathbf{v})$$
The proofs of (1) and (2) are left as Exercise 25.

Up to this point, composition seems to have no connection with isomorphisms. In fact, the two notions are closely related.

Theorem 5

Let V and W be finite dimensional vector spaces. The following conditions are equivalent for a linear transformation $T : V \to W$.
1. T is an isomorphism.
2. There exists a linear transformation $S : W \to V$ such that $ST = 1_V$ and $TS = 1_W$.
Moreover, in this case S is also an isomorphism and is uniquely determined by T: If $\mathbf{w}$ in W is written as $\mathbf{w} = T(\mathbf{v})$, then $S(\mathbf{w}) = \mathbf{v}$.

PROOF

(1) $\Rightarrow$ (2). If $B = \{\mathbf{e}_1, \dots, \mathbf{e}_n\}$ is a basis of V, then $D = \{T(\mathbf{e}_1), \dots, T(\mathbf{e}_n)\}$ is a basis of W by Theorem 1. Hence (using Theorem 3 §7.1), define $S : W \to V$ by
$$S[T(\mathbf{e}_i)] = \mathbf{e}_i \quad \text{for each } i \tag{$*$}$$
Since $\mathbf{e}_i = 1_V(\mathbf{e}_i)$, this gives $ST = 1_V$ by Theorem 2 §7.1. But applying T gives $T[S[T(\mathbf{e}_i)]] = T(\mathbf{e}_i)$ for each i, so $TS = 1_W$ (again by Theorem 2 §7.1, using the basis D of W).

---

3 Theorem 4 can be expressed by saying that vector spaces and linear transformations are an example of a category. In general a **category** consists of certain **objects** and, for any two objects X and Y, a set mor(X, Y). The elements α of mor(X, Y) are called **morphisms** from X to Y and are written $\alpha : X \to Y$. It is assumed that identity morphisms and composition are defined in such a way that Theorem 4 holds. Hence, in the category of vector spaces the objects are the vector spaces themselves and the morphisms are the linear transformations. Another example is the category of metric spaces, in which the objects are sets equipped with a distance function (called a metric), and the morphisms are continuous functions (with respect to the metric). The category of sets and functions is a very basic example. The study of categories is becoming increasingly important throughout mathematics.

(2) $\Rightarrow$ (1). If $T(\mathbf{v}) = T(\mathbf{v}_1)$, then $S[T(\mathbf{v})] = S[T(\mathbf{v}_1)]$. Because $ST = 1_V$, this reads $\mathbf{v} = \mathbf{v}_1$; that is, T is one-to-one. Given $\mathbf{w}$ in W, the fact that $TS = 1_W$ means that $\mathbf{w} = T[S(\mathbf{w})]$, so T is onto.

Finally, S is uniquely determined by the condition $ST = 1_V$ because this condition implies (*). S is an isomorphism because it carries the basis D to B. Finally, given $\mathbf{w}$ in W, write $\mathbf{w} = r_1 T(\mathbf{e}_1) + \cdots + r_n T(\mathbf{e}_n) = T(\mathbf{v})$, where $\mathbf{v} = r_1 \mathbf{e}_1 + \cdots + r_n \mathbf{e}_n$. Then $S(\mathbf{w}) = \mathbf{v}$ by (*).

Given an isomorphism $T : V \to W$, the unique isomorphism $S : W \to V$ satisfying condition (2) of Theorem 5 is called the **inverse** of T and is denoted by T^{-1}. Hence $T : V \to W$ and $T^{-1} : W \to V$ are related by the **fundamental identities**:

$$T^{-1}[T(\mathbf{v})] = \mathbf{v} \text{ for all } \mathbf{v} \text{ in } V \quad \text{and} \quad T[T^{-1}(\mathbf{w})] = \mathbf{w} \text{ for all } \mathbf{w} \text{ in } W$$

In other words, each of T and T^{-1} reverses the action of the other. In particular, equation (*) in the proof of Theorem 5 shows how to define T^{-1} using the image of a basis under the isomorphism T. Here is an example.

Example 6

Define $T : P_1 \to P_1$ by $T(a + bx) = (a - b) + ax$. Show that T has an inverse, and find the action of T^{-1}.

Solution The transformation T is linear (verify). Because $T(1) = 1 + x$ and $T(x) = -1$, T carries the basis $B = \{1, x\}$ to the basis $D = \{1 + x, -1\}$. Hence T is an isomorphism, and T^{-1} is defined by $T^{-1}(1 + x) = 1$ and $T^{-1}(-1) = x$. Because $a + bx = b(1 + x) + (b - a)(-1)$, we obtain

$$T^{-1}(a + bx) = bT^{-1}(1 + x) + (b - a)T^{-1}(-1) = b + (b - a)x.$$

Condition (2) in Theorem 5 characterizes the inverse of a linear transformation $T : V \to W$ as the (unique) transformation $S : W \to V$ that satisfies $ST = 1_V$ and $TS = 1_W$. This often determines the inverse.

Example 7

Define $T : \mathbb{R}^3 \to \mathbb{R}^3$ by $T(x, y, z) = (z, x, y)$. Show that $T^3 = 1_{\mathbb{R}^3}$, and hence find T^{-1}.

Solution $T^2(x, y, z) = T[T(x, y, z)] = T(z, x, y) = (y, z, x)$. Hence

$$T^3(x, y, z) = T[T^2(x, y, z)] = T(y, z, x) = (x, y, z)$$

This shows that $T^3 = 1_{\mathbb{R}^3}$, so $T(T^2) = 1_{\mathbb{R}^3} = (T^2)T$. Thus $T^{-1} = T^2$ by (2) of Theorem 5.

Example 8

Define $T : \mathbf{P}_n \to \mathbb{R}^{n+1}$ by $T(p) = (p(0), p(1), \ldots, p(n))$ for all p in $\mathbf{P}_n$. Show that T^{-1} exists.

Solution The verification that T is linear is left to the reader. If $T(p) = 0$, then $p(k) = 0$ for $k = 0, 1, \ldots, n$, so p has $n + 1$ distinct roots. Because p has degree at most n, this implies that $p = 0$ is the zero polynomial (Theorem 4 §6.5) and hence that T is one-to-one. But dim $\mathbf{P}_n = n + 1 = \dim \mathbb{R}^{n+1}$, so this means that T is also onto and hence is an isomorphism. Thus T^{-1} exists by Theorem 5. Note that we have not given an explicit description of the action of T^{-1}, we have merely shown that such a description exists. To give it requires some ingenuity; one method involves the Lagrange interpolation expansion (Theorem 3 §6.5).

Exercises 7.3

1. Verify that each of the following is an isomorphism (Theorem 3 is useful).
   (a) $T : \mathbb{R}^3 \to \mathbb{R}^3$; $T(x, y, z) = (x + y, y + z, z + x)$
   ◆(b) $T : \mathbb{R}^3 \to \mathbb{R}^3$; $T(x, y, z) = (x, x + y, x + y + z)$
   (c) $T : \mathbb{C} \to \mathbb{C}$; $T(z) = \bar{z}$
   ◆(d) $T : \mathbf{M}_{mn} \to \mathbf{M}_{mn}$; $T(X) = UXV$, U and V invertible
   (e) $T : \mathbf{P}_1 \to \mathbb{R}^2$; $T[p(x)] = [p(0), p(1)]$
   ◆(f) $T : V \to V$; $T(\mathbf{v}) = k\mathbf{v}$, $k \neq 0$ a fixed number, V any vector space
   (g) $T : \mathbf{M}_{22} \to \mathbb{R}^4$; $T\begin{bmatrix} a & b \\ c & d \end{bmatrix} = (a + b, d, c, a - b)$
   ◆(h) $T : \mathbf{M}_{mn} \to \mathbf{M}_{nm}$; $T(A) = A^T$

2. Show that $\{a + bx + cx^2, a_1 + b_1x + c_1x^2, a_2 + b_2x + c_2x^2\}$ is a basis of $\mathbf{P}_2$ if and only if $\{(a, b, c), (a_1, b_1, c_1), (a_2, b_2, c_2)\}$ is a basis of $\mathbb{R}^3$.

3. If V is any vector space, let V^n denote the space of all n-tuples $(\mathbf{v}_1, \mathbf{v}_2, \dots, \mathbf{v}_n)$, where each $\mathbf{v}_i$ lies in V. (This is a vector space with component-wise operations; see Exercise 17 §6.1.) If $C_j(A)$ denotes the jth column of the $m \times n$ matrix A, show that $T : \mathbf{M}_{mn} \to (\mathbb{R}^m)^n$ is an isomorphism if $T(A) = [C_1(A)\ C_2(A) \cdots C_n(A)]$. (Here $\mathbb{R}^m$ consists of columns.)

4. In each case, compute the action of ST and TS, and show that $ST \neq TS$.
   (a) $S : \mathbb{R}^2 \to \mathbb{R}^2$ with $S(x, y) = (y, x)$; $T : \mathbb{R}^2 \to \mathbb{R}^2$ with $T(x, y) = (x, 0)$
   ◆(b) $S : \mathbb{R}^3 \to \mathbb{R}^3$ with $S(x, y, z) = (x, 0, z)$; $T : \mathbb{R}^3 \to \mathbb{R}^3$ with $T(x, y, z) = (x + y, 0, y + z)$
   (c) $S : \mathbf{P}_2 \to \mathbf{P}_2$ with $S(p) = p(0) + p(1)x + p(2)x^2$; $T : \mathbf{P}_2 \to \mathbf{P}_2$ with $T(a + bx + cx^2) = b + cx + ax^2$
   ◆(d) $S : \mathbf{M}_{22} \to \mathbf{M}_{22}$ with $S\begin{bmatrix} a & b \\ c & d \end{bmatrix} = \begin{bmatrix} a & 0 \\ 0 & d \end{bmatrix}$;
   $T : \mathbf{M}_{22} \to \mathbf{M}_{22}$ with $T\begin{bmatrix} a & b \\ c & d \end{bmatrix} = \begin{bmatrix} c & a \\ d & b \end{bmatrix}$

5. In each case, show that the linear transformation T satisfies $T^2 = T$.
   (a) $T : \mathbb{R}^4 \to \mathbb{R}^4$; $T(x, y, z, w) = (x, 0, z, 0)$
   ◆(b) $T : \mathbb{R}^2 \to \mathbb{R}^2$; $T(x, y) = (x + y, 0)$
   (c) $T : \mathbf{P}_2 \to \mathbf{P}_2$; $T(a + bx + cx^2) = (a + b - c) + cx + cx^2$
   ◆(d) $T : \mathbf{M}_{22} \to \mathbf{M}_{22}$; $T\begin{bmatrix} a & b \\ c & d \end{bmatrix} = \frac{1}{2}\begin{bmatrix} a+c & b+d \\ a+c & b+d \end{bmatrix}$

6. Determine whether each of the following transformations T has an inverse and, if so, determine the action of T^{-1}.
   (a) $T : \mathbb{R}^3 \to \mathbb{R}^3$; $T(x, y, z) = (x + y, y + z, z + x)$
   ◆(b) $T : \mathbb{R}^4 \to \mathbb{R}^4$; $T(x, y, z, t) = (x + y, y + z, z + t, t + x)$
   (c) $T : \mathbf{M}_{22} \to \mathbf{M}_{22}$; $T\begin{bmatrix} a & b \\ c & d \end{bmatrix} = \begin{bmatrix} a-c & b-d \\ 2a-c & 2b-d \end{bmatrix}$
   ◆(d) $T : \mathbf{M}_{22} \to \mathbf{M}_{22}$; $T\begin{bmatrix} a & b \\ c & d \end{bmatrix} = \begin{bmatrix} a+2c & b+2d \\ 3c-a & 3d-b \end{bmatrix}$
   (e) $T : \mathbf{P}_2 \to \mathbb{R}^3$; $T(a + bx + cx^2) = (a - c, 2b, a + c)$
   ◆(f) $T : \mathbf{P}_2 \to \mathbb{R}^3$; $T(p) = [p(0), p(1), p(-1)]$

7. In each case, show that T is self-inverse: $T^{-1} = T$.
   (a) $T : \mathbb{R}^4 \to \mathbb{R}^4$; $T(x, y, z, w) = (x, -y, -z, w)$
   ◆(b) $T : \mathbb{R}^2 \to \mathbb{R}^2$; $T(x, y) = (ky - x, y)$, k any fixed number
   (c) $T : \mathbf{P}_n \to \mathbf{P}_n$; $T(p(x)) = p(3 - x)$
   ◆(d) $T : \mathbf{M}_{22} \to \mathbf{M}_{22}$; $T(X) = AX$ where $A = \frac{1}{4}\begin{bmatrix} 5 & -3 \\ 3 & -5 \end{bmatrix}$

8. In each case, show that $T^6 = 1_{\mathbb{R}^4}$ and so determine T^{-1}.
   (a) $T : \mathbb{R}^4 \to \mathbb{R}^4$; $T(x, y, z, w) = (-x, z, w, y)$
   ◆(b) $T : \mathbb{R}^4 \to \mathbb{R}^4$; $T(x, y, z, w) = (-y, x - y, z, -w)$

9. In each case, show that T is an isomorphism by defining T^{-1} explicitly.

(a) $T : \mathbf{P}_n \to \mathbf{P}_n$ is given by $T[p(x)] = p(x + 1)$.

♦(b) $T : \mathbf{M}_{nn} \to \mathbf{M}_{nn}$ is given by $T(A) = UA$ where U is invertible in $\mathbf{M}_{nn}$.

10. Given linear transformations $V \xrightarrow{T} W \xrightarrow{S} U$:

   (a) If S and T are both one-to-one, show that ST is one-to-one.

   ♦(b) If S and T are both onto, show that ST is onto.

11. Let $T : V \to W$ be a linear transformation.

   (a) If T is one-to-one and $TR = TR_1$ for transformations R and $R_1 : U \to V$, show that $R = R_1$.

   (b) If T is onto and $ST = S_1 T$ for transformations S and $S_1 : W \to U$, show that $S = S_1$.

12. Consider the linear transformations $V \xrightarrow{T} W \xrightarrow{R} U$.

   (a) Show that $\ker T \subseteq \ker RT$.

   (b) Show that $\operatorname{im} RT \subseteq \operatorname{im} R$.

13. Let $V \xrightarrow{T} U \xrightarrow{S} W$ be linear transformations.

   (a) If ST is one-to-one, show that T is one-to-one and that $\dim V \le \dim U$.

   ♦(b) If ST is onto, show that S is onto and that $\dim W \le \dim U$.

14. Let $T : V \to V$ be a linear transformation. Show that $T^2 = 1_V$ if and only if T is invertible and $T = T^{-1}$.

15. Let N be a nilpotent $n \times n$ matrix (that is, $N^k = 0$ for some k). Show that $T : \mathbf{M}_{nm} \to \mathbf{M}_{nm}$ is an isomorphism if $T(X) = X - NX$. [*Hint:* If X is in $\ker T$, show that $X = NX = N^2 X = \cdots$. Then use Theorem 3.]

♦16. Let $T : V \to W$ be a linear transformation, and let $\{\mathbf{e}_1, \ldots, \mathbf{e}_r, \mathbf{e}_{r+1}, \ldots, \mathbf{e}_n\}$ be any basis of V such that $\{\mathbf{e}_{r+1}, \ldots, \mathbf{e}_n\}$ is a basis of $\ker T$. Show that $\operatorname{im} T \cong \operatorname{span}\{\mathbf{e}_1, \ldots, \mathbf{e}_r\}$. [*Hint:* See Theorem 5 §7.2.]

17. Is every isomorphism $T : \mathbf{M}_{22} \to \mathbf{M}_{22}$ given (as in Example 3) by an invertible matrix U such that $T(X) = UX$ for all X in $\mathbf{M}_{22}$? Prove your answer.

18. Let $\mathbf{D}_n$ denote the space of all functions f from $\{1, 2, \ldots, n\}$ to $\mathbb{R}$ (see Exercise 33 §6.3). If $T : \mathbf{D}_n \to \mathbb{R}^n$ is defined by $T(f) = (f(1), f(2), \ldots, f(n))$, show that T is an isomorphism.

19. (a) Let V be the vector space of Exercise 3 §6.1. Find an isomorphism $T : V \to \mathbb{R}^1$.

   ♦(b) Let V be the vector space of Exercise 4 §6.1. Find an isomorphism $T : V \to \mathbb{R}^2$.

20. Let $V \xrightarrow{T} W \xrightarrow{S} V$ be linear transformations such that $ST = 1_V$. If $\dim V = \dim W = n$, show that $S = T^{-1}$ and $T = S^{-1}$. [*Hint:* Exercise 13 and Theorems 3, 4, and 5.]

21. Let $V \xrightarrow{T} W \xrightarrow{S} V$ be functions such that $TS = 1_W$ and $ST = 1_V$. If T is linear, show that S is also linear.

22. Let A and B be matrices of size $p \times m$ and $n \times q$. Assume that $mn = pq$. Define $R : \mathbf{M}_{mn} \to \mathbf{M}_{pq}$ by $R(X) = AXB$.

   (a) Show that $\mathbf{M}_{mn} \cong \mathbf{M}_{pq}$ by comparing dimensions.

   (b) Show that R is a linear transformation.

   (c) Show that if R is an isomorphism, then $m = p$ and $n = q$. [*Hint:* Show that $T : \mathbf{M}_{mn} \to \mathbf{M}_{pn}$ given by $T(X) = AX$ and $S : \mathbf{M}_{mn} \to \mathbf{M}_{mq}$ given by $S(X) = XB$ are both one-to-one, and use the dimension theorem.]

23. Let $T : V \to V$ be a linear transformation such that $T^2 = 0$ is the zero transformation.

   (a) If $V \ne \{\mathbf{0}\}$, show that T cannot be invertible.

   (b) If $R : V \to V$ is defined by $R(\mathbf{v}) = \mathbf{v} + T(\mathbf{v})$ for all $\mathbf{v}$ in V, show that R is linear and invertible.

24. Let V consist of all sequences $[x_0, x_1, x_2, \ldots)$ of numbers, and define vector operations
$$[x_0, x_1, \ldots) + [y_0, y_1, \ldots) = [x_0 + y_0, x_1 + y_1, \ldots)$$
$$r[x_0, x_1, \ldots) = [rx_0, rx_1, \ldots)$$

   (a) Show that V is a vector space of infinite dimension.

   (b) Define $T : V \to V$ and $S : V \to V$ by $T[x_0, x_1, \ldots) = [x_1, x_2, \ldots)$ and $S[x_0, x_1, \ldots) = [0, x_0, x_1, \ldots)$. Show that $TS = 1_V$, so TS is one-to-one and onto, but that T is not one-to-one and S is not onto.

25. Prove (1) and (2) of Theorem 4.

26. Define $T : \mathbf{P}_n \to \mathbf{P}_n$ by $T(p) = p(x) + xp'(x)$ for all p in $\mathbf{P}_n$.

   (a) Show that T is linear.

   ♦(b) Show that $\ker T = \{0\}$ and conclude that T is an isomorphism. [*Hint:* Write $p(x) = a_0 + a_1 x + \cdots + a_n x^n$ and compare coefficients if $p(x) = -xp'(x)$.]

(c) Conclude that each $q(x)$ in $\mathbf{P}_n$ has the form $q(x) = p(x) + xp'(x)$ for some unique polynomial $p(x)$.

(d) Does this remain valid if T is defined by $T[p(x)] = p(x) - xp'(x)$? Explain.

27. Let $T : V \to W$ be a linear transformation, where V and W are finite dimensional.

(a) Show that T is one-to-one if and only if there exists a linear transformation $S : W \to V$ with $ST = 1_V$. [*Hint:* If $\{e_1, \dots, e_n\}$ is a basis of V and T is one-to-one, show that W has a basis $\{T(e_1), \dots, T(e_n), f_{n+1}, \dots, f_{n+k}\}$ and use Theorems 2 and 3, §7.1.]

♦(b) Show that T is onto if and only if there exists a linear transformation $S : W \to V$ with $TS = 1_W$. [*Hint:* Let $\{e_1, \dots, e_r, \dots, e_n\}$ be a basis of V such that $\{e_{r+1}, \dots, e_n\}$ is a basis of $\ker T$. Use Theorem 5 §7.2 and Theorems 2 and 3, §7.1.]

28. Let S and T be linear transformations $V \to W$, where $\dim V = n$ and $\dim W = m$.

(a) Show that $\ker S = \ker T$ if and only if $T = RS$ for some isomorphism $R : W \to W$. [*Hint:* Let $\{e_1, \dots, e_r, \dots, e_n\}$ be a basis of V such that $\{e_{r+1}, \dots, e_n\}$ is a basis of $\ker S = \ker T$. Use Theorem 5 §7.2 to extend

$\{S(e_1), \dots, S(e_r)\}$ and $\{T(e_1), \dots, T(e_r)\}$ to bases of W.]

(b) Show that $\operatorname{im} S = \operatorname{im} T$ if and only if $T = SR$ for some isomorphism $R : V \to V$. [*Hint:* Show that $\dim(\ker S) = \dim(\ker T)$ and choose bases $\{e_1, \dots, e_r, \dots, e_n\}$ and $\{f_1, \dots, f_r, \dots, f_n\}$ of V where $\{e_{r+1}, \dots, e_n\}$ and $\{f_{r+1}, \dots, f_n\}$ are bases of $\ker S$ and $\ker T$, respectively. If $1 \le i \le r$, show that $S(e_i) = T(g_i)$ for some g_i in V, and prove that $\{g_1, \dots, g_r, f_{r+1}, \dots, f_n\}$ is a basis of V.]

♦29. If $T : V \to V$ is a linear transformation where $\dim V = n$, show that $TST = T$ for some isomorphism $S : V \to V$. [*Hint:* Let $\{e_1, \dots, e_r, e_{r+1}, \dots, e_n\}$ be as in Theorem 5 §7.2. Extend $\{T(e_1), \dots, T(e_r)\}$ to a basis of V, and use Theorem 1 and Theorems 2 and 3, §7.1.]

30. Let A and B denote $m \times n$ matrices. In each case show that (1) and (2) are equivalent.

(a) (1) A and B have the same null space.
    (2) $B = PA$ for some invertible $m \times m$ matrix P.

(b) (1) A and B have the same range.
    (2) $B = AQ$ for some invertible $n \times n$ matrix Q.

[*Hint:* Use Exercise 28.]

SECTION 7.4   More on Linear Recurrences[4]

In Section 3.4 we used diagonalization to study linear recurrences, and gave several examples. We now apply the theory of vector spaces and linear transformations to study the problem in more generality.

Consider the linear recurrence

$$x_{n+2} = 6x_n - x_{n+1} \quad \text{for } n \ge 0$$

If the initial values x_0 and x_1 are prescribed, this gives a sequence of numbers. For example, if $x_0 = 1$ and $x_1 = 1$ the sequence continues

$$x_2 = 5, \; x_3 = 1, \; x_4 = 29, \; x_5 = -23, \; x_6 = 197, \dots$$

as the reader can verify. Clearly, the entire sequence is uniquely determined by the recurrence and the two initial values. In this section we define a vector space structure on the set of *all* sequences, and study the subspace of those sequences that satisfy a particular recurrence.

Sequences will be considered entities in their own right, so it is useful to have a special notation for them. Let

$$[x_n) \quad \text{denote the sequence } x_0, \, x_1, \, x_2, \dots, x_n, \dots$$

---

4 This section requires only Sections 7.1–7.3.

Example 1

$[n)$	is the sequence $0, 1, 2, 3, \ldots$
$[n + 1)$	is the sequence $1, 2, 3, 4, \ldots$
$[2^n)$	is the sequence $1, 2, 2^2, 2^3, \ldots$
$[(-1)^n)$	is the sequence $1, -1, 1, -1, \ldots$
$[5)$	is the sequence $5, 5, 5, 5, \ldots$

Sequences of the form $[c)$ for a fixed number c will be referred to as **constant sequences**, and those of the form $[\lambda^n)$, λ some number, are **power sequences**.

Two sequences are regarded as **equal** when they are identical:

$$[x_n) = [y_n) \quad \text{means} \quad x_n = y_n \quad \text{for all } n = 0, 1, 2, \ldots$$

Addition and scalar multiplication of sequences are defined by

$$[x_n) + [y_n) = [x_n + y_n)$$
$$r[x_n) = [rx_n)$$

These operations are analogous to the addition and scalar multiplication in $\mathbb{R}^n$, and it is easy to check that the vector-space axioms are satisfied. The zero vector is the constant sequence $[0)$, and the negative of a sequence $[x_n)$ is given by $-[x_n) = [-x_n)$.

Now suppose k real numbers $r_0, r_1, \ldots, r_{k-1}$ are given, and consider the **linear recurrence relation** determined by these numbers.

$$x_{n+k} = r_0 x_n + r_1 x_{n+1} + \cdots + r_{k-1} x_{n+k-1} \tag{$*$}$$

When $r_0 \neq 0$, we say this recurrence has **length** k. (We shall usually assume that $r_0 \neq 0$; otherwise, we are essentially dealing with a recurrence of shorter length than k.) For example, the relation $x_{n+2} = 2x_n + x_{n+1}$ is of length 2.

A sequence $[x_n)$ is said to **satisfy** the relation $(*)$ if $(*)$ holds for all $n \geq 0$. Let V denote the set of all sequences that satisfy the relation. In symbols,

$$V = \{[x_n) \mid x_{n+k} = r_0 x_n + r_1 x_{n+1} + \cdots + r_{k-1} x_{n+k-1} \text{ holds for all } n \geq 0\}$$

It is easy to see that the constant sequence $[0)$ lies in V and that V is closed under addition and scalar multiplication of sequences. Hence V is vector space (being a subspace of the space of all sequences). The following important observation about V is needed (it was used implicitly earlier): If the first k terms of two sequences agree, then the sequences are identical. More formally,

Lemma

Let $[x_n)$ and $[y_n)$ denote two sequences in V. Then

$$[x_n) = [y_n) \quad \text{if and only if} \quad x_0 = y_0, \, x_1 = y_1, \ldots, x_{k-1} = y_{k-1}$$

PROOF

If $[x_n) = [y_n)$ then $x_n = y_n$ for *all* $n = 0, 1, 2, \ldots$. Conversely, if $x_i = y_i$ for all $i = 0, 1, \ldots, k - 1$, use the recurrence $(*)$ for $n = 0$.

$$x_k = r_0 x_0 + r_1 x_1 + \cdots + r_{k-1} x_{k-1} = r_0 y_0 + r_1 y_1 + \cdots + r_{k-1} y_{k-1} = y_k$$

Next the recurrence for $n = 1$ establishes $x_{k+1} = y_{k+1}$. The process continues to show that $x_{n+k} = y_{n+k}$ holds for *all* $n \geq 0$ by induction on n. Hence $[x_n) = [y_n)$.

This shows that a sequence in V is completely determined by its first k terms. In particular, given a k-tuple

$$\mathbf{v} = (v_0, v_1, \ldots, v_{k-1})$$

in $\mathbb{R}^k$, define $T(\mathbf{v})$ to be the sequence in V whose first k terms are $v_0, v_1, \ldots, v_{k-1}$. The rest of the sequence $T(\mathbf{v})$ is determined by the recurrence, so $T : \mathbb{R}^k \to V$ is a function. In fact, it is an isomorphism.

Theorem 1

Given real numbers $r_0, r_1, \ldots, r_{k-1}$, let

$$V = \{[x_n] \mid x_{n+k} = r_0 x_n + r_1 x_{n+1} + \cdots + r_{k-1}x_{n+k-1}, \text{ for all } n \geq 0\}$$

denote the vector space of all sequences satisfying the linear recurrence relation determined by $r_0, r_1, \ldots, r_{k-1}$. Then the function

$$T : \mathbb{R}^k \to V$$

defined above is an isomorphism. In particular:
1. $\dim V = k$.
2. If $\{\mathbf{v}_1, \ldots, \mathbf{v}_k\}$ is any basis of $\mathbb{R}^k$, then $\{T(\mathbf{v}_1), \ldots, T(\mathbf{v}_k)\}$ is a basis of V.

PROOF

(1) and (2) will follow from Theorems 1 and 2, §7.3 as soon as we show that T is an isomorphism. Given $\mathbf{v}$ and $\mathbf{w}$ in $\mathbb{R}^k$, write $\mathbf{v} = (v_0, v_1, \ldots, v_{k-1})$ and $\mathbf{w} = (w_0, w_1, \ldots, w_{k-1})$. The first k terms of $T(\mathbf{v})$ and $T(\mathbf{w})$ are $v_0, v_1, \ldots, v_{k-1}$ and $w_0, w_1, \ldots, w_{k-1}$, respectively, so the first k terms of $T(\mathbf{v}) + T(\mathbf{w})$ are $v_0 + w_0, v_1 + w_1, \ldots, v_{k-1} + w_{k-1}$. Because these terms agree with the first k terms of $T(\mathbf{v} + \mathbf{w})$, the lemma implies that $T(\mathbf{v} + \mathbf{w}) = T(\mathbf{v}) + T(\mathbf{w})$. The proof that $T(r\mathbf{v}) + rT(\mathbf{v})$ is similar, so T is linear.

Now let $[x_n]$ be any sequence in V, and let $\mathbf{v} = (x_0, x_1, \ldots, x_{k-1})$. Then the first k terms of $[x_n]$ and $T(\mathbf{v})$ agree, so $T(\mathbf{v}) = [x_n]$. Hence T is onto. Finally, if $T(\mathbf{v}) = [0]$ is the zero sequence, then the first k terms of $T(\mathbf{v})$ are all zero (*all* terms of $T(\mathbf{v})$ are zero!) so $\mathbf{v} = \mathbf{0}$. This means that $\ker T = \{\mathbf{0}\}$, so T is one-to-one.

Example 2

Show that the sequences $[1]$, $[n]$, and $[(-1)^n)$ are a basis of the space V of all solutions of the recurrence

$$x_{n+3} = -x_n + x_{n+1} + x_{n+2}.$$

Then find the solution satisfying $x_0 = 1, x_1 = 2, x_2 = 5$.

Solution

The verifications that these sequences satisfy the recurrence (and hence lie in V) are left to the reader. They are a basis because $[1] = T(1, 1, 1)$, $[n] = T(0, 1, 2)$, and $[(-1)^n)] = T(1, -1, 1)$; and $\{(1, 1, 1), (0, 1, 2), (1, -1, 1)\}$ is a basis of $\mathbb{R}^3$. Hence the sequence $[x_n]$ in V satisfying $x_0 = 1, x_1 = 2, x_2 = 5$ is a linear combination of this basis:

$$[x_n] = t_1[1] + t_2[n] + t_3[(-1)^n)]$$

The nth term is $x_n = t_1 + nt_2 + (-1)^n t_3$, so taking $n = 0, 1, 2$ gives

$$
\begin{aligned}
1 &= x_0 = t_1 + 0 + t_3 \\
2 &= x_1 = t_1 + t_2 - t_3 \\
5 &= x_2 = t_1 + 2t_2 + t_3
\end{aligned}
$$

This has the solution $t_1 = t_3 = \frac{1}{2}$, $t_2 = 2$, so $x_n = \frac{1}{2} + 2n + \frac{1}{2}(-1)^n$.

This technique clearly works for any linear recurrence of length k: Simply take your favourite basis $\{\mathbf{v}_1, \ldots, \mathbf{v}_k\}$ of $\mathbb{R}^k$—perhaps the standard basis—and compute $T(\mathbf{v}_1), \ldots, T(\mathbf{v}_k)$. This is a basis of V all right, but the nth term of $T(\mathbf{v}_i)$ is not usually given as an explicit function of n. (The basis in Example 2 was carefully chosen so that the nth terms of the three sequences were 1, n, and $(-1)^n$, respectively, each a simple function of n.)

However, it turns out that an explicit basis of V can be given in the general situation. Given the recurrence

$$x_{n+k} = r_0 x_n + r_1 x_{n+1} + \cdots + r_{k-1} x_{n+k-1} \tag{$*$}$$

the idea is to look for numbers λ such that the power sequence $[\lambda^n)$ satisfies $(*)$. This happens if and only if

$$\lambda^{n+k} = r_0 \lambda^n + r_1 \lambda^{n+1} + \cdots + r_{k-1} \lambda^{n+k-1}$$

holds for all $n \geq 0$. This is true just when the case $n = 0$ holds; that is,

$$\lambda^k = r_0 + r_1 \lambda + \cdots + r_{k-1} \lambda^{k-1}$$

The polynomial

$$p(x) = x^k - r_{k-1} x^{k-1} - \cdots - r_1 x - r_0$$

is called the polynomial **associated** with the linear recurrence $(*)$. Thus every root λ of $p(x)$ provides a sequence $[\lambda^n)$ satisfying $(*)$. If there are k distinct roots, the power sequences provide a basis. Incidentally, if $\lambda = 0$, the sequence $[\lambda^n)$ is $1, 0, 0, \ldots$; that is, we accept the convention that $0^0 = 1$.

Theorem 2

Let $r_0, r_1, \ldots, r_{k-1}$ be real numbers; let

$$V = \{[x_n) \mid x_{n+k} = r_0 x_n + \cdots + r_{k-1} x_{n+k-1} \quad \text{for all } n \geq 0\}$$

denote the vector space of all sequences satisfying the linear recurrence relation determined by $r_0, r_1, \ldots, r_{k-1}$; and let

$$p(x) = x^k - r_{k-1} x^{k-1} - \cdots - r_1 x - r_0$$

denote the polynomial associated with the recurrence relation. Then

1. $[\lambda^n)$ lies in V if and only if λ is a root of $p(x)$.
2. If $\lambda_1, \lambda_2, \ldots, \lambda_k$ are distinct real roots of $p(x)$, then $\{[\lambda_1^n), [\lambda_2^n), \ldots, [\lambda_k^n)\}$ is a basis of V.

PROOF

It remains to prove (2). But $[\lambda_i^n] = T(\mathbf{v}_i)$ where $\mathbf{v}_i = (1, \lambda_i, \lambda_i^2, \ldots, \lambda_i^{k-1})$, so (2) follows by Theorem 1, provided that $(\mathbf{v}_1, \mathbf{v}_2, \ldots, \mathbf{v}_n)$ is a basis of $\mathbb{R}^k$. This is true provided that the matrix

$$\begin{bmatrix} 1 & 1 & \cdots & 1 \\ \lambda_1 & \lambda_2 & \cdots & \lambda_k \\ \lambda_1^2 & \lambda_2^2 & \cdots & \lambda_k^2 \\ \vdots & \vdots & & \vdots \\ \lambda_1^{k-1} & \lambda_2^{k-1} & \cdots & \lambda_k^{k-1} \end{bmatrix}$$

is invertible. But this is the transpose of a Vandermonde matrix and so is invertible if the λ_i are distinct (Theorem 7 §3.2). This proves (2).

Example 3

Find the solution of $x_{n+2} = 2x_n + x_{n+1}$ that satisfies $x_0 = a$, $x_1 = b$.

Solution   The associated polynomial is $p(x) = x^2 - x - 2 = (x - 2)(x + 1)$. The roots are $\lambda_1 = 2$ and $\lambda_2 = -1$, so the sequences $[2^n]$ and $[(-1)^n]$ are a basis for the space of solutions by Theorem 2. Hence every solution $[x_n]$ is a linear combination

$$[x_n] = t_1[2^n] + t_2[(-1)^n]$$

This means that $x_n = t_1 2^n + t_2(-1)^n$ holds for $n = 0, 1, 2, \ldots$, so (taking $n = 0, 1$) $x_0 = a$ and $x_1 = b$ give

$$t_1 + t_2 = a$$
$$2t_1 - t_2 = b$$

These are easily solved: $t_1 = \frac{1}{3}(a + b)$ and $t_2 = \frac{1}{3}(2a - b)$, so

$$x_n = \tfrac{1}{3}[(a + b)2^n + (2a - b)(-1)^n]$$

If $p(x)$ is the polynomial associated with a linear recurrence relation of length k, and if $p(x)$ has k distinct roots $\lambda_1, \lambda_2, \ldots, \lambda_k$, then $p(x)$ factors completely:

$$p(x) = (x - \lambda_1)(x - \lambda_2) \cdots (x - \lambda_k)$$

Each root λ_i provides a sequence $[\lambda_i^n]$ satisfying the recurrence, and they are a basis of V by Theorem 2. In this case, each λ_i has multiplicity 1 as a root of $p(x)$. In general, a root λ has **multiplicity** m if $p(x) = (x - \lambda)^m q(x)$, where $q(\lambda) \neq 0$. In this case, there are fewer than k distinct roots and so fewer than k sequences $[\lambda_n]$ satisfying the recurrence. However, we can still obtain a basis because, if λ has multiplicity m (and $\lambda \neq 0$), it provides m linearly independent sequences that satisfy the recurrence. To prove this, it is convenient to give another way to describe the space V of all sequences satisfying a given linear recurrence relation.

Let $\mathbf{S}$ denote the vector space of *all* sequences and define a function

$$S : \mathbf{S} \to \mathbf{S} \quad \text{by} \quad S[x_n] = [x_{n+1}] = [x_1, x_2, x_3, \ldots]$$

S is clearly a linear transformation and is called the **shift operator** on $\mathbf{S}$. Note that powers of S shift the sequence further: $S^2[x_n] = S[x_{n+1}] = [x_{n+2}]$. In general,

$$S^k[x_n] = [x_{n+k}] = [x_k, x_{k+1}, \ldots] \quad \text{for each } k = 0, 1, 2, \ldots$$

But then a linear recurrence relation

$$x_{n+k} = r_0 x_n + r_1 x_{n+1} + \cdots + r_{k-1} x_{n+k-1} \quad \text{for } n = 0, 1, \ldots$$

can be written

$$S^k[x_n) = r_0[x_n) + r_1 S[x_n) + \cdots + r_{k-1} S^{k-1}[x_n) \qquad (*)$$

Now let $p(x) = x^k - r_{k-1} x^{k-1} - \cdots - r_1 x - r_0$ denote the polynomial associated with the recurrence relation. The set $\mathbf{L}[\mathbf{S}, \mathbf{S}]$ of all linear transformations from $\mathbf{S}$ to itself is a vector space (verify[5]) that is closed under composition. In particular,

$$p(S) = S^k - r_{k-1} S^{k-1} - \cdots - r_1 S - r_0$$

is a linear transformation called the **evaluation** of p at S. The point is that condition $(*)$ can be written as

$$p(S)[x_n) = 0$$

In other words, the space V of all sequences satisfying the recurrence relation is just $\ker[p(S)]$. This is the first assertion in the following theorem.

Theorem 3

Let $r_0, r_1, \ldots, r_{k-1}$ be real numbers, and let

$$V = \{[x_n) \mid x_{n+k} = r_0 x_n + r_1 x_{n+1} + \cdots + r_{k-1} x_{n+k-1} \quad \text{for all } n \geq 0\}$$

denote the space of all sequences satisfying the linear recurrence relation determined by $r_0, r_1, \ldots, r_{k-1}$. Let

$$p(x) = x^k - r_{k-1} x^{k-1} - \cdots - r_1 x - r_0$$

denote the corresponding polynomial. Then:
1. $V = \ker[p(S)]$, where S is the shift operator.
2. If $p(x) = (x - \lambda)^m q(x)$, where $\lambda \neq 0$ and $m > 1$, then the sequences

$$\{[\lambda^n), [n\lambda^n), [n^2\lambda^n), \ldots, [n^{m-1}\lambda^n)\}$$

all lie in V and are linearly independent.

PROOF

(Sketch) It remains to prove (2). If $\binom{n}{k} = \frac{n(n-1) \cdots (n-k+1)}{k!}$ denotes the binomial coefficient, the idea is to use (1) to show that the sequence $s_k = \left[\binom{n}{k}\lambda^n\right)$ is a solution for each $k = 0, 1, \ldots, m - 1$. Then (2) of Theorem 1 can be applied to show that $\{s_0, s_1, \ldots, s_{m-1}\}$ is linearly independent. Finally, the sequences $t_k = [n^k \lambda^n)$, $k = 0, 1, \ldots, m-1$, in the present theorem can be given by $t_k = \sum_{j=0}^{m-1} a_{kj} s_j$, where $A = [a_{ij}]$ is an invertible matrix. Then (2) follows. We omit the details.

This theorem combines with Theorem 2 to give a basis for V when $p(x)$ has k real roots (not necessarily distinct) none of which is zero. This last requirement means $r_0 \neq 0$, a condition that is unimportant in practice (see Remark 1 below).

---

5 See Exercises 19 and 20, §9.1.

Theorem 4

Let $r_0, r_1, \ldots, r_{k-1}$ be real numbers with $r_0 \neq 0$; let

$$V = \{[x_n] \mid x_{n+k} = r_0 x_n + r_1 x_{n+1} + \cdots + r_{k-1} x_{n+k-1} \text{ for all } n \geq 0\}$$

denote the space of all sequences satisfying the linear recurrence relation of length k determined by $r_0, \ldots, r_{k-1}$; and assume that the polynomial

$$p(x) = x^k - r_{k-1}x^{k-1} - \cdots - r_1 x - r_0$$

factors completely as

$$p(x) = (x - \lambda_1)^{m_1}(x - \lambda_2)^{m_2} \cdots (x - \lambda_p)^{m_p}$$

where $\lambda_1, \lambda_2, \ldots, \lambda_p$ are distinct real numbers and each $m_i \geq 1$. Then $\lambda_i \neq 0$ for each i, and

$$\{[\lambda_1^n), [n\lambda_1^n), \ldots, [n^{m_1-1}\lambda_1^n); [\lambda_2^n), [n\lambda_2^n), \ldots, [n^{m_2-1}\lambda_2^n); \ldots;$$
$$[\lambda_p^n), [n\lambda_p^n), \ldots, [n^{m_p-1}\lambda_p^n)\}$$

is a basis of V.

PROOF

There are $m_1 + m_2 + \cdots + m_p = k$ sequences in all so, because $\dim V = k$, it suffices to show that they are linearly independent. The assumption that $r_0 \neq 0$, implies that 0 is not a root of $p(x)$. Hence each $\lambda_i \neq 0$, so $\{[\lambda_i^n), [n\lambda_i^n), \ldots, [n^{m_i-1}\lambda_i^n)\}$ is linearly independent by Theorem 3.

The proof that the whole set of sequences is linearly independent is omitted.

Example 4

Find a basis for the space V of all sequences $[x_n)$ satisfying

$$x_{n+3} = -9x_n - 3x_{n+1} + 5x_{n+2}.$$

Solution   The associated polynomial is $p(x) = x^3 - 5x^2 + 3x + 9 = (x-3)^2(x+1)$. Hence 3 is a double root, so $[3^n)$ and $[n3^n)$ both lie in V by Theorem 3 (the reader should verify this). Similarly, $\lambda = -1$ is a root of multiplicity 1, so $[(-1)^n)$ lies in V. Hence $\{[3^n), [n3^n), [(-1)^n)\}$ is a basis by Theorem 4.

Remark 1   If $r_0 = 0$ [so $p(x)$ has 0 as a root], the recurrence reduces to one of shorter length. For example, consider

$$x_{n+4} = 0x_n + 0x_{n+1} + 3x_{n+2} + 2x_{n+3} \qquad (*)$$

If we set $y_n = x_{n+2}$, this recurrence becomes $y_{n+2} = 3y_n + 2y_{n+1}$, which has solutions $[3^n)$ and $[(-1)^n)$. These give the following solution to $(*)$:

$$[0, 0, 1, 3, 3^2, \ldots)$$
$$[0, 0, 1, -1, (-1)^2, \ldots)$$

In addition, it is easy to verify that

$$[1, \ 0, \ 0, \ 0, \ 0, \ldots)$$
$$[0, \ 1, \ 0, \ 0, \ 0, \ldots)$$

are also solutions to (∗). The space of all solutions of (∗) has dimension 4 (Theorem 1), so these sequences are a basis. This technique works whenever $r_0 = 0$.

Remark 2   Theorem 4 completely describes the space V of sequences that satisfy a linear recurrence relation for which the associated polynomial $p(x)$ has all real roots. However, in many cases of interest, $p(x)$ has complex roots that are not real. If $p(\mu) = 0$, μ complex, then $p(\bar{\mu}) = 0$ too ($\bar{\mu}$ the conjugate), and the main observation is that $[\mu^n + \bar{\mu}^n)$ and $[i(\mu^n - \bar{\mu}^n))$ are *real* solutions. Analogs of the preceding theorems can then be proved.

Exercises 7.4

1. Find a basis for the space V of sequences $[x_n)$ satisfying the following recurrences, and use it to find the sequence satisfying $x_0 = 1$, $x_1 = 2$, $x_2 = 1$.

   (a) $x_{n+3} = -2x_n + x_{n+1} + 2x_{n+2}$

   ◆(b) $x_{n+3} = -6x_n + 7x_{n+1}$

   (c) $x_{n+3} = -36x_n + 7x_{n+2}$

2. In each case, find a basis for the space V of all sequences $[x_n)$ satisfying the recurrence, and use it to find x_n if $x_0 = 1$, $x_1 = -1$, and $x_2 = 1$.

   (a) $x_{n+3} = x_n + x_{n+1} - x_{n-2}$

   ◆(b) $x_{n+3} = -2x_n + 3x_{n+1}$

   (c) $x_{n+3} = -4x_n + 3x_{n+2}$

   ◆(d) $x_{n+3} = x_n - 3x_{n+1} + 3x_{n+2}$

   (e) $x_{n+3} = 8x_n - 12x_{n+1} + 6x_{n+2}$

3. Find a basis for the space V of sequences $[x_n)$ satisfying each of the following recurrences.

   (a) $x_{n+2} = -a^2 x_n + 2ax_{n+1}$, $a \neq 0$

   ◆(b) $x_{n+2} = -abx_n + (a+b)x_{n+1}$, $(a \neq b)$

4. In each case, find a basis of V.

   (a) $V = \{[x_n) \mid x_{n+4} = 2x_{n+2} - x_{n+3} \text{ for } n \geq 0\}$

   ◆(b) $V = \{[x_n) \mid x_{n+4} = -x_{n+2} - 2x_{n+3} \text{ for } n \geq 0\}$

5. Suppose that $[x_n)$ satisfies a linear recurrence relation of length k. If

   $$\{\mathbf{e}_0 = (1, \ 0, \ldots, \ 0), \ \mathbf{e}_1 = (0, \ 1, \ldots, \ 0), \ldots,$$
   $$\mathbf{e}_{k-1} = (0, \ 0, \ldots, \ 1)\}$$

   is the standard basis of $\mathbb{R}^k$, show that $x_n = x_0 T(\mathbf{e}_0) + x_1 T(\mathbf{e}_1) + \cdots + x_{k-1} T(\mathbf{e}_{k-1})$ holds for all $n \geq k$. (Here T is as in Theorem 1.)

6. Show that the shift operator S is onto but not one-to-one. Find ker S.

◆7. Find a basis for the space V of all sequences $[x_n)$ satisfying $x_{n+2} = -x_n$.

8 Orthogonality

In Section 5.3 we introduced the dot product in $\mathbb{R}^n$ and extended the basic geometric notions of length and distance. A set $\{E_1, E_2, \ldots, E_m\}$ of nonzero vectors in $\mathbb{R}^n$ was called an **orthogonal set** if $E_i \bullet E_j = 0$ for all $i \neq j$, and it was proved that every orthogonal set is independent. In particular, it was observed that the expansion of a vector as a linear combination of orthogonal basis vectors is easy to obtain because formulas exist for the coefficients. Hence the orthogonal bases are the "nice" bases, and much of this chapter is devoted to extending results about bases to orthogonal bases. This results in some very powerful methods and theorems. Our first task is to show that every subspace of $\mathbb{R}^n$ *has* an orthogonal basis.

SECTION 8.1 Orthogonal Complements and Projections

If $\{\mathbf{v}_1, \ldots, \mathbf{v}_m\}$ is linearly independent in a general vector space, and if $\mathbf{v}_{m+1}$ is not in span$\{\mathbf{v}_1, \ldots, \mathbf{v}_m\}$, then $\{\mathbf{v}_1, \ldots, \mathbf{v}_m, \mathbf{v}_{m+1}\}$ is independent (Lemma 1 §6.4). Here is the analog for *orthogonal* sets in $\mathbb{R}^n$.

Lemma 1 Orthogonal Lemma

Let $\{E_1, E_2, \ldots, E_m\}$ be an orthogonal set in $\mathbb{R}^n$. Given X in $\mathbb{R}^n$, write

$$E_{m+1} = X - \frac{X \bullet E_1}{\|E_1\|^2} E_1 - \frac{X \bullet E_2}{\|E_2\|^2} E_2 - \cdots - \frac{X \bullet E_m}{\|E_m\|^2} E_m$$

Then:
1. E_{m+1} is orthogonal to each of $E_1, E_2, \ldots, E_m$.
2. If X is not in span$\{E_1, \ldots, E_m\}$, then $E_{m+1} \neq 0$ and $\{E_1, \ldots, E_m, E_{m+1}\}$ is an orthogonal set.

PROOF

For convenience, write $t_i = (X \bullet E_i) / \|E_i\|^2$ for each i. Given $1 \leq k \leq m$:

$$E_{m+1} \bullet E_k = (X - t_1 E_1 - \cdots - t_k E_k - \cdots - t_m E_m) \bullet E_k$$
$$= X \bullet E_k - t_1(E_1 \bullet E_k) - \cdots - t_k(E_k \bullet E_k) - \cdots - t_m(E_m \bullet E_k)$$

$$= X \cdot E_k - t_k \| E_k \|^2$$
$$= 0$$

This proves (1), and (2) follows because $E_{m+1} \neq 0$ if X is not in span$\{E_1, \dots, E_m\}$.

The orthogonal lemma has three important consequences for $\mathbb{R}^n$. The first is an extension for orthogonal sets of the fundamental fact that any independent set is part of a basis (Theorem 1 §6.4).

Theorem 1

Let U be a subspace of $\mathbb{R}^n$.
1. Every orthogonal subset $\{E_1, \dots, E_m\}$ in U is part of an orthogonal basis of U.
2. If $U \neq \{0\}$, it *has* an orthogonal basis.

PROOF

1. If span$\{E_1, \dots, E_m\} = U$, it is *already* a basis. Otherwise, there exists X in U outside span$\{E_1, \dots, E_m\}$. If E_{m+1} is as given in the orthogonal lemma, then E_{m+1} is in U and $\{E_1, \dots, E_m, E_{m+1}\}$ is orthogonal. If span$\{E_1, \dots, E_m, E_{m+1}\} = U$, we are done. Otherwise, the process continues to create larger and larger orthogonal subsets of U. They are all independent by Theorem 5 §5.3, so we have a basis when we reach a subset containing dim U vectors.
2. If $E \neq 0$ is in U, then $\{E\}$ is orthogonal, so (2) follows from (1).

We can improve upon (2) of Theorem 1. In fact, the second consequence of the orthogonal lemma is a procedure by which *any* basis $\{X_1, \dots, X_m\}$ of a subspace U of $\mathbb{R}^n$ can be systematically modified to yield an orthogonal basis $\{E_1, \dots, E_m\}$ of U. The E_i are constructed one at a time from the X_i.

To start the process, take $E_1 = X_1$. Then X_2 is not in span$\{E_1\}$ because $\{X_1, X_2\}$ is independent, so take

$$E_2 = X_2 - \frac{X_2 \cdot E_1}{\| E_1 \|^2} E_1$$

Thus $\{E_1, E_2\}$ is orthogonal by Lemma 1. Moreover, span$\{E_1, E_2\}$ = span$\{X_1, X_2\}$ (verify), so X_3 is not in span$\{E_1, E_2\}$. Hence $\{E_1, E_2, E_3\}$ is orthogonal where

$$E_3 = X_3 - \frac{X_3 \cdot E_1}{\| E_1 \|^2} E_1 - \frac{X_3 \cdot E_2}{\| E_2 \|^2} E_2$$

Again, span$\{E_1, E_2, E_3\}$ = span$\{X_1, X_2, X_3\}$, so X_4 is not in span$\{E_1, E_2, E_3\}$ and the process continues. At the mth iteration we construct an orthogonal set $\{E_1, \dots, E_m\}$ such that

$$\text{span}\{E_1, E_2, \dots, E_m\} = \text{span}\{X_1, X_2, \dots, X_m\} = U$$

Hence $\{E_1, E_2, \dots, E_m\}$ is the desired orthogonal basis of U. The procedure can be summarized as follows.

Theorem 2 Gram–Schmidt Orthogonalization Algorithm[1]

If $\{X_1, X_2, \ldots, X_m\}$ is any basis of a subspace U of $\mathbb{R}^n$, construct $E_1, E_2, \ldots, E_m$ in U successively as follows:

$$E_1 = X_1$$

$$E_2 = X_2 - \frac{X_2 \cdot E_1}{\|E_1\|^2} E_1$$

$$E_3 = X_3 - \frac{X_3 \cdot E_1}{\|E_1\|^2} E_1 - \frac{X_3 \cdot E_2}{\|E_2\|^2} E_2$$

$$\vdots$$

$$E_k = X_k - \frac{X_k \cdot E_1}{\|E_1\|^2} E_1 - \frac{X_k \cdot E_2}{\|E_2\|^2} E_2 - \cdots - \frac{X_k \cdot E_{k-1}}{\|E_{k-1}\|^2} E_{k-1}$$

for each $k = 2, 3, \ldots, m$. Then

1. $\{E_1, E_2, \ldots, E_m\}$ is an orthogonal basis of U.
2. $\text{span}\{E_1, E_2, \ldots, E_k\} = \text{span}\{X_1, X_2, \ldots, X_k\}$ for each $k = 1, 2, \ldots, m$.

Of course, the algorithm converts any basis of $\mathbb{R}^n$ itself into an orthogonal basis.

Example 1

Find an orthogonal basis of the row space of $A = \begin{bmatrix} 1 & 1 & -1 & -1 \\ 3 & 2 & 0 & 1 \\ 1 & 0 & 1 & 0 \end{bmatrix}$.

Solution Let X_1, X_2, X_3 denote the rows of A and observe that $\{X_1, X_2, X_3\}$ is linearly independent. Take $E_1 = X_1$. The algorithm gives

$$E_2 = X_2 - \frac{X_2 \cdot E_1}{\|E_1\|^2} E_1 = [3 \ \ 2 \ \ 0 \ \ 1] - \tfrac{4}{4}[1 \ \ 1 \ \ -1 \ \ -1] = [2 \ \ 1 \ \ 1 \ \ 2]$$

$$E_3 = X_3 - \frac{X_3 \cdot E_1}{\|E_1\|^2} E_1 - \frac{X_3 \cdot E_2}{\|E_2\|^2} E_2 = X_3 - \tfrac{0}{4} E_1 - \tfrac{3}{10} E_2 = \tfrac{1}{10}[4 \ \ -3 \ \ 7 \ \ -6]$$

Hence $\{[1 \ \ 1 \ \ -1 \ \ -1], [2 \ \ 1 \ \ 1 \ \ 2], \tfrac{1}{10}[4 \ \ -3 \ \ 7 \ \ -6]\}$ is the orthogonal basis provided by the algorithm. In hand calculations it may be convenient to eliminate fractions, so $\{[1 \ 1 \ -1 \ -1], [2 \ 1 \ 1 \ 2], [4 \ -3 \ 7 \ -6]\}$ is also an orthogonal basis for row A.

Observe that the vector $\dfrac{X \cdot E_i}{\|E_i\|^2} E_i$ is unchanged if a nonzero scalar multiple of E_i is used in place of E_i. Hence, if a newly constructed E_i is multiplied by a nonzero scalar at some stage of the Gram–Schmidt algorithm, the subsequent E's will be unchanged. This is useful in actual calculations.

---

1 Erhardt Schmidt (1876–1959) was a German mathematician who studied under the great David Hilbert and later developed the theory of Hilbert spaces. He first described the present algorithm in 1907. Jörgen Pederson Gram (1850–1916) was a Danish actuary.

Projections

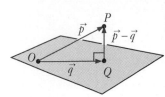

Suppose a point P and a plane through the origin O in $\mathbb{R}^3$ are given, and we want to find the point Q in the plane that is closest to P. Our geometric intuition assures us that such a point Q exists. If we let $\vec{p}$ be the position vector of P (that is, the vector from O to P), then what is required is to find the position vector $\vec{q}$ of Q (see the diagram). Again, our geometric insight assures us that, if $\vec{q}$ is chosen in such a way that $\vec{p} - \vec{q}$ is *perpendicular* to the plane, then $\vec{q}$ will be the vector we want.

Now we make two observations: first, that the set U of position vectors of points in the plane is a *subspace of* $\mathbb{R}^3$ (because the plane contains the origin), and second, that the condition that $\vec{p} - \vec{q}$ is perpendicular to the plane U means that $\vec{p} - \vec{q}$ is *orthogonal* to every vector in U. In these terms the whole discussion makes sense in $\mathbb{R}^n$. Furthermore, the orthogonal lemma provides exactly what is needed to find $\vec{q}$ in this more general setting.

If U is a subspace of $\mathbb{R}^n$, define the **orthogonal complement** $U^\perp$ of U (pronounced "U-perp") by

$$U^\perp = \{X \text{ in } \mathbb{R}^n \mid X \cdot Y = 0 \text{ for all } Y \text{ in } U\}.$$

It is easily verified that $U^\perp$ is a subspace of $\mathbb{R}^n$ and, if $U = \text{span}\{X_1, \dots, X_m\}$, that $U^\perp = \{X \text{ in } \mathbb{R}^n \mid X \cdot X_i = 0 \text{ for each } i = 1, 2, \dots, m\}$. (See Exercise 6.)

Example 2

Find $U^\perp$ if $U = \text{span}\{[1 \ -1 \ 2 \ 0], [1 \ 0 \ -2 \ 3]\}$ in $\mathbb{R}^4$.

Solution

$X = [x \ y \ z \ w]$ is in $U^\perp$ if and only if it is orthogonal to both $[1 \ -1 \ 2 \ 0]$ and $[1 \ 0 \ -2 \ 3]$; that is,

$$\begin{aligned} x - y + 2z \qquad &= 0 \\ x \qquad - 2z + 3w &= 0 \end{aligned}$$

Gaussian elimination gives $U^\perp = \text{span}\{[2 \ 4 \ 1 \ 0], [3 \ 3 \ 0 \ -1]\}$.

Now let $\{E_1, E_2, \dots, E_m\}$ be an orthogonal basis of a subspace U of $\mathbb{R}^n$. Given X in $\mathbb{R}^n$, consider the vector

$$P = \frac{X \cdot E_1}{\|E_1\|^2} E_1 + \frac{X \cdot E_2}{\|E_2\|^2} E_2 + \cdots + \frac{X \cdot E_m}{\|E_m\|^2} E_m \qquad (*)$$

Then P is in U, and $X - P$ is in $U^\perp$ (being orthogonal to each E_i by the orthogonal lemma). Moreover, we can show that P is *independent* of the choice of orthogonal basis in U. Indeed, suppose $\{E_1', \dots, E_m'\}$ is another orthogonal basis of U, and write

$$P' = \frac{X \cdot E_1'}{\|E_1'\|^2} E_1' + \frac{X \cdot E_2'}{\|E_2'\|^2} E_2' + \cdots + \frac{X \cdot E_m'}{\|E_m'\|^2} E_m'$$

Then, as before, P' is in U and $X - P'$ is in $U^\perp$. Write the vector $P - P'$ as follows:

$$P - P' = (X - P') - (X - P)$$

This vector is in U (because P and P' are in U) and also in $U^\perp$ (because $X - P$ and $X - P'$ are in $U^\perp$) and so must be zero (it is orthogonal to itself!). This means that $P = P'$, as asserted.

Hence, the vector P in equation $(*)$ depends only on X and the subspace U, and *not* on the choice of orthogonal basis $\{E_1, \dots, E_m\}$ of U used to compute it. Thus, we are entitled to make the following definition: Let U be a subspace of $\mathbb{R}^n$ with orthogonal basis $\{E_1, E_2, \dots, E_m\}$. If X is in $\mathbb{R}^n$, the vector

$$\text{proj}_U(X) = \frac{X \cdot E_1}{\|E_1\|^2} E_1 + \frac{X \cdot E_2}{\|E_2\|^2} E_2 + \cdots + \frac{X \cdot E_m}{\|E_m\|^2} E_m$$

is called the **orthogonal projection** of X on U. For the zero subspace $U = \{0\}$, we define $\text{proj}_{\{0\}}(X) = 0$. The preceding discussion proves (1) of the following theorem.

Theorem 3 Projection Theorem

If U is a subspace of $\mathbb{R}^n$ and X is in $\mathbb{R}^n$, write $P = \text{proj}_U(X)$. Then
1. P is in U and $X - P$ is in $U^\perp$.
2. P is the vector in U closest to X in the sense that

$$\|X - P\| < \|X - Y\| \quad \text{for all } Y \text{ in } U, \ Y \neq P$$

3. $\dim U + \dim U^\perp = n$.

PROOF

1. This is proved in the preceding discussion.
2. Write $X - Y = (X - P) + (P - Y)$. Then $P - Y$ is in U and so is orthogonal to $X - P$ by (1). Hence, the pythagorean theorem gives

$$\|X - Y\|^2 = \|X - P\|^2 + \|P - Y\|^2 > \|X - P\|^2$$

because $P - Y \neq 0$. This gives (2).
3. If $U = \{0\}$, then $U^\perp = \mathbb{R}^n$ and (3) holds. If $U^\perp = \{0\}$, then $U = \mathbb{R}^n$ by (1) and again (3) holds. So assume that U and $U^\perp$ are both nonzero and (by Theorem 1) let $\{E_1, \ldots, E_m\}$ and $\{F_1, \ldots, F_k\}$ be orthogonal bases of U and $U^\perp$, respectively. Then $\{E_1, \ldots, E_m, F_1, \ldots, F_k\}$ is orthogonal (by the definition of $U^\perp$), so it suffices to show that this set spans $\mathbb{R}^n$. But this follows from (1).

Example 3

Let $U = \text{span}\{X_1, X_2\}$ in $\mathbb{R}^4$ where $X_1 = [1\ 1\ 0\ 1]$ and $X_2 = [0\ 1\ 1\ 2]$. If $X = [3\ -1\ 0\ 2]$, find the vector in U closest to X and express X as the sum of a vector in U and a vector orthogonal to U.

Solution $\{X_1, X_2\}$ is independent but not orthogonal. The Gram–Schmidt process gives an orthogonal basis $\{E_1, E_2\}$ of U where $E_1 = X_1 = [1\ 1\ 0\ 1]$ and

$$E_2 = X_2 - \frac{X_2 \cdot E_1}{\|E_1\|^2} E_1 = X_2 - \tfrac{3}{3} E_1 = [-1\ 0\ 1\ 1]$$

Hence, we can compute the projection using $\{E_1, E_2\}$:

$$P = \text{proj}_U(X) = \frac{X \cdot E_1}{\|E_1\|^2} E_1 + \frac{X \cdot E_2}{\|E_2\|^2} E_2 = \tfrac{4}{3} E_1 + \tfrac{-1}{3} E_2 = \tfrac{1}{3}[5\ 4\ -1\ 3]$$

Thus, P is the vector in U closest to X, and $X - P = \tfrac{1}{3}[4\ -7\ 1\ 3]$ is orthogonal to every vector in U. (This can be verified by checking that it is orthogonal to the generators X_1 and X_2 of U.) The required decomposition of X is thus

$$X = P + (X - P) = \tfrac{1}{3}[5\ 4\ -1\ 3] + \tfrac{1}{3}[4\ -7\ 1\ 3].$$

Example 4

Find the point in the plane with equation $2x + y - z = 0$ that is closest to the point $P_0(2, -1, -3)$.

Solution

We write $\mathbb{R}^3$ as rows. The plane is the subspace U whose points $[x\ y\ z]$ satisfy $z = 2x + y$. Hence

$$U = \{[s\ t\ 2s + t] \mid s, r \text{ in } \mathbb{R}\} = \text{span}\{[0\ 1\ 1], [1\ 0\ 2]\}$$

The Gram–Schmidt process produces an orthogonal basis $\{\vec{e}_1, \vec{e}_2\}$ of U where $\vec{e}_1 = [0\ 1\ 1]$ and $\vec{e}_2 = [1\ -1\ 1]$. Hence, the vector in U closest to $\vec{v} = [2\ -1\ -3]$ is

$$\text{proj}_U(\vec{v}) = \frac{\vec{v} \cdot \vec{e}_1}{\|\vec{e}_1\|^2}\vec{e}_1 + \frac{\vec{v} \cdot \vec{e}_2}{\|\vec{e}_2\|^2}\vec{e}_2 = -2\vec{e}_1 + 0\vec{e}_2 = [0\ -2\ -2]$$

Thus, the point in U closest to $P_0(2, -1, -3)$ is $Q(0, -2, -2)$.

Example 5

If U is a subspace of $\mathbb{R}^n$, show that projection on U is a linear transformation. More precisely, show that $T : \mathbb{R}^n \to \mathbb{R}^n$ is a linear transformation where

$$T(X) = \text{proj}_U(X) \quad \text{for all } X \text{ in } \mathbb{R}^n.$$

Show further that $\ker T = U^\perp$ and $\text{im } T = U$.

Solution

Choose an orthonormal basis $\{E_1, E_2, \ldots, E_m\}$ of U. Then

$$T(X) = (X \cdot E_1)E_1 + (X \cdot E_2)E_2 + \cdots + (X \cdot E_m)E_m \qquad (*)$$

so axioms T1 and T2 hold because $(X + Y) \cdot E_i = X \cdot E_i + Y \cdot E_i$ and $(rX) \cdot E_i = r(X \cdot E_i)$ for each i. Hence T is linear.

We have $\text{im } T \subseteq U$ by $(*)$ because each E_i is in U; if X is in U, then $X = T(X)$ by $(*)$ and the expansion theorem applied to U (Theorem 6 §5.3). So $\text{im } T = U$.

If X is in $U^\perp$, then $X \cdot E_i = 0$ for each i (because E_i is in U), so $T(X) = 0$ by $(*)$. Hence $U^\perp \subseteq \ker T$. On the other hand, Theorem 3 shows that $X - T(X)$ is in $U^\perp$ for all X in $\mathbb{R}^n$. It follows that $\ker T \subseteq U^\perp$, and so $\ker T = U^\perp$.

Exercises 8.1

1. In each case, use the Gram–Schmidt algorithm to convert the given basis B of V into an orthogonal basis.

   (a) $V = \mathbb{R}^2$, $B = \{[1\ -1], [2\ 1]\}$

   ◆(b) $V = \mathbb{R}^2$, $B = \{[2\ 1], [1\ 2]\}$

   (c) $V = \mathbb{R}^3$, $B = \{[1\ -1\ 1], [1\ 0\ 1], [1\ 1\ 2]\}$

   ◆(d) $V = \mathbb{R}^3$, $B = \{[0\ 1\ 1], [1\ 1\ 1], [1\ -2\ 2]\}$

2. In each case, write X as the sum of a vector in U and a vector in $U^\perp$.

   (a) $X = [1\ 5\ 7]$, $U = \text{span}\{[1\ -2\ 3], [-1\ 1\ 1]\}$

   ◆(b) $X = [2\ 1\ 6]$, $U = \text{span}\{[3\ -1\ 2], [2\ 0\ -3]\}$

   (c) $X = [3\ 1\ 5\ 9]$,
   $U = \text{span}\{[1\ 0\ 1\ 1], [0\ 1\ -1\ 1], [-2\ 0\ 1\ 1]\}$

   ◆(d) $X = [2\ 0\ 1\ 6]$,
   $U = \text{span}\{[1\ 1\ 1\ 1], [1\ 1\ -1\ -1], [1\ -1\ 1\ -1]\}$

   (e) $X = [a\ b\ c\ d]$,
   $U = \text{span}\{[1\ 0\ 0\ 0], [0\ 1\ 0\ 0], [0\ 0\ 1\ 0]\}$

   ◆(f) $X = [a\ b\ c\ d]$,
   $U = \text{span}\{[1\ -1\ 2\ 0], [-1\ 1\ 1\ 1]\}$

3. Let $X = [1\ -2\ 1\ 6]$ in $\mathbb{R}^4$, and let $U = \text{span}\{[2\ 1\ 3\ -4], [1\ 2\ 0\ 1]\}$.

   ◆(a) Compute $\text{proj}_U(X)$.

(b) Show that $\{[1\ 0\ 2\ -3], [4\ 7\ 1\ 2]\}$ is another orthogonal basis of U.

♦(c) Use the basis in part (b) to compute $\text{proj}_U(X)$.

4. In each case, use the Gram–Schmidt algorithm to find an orthogonal basis of the subspace U, and find the vector in U closest to X.

   (a) $U = \text{span}\{[1\ 1\ 1], [0\ 1\ 1]\}$; $X = [-1\ 2\ 1]$
   ♦(b) $U = \text{span}\{[1\ -1\ 0], [-1\ 0\ 1]\}$; $X = [2\ 1\ 0]$
   (c) $U = \text{span}\{[1\ 0\ 1\ 0], [1\ 1\ 1\ 0], [1\ 1\ 0\ 0]\}$;
      $X = [2\ 0\ -1\ 3]$
   ♦(d) $U = \text{span}\{[1\ -1\ 0\ 1], [1\ 1\ 0\ 0], [1\ 1\ 0\ 1]\}$;
      $X = [2\ 0\ 3\ 1]$

5. Let $U = \text{span}\{V_1, V_2, \ldots, V_k\}$, V_i in $\mathbb{R}^n$, and let A be the $k \times n$ matrix with the V_i as rows.

   (a) Show that $U^\perp = \{X \mid X$ in $\mathbb{R}^n$, $AX^T = 0\}$.
   ♦(b) Use part (a) to find $U^\perp$ if
      $U = \text{span}\{[1\ -1\ 2\ 1], [1\ 0\ -1\ 1]\}$.

♦6. Let $U = \text{span}\{X_1, \ldots, X_m\}$ be a subspace of $\mathbb{R}^n$. Show that $U^\perp = \{X$ in $\mathbb{R}^n \mid X \bullet X_i = 0$ for each $i = 1, 2, \ldots, m\}$.

7. Let U be a subspace of $\mathbb{R}^n$. If X in $\mathbb{R}^n$ can be written in any way at all as $X = P + Q$ with P in U and Q in $U^\perp$, show that necessarily $P = \text{proj}_U(X)$.

8. Let U be a subspace of $\mathbb{R}^n$ and let X be a vector in $\mathbb{R}^n$. Using Exercise 7, or otherwise, show that X is in U if and only if $X = \text{proj}_U(X)$.

9. Let U be a subspace of $\mathbb{R}^n$.

   (a) Show that $U^\perp = \mathbb{R}^n$ if and only if $U = \{0\}$.
   (b) Show that $U^\perp = \{0\}$ if and only if $U = \mathbb{R}^n$.

♦10. If U is a subspace of $\mathbb{R}^n$, show that $\text{proj}_U(X) = X$ for all X in U.

11. If U is a subspace of $\mathbb{R}^n$, show that $X = \text{proj}_U(X) + \text{proj}_{U^\perp}(X)$ for all X in $\mathbb{R}^n$.

12. If $\{E_1, \ldots, E_n\}$ is an orthogonal basis of $\mathbb{R}^n$ and $U = \text{span}\{E_1, \ldots, E_m\}$, show that $U^\perp = \text{span}\{E_{m+1}, \ldots, E_n\}$.

13. If U is a subspace of $\mathbb{R}^n$, show that $U^{\perp\perp} = U$. [*Hint:* Show that $U \subseteq U^{\perp\perp}$, then use Theorem 3(3) twice.]

♦14. If U is a subspace of $\mathbb{R}^n$, show how to find an $n \times n$ matrix A such that $U = \{X \mid AX = 0\}$. [*Hint:* Exercise 13.]

15. Write $\mathbb{R}^n$ as rows. If A is an $n \times n$ matrix, write its null space as null $A = \{X$ in $\mathbb{R}^n \mid AX^T = 0\}$. Show that null $A = (\text{row } A)^\perp$.

16. Think of $\mathbb{R}^n$ as consisting of rows.

   (a) Let E be an $n \times n$ matrix, and let $U = \{XE \mid X$ in $\mathbb{R}^n\}$. Show that the following are equivalent.
      (i) $E^2 = E = E^T$ (E is a **projection matrix**).
      (ii) $(X - XE) \bullet (YE) = 0$ for all X and Y in $\mathbb{R}^n$.
      (iii) $\text{proj}_U(X) = XE$ for all X in $\mathbb{R}^n$.
      [*Hint:* For (ii) implies (iii): Write $X = XE + (X - XE)$ and use the uniqueness argument preceding the definition of $\text{proj}_U(X)$.
      For (iii) implies (ii): $X - XE$ is in $U^\perp$ for all X in $\mathbb{R}^n$.]

   (b) If E is a projection matrix, show that $I - E$ is also a projection matrix.

   (c) If $EF = 0 = FE$ and E and F are projection matrices, show that $E + F$ is also a projection matrix.

   ♦(d) If A is $m \times n$ and AA^T is invertible, show that $E = A^T(AA^T)^{-1}A$ is a projection matrix.

17. Let A be an $n \times n$ matrix of rank r. Show that there is an invertible $n \times n$ matrix U such that UA is a row-echelon matrix with the property that the first r rows are orthogonal. [*Hint:* Let R be the row-echelon form of A, and use the Gram–Schmidt process on the nonzero rows of R from the bottom up. Use Lemma 1 §2.4.]

18. Let A be an $(n - 1) \times n$ matrix with rows $X_1, X_2, \ldots, X_{n-1}$ and let A_i denote the $(n - 1) \times (n - 1)$ matrix obtained from A by deleting column i. Define the vector Y in $\mathbb{R}^n$ by $Y = [\det A_1\ -\det A_2\ \det A_3\ \cdots\ (-1)^{n+1} \det A_n]$ Show that:

   (a) $X_i \bullet Y = 0$ for all $i = 1, 2, \ldots, n - 1$. [*Hint:* Write $B_i = \begin{bmatrix} X_i \\ A \end{bmatrix}$ and show that $\det B_i = 0$.]

   (b) $Y \neq 0$ if and only if $\{X_1, X_2, \ldots, X_{n-1}\}$ is linearly independent. [*Hint:* If some det $A_i \neq 0$, the rows of A_i are linearly independent. Conversely, if the X_i are independent, consider $A = UR$ where R is in reduced row-echelon form.]

(c) If $\{X_1, X_2, \ldots, X_{n-1}\}$ is linearly independent, use Theorem 3(3) to show that all solutions to the system of $n - 1$ homogeneous equations
$$AX^T = 0$$
are given by tY, t a parameter.

SECTION 8.2 Orthogonal Diagonalization

Recall (Theorem 3 §5.5) that an $n \times n$ matrix A is diagonalizable if and only if it has n linearly independent eigenvectors. Moreover, the matrix P with these eigenvectors as columns is a diagonalizing matrix for A, that is,

$$P^{-1}AP \text{ is diagonal.}$$

As we have seen, the really nice bases of $\mathbb{R}^n$ are the orthogonal ones, so a natural question is: which $n \times n$ matrices have an *orthogonal* basis of eigenvectors. These turn out to be precisely the symmetric matrices, and this is the main result of this section.

If a matrix A has n orthogonal eigenvectors, they can (by normalizing) be taken to be orthonormal. The corresponding diagonalizing matrix P has orthonormal columns. Such matrices are very easy to invert.

Theorem 1

The following conditions are equivalent for an $n \times n$ matrix P.
1. P is invertible and $P^{-1} = P^T$.
2. The rows of P are orthonormal.
3. The columns of P are orthonormal.

PROOF

First recall that condition (1) is equivalent to $PP^T = I$ by the corollary to Theorem 5 §2.3. Let $X_1, X_2, \ldots, X_n$ denote the rows of P. Then X_j^T is the jth column of P^T, so the (i, j)-entry of PP^T is $X_i \cdot X_j$. Thus $PP^T = I$ means that $X_i \cdot X_j = 0$ if $i \neq j$ and $X_i \cdot X_j = 1$ if $i = j$, so condition (1) is equivalent to (2). The proof of the equivalence of (1) and (3) is similar.

An $n \times n$ matrix P is called an **orthogonal matrix**[2] if it satisfies one (and hence all) of the conditions in Theorem 1.

Example 1

The matrix $\begin{bmatrix} \cos\theta & -\sin\theta \\ \sin\theta & \cos\theta \end{bmatrix}$ is orthogonal for any angle θ.

Warning

It is not enough that the rows of a matrix A are merely orthogonal for A to be an orthogonal matrix.

2 In view of (2) and (3) of Theorem 1, *orthonormal matrix* might be a better name, but *orthogonal matrix* is standard.

Example 2

The matrix $\begin{bmatrix} 2 & 1 & 1 \\ -1 & 1 & 1 \\ 0 & -1 & 1 \end{bmatrix}$ has orthogonal rows but is not an orthogonal matrix.

However, if the rows are normalized, the resulting matrix $\begin{bmatrix} \frac{2}{\sqrt{6}} & \frac{1}{\sqrt{6}} & \frac{1}{\sqrt{6}} \\ \frac{-1}{\sqrt{3}} & \frac{1}{\sqrt{3}} & \frac{1}{\sqrt{3}} \\ 0 & \frac{-1}{\sqrt{2}} & \frac{1}{\sqrt{2}} \end{bmatrix}$ is

orthogonal (so the columns are also orthonormal as the reader can verify).

Example 3

If P and Q are orthogonal matrices, then PQ is also orthogonal, as is $P^{-1} = P^T$.

Solution P and Q are invertible, so PQ is also invertible and $(PQ)^{-1} = Q^{-1}P^{-1} = Q^T P^T = (PQ)^T$. Hence PQ is orthogonal. Similarly, $(P^{-1})^{-1} = P = (P^T)^T = (P^{-1})^T$ shows that P^{-1} is orthogonal.

An $n \times n$ matrix A is said to be **orthogonally diagonalizable** when an orthogonal matrix P can be found such that $P^{-1}AP = P^T AP$ is diagonal. This condition turns out to characterize the symmetric matrices.

Theorem 2 Principal Axis Theorem

The following conditions are equivalent for an $n \times n$ matrix A.
1. A has an orthonormal set of n eigenvectors.
2. A is orthogonally diagonalizable.
3. A is symmetric.

PROOF

(1) $\Leftrightarrow$ (2). This follows from the following observations: If $P = [X_1 \cdots X_n]$ is an $n \times n$ matrix, then P is orthogonal if and only if $\{X_1, \dots, X_n\}$ is an orthonormal set in $\mathbb{R}^n$; and $P^{-1}AP$ is diagonal if and only if $\{X_1, \dots, X_n\}$ consists of eigenvectors of A (see the proof of Theorem 3 §3.3).

(2) $\Rightarrow$ (3). If $P^T AP = D$ is diagonal, where $P^{-1} = P^T$, then $A = PDP^T$. Because $D^T = D$, this gives $A^T = P^{TT}D^T P^T = PDP^T = A$.

(3) $\Rightarrow$ (2). If A is an $n \times n$ symmetric matrix, we proceed by induction on n. If $n = 1$, A is already diagonal. If $n > 1$, assume that (3) $\Rightarrow$ (2) for $(n-1) \times (n-1)$ symmetric matrices. By Theorem 7 §5.5 let λ_1 be a (real) eigenvalue of A, and let $AX_1 = \lambda_1 X_1$, where $\|X_1\| = 1$. Use the Gram–Schmidt algorithm to find an orthonormal basis $\{X_1, X_2, \dots, X_n\}$ for $\mathbb{R}^n$, and let $P_1 = [X_1 \ X_2 \ \cdots \ X_n]$. Then P_1 is an orthogonal matrix and $P_1^T AP_1 = \begin{bmatrix} \lambda_1 & X \\ 0 & A_1 \end{bmatrix}$ in block form. But $P_1^T AP_1$ is symmetric (A is), so it follows that $X = 0$ and A_1 is symmetric. Then, by induction, there exists an $(n-1) \times (n-1)$ orthogonal matrix Q such that $Q^T A_1 Q = D_1$ is diagonal. Hence $P_2 = \begin{bmatrix} 1 & 0 \\ 0 & Q \end{bmatrix}$ is orthogonal and

$$(P_1P_2)^T A(P_1P_2) = P_2^T(P_1^T A P_1)P_2$$

$$= \begin{bmatrix} 1 & 0 \\ 0 & Q^T \end{bmatrix} \begin{bmatrix} \lambda_1 & 0 \\ 0 & A_1 \end{bmatrix} \begin{bmatrix} 1 & 0 \\ 0 & Q \end{bmatrix}$$

$$= \begin{bmatrix} \lambda_1 & 0 \\ 0 & D_1 \end{bmatrix}$$

is diagonal. Because P_1P_2 is orthogonal, this proves (2).

A set of orthonormal eigenvectors of a symmetric matrix A is called a set of **principal axes** for A. The name comes from geometry, and this is discussed in Section 8.9. Theorem 2 is also called the **real spectral theorem**, and the set of distinct eigenvalues is called the **spectrum** of the matrix. In full generality, the spectral theorem is a similar result for matrices with complex entries (Theorem 8 §8.6).

Example 4

Find an orthogonal matrix P such that $P^{-1}AP$ is diagonal, where $A = \begin{bmatrix} 1 & 0 & -1 \\ 0 & 1 & 2 \\ -1 & 2 & 5 \end{bmatrix}$.

Solution   The characteristic polynomial of A is

$$c_A(x) = \det \begin{bmatrix} x-1 & 0 & 1 \\ 0 & x-1 & -2 \\ 1 & -2 & x-5 \end{bmatrix} = x(x-1)(x-6)$$

Thus the eigenvalues are $\lambda = 0$, 1, and 6, and corresponding eigenvectors are

$$X_1 = \begin{bmatrix} 1 \\ -2 \\ 1 \end{bmatrix} \qquad X_2 = \begin{bmatrix} 2 \\ 1 \\ 0 \end{bmatrix} \qquad X_3 = \begin{bmatrix} -1 \\ 2 \\ 5 \end{bmatrix}$$

respectively. Moreover, by what appears to be remarkably good luck, these eigenvectors are *orthogonal*. We have $\|X_1\|^2 = 6$, $\|X_2\|^2 = 5$, and $\|X_3\|^2 = 30$, so

$$P = \begin{bmatrix} \frac{1}{\sqrt{6}}X_1 & \frac{1}{\sqrt{5}}X_2 & \frac{1}{\sqrt{30}}X_3 \end{bmatrix} = \frac{1}{\sqrt{30}}\begin{bmatrix} \sqrt{5} & 2\sqrt{6} & -1 \\ -2\sqrt{5} & \sqrt{6} & 2 \\ \sqrt{5} & 0 & 5 \end{bmatrix}$$

is orthogonal. Thus $P^{-1} = P^T$ and

$$P^TAP = \begin{bmatrix} 0 & 0 & 0 \\ 0 & 1 & 0 \\ 0 & 0 & 6 \end{bmatrix}$$

by the diagonalization algorithm.

Actually, the fact that the eigenvectors in Example 4 are orthogonal is no coincidence. Theorem 4 §5.5 guarantees they are linearly independent (they correspond to distinct eigenvalues); the fact that the matrix is symmetric implies that they are orthogonal. To prove this we need the following useful fact about symmetric matrices.

> **Theorem 3**
>
> If A is an $n \times n$ symmetric matrix, then
>
> $$(AX) \cdot Y = X \cdot (AY)$$
>
> for all columns X and Y in $\mathbb{R}^n$.

PROOF

Recall that $X \cdot Y = X^T Y$ for all columns X and Y. Because $A^T = A$, we get.

$$(AX) \cdot Y = (AX)^T Y = X^T A^T Y = X^T AY = X \cdot (AY).$$

> **Theorem 4**
>
> If A is a symmetric matrix, then eigenvectors of A corresponding to distinct eigenvalues are orthogonal.

PROOF

Let $AX = \lambda X$ and $AY = \mu Y$, where $\lambda \neq \mu$. Using Theorem 3, we compute

$$\lambda(X \cdot Y) = (\lambda X) \cdot Y = (AX) \cdot Y = X \cdot (AY) = X \cdot (\mu Y) = \mu(X \cdot Y)$$

Hence $(\lambda - \mu)(X \cdot Y) = 0$, and so $X \cdot Y = 0$ because $\lambda \neq \mu$.

Now the procedure for diagonalizing a symmetric $n \times n$ matrix is clear. Find the distinct eigenvalues (all real by Theorem 7 §5.5) and find orthonormal bases for each eigenspace (the Gram–Schmidt algorithm may be needed). Then the set of all these basis vectors is orthonormal (by Theorem 4) and contains n vectors.

Example 5

Orthogonally diagonalize $A = \begin{bmatrix} 8 & -2 & 2 \\ -2 & 5 & 4 \\ 2 & 4 & 5 \end{bmatrix}$.

Solution   The characteristic polynomial is

$$c_A(x) = \det \begin{bmatrix} x - 8 & 2 & -2 \\ 2 & x - 5 & -4 \\ -2 & -4 & x - 5 \end{bmatrix} = x(x - 9)^2$$

Hence the distinct eigenvalues are $\lambda = 0$ and 9 of multiplicities 1 and 2, respectively, so $\dim(E_0) = 1$ and $\dim(E_9) = 2$ by Theorem 6 §5.5 (A is diagonalizable, being symmetric). One eigenvector for $\lambda = 0$ is

$X_1 = \begin{bmatrix} 1 \\ 2 \\ -2 \end{bmatrix}$, so $E_0 = \text{span}\{X_1\}$. Gaussian elimination gives

$$E_9 = \text{span}\left\{ \begin{bmatrix} -2 \\ 1 \\ 0 \end{bmatrix}, \begin{bmatrix} 2 \\ 0 \\ 1 \end{bmatrix} \right\}$$

and these eigenvectors are both orthogonal to X_1 (as Theorem 4 guarantees) but not to each other. The Gram–Schmidt process gives an orthogonal basis $\{X_2, X_3\}$ of E_9 where $X_2 = \begin{bmatrix} -2 \\ 1 \\ 0 \end{bmatrix}$, $X_3 = \begin{bmatrix} 2 \\ 4 \\ 5 \end{bmatrix}$. Normalizing gives orthonormal eigenvectors

$$\left\{ \tfrac{1}{3} X_1, \ \tfrac{1}{\sqrt{5}} X_2, \ \tfrac{1}{3\sqrt{5}} X_3 \right\}$$

so

$$P = \begin{bmatrix} \tfrac{1}{3} X_1 & \tfrac{1}{\sqrt{5}} X_2 & \tfrac{1}{3\sqrt{5}} X_3 \end{bmatrix} = \frac{1}{3\sqrt{5}} \begin{bmatrix} \sqrt{5} & -6 & 2 \\ 2\sqrt{5} & 3 & 4 \\ -2\sqrt{5} & 0 & 5 \end{bmatrix}$$

is an orthogonal matrix such that $P^{-1}AP$ is diagonal.

It is worth noting that other, more convenient, matrices P exist. For example, $Y_2 = \begin{bmatrix} 2 \\ 1 \\ 2 \end{bmatrix}$ and $Y_3 = \begin{bmatrix} -2 \\ 2 \\ 1 \end{bmatrix}$ lie in E_9 and they are orthogonal. Moreover, they both have norm 3 (as does X_1), so

$$Q = \begin{bmatrix} \tfrac{1}{3} X_1 & \tfrac{1}{3} Y_2 & \tfrac{1}{3} Y_3 \end{bmatrix} = \frac{1}{3} \begin{bmatrix} 1 & 2 & -2 \\ 2 & 1 & 2 \\ -2 & 2 & 1 \end{bmatrix}$$

is a nicer orthogonal matrix with the property that $Q^{-1}AQ$ is diagonal.

If we are willing to replace *diagonal* by *upper triangular* in the principal axes theorem, we can weaken the requirement that A is symmetric to insisting only that A has real eigenvalues.

Theorem 5 Triangulation Theorem

If A is an $n \times n$ matrix with real eigenvalues, an orthogonal matrix P exists such that $P^T A P$ is upper triangular.[3]

PROOF

We modify the proof of Theorem 2. If $AX_1 = \lambda_1 X_1$ where $\|X_1\| = 1$, let $\{X_1, X_2, \ldots, X_n\}$ be an orthonormal basis of $\mathbb{R}^n$, and let $P_1 = [X_1 \ \cdots \ X_n]$. Then

---

3 There is also a lower triangular version.

P_1 is orthogonal and $P_1^T A P_1 = \begin{bmatrix} \lambda_1 & X \\ 0 & A_1 \end{bmatrix}$ in block form. By induction, let $Q^T A_1 Q = T_1$ be upper triangular where Q is orthogonal of size $(n-1) \times (n-1)$. Then $P_2 - \begin{bmatrix} 1 & 0 \\ 0 & Q \end{bmatrix}$ is orthogonal, so $P = P_1 P_2$ is also orthogonal and

$$P^T A P = \begin{bmatrix} \lambda_1 & XQ \\ 0 & T_1 \end{bmatrix}$$ is upper triangular.

The proof of Theorem 5 gives no way to construct the matrix P. However, an algorithm will be given in Section 9.4 where an improved version of Theorem 5 is presented.

As for a diagonal matrix, the eigenvalues of an upper triangular matrix are displayed along the main diagonal. Because, in Theorem 5, A and $P^T A P$ have the same determinant and trace, we obtain the following:

Corollary

If A is an $n \times n$ matrix with real eigenvalues $\lambda_1, \lambda_2, \dots, \lambda_n$ (possibly not all distinct), then $\det A = \lambda_1 \lambda_2 \cdots \lambda_n$ and $\operatorname{tr} A = \lambda_1 + \lambda_2 + \cdots + \lambda_n$.

This corollary remains true even if the eigenvalues are not real. In fact, a version of Theorem 5 holds for any $n \times n$ complex matrix (Schur's theorem). This is given in Section 8.6.

Exercises 8.2

1. Normalize the rows to make each of the following matrices orthogonal.

   (a) $\begin{bmatrix} 1 & 1 \\ -1 & 1 \end{bmatrix}$ ♦(b) $\begin{bmatrix} 3 & -4 \\ 4 & 3 \end{bmatrix}$

   (c) $\begin{bmatrix} 1 & 2 \\ -4 & 2 \end{bmatrix}$ ♦(d) $\begin{bmatrix} a & b \\ -b & a \end{bmatrix}$, $(a, b) \neq (0, 0)$

   (e) $\begin{bmatrix} \cos\theta & -\sin\theta & 0 \\ \sin\theta & \cos\theta & 0 \\ 0 & 0 & 2 \end{bmatrix}$ ♦(f) $\begin{bmatrix} 2 & 1 & -1 \\ 1 & -1 & 1 \\ 0 & 1 & 1 \end{bmatrix}$

   (g) $\begin{bmatrix} -1 & 2 & 2 \\ 2 & -1 & 2 \\ 2 & 2 & -1 \end{bmatrix}$ ♦(h) $\begin{bmatrix} 2 & 6 & -3 \\ 3 & 2 & 6 \\ -6 & 3 & 2 \end{bmatrix}$

♦2. If P is a triangular orthogonal matrix, show that P is diagonal and that all diagonal entries are 1 or -1.

3. If P is orthogonal, show that kP is orthogonal if and only if $k = 1$ or $k = -1$.

4. If the first two rows of an orthogonal matrix are $[\frac{1}{3}, \frac{2}{3}, \frac{2}{3}]$ and $[\frac{2}{3}, \frac{1}{3}, \frac{-2}{3}]$, find all possible third rows.

5. For each matrix A, find an orthogonal matrix P such that $P^{-1}AP$ is diagonal.

   (a) $A = \begin{bmatrix} 3 & 0 & 0 \\ 0 & 2 & 2 \\ 0 & 2 & 5 \end{bmatrix}$ ♦(b) $A = \begin{bmatrix} 3 & 0 & 7 \\ 0 & 5 & 0 \\ 7 & 0 & 3 \end{bmatrix}$

   (c) $A = \begin{bmatrix} 1 & 1 & 0 \\ 1 & 1 & 0 \\ 0 & 0 & 2 \end{bmatrix}$ ♦(d) $A = \begin{bmatrix} 5 & -2 & -4 \\ -2 & 8 & -2 \\ -4 & -2 & 5 \end{bmatrix}$

   (e) $A = \begin{bmatrix} 5 & 3 & 0 & 0 \\ 3 & 5 & 0 & 0 \\ 0 & 0 & 7 & 1 \\ 0 & 0 & 1 & 7 \end{bmatrix}$ ♦(f) $A = \begin{bmatrix} 3 & 5 & -1 & 1 \\ 5 & 3 & 1 & -1 \\ -1 & 1 & 3 & 5 \\ 1 & -1 & 5 & 3 \end{bmatrix}$

♦6. Consider $A = \begin{bmatrix} 0 & a & 0 \\ a & 0 & c \\ 0 & c & 0 \end{bmatrix}$ where one of $a, c \neq 0$.

   Show that $c_A(x) = x(x - k)(x + k)$, where

$k = \sqrt{a^2 + c^2}$ and find an orthogonal matrix P such that $P^{-1}AP$ is diagonal.

7. Consider $A = \begin{bmatrix} 0 & 0 & a \\ 0 & b & 0 \\ a & 0 & 0 \end{bmatrix}$. Show that $c_A(x) = (x - b)(x - a)(x + a)$ and find an orthogonal matrix P such that $P^{-1}AP$ is diagonal.

8. Given $A = \begin{bmatrix} b & a \\ a & b \end{bmatrix}$, show that $c_A(x) = (x - a - b)(x + a - b)$ and find an orthogonal matrix P such that $P^{-1}AP$ is diagonal.

9. Consider $A = \begin{bmatrix} b & 0 & a \\ 0 & b & 0 \\ a & 0 & b \end{bmatrix}$. Show that $c_A(x) = (x - b)(x - b - a)(x - b + a)$ and find an orthogonal matrix P such that $P^{-1}AP$ is diagonal.

10. Show that the following are equivalent for a symmetric matrix A.

    (a) A is orthogonal.   (b) $A^2 = I$.

    ♦(c) All eigenvalues of A are ± 1.
        [*Hint:* For (b) if and only if (c), use Theorem 2.]

11. We call matrices A and B **orthogonally similar** (and write $A \overset{\circ}{=} B$) if $B = P^T A P$ for an orthogonal matrix P.

    (a) Show that $A \overset{\circ}{=} A$ for all A; $A \overset{\circ}{=} B \Rightarrow B \overset{\circ}{=} A$; and $A \overset{\circ}{=} B$ and $B \overset{\circ}{=} C \Rightarrow A \overset{\circ}{=} C$.

    (b) Show that the following are equivalent for two symmetric matrices A and B.

        (i) A and B are similar.

        (ii) A and B are orthogonally similar.

        (iii) A and B have the same eigenvalues.

12. Assume that A and B are orthogonally similar (Exercise 11).

    (a) If A and B are invertible, show that A^{-1} and B^{-1} are orthogonally similar.

    ♦(b) Show that A^2 and B^2 are orthogonally similar.

    (c) Show that, if A is symmetric, so is B.

13. If A is symmetric, show that every eigenvalue of A is nonnegative if and only if $A = B^2$ for some symmetric matrix B.

♦14. Prove the converse of Theorem 3:
    If $(AX) \bullet Y = X \bullet (AY)$ for all n-columns X and Y, then A is symmetric.

15. Show that every eigenvalue of A is zero if and only if A is nilpotent ($A^k = 0$ for some $k \geq 1$).

16. If A has real eigenvalues, show that $A = B + C$ where B is symmetric and C is nilpotent. [*Hint:* Theorem 5.]

17. Let P be an orthogonal matrix.

    (a) Show that $\det P = 1$ or $\det P = -1$.

    ♦(b) Give 2×2 examples of P such that $\det P = 1$ and $\det P = -1$.

    (c) If $\det P = -1$, show that $I + P$ has no inverse. [*Hint:* $P^T(I + P) = (I + P)^T$.]

    (d) If P is $n \times n$ and $\det P \neq (-1)^n$, show that $I - P$ has no inverse. [*Hint:* $P^T(I - P) = -(I - P)^T$.]

18. We call a square matrix E a **projection matrix** if $E^2 = E = E^T$.

    (a) If E is a projection matrix, show that $P = I - 2E$ is orthogonal and symmetric.

    (b) If P is orthogonal and symmetric, show that $E = \frac{1}{2}(I - P)$ is a projection matrix.

    (c) If U is $m \times n$ and $U^T U = I$ (for example, a unit column in $\mathbb{R}^n$), show that $E = UU^T$ is a projection matrix.

19. A matrix that we obtain from the identity matrix by writing its rows in a different order is called a **permutation matrix**. Show that every permutation matrix is orthogonal.

♦20. If the rows $R_1, \ldots, R_n$ of the $n \times n$ matrix $A = [a_{ij}]$ are orthogonal, show that the (i, j)-entry of A^{-1} is $\dfrac{a_{ji}}{\| R_j \|^2}$.

21. (a) Let A be an $m \times n$ matrix. Show that the following are equivalent.

        (i) A has orthogonal rows.

        (ii) A can be factored as $A = DP$, where D is invertible and diagonal and P has orthonormal rows.

        (iii) AA^T is an invertible, diagonal matrix.

    (b) Show that an $n \times n$ matrix A has orthogonal rows if and only if A can be factored as $A = DP$, where P is orthogonal and D is diagonal and invertible.

22. Let A be a skew-symmetric matrix; that is, $A^T = -A$. Assume that A is an $n \times n$ matrix.

(a) Show that $I + A$ is invertible. [*Hint:* By Theorem 5 §2.3, it suffices to show that $(I + A)X = 0$, X in $\mathbb{R}^n$, implies $X = 0$. Compute $X \bullet X = X^T X$, and use the fact that $AX = -X$ and $A^2 X = X$.]

♦(b) Show that $P = (I - A)(I + A)^{-1}$ is orthogonal.

(c) Show that every orthogonal matrix P such that $I + P$ is invertible arises as in part (b) from some skew-symmetric matrix A. [*Hint:* Solve $P = (I - A)(I + A)^{-1}$ for A.]

23. Show that the following are equivalent for an $n \times n$ matrix P.

(a) P is orthogonal.

(b) $\|PX\| = \|X\|$ for all columns X in $\mathbb{R}^n$.

(c) $\|PX - PY\| = \|X - Y\|$ for all columns X and Y in $\mathbb{R}^n$.

(d) $(PX) \bullet (PY) = X \bullet Y$ for all columns X and Y in $\mathbb{R}^n$.
[*Hints:* For (c) $\Rightarrow$ (d), see Exercise 13(a) §5.3. For (d) $\Rightarrow$ (a), show that column i of P equals PE_i, where E_i is column i of the identity matrix.]

24. Show that every 2×2 orthogonal matrix has the form $\begin{bmatrix} \cos\theta & -\sin\theta \\ \sin\theta & \cos\theta \end{bmatrix}$ or $\begin{bmatrix} \cos\theta & \sin\theta \\ \sin\theta & -\cos\theta \end{bmatrix}$ for some angle θ. [*Hint:* If $a^2 + b^2 = 1$, then $a = \cos\theta$ and $b = \sin\theta$ for some angle θ.]

25. Use Theorem 5 to show that every symmetric matrix is orthogonally diagonalizable.

SECTION 8.3   Positive Definite Matrices

All the eigenvalues of any symmetric matrix are real; this section is about the case in which the eigenvalues are positive. These matrices, which arise whenever optimization (maximum and minimum) problems are encountered, have countless applications throughout science and engineering. They also arise in statistics (for example, in factor analysis used in the social sciences) and in geometry (see Section 8.9). We will encounter them again in Chapter 10 when describing all inner products in $\mathbb{R}^n$.

A square matrix is called **positive definite** if it is symmetric and all its eigenvalues are positive. Because these matrices are symmetric, the principal axis theorem plays a central role in the theory.

Theorem 1

If A is positive definite, then it is invertible and $\det A > 0$.

PROOF

If A is $n \times n$ and the eigenvalues are $\lambda_1, \lambda_2, \ldots, \lambda_n$, then $\det A = \lambda_1 \lambda_2 \cdots \lambda_n > 0$ by the principal axis theorem (or the corollary to Theorem 5 §8.2).

If X is a column in $\mathbb{R}^n$ and A is any real $n \times n$ matrix, we view the 1×1 matrix $X^T A X$ as a real number. With this convention, we have the following characterization of positive definite matrices.

Theorem 2

A symmetric matrix A is positive definite if and only if $X^T A X > 0$ for every column $X \neq 0$ in $\mathbb{R}^n$.

PROOF

A is symmetric so, by the principal axis theorem, let $P^TAP = D = \text{diag}(\lambda_1, \lambda_2, \ldots, \lambda_n)$ where $P^{-1} = P^T$ and the λ_i are the eigenvalues of A. Given X in $\mathbb{R}^n$, write $Y = P^TX = [y_1 \; y_2 \; \cdots \; y_n]^T$. Then

$$X^TAX = X^T(PDP^T)X = Y^TDY = \lambda_1 y_1^2 + \lambda_2 y_2^2 + \cdots + \lambda_n y_n^2 \qquad (*)$$

If A is positive definite and $X \neq 0$, then $X^TAX > 0$ by $(*)$ because some $y_j \neq 0$. Conversely, if $X^TAX > 0$ whenever $X \neq 0$, let $X = PE_j \neq 0$ where E_j is column j of I_n. Then $Y = E_j$, so $(*)$ reads $\lambda_j = X^TAX > 0$.

Example 1

If U is any invertible $n \times n$ matrix, show that $A = U^TU$ is positive definite.

Solution   If X is in $\mathbb{R}^n$ and $X \neq 0$, then

$$X^TAX = X^T(U^TU)X = (UX)^T(UX) = \|UX\|^2 > 0$$

because $UX \neq 0$ (U is invertible). Hence Theorem 2 applies.

It is remarkable that the converse to Example 1 is also true. In fact every positive definite matrix A can be factored as $A = U^TU$ where U is an upper triangular matrix with positive elements on the main diagonal. However, before verifying this, we introduce another concept that is central to any discussion of positive definite matrices.

If A is any $n \times n$ matrix, let $^{(r)}A$ denote the $r \times r$ submatrix of A in the upper left corner of A; that is $^{(r)}A$ is the matrix obtained from A by deleting the last $n - r$ rows and columns. The matrices $^{(1)}A, {}^{(2)}A, {}^{(3)}A, \ldots, {}^{(n)}A = A$ are called the **principal submatrices** of A.

Example 2

If $A = \begin{bmatrix} 10 & 5 & 2 \\ 5 & 3 & 2 \\ 2 & 2 & 3 \end{bmatrix}$ then $^{(1)}A = [10]$, $^{(2)}A = \begin{bmatrix} 10 & 5 \\ 5 & 3 \end{bmatrix}$ and $^{(3)}A = A$.

Lemma 1

If A is positive definite, so is each principal submatrix $^{(r)}A$ for $r = 1, 2, \ldots, n$.

PROOF

Write $A = \begin{bmatrix} {}^{(r)}A & P \\ Q & R \end{bmatrix}$ in block form. If $Y \neq 0$ in $\mathbb{R}^r$, write $X = \begin{bmatrix} Y \\ 0 \end{bmatrix}$ in $\mathbb{R}^n$. Then $X \neq 0$, so the fact that A is positive definite gives

$$0 < X^TAX = [Y^T \; 0]\begin{bmatrix} {}^{(r)}A & P \\ Q & R \end{bmatrix}\begin{bmatrix} Y \\ 0 \end{bmatrix} = Y^T({}^{(r)}A)Y$$

This shows that $^{(r)}A$ is positive definite by Theorem 2.[4]

---

[4] A similar argument shows that, if B is any matrix obtained from a positive definite matrix A by deleting certain rows and deleting the *same* columns, then B is also positive definite.

If A is positive definite, Lemma 1 and Theorem 1 show that $\det({}^{(r)}A) > 0$ for every r. This proves part of the following theorem which contains the converse to Example 1, and characterizes the positive definite matrices among the symmetric ones.

Theorem 3

The following conditions are equivalent for a symmetric $n \times n$ matrix A:
1. A is positive definite.
2. $\det({}^{(r)}A) > 0$ for each $r = 1, 2, \ldots, n$.
3. $A = U^T U$ where U is an upper triangular matrix with positive entries on the main diagonal.

Furthermore, the factorization in (3) is unique (called the **Cholesky Factorization**[5] of A).

PROOF

First, (3) $\Rightarrow$ (1) by Example 1, and (1) $\Rightarrow$ (2) by Lemma 1 and Theorem 1. (2) $\Rightarrow$ (3). Assume (2) and proceed by induction on n. If $n = 1$, then $A = [a]$ where $a > 0$ by (2), so take $U = [\sqrt{a}]$. If $n > 1$, write $B = {}^{(n-1)}A$. Then B is symmetric and satisfies (2) so, by induction, we have $B = U^T U$ as in (3) where U is of size $(n-1) \times (n-1)$. Then, as A is symmetric, it has block form

$$A = \begin{bmatrix} B & P \\ P^T & b \end{bmatrix}$$ where P is a column in $\mathbb{R}^{n-1}$ and b is in $\mathbb{R}$. If we write $X = (U^T)^{-1}P$

and $c = b - X^T X$, block multiplication gives

$$A = \begin{bmatrix} U^T U & P \\ P^T & b \end{bmatrix} = \begin{bmatrix} U^T & 0 \\ X^T & 1 \end{bmatrix} \begin{bmatrix} U & X \\ 0 & c \end{bmatrix}$$

as the reader can verify. Taking determinants and applying Theorem 5 §3.1 gives $\det A = \det(U^T) \det U \cdot c = c(\det U)^2$. Hence $c > 0$ because $\det A > 0$ by (2), so the above factorization can be written $A = \begin{bmatrix} U^T & 0 \\ X^T & \sqrt{c} \end{bmatrix} \begin{bmatrix} U & X \\ 0 & \sqrt{c} \end{bmatrix}$.

Since U has positive diagonal entries, this proves (3).

As to the uniqueness, suppose that $A = U^T U = U_1^T U_1$ are two Cholesky factorizations. Write $D = UU_1^{-1} = (U^T)^{-1}U_1^T$. Then D is both upper triangular, because $D = UU_1^{-1}$, and lower triangular, because $D = (U^T)^{-1}U_1^T$, and so is a diagonal matrix. Thus $U = DU_1$ and $U_1 = DU$, so it suffices to show that $D = I$. But eliminating U_1 gives $U = D^2 U$, so $D^2 = I$ because U is invertible. Since the diagonal entries of D are positive (this is true of U and U_1), it follows that $D = I$.

The remarkable thing is that the matrix U in the Cholesky factorization is easy to obtain from A using row operations. The key is that Step 1 of the following algorithm is possible for any positive definite matrix A. A proof of the algorithm is given following Example 3.

---

5 The factorization is sometimes defined as $A = LL^T$ where $L = U^T$ is lower triangular with positive diagonal entries.

Algorithm for the Cholesky Factorization

If A is a positive definite matrix, the Cholesky factorization $A = U^T U$ can be obtained as follows:

Step 1. Carry A to an upper triangular matrix U_1 with positive diagonal entries using row operations each of which adds a multiple of a row to a lower row.

Step 2. Obtain U from U_1 by dividing each row of U_1 by the square root of the diagonal entry in that row.

Example 3

Find the Cholesky factorization of $A = \begin{bmatrix} 10 & 5 & 2 \\ 5 & 3 & 2 \\ 2 & 2 & 3 \end{bmatrix}$.

Solution   The matrix A is positive definite by Theorem 3 because $\det{}^{(1)}A = 10 > 0$, $\det{}^{(2)}A = 5 > 0$ and $\det{}^{(3)}A = \det A = 3 > 0$. Hence Step 1 of the algorithm is carried out as follows:

$$A = \begin{bmatrix} 10 & 5 & 2 \\ 5 & 3 & 2 \\ 2 & 2 & 3 \end{bmatrix} \to \begin{bmatrix} 10 & 5 & 2 \\ 0 & \frac{1}{2} & 1 \\ 0 & 1 & \frac{13}{5} \end{bmatrix} \to \begin{bmatrix} 10 & 5 & 2 \\ 0 & \frac{1}{2} & 1 \\ 0 & 0 & \frac{3}{5} \end{bmatrix} = U_1$$

Now carry out Step 2 on U_1 to obtain $U = \begin{bmatrix} \sqrt{10} & \frac{5}{\sqrt{10}} & \frac{2}{\sqrt{10}} \\ 0 & \frac{1}{\sqrt{2}} & \sqrt{2} \\ 0 & 0 & \frac{\sqrt{3}}{\sqrt{5}} \end{bmatrix}$. The reader can

verify that $U^T U = A$.

PROOF OF THE CHOLESKY ALGORITHM

If A is positive definite, let $A = U^T U$ be the Cholesky factorization, and let $D = \text{diag}(d_1, \dots, d_n)$ be the common diagonal of U and U^T. Then $U^T D^{-1}$ is lower triangular with ones on the diagonal (call such matrices LT-1). Hence $L = (U^T D^{-1})^{-1}$ is also LT-1, and so $I_n \to L$ by a sequence of row operations each of which adds a multiple of a row to a lower row (verify; modify columns right to left). But then $A \to LA$ by the same sequence of row operations (see the discussion preceding Theorem 1 §2.4). Since $LA = [D(U^T)^{-1}][U^T U] = DU$ is upper triangular with positive entries on the diagonal, this shows that Step 1 of the algorithm is possible.

Turning to Step 2, let $A \to U_1$ as in Step 1 so that $U_1 = L_1 A$ where L_1 is LT-1. Since A is symmetric, we get

$$L_1 U_1^T = L_1(L_1 A)^T = L_1 A^T L_1^T = L_1 A L_1^T = U_1 L_1^T \qquad (*)$$

Let $D_1 = \text{diag}(e_1, \dots, e_n)$ denote the diagonal of U_1. Then $(*)$ gives $L_1(U_1^T D_1^{-1}) = U_1 L_1^T D_1^{-1}$. This is both upper triangular (right side) and LT-1 (left side), and so must equal I_n. In particular, $U_1^T D_1^{-1} = L_1^{-1}$. Now let $D_2 = \text{diag}(\sqrt{e_1}, \dots, \sqrt{e_n})$ so that $D_2^2 = D_1$. If we write $U = D_2^{-1} U_1$, we have

$$U^T U = (U_1^T D_2^{-1})(D_2^{-1} U_1) = U_1^T (D_2^2)^{-1} U_1 = (U_1^T D_1^{-1}) U_1 = (L_1^{-1}) U_1 = A$$

This proves Step 2 because $U = D_2^{-1} U_1$ is formed by dividing each row of U_1 by the square root of its diagonal entry (verify).

1. Find the Cholesky decomposition of each of the following matrices.

   (a) $\begin{bmatrix} 4 & 3 \\ 3 & 5 \end{bmatrix}$ ◆(b) $\begin{bmatrix} 2 & -1 \\ -1 & 1 \end{bmatrix}$

   (c) $\begin{bmatrix} 12 & 4 & 3 \\ 4 & 2 & -1 \\ 3 & -1 & 7 \end{bmatrix}$ ◆(d) $\begin{bmatrix} 20 & 4 & 5 \\ 4 & 2 & 3 \\ 5 & 3 & 5 \end{bmatrix}$

2. If A is positive definite, show that A^k is positive definite for all $k \geq 1$. Prove the converse if k is odd. What if k is even?

3. Find a symmetric matrix A such that A^2 is positive definite but A is not.

4. If A and B are positive definite, show that $A + B$ is positive definite.

5. If A and B are positive definite, show that $\begin{bmatrix} A & 0 \\ 0 & B \end{bmatrix}$ is positive definite.

◆6. If A is an $n \times n$ positive definite matrix and U is an $n \times m$ matrix of rank m, show that $U^T A U$ is positive definite.

7. If A is positive definite, show that each diagonal entry is positive.

8. Let A_0 be formed from A by deleting rows 2 and 4 and deleting columns 2 and 4. If A is positive definite, show that A_0 is positive definite.

9. If A is positive definite, show that $A = CC^T$ where C has orthogonal columns.

◆10. If A is positive definite, show that $A = C^2$ where C is positive definite.

11. Let A be a positive definite matrix. If a is a real number, show that aA is positive definite if and only if $a > 0$.

12. (a) Suppose an invertible matrix A can be factored in $\mathbf{M}_{nn}$ as $A = LDU$ where L is lower triangular with 1s on the diagonal, U is upper triangular with 1s on the diagonal, and D is diagonal with positive diagonal entries. Show that the factorization is unique: If $A = L_1 D_1 U_1$ is another such factorization, show that $L_1 = L$, $D_1 = D$, and $U_1 = U$.

    ◆(b) Show that a matrix A is positive definite if and only if A is symmetric and admits a factorization $A = LDU$ as in (a).

13. Let A be positive definite and write $d_r = \det {}^{(r)} A$ for each $r = 1, 2, \ldots, n$. If U is the upper triangular matrix obtained in step 1 of the algorithm, show that the diagonal elements $u_{11}, u_{22}, \ldots, u_{nn}$ of U are given by $u_{11} = d_1$, $u_{jj} = d_j / d_{j-1}$ if $j > 1$.
    [*Hint:* If $LA = U$ where L is lower triangular with 1s on the diagonal, use block multiplication to show that $\det {}^{(r)} A = \det {}^{(r)} U$ for each r.]

SECTION 8.4   QR-Factorization[6]

The main virtue of orthogonal matrices is that they can be inverted easily—simply take the transpose. This fact combines with the following theorem to give a useful way to simplify many matrix calculations as, for example, in approximation—see Section 8.7. The result is a matrix version of the Gram–Schmidt process resulting in a factorization of any matrix A with independent columns as the product $A = QR$ of a matrix Q with *orthonormal* columns and an invertible upper triangular matrix R. This is called (rather unimaginatively) a **QR-factorization** of A. It is important in applications, and is particularly useful in calculations because there are computer algorithms that accomplish the decomposition with good control over round-off error.

---

6 This section is not used elsewhere in this book.

Suppose $A = [C_1 \ C_2 \ \cdots \ C_n]$ is an $m \times n$ matrix with linearly independent columns $C_1, C_2, \ldots, C_n$. The Gram–Schmidt algorithm can be applied to these columns to provide orthogonal columns $F_1, F_2, \ldots, F_n$ where $F_1 = C_1$ and

$$F_k = C_k - \frac{C_k \bullet F_1}{\|F_1\|^2} F_1 - \frac{C_k \bullet F_2}{\|F_2\|^2} F_2 - \cdots - \frac{C_k \bullet F_{k-1}}{\|F_{k-1}\|^2} F_{k-1}$$

for each $k = 2, 3, \ldots, n$. Now write $Q_k = \dfrac{1}{\|F_k\|} F_k$ for each k. Then $Q_1, Q_2, \ldots, Q_n$ are orthonormal columns, and the above equation becomes

$$\|F_k\| Q_k = C_k - (C_k \bullet Q_1) Q_1 - (C_k \bullet Q_2) Q_2 - \cdots - (C_k \bullet Q_{k-1}) Q_{k-1}$$

Using these equations, express each C_k as a linear combination of the Q_i:

$$
\begin{aligned}
C_1 &= \|F_1\| Q_1 \\
C_2 &= (C_2 \bullet Q_1) Q_1 + \|F_2\| Q_2 \\
C_3 &= (C_3 \bullet Q_1) Q_1 + (C_3 \bullet Q_2) Q_2 + \|F_3\| Q_3 \\
&\ \ \vdots \qquad\qquad\qquad \vdots \\
C_n &= (C_n \bullet Q_1) Q_1 + (C_n \bullet Q_2) Q_2 + (C_n \bullet Q_3) Q_3 + \cdots + \|F_n\| Q_n
\end{aligned}
$$

Using block multiplication, these equations have a matrix form that gives the required factorization:

$$
\begin{aligned}
A &= [C_1 \ C_2 \ C_3 \ \cdots \ C_n] \\[2mm]
&= [Q_1 \ Q_2 \ Q_3 \ \cdots \ Q_n]
\begin{bmatrix}
\|F_1\| & C_2 \bullet Q_1 & C_3 \bullet Q_1 & \cdots & C_n \bullet Q_1 \\
0 & \|F_2\| & C_3 \bullet Q_2 & \cdots & C_n \bullet Q_2 \\
0 & 0 & \|F_3\| & \cdots & C_n \bullet Q_3 \\
\vdots & \vdots & \vdots & \ddots & \vdots \\
0 & 0 & 0 & \cdots & \|F_n\|
\end{bmatrix}
\end{aligned} \qquad (*)
$$

Here the first factor $Q = [Q_1 \ Q_2 \ Q_3 \ \cdots \ Q_n]$ has orthonormal columns, and the second factor is an $n \times n$ upper triangular matrix R with nonzero (in fact, positive) diagonal entries, and so is invertible. We record this in the following theorem.

Theorem 1 QR-Factorization

Every $m \times n$ matrix A with linearly independent columns has a QR-factorization $A = QR$. In fact, the upper triangular matrix R can be chosen with positive diagonal entries.

The matrices Q and R in Theorem 1 are almost uniquely determined by A; we return to this below.

Example 1

Find the QR-factorization of $A = \begin{bmatrix} 1 & 1 & 0 \\ -1 & 0 & 1 \\ 0 & 1 & 1 \\ 0 & 0 & 1 \end{bmatrix}$.

Solution   The columns of A are $C_1 = [1 \ -1 \ 0 \ 0]^T$, $C_2 = [1 \ 0 \ 1 \ 0]^T$, and $C_3 = [0 \ 1 \ 1 \ 1]^T$, and $\{C_1, C_2, C_3\}$ is independent. If we apply the Gram–Schmidt algorithm to these columns C_i, the result is

$$F_1 = C_1 \qquad\qquad = [1 \ -1 \ 0 \ 0]^T$$
$$F_2 = C_2 - \tfrac{1}{2}F_1 \qquad = [\tfrac{1}{2} \ \tfrac{1}{2} \ 1 \ 0]^T$$
$$F_3 = C_3 + \tfrac{1}{2}F_1 - F_2 = [0 \ 0 \ 0 \ 1]^T$$

Hence let $Q_j = \dfrac{1}{\|F_j\|} F_j$ for each j. Then equation $(*)$ preceding Theorem 1 gives $A = QR$ where

$$Q = [Q_1 \ Q_2 \ Q_3] = \begin{bmatrix} \frac{1}{\sqrt{2}} & \frac{1}{\sqrt{6}} & 0 \\ \frac{-1}{\sqrt{2}} & \frac{1}{\sqrt{6}} & 0 \\ 0 & \frac{2}{\sqrt{6}} & 0 \\ 0 & 0 & 1 \end{bmatrix} = \frac{1}{\sqrt{6}} \begin{bmatrix} \sqrt{3} & 1 & 0 \\ -\sqrt{3} & 1 & 0 \\ 0 & 2 & 0 \\ 0 & 0 & \sqrt{6} \end{bmatrix}$$

$$R = \begin{bmatrix} \|F_1\| & C_2 \bullet Q_1 & C_3 \bullet Q_1 \\ 0 & \|F_2\| & C_3 \bullet Q_2 \\ 0 & 0 & \|F_3\| \end{bmatrix} = \begin{bmatrix} \sqrt{2} & \frac{1}{\sqrt{2}} & \frac{-1}{\sqrt{2}} \\ 0 & \frac{\sqrt{3}}{\sqrt{2}} & \frac{\sqrt{3}}{\sqrt{2}} \\ 0 & 0 & 1 \end{bmatrix} = \frac{1}{\sqrt{2}} \begin{bmatrix} 2 & 1 & -1 \\ 0 & \sqrt{3} & \sqrt{3} \\ 0 & 0 & \sqrt{2} \end{bmatrix}$$

The reader can verify that indeed $A = QR$.

If a matrix A has independent rows and we apply QR-factorization to A^T, the result is:

Corollary

If A has independent rows, then A factors uniquely as $A = LP$ where P has orthonormal rows and L is an invertible lower triangular matrix with positive main diagonal entries.

Since a square matrix with orthonormal columns is orthogonal, we have

Theorem 2

Every square, invertible matrix A has factorizations $A = QR$ and $A = LP$ where Q and P are orthogonal, R is upper triangular and invertible, and L is lower triangular and invertible.

Remark   If an $m \times n$ matrix A has independent columns then $A^T A$ is invertible (by Theorem 4 §5.4), and it is often desirable to compute the inverse of $A^T A$ (see Section 8.7 for example). This is simplified if we have a QR-factorization of A (and is one of the main reasons for the importance of Theorem 1). For if $A = QR$ is such a factorization, then $Q^T Q = I_n$ because Q has orthonormal columns (verify), so we obtain

$$A^T A = R^T Q^T Q R = R^T R.$$

Hence computing $(A^TA)^{-1}$ amounts to finding R^{-1}, and this is a routine matter because R is upper triangular. Thus the difficulty in computing $(A^TA)^{-1}$ lies in obtaining the QR-factorization of A.

Uniqueness

The QR-factorization $A = QR$ in equation $(*)$ preceding Theorem 1 has the additional virtue that the diagonal entries of R are all positive. This, together with the following result, reveals the degree to which the QR-factorization is unique.

Lemma 1

Let $A = QR$ be a QR-factorization of the $m \times n$ matrix A. Then a QR-factorization $A = Q_1R_1$ in which R_1 has positive diagonal entries can be obtained as follows: Carry $R \to R_1$ by negating the rows in R with negative diagonal entries, and then obtain $Q \to Q_1$ by negating the corresponding columns of Q.

PROOF

Let D be the diagonal matrix obtained from I_n by negating the rows which, in R, have a negative diagonal entry. Then $R_1 = DR$ is obtained from R by negating the corresponding rows and so R_1 has positive diagonal entries.

Since $D^{-1} = D$, we have $R = DR_1$. Hence $A = Q(DR_1) = Q_1R_1$ is a QR-factorization where $Q_1 = QD$ is obtained from Q by negating the corresponding *columns* of Q.

Hence we can move from one QR-factorization $A = QR$ of a matrix A to another by negating rows in R and negating the corresponding columns in Q. The next theorem shows that the factorization with every diagonal entry of R positive is uniquely determined by A.

Theorem 3

Let A be an $m \times n$ matrix with independent columns. If $A = QR$ and $A = Q_1R_1$ are QR-factorizations of A in which R and R_1 have positive entries on the main diagonal, then $Q_1 = Q$ and $R_1 = R$.

PROOF

Write $Q = [C_1 \; C_2 \; \cdots \; C_n]$ and $Q_1 = [D_1 \; D_2 \; \cdots \; D_n]$ in terms of their columns, and observe first that $Q^TQ = I_n = Q_1^TQ_1$ because Q and Q_1 have orthonormal columns. Hence it suffices to show that $Q_1 = Q$ (then $R_1 = Q_1^TA = Q^TA = R$). Since $Q_1^TQ_1 = I_n$, the equation $QR = Q_1R_1$ gives $Q_1^TQ = R_1R^{-1}$; for convenience we write this matrix as

$$Q_1^TQ = R_1R^{-1} = [t_{ij}].$$

This matrix is upper triangular with positive diagonal elements (since this is true for R and R_1), so $t_{ii} > 0$ for each i and $t_{ij} = 0$ if $i > j$. On the other hand, the (i, j)-entry of

$Q_1^T Q$ is $D_i^T C_j = D_i \cdot C_j$, so we have $D_i \cdot C_j = t_{ij}$ for all i and j. But $\{D_1, D_2, \ldots, D_n\}$ is an orthonormal basis of $\mathbb{R}^n$, so the expansion theorem gives

$$C_j = (D_1 \cdot C_j)D_1 + (D_2 \cdot C_j)D_2 + \cdots + (D_n \cdot C_j)D_n = t_{1j}D_1 + t_{2j}D_2 + \cdots + t_{ij}D_i$$

because $D_i \cdot C_j = t_{ij} = 0$ if $i > j$. The first few equations here are

$$\begin{aligned}
C_1 &= t_{11}D_1 \\
C_2 &= t_{12}D_1 + t_{22}D_2 \\
C_3 &= t_{13}D_1 + t_{23}D_2 + t_{33}D_3 \\
C_4 &= t_{14}D_1 + t_{24}D_2 + t_{34}D_3 + t_{44}D_4 \\
&\vdots \qquad \qquad \vdots
\end{aligned}$$

The first of these equations gives $1 = \|C_1\| = \|t_{11}D_1\| = |t_{11}| \|D_1\| = t_{11}$, whence $C_1 = D_1$. But then $t_{12} = D_1 \cdot C_2 = C_1 \cdot C_2 = 0$, so the second equation becomes $C_2 = t_{22}D_2$. Now a similar argument gives $C_2 = D_2$, and then $t_{13} = 0$ and $t_{23} = 0$ follows in the same way. Hence $C_3 = t_{33}D_3$ and $C_3 = D_3$. Continue in this way to get $C_i = D_i$ for all i. This means that $Q_1 = Q$, which is what we wanted.

Exercises 8.4

1. In each case, factor A as $A = QR$, where R is invertible and upper triangular and P has orthonormal columns.

   (a) $A = \begin{bmatrix} 1 & -1 \\ -1 & 0 \end{bmatrix}$ ◆(b) $A = \begin{bmatrix} 2 & 1 \\ 1 & 1 \end{bmatrix}$

   (c) $A = \begin{bmatrix} 1 & 1 & 1 \\ 1 & 1 & 0 \\ 1 & 0 & 0 \\ 0 & 0 & 0 \end{bmatrix}$ ◆(d) $A = \begin{bmatrix} 1 & 1 & 0 \\ -1 & 0 & 1 \\ 0 & 1 & 1 \\ 1 & -1 & 0 \end{bmatrix}$

2. If $A = QR = Q_1R_1$ are two QR-factorizations of the invertible matrix A, show that $R_1 = DR$ and $Q_1 = QD$ for some diagonal matrix D with diagonal entries ± 1. [*Hint:* Consider $RR_1^{-1} = Q_1^{-1}Q$.]

SECTION 8.5   Computing Eigenvalues

In practice, the problem of finding eigenvalues and eigenvectors of a matrix is virtually never solved by finding the roots of the characteristic polynomial. This is difficult for large matrices; iterative methods are much better. Two of these will be described briefly in this section.

Recall that an eigenvalue λ of an $n \times n$ matrix A is called a **dominant eigenvalue** if λ has multiplicity 1, and

$$|\lambda| > |\mu| \quad \text{for all eigenvalues } \mu \neq \lambda.$$

Any corresponding eigenvector is called a **dominant eigenvector** of A. When such an eigenvalue exists, one technique for finding it is as follows: Let X_0 in $\mathbb{R}^n$ be a first approximation to a dominant eigenvector, and compute successive approximations $X_1, X_2, \ldots$ by

$$X_1 = AX_0 \qquad X_2 = AX_1 \qquad X_3 = AX_2 \qquad \cdots$$

In general, we define

$$X_{k+1} = AX_k \quad \text{for each } k \geq 0.$$

If the first estimate X_0 is good enough (see below), these vectors X_n will approximate dominant eigenvectors of A. This technique is called the **power method** (because $X_k = A^k X_0$ for each $k \geq 1$). Moreover, it can be used to approximate the dominant eigenvalue λ. Observe that if Z is any eigenvector corresponding to λ, then

$$\frac{Z \cdot (AZ)}{\|Z\|^2} = \frac{Z \cdot (\lambda Z)}{\|Z\|^2} = \lambda.$$

Because the vectors $X_1, X_2, \ldots, X_n, \ldots$ approximate dominant eigenvectors, we define the **Rayleigh quotients** as follows:

$$r_k = \frac{X_k \cdot AX_k}{\|X_k\|^2} = \frac{X_k \cdot X_{k+1}}{\|X_k\|^2} \quad \text{for } k \geq 1.$$

Then the numbers r_k approximate the dominant eigenvalue λ.

Example 1

Use the power method to approximate a dominant eigenvector and eigenvalue of $A = \begin{bmatrix} 1 & 1 \\ 2 & 0 \end{bmatrix}$.

Solution   The eigenvalues of A are 2 and -1, with eigenvectors $\begin{bmatrix} 1 \\ 1 \end{bmatrix}$ and $\begin{bmatrix} 1 \\ -2 \end{bmatrix}$. Take $X_0 = \begin{bmatrix} 1 \\ 0 \end{bmatrix}$ as the first approximation and compute $X_1, X_2, \ldots,$ successively, from $X_1 = AX_0$, $X_2 = AX_1, \ldots$. The result is

$$X_1 = \begin{bmatrix} 1 \\ 2 \end{bmatrix}, X_2 = \begin{bmatrix} 3 \\ 2 \end{bmatrix}, X_3 = \begin{bmatrix} 5 \\ 6 \end{bmatrix}, X_4 = \begin{bmatrix} 11 \\ 10 \end{bmatrix}, X_5 = \begin{bmatrix} 21 \\ 22 \end{bmatrix}, \ldots$$

These vectors are approaching scalar multiples of the dominant eigenvector $\begin{bmatrix} 1 \\ 1 \end{bmatrix}$. Moreover, the Rayleigh quotients are

$$r_1 = \tfrac{7}{5}, r_2 = \tfrac{27}{13}, r_3 = \tfrac{115}{61}, r_4 = \tfrac{451}{221}, \ldots$$

and these are approaching the dominant eigenvalue 2.

To see why the power method works, let $\lambda_1, \lambda_2, \ldots, \lambda_m$ be eigenvalues of A with λ_1 dominant and let $Y_1, Y_2, \ldots, Y_m$ be corresponding eigenvectors. What is required is that the first approximation X_0 be a linear combination of these eigenvectors:

$$X_0 = a_1 Y_1 + a_2 Y_2 + \cdots + a_m Y_m \quad \text{with } a_1 \neq 0$$

If $k \geq 1$, the fact that $A^k Y_i = \lambda_i^k Y_i$ for each i gives

$$X_k = a_1 \lambda_1^k Y_1 + a_2 \lambda_2^k Y_2 + \cdots + a_m \lambda_m^k Y_m \quad \text{for } k \geq 1$$

Hence

$$\frac{1}{\lambda_1^k} X_k = a_1 Y_1 + a_2 \left(\frac{\lambda_2}{\lambda_1}\right)^k Y_2 + \cdots + a_m \left(\frac{\lambda_m}{\lambda_1}\right)^k Y_m$$

The right side approaches $a_1 Y_1$ as k increases because λ_1 is dominant $\left(\left|\frac{\lambda_i}{\lambda_1}\right| < 1 \text{ for each } i > 1\right)$. Because $a_1 \neq 0$, this means that X_k approximates the dominant eigenvector $a_1 \lambda_1^k Y_1$.

The power method requires that the first approximation X_0 be a linear combination of eigenvectors. (In Example 1 the eigenvectors form a basis of $\mathbb{R}^2$.) But even in this case the method fails if $a_1 = 0$, where a_1 is the coefficient of the dominant eigenvector (try $X_0 = \begin{bmatrix} -1 \\ 2 \end{bmatrix}$ in Example 1). In general, the rate of convergence is quite slow if any of the ratios $\left|\frac{\lambda_i}{\lambda_1}\right|$ is near 1. Also, because the method requires repeated multiplications by A, it is not recommended unless these multiplications are easy to carry out (for example, if most of the entries of A are zero).

QR-Algorithm

A much better method depends on the factorization (using the Gram–Schmidt algorithm) of an invertible matrix A in the form

$$A = QR$$

where Q is orthogonal and R is invertible and upper triangular (see Theorem 2 §8.4). The **QR-algorithm** uses this repeatedly to create a sequence of matrices $A_1 = A, A_2, A_3, \ldots$ as follows:

1. Define $A_1 = A$ and factor it as $A_1 = Q_1 R_1$.
2. Define $A_2 = R_1 Q_1$ and factor it as $A_2 = Q_2 R_2$.
3. Define $A_3 = R_2 Q_2$ and factor it as $A_3 = Q_3 R_3$.

$$\vdots$$

In general, A_k is factored as $A_k = Q_k R_k$ and we define $A_{k+1} = R_k Q_k$. Then A_{k+1} is similar to A_k [in fact, $A_{k+1} = R_k Q_k = (Q_k^{-1} A_k) Q_k$], and hence each A_k has the same eigenvalues as A. If the eigenvalues of A are real and have distinct absolute values, the remarkable thing is that the sequence of matrices $A_1, A_2, A_3, \ldots$ converges to an upper triangular matrix with these eigenvalues on the main diagonal. [See below for the case of complex eigenvalues.]

Example 2

If $A = \begin{bmatrix} 1 & 1 \\ 2 & 0 \end{bmatrix}$ as in Example 1, use the QR-algorithm to approximate the eigenvalues.

Solution   The matrices A_1, A_2, and A_3 are as follows:

$$A_1 = \begin{bmatrix} 1 & 1 \\ 2 & 0 \end{bmatrix} = Q_1 R_1 \quad \text{where } Q_1 = \frac{1}{\sqrt{5}} \begin{bmatrix} 1 & 2 \\ 2 & -1 \end{bmatrix} \text{ and } R_1 = \frac{1}{\sqrt{5}} \begin{bmatrix} 5 & 1 \\ 0 & 2 \end{bmatrix}$$

$$A_2 = \tfrac{1}{5}\begin{bmatrix} 7 & 9 \\ 4 & -2 \end{bmatrix} = \begin{bmatrix} 1.4 & -1.8 \\ -0.8 & -0.4 \end{bmatrix} = Q_2 R_2$$

$$\text{where } Q_2 = \tfrac{1}{\sqrt{65}}\begin{bmatrix} 7 & 4 \\ 4 & -7 \end{bmatrix} \text{ and } R_2 = \tfrac{1}{\sqrt{65}}\begin{bmatrix} 13 & 11 \\ 0 & 10 \end{bmatrix}$$

$$A_3 = \tfrac{1}{13}\begin{bmatrix} 27 & -5 \\ 8 & -14 \end{bmatrix} = \begin{bmatrix} 2.08 & -0.38 \\ 0.62 & -1.08 \end{bmatrix}$$

This is converging to $\begin{bmatrix} 2 & * \\ 0 & -1 \end{bmatrix}$ and so is approximating the eigenvalues 2 and -1 on the main diagonal.

It is beyond the scope of this book to pursue a detailed discussion of these methods. The reader is referred to J. M. Wilkinson, *The Algebraic Eigenvalue Problem* (Oxford, England: Oxford University Press, 1965) or G. W. Stewart, *Introduction to Matrix Computations* (New York: Academic Press, 1973). We conclude with some remarks on the QR-algorithm.

Shifting Convergence is accelerated if, at stage k of the algorithm, a number s_k is chosen and $A_k - s_k I$ is factored in the form $Q_k R_k$ rather than A_k itself. Then

$$Q_k^{-1} A_k Q_k = Q_k^{-1}(Q_k R_k + s_k I)Q_k = R_k Q_k + s_k I$$

so we take $A_{k+1} = R_k Q_k + s_k I$. If the shifts s_k are carefully chosen, convergence can be greatly improved.

Preliminary Preparation A matrix such as

$$\begin{bmatrix} * & * & * & * & * \\ * & * & * & * & * \\ 0 & * & * & * & * \\ 0 & 0 & * & * & * \\ 0 & 0 & 0 & * & * \end{bmatrix}$$

is said to be in **upper Hessenberg** form, and the QR-factorizations of such matrices are greatly simplified. A series of orthogonal matrices $H_1, H_2, \ldots, H_m$ (called **Householder matrices**) can be easily constructed such that

$$B = H_m^T \cdots H_1^T A H_1 \cdots H_m$$

is in upper Hessenberg form. Then the QR-algorithm can be efficiently applied to B and, because B is similar to A, it produces the eigenvalues of A.

Complex Eigenvalues If some of the eigenvalues of a real matrix A are not real, the QR-algorithm converges to a block upper triangular matrix where the diagonal blocks are either 1×1 (the real eigenvalues) or 2×2 (each providing a pair of conjugate complex eigenvalues of A).

Exercises 8.5

1. In each case, find the exact eigenvalues and determine corresponding eigenvectors.
   Then start with $X_0 = \begin{bmatrix} 1 \\ 1 \end{bmatrix}$ and compute X_4 and r_3 using the power method.

   (a) $\begin{bmatrix} 2 & -4 \\ -3 & 3 \end{bmatrix}$ ◆(b) $\begin{bmatrix} 5 & 2 \\ -3 & -2 \end{bmatrix}$

   (c) $\begin{bmatrix} 1 & 2 \\ 2 & 1 \end{bmatrix}$ ◆(d) $\begin{bmatrix} 3 & 1 \\ 1 & 0 \end{bmatrix}$

2. In each case, find the exact eigenvalues and then approximate them using the QR-algorithm.

   (a) $\begin{bmatrix} 1 & 1 \\ 1 & 0 \end{bmatrix}$ ◆(b) $\begin{bmatrix} 3 & 1 \\ 1 & 0 \end{bmatrix}$

3. Apply the power method to $A = \begin{bmatrix} 0 & 1 \\ -1 & 0 \end{bmatrix}$,

   starting at $X_0 = \begin{bmatrix} 1 \\ 1 \end{bmatrix}$. Does it converge? Explain.

◆4. If A is symmetric, show that each matrix A_k in the QR-algorithm is also symmetric. Deduce that they converge to a diagonal matrix.

5. Apply the QR-algorithm to $A = \begin{bmatrix} 2 & -3 \\ 1 & -2 \end{bmatrix}$. Explain.

6. Given a matrix A, let A_k, Q_k, and R_k, $k \geq 1$, be the matrices constructed in the QR-algorithm. Show that $A_k = (Q_1 Q_2 \cdots Q_k)(R_k \cdots R_2 R_1)$ for each $k \geq 1$ and hence that this is a QR-factorization of A_k. [*Hint:* Show that $Q_k R_k = R_{k-1} Q_{k-1}$ for each $k \geq 2$, and use this equality to compute $(Q_1 Q_2 \cdots Q_k)(R_k \cdots R_2 R_1)$ "from the centre out." Use the fact that $(AB)^{n+1} = A(BA)^n B$ for any square matrices A and B.]

SECTION 8.6   Complex Matrices

If A is an $n \times n$ matrix, the characteristic polynomial $c_A(x)$ is a polynomial of degree n and the eigenvalues of A are just the roots of $c_A(x)$. In each of our examples these roots have been *real* numbers (in fact, the examples have been carefully chosen so this will be the case!); but it need not happen, even when the characteristic polynomial has real coefficients. For example, if $A = \begin{bmatrix} 0 & 1 \\ -1 & 0 \end{bmatrix}$, then $c_A(x) = x^2 + 1$

has roots i and $-i$, where i is the complex number satisfying $i^2 = -1$. Therefore, we have to deal with the possibility that the eigenvalues of a (real) square matrix might be complex numbers.

In fact, nearly everything in this book would remain true if the phrase *real number* were replaced by *complex number* wherever it occurs. Then we would deal with matrices with complex entries, systems of linear equations with complex coefficients (and complex solutions), determinants of complex matrices, and vector spaces with scalar multiplication by any complex number allowed. Moreover, the proofs of most theorems about (the real version of) these concepts extend easily to the complex case. It is not our intention here to give a full treatment of complex linear algebra. However, we will carry the theory far enough to give another proof that the eigenvalues of a real symmetric matrix A are real (Theorem 7 §5.5) and to prove the spectral theorem, an extension of the principal axis theorem (Theorem 2 §8.2).

The set of complex numbers is denoted $\mathbb{C}$. We will use only the most basic properties of these numbers (mainly conjugation and absolute values), and the reader can find this material in Appendix A.

If $n \geq 1$, we denote the set of all n-tuples of complex numbers by $\mathbb{C}^n$. As with $\mathbb{R}^n$, these n-tuples will be written either as column or row matrices and will be referred to as vectors. We define vector operations on $\mathbb{C}^n$ as follows:

$$[v_1 \quad v_2 \quad \cdots \quad v_n]^T + [w_1 \quad w_2 \quad \cdots \quad w_n]^T = [v_1 + w_1 \quad v_2 + w_2 \quad \cdots \quad v_n + w_n]^T$$
$$u[v_1 \quad v_2 \quad \cdots \quad v_n]^T = [uv_1 \quad uv_2 \quad \cdots \quad uv_n]^T \text{ for } u \text{ in } \mathbb{C}$$

With these definitions, $\mathbb{C}^n$ satisfies the axioms for a vector space (with complex scalars) given in Chapter 6. Thus we can speak of spanning sets for $\mathbb{C}^n$, of linearly independent subsets, and of bases. In all cases, the definitions are identical to the real case, except that the scalars are allowed to be complex numbers. In particular, the standard basis of $\mathbb{R}^n$ remains a basis of $\mathbb{C}^n$, called the **standard basis** of $\mathbb{C}^n$.

There is a natural generalization to $\mathbb{C}^n$ of the dot product in $\mathbb{R}^n$. Given $Z = [z_1 \ z_2 \ \cdots \ z_n]^T$ and $W = [w_1 \ w_2 \ \cdots \ w_n]^T$ in $\mathbb{C}^n$, define their **standard inner product** $\langle Z, W \rangle$ by

$$\langle Z, W \rangle = z_1\overline{w}_1 + z_2\overline{w}_2 + \cdots + z_n\overline{w}_n$$

where $\overline{w}$ is the conjugate of the complex number w. Clearly, if Z and W actually lie in $\mathbb{R}^n$, then $\langle Z, W \rangle = Z \cdot W$ is the usual dot product.

Example 1

If $Z = [2 \ 1-i \ 2i \ 3-i]^T$ and $W = [1-i \ -1 \ -i \ 3+2i]^T$, then

$$\langle Z, W \rangle = 2(1 + i) + (1 - i)(-1) + (2i)(i) + (3 - i)(3 - 2i) = 6 - 6i$$
$$\langle Z, Z \rangle = 2 \cdot 2 + (1 - i)(1 + i) + (2i)(-2i) + (3 - i)(3 + i) = 20$$

Note that $\langle Z, W \rangle$ is a complex number in general. However, if $W = Z = [z_1 \ z_2 \ \cdots \ z_n]^T$, the definition gives $\langle Z, Z \rangle = |z_1|^2 + \cdots + |z_n|^2$ which is a nonnegative real number, equal to 0 if and only if $Z = 0$. This explains the conjugation in the definition of $\langle Z, W \rangle$, and it gives (4) of the following theorem.

Theorem 1

Let Z, Z_1, W, and W_1 denote vectors in $\mathbb{C}^n$, and let λ denote a complex number.
1. $\langle Z + Z_1, W \rangle = \langle Z, W \rangle + \langle Z_1, W \rangle$ and $\langle Z, W + W_1 \rangle = \langle Z, W \rangle + \langle Z, W_1 \rangle$.
2. $\langle \lambda Z, W \rangle = \lambda \langle Z, W \rangle$ and $\langle Z, \lambda W \rangle = \overline{\lambda} \langle Z, W \rangle$.
3. $\langle Z, W \rangle = \overline{\langle W, Z \rangle}$.
4. $\langle Z, Z \rangle \geq 0$, and $\langle Z, Z \rangle = 0$ if and only if $Z = 0$.

PROOF

We leave (1) and (2) to the reader (Exercise 10), and (4) has already been proved. To prove (3), write $Z = [z_1 \ z_2 \ \cdots \ z_n]^T$ and $W = [w_1 \ w_2 \ \cdots \ w_n]^T$. Then

$$\overline{\langle W, Z \rangle} = \overline{(w_1\overline{z}_1 + \cdots + w_n\overline{z}_n)} = \overline{w}_1\overline{\overline{z}}_1 + \cdots + \overline{w}_n\overline{\overline{z}}_n$$
$$= z_1\overline{w}_1 + \cdots + z_n\overline{w}_n = \langle Z, W \rangle$$

As for the dot product on $\mathbb{R}^n$, property (4) enables us to define the **norm** or **length** $\|Z\|$ of a vector $Z = [z_1 \ z_2 \ \cdots \ z_n]$ in $\mathbb{C}^n$:

$$\|Z\| = \sqrt{\langle Z, Z \rangle} = \sqrt{|z_1|^2 + |z_2|^2 + \cdots + |z_n|^2}$$

The only properties of the norm function we shall need are the following (the proof is left to the reader):

Theorem 2

If Z is any vector in $\mathbb{C}^n$, then
1. $\|Z\| \geq 0$, and $\|Z\| = 0$ if and only if $Z = 0$.
2. $\|\lambda Z\| = |\lambda| \|Z\|$ for all complex numbers λ.

A vector U in $\mathbb{C}^n$ is called a **unit vector** if $\|U\| = 1$. Property (2) in Theorem 2 then shows that if $Z \neq 0$ is any nonzero vector in $\mathbb{C}^n$, then $U = \dfrac{1}{\|Z\|} Z$ is a unit vector.

Example 2

In $\mathbb{C}^4$, find a unit vector U that is a positive real multiple of $Z = [1 - i \ \ i \ \ 2 \ \ 3 + 4i]$.

Solution $\|Z\| = \sqrt{2 + 1 + 4 + 25} = \sqrt{32} = 4\sqrt{2}$, so take $U = \frac{1}{4\sqrt{2}} Z$.

A matrix $Z = [z_{ij}]$ is called a **complex matrix** if every entry z_{ij} is a complex number. The notion of conjugation for complex numbers extends to matrices as follows: Define the **conjugate** of $Z = [z_{ij}]$ to be the matrix

$$\overline{Z} = [\overline{z}_{ij}]$$

obtained from Z by conjugating every entry. Then (using Appendix A)

$$\overline{Z + W} = \overline{Z} + \overline{W} \quad \text{and} \quad \overline{ZW} = \overline{Z}\,\overline{W}$$

holds for all (complex) matrices of appropriate size.

Transposition of complex matrices is defined just as in the real case. The following notion is fundamental in the study of complex matrices. The **conjugate transpose** Z^* of a complex matrix Z is defined by

$$Z^* = (\overline{Z})^T = \overline{(Z^T)}$$

Observe that $Z^* = Z^T$ when Z is real.

Example 3

$$\begin{bmatrix} 3 & 1 - i & 2 + i \\ 2i & 5 + 2i & -i \end{bmatrix}^* = \begin{bmatrix} 3 & -2i \\ 1 + i & 5 - 2i \\ 2 - i & i \end{bmatrix}$$

The following properties of Z^* follow easily from the rules for transposition of real matrices and extend these rules to complex matrices. Note the conjugate in property (3).

Theorem 3

Let Z and W denote complex matrices, and let λ be a complex number.
1. $(Z^*)^* = Z$.
2. $(Z + W)^* = Z^* + W^*$.
3. $(\lambda Z)^* = \overline{\lambda} Z^*$.
4. $(ZW)^* = W^* Z^*$.

If A is a real symmetric matrix, it is clear that $A^* = A$. The complex matrices that satisfy this condition turn out to be the most natural generalization of the real symmetric matrices: A square complex matrix H is called **hermitian**[7] if $H^* = H$, equivalently $\bar{H} = H^T$. Hermitian matrices are easy to recognize because the entries on the main diagonal must be real, and the "reflection" of each nondiagonal entry in the main diagonal must be the conjugate of that entry.

Example 4

$$\begin{bmatrix} 3 & i & 2+i \\ -i & -2 & -7 \\ 2-i & -7 & 1 \end{bmatrix}$$ is hermitian, whereas $\begin{bmatrix} 1 & i \\ i & -2 \end{bmatrix}$ and $\begin{bmatrix} 1 & i \\ -i & i \end{bmatrix}$ are not.

The following gives a very useful characterization of hermitian matrices in terms of the standard inner product in $\mathbb{C}^n$.

Theorem 4

An $n \times n$ complex matrix H is hermitian if and only if

$$\langle HZ, W \rangle = \langle Z, HW \rangle$$

for all columns Z and W in $\mathbb{C}^n$.

PROOF

If H is hermitian, we have $H^T = \bar{H}$. If Z and W are columns in $\mathbb{C}^n$, then $\langle Z, W \rangle = Z^T \overline{W}$, so

$$\langle HZ, W \rangle = (HZ)^T \overline{W} = Z^T H^T \overline{W} = Z^T \bar{H}\overline{W} = Z^T (\overline{HW}) = \langle Z, HW \rangle.$$

To prove the converse, let E_j denote column j of the identity matrix. If $H = [h_{ij}]$, we have

$$\bar{h}_{ij} = \langle E_i, HE_j \rangle = \langle HE_i, E_j \rangle = h_{ji}.$$

Hence $\bar{H} = H^T$, so H is hermitian.

Let Z be an $n \times n$ complex matrix. As in the real case, a complex number λ is called an **eigenvalue** of Z if $ZX = \lambda X$ holds for some column $X \neq 0$ in $\mathbb{C}^n$. In this case X is called an **eigenvector** of Z corresponding to λ. The **characteristic polynomial** $c_Z(x)$ is defined by

$$c_Z(x) = \det(xI - Z).$$

This polynomial has complex coefficients (possibly nonreal). However, the proof of Theorem 2 §3.3 goes through to show that the eigenvalues of Z are the roots (possibly complex) of $c_Z(x)$.

---

7   The name hermitian honours Charles Hermite (1822–1901), a French mathematician who worked primarily in analysis and is remembered as the first to show that the number *e* from calculus is transcendental—that is, *e* is not the root of any polynomial with integer coefficients.

It is at this point that the advantage of working with complex numbers becomes apparent. The real numbers are incomplete in the sense that the characteristic polynomial of a real matrix may fail to have all its roots real. However, this difficulty does not occur for the complex numbers. The so-called fundamental theorem of algebra ensures that *every* polynomial of positive degree with complex coefficients has a complex root. Hence every square complex matrix has a (complex) eigenvalue. Indeed (Appendix A), $c_Z(x)$ factors completely as follows:

$$c_Z(x) = (x - \lambda_1)(x - \lambda_2) \cdots (x - \lambda_n)$$

where $\lambda_1, \lambda_2, \dots, \lambda_n$ are the eigenvalues of Z (with possible repetitions due to multiple roots).

The next result shows that, for hermitian matrices, the eigenvalues are actually real. Because symmetric real matrices are hermitian, it gives a proof of Theorem 7 §5.5. It also extends Theorem 4 §8.2, which asserts that eigenvectors of a symmetric real matrix corresponding to distinct eigenvalues are actually orthogonal. In the complex context, two columns Z and W in $\mathbb{C}^n$ are said to be **orthogonal** if $\langle Z, W \rangle = 0$.

Theorem 5

Let H denote a hermitian matrix.
1. The eigenvalues of H are real.
2. Eigenvectors of H corresponding to distinct eigenvalues are orthogonal.

PROOF

Let λ and μ be the eigenvalues of H with (nonzero) eigenvectors Z and W. Then $HZ = \lambda Z$ and $HW = \mu W$, so Theorem 4 gives

$$\lambda \langle Z, W \rangle = \langle \lambda Z, W \rangle = \langle HZ, W \rangle = \langle Z, HW \rangle = \langle Z, \mu W \rangle = \bar{\mu} \langle Z, W \rangle \qquad (*)$$

If $\mu = \lambda$ and $W = Z$, this becomes $\lambda \langle Z, Z \rangle = \bar{\lambda} \langle Z, Z \rangle$. Because $\langle Z, Z \rangle = \|Z\|^2 \neq 0$, this implies $\lambda = \bar{\lambda}$. Thus λ is real, proving (1). Similarly, μ is real, so equation $(*)$ gives $\lambda \langle Z, W \rangle = \mu \langle Z, W \rangle$. If $\lambda \neq \mu$, this implies $\langle Z, W \rangle = 0$, proving (2).

The principal axis theorem (Theorem 2 §8.2) asserts that every real symmetric matrix A is orthogonally diagonalizable—that is, $P^T A P$ is diagonal where P is an orthogonal matrix $(P^{-1} = P^T)$. The next theorem identifies the complex analogs of these orthogonal real matrices. As in the real case, a set of nonzero vectors $\{Z_1, Z_2, \dots, Z_m\}$ in $\mathbb{C}^n$ is called **orthogonal** if $\langle Z_i, Z_j \rangle = 0$ whenever $i \neq j$, and it is **orthonormal** if, in addition, $\|Z_i\| = 1$ for each i.

Theorem 6

The following are equivalent for an $n \times n$ complex matrix U.
1. $U^{-1} = U^*$.
2. The rows of U are an orthonormal set in $\mathbb{C}^n$.
3. The columns of U are an orthonormal set in $\mathbb{C}^n$.

The proof is a direct adaptation of the proof of Theorem 1 §8.2.

A square complex matrix U is called **unitary** if it satisfies the conditions in Theorem 6. Thus a real matrix is unitary if and only if it is orthogonal.

Example 5

The matrix $Z = \begin{bmatrix} 1+i & 1 \\ 1-i & i \end{bmatrix}$ has orthogonal columns, but the rows are not

orthogonal. Normalizing the columns gives the unitary matrix $\frac{1}{2}\begin{bmatrix} 1+i & \sqrt{2} \\ 1-i & \sqrt{2}i \end{bmatrix}$.

Given a real symmetric matrix A, the diagonalization algorithm in Section 3.3 leads to a procedure for finding an orthogonal matrix P such that $P^T AP$ is diagonal (see Example 4, §8.2). The following example illustrates Theorem 5 and shows that the technique works for complex matrices.

Example 6

Consider the hermitian matrix $H = \begin{bmatrix} 3 & 2+i \\ 2-i & 7 \end{bmatrix}$. Find the eigenvalues of H, find two orthonormal eigenvectors, and so find a unitary matrix U such that U^*HU is diagonal.

Solution   The characteristic polynomial of H is

$$c_H(x) = \det(xI - H) = \det\begin{bmatrix} x-3 & -2-i \\ -2+i & x-7 \end{bmatrix} = (x-2)(x-8)$$

Hence the eigenvalues are 2 and 8 (both real as expected), and corresponding eigenvectors are $\begin{bmatrix} 2+i \\ -1 \end{bmatrix}$ and $\begin{bmatrix} 1 \\ 2-i \end{bmatrix}$ (orthogonal as expected). Each has length

$\sqrt{6}$ so, as in the (real) diagonalization algorithm, let $U = \frac{1}{\sqrt{6}}\begin{bmatrix} 2+i & 1 \\ -1 & 2-i \end{bmatrix}$ be the unitary matrix with the normalized eigenvectors as columns. Then $U^*HU = \begin{bmatrix} 2 & 0 \\ 0 & 8 \end{bmatrix}$ is diagonal.

An $n \times n$ complex matrix Z is called **unitarily diagonalizable** if U^*ZU is diagonal for some unitary matrix U. As Example 6 suggests, we are going to prove that every hermitian matrix is unitarily diagonalizable. However, with only a little extra effort, we can get a very important theorem that has this result as an easy consequence.

A complex matrix is called **upper triangular** if every entry below the main diagonal is zero. We owe the following theorem to Issai Schur.[8]

---

8 Issai Schur (1875–1941) was a German mathematician who did fundamental work in the theory of representations of groups as matrices.

Theorem 7 Schur's Theorem

If Z is any $n \times n$ complex matrix, there exists a unitary matrix U such that

$$U^*ZU = T$$

is upper triangular. Moreover, the entries on the main diagonal of T are the eigenvalues $\lambda_1, \lambda_2, \ldots, \lambda_n$ of Z (including multiplicities).

PROOF

We use induction on n. If $n = 1$, Z is already upper triangular. If $n > 1$, assume the theorem is valid for $(n - 1) \times (n - 1)$ complex matrices. Let λ_1 be an eigenvalue of Z, and let Y_1 be an eigenvector with $\|Y_1\| = 1$. Then Y_1 is part of a basis of $\mathbb{C}^n$ (by the analog of Theorem 1 §6.4), so the (complex analog of the) Gram–Schmidt process provides $Y_2, \ldots, Y_n$ such that $\{Y_1, \ldots, Y_n\}$ is an orthonormal basis of $\mathbb{C}^n$.
If $U_1 = [Y_1 \ Y_2 \ \cdots \ Y_n]$ is the matrix with these vectors as its columns, then

$$U_1^* Z U_1 = \begin{bmatrix} \lambda_1 & X_1 \\ 0 & Z_1 \end{bmatrix}$$

in block form. Now apply induction to find a unitary $(n - 1) \times (n - 1)$ matrix W_1 such that $W_1^* Z_1 W_1 = T_1$ is upper triangular. Then $U_2 = \begin{bmatrix} 1 & 0 \\ 0 & W_1 \end{bmatrix}$ is a unitary $n \times n$ matrix. Hence $U = U_1 U_2$ is unitary (using Theorem 6), and

$$\begin{aligned} U^* Z U &= U_2^* (U_1^* Z U_1) U_2 \\ &= \begin{bmatrix} 1 & 0 \\ 0 & W_1^* \end{bmatrix} \begin{bmatrix} \lambda_1 & X_1 \\ 0 & Z_1 \end{bmatrix} \begin{bmatrix} 1 & 0 \\ 0 & W_1 \end{bmatrix} = \begin{bmatrix} \lambda_1 & X_1 W_1 \\ 0 & T_1 \end{bmatrix} \end{aligned}$$

is upper triangular. Finally, Z and $U^* Z U = T$ have the same eigenvalues by (the complex version of) Theorem 1 §5.5, and they are the diagonal entries of T because T is upper triangular.

The fact that similar matrices have the same traces and determinants gives the following consequence of Schur's theorem.

Corollary

Let Z be an $n \times n$ complex matrix, and let $\lambda_1, \lambda_2, \ldots, \lambda_n$ denote the eigenvalues of Z, including multiplicities. Then

$$\det Z = \lambda_1 \lambda_2 \cdots \lambda_n \quad \text{and} \quad \operatorname{tr} Z = \lambda_1 + \lambda_2 + \cdots + \lambda_n$$

Schur's theorem asserts that every complex matrix can be "unitarily triangularized." However, we cannot substitute "unitarily diagonalized" here. In fact, if $Z = \begin{bmatrix} 1 & 1 \\ 0 & 1 \end{bmatrix}$, there is no invertible complex matrix U at all such that $U^{-1} Z U$ is diagonal. However, the situation is much better for hermitian matrices.

Theorem 8 Spectral Theorem

If H is hermitian, there is a unitary matrix U such that U^*HU is diagonal.

PROOF

By Schur's theorem, let $U^*HU = T$ be upper triangular where U is unitary. Since H is hermitian, this gives

$$T^* = (U^*HU)^* = U^*H^*U^{**} = U^*HU = T$$

This means that T is both upper and lower triangular. Hence T is actually diagonal.

The principal axis theorem asserts that a real matrix A is symmetric if and only if it is orthogonally diagonalizable (that is, P^TAP is diagonal for some real orthogonal matrix P). Theorem 8 is the complex analog of half of this result. However, the converse is false for complex matrices: There exist unitarily diagonalizable matrices that are not hermitian.

Example 7

Show that the non-hermitian matrix $Z = \begin{bmatrix} 0 & 1 \\ -1 & 0 \end{bmatrix}$ is unitarily diagonalizable.

Solution   The characteristic polynomial is $c_Z(x) = x^2 + 1$. Hence the eigenvalues are i and $-i$, and it is easy to verify that $\begin{bmatrix} i \\ -1 \end{bmatrix}$ and $\begin{bmatrix} -1 \\ i \end{bmatrix}$ are corresponding eigenvectors. Moreover, these eigenvectors are orthogonal and both have length $\sqrt{2}$, so $U = \frac{1}{\sqrt{2}} \begin{bmatrix} i & -1 \\ -1 & i \end{bmatrix}$ is a unitary matrix such that $U^*ZU = \begin{bmatrix} i & 0 \\ 0 & -i \end{bmatrix}$ is diagonal.

There is a very simple way to characterize those complex matrices that are unitarily diagonalizable. To this end, an $n \times n$ complex matrix N is called **normal** if $NN^* = N^*N$. It is clear that every hermitian or unitary matrix is normal, as is the matrix $\begin{bmatrix} 0 & 1 \\ -1 & 0 \end{bmatrix}$ in Example 7. In fact we have the following result.

Theorem 9

An $n \times n$ complex matrix Z is unitarily diagonalizable if and only if Z is normal.

PROOF

Assume first that $U^*ZU = D$, where U is unitary and D is diagonal. Then $DD^* = D^*D$ as is easily verified. Because $DD^* = U^*(ZZ^*)U$ and $D^*D = U^*(Z^*Z)U$, it follows by cancellation that $ZZ^* = Z^*Z$. Conversely, assume Z is normal—that is, $ZZ^* = Z^*Z$. By Schur's theorem, let $U^*ZU = T$, where T is upper triangular and U is unitary. Then T is normal too:

$$TT^* = U^*(ZZ^*)U = U^*(Z^*Z)U = T^*T$$

Hence it suffices to show that a normal $n \times n$ upper triangular matrix T must be diagonal. We induct on n; it is clear if $n = 1$. If $n > 1$ and $T = [t_{ij}]$, then equating $(1, 1)$-entries in TT^* and T^*T gives

$$|t_{11}|^2 + |t_{12}|^2 + \cdots + |t_{1n}|^2 = |t_{11}|^2$$

This implies $t_{12} = t_{13} = \cdots = t_{1n} = 0$, so $T = \begin{bmatrix} t_{11} & 0 \\ 0 & T_1 \end{bmatrix}$ in block form. Hence

$T^* = \begin{bmatrix} \overline{t}_{11} & 0 \\ 0 & T_1^* \end{bmatrix}$ so $TT^* = T^*T$ implies $T_1 T_1^* = T_1 T_1^*$. Thus T_1 is diagonal by induction, and the proof is complete.

We conclude this section by using Schur's theorem (Theorem 7) to prove a famous theorem about matrices. Recall that the characteristic polynomial of a square matrix A is defined by $c_A(x) = \det(xI - A)$, and that the eigenvalues of A are just the roots of $c_A(x)$.

Theorem 10 Cayley–Hamilton Theorem[9]

If A is an $n \times n$ complex matrix, then $c_A(A) = 0$; that is, A is a root of its characteristic polynomial.

PROOF

If $p(x)$ is any polynomial with complex coefficients, then $p(P^{-1}AP) = P^{-1}p(A)P$. Hence, by Schur's theorem, we may assume that A is upper triangular. Then the eigenvalues $\lambda_1, \lambda_2, \ldots, \lambda_n$ of A appear along the main diagonal, so $c_A(x) = (x - \lambda_1)(x - \lambda_2)(x - \lambda_3)\cdots(x - \lambda_n)$. Thus

$$c_A(A) = (A - \lambda_1 I)(A - \lambda_2 I)(A - \lambda_3 I)\cdots(A - \lambda_n I).$$

Note that each matrix $A - \lambda_i I$ is upper triangular. Now observe:

1. $A - \lambda_1 I$ has zero first column because column 1 of A is $[\lambda_1 \ 0 \ 0 \ \cdots \ 0]^T$.
2. Then $(A - \lambda_1 I)(A - \lambda_2 I)$ has the first two columns zero because column 2 of $(A - \lambda_2 I)$ is $[b \ 0 \ 0 \ \cdots \ 0]^T$ for some constant b.
3. Next $(A - \lambda_1 I)(A - \lambda_2 I)(A - \lambda_3 I)$ has the first three columns zero because column 3 of $(A - \lambda_3 I)$ is $[c \ d \ 0 \ \cdots \ 0]^T$ for some constants c and d.

Continuing in this way we see that $(A - \lambda_1 I)(A - \lambda_2 I)(A - \lambda_3 I) \cdots (A - \lambda_n I)$ has all n columns zero; that is $c_A(A) = 0$.

Exercises 8.6

1. In each case, compute the norm of the complex vector.

   (a) $(1, 1 - i, -2, i)$

   ◆(b) $(1 - i, 1 + i, 1, -1)$

   (c) $(2 + i, 1 - i, 2, 0, -i)$

   ◆(d) $(-2, -i, 1 + i, 1 - i, 2i)$

2. In each case, determine whether the two vectors are orthogonal.

   (a) $(4, -3i, 2 + i), (i, 2, 2 - 4i)$

   ◆(b) $(i, -i, 2 + i), (i, i, 2 - i)$

   (c) $(1, 1, i, i), (1, i, -i, 1)$

   ◆(d) $(4 + 4i, 2 + i, 2i), (-1 + i, 2, 3 - 2i)$

---

9 Named after the English mathematician Arthur Cayley (1821–1895)—see page 27—and William Rowan Hamilton (1805–1865), an Irish mathematician famous for his work on physical dynamics.

3. A subset U of $\mathbb{C}^n$ is called a **complex subspace** of $\mathbb{C}^n$ if it contains 0 and if, given Z and W in U, both $Z + W$ and zZ lie in U (z any complex number). In each case, determine whether U is a complex subspace of $\mathbb{C}^3$.

  (a) $U = \{(w, \bar{w}, 0)\,|\, w \text{ in } \mathbb{C}\}$

  ♦(b) $U = \{(w, 2w, a)\,|\, w \text{ in } \mathbb{C}, a \text{ in } \mathbb{R}\}$

  (c) $U = \mathbb{R}^3$

  ♦(d) $U = \{(v + w, v - 2w, v)\,|\, v, w \text{ in } \mathbb{C}\}$

4. In each case, find a basis over $\mathbb{C}$, and determine the dimension of the complex subspace U of $\mathbb{C}^3$ (see the previous exercise).

  (a) $U = \{(w, v + w, v - iw)\,|\, v, w \text{ in } \mathbb{C}\}$

  ♦(b) $U = \{(iv + w, 0, 2v - w)\,|\, v, w \text{ in } \mathbb{C}\}$

  (c) $U = \{(u, v, w)\,|\, iu - 3v + (1 - i)w = 0;$
        $u, v, w \text{ in } \mathbb{C}\}$

  ♦(d) $U = \{(u, v, w)\,|\, 2u + (1 + i)v - iw = 0;$
        $u, v, w \text{ in } \mathbb{C}\}$

5. In each case, determine whether the given matrix is hermitian, unitary, or normal.

  (a) $\begin{bmatrix} 1 & -i \\ i & i \end{bmatrix}$
  ♦(b) $\begin{bmatrix} 2 & 3 \\ -3 & 2 \end{bmatrix}$

  (c) $\begin{bmatrix} 1 & i \\ -i & 2 \end{bmatrix}$
  ♦(d) $\begin{bmatrix} 1 & -i \\ i & -1 \end{bmatrix}$

  (e) $\frac{1}{\sqrt{2}} \cdot \begin{bmatrix} 1 & -1 \\ 1 & 1 \end{bmatrix}$
  ♦(f) $\begin{bmatrix} 1 & 1+i \\ 1+i & i \end{bmatrix}$

  (g) $\begin{bmatrix} 1+i & i \\ -i & -1+i \end{bmatrix}$
  ♦(h) $\frac{1}{\sqrt{2}|z|} \cdot \begin{bmatrix} z & z \\ \bar{z} & -\bar{z} \end{bmatrix}, z \neq 0$

6. Show that a matrix N is normal if and only if $\bar{N}N^T = N^T\bar{N}$.

7. Let $A = \begin{bmatrix} z & \bar{v} \\ v & w \end{bmatrix}$ where v, w, and z are complex numbers. Characterize in terms of v, w, and z when A is

  (a) hermitian

  (b) unitary

  (c) normal.

8. In each case, find a unitary matrix U such that U^*ZU is diagonal.

  (a) $Z = \begin{bmatrix} 1 & i \\ -i & 1 \end{bmatrix}$
  ♦(b) $Z = \begin{bmatrix} 4 & 3-i \\ 3+i & 1 \end{bmatrix}$

  (c) $Z = \begin{bmatrix} a & b \\ -b & a \end{bmatrix};$
  ♦(d) $Z = \begin{bmatrix} 2 & 1+i \\ 1-i & 3 \end{bmatrix}$

    a, b real

(e) $Z = \begin{bmatrix} 1 & 0 & 1+i \\ 0 & 2 & 0 \\ 1-i & 0 & 0 \end{bmatrix}$
♦(f) $Z = \begin{bmatrix} 1 & 0 & 0 \\ 0 & 1 & 1+i \\ 0 & 1-i & 2 \end{bmatrix}$

9. Show that $\langle ZX, Y \rangle = \langle X, Z^*Y \rangle$ holds for all $n \times n$ matrices Z and for all columns X and Y in $\mathbb{C}^n$.

10. (a) Prove (1) and (2) of Theorem 1.

  ♦(b) Prove Theorem 2.

  (c) Prove Theorem 3.

11. (a) Show that Z is hermitian if and only if $\bar{Z} = Z^T$.

  ♦(b) Show that the diagonal entries of any hermitian matrix are real.

12. (a) Show that every complex matrix Z can be written uniquely in the form $Z = A + iB$, where A and B are real matrices.

  (b) If $Z = A + iB$ as in (a), show that Z is hermitian if and only if A is symmetric, and B is skew-symmetric (that is, $B^T = -B$).

13. If Z is any complex $n \times n$ matrix, show that ZZ^* and $Z + Z^*$ are hermitian.

14. A complex matrix S is called **skew-hermitian** if $S^* = -S$.

  (a) Show that $Z - Z^*$ is skew-hermitian for any square complex matrix Z.

  ♦(b) If S is skew-hermitian, show that S^2 and iS are hermitian.

  (c) If S is skew-hermitian, show that the eigenvalues of S are pure imaginary ($i\lambda$ for real λ).

  ♦(d) Show that every $n \times n$ complex matrix Z can be written uniquely as $Z = H + S$, where H is hermitian and S is skew-hermitian.

15. Let U be a unitary matrix. Show that:

  (a) $\|UX\| = \|X\|$ for all columns X in $\mathbb{C}^n$.

  (b) $|\lambda| = 1$ for every eigenvalue λ of U.

16. (a) If Z is an invertible complex matrix, show that Z^* is invertible and that $(Z^*)^{-1} = (Z^{-1})^*$.

  ♦(b) Show that the inverse of a unitary matrix is again unitary.

  (c) If U is unitary, show that U^* is unitary.

17. Let Z be an $m \times n$ matrix such that $Z^*Z = I_n$ (for example, Z is a unit column in $\mathbb{C}^n$).

  (a) Show that $H = ZZ^*$ is hermitian and satisfies $H^2 = H$.

(b) Show that $U = I - 2ZZ^*$ is both unitary and hermitian (so $U^{-1} = U^* = U$).

18. (a) If N is normal, show that zN is also normal for all complex numbers z.

♦(b) Show that (a) fails if *normal* is replaced by *hermitian*.

19. Show that a real 2×2 normal matrix either is symmetric or has the form $\begin{bmatrix} a & b \\ -b & a \end{bmatrix}$.

20. If H is hermitian, show that all the coefficients of $c_H(x)$ are real numbers.

21. (a) If $A = \begin{bmatrix} 1 & 1 \\ 0 & 1 \end{bmatrix}$, show that $U^{-1}AU$ is not diagonal for any invertible complex matrix U.

♦(b) If $A = \begin{bmatrix} 0 & 1 \\ -1 & 0 \end{bmatrix}$, show that $U^{-1}AU$ is not upper triangular for any *real* invertible matrix U.

22. If Z is any $n \times n$ matrix, show that U^*ZU is lower triangular for some unitary matrix U.

23. If Z is a 3×3 matrix, show that $Z^2 = 0$ if and only if there exists a unitary matrix U such that U^*ZU has the form

$$\begin{bmatrix} 0 & 0 & u \\ 0 & 0 & v \\ 0 & 0 & 0 \end{bmatrix} \text{ or the form } \begin{bmatrix} 0 & u & v \\ 0 & 0 & 0 \\ 0 & 0 & 0 \end{bmatrix}.$$

24. If $Z^2 = Z$, show that rank $Z = $ tr Z. [*Hint:* Use Schur's theorem.]

SECTION 8.7   Best Approximation and Least Squares

A system of linear equations need not have a solution. However, even when no solution exists, it is often desirable to find a "best approximation" to a solution. In this section one definition of best approximation is given. Then it is shown that such approximations always exist, and a method for finding them is described. The result is then applied to least squares approximation of data, a subject introduced in Section 5.7.

Suppose A is an $m \times n$ matrix and B is a column in $\mathbb{R}^m$, and consider the system

$$AX = B$$

of m linear equations in n variables. This need not have a solution. However, given any column Z in $\mathbb{R}^n$, the distance $\|B - AZ\|$ is a measure of how far AZ is from B. Hence it is natural to ask whether there is a column Z in $\mathbb{R}^n$ that is as close as possible to a solution in the sense that

$$\|B - AZ\|$$

is the minimum value of $\|B - AX\|$ as X ranges over all columns in $\mathbb{R}^n$. Theorem 3 §8.1 (the projection theorem) answers this question in the affirmative. To see how, define

$$U = \{AX \mid X \text{ lies in } \mathbb{R}^n\}.$$

Then U is a subspace of $\mathbb{R}^m$, so we are to find AX in U as close as possible to B. The projection theorem guarantees a solution—call it AZ—and, in fact,

$$AZ = \text{proj}_U(B).$$

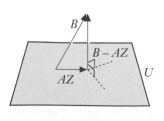

However, two computational problems are involved here. First, we need an orthogonal basis of U to compute $\text{proj}_U(B)$. Second, we end up with AZ rather than Z itself. So it is useful to find a way to compute Z directly.

The key observation is that $B - AZ$ is in $U^\perp$ (Theorem 3 §8.1) and so is orthogonal to every vector AX in U (see the diagram). Thus,

$$0 = (AX) \bullet (B - AZ) = (AX)^T(B - AZ) = X^TA^T(B - AZ)$$
$$= X \bullet [A^T(B - AZ)]$$

for all X in $\mathbb{R}^n$. In other words, the vector $A^T(B - AZ)$ in $\mathbb{R}^n$ is orthogonal to *every* vector in $\mathbb{R}^n$ and so must be zero. Hence Z satisfies

$$(A^TA)Z = A^TB.$$

This is a system of linear equations called the **normal equations** for Z. Note that this system can have more than one solution (see Exercise 5). However, the $n \times n$ matrix A^TA is invertible if (and only if) the columns of A are linearly independent (Theorem 4 §5.4); so, in this case, Z is uniquely determined and is given explicitly by

$$Z = (A^TA)^{-1}A^TB.$$

However, the most efficient way to find Z is to apply gaussian elimination to the normal equations.

This discussion is summarized in the following theorem.

Theorem 1　Best Approximation Theorem

Let A be an $m \times n$ matrix, let B be any column in $\mathbb{R}^m$, and consider the system

$$AX = B$$

of m equations in n variables. Any solution Z to the normal equations

$$(A^TA)Z = A^TB$$

is a best approximation to a solution to $AX = B$ in the sense that $\|B - AZ\|$ is the minimum value of $\|B - AX\|$ as X ranges over all columns in $\mathbb{R}^n$.

If the columns of A are linearly independent, then A^TA is invertible and Z is given uniquely by $Z = (A^TA)^{-1}A^TB$.

Note that if A is $n \times n$ and invertible, then

$$Z = (A^TA)^{-1}A^TB = A^{-1}B$$

is the solution to the system of equations, and $\|B - AZ\| = 0$. Hence $(A^TA)^{-1}A^T$ is playing the role of the inverse of the nonsquare matrix A in the case in which A has linearly independent columns. The matrix $A^T(AA^T)^{-1}$ plays a similar role when the rows of A are linearly independent. These are both special cases of the **generalized inverse** of a matrix A (see Exercise 7). However, we shall not pursue this topic here.

Example 1

The equations

$$\begin{aligned}
3x - y &= 4 \\
x + 2y &= 0 \\
2x + y &= 1
\end{aligned}$$

have no solution. Find the vector $Z = \begin{bmatrix} x_0 \\ y_0 \end{bmatrix}$ that best approximates a solution.

Solution　In this case,

$$A = \begin{bmatrix} 3 & -1 \\ 1 & 2 \\ 2 & 1 \end{bmatrix}, \quad \text{so} \quad A^TA = \begin{bmatrix} 3 & 1 & 2 \\ -1 & 2 & 1 \end{bmatrix}\begin{bmatrix} 3 & -1 \\ 1 & 2 \\ 2 & 1 \end{bmatrix} = \begin{bmatrix} 14 & 1 \\ 1 & 6 \end{bmatrix}$$

is invertible. The normal equations $(A^TA)Z = A^TB$ are

$$\begin{bmatrix} 14 & 1 \\ 1 & 6 \end{bmatrix} Z = \begin{bmatrix} 14 \\ -3 \end{bmatrix}, \quad \text{so} \quad Z = \tfrac{1}{83}\begin{bmatrix} 87 \\ -56 \end{bmatrix}.$$

Thus $x_0 = \tfrac{87}{83}$ and $y_0 = \tfrac{-56}{83}$. With these values of x and y, the left sides of the equations are

$$3x_0 - y_0 = \tfrac{317}{83} = 3.82$$
$$x_0 + 2y_0 = \tfrac{-25}{83} = -0.30$$
$$2x_0 + y_0 = \tfrac{118}{83} = 1.42$$

This is as close as possible to a solution.

Example 2

The average number g of goals per game scored by a hockey player seems to be related linearly to two factors: the number x_1 of years of experience and the number x_2 of goals in the preceding 10 games. The accompanying data were collected on four players. Find the linear function $g = a_0 + a_1x_1 + a_2x_2$ that best fits these data.

Solution   If the relationship is given by $g = r_0 + r_1x_1 + r_2x_2$, then the data can be described as follows:

g	x_1	x_2
0.8	5	3
0.8	3	4
0.6	1	5
0.4	2	1

$$\begin{bmatrix} 1 & 5 & 3 \\ 1 & 3 & 4 \\ 1 & 1 & 5 \\ 1 & 2 & 1 \end{bmatrix}\begin{bmatrix} r_0 \\ r_1 \\ r_2 \end{bmatrix} = \begin{bmatrix} 0.8 \\ 0.8 \\ 0.6 \\ 0.4 \end{bmatrix}$$

Using the notation in Theorem 1, we get

$$Z = (A^TA)^{-1}A^TB$$

$$= \tfrac{1}{42}\begin{bmatrix} 119 & -17 & -19 \\ -17 & 5 & 1 \\ -19 & 1 & 5 \end{bmatrix}\begin{bmatrix} 1 & 1 & 1 & 1 \\ 5 & 3 & 1 & 2 \\ 3 & 4 & 5 & 1 \end{bmatrix}\begin{bmatrix} 0.8 \\ 0.8 \\ 0.6 \\ 0.4 \end{bmatrix} = \begin{bmatrix} 0.14 \\ 0.09 \\ 0.08 \end{bmatrix}$$

Hence the best-fitting function is $g = 0.14 + 0.09x_1 + 0.08x_2$. The amount of computation would have been reduced if the normal equations had been constructed and then solved by gaussian elimination.

Least Squares Approximation

Theorem 1 applies directly to least squares approximation. This was treated in Section 5.7, though it was not possible to prove the main theorem (Theorem 2 §5.7) with the techniques then available.

Suppose that data are available giving pairs of corresponding values of the two variables x and y:

$$(x_1, y_1), (x_2, y_2), \ldots, (x_n, y_n)$$

Given such data pairs, assume for the moment that the variables x and y are related by a polynomial of degree m.

$$y = p(x) = r_0 + r_1x + \cdots + r_mx^m$$

Then for each x_i we have *two* values of the variable y, the observed value y_i and the computed value $p(x_i)$. The question now is this: Is it possible to choose the coefficients $r_0, r_1, \ldots, r_m$ in such a way that the $p(x_i)$ are as close as possible to the corresponding y_i? To apply Theorem 1, the following notation is convenient:

$$Y = \begin{bmatrix} y_1 \\ y_2 \\ \vdots \\ y_n \end{bmatrix} \qquad p(X) = \begin{bmatrix} p(x_1) \\ p(x_2) \\ \vdots \\ p(x_n) \end{bmatrix}$$

Then the problem takes the following form: Choose $r_0, r_1, \ldots, r_m$ such that

$$\|Y - p(X)\|^2 = [y_1 - p(x_1)]^2 + [y_2 - p(x_2)]^2 + \cdots + [y_n - p(x_n)]^2$$

is as small as possible. A polynomial $p(x)$ satisfying this condition is called a **least squares approximating polynomial** of degree m for the data pairs given. Now write

$$R = \begin{bmatrix} r_0 \\ r_1 \\ \vdots \\ r_m \end{bmatrix} \qquad M = \begin{bmatrix} 1 & x_1 & x_1^2 & \cdots & x_1^m \\ 1 & x_2 & x_2^2 & \cdots & x_2^m \\ \vdots & \vdots & \vdots & & \vdots \\ 1 & x_n & x_n^2 & \cdots & x_n^m \end{bmatrix}$$

Then $p(X)$ can be written

$$p(X) = \begin{bmatrix} p(x_1) \\ p(x_2) \\ \vdots \\ p(x_n) \end{bmatrix} = \begin{bmatrix} r_0 + r_1 x_1 + \cdots + r_m x_1^m \\ r_0 + r_1 x_2 + \cdots + r_m x_2^m \\ \vdots \\ r_0 + r_1 x_n + \cdots + r_m x_n^m \end{bmatrix} = MR$$

so we are to find a column R in $\mathbb{R}^{m+1}$ such that $\|Y - MR\|^2$ is as small as possible. In this form, Theorem 1 applies directly and gives the first part of Theorem 2.

Theorem 2

Let n data pairs $(x_1, y_1), (x_2, y_2), \ldots, (x_n, y_n)$ be given, and write

$$Y = \begin{bmatrix} y_1 \\ y_2 \\ \vdots \\ y_n \end{bmatrix} \qquad M = \begin{bmatrix} 1 & x_1 & x_1^2 & \cdots & x_1^m \\ 1 & x_2 & x_2^2 & \cdots & x_2^m \\ \vdots & \vdots & \vdots & & \vdots \\ 1 & x_n & x_n^2 & \cdots & x_n^m \end{bmatrix}$$

1. If $Z = \begin{bmatrix} z_0 & z_1 & \cdots & z_m \end{bmatrix}^T$ is any solution to the normal equations

$$(M^T M)Z = M^T Y$$

then the polynomial

$$\bar{p}(x) = z_0 + z_1 x + z_2 x^2 + \cdots + z_m x^m$$

is a least squares approximating polynomial of degree m for the given data pairs.

2. If at least $m + 1$ of the numbers $x_1, x_2, \ldots, x_n$ are distinct (so $n \geq m + 1$), the matrix $M^T M$ is invertible and Z is uniquely determined by

$$Z = (M^T M)^{-1} M^T Y$$

PROOF

It remains to prove (2), and for that we show that the columns of M are linearly independent (Theorem 4 §5.4). Suppose a linear combination of the columns vanishes:

$$r_0 \begin{bmatrix} 1 \\ 1 \\ \vdots \\ 1 \end{bmatrix} + r_1 \begin{bmatrix} x_1 \\ x_2 \\ \vdots \\ x_n \end{bmatrix} + \cdots + r_m \begin{bmatrix} x_1^m \\ x_2^m \\ \vdots \\ x_n^m \end{bmatrix} = \begin{bmatrix} 0 \\ 0 \\ \vdots \\ 0 \end{bmatrix}$$

If we write $q(x) = r_0 + r_1 x + \cdots + r_m x^m$, equating coefficients shows that $q(x_1) = q(x_2) = \cdots = q(x_n) = 0$. Hence $q(x)$ is a polynomial of degree m with at least $m + 1$ distinct roots, so $q(x)$ must be the zero polynomial (see Theorem 4 §6.5). Thus $r_0 = r_1 = \cdots = r_m = 0$ as required.

Several examples illustrating the use of this theorem were given in Section 5.7. The interested reader is referred to them.

There is an extension of Theorem 2 that should be mentioned. Given the data pairs $(x_1, y_1), \ldots, (x_n, y_n)$, that theorem shows how to find a polynomial

$$p(x) = r_0 + r_1 x + \cdots + r_m x^m$$

such that $\| Y - p(X) \|^2$ is as small as possible, where X and $p(X)$ are as before. Choosing the appropriate polynomial $p(x)$ amounts to choosing the coefficients $r_0, r_1, \ldots, r_m$, and the theorem gives a formula for the optimal choices. Now $p(x)$ is a linear combination of the functions $1, x, \ldots, x^m$, where the r_i are the coefficients, and this suggests applying the method to linear combinations of other functions. If $f_0(x), f_1(x), \ldots, f_m(x)$ are given functions, write

$$f(x) = r_0 f_0(x) + r_1 f_1(x) + \cdots + r_m f_m(x)$$

where $r_0, r_1, \ldots, r_m$ are real numbers. Then the more general question is whether $r_0, r_1, \ldots, r_m$ can be found such that $\| Y - f(X) \|^2$ is as small as possible, where now we write

$$f(X) = \begin{bmatrix} f(x_1) \\ f(x_2) \\ \vdots \\ f(x_n) \end{bmatrix}$$

The theorem follows.

Theorem 3

Let n data pairs $(x_1, y_1), (x_2, y_2), \ldots, (x_n, y_n)$ be given and suppose that $m + 1$ functions $f_0(x), f_1(x), \ldots, f_m(x)$ are specified. Write

$$Y = \begin{bmatrix} y_1 \\ y_2 \\ \vdots \\ y_n \end{bmatrix} \qquad M = \begin{bmatrix} f_0(x_1) & f_1(x_1) & \cdots & f_m(x_1) \\ f_0(x_2) & f_1(x_2) & \cdots & f_m(x_2) \\ \vdots & \vdots & & \vdots \\ f_0(x_n) & f_1(x_n) & \cdots & f_m(x_n) \end{bmatrix}$$

1. If $Z = [z_0 \ z_1 \ \cdots \ z_m]^T$ is any solution to the normal equations

$$(M^T M)Z = M^T Y$$

then

$$\overline{f}(x) = z_0 f_0(x) + z_1 f_1(x) + \cdots + z_m f_m(x)$$

is the best approximation for these data among all functions $f(x)$ of the form

$$f(x) = r_0 f_0(x) + r_1 f_1(x) + \cdots + r_m f_m(x) \quad r_i \text{ in } \mathbb{R}$$

in the sense that $\|Y - \overline{f}(X)\| \leq \|Y - f(X)\|$ holds for all choices of the r_i.

2. If $M^T M$ is invertible (that is, rank $M = m + 1$), then Z is uniquely determined by

$$Z = (M^T M)^{-1} M^T Y$$

PROOF

Observe that $f(X) = MR$, where $R = [r_0 \; r_1 \; \cdots \; r_m]^T$, so we are asked to choose R to minimize $\|Y - MR\|^2$. Theorem 1 applies as before.

The function $\overline{f}(x) = z_0 f_0(x) + z_1 f_1(x) + \cdots + z_m f_m(x)$ in Theorem 3 is called a **least squares approximating function** of the form $r_0 f_0(x) + \cdots + r_m f_m(x)$. This theorem contains Theorem 2 as a special case ($f_i(x) = x^i$ for each i), but there is no guarantee that $M^T M$ will be invertible in the general case if $m + 1$ of the x_i are distinct. Conditions for this to hold depend on the choice of the functions $f_0(x), f_1(x), \ldots, f_m(x)$, and are related to the *wronskian* of these functions, a condition that they are independent.

Example 3

Given the data pairs $(-1, 0)$, $(0, 1)$, and $(1, 4)$, find the least squares approximating function of the form $r_0 x + r_1 2^x$.

Solution   The functions are $f_0(x) = x$ and $f_1(x) = 2^x$, so the matrix M is

$$M = \begin{bmatrix} f_0(x_1) & f_1(x_1) \\ f_0(x_2) & f_1(x_2) \\ f_0(x_3) & f_1(x_3) \end{bmatrix} = \begin{bmatrix} -1 & 2^{-1} \\ 0 & 2^0 \\ 1 & 2^1 \end{bmatrix} = \tfrac{1}{2} \begin{bmatrix} -2 & 1 \\ 0 & 2 \\ 2 & 4 \end{bmatrix}$$

In this case $M^T M = \tfrac{1}{4} \begin{bmatrix} 8 & 6 \\ 6 & 21 \end{bmatrix}$ is invertible, so the normal equations

$$\tfrac{1}{4} \begin{bmatrix} 8 & 6 \\ 6 & 21 \end{bmatrix} Z = \begin{bmatrix} 4 \\ 9 \end{bmatrix} \quad \text{have a unique solution} \quad Z = \tfrac{1}{11} \begin{bmatrix} 10 \\ 16 \end{bmatrix}$$

Hence the best-fitting function of the form $r_0 x + r_1 2^x$ is $\overline{f}(x) = \tfrac{10}{11} x + \tfrac{16}{11} 2^x$.

$$\text{Note that } \overline{f}(X) = \begin{bmatrix} \overline{f}(-1) \\ \overline{f}(0) \\ \overline{f}(1) \end{bmatrix} = \begin{bmatrix} \frac{-2}{11} \\ \frac{16}{11} \\ \frac{42}{11} \end{bmatrix}, \text{ compared with } Y = \begin{bmatrix} 0 \\ 1 \\ 4 \end{bmatrix}.$$

Exercises 8.7

1. Find the best approximation to a solution of each of the following systems of equations.

   (a) $x + y - z = 5$
   $2x - y + 6z = 1$
   $3x + 2y - z = 6$
   $-x + 4y + z = 0$

   ◆(b) $3x + y + z = 6$
   $2x + 3y - z = 1$
   $2x - y + z = 0$
   $3x - 3y + 3z = 8$

2. Find a least squares approximating function of the form $r_0 x + r_1 x^2 + r_2 2^x$ for each of the following sets of data pairs.

   (a) $(-1, 1), (0, 3), (1, 1), (2, 0)$

   ◆(b) $(0, 1), (1, 1), (2, 5), (3, 10)$

3. Find the least squares approximating function of the form $r_0 + r_1 x^2 + r_2 \sin \frac{\pi x}{2}$ for each of the following sets of data pairs.

   (a) $(0, 3), (1, 0), (1, -1), (-1, 2)$

   ◆(b) $(-1, \frac{1}{2}), (0, 1), (2, 5), (3, 9)$

◆4. The yield y of wheat in bushels per acre appears to be a linear function of the number of days x_1 of sunshine, the number of inches x_2 of rain, and the number of pounds x_3 of fertilizer applied per acre. Find the best fit to the data in the table by an equation of the form $y = r_0 + r_1 x_1 + r_2 x_2 + r_3 x_3$. [*Hint:* If a calculator for inverting $A^T A$ is not available, the inverse is given in the answer.]

y	x_1	x_2	x_3
28	50	18	10
30	40	20	16
21	35	14	10
23	40	12	12
23	30	16	14

5. Let A be any $m \times n$ matrix and write $K = \{X \mid A^T A X = 0\}$. Let B be an m-column. Show that, if Z is an n-column such that $\|B - AZ\|$ is minimal, then *all* such vectors have the form $Z + X$ for some X in K. [*Hint:* $\|B - AY\|$ is minimal if and only if $A^T A Y = A^T B$.]

6. Given the situation in Theorem 3, write

   $$f(x) = r_0 f_0(x) + r_1 f_1(x) + \cdots + r_m f_m(x)$$

   Suppose that $f(x)$ has at most k roots for any choice of the coefficients $r_0, r_1, \ldots, r_m$, not all zero.

   (a) Show that $M^T M$ is invertible if at least $k + 1$ of the x_i are distinct.

   ◆(b) If at least two of the x_i are distinct, show that there is always a best approximation of the form $r_0 + r_1 e^x$.

   (c) If at least three of the x_i are distinct, show that there is always a best approximation of the form $r_0 + r_1 x + r_2 e^x$. [Calculus is needed.]

7. If A is an $m \times n$ matrix, it can be proved that there exists a unique $n \times m$ matrix $A^\#$ satisfying the following four conditions: $A A^\# A = A$; $A^\# A A^\# = A^\#$; $A A^\#$ and $A^\# A$ are symmetric. The matrix $A^\#$ is called the **generalized inverse** of A, or the **Moore–Penrose** inverse.

   (a) If A is square and invertible, show that $A^\# = A^{-1}$.

   (b) If rank $A = m$, show that $A^\# = A^T (A A^T)^{-1}$.

   (c) If rank $A = n$, show that $A^\# = (A^T A)^{-1} A^T$.

SECTION 8.8   Finite Fields and Linear Codes

For centuries mankind has been using codes to transmit messages. In many cases, for example transmitting financial, medical, or military information, the message is disguised in such a way that it cannot be understood by an intruder who intercepts it, but can be easily "decoded" by the intended receiver. This subject is called *cryptography* and, while intriguing, is not our focus here. Instead, we investigate methods for detecting and correcting errors in the transmission of the message.

The stunning photos from the planet Saturn sent by the space probe are a very good example of how successful these methods can be. These messages are subject to "noise" such as solar interference which causes errors in the message. The signal is received on Earth with errors that must be detected and corrected before the high-quality pictures can be printed. This is done using error-correcting codes. To see how, we first discuss a system of adding and multiplying integers while ignoring multiples of a fixed integer.

Modular Arithmetic

We work in the set $\mathbb{Z} = \{0, \pm 1, \pm 2, \pm 3, \ldots\}$ of **integers**, that is the set of whole numbers. Everyone is familiar with the process of "long division" from arithmetic. For example, we can divide an integer a by 5 and leave a remainder "modulo 5" in the set $\{0, 1, 2, 3, 4\}$. As an illustration

$$19 = 3 \cdot 5 + 4,$$

so the remainder of 19 modulo 5 is 4. Similarly, the remainder of 137 modulo 5 is 2 because $137 = 27 \cdot 5 + 2$. This works even for negative integers: For example,

$$-17 = (-4) \cdot 5 + 3,$$

so the remainder of -17 modulo 5 is 3.

This process is called the **division algorithm**. More formally, let $n \geq 2$ denote a positive integer. Then every integer a can be written uniquely in the form

$$a = qn + r \quad \text{where } q \text{ and } r \text{ are integers and } 0 \leq r \leq n - 1.$$

Here q is called the **quotient** of a **modulo** n, and r is called the **remainder** of a **modulo** n. We refer to n as the **modulus**. Thus, if $n = 6$, the fact that $134 = 22 \cdot 6 + 2$ means that 134 has quotient 22 and remainder 2 modulo 6.

Our interest here is in the set of *all* possible remainders modulo n. This set is denoted

$$\mathbb{Z}_n = \{0, 1, 2, 3, \ldots, n - 1\}$$

and is called the set of **integers modulo** n. Thus every integer is uniquely represented in $\mathbb{Z}_n$ by its remainder modulo n.

We are going to show how to do arithmetic in $\mathbb{Z}_n$ by adding and multiplying modulo n. That is, we add or multiply two numbers in $\mathbb{Z}_n$ by calculating the usual sum or product in $\mathbb{Z}$ and taking the remainder modulo n. It is proved in books on abstract algebra that the usual laws of arithmetic hold in $\mathbb{Z}_n$ for any modulus $n \geq 2$. This seems remarkable until we remember that these laws are true for ordinary addition and multiplication and all we are doing is reducing modulo n.

Consider the case $n = 6$, so that $\mathbb{Z}_6 = \{0, 1, 2, 3, 4, 5\}$. Then $2 + 5 = 1$ in $\mathbb{Z}_6$ because 7 leaves a remainder of 1 when divided by 6. Similarly, $2 \cdot 5 = 4$ in $\mathbb{Z}_6$, while $3 + 5 = 2$, and $3 + 3 = 0$. In this way we can fill in the addition and multiplication tables for $\mathbb{Z}_6$; the result is:

TABLES FOR $\mathbb{Z}_6$

+	0	1	2	3	4	5
0	0	1	2	3	4	5
1	1	2	3	4	5	0
2	2	3	4	5	0	1
3	3	4	5	0	1	2
4	4	5	0	1	2	3
5	5	0	1	2	3	4

×	0	1	2	3	4	5
0	0	0	0	0	0	0
1	0	1	2	3	4	5
2	0	2	4	0	2	4
3	0	3	0	3	0	3
4	0	4	2	0	4	2
5	0	5	4	3	2	1

Calculations in $\mathbb{Z}_6$ are carried out much as in $\mathbb{Z}$. As an illustration, consider the "distributive law" $a(b + c) = ab + ac$ familiar from ordinary arithmetic. This holds for all a, b, and c in $\mathbb{Z}_6$; we verify a particular case:

$$3(5 + 4) = 3 \cdot 5 + 3 \cdot 4 \quad \text{in } \mathbb{Z}_6.$$

In fact, the left side is $3(5 + 4) = 3 \cdot 3 = 3$, and the right side is $3 \cdot 5 + 3 \cdot 4 = 3 + 0 = 3$ too. Hence doing arithmetic in $\mathbb{Z}_6$ is familiar. However, there are differences. For example, $3 \cdot 4 = 0$ in $\mathbb{Z}_6$, in contrast to the fact that $a \cdot b = 0$ in $\mathbb{Z}$ can only happen when either $a = 0$ or $b = 0$.

Note that we will make statements like $-30 = 19$ in $\mathbb{Z}_7$; it means that -30 and 19 leave the same remainder—5—when divided by 7, and so are equal in $\mathbb{Z}_7$; they both equal 5. In general, if $n \geq 2$ is any modulus, the operative fact is that

$$a = b \text{ in } \mathbb{Z}_n \quad \text{if and only if} \quad a - b \text{ is a multiple of } n.$$

In this case we say that a and b are **equal modulo** n, and write $a = b \pmod{n}$.

Arithmetic in $\mathbb{Z}_n$ is, in a sense, simpler than that for the real numbers. For example, consider negatives. Given the element 8 in $\mathbb{Z}_{17}$, what is -8? The answer lies in the observation that $8 + 9 = 0$ in $\mathbb{Z}_{17}$, so $-8 = 9$ (and $-9 = 8$). In the same way, finding negatives is not difficult in $\mathbb{Z}_n$ for any modulus n.

Check-Digit Codes

0 78787 73104 2

Everyone has seen the bar codes on products at the supermarket (see the diagram). The stripes on the bar are scanned by a laser to determine a sequence of twelve numbers

$$\mathbf{v} = [d_1, d_2, \ldots, d_{11}, d_{12}] \quad \text{where } 0 \leq d_i \leq 9 \text{ for each } i.$$

This is called the **Universal Product Code** (UPC) for the product, and the integers d_i are called *digits*. The first eleven digits of $\mathbf{v}$ provide information to the store about the product being sold. The last digit d_{12} in the sequence is called a *check-digit* and is chosen so that the dot product

$$\mathbf{v} \cdot \mathbf{c} = 0 \quad \text{in } \mathbb{Z}_{10}$$

where $\mathbf{c} = [3, 1, 3, 1, 3, 1, 3, 1, 3, 1, 3, 1]$ is called a *check vector*. Here we are regarding the digits of $\mathbf{v}$ and $\mathbf{c}$ as elements in $\mathbb{Z}_{10}$, and the dot product is defined (as in $\mathbb{R}^{12}$) by multiplying corresponding digits in $\mathbb{Z}_{10}$ and adding the results.

In practice, if a manufacturer wants to assign a UPC vector $\mathbf{v}$ to a product, he first chooses the first eleven digits from $\mathbb{Z}_{10}$ to describe the product, and then chooses the check-digit so that $\mathbf{v} \cdot \mathbf{c} = 0$ in $\mathbb{Z}_{10}$. For example, if $\mathbf{v}$ has the form $\mathbf{v} = [2, 7, 1, 0, 4, 5, 3, 0, 1, 0, 3, d]$, then $\mathbf{v} \cdot \mathbf{c} = 4 + d$ in $\mathbb{Z}_{10}$ (verify), so insisting that $\mathbf{v} \cdot \mathbf{c} = 0$ in $\mathbb{Z}_{10}$ requires that $d = 6$.

The UPC can detect single digit errors and some adjacent interchanges. Suppose that a UPC vector $\mathbf{v}$ is read by the laser scanner as $\mathbf{v}'$, and an error occurs. Then the scanner computes $\mathbf{v}' \cdot \mathbf{c}$ and the error will be detected if $\mathbf{v}' \cdot \mathbf{c} \neq 0$ in $\mathbb{Z}_{10}$. Of course it could happen that $\mathbf{v}' \cdot \mathbf{c} = 0$, so the error would *not* be detected. But this does not happen for errors to a single digit, or for most errors where two adjacent digits are interchanged.

Suppose that an error is made in the first digit of $\mathbf{v} = [d_1, d_2, d_3, \ldots]$, and it is scanned as $\mathbf{v}' = [a, d_2, d_3, \ldots]$ where $a \neq d_1$. Then $\mathbf{v}' = \mathbf{v} + [a - d_1, 0, 0, \ldots]$, so

$$\mathbf{v}' \cdot \mathbf{c} = \mathbf{v} \cdot \mathbf{c} + 3(a - d_1) = 3(a - d_1) \neq 0 \quad \text{in } \mathbb{Z}_{10}$$

because $3x \neq 0$ in $\mathbb{Z}_{10}$ whenever $x \neq 0$ in $\mathbb{Z}_{10}$ (verify). This means that the error is detected. This argument works if a single error occurs in any odd digit of $\mathbf{v}$, and an analogous proof works for single even digit errors.

Similarly, the UPC can detect most errors where two adjacent (nonequal) digits are interchanged. Suppose the vector $\mathbf{v} = [a, b, d_3, d_4, \ldots]$ is read incorrectly as $\mathbf{v}' = [b, a, d_3, d_4, \ldots]$, where $a \neq b$ in $\mathbb{Z}_{10}$. Then $\mathbf{v}' = \mathbf{v} + [b - a, a - b, 0, 0, \ldots]$ so

$$\mathbf{v}' \cdot \mathbf{c} = \mathbf{v} \cdot \mathbf{c} + (3(a - b) + (b - a)) = 2(a - b) \quad \text{in } \mathbb{Z}_{10}.$$

Hence the error will be detected if $2(a - b) \neq 0$ in $\mathbb{Z}_{10}$. But $2x = 0$ in $\mathbb{Z}_{10}$ can only happen if $x = 0$ or $x = 5$, so the error will be detected if $a - b$ is not 5 in $\mathbb{Z}_{10}$. A similar argument works if any two adjacent digits are interchanged. Hence *all* adjacent interchange errors will be detected if the UPC codes used have the property that no nonequal adjacent digits differ by 5.

There are other common examples of check-digit codes. The **International Standard Book Number** (ISBN) is a ten-digit vector $\mathbf{v}$ assigned to each book. However, in this case the digits of $\mathbf{v}$ are drawn from $\mathbb{Z}_{11}$ (rather than $\mathbb{Z}_{10}$ as in the UPC code). Again, the first nine digits give publication information about the book, and the tenth digit is a check-digit, chosen so that $\mathbf{v} \cdot \mathbf{c} = 0$ in $\mathbb{Z}_{11}$. In this case, however, the check vector is $\mathbf{c} = [10, 9, 8, 7, 6, 5, 4, 3, 2, 1]$—with digits drawn from $\mathbb{Z}_{11}$. The ISBN code detects all single errors and all adjacent transposition errors.

In a similar way, the number on your credit card, and the number on your bank account are check-digit codes. The check vector is different in each case, and more elaborate schemes are used to detect errors.

Finite Fields

In our study of linear algebra so far the scalars have been real (possibly complex) numbers. The set $\mathbb{R}$ of real numbers has the property that it is closed under addition and multiplication, that the usual laws of arithmetic hold, and that every nonzero real number has an inverse in $\mathbb{R}$. Such a system is called a **field**. Hence the real numbers $\mathbb{R}$ form a field, as does the set $\mathbb{C}$ of complex numbers. Another example is the set $\mathbb{Q}$ of all rational numbers (fractions); however the set $\mathbb{Z}$ of integers is *not* a field—for example, 2 has no inverse *in* the set $\mathbb{Z}$.

Our motivation for isolating the concept of a field is that nearly everything we have done remains valid if the scalars are restricted to some field: The gaussian algorithm can be used to solve systems of linear equations with coefficients in the field; a square matrix with entries from the field is invertible if and only if its determinant is nonzero; the matrix inversion algorithm works in the same way; and so on. The reason is that the field has all the properties used in the proofs of these results, so all the theorems remain valid.

It turns out that there are *finite* fields—that is, finite sets that satisfy the usual laws of arithmetic and in which every nonzero element has an inverse. If $n \geq 2$ is an integer, the modular system $\mathbb{Z}_n$ certainly satisfies the basic laws of arithmetic, but it need not be a field. For example we have $2 \cdot 3 = 0$ in $\mathbb{Z}_6$ so 3 has no inverse in $\mathbb{Z}_6$ (if $3a = 1$ then $2 = 2 \cdot 1 = 2(3a) = 0a = 0$ in $\mathbb{Z}_6$, a contradiction). The problem is that $6 = 2 \cdot 3$ can be properly factored in $\mathbb{Z}$.

More generally, an integer $p \geq 2$ is called a **prime** if p *cannot* be factored as $p = ab$ where a and b are positive integers and neither a nor b equals 1. Thus the first few primes are 2, 3, 5, 7, 11, 13, 17, …. If $n \geq 2$ is not a prime and $n = ab$ where $2 \leq a$ and $b \leq n - 1$, then $ab = 0$ in $\mathbb{Z}_n$ and it follows (as above in the case $n = 6$) that b cannot have an inverse in $\mathbb{Z}_n$, and hence that $\mathbb{Z}_n$ is not a field. In other words, if $\mathbb{Z}_n$ is a field, then n must be a prime. Surprisingly, the converse is true:

Theorem 1

If p is a prime, then $\mathbb{Z}_p$ is a field using addition and multiplication modulo p.

The proof can be found in books on abstract algebra.[10] If p is a prime, the field $\mathbb{Z}_p$ is called the **field of integers modulo** p.

For example, consider the case $n = 5$. Then $\mathbb{Z}_5 = \{0, 1, 2, 3, 4\}$ and the addition and multiplication tables are:

+	0	1	2	3	4
0	0	1	2	3	4
1	1	2	3	4	0
2	2	3	4	0	1
3	3	4	0	1	2
4	4	0	1	2	3

×	0	1	2	3	4
0	0	0	0	0	0
1	0	1	2	3	4
2	0	2	4	1	3
3	0	3	1	4	2
4	0	4	3	2	1

Hence 1 and 4 are self-inverse in $\mathbb{Z}_5$, and 2 and 3 are inverses of each other, so $\mathbb{Z}_5$ is indeed a field. Here is another important example.

Example 1

If $p = 2$, then $\mathbb{Z}_2 = \{0, 1\}$ is a field with addition and multiplication modulo 2 given by the tables

+	0	1
0	0	1
1	1	0

and

×	0	1
0	0	0
1	0	1

This is binary arithmetic, the basic algebra of computers.

While it is routine to find negatives of elements of $\mathbb{Z}_p$, it is a bit more difficult to find inverses in $\mathbb{Z}_p$. For example, how does one find 14^{-1} in $\mathbb{Z}_{17}$? Since we want $14^{-1} \cdot 14 = 1$ in $\mathbb{Z}_{17}$, we are looking for an integer a with the property that $a \cdot 14 = 1$ modulo 17. Of course we can try all possibilities (there are only 17 of them!), and the result is $a = 11$ (verify). However this method is of little use for large primes p, and it is a comfort to know that there is a systematic procedure (called the **euclidean algorithm**) for finding inverses in $\mathbb{Z}_p$ for any prime p. Furthermore, this algorithm is easy to program for a computer. To illustrate the method, let us once again find the inverse of 14 in $\mathbb{Z}_{17}$.

---

10 For example W. K. Nicholson, *Introduction to Abstract Algebra*, 2nd ed., (New York: Wiley, 1999).

Example 2

Solution

Find the inverse of 14 in $\mathbb{Z}_{17}$.

The idea is to first divide $p = 17$ by 14:

$$17 = 1 \cdot 14 + 3.$$

Now divide 14 by the new remainder 3 to get

$$14 = 4 \cdot 3 + 2,$$

and then divide 3 by the new remainder 2 to get

$$3 = 1 \cdot 2 + 1.$$

It is a theorem of number theory that, because 17 is a prime, this procedure will always lead to a remainder of 1. At this point we eliminate remainders in these equations from the bottom up:

$$
\begin{aligned}
1 &= 3 - 1 \cdot 2 \\
&= 3 - 1 \cdot (14 - 4 \cdot 3) = 5 \cdot 3 - 1 \cdot 14 \\
&= 5 \cdot (17 - 1 \cdot 14) - 1 \cdot 14 = 5 \cdot 17 - 6 \cdot 14.
\end{aligned}
$$

Hence $(-6) \cdot 14 = 1$ in $\mathbb{Z}_{17}$, that is $11 \cdot 14 = 1$. So $14^{-1} = 11$ in $\mathbb{Z}_{17}$.

As mentioned above, nearly everything we have done with matrices over the field of real numbers can be done in the same way for matrices with entries from $\mathbb{Z}_p$. We illustrate this with one example. Again the reader is referred to books on abstract algebra.

Example 3

Solution

Determine if the matrix $A = \begin{bmatrix} 1 & 4 \\ 6 & 5 \end{bmatrix}$ from $\mathbb{Z}_7$ is invertible and, if so, find its inverse.

Working in $\mathbb{Z}_7$ we have $\det A = 1 \cdot 5 - 6 \cdot 4 = 5 - 3 = 2 \neq 0$ in $\mathbb{Z}_7$, so A is invertible. Hence Example 4 §2.3 gives $A^{-1} = 2^{-1} \begin{bmatrix} 5 & -4 \\ -6 & 1 \end{bmatrix}$. Note that $2^{-1} = 4$ in $\mathbb{Z}_7$ (because $2 \cdot 4 = 1$ in $\mathbb{Z}_7$). Note also that $-4 = 3$ and $-6 = 1$ in $\mathbb{Z}_7$, so finally $A^{-1} = 4 \begin{bmatrix} 5 & 3 \\ 1 & 1 \end{bmatrix} = \begin{bmatrix} 6 & 5 \\ 4 & 4 \end{bmatrix}$. The reader can verify that indeed $\begin{bmatrix} 1 & 4 \\ 6 & 5 \end{bmatrix}\begin{bmatrix} 6 & 5 \\ 4 & 4 \end{bmatrix} = \begin{bmatrix} 1 & 0 \\ 0 & 1 \end{bmatrix}$ in $\mathbb{Z}_7$.

While we shall not use them, there are finite fields other than $\mathbb{Z}_p$ for the various primes p. Let F be a finite field, and let 1 denote the unity element of F (that is, $1a = a$ for all a in F). If $k \geq 1$ is an integer, let $k1 = 1 + 1 + \cdots + 1$ (k summands). Then the elements $1, 21, 31, \ldots$ cannot be all distinct (F is finite), say $a1 = b1$ for integers $a > b \geq 1$. If $k = a - b$, this means that $k1 = 0$ where $k \geq 1$. Now let

$$p \text{ be the smallest positive integer such that } p1 = 0 \text{ in } F.$$

We claim that p is a prime. Indeed, if $p = km$ where $1 < k < p$ and $1 < m < p$ are integers, then $0 = p\mathbf{1} = (k\mathbf{1})(m\mathbf{1})$ so either $k\mathbf{1} = 0$ or $m\mathbf{1} = 0$,[11] contrary to the minimality of p. Hence p is a prime, called the **characteristic** of F.

We now claim that the subset $\{0, \mathbf{1}, 2\mathbf{1}, 3\mathbf{1}, \ldots, (p-1)\mathbf{1}\}$ of F is a copy of the field $\mathbb{Z}_p$. In fact, define a function

$$T : \mathbb{Z}_p \to \{0, \mathbf{1}, 2\mathbf{1}, 3\mathbf{1}, \ldots, (p-1)\mathbf{1}\} \quad \text{by} \quad T(a) = a\mathbf{1} \text{ for all } a \text{ in } \mathbb{Z}_p.$$

Note that this makes sense only if $a = b$ in $\mathbb{Z}_p$ implies that $a\mathbf{1} = b\mathbf{1}$. But $a = b$ in $\mathbb{Z}_p$ means that $a - b = kp$ for some integer k, so $a\mathbf{1} - b\mathbf{1} = (a - b)\mathbf{1} = kp\mathbf{1} = 0$ in F, as required. So T makes sense (we say that T is *well defined*) and it is routine to verify that T is one-to-one and onto, and that the correspondence $a \leftrightarrow T(a) = a\mathbf{1}$ preserves addition and multiplication:

$$a + b \leftrightarrow a\mathbf{1} + b\mathbf{1} \quad \text{and} \quad ab \leftrightarrow a\mathbf{1}\, b\mathbf{1}.$$

In other words, $\{0, \mathbf{1}, 2\mathbf{1}, 3\mathbf{1}, \ldots, (p-1)\mathbf{1}\}$ is a copy of $\mathbb{Z}_p$, and it is customary to *identify* it with $\mathbb{Z}_p$. This being done, we have

$$\mathbb{Z}_p \subseteq F.$$

In this case, if a is in $\mathbb{Z}_p$ and v is in F, the product av is again in F. Using this as the scalar multiplication, F becomes a vector space over $\mathbb{Z}_p$ (with the addition of F), say with basis $\{f_1, f_2, \ldots, f_n\}$. Then each element of F has a unique representation as a linear combination $a_1 f_1 + a_2 f_2 + \cdots + a_n f_n$ where the a_i are in $\mathbb{Z}_p$. Since there are p choices for each of the a_i, it follows that the finite field F has exactly p^n elements.

Surprisingly, the converse is true: For every prime p and every integer $n \geq 1$, there *exists* a field with exactly p^n elements, and this field is *unique*.[12] It is called the **Galois field** of order p^n, and is denoted $GF(p^n)$. Note that the above argument shows that we may assume that $\mathbb{Z}_p \subseteq GF(p^n)$.

Error Correcting Codes

Coding theory is concerned with the transmission of information over a *channel* that is affected by *noise*. The noise causes errors, so the aim of the theory is to find ways to detect such errors and correct at least some of them. General coding theory originated with the work of Claude Shannon (1916–2001) who showed that information can be transmitted at near optimal rates with arbitrarily small chance of error.

Let F denote a finite field and, if $n \geq 1$, let F^n denote the F-vector space of $n \times 1$ row matrices over F with the usual componentwise addition and scalar multiplication. In this context, the rows in F^n are called **words** (or n-**words**) and, as the name implies, will be written as $[a \ b \ c \ d] = abcd$. The individual components of a word are called its **digits**. A nonempty subset C of F^n is called a **code** (or an n-**code**), and the elements in C are called **code words**. If $F = \mathbb{Z}_2$, these are called **binary** codes.

If a code word $\mathbf{w}$ is transmitted and an error occurs, the resulting word $\mathbf{v}$ is decoded as the code word "closest" to $\mathbf{v}$ in F^n. To make sense of what "closest" means, we need a distance function on F^n analogous to that in $\mathbb{R}^n$ (see Theorem 3 §5.3).

---

11 If $ab = 0$ in a field F, then either $a = 0$ or $b = 0$ (because if $a \neq 0$ it has an inverse in F).

12 See, for example, W. K. Nicholson, *Introduction to Abstract Algebra*, 2nd ed., (New York: Wiley, 1999).

The usual definition in $\mathbb{R}^n$ does not work in this situation. For example, if $\mathbf{w} = 1111$ in $\mathbb{Z}_2^4$, then the square of the distance of $\mathbf{w}$ from $\mathbf{0}$ is $(1-0)^2 + (1-0)^2 + (1-0)^2 + (1-0)^2 = 0$, even though $\mathbf{w} \neq \mathbf{0}$.

However there is a satisfactory notion of distance in F^n due to Richard Hamming (1915–1998). Given a word $\mathbf{w} = a_1 a_2 \cdots a_n$ in F^n, we first define the **Hamming weight** $wt(\mathbf{w})$ to be the number of nonzero digits in $\mathbf{w}$:

$$wt(\mathbf{w}) = |\{i \,|\, a_i \neq 0\}|$$

Clearly, $0 \leq wt(\mathbf{w}) \leq n$ for every word $\mathbf{w}$ in F^n. Given another word $\mathbf{v} = b_1 b_2 \cdots b_n$ in F^n, the **Hamming distance** $d(\mathbf{v}, \mathbf{w})$ between $\mathbf{v}$ and $\mathbf{w}$ is defined by

$$d(\mathbf{v}, \mathbf{w}) = wt(\mathbf{v} - \mathbf{w}) = |\{i \,|\, b_i \neq a_i\}|.$$

In other words, $d(\mathbf{v}, \mathbf{w})$ is the number of places at which the digits of $\mathbf{v}$ and $\mathbf{w}$ differ. The next result justifies using the term *distance* for this function d.

Theorem 2

Let $\mathbf{u}$, $\mathbf{v}$, and $\mathbf{w}$ denote words in F^n. Then:
1. $d(\mathbf{v}, \mathbf{w}) \geq 0$.
2. $d(\mathbf{v}, \mathbf{w}) = 0$ *if and only if* $\mathbf{v} = \mathbf{w}$.
3. $d(\mathbf{v}, \mathbf{w}) = d(\mathbf{w}, \mathbf{v})$.
4. $d(\mathbf{v}, \mathbf{w}) \leq d(\mathbf{v}, \mathbf{u}) + d(\mathbf{u}, \mathbf{w})$.

PROOF

(1) and (3) are clear, and (2) follows because $wt(\mathbf{v}) = 0$ if and only if $\mathbf{v} = \mathbf{0}$. To prove (4), write $\mathbf{x} = \mathbf{v} - \mathbf{u}$ and $\mathbf{y} = \mathbf{u} - \mathbf{w}$. Then (4) reads $wt(\mathbf{x} + \mathbf{y}) \leq wt(\mathbf{x}) + wt(\mathbf{y})$. If $\mathbf{x} = a_1 a_2 \cdots a_n$ and $\mathbf{y} = b_1 b_2 \cdots b_n$, this follows if $a_i + b_i \neq 0$ implies that either $a_i \neq 0$ or $b_i \neq 0$. But this is clear.

Given a word $\mathbf{w}$ in F^n and a real number $r > 0$, define the **sphere** $S_r(\mathbf{w})$ of radius r (or simply the r-**sphere**) about $\mathbf{w}$ as follows:

$$S_r(\mathbf{w}) = \{\mathbf{x} \in F^n \,|\, d(\mathbf{w}, \mathbf{x}) \leq r\}.$$

Using this we can describe one of the most useful decoding methods.

Nearest Neighbour Decoding

Let C be an n-code, and suppose a word v is transmitted and w is received. Then w is decoded as the code word in C closest to it. (If there is a tie, choose arbitrarily.)

Using this method, we can describe how to construct a code C that can detect (or correct) t errors. Suppose a code word $\mathbf{c}$ is transmitted and a word $\mathbf{w}$ is received with s errors where $1 \leq s \leq t$. Then s is the number of places at which the $\mathbf{c}$- and $\mathbf{w}$-digits differ, that is $s = d(\mathbf{c}, \mathbf{w})$. Hence $S_t(\mathbf{c})$ consists of all possible received words where at most t errors have occurred.

Assume first that C has the property that no code word lies in the t-sphere of another code word. Because $\mathbf{w}$ is in $S_t(\mathbf{c})$ and $\mathbf{w} \neq \mathbf{c}$, this means that $\mathbf{w}$ is not a code word and the error has been detected. If we strengthen the assumption on C to require that the t-spheres about code words are pairwise disjoint, then $\mathbf{w}$ belongs to a unique sphere (the one about $\mathbf{c}$), and so $\mathbf{w}$ will be correctly decoded as $\mathbf{c}$.

To describe when this happens, let C be an n-code. The **minimum distance** d of C is defined to be the smallest distance between two distinct code words in C; that is,

$$d = \min\{d(\mathbf{v},\mathbf{w}) \mid \mathbf{v} \text{ and } \mathbf{w} \text{ in } C; \mathbf{v} \neq \mathbf{w}\}.$$

Theorem 3

Let C be an n-code with minimum distance d. Assume that nearest neighbour decoding is used. Then:
1. If $t < d$, then C can detect t errors.[13]
2. If $2t < d$, then C can correct t errors.

PROOF

1. Let $\mathbf{c}$ be a code word in C. If $\mathbf{w} \in S_t(\mathbf{c})$, then $d(\mathbf{w},\mathbf{c}) \leq t < d$ by hypothesis. Thus the t-sphere $S_t(\mathbf{c})$ contains no other code word, so C can detect t errors by the preceding discussion.
2. If $2t < d$, it suffices (again by the preceding discussion) to show that the t-spheres about distinct code words are pairwise disjoint. But if $\mathbf{c} \neq \mathbf{c}'$ are code words in C and $\mathbf{w}$ is in $S_t(\mathbf{c}') \cap S_t(\mathbf{c})$, then Theorem 2 gives

$$d(\mathbf{c},\mathbf{c}') \leq d(\mathbf{c},\mathbf{w}) + d(\mathbf{w},\mathbf{c}') \leq t + t = 2t < d$$

by hypothesis, contradicting the minimality of d.

Example 4

If $F = \mathbb{Z}_3 = \{0, 1, 2\}$, the 6-code $\{111111, 111222, 222111\}$ has minimum distance 3 and so can detect 2 errors and correct 1 error.

Let $\mathbf{c}$ be any word in F^n. A word $\mathbf{w}$ satisfies $d(\mathbf{w}, \mathbf{c}) = r$ if and only if $\mathbf{w}$ and $\mathbf{c}$ differ in exactly r digits. If $|F| = q$, there are exactly $\binom{n}{r}(q-1)^r$ such words where $\binom{n}{r}$ is the binomial coefficient. Indeed, choose the r places where they differ in $\binom{n}{r}$ ways, and then fill those places in $\mathbf{w}$ in $(q-1)^r$ ways. It follows that the number of words in the t-sphere about $\mathbf{c}$ is

$$|S_t(\mathbf{c})| = \binom{n}{0} + \binom{n}{1}(q-1) + \binom{n}{2}(q-1)^2 + \cdots + \binom{n}{t}(q-1)^t = \sum_{i=0}^{t}\binom{n}{i}(q-1)^i.$$

This leads to a useful bound on the size of error-correcting codes.

Theorem 4   Hamming Bound

Let C be an n-code over a field F that can correct t errors using nearest neighbour decoding. If $|F| = q$, then

$$|C| \leq \frac{q^n}{\sum_{i=0}^{t}\binom{n}{i}(q-1)^i}.$$

---

13 If C can detect (or correct) t or fewer errors, we simply say that C detects (corrects) t errors.

PROOF

Write $k = \sum_{i=0}^{t} \binom{n}{i}(q-1)^i$. The t-spheres centred at distinct code words each contain k words, and there are $|C|$ of them. Moreover they are pairwise disjoint because the code corrects t errors (see the discussion preceding Theorem 3). Hence they contain $k \cdot |C|$ distinct words, and so $k \cdot |C| \le |F^n| = q^n$, proving the theorem.

A code is called **perfect** if there is equality in the Hamming bound; equivalently, if every word in F^n lies in exactly one t-sphere about a code word. For example, if $F = \mathbb{Z}_2$, $n = 3$ and $t = 1$, then $q = 2$ and $\binom{3}{0} + \binom{3}{1} = 4$, so the Hamming bound is $\frac{2^3}{4} = 2$. The 3-code $C = \{000, 111\}$ has minimum distance 3 and so can correct 1 error by Theorem 3. Hence C is perfect.

Linear Codes

Up to this point we have been regarding *any* nonempty subset of the F-vector space F^n as a code. However many important codes are actually subspaces. A subspace $C \subseteq F^n$ of dimension $k \ge 1$ over F is called an (n, k)-**linear code**, or simply an (n, k)-**code**. We do not regard the zero subspace ($k = 0$) as a code.

Example 5

If $F = \mathbb{Z}_2$ and $n \ge 2$, the n-**parity-check code** is constructed as follows: An extra digit is added to each word in F^{n-1} to make the number of 1's in the resulting word even (we say such words have **even parity**). The resulting $(n, n-1)$-code is linear because the sum of two words of even parity again has even parity.

Many of the properties of general codes take a simpler form for linear codes. The following result gives a much easier way to find the minimal distance of a linear code, and sharpens the results in Theorem 3.

Theorem 5

Let C be an (n, k)-code with minimum distance d over a finite field F, and use nearest neighbour decoding.
  1. $d = \min\{wt(\mathbf{w}) \,|\, \mathbf{0} \ne \mathbf{w} \text{ in } C\}$.
  2. C can detect $t \ge 1$ errors if and only if $t < d$.
  3. C can correct $t \ge 1$ errors if and only if $2t < d$.
  4. If C can correct $t \ge 1$ errors and $|F| = q$, then
  $$\binom{n}{0} + \binom{n}{1}(q-1) + \binom{n}{2}(q-1)^2 + \cdots + \binom{n}{t}(q-1)^t \le q^{n-k}.$$

PROOF

  1. Write $d' = \min\{wt(\mathbf{w}) \,|\, \mathbf{0} \ne \mathbf{w} \text{ in } C\}$. If $\mathbf{v} \ne \mathbf{w}$ are words in C, then $d(\mathbf{v}, \mathbf{w}) = wt(\mathbf{v} - \mathbf{w}) \ge d'$ because $\mathbf{v} - \mathbf{w}$ is in the subspace C. Hence $d \ge d'$. Conversely, given $\mathbf{w} \ne \mathbf{0}$ in C then, since $\mathbf{0}$ is in C, we have $wt(\mathbf{w}) = d(\mathbf{w}, \mathbf{0}) \ge d$ by the definition of d. Hence $d' \ge d$ and (1) is proved.
  2. Assume that C can detect t errors. Given $\mathbf{w} \ne \mathbf{0}$ in C, the t-sphere about $\mathbf{w}$

contains no other code word (see the discussion preceding Theorem 3). In particular, it does not contain the code word $\mathbf{0}$, so $t < d(\mathbf{w}, \mathbf{0}) = wt(\mathbf{w})$. Hence $t < d$ by (1). The converse is part of Theorem 3.

3. We require a result of interest in itself.

Claim. Suppose $\mathbf{c}$ in C has $wt(\mathbf{c}) \leq 2t$. Then $S_t(\mathbf{0}) \cap S_t(\mathbf{c})$ is nonempty.

Proof. If $wt(\mathbf{c}) \leq t$, then $\mathbf{c}$ itself is in $S_t(\mathbf{0}) \cap S_t(\mathbf{c})$. So assume $t < wt(\mathbf{c}) \leq 2t$. Then $\mathbf{c}$ has more than t nonzero digits, so we can form a new word $\mathbf{w}$ by changing exactly t of these nonzero digits to zero. Then $d(\mathbf{w}, \mathbf{c}) = t$, so $\mathbf{w}$ is in $S_t(\mathbf{c})$. But $wt(\mathbf{w}) = wt(\mathbf{c}) - t \leq t$, so $\mathbf{w}$ is also in $S_t(\mathbf{0})$. Hence $\mathbf{w}$ is in $S_t(\mathbf{0}) \cap S_t(\mathbf{c})$, proving the Claim.

If C corrects t errors, the t-spheres about code words are pairwise disjoint (see the discussion preceding Theorem 3). Hence the claim shows that $wt(\mathbf{c}) > 2t$ for all $\mathbf{c} \neq \mathbf{0}$ in C, from which $d > 2t$ by (1). The other inequality comes from Theorem 3.

4. We have $|C| = q^k$ because $\dim_F C = k$, so this assertion restates Theorem 4.

Example 6

If $F = \mathbb{Z}_2$, then

$$C = \{0000000, 0101010, 1010101, 1110000,$$
$$1011010, 0100101, 0001111, 1111111\}$$

is a $(7, 3)$-code; in fact $C = \text{span}\{0101010, 1010101, 1110000\}$. The minimum distance for C is 3, the minimum weight of a nonzero word in C.

Matrix Generators

Given a linear n-code C over a finite field F, the way encoding works in practice is as follows. A message stream is blocked off into segments of length $k \leq n$ called **messages**. Each message $\mathbf{u}$ in F^k is encoded as a code word, the code word is transmitted, the receiver decodes the received word as the nearest code word, and then re-creates the original message. A fast and convenient method is needed to encode the incoming messages, to decode the received word after transmission (with or without error), and finally to retrieve messages from code words. All this can be achieved for any linear code using matrix multiplication.

Let G denote a $k \times n$ matrix over a finite field F, and encode each message $\mathbf{u}$ in F^k as a word $\mathbf{u}G$ in F^n using matrix multiplication (thinking of words as rows). This amounts to saying that the set of code words is the subspace $C = \{\mathbf{u}G \mid \mathbf{u} \text{ in } F^k\}$ of F^n. This subspace need not have dimension k for every $k \times n$ matrix G. But, if $\{\mathbf{e}_1, \mathbf{e}_2, \dots, \mathbf{e}_k\}$ is the standard basis of F^k, then $\mathbf{e}_i G$ is row i of G for each I and $\{\mathbf{e}_1 G, \mathbf{e}_2 G, \dots, \mathbf{e}_k G\}$ spans C. Hence $\dim C = k$ if and only if the rows of G are independent in F^n, and these matrices turn out to be exactly the ones we need. For reference, we state their main properties in Lemma 1 below.

Lemma 1

The following are equivalent for a $k \times n$ matrix G over a finite field F:
1. $\text{rank } G = k$.
2. The columns of G span F^k.
3. The rows of G are independent in F^n.
4. The system $GX = B$ is consistent for every column B in $\mathbb{R}^k$.
5. $GK = I_k$ for some $n \times k$ matrix K.

Clearly each nonzero code word in C has weight at least 3, so C has minimum distance $d = 3$. Hence C can detect two errors and correct one error by Theorem 5. The dual code has minimum distance 4 and so can detect 3 errors and correct 1 error.

Exercises 8.8

1. If $\mathbf{v} = [1, 0, 5, 6, 3, 4, 0, 0, 5, 8, 2, d]$ is a UPC vector for an article, find the check-digit d.

2. (a) Show that if $3a = 0$ in $\mathbb{Z}_{10}$, then necessarily $a = 0$ in $\mathbb{Z}_{10}$.

   ◆(b) Show that $2a = 0$ in $\mathbb{Z}_{10}$ holds in $\mathbb{Z}_{10}$ if and only if $a = 0$ or $a = 5$.

3. Find the inverse of: (a) 8 in $\mathbb{Z}_{13}$; (b) 11 in $\mathbb{Z}_{19}$.

4. If $ab = 0$ in a field F, show that either $a = 0$ or $b = 0$.

5. Show that the entries of the last column of the multiplication table of $\mathbb{Z}_n$ are $0, n-1, n-2, \ldots, 2, 1$ in that order.

6. In each case show that the matrix A is invertible over the given field, and find A^{-1}.

   (a) $A = \begin{bmatrix} 1 & 4 \\ 2 & 1 \end{bmatrix}$ over $\mathbb{Z}_5$. ◆(b) $A = \begin{bmatrix} 5 & 6 \\ 4 & 3 \end{bmatrix}$ over $\mathbb{Z}_7$.

7. Consider the linear system $\begin{aligned} 3x + y + 4z &= 3 \\ 4x + 3y + z &= 1 \end{aligned}$. In each case solve the system by reducing the augmented matrix to reduced row-echelon form over the given field:

   (a) $\mathbb{Z}_5$.   ◆(b) $\mathbb{Z}_7$.

8. Let K be a vector space over $\mathbb{Z}_2$ with basis $\{1, t\}$, so $K = \{a + bt \mid a, b \text{ in } \mathbb{Z}_2\}$. It is known that K becomes a field of four elements if we define $t^2 = 1 + t$. Write down the multiplication table of K.

9. Let K be a vector space over $\mathbb{Z}_3$ with basis $\{1, t\}$, so $K = \{a + bt \mid a, b \text{ in } \mathbb{Z}_3\}$. It is known that K becomes a field of nine elements if we define $t^2 = -1 + 2t$ in $\mathbb{Z}_3$. In each case find the inverse of the element x of K:

   (a) $x = 1 + 2t$.

   ◆(b) $x = 1 + t$.
   [*Hint:* How would you find the inverse in the field $\mathbb{C}$ of complex numbers?]

10. How many errors can be detected or corrected by each of the following binary linear codes?

    (a) $C = \{0000000, 0011110, 0100111, 0111001, 1001011, 1010101, 1101100, 1110010\}$

    ◆(b) $C = \{0000000000, 0010011111, 0101100111, 0111111000, 1001110001, 1011101110, 1100010110, 1110001001\}$

11. (a) If a binary linear $(n, 2)$-code corrects one error, show that $n \geq 5$. [*Hint:* Hamming bound.]

    ◆(b) Find a $(5, 2)$-code that corrects one error.

12. (a) If a binary linear $(n, 3)$-code corrects two errors, show that $n \geq 9$. [*Hint:* Hamming bound.]

    ◆(b) If $G = \begin{bmatrix} 1 & 0 & 0 & 1 & 1 & 1 & 1 & 0 & 0 & 0 \\ 0 & 1 & 0 & 1 & 1 & 0 & 0 & 1 & 1 & 0 \\ 0 & 0 & 1 & 1 & 0 & 1 & 0 & 1 & 1 & 1 \end{bmatrix}$ show that the binary $(10, 3)$-code generated by G corrects two errors. [It can be shown that no binary $(9, 3)$-code corrects two errors.]

13. (a) Show that no binary linear $(4, 2)$-code can correct single errors.

    ◆(b) Find a binary linear $(5, 2)$-code that can correct one error.

14. Find the standard generator matrix G and the parity-check matrix H for each of the following systematic codes:

    (a) $\{00000, 11111\}$ over $\mathbb{Z}_2$.

    ◆(b) Any systematic $(n, 1)$-code where $n \geq 2$.

    (c) The code in Exercise 10(a).

    (d) The code in Exercise 10(b).

15. Let $\mathbf{c}$ be a word in F^n. Show that $S_t(\mathbf{c}) = \mathbf{c} + S_t(\mathbf{0})$, where we write $\mathbf{c} + S_t(\mathbf{0}) = \{\mathbf{c} + \mathbf{v} \mid \mathbf{v} \text{ in } S_t(\mathbf{0})\}$.

16. If a (n, k)-code has two standard generator matrices G and G_1, show that $G = G_1$.

17. Let C be a binary linear n-code (over $\mathbb{Z}_2$). Show that either each word in C has even weight, or half the words in C have even weight and half have odd weight. [*Hint:* The dimension theorem.]

SECTION 8.9   An Application to Quadratic Forms[16]

An expression like $x_1^2 + x_2^2 + x_3^2 - 2x_1x_3 + x_2x_3$ is called a quadratic form in the variables x_1, x_2, and x_3. In this section we show that a change of variables can always be made so that the quadratic form, when expressed in terms of the new variables y_1, y_2, and y_3, has no cross terms y_1y_2, y_1y_3, or y_2y_3. Moreover, we do this for forms involving any finite number of variables using orthogonal diagonalization. This has far-reaching applications; quadratic forms arise in such diverse areas as statistics, physics, the theory of functions of several variables, number theory, and geometry.

A **quadratic form** q in the n variables $x_1, x_2, \ldots, x_n$ is a linear combination of terms $x_1^2, x_2^2, \ldots, x_n^2$ and cross terms $x_1x_2, x_1x_3, x_2x_3 \ldots$. If $n = 3$, q has the form

$$q = a_{11}x_1^2 + a_{22}x_2^2 + a_{33}x_3^2 + a_{12}x_1x_2 + a_{21}x_2x_1$$
$$+ a_{13}x_1x_3 + a_{31}x_3x_1 + a_{23}x_2x_3 + a_{33}x_3x_2$$

In general

$$q = a_{11}x_1^2 + a_{22}x_2^2 + \cdots + a_{nn}x_n^2 + a_{12}x_1x_2 + a_{13}x_1x_3 + \cdots$$

This sum can be written compactly as a matrix product

$$q = q(X) = X^T A X$$

where $X = [x_1 \ \cdots \ x_n]^T$ and $A = [a_{ij}]$ is a real $n \times n$ matrix. Note that if $i \neq j$, two separate terms $a_{ij}x_ix_j$ and $a_{ji}x_jx_i$ are listed, each of which involves x_ix_j, and they can (rather cleverly) be replaced by

$$\tfrac{1}{2}(a_{ij} + a_{ji})x_ix_j \quad \text{and} \quad \tfrac{1}{2}(a_{ij} + a_{ji})x_jx_i$$

respectively, *without altering the sum*. Hence there is no loss of generality in assuming that x_ix_j and x_jx_i have the same coefficient in the sum for q. In other words, *we may assume that A is symmetric*.

Example 1

Write $q = x_1^2 + 3x_3^2 + 2x_1x_2 - x_1x_3$ in the form $q(X) = X^T A X$, where A is a symmetric 3×3 matrix.

Solution The cross terms are $2x_1x_2 = x_1x_2 + x_2x_1$ and $-x_1x_3 = -\tfrac{1}{2}x_1x_3 - \tfrac{1}{2}x_3x_1$. Of course, x_2x_3 and x_3x_2 both have coefficient zero, as does x_2^2. Hence

$$q(X) = [x_1 \ \ x_2 \ \ x_3] \begin{bmatrix} 1 & 1 & -\tfrac{1}{2} \\ 1 & 0 & 0 \\ -\tfrac{1}{2} & 0 & 3 \end{bmatrix} \begin{bmatrix} x_1 \\ x_2 \\ x_3 \end{bmatrix}$$

is the required form (verify).

We shall assume from now on that all quadratic forms are given by

$$q(X) = X^T A X$$

where A is symmetric. Given such a form, the problem is to find new variables $y_1, y_2, \ldots, y_n$, related to $x_1, x_2, \ldots, x_n$, with the property that, when q is expressed in terms of $y_1, y_2, \ldots, y_n$, there are no cross terms. If we write

---

16 This section requires only Section 8.2.

$$Y = [y_1 \; y_2 \; \cdots \; y_n]^T$$

this amounts to asking that $q = Y^T D Y$ where D is diagonal. It turns out that this can always be accomplished and, not surprisingly, that D is the matrix obtained when the symmetric matrix A is diagonalized. In fact, as Theorem 2 §8.2 shows, a matrix P can be found that is orthogonal (that is, $P^{-1} = P^T$) and diagonalizes A:

$$P^T A P = D = \begin{bmatrix} \lambda_1 & 0 & \cdots & 0 \\ 0 & \lambda_2 & \cdots & 0 \\ \vdots & \vdots & & \vdots \\ 0 & 0 & \cdots & \lambda_n \end{bmatrix}$$

The diagonal entries $\lambda_1, \lambda_2, \ldots, \lambda_n$ are the (not necessarily distinct) eigenvalues of A, repeated according to their multiplicities in $c_A(x)$, and the columns of P are the corresponding (orthonormal) eigenvectors of A. As A is symmetric, the λ_i are real by Theorem 7 §5.5.

Now define new variables Y by the equations

$$X = PY \quad \text{equivalently} \quad Y = P^T X$$

Then substitution in $q(X) = X^T A X$ gives

$$q = (PY)^T A (PY) = Y^T (P^T A P) Y = Y^T D Y = \lambda_1 y_1^2 + \lambda_2 y_2^2 + \cdots + \lambda_n y_n^2$$

Hence this change of variables produces the desired simplification in q.

Theorem 1 Diagonalization Theorem

Let $q = X^T A X$ be a quadratic form in the variables $x_1, x_2, \ldots, x_n$, where $X = [x_1 \; x_2 \; \cdots \; x_n]^T$ and A is a symmetric $n \times n$ matrix. Let P be an orthogonal matrix such that $P^T A P$ is diagonal, and define new variables $Y = [y_1 \; y_2 \; \cdots \; y_n]^T$ by

$$X = PY \quad \text{equivalently} \quad Y = P^T X$$

If q is expressed in terms of these new variables $y_1, y_2, \ldots, y_n$, the result is

$$q = \lambda_1 y_1^2 + \lambda_2 y_2^2 + \cdots + \lambda_n y_n^2$$

where $\lambda_1, \lambda_2, \ldots, \lambda_n$ are the eigenvalues of A repeated according to their multiplicities.

Let $q = X^T A X$ be a quadratic form where A is a symmetric matrix and let $\lambda_1, \ldots, \lambda_n$ be the (real) eigenvalues of A repeated according to their multiplicities. A corresponding set $\{F_1, \ldots, F_n\}$ of orthonormal eigenvectors for A is called a set **principal axes** for the quadratic form q. (The reason for the name will become clear later.) The orthogonal matrix P in Theorem 1 is given as $P = [F_1 \; \cdots \; F_n]$, so the variables X and Y are related by

$$X = PY = [F_1 \; F_2 \; \cdots \; F_n] \begin{bmatrix} y_1 \\ y_2 \\ \vdots \\ y_n \end{bmatrix} = y_1 F_1 + y_2 F_2 + \cdots + y_n F_n.$$

Thus the new variables y_i are the coefficients when X is expanded in terms of the orthonormal basis $\{F_1, \ldots, F_n\}$ of $\mathbb{R}^n$. In particular, the coefficients y_i are given by $y_i = X \cdot F_i$ by the expansion theorem (Theorem 6 §5.3). Hence q itself is easily computed from the eigenvalues λ_i and the principal axes F_i:

$$q = q(X) = \lambda_1(X \cdot F_1)^2 + \cdots + \lambda_n(X \cdot F_n)^2.$$

Example 2

Find new variables $y_1, y_2, y_3,$ and y_4 such that

$$q = 3(x_1^2 + x_2^2 + x_3^2 + x_4^2) + 2x_1x_2 - 10x_1x_3 + 10x_1x_4 + 10x_2x_3$$
$$- 10x_2x_4 + 2x_3x_4$$

has diagonal form, and find the corresponding principal axes.

Solution

The form can be written as $q = X^TAX$, where

$$X = \begin{bmatrix} x_1 \\ x_2 \\ x_3 \\ x_4 \end{bmatrix} \qquad A = \begin{bmatrix} 3 & 1 & -5 & 5 \\ 1 & 3 & 5 & -5 \\ -5 & 5 & 3 & 1 \\ 5 & -5 & 1 & 3 \end{bmatrix}$$

A routine calculation yields

$$c_A(x) = \det(xI - A) = (x - 12)(x + 8)(x - 4)^2$$

so the eigenvalues are $\lambda_1 = 12$, $\lambda_2 = -8$, and $\lambda_3 = \lambda_4 = 4$. The corresponding orthonormal eigenvectors are the principal axes:

$$F_1 = \tfrac{1}{2}\begin{bmatrix} 1 \\ -1 \\ -1 \\ 1 \end{bmatrix} \quad F_2 = \tfrac{1}{2}\begin{bmatrix} 1 \\ -1 \\ 1 \\ -1 \end{bmatrix} \quad F_3 = \tfrac{1}{2}\begin{bmatrix} 1 \\ 1 \\ 1 \\ 1 \end{bmatrix} \quad F_4 = \tfrac{1}{2}\begin{bmatrix} 1 \\ 1 \\ -1 \\ -1 \end{bmatrix}$$

The matrix

$$P = [F_1 \ \ F_2 \ \ F_3 \ \ F_4] = \tfrac{1}{2}\begin{bmatrix} 1 & 1 & 1 & 1 \\ -1 & -1 & 1 & 1 \\ -1 & 1 & 1 & -1 \\ 1 & -1 & 1 & -1 \end{bmatrix}$$

is thus orthogonal, and $P^{-1}AP = P^TAP$ is diagonal. Hence the new variables Y and the old variables X are related by $Y = P^TX$ and $X = PY$. Explicitly,

$$y_1 = \tfrac{1}{2}(x_1 - x_2 - x_3 + x_4) \qquad x_1 = \tfrac{1}{2}(y_1 + y_2 + y_3 + y_4)$$
$$y_2 = \tfrac{1}{2}(x_1 - x_2 + x_3 - x_4) \qquad x_2 = \tfrac{1}{2}(-y_1 - y_2 + y_3 + y_4)$$
$$y_3 = \tfrac{1}{2}(x_1 + x_2 + x_3 + x_4) \qquad x_3 = \tfrac{1}{2}(-y_1 + y_2 + y_3 - y_4)$$
$$y_4 = \tfrac{1}{2}(x_1 + x_2 - x_3 - x_4) \qquad x_4 = \tfrac{1}{2}(y_1 - y_2 + y_3 - y_4)$$

If these x_i are substituted in the original expression for q, the result is

$$q = 12y_1^2 - 8y_2^2 + 4y_3^2 + 4y_4^2$$

This is the required diagonal form.

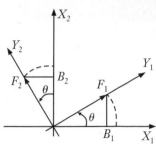

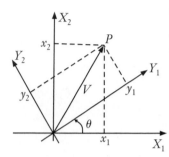

It is instructive to look at the case of quadratic forms in two variables x_1 and x_2. Then the principal axes can always be found by rotating the X_1 and X_2 axes counterclockwise about the origin through an angle θ to produce the Y_1 and Y_2 axes as in the diagram. To conform with Theorem 1, we write ordered pairs as columns. Then the rotation carries the X_1 and X_2 coordinate vectors $B_1 = \begin{bmatrix} 1 \\ 0 \end{bmatrix}$ and $B_2 = \begin{bmatrix} 0 \\ 1 \end{bmatrix}$ to the Y_1 and Y_2 coordinate vectors F_1 and F_2. Simple trigonometry gives

$$F_1 = \begin{bmatrix} \cos\theta \\ \sin\theta \end{bmatrix} \quad \text{and} \quad F_2 = \begin{bmatrix} -\sin\theta \\ \cos\theta \end{bmatrix} \tag{$*$}$$

These can be used to determine the Y_1 and Y_2 coordinates y_1, y_2 of a point P from the X_1 and X_2 coordinates x_1, x_2 of P. If V is the position vector of P as in the second diagram, then $V = x_1 B_1 + x_2 B_2$ in the original X_1 and X_2 coordinate system, and $V = y_1 F_1 + y_2 F_2$ in the new Y_1 and Y_2 system. Hence

$$\begin{aligned} \begin{bmatrix} x_1 \\ x_2 \end{bmatrix} = V = y_1 F_1 + y_2 F_2 &= y_1 \begin{bmatrix} \cos\theta \\ \sin\theta \end{bmatrix} + y_2 \begin{bmatrix} -\sin\theta \\ \cos\theta \end{bmatrix} \\ &= \begin{bmatrix} \cos\theta & -\sin\theta \\ \sin\theta & \cos\theta \end{bmatrix} \begin{bmatrix} y_1 \\ y_2 \end{bmatrix} \end{aligned} \tag{$**$}$$

is the change-of-variables formula for the rotation. Note that the matrix $P = \begin{bmatrix} \cos\theta & -\sin\theta \\ \sin\theta & \cos\theta \end{bmatrix}$ is orthogonal, so equation ($**$) takes the form $X = PY$ as in Theorem 1. Note that $P = R_\theta$ in the notation of Section 2.5.

We can now completely analyze quadratic forms in two variables. Consider the graph of the equation $rx^2 + sy^2 = 1$. Call the graph an **ellipse** if $rs > 0$ and an **hyperbola** if $rs < 0$. Theorem 2 asserts that every equation $ax^2 + bxy + cy^2 = 1$ can be transformed into such a diagonal form by a rotation, and also gives a simple way of deciding which conic it is.

Theorem 2

Consider the quadratic form $q = ax_1^2 + bx_1x_2 + cx_2^2$ where a, b, and c are not all zero.
1. There is a counterclockwise rotation of the coordinate axes about the origin such that, in the new coordinate system, q has no cross term.
2. The graph of the equation $ax_1^2 + bx_1x_2 + cx_2^2 = 1$ is an ellipse if $b^2 - 4ac < 0$ and an hyperbola if $b^2 - 4ac > 0$.

PROOF

If $b = 0$, q *already* has no cross term and (1) and (2) are clear. So assume $b \neq 0$. The matrix $A = \begin{bmatrix} a & \frac{1}{2}b \\ \frac{1}{2}b & c \end{bmatrix}$ of q has characteristic polynomial $c_A(x) = x^2 - (a+c)x - \frac{1}{4}(b^2 - 4ac)$. If we write $d = \sqrt{b^2 + (a-c)^2}$ for convenience; then the quadratic formula gives the eigenvalues

$$\lambda_1 = \tfrac{1}{2}[a + c - d] \quad \text{and} \quad \lambda_2 = \tfrac{1}{2}[a + c + d]$$

with corresponding principal axes

$$F_1 = \frac{1}{\sqrt{b^2 + (a - c - d)^2}} \begin{bmatrix} a - c - d \\ b \end{bmatrix} \quad \text{and}$$

$$F_2 = \frac{1}{\sqrt{b^2 + (a - c - d)^2}} \begin{bmatrix} -b \\ a - c - d \end{bmatrix}$$

as the reader can verify. These agree with equation (∗) if θ is an angle such that

$$\cos \theta = \frac{a - c - d}{\sqrt{b^2 + (a - c - d)^2}} \quad \text{and} \quad \sin \theta = \frac{b}{\sqrt{b^2 + (a - c - d)^2}}$$

Then $P = [F_1 \ F_2] = \begin{bmatrix} \cos \theta & -\sin \theta \\ \sin \theta & \cos \theta \end{bmatrix}$ diagonalizes A and equation (∗∗) becomes the formula $X = PY$ in Theorem 1. This proves (1).

Finally, A is similar to $\begin{bmatrix} \lambda_1 & 0 \\ 0 & \lambda_2 \end{bmatrix}$, so $\lambda_1 \lambda_2 = \det A = \frac{1}{4}(4ac - b^2)$. Hence the graph of $\lambda_1 y_1^2 + \lambda_2 y_2^2 = 1$ is an ellipse if $b^2 < 4ac$ and an hyperbola if $b^2 > 4ac$. This proves (2).

Example 3

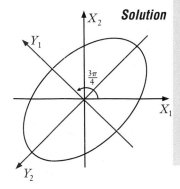

Solution

Consider the equation $x^2 + xy + y^2 = 1$. Find a rotation so that the equation has no cross term.

Here $a = b = c = 1$ in the notation of Theorem 2, so $\cos \theta = \frac{-1}{\sqrt{2}}$ and $\sin \theta = \frac{1}{\sqrt{2}}$. Hence $\theta = \frac{3\pi}{4}$ will do it. The new variables are $y_1 = \frac{1}{\sqrt{2}}(x_2 - x_1)$ and $y_2 = \frac{-1}{\sqrt{2}}(x_2 + x_1)$ by (∗∗), and the equation becomes $y_1^2 + 3y_2^2 = 2$. The angle θ has been chosen such that the new Y_1 and Y_2 axes are the axes of symmetry of the ellipse (see the diagram). The eigenvectors $F_1 = \frac{1}{\sqrt{2}}\begin{bmatrix} -1 \\ 1 \end{bmatrix}$ and $F_2 = \frac{1}{\sqrt{2}}\begin{bmatrix} -1 \\ -1 \end{bmatrix}$ point along these axes of symmetry, and this is the reason for the name *principal axes*.

The determinant of any orthogonal matrix P is either 1 or −1 (because $PP^T = I$). The orthogonal matrices $\begin{bmatrix} \cos \theta & -\sin \theta \\ \sin \theta & \cos \theta \end{bmatrix}$ arising from rotations all have determinant 1. More generally, given any quadratic form $q = X^T A X$, the orthogonal matrix P such that $P^T A P$ is diagonal can always be chosen so that $\det P = 1$ by interchanging two eigenvalues (and hence the corresponding columns of P). It is shown in Theorem 4 §10.4 that orthogonal 2×2 matrices with determinant 1 correspond to rotations. Similarly, orthogonal 3×3 matrices with determinant 1 correspond to rotations about a line through the origin. This extends Theorem 2: Every quadratic form in two or three variables can be diagonalized by a rotation of the coordinate system.

Congruence

We return to the study of quadratic forms in general.

Theorem 3

If $q(X) = X^T A X$ is a quadratic form given by a symmetric matrix A, then A is uniquely determined by q.

PROOF

Let $q(X) = X^T B X$ for all X where $B^T = B$. If $C = A - B$, then $X^T C X = 0$ for all X, and we must show that $C = 0$. Given Y in $\mathbb{R}^n$,

$$0 = (X + Y)^T C(X + Y) = X^T C X + X^T C Y + Y^T C X + Y^T C Y$$
$$= X^T C Y + Y^T C X$$

But $Y^T C X = (X^T C Y)^T = X^T C Y$ (it is 1×1). Hence $X^T C Y = 0$ for all X and Y in $\mathbb{R}^n$. If E_j is column j of I_n, then the (i, j)-entry of C is $E_i^T C E_j = 0$. Thus $C = 0$.

Hence we can speak of *the* symmetric matrix of a quadratic form.

A quadratic form q can be diagonalized in several ways. For example, if

$$q = 2x_1^2 - 4x_1 x_2 + 5x_2^2, \quad A = \begin{bmatrix} 2 & -1 \\ -2 & 5 \end{bmatrix} \text{ so Theorem 1 gives}$$

$$q = 6y_1^2 + y_2^2 \quad \text{where} \quad Y = P^T X, \; P = \frac{1}{\sqrt{5}} \begin{bmatrix} 1 & 2 \\ -2 & 1 \end{bmatrix}$$

On the other hand, $q = 2(x_1 - x_2)^2 + 3x_2^2$, so

$$q = 2z_1^2 + 3z_2^2 \quad \text{where} \quad Z = QX, \; Q = \begin{bmatrix} 1 & -1 \\ 0 & 1 \end{bmatrix}$$

The question arises: How are these changes of variables related and what properties do they share?

Given a quadratic form $q = q(X) = X^T A X$, suppose new variables Y are given by an invertible matrix U:

$$Y = U^{-1} X \quad \text{equivalently} \quad X = UY$$

In terms of these new variables, q takes the form

$$q = q(Y) = (UY)^T A(UY) = Y^T (U^T A U) Y$$

That is, q has matrix $U^T A U$ with respect to the new variables Y. Hence, to study changes of variables in quadratic forms, we study the following relationship on matrices: Two $n \times n$ matrices A and B are called **congruent**, written $A \overset{c}{\sim} B$, if $B = U^T A U$ for some invertible matrix U. Here are some properties of congruence:

1. $A \overset{c}{\sim} A$ for all A.
2. If $A \overset{c}{\sim} B$, then $B \overset{c}{\sim} A$.
3. If $A \overset{c}{\sim} B$ and $B \overset{c}{\sim} C$, then $A \overset{c}{\sim} C$.
4. If $A \overset{c}{\sim} B$, then A is symmetric if and only if B is symmetric.
5. If $A \overset{c}{\sim} B$, then rank $A =$ rank B.

The converse to (5) can fail even for symmetric matrices.

Example 4

> The symmetric matrices $A = \begin{bmatrix} 1 & 0 \\ 0 & 1 \end{bmatrix}$ and $B = \begin{bmatrix} 1 & 0 \\ 0 & -1 \end{bmatrix}$ have the same rank but are not congruent. Indeed, if $A \cong B$, an invertible matrix U exists such that $B = U^T A U = U^T U$. But then $-1 = \det B = (\det U)^2$, a contradiction because we are working with real matrices.

The key distinction between A and B in Example 4 is that A has two positive eigenvalues (counting multiplicities) whereas B has only one.

Theorem 4 Sylvester's Law of Inertia

If $A \cong B$, then A and B have the same number of positive eigenvalues, counting multiplicities.

The proof is given at the end of this section.

The **index** of a symmetric matrix A is the number of positive eigenvalues of A. If $q = q(X) = X^T A X$ is a quadratic form, the **index** and **rank** of q are defined to be, respectively, the index and rank of the matrix A. As we saw before, if the variables expressing a quadratic form q are changed, the new matrix is congruent to the old one. Hence the index and rank depend only on q and not on the way it is expressed.

Now let $q = q(X) = X^T A X$ be any quadratic form in n variables, of index k and rank r, where A is symmetric. We claim that new variables Z can be found so that q is **completely diagonalized**—that is,

$$q(Z) = z_1^2 + \cdots + z_k^2 - z_{k+1}^2 - \cdots - z_r^2$$

If $k \le r \le n$, let $D_n(k, r)$ denote the $n \times n$ diagonal matrix whose main diagonal consists of k ones, followed by $r - k$ minus ones, followed by $n - r$ zeros. Then we seek new variables Z such that

$$q(Z) = Z^T D_n(k, r) Z$$

To determine Z, first diagonalize A as follows: Find an orthogonal matrix P_0 such that

$$P_0^T A P_0 = D = \operatorname{diag}(\lambda_1, \lambda_2, \ldots, \lambda_r, 0, \ldots, 0)$$

is diagonal with the nonzero eigenvalues $\lambda_1, \lambda_2, \ldots, \lambda_r$ of A on the main diagonal (followed by $n - r$ zeros). By reordering the columns of P_0, if necessary, we may assume that $\lambda_1, \ldots, \lambda_k$ are positive and $\lambda_{k+1}, \ldots, \lambda_r$ are negative. This being the case, let D_0 be the $n \times n$ diagonal matrix

$$D_0 = \operatorname{diag}\left(\frac{1}{\sqrt{\lambda_1}}, \ldots, \frac{1}{\sqrt{\lambda_k}}, \frac{1}{\sqrt{-\lambda_{k+1}}}, \ldots, \frac{1}{\sqrt{-\lambda_r}}, 1, \ldots, 1 \right)$$

Then $D_0^T D D_0 = D_n(k, r)$, so if new variables Z are given by $X = (P_0 D_0) Z$, we obtain

$$q(Z) = Z^T D_n(k, r) Z = z_1^2 + \cdots + z_k^2 - z_{k+1}^2 - \cdots - z_r^2$$

as required. Note that the change-of-variables matrix $P_0 D_0$ from Z to X has orthogonal columns (in fact, scalar multiples of the columns of P_0).

Example 5

Completely diagonalize the quadratic form q in Example 2 and find the index and rank.

Solution

In the notation of Example 2, the eigenvalues of the matrix A of q are 12, -8, 4, 4; so the index is 3 and the rank is 4. Moreover, the corresponding orthogonal eigenvectors are F_1, F_2, F_3, and F_4. Hence $P_0 = [F_1 \ F_3 \ F_4 \ F_2]$ is orthogonal and

$$P_0^T A P_0 = \text{diag}(12, \ 4, \ 4, \ -8)$$

As before, take $D_0 = \text{diag}(\frac{1}{\sqrt{12}}, \frac{1}{2}, \frac{1}{2}, \frac{1}{\sqrt{8}})$ and define the new variables Z by $X = (P_0 D_0)Z$. Hence the new variables are given by $Z = D_0^{-1} P_0^T X$. The result is

$$z_1 = \sqrt{3}(x_1 - x_2 - x_3 + x_4)$$
$$z_2 = x_1 + x_2 + x_3 + x_4$$
$$z_3 = x_1 + x_2 - x_3 - x_4$$
$$z_4 = \sqrt{2}(x_1 - x_2 + x_3 - x_4)$$

This discussion gives the following information about symmetric matrices.

Theorem 5

Let A and B be symmetric $n \times n$ matrices, and let $0 \le k \le r \le n$.
1. A has index k and rank r if and only if $A \stackrel{c}{\sim} D_n(k, r)$.
2. $A \stackrel{c}{\sim} B$ if and only if they have the same rank and index.

PROOF

1. If A has index k and rank r, take $U = P_0 D_0$ where P_0 and D_0 are as described prior to Example 5. Then $U^T A U = D_n(k, r)$. The converse is true because $D_n(k, r)$ has index k and rank r (using Theorem 4).
2. If A and B both have index k and rank r, then $A \stackrel{c}{\sim} D_n(k, r) \stackrel{c}{\sim} B$ by (1). The converse was given earlier.

PROOF OF THEOREM 4

By Theorem 1, $A \stackrel{c}{\sim} D_1$ and $B \stackrel{c}{\sim} D_2$ where D_1 and D_2 are diagonal and have the same eigenvalues as A and B, respectively. We have $D_1 \stackrel{c}{\sim} D_2$, (because $A \stackrel{c}{\sim} B$), so we may assume that A and B are both diagonal. Consider the quadratic form $q(X) = X^T A X$. If A has k positive eigenvalues, q has the form

$$q(X) = a_1 x_1^2 + \cdots + a_k x_k^2 - a_{k+1} x_{k+1}^2 - \cdots - a_r x_r^2, \quad a_i > 0$$

where $r = \text{rank } A = \text{rank } B$. The subspace $W_1 = \{X \mid x_{k+1} = \cdots = x_r = 0\}$ of $\mathbb{R}^n$ has dimensions $n - r + k$ and satisfies $q(X) > 0$ for all $X \ne 0$ in W_1.

On the other hand, if $B = U^T A U$, define new variables Y by $X = UY$. If B has k' positive eigenvalues, q has the form

$$q(X) = b_1 y_1^2 + \cdots + b_{k'} y_{k'}^2 - b_{k'+1} y_{k'+1}^2 - \cdots - b_r y_r^2, \quad b_i > 0$$

Let $F_1, \ldots, F_n$ denote the columns of U. They are a basis of $\mathbb{R}^n$ and

$$X = UY = [F_1 \;\cdots\; F_n] \begin{bmatrix} y_1 \\ \vdots \\ y_n \end{bmatrix} = y_1 F_1 + \cdots + y_n F_n$$

Hence the subspace $W_2 = \text{span}\{F_{k'+1}, \ldots, F_r\}$ satisfies $q(X) < 0$ for all $X \neq 0$ in W_2. Note that $\dim W_2 = r - k'$. It follows that W_1 and W_2 have only the zero vector in common. Hence, if B_1 and B_2 are bases of W_1 and W_2, respectively, then (Exercise 31 §6.3) $B_1 \cup B_2$ is an independent set of $(n - r + k) + (r - k') = n + k - k'$ vectors in $\mathbb{R}^n$. This implies that $k \leq k'$, and a similar argument shows $k' \leq k$.

Exercises 8.9

1. In each case, find a symmetric matrix A such that $q = X^T B X$ takes the form $q = X^T A X$.

   (a) $B = \begin{bmatrix} 1 & 1 \\ 0 & 1 \end{bmatrix}$ ◆(b) $B = \begin{bmatrix} 1 & 1 \\ -1 & 2 \end{bmatrix}$

   (c) $B = \begin{bmatrix} 1 & 0 & 1 \\ 1 & 1 & 0 \\ 0 & 1 & 1 \end{bmatrix}$ ◆(d) $B = \begin{bmatrix} 1 & 2 & -1 \\ 4 & 1 & 0 \\ 5 & -2 & 3 \end{bmatrix}$

2. In each case, find a change of variables that will diagonalize the quadratic form q. Determine the index and rank of q.

   (a) $q = x_1^2 + 2x_1 x_2 + x_2^2$

   ◆(b) $q = x_1^2 + 4x_1 x_2 + x_2^2$

   (c) $q = x_1^2 + x_2^2 + x_3^2 - 4(x_1 x_2 + x_1 x_3 + x_2 x_3)$

   ◆(d) $q = 7x_1^2 + x_2^2 + x_3^2 + 8x_1 x_2 + 8x_1 x_3 - 16x_2 x_3$

   (e) $q = 2(x_1^2 + x_2^2 + x_3^2 - x_1 x_2 + x_1 x_3 - x_2 x_3)$

   ◆(f) $q = 5x_1^2 + 8x_2^2 + 5x_3^2 - 4(x_1 x_2 + 2x_1 x_3 + x_2 x_3)$

   (g) $q = x_1^2 - x_3^2 - 4x_1 x_2 + 4x_2 x_3$

   ◆(h) $q = x_1^2 + x_3^2 - 2x_1 x_2 + 2x_2 x_3$

3. For each of the following, write the equation in terms of new variables so that it is in standard position, and identify the curve.

   (a) $xy = 1$

   ◆(b) $3x^2 - 4xy = 2$

   (c) $6x^2 + 6xy - 2y^2 = 5$

   ◆(d) $2x^2 + 4xy + 5y^2 = 1$

4. Consider the equation $ax^2 + bxy + cy^2 = d$, where $b \neq 0$. Introduce new variables x_1 and y_1 by rotating the axes counterclockwise through an

angle θ. Show that the resulting equation has no $x_1 y_1$-term if θ is given by

$$\cos 2\theta = \frac{a - c}{\sqrt{b^2 + (a - c)^2}},$$

$$\sin 2\theta = \frac{b}{\sqrt{b^2 + (a - c)^2}}$$

   [*Hint:* Use equation $(**)$ in Example 2 to get x and y in terms of x_1 and y_1, and substitute.]

5. Prove properties (1)–(5) preceding Example 4.

6. If $A \overset{c}{\cong} B$ show that A is invertible if and only if B is invertible.

7. If $X = [x_1 \;\cdots\; x_n]^T$ is a column of variables, $A = A^T$ is $n \times n$, B is $1 \times n$, and c is a constant, $X^T A X + B X = c$ is called a **quadratic equation** in the variables x_i.

   (a) Show that new variables $y_1, \ldots, y_n$ can be found such that the equation takes the form
   $$\lambda_1 y_1^2 + \cdots + \lambda_r y_r^2 + k_1 y_1 + \cdots + k_n y_n = c.$$

   ◆(b) Write
   $$x_1^2 + 3x_2^2 + 3x_3^2 + 4x_1 x_2 - 4x_1 x_3 + 5x_1 - 6x_3 = 7$$
   in this form and find variables y_1, y_2, y_3 as in (a).

8. Given a symmetric matrix A, define $q_A(X) = X^T A X$. Show that $B \overset{c}{\cong} A$ if and only if B is symmetric and there is an invertible matrix U such that $q_B(X) = q_A(UX)$ for all X. [*Hint:* Theorem 3.]

9. Let $q(X) = X^T A X$ be a quadratic form, $A = A^T$.

   (a) Show that $q(X) > 0$ for all $X \neq 0$, if and only if A is positive definite (all eigenvalues are positive). In this case, q is called **positive definite**.

(b) Show that new variables Y can be found such that $q = \|Y\|^2$ and $Y = UX$ where U is upper triangular with positive diagonal entries. [*Hint:* Theorem 3 §8.3.]

10. A **bilinear form** β on $\mathbb{R}^n$ is a function that assigns to every pair X, Y of columns in $\mathbb{R}^n$ a number $\beta(X, Y)$ in such a way that

$$\beta(rX + sY, Z) = r\beta(X, Z) + s\beta(Y, Z)$$

$$\beta(X, rY + sZ) = r\beta(X, Y) + s\beta(X, Z)$$

for all X, Y, Z in $\mathbb{R}^n$ and r, s in $\mathbb{R}$.
If $\beta(X, Y) = \beta(Y, X)$ for all X, Y, β is called **symmetric**.

(a) If β is a bilinear form, show that an $n \times n$ matrix A exists such that $\beta(X, Y) = X^T AY$ for all X, Y.

(b) Show that A is uniquely determined by β.

(c) Show that β is symmetric if and only if $A = A^T$.

SECTION 8.10   An Application to Systems of Differential Equations[17]

Solving a variety of problems, particularly in science and engineering, comes down to solving a differential equation or a system of such equations. In this section, vector spaces and matrix multiplication will be used to describe systems of differential equations, and diagonalization will be used to solve such systems. Of course, our methods really are only a first step into the vast theory of differential equations but, at least for linear systems of first-order differential equations, the techniques do solve the problem and provide a basis for further work.

If f is a function of a real variable x, and if f' and f'' denote the first and second derivatives of f, then equations of the form

$$f' + af = 0 \quad \text{or} \quad f'' + af' + bf = 0 \quad (a \text{ and } b \text{ numbers})$$

are called **differential equations of order 1 and 2**, respectively. One approach to such equations is to reduce those of order greater than one to *systems* of first-order equations. Then matrix diagonalization techniques can be applied. We treat such systems in this section.

The general problem is to find differentiable functions $f_1, f_2, \ldots, f_n$ that satisfy a system of equations of the form

$$f_1' = a_{11}f_1 + a_{12}f_2 + \cdots + a_{1n}f_n$$
$$f_2' = a_{21}f_1 + a_{22}f_2 + \cdots + a_{2n}f_n$$
$$\vdots \qquad \vdots \qquad \vdots \qquad \qquad \vdots$$
$$f_n' = a_{n1}f_1 + a_{n2}f_2 + \cdots + a_{nn}f_n$$

where the a_{ij} are constants. This is called a **linear system of differential equations**. The first step is to put it in matrix form. Write

$$\mathbf{f} = \begin{bmatrix} f_1 \\ f_2 \\ \vdots \\ f_n \end{bmatrix} \qquad \mathbf{f}' = \begin{bmatrix} f_1' \\ f_2' \\ \vdots \\ f_n' \end{bmatrix} \qquad A = \begin{bmatrix} a_{11} & a_{12} & \cdots & a_{1n} \\ a_{21} & a_{22} & \cdots & a_{2n} \\ \vdots & \vdots & & \vdots \\ a_{n1} & a_{n2} & \cdots & a_{nn} \end{bmatrix}$$

Then the system can be written compactly as

$$\mathbf{f}' = A\mathbf{f}$$

and, given the matrix A, the problem is to find a column $\mathbf{f}$ of differentiable functions that satisfies this condition.

---

17 This section requires only Chapter 6.

Linear algebra enters into this as follows: These columns of functions become a vector space if matrix addition and scalar multiplication are used:

$$\begin{bmatrix} f_1 \\ f_2 \\ \vdots \\ f_n \end{bmatrix} + \begin{bmatrix} g_1 \\ g_2 \\ \vdots \\ g_n \end{bmatrix} = \begin{bmatrix} f_1 + g_1 \\ f_2 + g_2 \\ \vdots \\ f_n + g_n \end{bmatrix} \qquad a \begin{bmatrix} f_1 \\ f_2 \\ \vdots \\ f_n \end{bmatrix} = \begin{bmatrix} af_1 \\ af_2 \\ \vdots \\ af_n \end{bmatrix}$$

Of course, addition $f_i + g_i$ and scalar multiplication af_i of the individual functions are defined pointwise (as in Example 6 §6.1). That is, the actions of $f_i + g_i$ and af_i are given by

$$(f_i + g_i)(x) = f_i(x) + g_i(x) \quad \text{and} \quad (af_i)(x) = af_i(x)$$

for all x. With these definitions, the set of all n-columns of functions becomes a vector space, as the reader can verify. The zero vector and the negative of a vector are

$$\mathbf{0} = \begin{bmatrix} 0 \\ 0 \\ \vdots \\ 0 \end{bmatrix} \quad \text{and} \quad - \begin{bmatrix} f_1 \\ f_2 \\ \vdots \\ f_n \end{bmatrix} = \begin{bmatrix} -f_1 \\ -f_2 \\ \vdots \\ -f_n \end{bmatrix}$$

just as for matrices. This vector space will be denoted F^n.

Our concern here is not for F^n but for those columns of functions in it that satisfy the linear system:

$$U = \{\mathbf{f} \mid \mathbf{f} \text{ lies in } F^n \text{ and } \mathbf{f}' = A\mathbf{f}\}$$

Now recall the following three basic facts about differentiation:

$$\mathbf{0}' = \mathbf{0} \qquad (\mathbf{f} + \mathbf{g})' = \mathbf{f}' + \mathbf{g}' \qquad (c\mathbf{f})' = c\mathbf{f}'$$

Hence U is a subspace of F^n. The problem now is to find a convenient basis for U.

The case $n = 1$ has been discussed earlier. Here the problem is to find all functions f satisfying

$$f' = af \quad a \text{ a constant}$$

and it follows from Theorem 1 §6.6 that the space of solutions has dimension 1 and that $\{e^{ax}\}$ is a basis. If the matrix A is diagonalizable, this case can be used in the general situation. The following example provides an illustration.

Example 1

Find a solution to the system

$$\begin{aligned} f_1' &= f_1 + 3f_2 \\ f_2' &= 2f_1 + 2f_2 \end{aligned}$$

that satisfies $f_1(0) = 0$, $f_2(0) = 5$.

Solution   This is $\mathbf{f}' = A\mathbf{f}$, where $\mathbf{f} = \begin{bmatrix} f_1 \\ f_2 \end{bmatrix}$ and $A = \begin{bmatrix} 1 & 3 \\ 2 & 2 \end{bmatrix}$. The reader can verify that $c_A(x) = (x - 4)(x + 1)$, and that $X_1 = \begin{bmatrix} 1 \\ 1 \end{bmatrix}$ and $X_2 = \begin{bmatrix} 3 \\ -2 \end{bmatrix}$ are eigenvectors corresponding to the eigenvalues 4 and -1, respectively. Hence the diagonalization algorithm gives $P^{-1}AP = \begin{bmatrix} 4 & 0 \\ 0 & -1 \end{bmatrix}$, where

$P = [X_1 \ X_2] = \begin{bmatrix} 1 & 3 \\ 1 & -2 \end{bmatrix}$. Now consider new functions g_1 and g_2 given by $\mathbf{f} = P\mathbf{g}$ (equivalently, $\mathbf{g} = P^{-1}\mathbf{f}$). Hence

$$\begin{bmatrix} f_1 \\ f_2 \end{bmatrix} = \begin{bmatrix} 1 & 3 \\ 1 & -2 \end{bmatrix}\begin{bmatrix} g_1 \\ g_2 \end{bmatrix} \quad \text{that is,} \quad \begin{matrix} f_1 = g_1 + 3g_2 \\ f_2 = g_1 - 2g_2 \end{matrix}$$

Then $f_1' = g_1' + 3g_2'$ and $f_2' = g_1' - 2g_2'$ so that

$$\mathbf{f}' = \begin{bmatrix} f_1' \\ f_2' \end{bmatrix} = \begin{bmatrix} 1 & 3 \\ 1 & -2 \end{bmatrix}\begin{bmatrix} g_1' \\ g_2' \end{bmatrix} = P\mathbf{g}'$$

If this is substituted in $\mathbf{f}' = A\mathbf{f}$, the result is $P\mathbf{g}' = AP\mathbf{g}$, whence

$$\mathbf{g}' = P^{-1}AP\mathbf{g}$$

But this means that

$$\begin{bmatrix} g_1' \\ g_2' \end{bmatrix} = \begin{bmatrix} 4 & 0 \\ 0 & -1 \end{bmatrix}\begin{bmatrix} g_1 \\ g_2 \end{bmatrix}, \quad \text{so} \quad \begin{matrix} g_1' = 4g_1 \\ g_2' = -g_2 \end{matrix}$$

Then the case $n = 1$ gives $g_1(x) = ce^{4x}$, $g_2(x) = de^{-x}$, where c and d are constants. Finally, then,

$$\begin{bmatrix} f_1(x) \\ f_2(x) \end{bmatrix} = P\begin{bmatrix} g_1(x) \\ g_2(x) \end{bmatrix} = \begin{bmatrix} 1 & 3 \\ 1 & -2 \end{bmatrix}\begin{bmatrix} ce^{4x} \\ de^{-x} \end{bmatrix} = \begin{bmatrix} ce^{4x} + 3de^{-x} \\ ce^{4x} - 2de^{-x} \end{bmatrix}$$

so the *general solution* is

$$\begin{matrix} f_1(x) = ce^{4x} + 3de^{-x} \\ f_2(x) = ce^{4x} - 2de^{-x} \end{matrix} \quad c \text{ and } d \text{ constants.}$$

It is worth observing that this can be written in matrix form as

$$\begin{bmatrix} f_1(x) \\ f_2(x) \end{bmatrix} = c\begin{bmatrix} 1 \\ 1 \end{bmatrix}e^{4x} + d\begin{bmatrix} 3 \\ -2 \end{bmatrix}e^{-x}$$

That is,

$$\mathbf{f}(x) = cX_1 e^{4x} + dX_2 e^{-x}$$

This form of the solution works more generally, as will be shown.

Finally, the requirement in this example that $f_1(0) = 0$ and $f_2(0) = 5$ determines the constants c and d:

$$0 = f_1(0) = ce^0 + 3de^0 = c + 3d$$
$$5 = f_2(0) = ce^0 - 2de^0 = c - 2d$$

These equations give $c = 3$ and $d = -1$, so

$$f_1(x) = 3e^{4x} - 3e^{-x}$$
$$f_2(x) = 3e^{4x} + 2e^{-x}$$

satisfy all the requirements.

The technique of this example works in general.

Theorem 1

Consider a linear system

$$\mathbf{f}' = A\mathbf{f}$$

of differential equations, where A is an $n \times n$ diagonalizable matrix. Let $P^{-1}AP$ be diagonal, where P is given in terms of its columns

$$P = [X_1 \ X_2 \ \cdots \ X_n]$$

and $\{X_1, X_2, \ldots, X_n\}$ are independent eigenvectors of A. If X_i corresponds to the eigenvalue λ_i for each i, then

$$\{X_1 e^{\lambda_1 x}, X_2 e^{\lambda_2 x}, \ldots, X_n e^{\lambda_n x}\}$$

is a basis for the space of solutions of $\mathbf{f}' = A\mathbf{f}$.

PROOF

Such X_i exist by virtue of the assumption that A is diagonalizable, and their independence guarantees that P is invertible. Let

$$P^{-1}AP = \begin{bmatrix} \lambda_1 & 0 & \cdots & 0 \\ 0 & \lambda_2 & \cdots & 0 \\ \vdots & \vdots & & \vdots \\ 0 & 0 & \cdots & \lambda_n \end{bmatrix}$$

where $\lambda_1, \lambda_2, \ldots, \lambda_n$ are the (not necessarily distinct) eigenvalues of A. As in the example, define a new column of functions $\mathbf{g} = \begin{bmatrix} g_1 \\ g_2 \\ \vdots \\ g_n \end{bmatrix}$ by $\mathbf{g} = P^{-1}\mathbf{f}$; equivalently,

$\mathbf{f} = P\mathbf{g}$. If f_i is the ith component of $\mathbf{f}$ and $P = [p_{ij}]$, this gives

$$f_i = p_{i1}g_1 + p_{i2}g_2 + \cdots + p_{in}g_n.$$

Differentiation preserves this relationship:

$$f_i' = p_{i1}g_1' + p_{i2}g_2' + \cdots + p_{in}g_n'$$

so $\mathbf{f}' = P\mathbf{g}'$. Substituting this into $\mathbf{f}' = A\mathbf{f}$ gives $P\mathbf{g}' = AP\mathbf{g}$. But then multiplication by P^{-1} gives $\mathbf{g}' = P^{-1}AP\mathbf{g}$, so the original system of equations for $\mathbf{f}$ becomes much simpler in terms of $\mathbf{g}$:

$$\begin{bmatrix} g_1' \\ g_2' \\ \vdots \\ g_n' \end{bmatrix} = \begin{bmatrix} \lambda_1 & 0 & \cdots & 0 \\ 0 & \lambda_2 & \cdots & 0 \\ \vdots & \vdots & & \vdots \\ 0 & 0 & \cdots & \lambda_n \end{bmatrix} \begin{bmatrix} g_1 \\ g_2 \\ \vdots \\ g_n \end{bmatrix}$$

Hence $g_i' = \lambda_i g_i$ holds for each i, and Theorem 1 §6.6 implies that the only solutions are

$$g_i(x) = c_i e^{\lambda_i x} \quad c_i \text{ some constant.}$$

Then the relationship $\mathbf{f} = P\mathbf{g}$ gives the functions $f_1, f_2, \ldots, f_n$ as follows:

$$\mathbf{f}(x) = [X_1 \ X_2 \ \cdots \ X_n] \begin{bmatrix} c_1 e^{\lambda_1 x} \\ c_2 e^{\lambda_2 x} \\ \vdots \\ c_n e^{\lambda_n x} \end{bmatrix} = c_1 X_1 e^{\lambda_1 x} + c_2 X_2 e^{\lambda_2 x} + \cdots + c_n X_n e^{\lambda_n x}$$

Hence the columns $\{X_1 e^{\lambda_1 x}, X_2 e^{\lambda_2 x}, \ldots, X_n e^{\lambda_n x}\}$ span the space U of solutions. They are independent because, if

$$c_1 e^{\lambda_1 x} X_1 + c_2 e^{\lambda_2 x} X_2 + \cdots + c_n e^{\lambda_n x} X_n = 0$$

in F^n, then the equation must hold for all x. In particular, taking $x = 0$ gives $c_1 X_1 + \cdots + c_n X_1 = 0$, so the independence of the X_i gives $c_1 = \cdots = c_n = 0$.

The theorem shows that *every* solution to $\mathbf{f}' = A\mathbf{f}$ is a linear combination

$$\mathbf{f}(x) = c_1 X_1 e^{\lambda_1 x} + c_2 X_2 e^{\lambda_2 x} + \cdots + c_n X_n e^{\lambda_n x}$$

where the coefficients c_i are arbitrary. Hence this is called the **general solution** to the system of differential equations. In most cases the solution functions $f_i(x)$ are required to satisfy boundary conditions, often of the form $f_i(a) = b_i$, where $a, b_1, \ldots, b_n$ are prescribed numbers. These conditions determine the constants c_i. The following example illustrates this and displays a situation where one eigenvalue has multiplicity greater than 1.

Example 2

Find the general solution to the system

$$\begin{aligned} f_1' &= 5f_1 + 8f_2 + 16f_3 \\ f_2' &= 4f_1 + f_2 + 8f_3 \\ f_3' &= -4f_1 - 4f_2 - 11f_3 \end{aligned}$$

Then find a solution satisfying the boundary conditions $f_1(0) = f_2(0) = f_3(0) = 1$.

Solution   The system has the form $\mathbf{f}' = A\mathbf{f}$, where $A = \begin{bmatrix} 5 & 8 & 16 \\ 4 & 1 & 8 \\ -4 & -4 & -11 \end{bmatrix}$. Then

$c_A(x) = (x + 3)^2(x - 1)$ and independent eigenvectors corresponding to the eigenvalues -3, -3, and 1 are, respectively,

$$X_1 = \begin{bmatrix} -1 \\ 1 \\ 0 \end{bmatrix} \qquad X_2 = \begin{bmatrix} -2 \\ 0 \\ 1 \end{bmatrix} \qquad X_3 = \begin{bmatrix} 2 \\ 1 \\ -1 \end{bmatrix}$$

Hence $\{X_1 e^{-3x}, X_2 e^{-3x}, X_3 e^x\}$ spans the space of solutions, so the general solution is

$$\mathbf{f}(x) = c_1 \begin{bmatrix} -1 \\ 1 \\ 0 \end{bmatrix} e^{-3x} + c_2 \begin{bmatrix} -2 \\ 0 \\ 1 \end{bmatrix} e^{-3x} + c_3 \begin{bmatrix} 2 \\ 1 \\ -1 \end{bmatrix} e^x, \quad c_i \text{ constants.}$$

The boundary conditions $f_1(0) = f_2(0) = f_3(0) = 1$ determine the constants c_i.

$$
\begin{bmatrix} 1 \\ 1 \\ 1 \end{bmatrix} = \mathbf{f}(0) = c_1 \begin{bmatrix} -1 \\ 1 \\ 0 \end{bmatrix} + c_2 \begin{bmatrix} -2 \\ 0 \\ 1 \end{bmatrix} + c_3 \begin{bmatrix} 2 \\ 1 \\ -1 \end{bmatrix}
$$

$$
= \begin{bmatrix} -1 & -2 & 2 \\ 1 & 0 & 1 \\ 0 & 1 & -1 \end{bmatrix} \begin{bmatrix} c_1 \\ c_2 \\ c_3 \end{bmatrix}
$$

The solution is $c_1 = -3$, $c_2 = 5$, $c_3 = 4$, so the required specific solution is

$$
f_1(x) = -7e^{-3x} + 8e^x
$$
$$
f_2(x) = -3e^{-3x} + 4e^x
$$
$$
f_3(x) = 5e^{-3x} - 4e^x
$$

The foregoing analysis fails if A is not diagonalizable, a situation that will not be treated in this book.

Exercises 8.10

1. Use Theorem 1 to find the general solution to each of the following systems. Then find a specific solution satisfying the given boundary condition.

   (a) $f_1' = 2f_1 + 4f_2$, $f_1(0) = 0$
   $f_2' = 3f_1 + 3f_2$, $f_2(0) = 1$

   ◆(b) $f_1' = -f_1 + 5f_2$, $f_1(0) = 1$
   $f_2' = f_1 + 3f_2$, $f_2(0) = -1$

   (c) $f_1' = \qquad 4f_2 + 4f_3$
   $f_2' = f_1 + f_2 - 2f_3$
   $f_3' = -f_1 + f_2 + 4f_3$
   $f_1(0) = f_2(0) = f_3(0) = 1$

   ◆(d) $f_1' = 2f_1 + f_2 + 2f_3$
   $f_2' = 2f_1 + 2f_2 - 2f_3$
   $f_3' = 3f_1 + f_2 + f_3$
   $f_1(0) = f_2(0) = f_3(0) = 1$

2. (a) Show that $e^{\lambda x}$, $e^{\mu x}$, and $e^{\delta x}$ are linearly independent functions if λ, μ, and δ are distinct. [*Hint:* If $re^{\lambda x} + se^{\mu x} + te^{\delta x} = 0$, differentiate twice and use Theorem 7 §3.2.]

   (b) Generalize part (a) to $\{e^{\lambda_1 x}, e^{\lambda_2 x}, \ldots, e^{\lambda_n x}\}$.

Change of Basis

If A is an $m \times n$ matrix, the corresponding **matrix transformation** $T_A : \mathbb{R}^n \to \mathbb{R}^m$ is defined by

$$T_A(X) = AX \quad \text{for all columns } X \text{ in } \mathbb{R}^n.$$

It was shown in Theorem 2 §2.5 that every linear transformation $T : \mathbb{R}^n \to \mathbb{R}^m$ is a matrix transformation; that is, $T = T_A$ for some $m \times n$ matrix A. Furthermore, the matrix A is uniquely determined by T. In fact A is given in terms of its columns by

$$A = [T(E_1)\, T(E_2) \cdots T(E_n)]$$

where $\{E_1, E_2, \dots, E_n\}$ is the standard basis of $\mathbb{R}^n$.

In this chapter we show how to associate a matrix with any linear transformation $T : V \to W$ where V and W are finite-dimensional vector spaces, and we describe how the matrix can be used to compute $T(\mathbf{v})$ for any $\mathbf{v}$ in V. The matrix depends on the choice of a basis B in V and a basis D in W, and much of the chapter is involved in how to choose the bases B and D so that the matrix takes a simple form. This is particularly important when $W = V$ (and $D = B$), and this case leads to some of the most important theorems in linear algebra.

SECTION 9.1 The Matrix of a Linear Transformation

Let $T : V \to W$ be a linear transformation where $\dim V = n$ and $\dim V = m$. The aim in this section is to describe the action of T as multiplication by an $m \times n$ matrix A. The idea is to convert a vector $\mathbf{v}$ in V into a column in $\mathbb{R}^n$, multiply that column by A to get a column in $\mathbb{R}^m$, and convert this column back to get $T(\mathbf{v})$ in W.

Converting vectors to columns is a simple matter, but one small change is needed. Up to now the *order* of the vectors in a basis has been of no importance. However, in this section, we shall speak of an **ordered basis** $\{\mathbf{e}_1, \mathbf{e}_2, \dots, \mathbf{e}_n\}$, which is just a basis where the order in which the vectors are listed is taken into account. Hence $\{\mathbf{e}_2, \mathbf{e}_1, \mathbf{e}_3\}$ is a different *ordered* basis from $\{\mathbf{e}_1, \mathbf{e}_2, \mathbf{e}_3\}$.

If $B = \{\mathbf{e}_1, \mathbf{e}_2, \dots, \mathbf{e}_n\}$ is an ordered basis in a vector space V, and if

$$\mathbf{v} = v_1\mathbf{e}_1 + v_2\mathbf{e}_2 + \cdots + v_n\mathbf{e}_n$$

is a vector in V, then the (uniquely determined) numbers $v_1, v_2, \ldots, v_n$ are called the **coordinates** of $\mathbf{v}$ with respect to the basis B. The **coordinate vector** of $\mathbf{v}$ with respect to B is defined to be

$$C_B(\mathbf{v}) = \begin{bmatrix} v_1 \\ v_2 \\ \vdots \\ v_n \end{bmatrix} = [v_1 \ \ v_2 \ \ \cdots \ \ v_n]^T$$

The reason for writing $C_B(\mathbf{v})$ as a column instead of a row will become clear later.

Example 1

> The coordinate vector for $\mathbf{v} = (2, 1, 3)$ with respect to the ordered basis $B = \{(1, 1, 0), (1, 0, 1), (0, 1, 1)\}$ of $\mathbb{R}^3$ is $C_B(\mathbf{v}) = [0 \ \ 2 \ \ 1]^T$ because
>
> $$\mathbf{v} = (2, 1, 3) = 0(1, 1, 0) + 2(1, 0, 1) + 1(0, 1, 1).$$

Theorem 1

> If V has dimension n and B is any ordered basis of V, the coordinate transformation $C_B : V \to \mathbb{R}^n$ is an isomorphism.

PROOF

> The verification that C_B is linear is Exercise 13. If $B = \{\mathbf{e}_1, \ldots, \mathbf{e}_n\}$, then $C_B(\mathbf{e}_j)$ is column j of the identity matrix. Hence C_B carries B to the standard basis of $\mathbb{R}^n$, so it is an isomorphism by Theorem 1 §7.3.

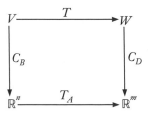

Now let $T : V \to W$ be a linear transformation where $\dim V = n$ and $\dim W = m$, and let $B = \{\mathbf{e}_1, \mathbf{e}_2, \ldots, \mathbf{e}_n\}$ and D be ordered bases of V and W, respectively. Then $C_B : V \to \mathbb{R}^n$ and $C_D : W \to \mathbb{R}^m$ are isomorphisms and we have the situation shown in the diagram where A is an $m \times n$ matrix (to be determined). Indeed, the composite $C_D T C_B^{-1} : \mathbb{R}^n \to \mathbb{R}^m$ is a linear transformation, so Theorem 2 §2.5 shows that a unique $m \times n$ matrix A exists such that

$$C_D T C_B^{-1} = T_A, \quad \text{equivalently} \quad C_D T = T_A C_B$$

T_A acts by left multiplication by A, so this condition is

$$C_D[T(\mathbf{v})] = A C_B(\mathbf{v}) \quad \text{for all } \mathbf{v} \text{ in } V$$

This requirement completely determines A. Indeed, the fact that $C_B(\mathbf{e}_j)$ is column j of the identity matrix gives

$$\text{column } j \text{ of } A = A C_B(\mathbf{e}_j) = C_D[T(\mathbf{e}_j)]$$

for all j. Hence, in terms of its columns,

$$A = [C_D[T(\mathbf{e}_1)] \ \ C_D[T(\mathbf{e}_2)] \ \ \cdots \ \ C_D[T(\mathbf{e}_n)]].$$

This is called the **matrix of T corresponding to the ordered bases B and D**, and we use the following notation:

$$M_{DB}(T) = [C_D[T(\mathbf{e}_1)] \ \ C_D[T(\mathbf{e}_2)] \ \ \cdots \ \ C_D[T(\mathbf{e}_n)]]$$

This discussion is summarized in the following important theorem.

Theorem 2

Let $T : V \to W$ be a linear transformation where $\dim V = n$ and $\dim W = m$, and let $B = \{\mathbf{e}_1, \ldots, \mathbf{e}_n\}$ and D be ordered bases of V and W, respectively. Then the matrix $M_{DB}(T)$ just given is the unique $m \times n$ matrix A that satisfies

$$C_D T = T_A C_B$$

Hence the defining property of $M_{DB}(T)$ is

$$C_D[T(\mathbf{v})] = M_{DB}(T)C_B(\mathbf{v}) \quad \text{for all } \mathbf{v} \text{ in } V$$

The fact that $T = C_D^{-1} T_A C_B$ means that the action of T on a vector $\mathbf{v}$ in V can be performed by first taking coordinates (that is, applying C_B to $\mathbf{v}$), then multiplying by A (applying T_A), and finally converting the resulting m-tuple back to a vector in W (applying C_D^{-1}).

Example 2

Define $T : \mathbf{P}_2 \to \mathbb{R}^2$ by $T(a + bx + cx^2) = (a + c, b - a - c)$ for all polynomials $a + bx + cx^2$. If $B = \{\mathbf{e}_1, \mathbf{e}_2, \mathbf{e}_3\}$ and $D = \{\mathbf{f}_1, \mathbf{f}_2\}$ where

$$\mathbf{e}_1 = 1, \ \mathbf{e}_2 = x, \ \mathbf{e}_3 = x^2 \quad \text{and} \quad \mathbf{f}_1 = (1, 0), \ \mathbf{f}_2 = (0, 1)$$

compute $M_{DB}(T)$ and verify Theorem 2.

Solution   We have $T(\mathbf{e}_1) = \mathbf{f}_1 - \mathbf{f}_2$, $T(\mathbf{e}_2) = \mathbf{f}_2$, and $T(\mathbf{e}_3) = \mathbf{f}_1 - \mathbf{f}_2$. Hence

$$M_{DB}(T) = \begin{bmatrix} C_D[T(\mathbf{e}_1)] & C_D[T(\mathbf{e}_2)] & C_D[T(\mathbf{e}_3)] \end{bmatrix} = \begin{bmatrix} 1 & 0 & 1 \\ -1 & 1 & -1 \end{bmatrix}.$$

If $\mathbf{v} = a + bx + cx^2 = a\mathbf{e}_1 + b\mathbf{e}_2 + c\mathbf{e}_3$, then $T(\mathbf{v}) = (a + c)\mathbf{f}_1 + (b - a - c)\mathbf{f}_2$, so

$$C_D[T(\mathbf{v})] = \begin{bmatrix} a + c \\ b - a - c \end{bmatrix} = \begin{bmatrix} 1 & 0 & 1 \\ -1 & 1 & -1 \end{bmatrix} \begin{bmatrix} a \\ b \\ c \end{bmatrix} = M_{DB}(T)C_B(\mathbf{v})$$

as Theorem 2 asserts.

Example 3

Suppose that $T : \mathbf{P}_2 \to \mathbb{R}^2$ has matrix $M_{DB}(T) = \begin{bmatrix} 5 & 2 & -1 \\ 3 & 0 & 4 \end{bmatrix}$ where $B = \{1, x, x^2\}$ and $D = \{(1, 0), (1, 1)\}$. Find $T(\mathbf{v})$ where $\mathbf{v} = a + bx + cx^2$.

Solution   The idea is to compute $C_D[T(\mathbf{v})]$ first and then get $T(\mathbf{v})$:

$$C_D[T(\mathbf{v})] = M_{DB}(T)C_B(\mathbf{v}) = \begin{bmatrix} 5 & 2 & -1 \\ 3 & 0 & 4 \end{bmatrix} \begin{bmatrix} a \\ b \\ c \end{bmatrix} = \begin{bmatrix} 5a + 2b - c \\ 3a + 4c \end{bmatrix}$$

Hence $T(\mathbf{v}) = (5a + 2b - c)(1, 0) + (3a + 4c)(1, 1) = (8a + 2b + 3c, 3a + 4c)$.

The next two examples will be referred to later.

Example 4

Let A be an $m \times n$ matrix, and let $T_A : \mathbb{R}^n \to \mathbb{R}^m$ be the matrix transformation induced by $A : T_A(X) = AX$ for all columns X in $\mathbb{R}^n$. If B and D are the standard bases of $\mathbb{R}^n$ and $\mathbb{R}^m$, respectively (in the usual order), then

$$M_{DB}(T_A) = A$$

In other words, the matrix of T_A corresponding to the standard bases is A itself.

Solution Write $B = \{E_1, \dots, E_n\}$. Because D is the standard basis of $\mathbb{R}^m$, it is easy to verify that $C_D(Y) = Y$ for all columns Y in $\mathbb{R}^m$. Hence

$$M_{DB}(T_A) = [T_A(E_1)\ \ T_A(E_2)\ \ \cdots\ \ T_A(E_n)] = [AE_1\ \ AE_2\ \ \cdots\ \ AE_n] = A$$

because AE_j is the jth column of A.

Example 5

Let V and W have ordered bases B and D, respectively. Let $\dim V = n$.
1. The identity transformation $1_V : V \to V$ has matrix $M_{BB}(1_V) = I_n$.
2. The zero transformation $0 : V \to W$ has matrix $M_{DB}(0) = 0$.

The first result in Example 5 is false if the two bases of V are not equal. In fact, if B is the standard basis of $\mathbb{R}^n$, then the basis D of $\mathbb{R}^n$ can be chosen so that $M_{BD}(1_{\mathbb{R}^n})$ turns out to be any invertible matrix at all (Exercise 14).

The next two theorems show that composition of linear transformations is compatible with multiplication of the corresponding matrices.

Theorem 3

Let $V \xrightarrow{T} W \xrightarrow{S} U$, be linear transformations and let B, D, and E be finite ordered bases of V, W, and U, respectively. Then

$$M_{EB}(ST) = M_{ED}(S) \bullet M_{DB}(T)$$

PROOF

We use the property in Theorem 2 three times. If $\mathbf{v}$ is in V,

$$M_{ED}(S)M_{DB}(T)C_B(\mathbf{v}) = M_{ED}(S)C_D[T(\mathbf{v})] = C_E[ST(\mathbf{v})] = M_{EB}(ST)C_B(\mathbf{v})$$

If $B = \{\mathbf{e}_1, \dots, \mathbf{e}_n\}$, then $C_B(\mathbf{e}_j)$ is column j of I_n. Hence taking $\mathbf{v} = \mathbf{e}_j$ shows that $M_{ED}(S)M_{DB}(T)$ and $M_{EB}(ST)$ have equal jth columns.

Theorem 4

Let $T : V \to W$ be a linear transformation, where $\dim V = \dim W = n$. The following are equivalent.
1. T is an isomorphism.
2. $M_{DB}(T)$ is invertible for all ordered bases B and D of V and W.
3. $M_{DB}(T)$ is invertible for some pair of ordered bases B and D of V and W.
When this is the case, $M_{DB}(T)^{-1} = M_{BD}(T^{-1})$.

PROOF

$(1) \Rightarrow (2)$. We have $V \overset{T}{\to} W \overset{T^{-1}}{\to} V$, so Theorem 3 and Example 5 give

$$M_{BD}(T^{-1})M_{DB}(T) = M_{BB}(T^{-1}T) = M_{BB}(1_V) = I_n$$

Similarly, $M_{DB}(T)M_{BD}(T^{-1}) = I_n$, proving (2) (and the last statement in the theorem).

$(2) \Rightarrow (3)$. This is clear.

$(3) \Rightarrow (1)$. Let $M_{DB}(T)$ be invertible for some B and D and, for convenience, write $A = M_{DB}(T)$. Then $C_D T = T_A C_B$ by Theorem 2. Hence $T = C_D^{-1} T_A C_B$ is a composite of isomorphisms by Theorem 1 (note that $(T_A)^{-1} = T_{A^{-1}}$) and so is itself an isomorphism.

In view of Theorem 4, it is not surprising that there is a connection between the rank of a linear transformation (see Section 7.2) and the rank of the corresponding matrices.

Theorem 5

Let $T : V \to W$ be a linear transformation where $\dim V = n$ and $\dim W = m$. If B and D are any ordered bases of V and W, then rank $T = \text{rank}[M_{DB}(T)]$.

PROOF

Write $A = M_{DB}(T)$ for convenience. The column space of A is $U = \{AX \mid X \text{ in } \mathbb{R}^n\}$. Hence rank $A = \dim U$ and so, because rank $T = \dim(\text{im } T)$, it suffices to find an isomorphism $S : \text{im } T \to U$. Now every vector in im T has the form $T(\mathbf{v})$, $\mathbf{v}$ in V. By Theorem 2, $C_D[T(\mathbf{v})] = AC_B(\mathbf{v})$ lies in U. So define $S : \text{im } T \to U$ by

$$S[T(\mathbf{v})] = C_D[T(\mathbf{v})] \quad \text{for all vectors } T(\mathbf{v}) \text{ in im } T$$

The fact that C_D is linear and one-to-one implies immediately that S is linear and one-to-one. To see that S is onto, let AX be any member of U, X in $\mathbb{R}^n$. Then $X = C_B(\mathbf{v})$ for some $\mathbf{v}$ in V because C_B is onto. Hence $AX = AC_B(\mathbf{v}) = C_D[T(\mathbf{v})] = S[T(\mathbf{v})]$, so S is onto. This means that S is an isomorphism.

We conclude with an example showing that the matrix of a linear transformation can be made very simple by a careful choice of the two bases.

Example 6

Let $T : V \to W$ be a linear transformation where $\dim V = n$ and $\dim W = m$. Choose an ordered basis $B = \{\mathbf{e}_1, \ldots, \mathbf{e}_r, \mathbf{e}_{r+1}, \ldots, \mathbf{e}_n\}$ of V in which $\{\mathbf{e}_{r+1}, \ldots, \mathbf{e}_n\}$ is a basis of ker T, possibly empty. Then $\{T(\mathbf{e}_1), \ldots, T(\mathbf{e}_r)\}$ is a basis of im T by Theorem 5 §7.2, so extend it to an ordered basis

$D = \{T(\mathbf{e}_1), \ldots, T(\mathbf{e}_r), \mathbf{f}_{r+1}, \ldots, \mathbf{f}_m\}$ of W. Because $T(\mathbf{e}_{r+1}) = \cdots = T(\mathbf{e}_n) = \mathbf{0}$, we have

$$M_{DB}(T) = \begin{bmatrix} C_D[T(\mathbf{e}_1)] & \cdots & C_D[T(\mathbf{e}_r)] & C_D[T(\mathbf{e}_{r+1})] & \cdots & C_D[T(\mathbf{e}_n)] \end{bmatrix} = \begin{bmatrix} I_r & 0 \\ 0 & 0 \end{bmatrix}.$$

Incidentally, this shows that rank $T = r$ by Theorem 5.

Exercises 9.1

1. In each case, find the coordinates of **v** with respect to the basis B of the vector space V.

   (a) $V = \mathbf{P}_2$, $\mathbf{v} = 2x^2 + x - 1$, $B = \{x + 1, x^2, 3\}$

   ◆(b) $V = \mathbf{P}_2$, $\mathbf{v} = ax^2 + bx + c$, $B = \{x^2, x + 1, x + 2\}$

   (c) $V = \mathbb{R}^3$, $\mathbf{v} = (1, -1, 2)$,
   $B = \{(1, -1, 0), (1, 1, 1), (0, 1, 1)\}$

   ◆(d) $V = \mathbb{R}^3$, $\mathbf{v} = (a, b, c)$,
   $B = \{(1, -1, 2), (1, 1, -1), (0, 0, 1)\}$

   (e) $V = \mathbf{M}_{22}$, $\mathbf{v} = \begin{bmatrix} 1 & 2 \\ -1 & 0 \end{bmatrix}$,
   $B = \left\{ \begin{bmatrix} 1 & 1 \\ 0 & 0 \end{bmatrix}, \begin{bmatrix} 1 & 0 \\ 1 & 0 \end{bmatrix}, \begin{bmatrix} 0 & 0 \\ 1 & 1 \end{bmatrix}, \begin{bmatrix} 1 & 0 \\ 0 & 1 \end{bmatrix} \right\}$

2. Suppose $T : \mathbf{P}_2 \to \mathbb{R}^2$ is a linear transformation. If $B = \{1, x, x^2\}$ and $D = \{(1, 1), (0, 1)\}$, find the action of T given:

   (a) $M_{DB}(T) = \begin{bmatrix} 1 & 2 & -1 \\ -1 & 0 & 1 \end{bmatrix}$

   ◆(b) $M_{DB}(T) = \begin{bmatrix} 2 & 1 & 3 \\ -1 & 0 & -2 \end{bmatrix}$

3. In each case, find the matrix of $T : V \to W$ corresponding to the bases B and D of V and W, respectively.

   (a) $T : \mathbf{M}_{22} \to \mathbb{R}$, $T(A) = \text{tr } A$;
   $B = \left\{ \begin{bmatrix} 1 & 0 \\ 0 & 0 \end{bmatrix}, \begin{bmatrix} 0 & 1 \\ 0 & 0 \end{bmatrix}, \begin{bmatrix} 0 & 0 \\ 1 & 0 \end{bmatrix}, \begin{bmatrix} 0 & 0 \\ 0 & 1 \end{bmatrix} \right\}$,
   $D = \{1\}$

   ◆(b) $T : \mathbf{M}_{22} \to \mathbf{M}_{22}$, $T(A) = A^T$;
   $B = D = \left\{ \begin{bmatrix} 1 & 0 \\ 0 & 0 \end{bmatrix}, \begin{bmatrix} 0 & 1 \\ 0 & 0 \end{bmatrix}, \begin{bmatrix} 0 & 0 \\ 1 & 0 \end{bmatrix}, \begin{bmatrix} 0 & 0 \\ 0 & 1 \end{bmatrix} \right\}$

   (c) $T : \mathbf{P}_2 \to \mathbf{P}_3$, $T[p(x)] = xp(x)$; $B = \{1, x, x^2\}$, $D = \{1, x, x^2, x^3\}$

   ◆(d) $T : \mathbf{P}_2 \to \mathbf{P}_2$, $T[p(x)] = p(x + 1)$;
   $B = D = \{1, x, x^2\}$

4. In each case, find the matrix of $T : V \to W$ corresponding to the bases B and D, respectively, and use it to compute $C_D[T(\mathbf{v})]$, and hence $T(\mathbf{v})$.

   (a) $T : \mathbb{R}^3 \to \mathbb{R}^4$,
   $T(x, y, z) = (x + z, 2z, y - z, x + 2y)$; B and D standard; $\mathbf{v} = (1, -1, 3)$

   ◆(b) $T : \mathbb{R}^2 \to \mathbb{R}^4$, $T(x, y) = (2x - y, 3x + 2y, 4y, x)$; $B = \{(1, 1), (1, 0)\}$, D standard; $\mathbf{v} = (a, b)$

   (c) $T : \mathbf{P}_2 \to \mathbb{R}^2$, $T(a + bx + cx^2) = (a + c, 2b)$;
   $B = \{1, x, x^2\}$, $D = \{(1, 0), (1, -1)\}$;
   $\mathbf{v} = a + bx + cx^2$

   ◆(d) $T : \mathbf{P}_2 \to \mathbb{R}^2$, $T(a + bx + cx^2) = (a + b, c)$;
   $B = \{1, x, x^2\}$, $D = \{(1, -1), (1, 1)\}$;
   $\mathbf{v} = a + bx + cx^2$

   (e) $T : \mathbf{M}_{22} \to \mathbb{R}$, $T\begin{bmatrix} a & b \\ c & d \end{bmatrix} = a + b + c + d$;
   $B = \left\{ \begin{bmatrix} 1 & 0 \\ 0 & 0 \end{bmatrix}, \begin{bmatrix} 0 & 1 \\ 0 & 0 \end{bmatrix}, \begin{bmatrix} 0 & 0 \\ 1 & 0 \end{bmatrix}, \begin{bmatrix} 0 & 0 \\ 0 & 1 \end{bmatrix} \right\}$,
   $D = \{1\}$; $\mathbf{v} = \begin{bmatrix} a & b \\ c & d \end{bmatrix}$

   ◆(f) $T : \mathbf{M}_{22} \to \mathbf{M}_{22}$, $T\begin{bmatrix} a & b \\ c & d \end{bmatrix} = \begin{bmatrix} a & b + c \\ b + c & d \end{bmatrix}$;
   $B = D = \left\{ \begin{bmatrix} 1 & 0 \\ 0 & 0 \end{bmatrix}, \begin{bmatrix} 0 & 1 \\ 0 & 0 \end{bmatrix}, \begin{bmatrix} 0 & 0 \\ 1 & 0 \end{bmatrix}, \begin{bmatrix} 0 & 0 \\ 0 & 1 \end{bmatrix} \right\}$;
   $\mathbf{v} = \begin{bmatrix} a & b \\ c & d \end{bmatrix}$

5. In each case, verify Theorem 3. Use the standard basis in $\mathbb{R}^n$ and $\{1, x, x^2\}$ in $\mathbf{P}_2$.

   (a) $\mathbb{R}^3 \xrightarrow{T} \mathbb{R}^2 \xrightarrow{S} \mathbb{R}^4$; $T(a, b, c) = (a + b, b - c)$,
   $S(a, b) = (a, b - 2a, 3b, a + b)$

   ◆(b) $\mathbb{R}^3 \xrightarrow{T} \mathbb{R}^4 \xrightarrow{S} \mathbb{R}^2$;
   $T(a, b, c) = (a + b, c + b, a + c, b - a)$,
   $S(a, b, c, d) = (a + b, c - d)$

   (c) $\mathbf{P}_2 \xrightarrow{T} \mathbb{R}^3 \xrightarrow{S} \mathbf{P}_2$;
   $T(a + bx + cx^2) = (a, b - c, c - a)$,
   $S(a, b, c) = b + cx + (a - c)x^2$

   ◆(d) $\mathbb{R}^3 \xrightarrow{T} \mathbf{P}_2 \xrightarrow{S} \mathbb{R}^2$;
   $T(a, b, c) = (a - b) + (c - a)x + bx^2$,
   $S(a + bx + cx^2) = (a - b, c)$

6. Verify Theorem 3 for $\mathbf{M}_{22} \xrightarrow{T} \mathbf{M}_{22} \xrightarrow{S} \mathbf{P}_2$, where
   $T(A) = A^T$ and $S\begin{bmatrix} a & b \\ c & d \end{bmatrix} = b + (a + d)x + cx^2$.
   Use the bases
   $B = D = \left\{ \begin{bmatrix} 1 & 0 \\ 0 & 0 \end{bmatrix}, \begin{bmatrix} 0 & 1 \\ 0 & 0 \end{bmatrix}, \begin{bmatrix} 0 & 0 \\ 1 & 0 \end{bmatrix}, \begin{bmatrix} 0 & 0 \\ 0 & 1 \end{bmatrix} \right\}$
   and $E = \{1, x, x^2\}$.

7. In each case, find T^{-1} and verify that $M_{DB}(T)^{-1} = M_{BD}(T^{-1})$.

(a) $T: \mathbb{R}^2 \to \mathbb{R}^2$, $T(a, b) = (a + 2b, 2a + 5b)$;
$B = D =$ standard

♦(b) $T: \mathbb{R}^3 \to \mathbb{R}^3$, $T(a, b, c) = (b + c, a + c, a + b)$;
$B = D =$ standard

(c) $T: \mathbf{P}_2 \to \mathbb{R}^3$,
$T(a + bx + cx^2) = (a - c, b, 2a - c)$;
$B = \{1, x, x^2\}$, $D =$ standard

♦(d) $T: \mathbf{P}_2 \to \mathbb{R}^3$,
$T(a + bx + cx^2) = (a + b + c, b + c, c)$;
$B = \{1, x, x^2\}$, $D =$ standard

8. In each case, show that $M_{DB}(T)$ is invertible and use the fact that $M_{BD}(T^{-1}) = [M_{BD}(T)]^{-1}$ to determine the action of T^{-1}.

(a) $T: \mathbf{P}_2 \to \mathbb{R}^3$, $T(a + bx + cx^2) = (a + c, c, b - c)$;
$B = \{1, x, x^2\}$, $D =$ standard

♦(b) $T: \mathbf{M}_{22} \to \mathbb{R}^4$,

$$T\begin{bmatrix} a & b \\ c & d \end{bmatrix} = (a + b + c, b + c, c, d);$$

$$B = \left\{ \begin{bmatrix} 1 & 0 \\ 0 & 0 \end{bmatrix}, \begin{bmatrix} 0 & 1 \\ 0 & 0 \end{bmatrix}, \begin{bmatrix} 0 & 0 \\ 1 & 0 \end{bmatrix}, \begin{bmatrix} 0 & 0 \\ 0 & 1 \end{bmatrix} \right\},$$

$D =$ standard

9. Let $D: \mathbf{P}_3 \to \mathbf{P}_2$ be the differentiation map given by $D[p(x)] = p'(x)$. Find the matrix of D corresponding to the bases $B = \{1, x, x^2, x^3\}$ and $E = \{1, x, x^2\}$, and use it to compute $D(a + bx + cx^2 + dx^3)$.

10. Use Theorem 4 to show that $T: V \to V$ is not an isomorphism if $\ker T \neq 0$ (assume $\dim V = n$). [*Hint:* Choose any ordered basis B containing a vector in $\ker T$.]

11. Let $T: V \to \mathbb{R}$ be a linear transformation, and let $D = \{1\}$ be the basis of $\mathbb{R}$. Given any ordered basis $B = \{\mathbf{e}_1, \dots, \mathbf{e}_n\}$ of V, show that $M_{DB}(T) = [T(\mathbf{e}_1) \cdots T(\mathbf{e}_n)]$.

♦12. Let $T: V \to W$ be an isomorphism, let $B = \{\mathbf{e}_1, \dots, \mathbf{e}_n\}$ be an ordered basis of V, and let $D = \{T(\mathbf{e}_1), \dots, T(\mathbf{e}_n)\}$. Show that $M_{DB}(T) = I_n$ —the $n \times n$ identity matrix.

13. Complete the proof of Theorem 1.

14. Let U be any invertible $n \times n$ matrix, and let $D = \{\mathbf{f}_1, \mathbf{f}_2, \dots, \mathbf{f}_n\}$ where $\mathbf{f}_j$ is column j of U. Show that $M_{BD}(1_{\mathbb{R}^n}) = U$ when B is the standard basis of $\mathbb{R}^n$.

15. Let B be an ordered basis of the n-dimensional space V and let $C_B: V \to \mathbb{R}^n$ be the coordinate transformation. If D is the standard basis of $\mathbb{R}^n$, show that $M_{DB}(C_B) = I_n$.

16. Let $T: \mathbf{P}_2 \to \mathbb{R}^3$ be defined by $T(p) = (p(0), p(1), p(2))$ for all p in $\mathbf{P}_2$. Let $B = \{1, x, x^2\}$ and $D = \{(1, 0, 0), (0, 1, 0), (0, 0, 1)\}$.

(a) Show that $M_{DB}(T) = \begin{bmatrix} 1 & 0 & 0 \\ 1 & 1 & 1 \\ 1 & 2 & 4 \end{bmatrix}$ and conclude that T is an isomorphism.

♦(b) Generalize to $T: \mathbf{P}_n \to \mathbb{R}^{n+1}$ where $T(p) = (p(a_0), p(a_1), \dots, p(a_n))$ and $a_0, a_1, \dots, a_n$ are distinct real numbers. [*Hint:* Theorem 7 §3.2.]

17. Let $T: \mathbf{P}_n \to \mathbf{P}_n$ be defined by $T[p(x)] = p(x) + xp'(x)$, where $p'(x)$ denotes the derivative. Show that T is an isomorphism by finding $M_{BB}(T)$ when $B = \{1, x, x^2, \dots, x^n\}$.

18. If k is any number, define $T_k: \mathbf{M}_{22} \to \mathbf{M}_{22}$ by $T_k(A) = A + kA^T$.

(a) If $B = \left\{ \begin{bmatrix} 1 & 0 \\ 0 & 0 \end{bmatrix}, \begin{bmatrix} 0 & 0 \\ 0 & 1 \end{bmatrix}, \begin{bmatrix} 0 & 1 \\ 1 & 0 \end{bmatrix}, \begin{bmatrix} 0 & 1 \\ -1 & 0 \end{bmatrix} \right\}$ find $M_{BB}(T_k)$, and conclude that T_k is invertible if $k \neq 1$ and $k \neq -1$.

(b) Repeat for $T_k: \mathbf{M}_{33} \to \mathbf{M}_{33}$. Can you generalize?

The remaining exercises require the following definitions. If V and W are vector spaces, the set of all linear transformations from V to W will be denoted by

$$\mathbf{L}(V, W) = \{T \mid T: V \to W \text{ is a linear transformation}\}$$

Given S and T in $\mathbf{L}(V, W)$ and a in $\mathbb{R}$, define $S + T: V \to W$ and $aT: V \to W$ by

$$(S + T)(\mathbf{v}) = S(\mathbf{v}) + T(\mathbf{v}) \quad \text{for all } \mathbf{v} \text{ in } V$$
$$(aT)(\mathbf{v}) = aT(\mathbf{v}) \qquad \text{for all } \mathbf{v} \text{ in } V$$

19. Show that $\mathbf{L}(V, W)$ is a vector space.

20. Show that the following properties hold provided that the transformations link together in such a way that all the operations are defined.

(a) $R(ST) = (RS)T$

(b) $1_W T = T = T1_V$

(c) $R(S + T) = RS + RT$

♦(d) $(S + T)R = SR + TR$

(e) $(aS)T = a(ST) = S(aT)$

21. Given S and T in $\mathbf{L}(V, W)$, show that:

(a) $\ker S \cap \ker T \subseteq \ker(S + T)$

♦(b) $\operatorname{im}(S + T) \subseteq \operatorname{im} S + \operatorname{im} T$

22. Le
    sul

   (a)

 ◆(b)

   (c)

23. De
    ea
    gi
    R

24. Le
    is
    S_v

   (a)

   (b)

25. L
    B
    de

   (a)

 ◆(b)

   (c)

26. L
    or

Unfortunately, not every linear operator $T : V \to V$ is reducible. In fact, the linear operator in Example 4 has *no* invariant subspaces except 0 and V. On the other hand, one might expect that this is the only type of nonreducible operator; that is, if the operator *has* an invariant subspace that is not 0 or V, then *some* invariant complement must exist. The next example shows that even this is not valid.

Example 11

Consider the operator $T : \mathbb{R}^2 \to \mathbb{R}^2$ given by $T\begin{bmatrix} a \\ b \end{bmatrix} = \begin{bmatrix} a+b \\ b \end{bmatrix}$. Show that $U_1 = \mathbb{R}\begin{bmatrix} 1 \\ 0 \end{bmatrix}$ is T-invariant but that U_1 has no T-invariant complement in $\mathbb{R}^2$.

Solution   Because $U_1 = \text{span}\left\{\begin{bmatrix} 1 \\ 0 \end{bmatrix}\right\}$ and $T\begin{bmatrix} 1 \\ 0 \end{bmatrix} = \begin{bmatrix} 1 \\ 0 \end{bmatrix}$, it follows (by Example 3) that U_1 is T-invariant. Now assume, if possible, that U_1 has a T-invariant complement U_2 in $\mathbb{R}^2$. Then $U_1 \oplus U_2 = \mathbb{R}^2$ and $T(U_2) \subseteq U_2$. Theorem 6 gives

$$2 = \dim \mathbb{R}^2 = \dim U_1 + \dim U_2 = 1 + \dim U_2$$

so $\dim U_2 = 1$. Let $U_2 = \mathbb{R}\begin{bmatrix} p \\ q \end{bmatrix}$. Then $\begin{bmatrix} p \\ q \end{bmatrix}$ is not in U_1 (because $U_1 \cap U_2 = 0$) and hence $q \neq 0$. On the other hand, $T\begin{bmatrix} p \\ q \end{bmatrix}$ lies in $U_2 = \mathbb{R}\begin{bmatrix} p \\ q \end{bmatrix}$, say

$$\begin{bmatrix} p+q \\ q \end{bmatrix} = T\begin{bmatrix} p \\ q \end{bmatrix} = \lambda \begin{bmatrix} p \\ q \end{bmatrix} \quad \text{for some } \lambda \text{ } in \text{ } \mathbb{R}$$

Hence $p + q = \lambda p$ and $q = \lambda q$. Because $q \neq 0$, the second of these equations implies that $\lambda = 1$, whence the first implies that $q = 0$, a contradiction. So a T-invariant complement of U_1 does not exist.

Exercises 9.3

1. Let T be a linear operator on V. If U and U_1 are T-invariant, show that $U \cap U_1$ and $U + U_1$ are also T-invariant.

2. If $T : V \to V$ is any linear operator, show that $\ker T$ and $\text{im } T$ are T-invariant subspaces.

3. Let S and T be linear operators on V and assume that $ST = TS$.
   (a) Show that $\text{im } S$ and $\ker S$ are T-invariant.
   ◆(b) If U is T-invariant, show that $S(U)$ is T-invariant.

4. Let $T : V \to V$ be a linear operator. Given $\mathbf{e}$ in V, let U denote the set of vectors in V that lie in every T-invariant subspace that contains $\mathbf{e}$.
   (a) Show that U is a T-invariant subspace of V containing $\mathbf{e}$.

   (b) Show that U is contained in every T-invariant subspace of V that contains $\mathbf{e}$.

5. (a) Show that every subspace is T-invariant if T is a scalar operator (see Example 1 §7.1).
   (b) Conversely, if every subspace is T-invariant, show that T is scalar.

◆6. Show that the only subspaces of V that are T-invariant for every operator $T : V \to V$ are 0 and V. Assume that V is finite dimensional. [*Hint:* Theorem 3 §7.1.]

7. Suppose that $T : V \to V$ is a linear operator and that U is a T-invariant subspace of V. If S is an invertible operator, put $T' = STS^{-1}$. Show that $S(U)$ is a T'-invariant subspace.

8. In each case, show that U is T-invariant, use it to find a block upper triangular matrix for T, and use that to compute $c_T(x)$.

   (a) $T : \mathbf{P}_2 \to \mathbf{P}_2$, $T(a + bx + cx^2) = (-a + 2b + c)$
   $+ (a + 3b + c)x + (a + 4b)x^2$, $U = \text{span}\{1, x + x^2\}$

   ◆(b) $T : \mathbf{P}_2 \to \mathbf{P}_2$, $T(a + bx + cx^2) = (5a - 2b + c)$
   $+ (5a - b + c)x + (a + 2c)x^2$,
   $U = \text{span}\{1 - 2x^2, x + x^2\}$

9. In each case, show that $T_A : \mathbb{R}^2 \to \mathbb{R}^2$ has no invariant subspaces except 0 and $\mathbb{R}^2$.

   (a) $A = \begin{bmatrix} 1 & 2 \\ -1 & -1 \end{bmatrix}$

   ◆(b) $A = \begin{bmatrix} \cos\theta & -\sin\theta \\ \sin\theta & \cos\theta \end{bmatrix}$, $0 < \theta < \pi$

10. In each case, show that $V = U \oplus W$.

    (a) $V = \mathbb{R}^4$, $U = \text{span}\{(1, 1, 0, 0), (0, 1, 1, 0)\}$,
    $W = \text{span}\{(0, 1, 0, 1), (0, 0, 1, 1)\}$

    ◆(b) $V = \mathbb{R}^4$, $U = \{(a, a, b, b) \mid a, b \text{ in } \mathbb{R}\}$,
    $W = \{(c, d, c, -d) \mid c, d \text{ in } \mathbb{R}\}$

    (c) $V = \mathbf{P}_3$, $U = \{a + bx \mid a, b \text{ in } \mathbb{R}\}$,
    $W = \{ax^2 + bx^3 \mid a, b \text{ in } \mathbb{R}\}$

    (d) $V = \mathbf{M}_{22}$, $U = \left\{ \begin{bmatrix} a & a \\ b & b \end{bmatrix} \middle| a, b \text{ in } \mathbb{R} \right\}$,

    $W = \left\{ \begin{bmatrix} a & b \\ -a & b \end{bmatrix} \middle| a, b \text{ in } \mathbb{R} \right\}$

11. Let $U = \text{span}\{(1, 0, 0, 0), (0, 1, 0, 0)\}$ in $\mathbb{R}^4$. Show that $\mathbb{R}^4 = U \oplus W_1$ and $\mathbb{R}^4 = U \oplus W_2$, where $W_1 = \text{span}\{(0, 0, 1, 0), (0, 0, 0, 1)\}$ and $W_2 = \text{span}\{(1, 1, 1, 1), (1, 1, 1, -1)\}$.

12. Let U be a subspace of V, and suppose that $V = U \oplus W_1$ and $V = U \oplus W_2$ hold for subspaces W_1 and W_2. Show that $\dim W_1 = \dim W_2$.

13. If U and W denote the subspaces of even and odd polynomials in $\mathbf{P}_n$, respectively, show that $\mathbf{P}_n = U \oplus W$. (See Exercise 34 §6.3.) [*Hint:* $f(x) + f(-x)$ is even.]

◆14. Let E be a 2×2 matrix such that $E^2 = E$. Show that $\mathbf{M}_{22} = U \oplus W$, where $U = \{A \mid AE = A\}$ and $W = \{B \mid BE = 0\}$. [*Hint:* XE lies in U for every matrix X.]

15. If U and W are the subspaces of symmetric and skew-symmetric $n \times n$ matrices, show that $\mathbf{M}_{nn} = U \oplus W$. [*Hint:* $X + X^T$ is symmetric.]

16. Let $V \xrightarrow{T} W \xrightarrow{S} V$ be linear transformations, and assume that $\dim V$ and $\dim W$ are finite.

    (a) If $ST = 1_V$, show that $W = \text{im } T \oplus \text{ker } S$. [*Hint:* Given $\mathbf{w}$ in W, show that $\mathbf{w} - TS(\mathbf{w})$ lies in $\text{ker } S$.]

    (b) Illustrate with $\mathbb{R}^2 \xrightarrow{T} \mathbb{R}^3 \xrightarrow{S} \mathbb{R}^2$ where $T(x, y) = (x, y, 0)$ and $S(x, y, z) = (x, y)$.

17. Let V be a finite dimensional vector space, and let U and W be subspaces such that $U \cap W = \{\mathbf{0}\}$. If $\dim U + \dim W = \dim V$, show that $V = U \oplus W$.

18. Let $A = \begin{bmatrix} 0 & 1 \\ 0 & 0 \end{bmatrix}$ and consider $T_A : \mathbb{R}^2 \to \mathbb{R}^2$.

    (a) Show that the only eigenvalue of T_A is $\lambda = 0$.

    ◆(b) Show that $\ker(T_A) = \mathbb{R}\begin{bmatrix} 1 \\ 0 \end{bmatrix}$ is the unique T_A-invariant subspace of $\mathbb{R}^2$ (except for 0 and $\mathbb{R}^2$).

19. If $A = \begin{bmatrix} 2 & -5 & 0 & 0 \\ 1 & -2 & 0 & 0 \\ 0 & 0 & -1 & -2 \\ 0 & 0 & 1 & 1 \end{bmatrix}$, show that $T_A : \mathbb{R}^4 \to \mathbb{R}^4$

    has two-dimensional T-invariant subspaces U and W such that $\mathbb{R}^4 = U \oplus W$, but A has no real eigenvalue.

◆20. Let $T : V \to V$ be a linear operator where $\dim V = n$. If U is a T-invariant subspace of V, let $T_1 : U \to U$ denote the restriction of T to U (so $T_1(\mathbf{u}) = T(\mathbf{u})$ for all $\mathbf{u}$ in U). Show that $c_T(x) = c_{T_1}(x) \cdot q(x)$ for some polynomial $q(x)$. [*Hint:* Theorem 1.]

21. Let $T : V \to V$ be a linear operator where $\dim V = n$. Show that V has a basis of eigenvectors if and only if V has a basis B such that $M_B(T)$ is diagonal.

22. In each case, show that $T^2 = 1$ and find (as in Example 10) an ordered basis B such that $M_B(T)$ has the given block form.

    (a) $T : \mathbf{M}_{22} \to \mathbf{M}_{22}$ where $T(A) = A^T$,
    $$M_B(T) = \begin{bmatrix} I_3 & 0 \\ 0 & -1 \end{bmatrix}$$

    ◆(b) $T : \mathbf{P}_3 \to \mathbf{P}_3$ where $T[p(x)] = p(-x)$,
    $$M_B(T) = \begin{bmatrix} I_2 & 0 \\ 0 & -I_2 \end{bmatrix}$$

(c) $T : \mathbb{C} \to \mathbb{C}$ where $T(a + bi) = a - bi$,

$$M_B(T) = \begin{bmatrix} 1 & 0 \\ 0 & -1 \end{bmatrix}$$

◆(d) $T : \mathbb{R}^3 \to \mathbb{R}^3$ where $T(a, b, c) =$

$(-a + 2b + c, b + c, -c)$, $M_B(T) = \begin{bmatrix} 1 & 0 \\ 0 & -I_2 \end{bmatrix}$

(e) $T : V \to V$ where $T(\mathbf{v}) = -\mathbf{v}$, $\dim V = n$,
$M_B(T) = -I_n$

23. Let U and W denote subspaces of a vector space V.

   (a) If $V = U \oplus W$, define $T : V \to V$ by $T(\mathbf{v}) = \mathbf{w}$ where $\mathbf{v}$ is written (uniquely) as $\mathbf{v} = \mathbf{u} + \mathbf{w}$ with $\mathbf{u}$ in U and $\mathbf{w}$ in W. Show that T is a linear transformation, $U = \ker T$, $W = \operatorname{im} T$, and $T^2 = T$.

   (b) Conversely, if $T : V \to V$ is a linear transformation such that $T^2 = T$, show that $V = \ker T \oplus \operatorname{im} T$. [*Hint*: $\mathbf{v} - T(\mathbf{v})$ lies in $\ker T$ for all $\mathbf{v}$ in V.]

24. Let $T : V \to V$ be a linear operator satisfying $T^2 = T$ (such operators are called **idempotents**). Define $U_1 = \{\mathbf{v} \mid T(\mathbf{v}) = \mathbf{v}\}$ and $U_2 = \ker T = \{\mathbf{v} \mid T(\mathbf{v}) = \mathbf{0}\}$.

   (a) Show that $V = U_1 \oplus U_2$.

   (b) If $\dim V = n$, find a basis B of V such that
   $$M_B(T) = \begin{bmatrix} I_r & 0 \\ 0 & 0 \end{bmatrix}, \text{ where } r = \operatorname{rank} T.$$

   (c) If A is an $n \times n$ matrix such that $A^2 = A$, show that A is similar to $\begin{bmatrix} I_r & 0 \\ 0 & 0 \end{bmatrix}$, where $r = \operatorname{rank} A$. [*Hint*: Example 10.]

25. In each case, show that $T^2 = T$ and find (as in the preceding exercise) an ordered basis B such that $M_B(T)$ has the form given (0_k is the $k \times k$ zero matrix).

   (a) $T : \mathbf{P}_2 \to \mathbf{P}_2$ where $T(a + bx + cx^2) =$

   $(a - b + c)(1 + x + x^2)$, $M_B(T) = \begin{bmatrix} 1 & 0 \\ 0 & 0_2 \end{bmatrix}$

   ◆(b) $T : \mathbb{R}^3 \to \mathbb{R}^3$ where $T(a, b, c) =$

   $(a + 2b, 0, 4b + c)$, $M_B(T) = \begin{bmatrix} I_2 & 0 \\ 0 & 0 \end{bmatrix}$

   (c) $T : \mathbf{M}_{22} \to \mathbf{M}_{22}$ where $T \begin{bmatrix} a & b \\ c & d \end{bmatrix} =$

   $\begin{bmatrix} -5 & -15 \\ 2 & 6 \end{bmatrix} \begin{bmatrix} a & b \\ c & d \end{bmatrix}$, $M_B(T) = \begin{bmatrix} I_2 & 0 \\ 0 & 0_2 \end{bmatrix}$

26. Let $T : V \to V$ be an operator satisfying $T^2 = cT$, $c \neq 0$.

   (a) Show that $V = U \oplus \ker T$, where $U = \{\mathbf{u} \mid T(\mathbf{u}) = c\mathbf{u}\}$. [*Hint*: Compute $T(\mathbf{v} - \frac{1}{c}T(\mathbf{v}))$.]

   (b) If $\dim V = n$, show that V has a basis B such that $M_B(T) = \begin{bmatrix} cI_r & 0 \\ 0 & 0 \end{bmatrix}$, where $r = \operatorname{rank} T$.

   (c) If A is any $n \times n$ matrix of rank r such that $A^2 = cA$, $c \neq 0$, show that A is similar to $\begin{bmatrix} cI_r & 0 \\ 0 & 0 \end{bmatrix}$.

27. Let $T : V \to V$ be an operator such that $T^2 = c^2$, $c \neq 0$.

   (a) Show that $V = U_1 \oplus U_2$, where $U_1 = \{\mathbf{v} \mid T(\mathbf{v}) = c\mathbf{v}\}$ and $U_2 = \{\mathbf{v} \mid T(\mathbf{v}) = -c\mathbf{v}\}$. [*Hint*: $\mathbf{v} = \frac{1}{2c}\{[T(\mathbf{v}) + c\mathbf{v}] - [T(\mathbf{v}) - c\mathbf{v}]\}$.]

   (b) If $\dim V = n$, show that V has a basis B such that $M_B(T) = \begin{bmatrix} cI_k & 0 \\ 0 & -cI_{n-k} \end{bmatrix}$ for some k.

   (c) If A is an $n \times n$ matrix such that $A^2 = c^2I$, $c \neq 0$, show that A is similar to $\begin{bmatrix} cI_k & 0 \\ 0 & -cI_{n-k} \end{bmatrix}$ for some k.

28. If P is a fixed $n \times n$ matrix, define $T : \mathbf{M}_{nn} \to \mathbf{M}_{nn}$ by $T(A) = PA$. Let U_j denote the subspace of $\mathbf{M}_{nn}$ consisting of all matrices with all columns zero except possibly column j.

   (a) Show that each U_j is T-invariant.

   (b) Show that $\mathbf{M}_{nn}$ has a basis B such that $M_B(T)$ is block diagonal with each block on the diagonal equal to P.

29. Let V be a vector space. If $f : V \to \mathbb{R}$ is a linear transformation and $\mathbf{z}$ is a vector in V, define $T_{f, \mathbf{z}} : V \to V$ by $T_{f, \mathbf{z}}(\mathbf{v}) = f(\mathbf{v})\mathbf{z}$ for all $\mathbf{v}$ in V. Assume that $f \neq 0$ and $\mathbf{z} \neq \mathbf{0}$.

   (a) Show that $T_{f, \mathbf{z}}$ is a linear operator of rank 1.

   ◆(b) If $f \neq 0$, show that $T_{f, \mathbf{z}}$ is an idempotent if and only if $f(\mathbf{z}) = 1$. (Recall that $T : V \to V$ is called an idempotent if $T^2 = T$.)

   (c) Show that every idempotent $T : V \to V$ of rank 1 has the form $T = T_{f, \mathbf{z}}$ for some $f : V \to \mathbb{R}$ and some $\mathbf{z}$ in V with $f(\mathbf{z}) = 1$. [*Hint*: Write $\operatorname{im} T = \mathbb{R}\mathbf{z}$ and show that $T(\mathbf{z}) = \mathbf{z}$. Then use Exercise 23.]

30. Let U be a fixed $n \times n$ matrix, and consider the operator $T : \mathbf{M}_{nn} \to \mathbf{M}_{nn}$ given by $T(A) = UA$.

    (a) Show that λ is an eigenvalue of T if and only if it is an eigenvalue of U.

    ◆(b) If λ is an eigenvalue of T, show that $E_\lambda(T)$ consists of all matrices whose columns lie in $E_\lambda(U)$:

    $$E_\lambda(T) = \{[P_1\ P_2\ \cdots\ P_n] \mid P_i \text{ in } E_\lambda(U) \text{ for each } i\}$$

    (c) Show that if $\dim[E_\lambda(U)] = d$, then $\dim[E_\lambda(T)] = nd$. [*Hint*: If $B = \{E_1, \dots, E_d\}$ is a basis of $E_\lambda(U)$, consider the set of all matrices with one column from B and the other columns zero.]

31. Let $T: V \to V$ be a linear operator where V is finite dimensional. If $U \subseteq V$ is a subspace, let $\bar{U} = \{\mathbf{u}_0 + T(\mathbf{u}_1) + T^2(\mathbf{u}_2) + \cdots + T^k(\mathbf{u}_k) \mid \mathbf{u}_i$ in $U, k \geq 0\}$. Show that $\bar{U}$ is the smallest T-invariant subspace containing U (that is, it is T-invariant, contains U, and is contained in every such subspace).

32. Let $U_1, \dots, U_m$ be subspaces of V and assume that $V = U_1 + \cdots + U_m$; that is, every $\mathbf{v}$ in V can be written (in at least one way) in the form $\mathbf{v} = \mathbf{u}_1 + \cdots + \mathbf{u}_m$, $\mathbf{u}_i$ in U_i. Show that the following conditions are equivalent.

    (i) If $\mathbf{u}_1 + \cdots + \mathbf{u}_m = \mathbf{0}$, $\mathbf{u}_i$ in U_i, then $\mathbf{u}_i = \mathbf{0}$ for each i.

    (ii) If $\mathbf{u}_1 + \cdots + \mathbf{u}_m = \mathbf{u}'_1 + \cdots + \mathbf{u}'_m$, $\mathbf{u}_i$ and $\mathbf{u}'_i$ in U_i, then $\mathbf{u}_i = \mathbf{u}'_i$ for each i.

    (iii) $U_i \cap (U_1 + \cdots + U_{i-1} + U_{i+1} + \cdots + U_m) = 0$ for each $i = 1, 2, \dots, m$.

    (iv) $U_i \cap (U_{i+1} + \cdots + U_m) = 0$ for each $i = 1, 2, \dots, m - 1$.

    When these conditions are satisfied, we say that V is the **direct sum** of the subspaces U_i, and write $V = U_1 \oplus U_2 \oplus \cdots \oplus U_m$.

33. (a) Let B be a basis of V and let $B = B_1 \cup B_2 \cup \cdots \cup B_m$ where the B_i are pairwise disjoint, nonempty subsets of B. If $U_i = \text{span } B_i$ for each i, show that $V = U_1 \oplus U_2 \oplus \cdots \oplus U_m$ (preceding exercise).

    (b) Conversely if $V = U_1 \oplus \cdots \oplus U_m$ and B_i is a basis of U_i for each i, show that $B = B_1 \cup \cdots \cup B_m$ is a basis of V as in (a).

SECTION 9.4 Block Triangular Form

We have shown (Theorem 5 §8.2) that any $n \times n$ matrix A with every eigenvalue real is similar to an upper triangular matrix U. The following theorem shows that U can be chosen in a special way.

Theorem 1 Block Triangulation Theorem

Let A be an $n \times n$ matrix with real eigenvalues and let

$$c_A(x) = (x - \lambda_1)^{m_1}(x - \lambda_2)^{m_2} \cdots (x - \lambda_k)^{m_k}$$

where $\lambda_1, \lambda_2, \dots, \lambda_k$ are the distinct eigenvalues of A. Then an invertible matrix P exists such that

$$P^{-1}AP = \begin{bmatrix} U_1 & 0 & 0 & \cdots & 0 \\ 0 & U_2 & 0 & \cdots & 0 \\ 0 & 0 & U_3 & \cdots & 0 \\ \vdots & \vdots & \vdots & & \vdots \\ 0 & 0 & 0 & \cdots & U_k \end{bmatrix}$$

where U_i is an $m_i \times m_i$ upper triangular matrix with every entry on the main diagonal equal to λ_i.

The proof is given at the end of this section. For now, we focus on a method for finding the matrix P. The key concept is as follows.

If A is as in Theorem 1, the **generalized eigenspace** $G_{\lambda_i}(A)$ is defined by

$$G_{\lambda_i}(A) = \text{null}[(\lambda_i I - A)^{m_i}].$$

Observe that $E_{\lambda_i}(A) = \text{null}(\lambda_i I - A)$ is a subspace of $G_{\lambda_i}(A)$. We need three technical results.

Lemma 1

Using the notation of Theorem 1, we have $\dim[G_{\lambda_i}(A)] = m_i$.

PROOF

Write $A_i = (\lambda_i I - A)^{m_i}$ for convenience and let P be as in Theorem 1. The spaces $G_{\lambda_i}(A) = \text{null}(A_i)$ and $\text{null}(P^{-1}A_iP)$ are isomorphic via $X \leftrightarrow P^{-1}X$, so we show that $\dim[\text{null}(P^{-1}A_iP)] = m_i$. Now $P^{-1}A_iP = (\lambda_i I - P^{-1}AP)^{m_i}$; in block form this is

$$P^{-1}A_iP = \begin{bmatrix} \lambda_i I - U_1 & 0 & \cdots & 0 \\ 0 & \lambda_i I - U_2 & \cdots & 0 \\ \vdots & \vdots & & \vdots \\ 0 & 0 & \cdots & \lambda_i I - U_k \end{bmatrix}^{m_i}$$

$$= \begin{bmatrix} (\lambda_i I - U_1)^{m_i} & 0 & \cdots & 0 \\ 0 & (\lambda_i I - U_2)^{m_i} & \cdots & 0 \\ \vdots & \vdots & & \vdots \\ 0 & 0 & \cdots & (\lambda_i I - U_k)^{m_i} \end{bmatrix}$$

The matrix $(\lambda_i I - U_j)^{m_i}$ is invertible if $j \neq i$ and zero if $j = i$ (because then U_i is an $m_i \times m_i$ upper triangular matrix with each entry on the main diagonal equal to λ_i). It follows that $m_i = \dim[\text{null}(P^{-1}A_iP)]$, as required.

Lemma 2

If P is as in Theorem 1, denote the columns of P as follows:

$$P_{11}, P_{12}, \dots, P_{1m_1}; P_{21}, P_{22}, \dots, P_{2m_2}; \dots; P_{k1}, P_{k2}, \dots, P_{km_k}$$

Then $\{P_{i1}, P_{i2}, \dots, P_{im_i}\}$ is a basis of $G_{\lambda_i}(A)$.

PROOF

It suffices by Lemma 1 to show that each P_{ij} is *in* $G_{\lambda_i}(A)$. Write the matrix in Theorem 1 as $P^{-1}AP = \text{diag}(U_1, U_2, \dots, U_k)$. Then

$$AP = P \,\text{diag}(U_1, U_2, \dots, U_k)$$

Comparing columns gives, successively:

$$AP_{11} = \lambda_1 P_{11}, \qquad\qquad \text{so } (\lambda_1 I - A)P_{11} = 0$$
$$AP_{12} = uP_{11} + \lambda_1 P_{12}, \qquad \text{so } (\lambda_1 I - A)^2 P_{12} = 0$$
$$AP_{13} = wP_{11} + vP_{12} + \lambda_1 P_{13}, \quad \text{so } (\lambda_1 I - A)^3 P_{13} = 0$$
$$\vdots \qquad\qquad\qquad\qquad\qquad \vdots \qquad\qquad \vdots$$

where u, v, w are in $\mathbb{R}$. In general, $(\lambda_1 I - A)^j P_{1j} = 0$ for $j = 1, 2, \dots, m_1$, so P_{1j} is in $G_{\lambda_1}(A)$. Similarly, P_{ij} is in $G_{\lambda_i}(A)$ for each i and j.

Lemma 3

If B_i is any basis of $G_{\lambda_i}(A)$, then $B = B_1 \cup B_2 \cup \cdots \cup B_k$ is a basis of $\mathbb{R}^n$.

PROOF

It suffices by Lemma 1 to show that B is independent. If a linear combination from B vanishes, let X_i be the sum of the terms from B_i. Then $X_1 + \cdots + X_k = 0$. But $X_i = \sum_j r_{ij} P_{ij}$ by Lemma 2, so $\sum_{i,j} r_{ij} P_{ij} = 0$. Hence each $X_i = 0$, so each coefficient in X_i is zero.

Lemma 2 suggests an algorithm for finding the matrix P in Theorem 1. Observe that there is an ascending chain of subspaces leading from $E_{\lambda_i}(A)$ to $G_{\lambda_i}(A)$:

$$E_{\lambda_i}(A) = \text{null}[(\lambda_i I - A)] \subseteq \text{null}[(\lambda_i I - A)^2] \subseteq \cdots \subseteq \text{null}[(\lambda_i I - A)^{m_i}] = G_{\lambda_i}(A)$$

We construct a basis for $G_{\lambda_i}(A)$ by climbing up this chain.

Triangulation Algorithm

Suppose A has characteristic polynomial

$$c_A(x) = (x - \lambda_1)^{m_1}(x - \lambda_2)^{m_2} \cdots (x - \lambda_k)^{m_k}$$

1. Choose a basis of $\text{null}[(\lambda_1 I - A)]$; enlarge it by adding vectors (possibly none) to a basis of $\text{null}[(\lambda_1 I - A)^2]$; enlarge that to a basis of $\text{null}[(\lambda_1 I - A)^3]$, and so on. Continue to obtain an ordered basis $\{P_{11}, P_{12}, \dots, P_{1m_1}\}$ of $G_{\lambda_1}(A)$.
2. As in (1) choose a basis $\{P_{i1}, P_{i2}, \dots, P_{im_i}\}$ of $G_{\lambda_i}(A)$ for each i.
3. Let $P = [P_{11}P_{12} \cdots P_{1m_1}; P_{21}P_{22} \cdots P_{2m_2}; \cdots; P_{k_1}P_{k_2} \cdots P_{km_k}]$ be the matrix with these basis vectors (in order) as columns.

Then $P^{-1}AP = \text{diag}(U_1, U_2, \dots, U_k)$ as in Theorem 1.

PROOF

Lemma 3 guarantees that $B = \{P_{11}, \dots, P_{km_k}\}$ is a basis of $\mathbb{R}^n$, and Theorem 4 §9.2 shows that $P^{-1}AP = M_B(T_A)$. Now $G_{\lambda_i}(A)$ is T_A-invariant for each i because

$$(\lambda_i I - A)^{m_i} X = 0 \quad \text{implies} \quad (\lambda_i I - A)^{m_i}(AX) = A(\lambda_i I - A)^{m_i} X = 0$$

By Theorem 7 §9.3 (and induction), we have $P^{-1}AP = M_B(T_A) = \text{diag}(U_1, U_2, \ldots, U_k)$ where U_i is the matrix of the restriction of T_A to $G_{\lambda_i}(A)$, and it remains to show that U_i has the desired upper triangular form. Given s, let P_{ij} be a basis vector in $\text{null}[(\lambda_i I - A)^{s+1}]$. Then $(\lambda_i I - A)P_{ij}$ is in $\text{null}[(\lambda_i I - A)^s]$, and therefore is a linear combination of the basis vectors P_{it} coming *before* P_{ij}. Hence

$$T_A(P_{ij}) = AP_{ij} = \lambda_i P_{ij} - (\lambda_i I - A)P_{ij}$$

shows that the column of U_i corresponding to P_{ij} has λ_i on the main diagonal and zeros below the main diagonal. This is what we wanted.

Example 1

If $A = \begin{bmatrix} 2 & 0 & 0 & 1 \\ 0 & 2 & 0 & -1 \\ -1 & 1 & 2 & 0 \\ 0 & 0 & 0 & 2 \end{bmatrix}$, find P such that $P^{-1}AP$ is block triangular.

Solution $c_A(x) = \det[xI - A] = (x - 2)^4$, so $\lambda_1 = 2$ is the only eigenvalue and we are in the case $k = 1$ of Theorem 1. Compute:

$$(2I - A) = \begin{bmatrix} 0 & 0 & 0 & -1 \\ 0 & 0 & 0 & 1 \\ 1 & -1 & 0 & 0 \\ 0 & 0 & 0 & 0 \end{bmatrix} \qquad (2I - A)^2 = \begin{bmatrix} 0 & 0 & 0 & 0 \\ 0 & 0 & 0 & 0 \\ 0 & 0 & 0 & -2 \\ 0 & 0 & 0 & 0 \end{bmatrix} \qquad (2I - A)^3 = 0$$

By gaussian elimination find a basis $\{P_{11}, P_{12}\}$ of $\text{null}(2I - A)$; then extend in any way to a basis $\{P_{11}, P_{12}, P_{13}\}$ of $\text{null}[(2I - A)^2]$; and finally get a basis $\{P_{11}, P_{12}, P_{13}, P_{14}\}$ of $\text{null}[(2I - A)^3] = \mathbb{R}^4$. One choice is

$$P_{11} = \begin{bmatrix} 1 \\ 1 \\ 0 \\ 0 \end{bmatrix} \qquad P_{12} = \begin{bmatrix} 0 \\ 0 \\ 1 \\ 0 \end{bmatrix} \qquad P_{13} = \begin{bmatrix} 0 \\ 1 \\ 0 \\ 0 \end{bmatrix} \qquad P_{14} = \begin{bmatrix} 0 \\ 0 \\ 0 \\ 1 \end{bmatrix}$$

Hence $P = [P_{11} \ P_{12} \ P_{13} \ P_{14}] = \begin{bmatrix} 1 & 0 & 0 & 0 \\ 1 & 0 & 1 & 0 \\ 0 & 1 & 0 & 0 \\ 0 & 0 & 0 & 1 \end{bmatrix}$ gives $P^{-1}AP = \begin{bmatrix} 2 & 0 & 0 & 1 \\ 0 & 2 & 1 & 0 \\ 0 & 0 & 2 & -2 \\ 0 & 0 & 0 & 2 \end{bmatrix}$

Example 2

If $A = \begin{bmatrix} 2 & 0 & 1 & 1 \\ 3 & 5 & 4 & 1 \\ -4 & -3 & -3 & -1 \\ 1 & 0 & 1 & 2 \end{bmatrix}$, find P such that $P^{-1}AP$ is block triangular.

Solution   The eigenvalues are $\lambda_1 = 1$ and $\lambda_2 = 2$ because

$$c_A(x) = \begin{vmatrix} x - 2 & 0 & -1 & -1 \\ -3 & x - 5 & -4 & -1 \\ 4 & 3 & x + 3 & 1 \\ -1 & 0 & -1 & x - 2 \end{vmatrix} = \begin{vmatrix} x - 1 & 0 & 0 & -x + 1 \\ -3 & x - 5 & -4 & -1 \\ 4 & 3 & x + 3 & 1 \\ -1 & 0 & -1 & x - 2 \end{vmatrix}$$

$$
\begin{aligned}
&= \begin{vmatrix} x-1 & 0 & 0 & 0 \\ -3 & x-5 & -4 & -4 \\ 4 & 3 & x+3 & 5 \\ -1 & 0 & -1 & x-3 \end{vmatrix} = (x-1)\begin{vmatrix} x-5 & -4 & -4 \\ 3 & x+3 & 5 \\ 0 & -1 & x-3 \end{vmatrix} \\
&= (x-1)\begin{vmatrix} x-5 & -4 & 0 \\ 3 & x+3 & -x+2 \\ 0 & -1 & x-2 \end{vmatrix} = (x-1)\begin{vmatrix} x-5 & -4 & 0 \\ 3 & x+2 & 0 \\ 0 & -1 & x-2 \end{vmatrix} \\
&= (x-1)(x-2)\begin{vmatrix} x-5 & -4 \\ 3 & x+2 \end{vmatrix} = (x-1)^2(x-2)^2
\end{aligned}
$$

By solving equations, we find $\mathrm{null}(I-A) = \mathrm{span}\{P_{11}\}$ and $\mathrm{null}(I-A)^2 = \mathrm{span}\{P_{11}, P_{12}\}$ where

$$
P_{11} = \begin{bmatrix} 1 \\ 1 \\ -2 \\ 1 \end{bmatrix} \quad \text{and} \quad P_{12} = \begin{bmatrix} 0 \\ 3 \\ -4 \\ 1 \end{bmatrix}
$$

Since $\lambda_1 = 1$ has multiplicity 2 as a root of $c_A(x)$, $\dim G_{\lambda_1}(A) = 2$ by Lemma 1. Since P_{11} and P_{12} both lie in $G_{\lambda_1}(A)$, we have $G_{\lambda_1}(A) = \mathrm{span}\{P_{11}, P_{12}\}$. Turning to $\lambda_2 = 2$, we find that $\mathrm{null}(2I-A) = \mathrm{span}\{P_{21}\}$ and $\mathrm{null}[(2I-A)^2] = \mathrm{span}\{P_{21}, P_{22}\}$ where

$$
P_{21} = \begin{bmatrix} 1 \\ 0 \\ -1 \\ 1 \end{bmatrix} \quad \text{and} \quad P_{22} = \begin{bmatrix} 0 \\ -4 \\ 3 \\ 0 \end{bmatrix}
$$

Again, $\dim G_{\lambda_2}(A) = 2$ as λ_2 has multiplicity 2, so $G_{\lambda_2}(A) = \mathrm{span}\{P_{21}, P_{22}\}$.

Hence $P = \begin{bmatrix} 1 & 0 & 1 & 0 \\ 1 & 3 & 0 & -4 \\ -2 & -4 & -1 & 3 \\ 1 & 1 & 1 & 0 \end{bmatrix}$ gives $P^{-1}AP = \begin{bmatrix} 1 & -3 & 0 & 0 \\ 0 & 1 & 0 & 0 \\ 0 & 0 & 2 & 3 \\ 0 & 0 & 0 & 2 \end{bmatrix}$.

If $p(x)$ is a polynomial and A is an $n \times n$ matrix, then $p(A)$ is also an $n \times n$ matrix if we interpret $A^0 = I_n$. For example, if $p(x) = x^2 - 2x + 3$, then $p(A) = A^2 - 2A + 3I$. Theorem 1 gives another proof of the Cayley–Hamilton theorem (see also Theorem 10 §8.6). As before, let $c_A(x)$ denote the characteristic polynomial of A.

Theorem 2 Cayley–Hamilton Theorem

If A is a square matrix with real eigenvalues, then $c_A(A) = 0$.

PROOF

Write $c_A(x) = (x-\lambda_1)^{m_1} \cdots (x-\lambda_k)^{m_k}$, and let $D = P^{-1}AP = \mathrm{diag}(U_1, U_2, \ldots, U_k)$ as in Theorem 1. Then $c_A(U_i) = 0$ for each i because $(U_i - \lambda_i I)^{m_i} = 0$. Hence

$$
\begin{aligned}
P^{-1}c_A(A)P = c_A(D) &= c_A[\mathrm{diag}(U_1, \ldots, U_k)] \\
&= \mathrm{diag}[c_A(U_1), \ldots, c_A(U_k)] \\
&= 0
\end{aligned}
$$

It follows that $c_A(A) = 0$.

Example 3

If $A = \begin{bmatrix} 1 & 3 \\ -1 & 2 \end{bmatrix}$, then $c_A(x) = \det \begin{bmatrix} x-1 & -3 \\ 1 & x-2 \end{bmatrix} = x^2 - 3x + 5$. Then

$$c_A(A) = A^2 - 3A + 5I_2 = \begin{bmatrix} -2 & 9 \\ -3 & 1 \end{bmatrix} - \begin{bmatrix} 3 & 9 \\ -3 & 6 \end{bmatrix} + \begin{bmatrix} 5 & 0 \\ 0 & 5 \end{bmatrix} = \begin{bmatrix} 0 & 0 \\ 0 & 0 \end{bmatrix}.$$

Theorem 1 will be refined even further in the next section.

Proof of Theorem 1

The proof of Theorem 1 requires the following fact, the proof of which we leave to the reader.

Lemma 4

If $\{\mathbf{v}_1, \mathbf{v}_2, \dots, \mathbf{v}_n\}$ is a basis of a vector space V, so also is $\{\mathbf{v}_1 + s\mathbf{v}_2, \mathbf{v}_2, \dots, \mathbf{v}_n\}$ for any scalar s.

PROOF OF THEOREM 1

Let A be as in Theorem 1, and let $T = T_A \colon \mathbb{R}^n \to \mathbb{R}^n$ be the matrix transformation induced by A. For convenience, call a matrix a λ-m-ut matrix if it is an $m \times m$ upper triangular matrix and every diagonal entry equals λ. Then we must find a basis B of $\mathbb{R}^n$ such that $M_B(T) = \operatorname{diag}(U_1, U_2, \dots, U_k)$ where U_i is a λ_i-m_i-ut matrix for each i. We proceed by induction on n. If $n = 1$, take $B = \{\mathbf{v}\}$ where $\mathbf{v}$ is any eigenvector of T. If $n > 1$, let $\mathbf{v}_1$ be a λ_1-eigenvector of T, and let $B_0 = \{\mathbf{v}_1, \mathbf{w}_1, \dots, \mathbf{w}_{n-1}\}$ be any basis of $\mathbb{R}^n$ containing $\mathbf{v}_1$. Then (see Lemma 2 §5.5)

$$M_{B_0}(T) = \begin{bmatrix} \lambda_1 & X \\ 0 & A_1 \end{bmatrix}$$

in block form where A_1 is $(n-1) \times (n-1)$. Moreover, A and $M_{B_0}(T)$ are similar, so

$$c_A(x) = c_{M_{B_0}(T)}(x) = (x - \lambda_1)c_{A_1}(x)$$

Hence $c_{A_1}(x) = (x - \lambda_1)^{m_1 - 1}(x - \lambda_2)^{m_2} \cdots (x - \lambda_k)^{m_k}$, so (by induction) let

$$Q^{-1}A_1Q = \operatorname{diag}(Z_1, U_2, \dots, U_k)$$

where Z_1 is a λ_1-$(m_1 - 1)$-ut matrix and U_i is a λ_i-m_i-ut matrix for each $i > 1$. If $P = \begin{bmatrix} 1 & 0 \\ 0 & Q \end{bmatrix}$, then $P^{-1}M_{B_0}(T)P = \begin{bmatrix} \lambda_1 & XQ \\ 0 & Q^{-1}A_1Q \end{bmatrix} = A'$, say. Hence $A' \sim M_{B_0}(T) \sim A$ so, by Theorem 4(1) §9.2 there is a basis B_1 of $\mathbb{R}^n$ such that $M_{B_1}(T) = M_{B_1}(T_A) = A'$. Thus $M_{B_1}(T)$ has the form

$$M_{B_1}(T) = \left[\begin{array}{c|c|ccccc} \lambda_1 & X_1 & & & Y & & \\ \hline 0 & Z_1 & 0 & 0 & \cdots & 0 \\ \hline & & U_2 & 0 & \cdots & 0 \\ & 0 & 0 & U_3 & \cdots & 0 \\ & & \vdots & \vdots & & \vdots \\ & & 0 & 0 & \cdots & U_k \end{array}\right] \qquad (*)$$

If we write $U_1 = \begin{bmatrix} \lambda_1 & X_1 \\ 0 & Z_1 \end{bmatrix}$, the basis B_1 fulfills our needs except that the row matrix Y may not be zero.

We remedy this defect as follows. Observe that the first vector in B_1 is a λ_1 eigenvector of T, which we continue to denote as $\mathbf{v}_1$. The idea is to add suitable scalar multiples of $\mathbf{v}_1$ to the other vectors in B_1. This results in a new basis by Lemma 4, and the multiples can be chosen so that the new matrix of T is the same as $(*)$ except that $Y = 0$. Let $\{\mathbf{w}_1, \ldots, \mathbf{w}_{m_2}\}$ be the vectors in B_1 corresponding to λ_2 (giving rise to U_2 in $(*)$). Write

$$U_2 = \begin{bmatrix} \lambda_2 & u_{12} & u_{13} & \cdots & u_{1m_2} \\ 0 & \lambda_2 & u_{23} & \cdots & u_{2m_2} \\ 0 & 0 & \lambda_2 & \cdots & u_{3m_2} \\ \vdots & \vdots & \vdots & & \vdots \\ 0 & 0 & 0 & \cdots & \lambda_2 \end{bmatrix} \quad \text{and} \quad Y = [y_1 \;\; y_2 \;\; \cdots \;\; y_{m_2}]$$

We first replace $\mathbf{w}_1$ by $\mathbf{w}_1' = \mathbf{w}_1 + s\mathbf{v}_1$ where s is to be determined. Then $(*)$ gives

$$\begin{aligned} T(\mathbf{w}_1') &= T(\mathbf{w}_1) + sT(\mathbf{v}_1) \\ &= (y_1\mathbf{v}_1 + \lambda_2\mathbf{w}_1) + s\lambda_1\mathbf{v}_1 \\ &= y_1\mathbf{v}_1 + \lambda_2(\mathbf{w}_1' - s\mathbf{v}_1) + s\lambda_1\mathbf{v}_1 \\ &= \lambda_2\mathbf{w}_1' + [y_1 - s(\lambda_2 - \lambda_1)]\mathbf{v}_1 \end{aligned}$$

Because $\lambda_2 \neq \lambda_1$ we can choose s such that $T(\mathbf{w}_1') = \lambda_2\mathbf{w}_1'$. Similarly, let $\mathbf{w}_2' = \mathbf{w}_2 + t\mathbf{v}_1$ where t is to be chosen. Then, as before,

$$\begin{aligned} T(\mathbf{w}_2') &= T(\mathbf{w}_2) + tT(\mathbf{v}_1) \\ &= (y_2\mathbf{v}_1 + u_{12}\mathbf{w}_1 + \lambda_2\mathbf{w}_2) + t\lambda_1\mathbf{v}_1 \\ &= u_{12}\mathbf{w}_1' + \lambda_2\mathbf{w}_2' + [(y_2 - u_{12}s) - t(\lambda_2 - \lambda_1)]\mathbf{v}_1 \end{aligned}$$

Again, t can be chosen so that $T(\mathbf{w}_2') = u_{12}\mathbf{w}_1' + \lambda_2\mathbf{w}_2'$. Continue in this way to eliminate $y_1, \ldots, y_{m_2}$. This procedure also works for $\lambda_3, \lambda_4, \ldots$ and so produces a new basis B such that $M_B(T)$ is as in $(*)$ but with $Y = 0$.

Exercises 9.4

1. In each case, find a matrix P such that $P^{-1}AP$ is in block triangular form as in Theorem 1.

(a) $\begin{bmatrix} 2 & 3 & 2 \\ -1 & -1 & -1 \\ 1 & 2 & 2 \end{bmatrix}$

♦(b) $\begin{bmatrix} -5 & 3 & 1 \\ -4 & 2 & 1 \\ -4 & 3 & 0 \end{bmatrix}$

(c) $\begin{bmatrix} 0 & 1 & 1 \\ 2 & 3 & 6 \\ -1 & -1 & -2 \end{bmatrix}$

♦(d) $\begin{bmatrix} -3 & -1 & 0 \\ 4 & -1 & 3 \\ 4 & -2 & 4 \end{bmatrix}$

(e) $\begin{bmatrix} -1 & -1 & -1 & 0 \\ 3 & 2 & 3 & -1 \\ 2 & 1 & 3 & -1 \\ 2 & 1 & 4 & -2 \end{bmatrix}$

♦(f) $\begin{bmatrix} -3 & 6 & 3 & 2 \\ -2 & 3 & 2 & 2 \\ -1 & 3 & 0 & 1 \\ -1 & 1 & 2 & 0 \end{bmatrix}$

2. Show that the following conditions are equivalent for a linear operator T on a finite dimensional space V.

(1) $M_B(T)$ is upper triangular for some ordered basis B of E.

(2) A basis $\{e_1, \ldots, e_n\}$ of V exists such that, for each i, $T(e_i)$ is a linear combination of $e_1, \ldots, e_i$.

(3) There exist T-invariant subspaces $V_1 \subseteq V_2 \subseteq \cdots \subseteq V_n = V$ such that $\dim V_i = i$ for each i.

3. If A is an $n \times n$ invertible matrix, show that $A^{-1} = r_0 I + r_1 A + \cdots + r_{n-1}A^{n-1}$ for some scalars $r_0, r_1, \ldots, r_{n-1}$. [*Hint:* Cayley–Hamilton theorem.]

♦4. If $T : V \to V$ is a linear operator where V is finite dimensional, show that $c_T(T) = 0$. [*Hint:* Exercise 26 §9.1.]

5. Define $T : \mathbf{P} \to \mathbf{P}$ by $T[\,p(x)] = xp(x)$. Show that:
   (a) T is linear and $f(T)[\,p(x)] = f(x)p(x)$ for all polynomials $f(x)$.
   (b) Conclude that $f(T) \neq 0$ for all nonzero polynomials $f(x)$.

SECTION 9.5   Jordan Canonical Form

Two $m \times n$ matrices A and B are called row-equivalent if A can be carried to B using row operations and, equivalently, if $B = UA$ for some invertible matrix U. We know (Theorem 4 §2.4) that each $m \times n$ matrix is row-equivalent to a unique matrix in reduced row-echelon form, and we say that these reduced row-echelon matrices are *canonical forms* for $m \times n$ matrices using row operations. If we allow column operations as well, then $A \to UAV = \begin{bmatrix} I_r & 0 \\ 0 & 0 \end{bmatrix}$ for invertible U and V, and the canonical forms are the matrices $\begin{bmatrix} I_r & 0 \\ 0 & 0 \end{bmatrix}$ where r is the rank (this is the Smith normal form and is discussed in Section 2.4). In this section, we discover the canonical forms for square matrices under similarity $A \to P^{-1}AP$.

If A is an $n \times n$ matrix with distinct real eigenvalues $\lambda_1, \lambda_2, \ldots, \lambda_k$, we saw in Theorem 1 §9.4 that A is similar to a block triangular matrix; more precisely, an invertible matrix P exists such that

$$P^{-1}AP = \begin{bmatrix} U_1 & 0 & \cdots & 0 \\ 0 & U_2 & \cdots & 0 \\ \vdots & \vdots & \ddots & \vdots \\ 0 & 0 & \cdots & U_k \end{bmatrix} = \mathrm{diag}(U_1, U_2, \ldots, U_k) \qquad (*)$$

where, for each i, U_i is block, upper triangular, with λ_i repeated on the main diagonal. The Jordan canonical form is a refinement of this theorem. The proof we gave of $(*)$ is matrix theoretic because we wanted to give an algorithm for actually finding the matrix P. However, we are going to employ abstract methods here. Consequently, we reformulate Theorem 1 §9.4 as follows:

Theorem 1

Let $T : V \to V$ be a linear operator where $\dim V = n$. Assume that $\lambda_1, \lambda_2, \ldots, \lambda_k$ are the distinct eigenvalues of T, and that the λ_i are all real. Then there exists a basis E of V such that $M_E(T) = \mathrm{diag}(U_1, U_2, \ldots, U_k)$ where, for each i, U_i is square, upper triangular, with λ_i repeated on the main diagonal.

PROOF

Choose any basis $B = \{\mathbf{b}_1, \mathbf{b}_2, \ldots, \mathbf{b}_n\}$ of V and write $A = M_B(T)$. Since A has the same eigenvalues as T, Theorem 1 §9.4 shows that an invertible matrix P exists such that $P^{-1}AP = \text{diag}(U_1, U_2, \ldots, U_k)$ where the U_i are as in the statement of the Theorem. If P_j denotes column j of P and $C_B : V \to \mathbb{R}^n$ is the coordinate isomorphism, let $\mathbf{e}_j = C_B^{-1}(P_j)$ for each j. Then $E = \{\mathbf{e}_1, \mathbf{e}_2, \ldots, \mathbf{e}_n\}$ is a basis of E and $C_B(\mathbf{e}_j) = P_j$ for each j. This means that $P_{B \leftarrow E} = [C_B(\mathbf{e}_j)] = [P_j] = P$, and hence (by Theorem 2 §9.2) that $P_{E \leftarrow B} = P^{-1}$. With this, column j of $M_E(T)$ is

$$C_E(T(\mathbf{e}_j)) = P_{E \leftarrow B}C_B(T(\mathbf{e}_j)) = P^{-1}M_B(T)C_B(\mathbf{e}_j) = P^{-1}AP_j$$

for all j. Hence

$$M_E(T) = [C_E(T(\mathbf{e}_j))] = [P^{-1}AP_j] = P^{-1}A[P_j] = P^{-1}AP = \text{diag}(U_1, U_2, \ldots, U_k),$$

as required.

If λ is a number, matrices of the form

$$J_1(\lambda) = [\lambda], \qquad J_2(\lambda) = \begin{bmatrix} \lambda & 1 \\ 0 & \lambda \end{bmatrix}, \qquad J_3(\lambda) = \begin{bmatrix} \lambda & 1 & 0 \\ 0 & \lambda & 1 \\ 0 & 0 & \lambda \end{bmatrix},$$

$$J_4(\lambda) = \begin{bmatrix} \lambda & 1 & 0 & 0 \\ 0 & \lambda & 1 & 0 \\ 0 & 0 & \lambda & 1 \\ 0 & 0 & 0 & \lambda \end{bmatrix}, \ldots$$

are called **Jordan blocks** corresponding to λ. We are going to show that Theorem 1 holds with each block U_i replaced by Jordan blocks corresponding to eigenvalues. It turns out that the whole thing hinges on the case $\lambda = 0$. An operator T is called **nilpotent** if $T^m = 0$ for some $m \geq 1$, and in this case $\lambda = 0$ for every eigenvalue λ of T. Moreover, the converse holds by Theorem 1 §9.4. Hence the following lemma is crucial.

Lemma 1

Let $T : V \to V$ be a linear operator where $\dim V = n$, and assume that T is nilpotent; that is, $T^m = 0$ for some $m \geq 1$. Then V has a basis B such that

$$M_B(T) = \text{diag}(J_1, J_2, \ldots, J_k)$$

where each J_i is a Jordan block corresponding to $\lambda = 0$.[1]

A proof is given at the end of this section.

Theorem 2   Real Jordan Canonical Form

Let $T : V \to V$ be a linear operator where $\dim V = n$, and assume that $\lambda_1, \lambda_2, \ldots, \lambda_m$ are the distinct eigenvalues of T, and that the λ_i are all real. Then there exists a basis E of V such that

$$M_E(T) = \text{diag}(J_1, J_2, \ldots, J_m)$$

where each J_i is a Jordan block corresponding to some λ_i.

---

1   The converse is true too: If $M_B(T)$ has this form for some basis B of V, then T is nilpotent.

PROOF

Let $E = \{\mathbf{e}_1, \mathbf{e}_2, \ldots, \mathbf{e}_n\}$ be a basis of V as in Theorem 1, and assume that U_i is an $n_i \times n_i$ matrix for each i. Let

$$E_1 = \{\mathbf{e}_1, \ldots, \mathbf{e}_{n_1}\}, \quad E_2 = \{\mathbf{e}_{n_1+1}, \ldots, \mathbf{e}_{n_2}\}, \quad \ldots, \quad E_k = \{\mathbf{e}_{n_{k-1}+1}, \ldots, \mathbf{e}_{n_k}\}$$

where $n_k = n$, and define $V_i = \mathrm{span}\{E_i\}$ for each i. Because $M_E(T) = \mathrm{diag}(U_1, U_2, \ldots, U_m)$ is block diagonal, it follows that each V_i is T-invariant and $M_{E_i}(T) = U_i$ for each i. Let U_i have λ_i repeated along the main diagonal, and consider the restriction $T : V_i \to V_i$. Then $M_{E_i}(T - \lambda_i I_{n_i})$ is a nilpotent matrix, and hence $T - \lambda_i I_{n_i}$ is a nilpotent operator on V_i. Hence Lemma 1 shows that V_i has a basis B_i such that $M_{B_i}(T - \lambda_i I_{n_i}) = \mathrm{diag}(K_1, K_2, \ldots, K_{t_i})$ where each K_i is a Jordan block corresponding to $\lambda = 0$. Hence

$$\begin{aligned} M_{B_i}(T) &= M_{B_i}(\lambda_i I_{n_i}) + M_{B_i}(T - \lambda_i I_{n_i}) \\ &= \lambda_i I_{n_i} + \mathrm{diag}(K_1, K_2, \ldots, K_{t_i}) = \mathrm{diag}(J_1, J_2, \ldots, J_{t_i}) \end{aligned}$$

where $J_i = \lambda_i I_{f_i} + K_i$ is a Jordan block corresponding to λ_i (where K_i is $f_i \times f_i$). Finally, $B = B_1 \cup B_2 \cup \cdots \cup B_k$ is a basis of V with respect to which T has the desired matrix.

Corollary

If A is an $n \times n$ matrix with real eigenvalues, an invertible matrix P exists such that $P^{-1}AP = \mathrm{diag}(J_1, J_2, \ldots, J_k)$ where each J_i is a Jordan block corresponding to an eigenvalue λ_i.

PROOF

Apply Theorem 2 to the matrix transformation $T_A : \mathbb{R}^n \to \mathbb{R}^n$ to find a basis B of $\mathbb{R}^n$ such that $M_B(T_A)$ has the desired form. If P is the (invertible) $n \times n$ matrix with the vectors of B as its columns, then $P^{-1}AP = M_B(T_A)$ by Theorem 4 §9.2.

Of course if we work over the field $\mathbb{C}$ of complex numbers rather than $\mathbb{R}$, the characteristic polynomial of a (complex) matrix A splits completely as a product of linear factors. The proof of Theorem 2 goes through to give

Camille Jordan. Photo
© Corbis.

Theorem 3 Jordan Canonical Form[2]

Let $T : V \to V$ be a linear operator on a complex vector space V where $\dim V = n$, and assume that $\lambda_1, \lambda_2, \ldots, \lambda_m$ are the distinct eigenvalues of T. Then there exists a basis E of V such that

$$M_E(T) = \mathrm{diag}(J_1, J_2, \ldots, J_m)$$

where each J_i is a Jordan block corresponding to some λ_i.

Except for the order of the Jordan blocks J_i, the Jordan canonical form is uniquely determined by the operator T. That is, for each eigenvalue λ the number and size

---

2 This was first proved in 1870 by the French mathematician Camille Jordan (1838–1922) in his monumental *Traité des substitutions et des équations algébriques*.

of the Jordan blocks corresponding to λ is uniquely determined. Thus, for example, two matrices (or two operators) are similar if and only if they have the same Jordan canonical form. We omit the proof of uniqueness; it is best presented using modules in a course on abstract algebra.

Proof of Lemma 1

Lemma 1

Let $T : V \rightarrow V$ be a linear operator where dim $V = n$, and assume that T is nilpotent; that is, $T^m = 0$ for some $m \geq 1$. Then V has a basis B such that

$$M_B(T) = \text{diag}(J_1, J_2, \ldots, J_k)$$

where each $J_i = J_{n_i}(0)$ is a Jordan block corresponding to $\lambda = 0$.

PROOF

The proof proceeds by induction on n. If $n = 1$, then T is a scalar operator, and so $T = 0$ and the lemma holds. If $n \geq 1$, we may assume that $T \neq 0$, so $m \geq 1$ and we may assume that m is chosen such that $T^m = 0$, but $T^{m-1} \neq 0$, say $T^{m-1}\mathbf{u} \neq \mathbf{0}$ for some $\mathbf{u}$ in V. [3]

Claim. $\{\mathbf{u}, T\mathbf{u}, T^2\mathbf{u}, \ldots, T^{m-1}\mathbf{u}\}$ is independent.
Proof. Suppose $a_0\mathbf{u} + a_1 T\mathbf{u} + a_2 T^2\mathbf{u} + \cdots + a_{m-1}T^{m-1}\mathbf{u} = \mathbf{0}$ where each a_i is in $\mathbb{R}$. Since $T^m = 0$, applying T^{m-1} gives $\mathbf{0} = T^{m-1}\mathbf{0} = a_0 T^{m-1}\mathbf{u}$, whence $a_0 = 0$. Hence $a_1 T\mathbf{u} + a_2 T^2\mathbf{u} + \cdots + a_{m-1}T^{m-1}\mathbf{u} = \mathbf{0}$ and applying T^{m-2} gives $a_1 = 0$ in the same way. Continue in this fashion to obtain $a_i = 0$ for each i. This proves the Claim.

Now define $P = \text{span}\{\mathbf{u}, T\mathbf{u}, T^2\mathbf{u}, \ldots, T^{m-1}\mathbf{u}\}$. Then P is a T-invariant subspace (because $T^m = 0$), and $T : P \rightarrow P$ is nilpotent with matrix $M_B(T) = J_m(0)$ where $B = \{\mathbf{u}, T\mathbf{u}, T^2\mathbf{u}, \ldots, T^{m-1}\mathbf{u}\}$. Hence we are done, by induction, if $V = P \oplus Q$ where Q is T-invariant (then dim $Q = n - \text{dim } P < n$ because $P \neq 0$, and $T : Q \rightarrow Q$ is nilpotent). With this in mind, choose a T-invariant subspace Q of maximal dimension such that $P \cap Q = \{\mathbf{0}\}$. [4] We assume that $V \neq P \oplus Q$ and look for a contradiction.

Choose $\mathbf{x} \in V$ such that $\mathbf{x} \notin P \oplus Q$. Then $T^m\mathbf{x} = \mathbf{0} \in P \oplus Q$ while $T^0\mathbf{x} = \mathbf{x} \notin P \oplus Q$. Hence there exists k, $1 \leq k \leq m$, such that $T^k\mathbf{x} \in P \oplus Q$ but $T^{k-1}\mathbf{x} \notin P \oplus Q$. Write $\mathbf{v} = T^{k-1}\mathbf{x}$, so that

$$\mathbf{v} \notin P \oplus Q \quad \text{and} \quad T\mathbf{v} \in P \oplus Q$$

Let $T\mathbf{v} = \mathbf{p} + \mathbf{q}$ with $\mathbf{p}$ in P and $\mathbf{q}$ in Q. Then $\mathbf{0} = T^{m-1}(T\mathbf{v}) = T^{m-1}\mathbf{p} + T^{m-1}\mathbf{q}$ so, since P and Q are T-invariant, $T^{m-1}\mathbf{p} = -T^{m-1}\mathbf{q} \in P \cap Q = \{\mathbf{0}\}$. Hence

$$T^{m-1}\mathbf{p} = \mathbf{0}.$$

Since $\mathbf{p} \in P$ we have $\mathbf{p} = a_0\mathbf{u} + a_1 T\mathbf{u} + a_2 T^2\mathbf{u} + \cdots + a_{m-1}T^{m-1}\mathbf{u}$ for $a_i \in \mathbb{R}$. Since $T^m = 0$, applying T^{m-1} gives $\mathbf{0} = T^{m-1}\mathbf{p} = a_0 T^{m-1}\mathbf{u}$, whence $a_0 = 0$. Thus $\mathbf{p} = T(\mathbf{p}_1)$ where $\mathbf{p}_1 = a_1\mathbf{u} + a_2 T\mathbf{u} + \cdots + a_{m-1}T^{m-2}\mathbf{u} \in P$. If we write $\mathbf{v}_1 = \mathbf{v} - \mathbf{p}_1$ we have

---

3 If $S : V \rightarrow V$ is an operator, we abbreviate $S(\mathbf{u})$ by $S\mathbf{u}$ for simplicity.
4 Observe that there *is* at least one such subspace: $Q = \{\mathbf{0}\}$.

$$T(\mathbf{v}_1) = T(\mathbf{v} - \mathbf{p}_1) = T\mathbf{v} - \mathbf{p} = \mathbf{q} \in Q.$$

Since $T(Q) \subseteq Q$, it follows that $T(Q + \mathbb{R}\mathbf{v}_1) \subseteq Q \subseteq Q + \mathbb{R}\mathbf{v}_1$. Moreover $\mathbf{v}_1 \notin Q$ (otherwise $\mathbf{v} = \mathbf{v}_1 + \mathbf{p}_1 \in P \oplus Q$, a contradiction). Hence $Q \subset Q + \mathbb{R}\mathbf{v}_1$ so, by the maximality of Q, we must have $(Q + \mathbb{R}\mathbf{v}_1) \cap P \neq \{\mathbf{0}\}$, say

$$\mathbf{0} \neq \mathbf{p}_2 = \mathbf{q}_1 + a\mathbf{v}_1 \quad \text{where} \quad \mathbf{p}_2 \in P, \quad \mathbf{q}_1 \in Q, \quad \text{and} \quad a \in \mathbb{R}.$$

Thus $a\mathbf{v}_1 = \mathbf{p}_2 - \mathbf{q}_1 \in P \oplus Q$. But since $\mathbf{v}_1 = \mathbf{v} - \mathbf{p}_1$ we have

$$a\mathbf{v} = a\mathbf{v}_1 + a\mathbf{p}_1 \in (P \oplus Q) + P = P \oplus Q.$$

Since $\mathbf{v} \notin P \oplus Q$, this implies that $a = 0$. But then $\mathbf{p}_2 = \mathbf{q}_1 \in P \cap Q = \{\mathbf{0}\}$ —a contradiction. This completes the proof.

Exercises 9.5

1. By direct computation, show that there is no invertible complex matrix C such that

$$C^{-1}\begin{bmatrix} 1 & 1 & 0 \\ 0 & 1 & 1 \\ 0 & 0 & 1 \end{bmatrix} C = \begin{bmatrix} 1 & 1 & 0 \\ 0 & 1 & 0 \\ 0 & 0 & 1 \end{bmatrix}.$$

♦2. Show that $\begin{bmatrix} a & 1 & 0 \\ 0 & a & 0 \\ 0 & 0 & b \end{bmatrix}$ is similar to $\begin{bmatrix} b & 0 & 0 \\ 0 & a & 1 \\ 0 & 0 & a \end{bmatrix}$.

3. (a) Show that every complex matrix is similar to its transpose.

   (b) Show every real matrix is similar to its transpose.

   [*Hint:* Show that $J_k(0)Q = Q[J_k(0)]^T$ where Q is the $k \times k$ matrix with 1s down the "counter diagonal", that is from the $(1, k)$-position to the $(k, 1)$-position.]

10 Inner Product Spaces

SECTION 10.1   Inner Products and Norms

The dot product was introduced in $\mathbb{R}^n$ to provide a natural generalization of the geometrical notions of length and orthogonality that were so important in Chapter 4. The plan in this chapter is to define an *inner product* on an arbitrary vector space V (of which the dot product is an example in $\mathbb{R}^n$) and use it to introduce these concepts in V.

An **inner product** on a real vector space V is a function that assigns a number $\langle \mathbf{v}, \mathbf{w} \rangle$ to every pair $\mathbf{v}$, $\mathbf{w}$ of vectors in V in such a way that the following axioms are satisfied.

P1. $\langle \mathbf{v}, \mathbf{w} \rangle$ is a real number for all $\mathbf{v}$ and $\mathbf{w}$ in V.

P2. $\langle \mathbf{v}, \mathbf{w} \rangle = \langle \mathbf{w}, \mathbf{v} \rangle$ for all $\mathbf{v}$ and $\mathbf{w}$ in V.

P3. $\langle \mathbf{v} + \mathbf{w}, \mathbf{u} \rangle = \langle \mathbf{v}, \mathbf{u} \rangle + \langle \mathbf{w}, \mathbf{u} \rangle$ for all $\mathbf{u}$, $\mathbf{v}$, and $\mathbf{w}$ in V.

P4. $\langle r\mathbf{v}, \mathbf{w} \rangle = r\langle \mathbf{v}, \mathbf{w} \rangle$ for all $\mathbf{v}$ and $\mathbf{w}$ in V and all r in $\mathbb{R}$.

P5. $\langle \mathbf{v}, \mathbf{v} \rangle > 0$ for all $\mathbf{v} \neq \mathbf{0}$ in V.

A vector space V with an inner product $\langle \ , \ \rangle$ will be called an **inner product space**.

Example 1

$\mathbb{R}^n$ is an inner product space with the dot product as inner product: $\langle X, Y \rangle = X \cdot Y$. This is also called the **euclidean** inner product, and $\mathbb{R}^n$ equipped with the dot product is called **euclidean n-space**.

Example 2

If A and B are $m \times n$ matrices, define $\langle A, B \rangle = \text{tr}(AB^T)$ where $\text{tr}(X)$ is the trace of the square matrix X. Show that $\langle \ , \ \rangle$ is an inner product in $\mathbf{M}_{mn}$.

Solution   We must verify that axioms P1–P5 are valid; checking P1–P4 is left as Exercise 19. If $R_1, R_2, \dots, R_n$ denote the rows of A, the (i, j)-entry of AA^T is $R_i \cdot R_j$, so

$$\langle A, A \rangle = \text{tr}(AA^T) = R_1 \cdot R_1 + R_2 \cdot R_2 + \cdots + R_n \cdot R_n$$

But $R_j \cdot R_j$ is the sum of the squares of the entries of R_j, so this shows that $\langle A, A \rangle$ is the sum of the squares of all nm entries of A. Axiom P5 follows.

The next example is important in analysis.

Example 3[1]

Let $\mathbf{C}[a, b]$ denote the vector space of **continuous functions** from $[a, b]$ to $\mathbb{R}$, a subspace of $\mathbf{F}[a, b]$. Show that

$$\langle f, g \rangle = \int_a^b f(x)g(x)\,dx$$

defines an inner product on $\mathbf{C}[a, b]$.

Solution

Axioms P1 and P2 are clear. As to axiom P4,

$$\langle rf, g \rangle = \int_a^b rf(x)g(x)\,dx = r\int_a^b f(x)g(x)\,dx = r\langle f, g \rangle$$

Axiom P3 is similar. Finally, theorems of calculus show that $\langle f, f \rangle = \int_a^b f(x)^2\,dx \geq 0$ and, if f is continuous, that this is zero if and only if $f(x)$ is the zero function. This gives axiom P5.

If $\mathbf{v}$ is any vector, then, using axiom P3, we get

$$\langle \mathbf{0}, \mathbf{v} \rangle = \langle \mathbf{0} + \mathbf{0}, \mathbf{v} \rangle = \langle \mathbf{0}, \mathbf{v} \rangle + \langle \mathbf{0}, \mathbf{v} \rangle$$

and it follows that the number $\langle \mathbf{0}, \mathbf{v} \rangle$ must be zero. This observation is recorded for reference in the following theorem, along with several other properties of inner products. The other proofs are left as Exercise 20.

Theorem 1

Let $\langle\ ,\ \rangle$ be an inner product on a space V; let $\mathbf{v}$, $\mathbf{u}$, and $\mathbf{w}$ denote vectors in V; and let r denote a real number.
1. $\langle \mathbf{u}, \mathbf{v} + \mathbf{w} \rangle = \langle \mathbf{u}, \mathbf{v} \rangle + \langle \mathbf{u}, \mathbf{w} \rangle$.
2. $\langle \mathbf{v}, r\mathbf{w} \rangle = r\langle \mathbf{v}, \mathbf{w} \rangle = \langle r\mathbf{v}, \mathbf{w} \rangle$.
3. $\langle \mathbf{v}, \mathbf{0} \rangle = 0 = \langle \mathbf{0}, \mathbf{v} \rangle$.
4. $\langle \mathbf{v}, \mathbf{v} \rangle = 0$ if and only if $\mathbf{v} = \mathbf{0}$.

If $\langle\ ,\ \rangle$ is an inner product on a space V, then, given $\mathbf{u}$, $\mathbf{v}$, and $\mathbf{w}$ in V,

$$\langle r\mathbf{u} + s\mathbf{v}, \mathbf{w} \rangle = \langle r\mathbf{u}, \mathbf{w} \rangle + \langle s\mathbf{v}, \mathbf{w} \rangle = r\langle \mathbf{u}, \mathbf{w} \rangle + s\langle \mathbf{v}, \mathbf{w} \rangle$$

for all r and s in $\mathbb{R}$ by axioms P3 and P4. Moreover, there is nothing special about the fact that there are two terms in the linear combination or that it is in the first component:

$$\langle r_1\mathbf{v}_1 + r_2\mathbf{v}_2 + \cdots + r_n\mathbf{v}_n, \mathbf{w} \rangle = r_1\langle \mathbf{v}_1, \mathbf{w} \rangle + r_2\langle \mathbf{v}_2, \mathbf{w} \rangle + \cdots + r_n\langle \mathbf{v}_n, \mathbf{w} \rangle$$

and

$$\langle \mathbf{v}, s_1\mathbf{w}_1 + s_2\mathbf{w}_2 + \cdots + s_m\mathbf{w}_m \rangle = s_1\langle \mathbf{v}, \mathbf{w}_1 \rangle + s_2\langle \mathbf{v}, \mathbf{w}_2 \rangle + \cdots + s_m\langle \mathbf{v}, \mathbf{w}_m \rangle$$

hold for all r_i and s_i in $\mathbb{R}$ and all $\mathbf{v}$, $\mathbf{w}$, $\mathbf{v}_i$, and $\mathbf{w}_j$ in V. These results are described by saying that inner products "preserve" linear combinations. For example,

$$\begin{aligned}
\langle 2\mathbf{u} - \mathbf{v}, 3\mathbf{u} + 2\mathbf{v} \rangle &= \langle 2\mathbf{u}, 3\mathbf{u} \rangle + \langle 2\mathbf{u}, 2\mathbf{v} \rangle + \langle -\mathbf{v}, 3\mathbf{u} \rangle + \langle -\mathbf{v}, 2\mathbf{v} \rangle \\
&= 6\langle \mathbf{u}, \mathbf{u} \rangle + 4\langle \mathbf{u}, \mathbf{v} \rangle - 3\langle \mathbf{v}, \mathbf{u} \rangle - 2\langle \mathbf{v}, \mathbf{v} \rangle \\
&= 6\langle \mathbf{u}, \mathbf{u} \rangle + \langle \mathbf{u}, \mathbf{v} \rangle - 2\langle \mathbf{v}, \mathbf{v} \rangle
\end{aligned}$$

[1] This example (and others later that refer to it) can be omitted with no loss of continuity by students with no calculus background.

If A is a symmetric $n \times n$ matrix and X and Y are columns in $\mathbb{R}^n$, we regard the 1×1 matrix $X^T A Y$ as a number. If we write

$$\langle X, Y \rangle = X^T A Y \quad \text{for all columns } X, Y \text{ in } \mathbb{R}^n$$

then axioms P1–P4 follow from matrix arithmetic (only P2 requires that A is symmetric). Axiom P5 reads

$$X^T A X > 0 \quad \text{for all columns } X \neq 0 \text{ in } \mathbb{R}^n$$

and this condition characterizes the positive definite matrices (Theorem 2 §8.3). This proves the first assertion in the next theorem.

Theorem 2

If A is any $n \times n$ positive definite matrix, then

$$\langle X, Y \rangle = X^T A Y \quad \text{for all columns } X, Y \text{ in } \mathbb{R}^n$$

defines an inner product on $\mathbb{R}^n$, and every inner product arises in this way.

PROOF

Given an inner product $\langle \, , \, \rangle$ on $\mathbb{R}^n$, let $\{E_1, E_2, \dots, E_n\}$ be the standard basis of $\mathbb{R}^n$. If $X = \sum_{i=1}^{n} x_i E_i$ and $Y = \sum_{j=1}^{n} y_j E_j$ are two vectors in $\mathbb{R}^n$, compute $\langle X, Y \rangle$ by adding the inner product of each term $x_i E_i$ to each term $y_j E_j$. The result is a double sum.

$$\langle X, Y \rangle = \sum_{i=1}^{n} \sum_{j=1}^{n} \langle x_i E_i, y_j E_j \rangle = \sum_{i=1}^{n} \sum_{j=1}^{n} x_i \langle E_i, E_j \rangle y_j$$

As the reader can verify, let this is a matrix product:

$$\langle X, Y \rangle = [x_1 \;\; x_2 \;\; \cdots \;\; x_n] \begin{bmatrix} \langle E_1, E_1 \rangle & \langle E_1, E_2 \rangle & \cdots & \langle E_1, E_n \rangle \\ \langle E_2, E_1 \rangle & \langle E_2, E_2 \rangle & \cdots & \langle E_2, E_n \rangle \\ \vdots & \vdots & \cdots & \vdots \\ \langle E_n, E_1 \rangle & \langle E_n, E_2 \rangle & \cdots & \langle E_n, E_n \rangle \end{bmatrix} \begin{bmatrix} y_1 \\ y_2 \\ \vdots \\ y_n \end{bmatrix}$$

Hence $\langle X, Y \rangle = X^T A Y$, where A is the $n \times n$ matrix whose (i, j)-entry is $\langle E_i, E_j \rangle$. The fact that $\langle E_i, E_j \rangle = \langle E_j, E_i \rangle$ shows that A is symmetric. Finally, A is positive definite by Theorem 2 §8.3.

Thus, just as every linear operator $\mathbb{R}^n \to \mathbb{R}^n$ corresponds to an $n \times n$ matrix, every inner product on $\mathbb{R}^n$ corresponds to a positive definite $n \times n$ matrix. In particular, the dot product corresponds to the identity matrix I_n.

Remark   If we refer to the inner product space $\mathbb{R}^n$ without specifying the inner product, we mean that the dot product is to be used.

Example 4

Let the inner product $\langle \, , \, \rangle$ be defined on $\mathbb{R}^2$ by

$$\left\langle \begin{bmatrix} v_1 \\ v_2 \end{bmatrix}, \begin{bmatrix} w_1 \\ w_2 \end{bmatrix} \right\rangle = 2v_1 w_1 - v_1 w_2 - v_2 w_1 + v_2 w_2$$

Find a symmetric 2×2 matrix A such that $\langle X, Y \rangle = X^T A Y$ for all X, Y in $\mathbb{R}^2$.

Solution   The (i, j)-entry of the matrix A is the coefficient of $v_i w_j$ in the expression, so $A = \begin{bmatrix} 2 & -1 \\ -1 & 1 \end{bmatrix}$. Incidentally, if $X = \begin{bmatrix} x \\ y \end{bmatrix}$, then

$$\langle X, X \rangle = 2x^2 - 2xy + y^2 = x^2 + (x - y)^2 \geq 0$$

for all X, and $\langle X, X \rangle = 0$ implies $X = 0$. Hence $\langle \, , \, \rangle$ is indeed an inner product, so A is positive definite.

Let $\langle \, , \, \rangle$ be an inner product on $\mathbb{R}^n$ given as in Theorem 2 by a positive definite matrix A. If $X = [x_1 \ x_2 \ \cdots \ x_n]^T$, then $\langle X, X \rangle = X^T A X$ is an expression in the variables $x_1, x_2, \ldots, x_n$ called a **quadratic form**. These are studied in detail in Section 8.9.

Norms and Distance

As in $\mathbb{R}^n$, if $\langle \, , \, \rangle$ is an inner product on a space V, the **norm**[2] $\|\mathbf{v}\|$ of a vector $\mathbf{v}$ in V is defined by

$$\|\mathbf{v}\| = \sqrt{\langle \mathbf{v}, \mathbf{v} \rangle}$$

Note that axiom P5 guarantees that $\langle \mathbf{v}, \mathbf{v} \rangle \geq 0$, so $\|\mathbf{v}\|$ is a real number. We define the **distance** between vectors $\mathbf{v}$ and $\mathbf{w}$ in an inner product space V to be

$$d(\mathbf{v}, \mathbf{w}) = \|\mathbf{v} - \mathbf{w}\|$$

Example 5

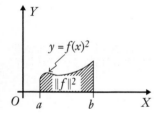

The norm of a continuous function $f = f(x)$ in $\mathbf{C}[a, b]$ (with the inner product from Example 3) is given by

$$\|f\| = \sqrt{\int_a^b f(x)^2 \, dx}$$

Hence $\|f\|^2$ is the area beneath the graph of $y = f(x)^2$ between $x = a$ and $x = b$ (see the diagram).

Example 6

Show that $\langle \mathbf{u} + \mathbf{v}, \mathbf{u} - \mathbf{v} \rangle = \|\mathbf{u}\|^2 - \|\mathbf{v}\|^2$ in any inner product space.

Solution
$$\langle \mathbf{u} + \mathbf{v}, \mathbf{u} - \mathbf{v} \rangle = \langle \mathbf{u}, \mathbf{u} \rangle - \langle \mathbf{u}, \mathbf{v} \rangle + \langle \mathbf{v}, \mathbf{u} \rangle - \langle \mathbf{v}, \mathbf{v} \rangle$$
$$= \|\mathbf{u}\|^2 - \langle \mathbf{u}, \mathbf{v} \rangle + \langle \mathbf{u}, \mathbf{v} \rangle - \|\mathbf{v}\|^2$$
$$= \|\mathbf{u}\|^2 - \|\mathbf{v}\|^2$$

A vector $\mathbf{v}$ in an inner product space V is called a **unit vector** if $\|\mathbf{v}\| = 1$. The set of all unit vectors in V is called the **unit ball** in V. For example, if $V = \mathbb{R}^2$ (with the dot product) and $\mathbf{v} = (x, y)$, then

$$\|\mathbf{v}\| = 1 \quad \text{if and only if} \quad x^2 + y^2 = 1$$

Hence the unit ball in $\mathbb{R}^2$ is the **unit circle** $x^2 + y^2 = 1$ with centre at the origin and radius 1. However, the shape of the unit ball varies with the choice of inner product.

---

2 If the dot product is used in $\mathbb{R}^n$, the norm $\|X\|$ of a vector X is usually called the **length** of X.

Example 7

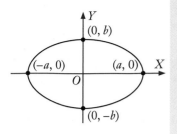

If $a > 0$ and $b > 0$, define an inner product on $\mathbb{R}^2$ by

$$\langle \mathbf{v}, \mathbf{w} \rangle = \frac{xx_1}{a^2} + \frac{yy_1}{b^2}$$

where $\mathbf{v} = (x, y)$ and $\mathbf{w} = (x_1, y_1)$. The reader can verify (Exercise 5) that this is indeed an inner product. In this case

$$\|\mathbf{v}\| = 1 \quad \text{if and only if} \quad \frac{x^2}{a^2} + \frac{y^2}{b^2} = 1$$

so the unit ball is the ellipse shown in the diagram.

Example 7 graphically illustrates the fact that norms and distances in an inner product space V vary with the choice of inner product in V.

The proof of the next result is left to the reader.

Theorem 3

If $\mathbf{v} \neq \mathbf{0}$ is any vector in an inner product space V, then $\dfrac{1}{\|\mathbf{v}\|}\mathbf{v}$ is the unique unit vector that is a positive multiple of $\mathbf{v}$.

The next theorem reveals an important and useful fact about the relationship between norms and inner products. The proof of the Cauchy inequality for $\mathbb{R}^n$ (Theorem 2 §5.3) goes through in the general context.

Theorem 4 Schwarz Inequality[3]

If $\mathbf{v}$ and $\mathbf{w}$ are two vectors in an inner product space V, then

$$\langle \mathbf{v}, \mathbf{w} \rangle^2 \leq \|\mathbf{v}\|^2 \|\mathbf{w}\|^2$$

Moreover, equality occurs if and only if one of $\mathbf{v}$ and $\mathbf{w}$ is a scalar multiple of the other.

Example 8

If f and g are continuous functions on the interval $[a, b]$, then (see Example 3)

$$\left\{ \int_a^b f(x)g(x)\, dx \right\}^2 \leq \int_a^b f(x)^2\, dx \int_a^b g(x)^2\, dx$$

Another famous inequality, the so-called *triangle inequality*, also comes from the Schwarz inequality. It is included in the following list of basic properties of the norm of a vector.

---

3 Herman Amandus Schwarz (1843–1921) was a German mathematician at the University of Berlin. He had strong geometric intuition, which he applied with great ingenuity to particular problems. A version of the inequality appeared in 1885.

Theorem 5

If V is an inner product space, the norm $\|\cdot\|$ has the following properties.
1. $\|\mathbf{v}\| \geq 0$ for every vector $\mathbf{v}$ in V.
2. $\|\mathbf{v}\| = 0$ if and only if $\mathbf{v} = \mathbf{0}$.
3. $\|r\mathbf{v}\| = |r|\|\mathbf{v}\|$ for every $\mathbf{v}$ in V and every r in $\mathbb{R}$.
4. $\|\mathbf{v} + \mathbf{w}\| \leq \|\mathbf{v}\| + \|\mathbf{w}\|$ for all $\mathbf{v}$ and $\mathbf{w}$ in V (**triangle inequality**).

PROOF

Because $\|\mathbf{v}\| = \sqrt{\langle \mathbf{v}, \mathbf{v} \rangle}$, properties (1) and (2) follow immediately from (3) and (4) of Theorem 1. As to (3), compute

$$\|r\mathbf{v}\|^2 = \langle r\mathbf{v}, r\mathbf{v} \rangle = r^2 \langle \mathbf{v}, \mathbf{v} \rangle = r^2 \|\mathbf{v}\|^2$$

Hence (3) follows by taking positive square roots. Finally, the fact that $\langle \mathbf{v}, \mathbf{w} \rangle \leq \|\mathbf{v}\|\|\mathbf{w}\|$ by the Schwarz inequality gives

$$\begin{aligned}
\|\mathbf{v} + \mathbf{w}\|^2 = \langle \mathbf{v} + \mathbf{w}, \mathbf{v} + \mathbf{w} \rangle &= \|\mathbf{v}\|^2 + 2\langle \mathbf{v}, \mathbf{w} \rangle + \|\mathbf{w}\|^2 \\
&\leq \|\mathbf{v}\|^2 + 2\|\mathbf{v}\|\|\mathbf{w}\| + \|\mathbf{w}\|^2 \\
&= (\|\mathbf{v}\| + \|\mathbf{w}\|)^2
\end{aligned}$$

Hence (4) follows by taking positive square roots.

It is worth noting that the usual triangle inequality for absolute values, $|r + s| \leq |r| + |s|$ for all real numbers r and s, is a special case of (4) where $V = \mathbb{R} = \mathbb{R}^1$ and the dot product $\langle r, s \rangle = rs$ is used.

In many calculations in an inner product space, it is required to show that some vector $\mathbf{v}$ is zero. This is often accomplished most easily by showing that its norm $\|\mathbf{v}\|$ is zero. Here is an example.

Example 9

Let $\{\mathbf{v}_1, \ldots, \mathbf{v}_n\}$ be a spanning set for an inner product space V. If $\mathbf{v}$ in V satisfies $\langle \mathbf{v}, \mathbf{v}_i \rangle = 0$ for each $i = 1, 2, \ldots, n$, show that $\mathbf{v} = \mathbf{0}$.

Solution   Write $\mathbf{v} = r_1\mathbf{v}_1 + \cdots + r_n\mathbf{v}_n$, r_i in $\mathbb{R}$. To show that $\mathbf{v} = \mathbf{0}$, we show that $\|\mathbf{v}\|^2 = \langle \mathbf{v}, \mathbf{v} \rangle = 0$. Compute:

$$\langle \mathbf{v}, \mathbf{v} \rangle = \langle \mathbf{v}, r_1\mathbf{v}_1 + \cdots + r_n\mathbf{v}_n \rangle = r_1 \langle \mathbf{v}, \mathbf{v}_1 \rangle + \cdots + r_n \langle \mathbf{v}, \mathbf{v}_n \rangle = 0$$

by hypothesis, and the result follows.

The norm properties in Theorem 5 translate to the following properties of distance familiar from geometry. The proof is Exercise 21.

Theorem 6

Let V be an inner product space.
1. $d(\mathbf{v}, \mathbf{w}) \geq 0$ for all $\mathbf{v}, \mathbf{w}$ in V.
2. $d(\mathbf{v}, \mathbf{w}) = 0$ if and only if $\mathbf{v} = \mathbf{w}$.
3. $d(\mathbf{v}, \mathbf{w}) = d(\mathbf{w}, \mathbf{v})$ for all $\mathbf{v}$ and $\mathbf{w}$ in V.
4. $d(\mathbf{v}, \mathbf{w}) \leq d(\mathbf{v}, \mathbf{u}) + d(\mathbf{u}, \mathbf{w})$ for all $\mathbf{v}, \mathbf{u}$, and $\mathbf{w}$ in V.

Exercises 10.1

1. In each case, determine which of axioms P1–P5 fail to hold.

   (a) $V = \mathbb{R}^2$, $\langle (x_1, y_1), (x_2, y_2) \rangle = x_1 y_1 x_2 y_2$

   ◆(b) $V = \mathbb{R}^3$, $\langle (x_1, x_2, x_3), (y_1, y_2, y_3) \rangle$
   $$= x_1 y_1 - x_2 y_2 + x_3 y_3$$

   (c) $V = \mathbb{R}$, $\langle z, w \rangle = z\overline{w}$, where $\overline{w}$ is complex conjugation

   ◆(d) $V = \mathbf{P}_3$, $\langle p(x), q(x) \rangle = p(1)q(1)$

   (e) $V = \mathbf{M}_{22}$, $\langle A, B \rangle = \det(AB)$

   ◆(f) $V = \mathbf{F}[0, 1]$, $\langle f, g \rangle = f(1)g(0) + f(0)g(1)$

2. Verify that the dot product on $\mathbb{R}^n$ satisfies axioms P1–P5.

3. In each case, find a scalar multiple of $\mathbf{v}$ that is a unit vector.

   (a) $\mathbf{v} = f$ in $\mathbf{C}[0, 1]$ where $f(x) = x^2$

   ◆(b) $\mathbf{v} = f$ in $\mathbf{C}[-\pi, \pi]$ where $f(x) = \cos x$

   (c) $\mathbf{v} = \begin{bmatrix} 1 \\ 3 \end{bmatrix}$ in $\mathbb{R}^2$ $\left(\langle \mathbf{v}, \mathbf{w} \rangle = \mathbf{v}^T \begin{bmatrix} 1 & 1 \\ 1 & 2 \end{bmatrix} \mathbf{w} \right)$

   ◆(d) $\mathbf{v} = \begin{bmatrix} 3 \\ -1 \end{bmatrix}$ in $\mathbb{R}^2$ $\left(\langle \mathbf{v}, \mathbf{w} \rangle = \mathbf{v}^T \begin{bmatrix} 1 & -1 \\ -1 & 2 \end{bmatrix} \mathbf{w} \right)$

4. In each case, find the distance between $\mathbf{u}$ and $\mathbf{v}$.

   (a) $\mathbf{u} = (3, -1, 2, 0)$, $\mathbf{v} = (1, 1, 1, 3)$

   ◆(b) $\mathbf{u} = (1, 2, -1, 2)$, $\mathbf{v} = (2, 1, -1, 3)$

   (c) $\mathbf{u} = f$, $\mathbf{v} = g$ in $\mathbf{C}[0, 1]$ where $f(x) = x^2$ and $g(x) = 1 - x$

   ◆(d) $\mathbf{u} = f$, $\mathbf{v} = g$ in $\mathbf{C}[-\pi, \pi]$ where $f(x) = 1$ and $g(x) = \cos x$

5. Let $a_1, a_2, \ldots, a_n$ be positive numbers. Given $\mathbf{v} = (v_1, v_2, \ldots, v_n)$ and $\mathbf{w} = (w_1, w_2, \ldots, w_n)$, define $\langle \mathbf{v}, \mathbf{w} \rangle = a_1 v_1 w_1 + \cdots + a_n v_n w_n$. Show that this is an inner product on $\mathbb{R}^n$.

6. If $\{\mathbf{b}_1, \ldots, \mathbf{b}_n\}$ is a basis of V and if $\mathbf{v} = v_1 \mathbf{b}_1 + \cdots + v_n \mathbf{b}_n$ and $\mathbf{w} = w_1 \mathbf{b}_1 + \cdots + w_n \mathbf{b}_n$ are vectors in V, define $\langle \mathbf{v}, \mathbf{w} \rangle = v_1 w_1 + \cdots + v_n w_n$. Show that this is an inner product on V.

7. If $p = p(x)$ and $q = q(x)$ are polynomials in $\mathbf{P}_n$, define
   $$\langle p, q \rangle = p(0)q(0) + p(1)q(1) + \cdots + p(n)q(n)$$
   Show that this is an inner product on $\mathbf{P}_n$. [*Hint for P5:* Theorem 4 §6.5.]

◆8. Let $\mathbf{D}_n$ denote the space of all functions from the set $\{1, 2, 3, \ldots, n\}$ to $\mathbb{R}$ with pointwise addition and scalar multiplication (see Example 6 §6.1). Show that $\langle \, , \, \rangle$ is an inner product on $\mathbf{D}_n$ if $\langle f, g \rangle = f(1)g(1) + f(2)g(2) + \cdots + f(n)g(n)$.

9. Let $\text{re}(z)$ denote the real part of the complex number z. Show that $\langle \, , \, \rangle$ is an inner product on $\mathbb{C}$ if $\langle z, w \rangle = \text{re}(z\overline{w})$.

10. If $T : V \to V$ is an isomorphism of the inner product space V, show that
    $\langle \mathbf{v}, \mathbf{w} \rangle_1 = \langle T(\mathbf{v}), T(\mathbf{w}) \rangle$ defines a new inner product $\langle \, , \, \rangle_1$ on V.

11. Show that every inner product $\langle \, , \, \rangle$ on $\mathbb{R}^n$ has the form $\langle X, Y \rangle = (UX) \bullet (UY)$ for some upper triangular matrix U with positive diagonal entries. [*Hint:* Theorem 3 §8.3.]

12. In each case, show that $\langle \mathbf{v}, \mathbf{w} \rangle = \mathbf{v}^T A \mathbf{w}$ defines an inner product on $\mathbb{R}^2$ and hence show that A is positive definite.

    (a) $A = \begin{bmatrix} 2 & 1 \\ 1 & 1 \end{bmatrix}$ 　 ◆(b) $A = \begin{bmatrix} 5 & -3 \\ -3 & 2 \end{bmatrix}$

    (c) $A = \begin{bmatrix} 3 & 2 \\ 2 & 3 \end{bmatrix}$ 　 ◆(d) $A = \begin{bmatrix} 3 & 4 \\ 4 & 6 \end{bmatrix}$

13. In each case, find a symmetric matrix A such that $\langle \mathbf{v}, \mathbf{w} \rangle = \mathbf{v}^T A \mathbf{w}$.

    (a) $\left\langle \begin{bmatrix} v_1 \\ v_2 \end{bmatrix}, \begin{bmatrix} w_1 \\ w_2 \end{bmatrix} \right\rangle = v_1 w_1 + 2v_1 w_2 + 2v_2 w_1 + 5v_2 w_2$

    ◆(b) $\left\langle \begin{bmatrix} v_1 \\ v_2 \end{bmatrix}, \begin{bmatrix} w_1 \\ w_2 \end{bmatrix} \right\rangle = v_1 w_1 - v_1 w_2 - v_2 w_1 + 2v_2 w_2$

    (c) $\left\langle \begin{bmatrix} v_1 \\ v_2 \\ v_3 \end{bmatrix}, \begin{bmatrix} w_1 \\ w_2 \\ w_3 \end{bmatrix} \right\rangle = 2v_1 w_1 + v_2 w_2 + v_3 w_3 - v_1 w_2$
    $$- v_2 w_1 + v_2 w_3 + v_3 w_2$$

    ◆(d) $\left\langle \begin{bmatrix} v_1 \\ v_2 \\ v_3 \end{bmatrix}, \begin{bmatrix} w_1 \\ w_2 \\ w_3 \end{bmatrix} \right\rangle = v_1 w_1 + 2v_2 w_2 + 5v_3 w_3$
    $$- 2v_1 w_3 - 2v_3 w_1$$

14. If A is symmetric and $X^T A X = 0$ for all columns X in $\mathbb{R}^n$, show that $A = 0$. [*Hint:* Consider $\langle X + Y, X + Y \rangle$ where $\langle X, Y \rangle = X^T A Y$.]

15. Show that the sum of two inner products on V is again an inner product.

16. Let $\|\mathbf{u}\| = 1$, $\|\mathbf{v}\| = 2$, $\|\mathbf{w}\| = \sqrt{3}$, $\langle \mathbf{u}, \mathbf{v} \rangle = -1$, $\langle \mathbf{u}, \mathbf{w} \rangle = 0$ and $\langle \mathbf{v}, \mathbf{w} \rangle = 3$. Compute:

   (a) $\langle \mathbf{v} + \mathbf{w}, \ 2\mathbf{u} - \mathbf{v} \rangle$

   ◆(b) $\langle \mathbf{u} - 2\mathbf{v} - \mathbf{w}, \ 3\mathbf{w} - \mathbf{v} \rangle$

17. Given the data in Exercise 16, show that $\mathbf{u} + \mathbf{v} = \mathbf{w}$.

18. Show that no vectors exist such that $\|\mathbf{u}\| = 1$, $\|\mathbf{v}\| = 2$, and $\langle \mathbf{u}, \mathbf{v} \rangle = -3$.

19. Complete Example 2.

◆20. Prove Theorem 1.

21. Prove Theorem 6.

22. Let $\mathbf{u}$ and $\mathbf{v}$ be vectors in an inner product space V.

   (a) Expand $\langle 2\mathbf{u} - 7\mathbf{v}, \ 3\mathbf{u} + 5\mathbf{v} \rangle$.

   ◆(b) Expand $\langle 3\mathbf{u} - 4\mathbf{v}, \ 5\mathbf{u} + \mathbf{v} \rangle$.

   (c) Show that $\|\mathbf{u} + \mathbf{v}\|^2 = \|\mathbf{u}\|^2 + 2\langle \mathbf{u}, \mathbf{v} \rangle + \|\mathbf{v}\|^2$.

   (d) Show that $\|\mathbf{u} - \mathbf{v}\|^2 = \|\mathbf{u}\|^2 - 2\langle \mathbf{u}, \mathbf{v} \rangle + \|\mathbf{v}\|^2$.

23. Show that $\|\mathbf{v}\|^2 + \|\mathbf{w}\|^2 = \frac{1}{2}\{\|\mathbf{v} + \mathbf{w}\|^2 + \|\mathbf{v} - \mathbf{w}\|^2\}$ for any $\mathbf{v}$ and $\mathbf{w}$ in an inner product space.

24. Let $\langle \ , \ \rangle$ be an inner product on a vector space V. Show that the corresponding distance function is translation invariant. That is, show that $d(\mathbf{v}, \mathbf{w}) = d(\mathbf{v} + \mathbf{u}, \mathbf{w} + \mathbf{u})$ for all $\mathbf{v}$, $\mathbf{w}$, and $\mathbf{u}$ in V.

25. (a) Show that $\langle \mathbf{u}, \mathbf{v} \rangle = \frac{1}{4}[\|\mathbf{u} + \mathbf{v}\|^2 - \|\mathbf{u} - \mathbf{v}\|^2]$ for all $\mathbf{u}$, $\mathbf{v}$ in an inner product space V.

   (b) If $\langle \ , \ \rangle$ and $\langle \ , \ \rangle'$ are two inner products on V that have equal associated norm functions, show that $\langle \mathbf{u}, \mathbf{v} \rangle = \langle \mathbf{u}, \mathbf{v} \rangle'$ holds for all $\mathbf{u}$ and $\mathbf{v}$.

26. Let $\mathbf{v}$ denote a vector in an inner product space V.

   (a) Show that $W = \{\mathbf{w} \mid \mathbf{w} \text{ in } V, \ \langle \mathbf{v}, \mathbf{w} \rangle = 0\}$ is a subspace of V.

   ◆(b) If $V = \mathbb{R}^3$ with the dot product, and if $\mathbf{v} = (1, -1, 2)$, find a basis for W.

27. Given vectors $\mathbf{w}_1, \mathbf{w}_2, \ldots, \mathbf{w}_n$ and $\mathbf{v}$, assume that $\langle \mathbf{v}, \mathbf{w}_i \rangle = 0$ for each i. Show that $\langle \mathbf{v}, \mathbf{w} \rangle = 0$ for all $\mathbf{w}$ in $\text{span}\{\mathbf{w}_1, \mathbf{w}_2, \ldots, \mathbf{w}_n\}$.

◆28. If $V = \text{span}\{\mathbf{v}_1, \mathbf{v}_2, \ldots, \mathbf{v}_n\}$ and $\langle \mathbf{v}, \mathbf{v}_i \rangle = \langle \mathbf{w}, \mathbf{v}_i \rangle$ holds for each i. Show that $\mathbf{v} = \mathbf{w}$.

29. Use the Schwarz inequality in an inner product space to show that:

   (a) If $\|\mathbf{u}\| \le 1$, then $\langle \mathbf{u}, \mathbf{v} \rangle^2 \le \|\mathbf{v}\|^2$ for all $\mathbf{v}$ in V.

   ◆(b) $(x \cos \theta + y \sin \theta)^2 \le x^2 + y^2$ for all real x, y, and θ.

   (c) $\|r_1\mathbf{v}_1 + \cdots + r_n\mathbf{v}_n\|^2 \le \{r_1\|\mathbf{v}_1\| + \cdots + r_n\|\mathbf{v}_n\|\}^2$ for all vectors $\mathbf{v}_i$, and all $r_i > 0$ in $\mathbb{R}$.

30. If A is a $2 \times n$ matrix, let $\mathbf{u}$ and $\mathbf{v}$ denote the rows of A.

   (a) Show that $AA^T = \begin{bmatrix} \|\mathbf{u}\|^2 & \mathbf{u} \cdot \mathbf{v} \\ \mathbf{u} \cdot \mathbf{v} & \|\mathbf{v}\|^2 \end{bmatrix}$

   (b) Show that $\det(AA^T) \ge 0$.

31. (a) If $\mathbf{v}$ and $\mathbf{w}$ are nonzero vectors in an inner product space V, show that $-1 \le \dfrac{\langle \mathbf{v}, \mathbf{w} \rangle}{\|\mathbf{v}\|\|\mathbf{w}\|} \le 1$, and hence that a unique angle θ exists such that $\dfrac{\langle \mathbf{v}, \mathbf{w} \rangle}{\|\mathbf{v}\|\|\mathbf{w}\|} = \cos \theta$ and $0 \le \theta \le \pi$. This angle θ is called the **angle between** $\mathbf{v}$ and $\mathbf{w}$.

   (b) Find the angle between $\mathbf{v} = (1, 2, -1, 1, 3)$ and $\mathbf{w} = (2, 1, 0, 2, 0)$ in $\mathbb{R}^5$ with the dot product.

   (c) If θ is the angle between $\mathbf{v}$ and $\mathbf{w}$, show that the **law of cosines** is valid:
   $\|\mathbf{v} - \mathbf{w}\|^2 = \|\mathbf{v}\|^2 + \|\mathbf{w}\|^2 - 2\|\mathbf{v}\|\|\mathbf{w}\|\cos \theta$.

32. Let $\{\mathbf{e}_1, \mathbf{e}_2, \ldots, \mathbf{e}_n\}$ be an orthogonal basis of V (see the beginning of the next section), and let θ_i be the angle between $\mathbf{v} \ne \mathbf{0}$ and $\mathbf{e}_i$ for each i (preceding exercise). Show that $\cos^2\theta_1 + \cos^2\theta_2 + \cdots + \cos^2\theta_n = 1$.

33. If $V = \mathbb{R}^2$, define $\|(x, y)\| = |x| + |y|$.

   (a) Show that $\|\cdot\|$ satisfies the conditions in Theorem 5.

   (b) Show that $\|\cdot\|$ does not arise from an inner product on $\mathbb{R}^2$. [*Hint:* If it did, use Theorem 2 to find numbers a, b, and c such that $\|(x, y)\|^2 = ax^2 + bxy + cy^2$ for all x and y.]

SECTION 10.2   Orthogonal Sets of Vectors

The idea that two lines can be perpendicular is fundamental in geometry, and this section is devoted to introducing this notion into a general inner product space V. To motivate the definition, recall that two nonzero geometric vectors X and Y in $\mathbb{R}^n$

are perpendicular (or orthogonal) if and only if $X \cdot Y = 0$. In general, two vectors $\mathbf{v}$ and $\mathbf{w}$ in an inner product space V are said to be **orthogonal** if

$$\langle \mathbf{v}, \mathbf{w} \rangle = 0$$

A set $\{\mathbf{e}_1, \mathbf{e}_2, \dots, \mathbf{e}_n\}$ of vectors is called an **orthogonal set of vectors** if

1. Each $\mathbf{e}_i \neq \mathbf{0}$.
2. $\langle \mathbf{e}_i, \mathbf{e}_j \rangle = 0$ for all $i \neq j$.

If, in addition, $\|\mathbf{e}_i\| = 1$ for each i, the set $\{\mathbf{e}_1, \mathbf{e}_2, \dots, \mathbf{e}_n\}$ is called an **orthonormal** set.

Example 1

$\{\sin x, \cos x\}$ is orthogonal in $\mathbf{C}[-\pi, \pi]$ because

$$\int_{-\pi}^{\pi} \sin x \cos x \, dx = [-\tfrac{1}{4} \cos 2x]_{-\pi}^{\pi} = 0$$

The first result about orthogonal sets extends Pythagoras' theorem in $\mathbb{R}^n$ (Theorem 4 §5.3) and the same proof works.

Theorem 1 Pythagoras' Theorem

If $\{\mathbf{e}_1, \mathbf{e}_2, \dots, \mathbf{e}_n\}$ is an orthogonal set of vectors, then

$$\|\mathbf{e}_1 + \mathbf{e}_2 + \dots + \mathbf{e}_n\|^2 = \|\mathbf{e}_1\|^2 + \|\mathbf{e}_2\|^2 + \dots + \|\mathbf{e}_n\|^2$$

The proof of the next result is left to the reader.

Theorem 2

Let $\{\mathbf{e}_1, \mathbf{e}_2, \dots, \mathbf{e}_n\}$ be an orthogonal set of vectors.
1. $\{r_1\mathbf{e}_1, r_2\mathbf{e}_2, \dots, r_n\mathbf{e}_n\}$ is also orthogonal for any $r_i \neq 0$ in $\mathbb{R}$.

2. $\left\{ \dfrac{1}{\|\mathbf{e}_1\|}\mathbf{e}_1, \dfrac{1}{\|\mathbf{e}_2\|}\mathbf{e}_2, \dots, \dfrac{1}{\|\mathbf{e}_n\|}\mathbf{e}_n \right\}$ is an orthonormal set.

As before, the process of passing from an orthogonal set to an orthonormal one is called **normalizing** the orthogonal set. The proof of Theorem 5 §5.3 goes through to give

Theorem 3

Every orthogonal set of vectors is linearly independent.

Example 2

Show that $\left\{ \begin{bmatrix} 2 \\ -1 \\ 0 \end{bmatrix}, \begin{bmatrix} 0 \\ 1 \\ 1 \end{bmatrix}, \begin{bmatrix} 0 \\ -1 \\ 2 \end{bmatrix} \right\}$ is an orthogonal basis of $\mathbb{R}^3$ with inner product

$\langle \mathbf{v}, \mathbf{w} \rangle = \mathbf{v}^T A \mathbf{w}$, where $A = \begin{bmatrix} 1 & 1 & 0 \\ 1 & 2 & 0 \\ 0 & 0 & 1 \end{bmatrix}$.

Solution We have

$$\left\langle \begin{bmatrix} 2 \\ -1 \\ 0 \end{bmatrix}, \begin{bmatrix} 0 \\ 1 \\ 1 \end{bmatrix} \right\rangle = \begin{bmatrix} 2 & -1 & 0 \end{bmatrix} \begin{bmatrix} 1 & 1 & 0 \\ 1 & 2 & 0 \\ 0 & 0 & 1 \end{bmatrix} \begin{bmatrix} 0 \\ 1 \\ 1 \end{bmatrix} = \begin{bmatrix} 1 & 0 & 0 \end{bmatrix} \begin{bmatrix} 0 \\ 1 \\ 1 \end{bmatrix} = 0$$

and the reader can verify that the other pairs are orthogonal too. Hence the set is orthogonal, so it is linearly independent by Theorem 3. Because $\dim \mathbb{R}^3 = 3$, it is a basis.

The proof of Theorem 6 §5.3 generalizes to give the following:

Theorem 4 Expansion Theorem

Let $\{e_1, e_2, \dots, e_n\}$ be an orthogonal basis of an inner product space V. If v is any vector in V, then

$$v = \frac{\langle v, e_1 \rangle}{\|e_1\|^2} e_1 + \frac{\langle v, e_2 \rangle}{\|e_2\|^2} e_2 + \dots + \frac{\langle v, e_n \rangle}{\|e_n\|^2} e_n$$

is the expansion of v as a linear combination of the basis vectors.

The coefficients $\dfrac{\langle v, e_1 \rangle}{\|e_1\|^2}, \dfrac{\langle v, e_2 \rangle}{\|e_2\|^2}, \dots, \dfrac{\langle v, e_n \rangle}{\|e_n\|^2}$ in the expansion theorem are sometimes called the **Fourier coefficients** of v with respect to the orthogonal basis $\{e_1, e_2, \dots, e_n\}$. This is in honour of the French mathematician J. B. J. Fourier (1768–1830). His original work was with a particular orthogonal set in the space $C[a, b]$, about which there will be more to say in Section 10.5.

Example 3

If $a_0, a_1, \dots, a_n$ are distinct numbers and $p(x)$ and $q(x)$ are in P_n, define

$$\langle p(x), q(x) \rangle = p(a_0)q(a_0) + p(a_1)q(a_1) + \dots + p(a_n)q(a_n)$$

This is an inner product on P_n. (Axioms P1–P4 are routinely verified, and P5 holds because 0 is the only polynomial of degree n with $n + 1$ distinct roots. See Theorem 4 §6.5.) Recall that the **Lagrange polynomials** $\delta_0(x), \delta_1(x), \dots, \delta_n(x)$ relative to the numbers $a_0, a_1, \dots, a_n$ are defined as follows (see Section 6.5):

$$\delta_k(x) = \frac{\prod_{i \neq k}(x - a_i)}{\prod_{i \neq k}(a_k - a_i)} \qquad k = 0, 1, 2, \dots, n$$

where $\prod_{i \neq k}(x - a_i)$ means the product of all the terms $(x - a_0), (x - a_1), (x - a_2), \dots, (x - a_n)$ except that the kth term is omitted. Then $\{\delta_0(x), \delta_1(x), \dots, \delta_n(x)\}$ is orthonormal with respect to $\langle \, , \, \rangle$ because $\delta_k(a_i) = 0$ if $i \neq k$ and $\delta_k(a_k) = 1$. These facts also show that $\langle p(x), \delta_k(x) \rangle = p(a_k)$, so the expansion theorem gives

$$p(x) = p(a_0)\delta_0(x) + p(a_1)\delta_1(x) + \dots + p(a_n)\delta_n(x)$$

for each $p(x)$ in P_n. This is the **Lagrange interpolation expansion** of $p(x)$, Theorem 3 §6.5, which is important in numerical integration.

Orthogonal Lemma

Let $\{\mathbf{e}_1, \mathbf{e}_2, \dots, \mathbf{e}_m\}$ be an orthogonal set of vectors in an inner product space V, and let $\mathbf{v}$ be any vector *not* in span$\{\mathbf{e}_1, \mathbf{e}_2, \dots, \mathbf{e}_m\}$. Define

$$\mathbf{e}_{m+1} = \mathbf{v} - \frac{\langle \mathbf{v}, \mathbf{e}_1 \rangle}{\|\mathbf{e}_1\|^2}\mathbf{e}_1 - \frac{\langle \mathbf{v}, \mathbf{e}_2 \rangle}{\|\mathbf{e}_2\|^2}\mathbf{e}_2 - \cdots - \frac{\langle \mathbf{v}, \mathbf{e}_m \rangle}{\|\mathbf{e}_m\|^2}\mathbf{e}_m$$

Then $\{\mathbf{e}_1, \mathbf{e}_2, \dots, \mathbf{e}_m, \mathbf{e}_{m+1}\}$ is an orthogonal set of vectors.

The proof of this result (and the next) is the same as for the dot product in $\mathbb{R}^n$ (Lemma 1 and Theorem 2 in Section 8.1).

Theorem 5 Gram–Schmidt Orthogonalization Algorithm

Let V be an inner product space and let $\{\mathbf{v}_1, \mathbf{v}_2, \dots, \mathbf{v}_n\}$ be any basis of V. Define vectors $\mathbf{e}_1, \mathbf{e}_2, \dots, \mathbf{e}_n$ in V successively as follows:

$$\mathbf{e}_1 = \mathbf{v}_1$$
$$\mathbf{e}_2 = \mathbf{v}_2 - \frac{\langle \mathbf{v}_2, \mathbf{e}_1 \rangle}{\|\mathbf{e}_1\|^2}\mathbf{e}_1$$
$$\mathbf{e}_3 = \mathbf{v}_3 - \frac{\langle \mathbf{v}_3, \mathbf{e}_1 \rangle}{\|\mathbf{e}_1\|^2}\mathbf{e}_1 - \frac{\langle \mathbf{v}_3, \mathbf{e}_2 \rangle}{\|\mathbf{e}_2\|^2}\mathbf{e}_2$$
$$\vdots$$
$$\mathbf{e}_k = \mathbf{v}_k - \frac{\langle \mathbf{v}_k, \mathbf{e}_1 \rangle}{\|\mathbf{e}_1\|^2}\mathbf{e}_1 - \frac{\langle \mathbf{v}_k, \mathbf{e}_2 \rangle}{\|\mathbf{e}_2\|^2}\mathbf{e}_2 - \cdots - \frac{\langle \mathbf{v}_k, \mathbf{e}_{k-1} \rangle}{\|\mathbf{e}_{k-1}\|^2}\mathbf{e}_{k-1}$$

for each $k = 2, 3, \dots, n$. Then
1. $\{\mathbf{e}_1, \mathbf{e}_2, \dots, \mathbf{e}_n\}$ is an orthogonal basis of V.
2. span$\{\mathbf{e}_1, \mathbf{e}_2, \dots, \mathbf{e}_k\}$ = span$\{\mathbf{v}_1, \mathbf{v}_2, \dots, \mathbf{v}_k\}$ holds for each $k = 1, 2, \dots, n$.

The purpose of the Gram–Schmidt algorithm is to convert a basis of an inner product space into an *orthogonal* basis. In particular, Theorem 5 shows that every finite dimensional inner product space *has* an orthogonal basis.

Example 4

Consider $V = \mathbf{P}_3$ with the inner product $\langle p, q \rangle = \int_{-1}^{1} p(x)q(x)\,dx$. If the Gram–Schmidt algorithm is applied to the basis $\{1, x, x^2, x^3\}$, show that the result is the orthogonal basis

$$\{1, x, \tfrac{1}{3}(3x^2 - 1), \tfrac{1}{5}(5x^3 - 3x)\}.$$

Solution Take $\mathbf{e}_1 = 1$. Then the algorithm gives

$$\mathbf{e}_2 = x - \frac{\langle x, \mathbf{e}_1 \rangle}{\|\mathbf{e}_1\|^2}\mathbf{e}_1 = x - \frac{0}{2}\mathbf{e}_1 = x$$

$$\mathbf{e}_3 = x^2 - \frac{\langle x^2, \mathbf{e}_1 \rangle}{\|\mathbf{e}_1\|^2}\mathbf{e}_1 - \frac{\langle x^2, \mathbf{e}_2 \rangle}{\|\mathbf{e}_2\|^2}\mathbf{e}_2$$
$$= x^2 - \frac{\frac{2}{3}}{2}1 - \frac{0}{\frac{2}{3}}x$$
$$= \tfrac{1}{3}(3x^2 - 1)$$

The verification that $\mathbf{e}_4 = \tfrac{1}{5}(5x^3 - 3x)$ is omitted.

The polynomials in Example 4 are such that the leading coefficient is 1 in each case. In other contexts (the study of differential equations, for example), it is customary to take multiples of these polynomials $p(x)$ such that $p(1) = 1$. The resulting orthogonal basis of $\mathbf{P}_3$ is

$$\{1, \ x, \ \tfrac{1}{2}(3x^2 - 1), \ \tfrac{1}{2}(5x^3 - 3x^2)\}$$

and these are the first four **Legendre polynomials**, so called to honour the French mathematician A. M. Legendre (1752–1833). They are important in the study of differential equations.

The orthogonal complement of a subspace U of $\mathbb{R}^n$ was defined (in Chapter 8) to be the set of all vectors in $\mathbb{R}^n$ that are orthogonal to every vector in U. This notion has a natural extension in an arbitrary inner product space. Let U be a subspace of an inner product space V. As in $\mathbb{R}^n$, the **orthogonal complement** $U^\perp$ of U in V is defined by

$$U^\perp = \{\mathbf{v} \mid \mathbf{v} \text{ in } V, \ \langle \mathbf{v}, \mathbf{u} \rangle = 0 \text{ for all } \mathbf{u} \text{ in } U\}.$$

Theorem 6

Let U be a finite dimensional subspace of an inner product space V.
1. $U^\perp$ is a subspace of V and $V = U \oplus U^\perp$.
2. If $\dim V = n$, then $\dim U + \dim U^\perp = n$.
3. If $\dim V = n$, then $U^{\perp\perp} = U$.

PROOF

1. $U^\perp$ is a subspace using Theorem 1 §10.1. If $\mathbf{v}$ is in $U \cap U^\perp$ then $\langle \mathbf{v}, \mathbf{v} \rangle = 0$. Hence $U \cap U^\perp = \{\mathbf{0}\}$, and $U + U^\perp = V$ follows from the orthogonal lemma.
2. This follows from Theorem 6 §9.3.
3. $U \subseteq U^{\perp\perp}$ always holds (verify) and $\dim U^{\perp\perp} = n - \dim U^\perp = \dim U$, where (2) was used twice.

We digress briefly and consider a subspace U of an arbitrary vector space V. As in Section 9.3, If W is any complement of U in V—that is, $V = U \oplus W$—each vector $\mathbf{v}$ in V has a *unique* representation as a sum $\mathbf{v} = \mathbf{u} + \mathbf{w}$ where $\mathbf{u}$ is in U and $\mathbf{w}$ is in W. Hence we may define a function $T : V \to V$ as follows:

$$T(\mathbf{v}) = \mathbf{u} \quad \text{where } \mathbf{v} = \mathbf{u} + \mathbf{w}, \ \mathbf{u} \text{ in } U, \ \mathbf{w} \text{ in } W.$$

Thus, to compute $T(\mathbf{v})$, express $\mathbf{v}$ in any way at all as the sum of a vector $\mathbf{u}$ in U and a vector in W; then $T(\mathbf{v}) = \mathbf{u}$.

This function T is a linear operator on V. Indeed, if $\mathbf{v}_1 = \mathbf{u}_1 + \mathbf{w}_1$ where $\mathbf{u}_1$ is in U and $\mathbf{w}_1$ is in W, then $\mathbf{v} + \mathbf{v}_1 = (\mathbf{u} + \mathbf{u}_1) + (\mathbf{w} + \mathbf{w}_1)$, so

$$T(\mathbf{v} + \mathbf{v}_1) = \mathbf{u} + \mathbf{u}_1 = T(\mathbf{v}) + T(\mathbf{v}_1)$$

because $\mathbf{u} + \mathbf{u}_1$ is in U and $\mathbf{w} + \mathbf{w}_1$ is in W. Similarly, $T(a\mathbf{v}) = aT(\mathbf{v})$ for all a in $\mathbb{R}$, so T is a linear operator. Furthermore, $\text{im } T = U$ and $\ker T = W$ as the reader can verify, and T is called the **projection on U with kernel W**.

If U is a subspace of V, there are many projections on U, one for each complementary subspace W with $V = U \oplus W$. If V is an *inner product space*, we single out one for special attention. Let U be a finite dimensional subspace of an inner product space V. The projection on U with kernel $U^{\perp}$ is called the **orthogonal projection** on U (or simply the **projection** on U) and is denoted $\text{proj}_U : V \to V$.

Theorem 7 Projection Theorem

Let U be a finite dimensional subspace of an inner product space V and let $\mathbf{v}$ be a vector in V.
1. $\text{proj}_U : V \to V$ is a linear operator with image U and kernel $U^{\perp}$.
2. $\text{proj}_U(\mathbf{v})$ is in U and $\mathbf{v} - \text{proj}_U(\mathbf{v})$ is in $U^{\perp}$.
3. If $\{\mathbf{e}_1, \mathbf{e}_2, \dots, \mathbf{e}_m\}$ is any orthogonal basis of U, then

$$\text{proj}_U(\mathbf{v}) = \frac{\langle \mathbf{v}, \mathbf{e}_1 \rangle}{\|\mathbf{e}_1\|^2}\mathbf{e}_1 + \frac{\langle \mathbf{v}, \mathbf{e}_2 \rangle}{\|\mathbf{e}_2\|^2}\mathbf{e}_2 + \cdots + \frac{\langle \mathbf{v}, \mathbf{e}_m \rangle}{\|\mathbf{e}_m\|^2}\mathbf{e}_m$$

PROOF

Only (3) remains to be proved. If $\{\mathbf{e}_{m+1}, \dots, \mathbf{e}_n\}$ is an orthogonal basis of $U^{\perp}$ (it exists by Theorem 5), then $\{\mathbf{e}_1, \dots, \mathbf{e}_m, \mathbf{e}_{m+1}, \dots, \mathbf{e}_n\}$ is an orthogonal basis of V (Theorem 5 §9.3) and the expansion theorem gives

$$\text{proj}_U(\mathbf{v}) = \frac{\langle \mathbf{v}, \mathbf{e}_1 \rangle}{\|\mathbf{e}_1\|^2}\mathbf{e}_1 + \cdots + \frac{\langle \mathbf{v}, \mathbf{e}_m \rangle}{\|\mathbf{e}_m\|^2}\mathbf{e}_m + \frac{\langle \mathbf{v}, \mathbf{e}_{m+1} \rangle}{\|\mathbf{e}_{m+1}\|^2}\mathbf{e}_{m+1} + \cdots + \frac{\langle \mathbf{v}, \mathbf{e}_n \rangle}{\|\mathbf{e}_n\|^2}\mathbf{e}_n$$

Then (3) follows because the first m terms are in U and the last $n - m$ terms are in $U^{\perp}$.

Example 5

Let U be a subspace of the finite dimensional inner product space V. Show that $\text{proj}_{U^{\perp}}(\mathbf{v}) = \mathbf{v} - \text{proj}_U(\mathbf{v})$ for all $\mathbf{v}$ in V.

Solution 1   The proof of Theorem 7 shows that $\mathbf{v} = \text{proj}_U(\mathbf{v}) + \text{proj}_{U^{\perp}}(\mathbf{v})$.

Solution 2   We have $V = U^{\perp} \oplus U^{\perp\perp}$ by Theorem 6. If we write $\mathbf{p} = \text{proj}_U(\mathbf{v})$, then $\mathbf{v} = (\mathbf{v} - \mathbf{p}) + \mathbf{p}$ where $\mathbf{v} - \mathbf{p}$ is in $U^{\perp}$ and $\mathbf{p}$ is in $U = U^{\perp\perp}$ by Theorem 7. Hence $\text{proj}_{U^{\perp}}(\mathbf{v}) = \mathbf{v} - \mathbf{p}$.

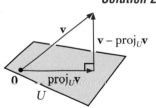

The vectors $\mathbf{v}$, $\text{proj}_U(\mathbf{v})$, and $\mathbf{v} - \text{proj}_U(\mathbf{v})$ in Theorem 7 can be visualized geometrically as in the diagram. This suggests that $\text{proj}_U(\mathbf{v})$ is the vector in U closest to $\mathbf{v}$. This is, in fact, the case.

Theorem 8 Approximation Theorem

Let U be a finite dimensional subspace of an inner product space V. If $\mathbf{v}$ is any vector in V, then $\text{proj}_U(\mathbf{v})$ is the vector in U that is closest to $\mathbf{v}$. Here *closest* means that

$$\|\mathbf{v} - \text{proj}_U(\mathbf{v})\| < \|\mathbf{v} - \mathbf{u}\|$$

for all $\mathbf{u}$ in U, $\mathbf{u} \neq \text{proj}_U(\mathbf{v})$.

PROOF

Write $\mathbf{p} = \text{proj}_U(\mathbf{v})$, and consider $\mathbf{v} - \mathbf{u} = (\mathbf{v} - \mathbf{p}) + (\mathbf{p} - \mathbf{u})$. Because $\mathbf{v} - \mathbf{p}$ is in $U^\perp$ and $\mathbf{p} - \mathbf{u}$ is in U, Pythagoras' theorem gives

$$\| \mathbf{v} - \mathbf{u} \|^2 = \| \mathbf{v} - \mathbf{p} \|^2 + \| \mathbf{p} - \mathbf{u} \|^2 > \| \mathbf{v} - \mathbf{p} \|^2$$

because $\mathbf{p} - \mathbf{u} \neq \mathbf{0}$. The result follows.

Example 6

Consider the space $\mathbf{C}[-1, 1]$ of real-valued continuous functions on the interval $[-1, 1]$ with inner product $\langle f, g \rangle = \int_{-1}^{1} f(x)g(x)\, dx$. Find the polynomial $p = p(x)$ of degree at most 2 that best approximates the absolute-value function f given by $f(x) = |x|$.

Solution

Here we want the vector p in the subspace $U = \mathbf{P}_2$ of $\mathbf{C}[-1, 1]$ that is closest to f. In Example 4 the Gram–Schmidt algorithm was applied to give an orthogonal basis $\{\mathbf{e}_1 = 1, \mathbf{e}_2 = x, \mathbf{e}_3 = 3x^2 - 1\}$ of $\mathbf{P}_2$ (where, for convenience, we have changed $\mathbf{e}_3$ by a numerical factor). Hence the required polynomial is

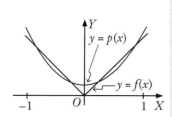

$$
\begin{aligned}
p &= \text{proj}_{\mathbf{P}_2}(f) \\
&= \frac{\langle f, \mathbf{e}_1 \rangle}{\| \mathbf{e}_1 \|^2} \mathbf{e}_1 + \frac{\langle f, \mathbf{e}_2 \rangle}{\| \mathbf{e}_2 \|^2} \mathbf{e}_2 + \frac{\langle f, \mathbf{e}_3 \rangle}{\| \mathbf{e}_3 \|^2} \mathbf{e}_3 \\
&= \tfrac{1}{2}\mathbf{e}_1 + 0\mathbf{e}_2 + \frac{\frac{1}{2}}{\frac{8}{5}}\mathbf{e}_3 \\
&= \tfrac{3}{16}(5x^2 + 1)
\end{aligned}
$$

The graphs of $p(x)$ and $f(x)$ are given in the diagram.

If polynomials of degree at most n are allowed in Example 6, the polynomial in $\mathbf{P}_n$ is $\text{proj}_{\mathbf{P}_n}(f)$, and it is calculated in the same way. Because the subspaces $\mathbf{P}_n$ get larger as n increases, it turns out that the approximating polynomials $\text{proj}_{\mathbf{P}_n}(f)$ get closer and closer to f. In fact, solving many practical problems comes down to approximating some interesting vector $\mathbf{v}$ (often a function) in an infinite dimensional inner product space V by vectors in finite dimensional subspaces (which can be computed). If $U_1 \subseteq U_2$ are finite dimensional subspaces of V, then

$$\| \mathbf{v} - \text{proj}_{U_2}(\mathbf{v}) \| \leq \| \mathbf{v} - \text{proj}_{U_1}(\mathbf{v}) \|$$

by Theorem 8 (because $\text{proj}_{U_1}(\mathbf{v})$ lies in U_1 and hence in U_2). Thus $\text{proj}_{U_2}(\mathbf{v})$ is a better approximation to $\mathbf{v}$ than $\text{proj}_{U_1}(\mathbf{v})$. Hence a general method in approximation theory might be described as follows: Given $\mathbf{v}$, use it to construct a sequence of finite dimensional subspaces

$$U_1 \subseteq U_2 \subseteq U_3 \subseteq \cdots$$

of V in such a way that $\| \mathbf{v} - \text{proj}_{U_k}(\mathbf{v}) \|$ approaches zero as k increases. Then $\text{proj}_{U_k}(\mathbf{v})$ is a suitable approximation to $\mathbf{v}$ if k is large enough. For more information, the interested reader may wish to consult *Interpolation and Approximation* by Philip J. Davis (New York: Blaisdell, 1963).

Exercises 10.2

Use the dot product in $\mathbb{R}^n$ unless otherwise instructed.

1. In each case, verify that B is an orthogonal basis of V with the given inner product and use the expansion theorem to express $\mathbf{v}$ as a linear combination of the basis vectors.

   (a) $\mathbf{v} = \begin{bmatrix} a \\ b \end{bmatrix}$, $B = \left\{ \begin{bmatrix} 1 \\ -1 \end{bmatrix}, \begin{bmatrix} 1 \\ 0 \end{bmatrix} \right\}$, $V = \mathbb{R}^2$,

   $\langle \mathbf{v}, \mathbf{w} \rangle = \mathbf{v}^T A \mathbf{w}$ where $A = \begin{bmatrix} 2 & 2 \\ 2 & 5 \end{bmatrix}$

   ◆(b) $\mathbf{v} = \begin{bmatrix} a \\ b \\ c \end{bmatrix}$, $B = \left\{ \begin{bmatrix} 1 \\ 1 \\ 1 \end{bmatrix}, \begin{bmatrix} -1 \\ 0 \\ 1 \end{bmatrix}, \begin{bmatrix} 1 \\ -6 \\ 1 \end{bmatrix} \right\}$,

   $V = \mathbb{R}^3$, $\langle \mathbf{v}, \mathbf{w} \rangle = \mathbf{v}^T A \mathbf{w}$ where $A = \begin{bmatrix} 2 & 0 & 1 \\ 0 & 1 & 0 \\ 1 & 0 & 2 \end{bmatrix}$

   (c) $\mathbf{v} = a + bx + cx^2$, $B = \{1, x, 2 - 3x^2\}$, $V = \mathbf{P}_2$,
   $\langle p, q \rangle = p(0)q(0) + p(1)q(1) + p(-1)q(-1)$

   ◆(d) $\mathbf{v} = \begin{bmatrix} a & b \\ c & d \end{bmatrix}$

   $B = \left\{ \begin{bmatrix} 1 & 0 \\ 0 & 1 \end{bmatrix}, \begin{bmatrix} 1 & 0 \\ 0 & -1 \end{bmatrix}, \begin{bmatrix} 0 & 1 \\ 1 & 0 \end{bmatrix}, \begin{bmatrix} 0 & 1 \\ -1 & 0 \end{bmatrix} \right\}$,

   $V = \mathbf{M}_{22}$, $\langle X, Y \rangle = \text{tr}(XY^T)$

2. Let $\mathbb{R}^3$ have the inner product
   $\langle (x, y, z), (x', y', z') \rangle = 2xx' + yy' + 3zz'$. In each case, use the Gram–Schmidt algorithm to transform B into an orthogonal basis.

   (a) $B = \{(1, 1, 0), (1, 0, 1), (0, 1, 1)\}$
   ◆(b) $B = \{(1, 1, 1), (1, -1, 1), (1, 1, 0)\}$

3. Let $\mathbf{M}_{22}$ have the inner product $\langle X, Y \rangle = \text{tr}(XY^T)$. In each case, use the Gram–Schmidt algorithm to transform B into an orthogonal basis.

   (a) $B = \left\{ \begin{bmatrix} 1 & 1 \\ 0 & 0 \end{bmatrix}, \begin{bmatrix} 1 & 0 \\ 1 & 0 \end{bmatrix}, \begin{bmatrix} 0 & 1 \\ 0 & 1 \end{bmatrix}, \begin{bmatrix} 1 & 0 \\ 0 & 1 \end{bmatrix} \right\}$

   ◆(b) $B = \left\{ \begin{bmatrix} 1 & 1 \\ 0 & 1 \end{bmatrix}, \begin{bmatrix} 1 & 0 \\ 1 & 1 \end{bmatrix}, \begin{bmatrix} 1 & 0 \\ 0 & 1 \end{bmatrix}, \begin{bmatrix} 1 & 0 \\ 0 & 0 \end{bmatrix} \right\}$

4. In each case, use the Gram–Schmidt process to convert the basis $B = \{1, x, x^2\}$ into an orthogonal basis of $\mathbf{P}_2$.

   (a) $\langle p, q \rangle = p(0)q(0) + p(1)q(1) + p(2)q(2)$
   ◆(b) $\langle p, q \rangle = \int_0^2 p(x)q(x)\, dx$

5. Show that $\{1, x - \frac{1}{2}, x^2 - x + \frac{1}{6}\}$ is an orthogonal basis of $\mathbf{P}_2$ with the inner product $\langle p, q \rangle = \int_0^1 p(x)q(x)\, dx$, and find the corresponding orthonormal basis.

6. In each case find $U^\perp$ and compute $\dim U$ and $\dim U^\perp$.

   (a) $U = \text{span}\{[1\ 1\ 2\ 0], [3\ -1\ 2\ 1], [1\ -3\ -2\ 1]\}$ in $\mathbb{R}^4$
   ◆(b) $U = \text{span}\{[1\ 1\ 0\ 0]\}$ in $\mathbb{R}^4$
   (c) $U = \text{span}\{1, x\}$ in $\mathbf{P}_2$ with
   $\langle p, q \rangle = p(0)q(0) + p(1)q(1) + p(2)q(2)$
   ◆(d) $U = \text{span}\{x\}$ in $\mathbf{P}_2$ with
   $\langle p, q \rangle = \int_0^1 p(x)q(x)\, dx$

   (e) $U = \text{span}\left\{ \begin{bmatrix} 1 & 0 \\ 0 & 1 \end{bmatrix}, \begin{bmatrix} 1 & 1 \\ 0 & 0 \end{bmatrix} \right\}$ in $\mathbf{M}_{22}$
   with $\langle X, Y \rangle = \text{tr}(XY^T)$

   ◆(f) $U = \text{span}\left\{ \begin{bmatrix} 1 & 1 \\ 0 & 0 \end{bmatrix}, \begin{bmatrix} 1 & 0 \\ 1 & 0 \end{bmatrix}, \begin{bmatrix} 1 & 0 \\ 1 & 1 \end{bmatrix} \right\}$ in $\mathbf{M}_{22}$
   with $\langle X, Y \rangle = \text{tr}(XY^T)$

7. Let $\langle X, Y \rangle = \text{tr}(XY^T)$ in $\mathbf{M}_{22}$. In each case find the matrix in U closest to A.

   (a) $U = \text{span}\left\{ \begin{bmatrix} 1 & 0 \\ 0 & 1 \end{bmatrix}, \begin{bmatrix} 1 & 1 \\ 1 & 1 \end{bmatrix} \right\}$; $A = \begin{bmatrix} 1 & -1 \\ 2 & 3 \end{bmatrix}$

   ◆(b) $U = \text{span}\left\{ \begin{bmatrix} 1 & 0 \\ 0 & 1 \end{bmatrix}, \begin{bmatrix} 1 & 1 \\ 1 & -1 \end{bmatrix}, \begin{bmatrix} 1 & 1 \\ 0 & 0 \end{bmatrix} \right\}$;

   $A = \begin{bmatrix} 2 & 1 \\ 3 & 2 \end{bmatrix}$

8. Let $\langle p(x), q(x) \rangle = p(0)q(0) + p(1)q(1) + p(2)q(2)$ in $\mathbf{P}_2$. In each case find the polynomial in U closest to $f(x)$.

   (a) $U = \text{span}\{1 + x, x^2\}$, $f(x) = 1 + x^2$
   ◆(b) $U = \text{span}\{1, 1 + x^2\}$; $f(x) = x$

9. Using the inner product $\langle p, q \rangle = \int_0^1 p(x)q(x)\, dx$ on $\mathbf{P}_2$, write $\mathbf{v}$ as the sum of a vector in U and a vector in $U^\perp$.

   (a) $\mathbf{v} = x^2$, $U = \text{span}\{x + 1, 9x - 5\}$
   ◆(b) $\mathbf{v} = x^2 + 1$, $U = \text{span}\{1, 2x - 1\}$

10. (a) Show that $\{\mathbf{u}, \mathbf{v}\}$ is orthogonal if and only if
$\|\mathbf{u} + \mathbf{v}\|^2 = \|\mathbf{u}\|^2 + \|\mathbf{v}\|^2$.

 (b) If $\mathbf{u} = \mathbf{v} = (1, 1)$ and $\mathbf{w} = (-1, 0)$, show that
$\|\mathbf{u} + \mathbf{v} + \mathbf{w}\|^2 = \|\mathbf{u}\|^2 + \|\mathbf{v}\|^2 + \|\mathbf{w}\|^2$ but
$\{\mathbf{u}, \mathbf{v}, \mathbf{w}\}$ is *not* orthogonal. Hence the converse to Pythagoras' theorem need not hold for more than two vectors.

11. Let $\mathbf{v}$ and $\mathbf{w}$ be vectors in an inner product space V. Show that:

 (a) $\mathbf{v}$ is orthogonal to $\mathbf{w}$ if and only if
$\|\mathbf{v} + \mathbf{w}\| = \|\mathbf{v} - \mathbf{w}\|$.

 (b) $\mathbf{v} + \mathbf{w}$ and $\mathbf{v} - \mathbf{w}$ are orthogonal if and only if $\|\mathbf{v}\| = \|\mathbf{w}\|$.

12. Let U and W be subspaces of an n-dimensional inner product space V. If $\dim U + \dim W = n$ and $\langle \mathbf{u}, \mathbf{w}\rangle = 0$ for all $\mathbf{u}$ in U and $\mathbf{w}$ in W, show that $U^\perp = W$.

13. If U and W are subspaces of an inner product space, show that $(U + W)^\perp = U^\perp \cap W^\perp$.

14. If X is any set of vectors in an inner product space V, define
$X^\perp = \{\mathbf{v} \mid \mathbf{v}$ in $V, \langle \mathbf{v}, \mathbf{x}\rangle = 0$ for all $\mathbf{x}$ in $X\}$.

 (a) Show that $X^\perp$ is a subspace of V.

 (b) If $U = \text{span}\{\mathbf{u}_1, \mathbf{u}_2, \ldots, \mathbf{u}_m\}$, show that $U^\perp = \{\mathbf{u}_1, \ldots, \mathbf{u}_m\}^\perp$.

 (c) If $X \subseteq Y$, show that $Y^\perp \subseteq X^\perp$.

 (d) Show that $X^\perp \cap Y^\perp = (X \cup Y)^\perp$.

15. If $\dim V = n$ and $\mathbf{w} \neq \mathbf{0}$ in V, show that $\dim \{\mathbf{v} \mid \mathbf{v}$ in $V, \langle \mathbf{v}, \mathbf{w}\rangle = 0\} = n - 1$.

16. If the Gram–Schmidt process is used on an orthogonal basis $\{\mathbf{v}_1, \ldots, \mathbf{v}_n\}$ of V, show that $\mathbf{e}_k = \mathbf{v}_k$ holds for each $k = 1, 2, \ldots, n$. That is, show that the algorithm reproduces the same basis.

17. If $\{\mathbf{e}_1, \mathbf{e}_2, \ldots, \mathbf{e}_{n-1}\}$ is orthonormal in an inner product space of dimension n, prove that there are exactly two vectors $\mathbf{e}_n$ such that $\{\mathbf{e}_1, \mathbf{e}_2, \ldots, \mathbf{e}_{n-1}, \mathbf{e}_n\}$ is an orthonormal basis.

18. Let U be a finite dimensional subspace of an inner product space V, and let $\mathbf{v}$ be a vector in V.

 (a) Show that $\mathbf{v}$ lies in U if and only if $\mathbf{v} = \text{proj}_U(\mathbf{v})$.

 (b) If $V = \mathbb{R}^3$, show that $(-5, 4, -3)$ lies in $\text{span}\{(3, -2, 5), (-1, 1, 1)\}$ but that $(-1, 0, 2)$ does not.

19. Let $\vec{n} \neq \vec{0}$ and $\vec{w} \neq \vec{0}$ be nonparallel vectors in $\mathbb{R}^3$ (as in Chapter 4).

 (a) Show that $\left\{\vec{n}, \vec{n} \times \vec{w}, \vec{w} - \dfrac{\vec{n} \cdot \vec{w}}{\|\vec{n}\|^2} \vec{n}\right\}$ is an orthogonal basis of $\mathbb{R}^3$.

 (b) Show that $\text{span}\left\{\vec{n} \times \vec{w}, \vec{w} - \dfrac{\vec{n} \cdot \vec{w}}{\|\vec{n}\|^2} \vec{n}\right\}$ is the plane through the origin with normal $\vec{n}$.

20. Let $E = \{\mathbf{e}_1, \mathbf{e}_2, \ldots, \mathbf{e}_n\}$ be an orthonormal basis of V.

 (a) Show that $\langle \mathbf{v}, \mathbf{w}\rangle = C_E(\mathbf{v}) \cdot C_E(\mathbf{w})$ for all $\mathbf{v}, \mathbf{w}$ in V.

 (b) If $P = [p_{ij}]$ is an $n \times n$ matrix, define $\mathbf{b}_i = p_{i1}\mathbf{e}_1 + \cdots + p_{in}\mathbf{e}_n$ for each i. Show that $B = \{\mathbf{b}_1, \mathbf{b}_2, \ldots, \mathbf{b}_n\}$ is an orthonormal basis if and only if P is an orthogonal matrix.

21. Let $\{\mathbf{e}_1, \ldots, \mathbf{e}_n\}$ be an orthogonal basis of V. If $\mathbf{v}$ and $\mathbf{w}$ are in V, show that
$$\langle \mathbf{v}, \mathbf{w}\rangle = \frac{\langle \mathbf{v}, \mathbf{e}_1\rangle\langle \mathbf{w}, \mathbf{e}_1\rangle}{\|\mathbf{e}_1\|^2} + \cdots + \frac{\langle \mathbf{v}, \mathbf{e}_n\rangle\langle \mathbf{w}, \mathbf{e}_n\rangle}{\|\mathbf{e}_n\|^2}$$

22. Let $\{\mathbf{e}_1, \ldots, \mathbf{e}_n\}$ be an orthonormal basis of V, and let $\mathbf{v} = v_1\mathbf{e}_1 + \cdots + v_n\mathbf{e}_n$ and $\mathbf{w} = w_1\mathbf{e}_1 + \cdots + w_n\mathbf{e}_n$. Show that $\langle \mathbf{v}, \mathbf{w}\rangle = v_1w_1 + \cdots + v_nw_n$ and $\|\mathbf{v}\|^2 = v_1^2 + \cdots + v_n^2$ (**Parseval's formula**).

23. Let $\mathbf{v}$ be a vector in an inner product space V.

 (a) Show that $\|\mathbf{v}\| \geq \|\text{proj}_U(\mathbf{v})\|$ holds for all finite dimensional subspaces U. [*Hint:* Pythagoras' theorem.]

 (b) If $\{\mathbf{e}_1, \mathbf{e}_2, \ldots, \mathbf{e}_m\}$ is any orthogonal set in V, prove **Bessel's inequality**:
$$\frac{\langle \mathbf{v}, \mathbf{e}_1\rangle^2}{\|\mathbf{e}_1\|^2} + \cdots + \frac{\langle \mathbf{v}, \mathbf{e}_m\rangle^2}{\|\mathbf{e}_m\|^2} \leq \|\mathbf{v}\|^2$$

24. (a) Let S denote a set of vectors in a finite dimensional inner product space V, and suppose that $\langle \mathbf{u}, \mathbf{v}\rangle = 0$ for all $\mathbf{u}$ in S implies $\mathbf{v} = \mathbf{0}$. Show that $V = \text{span } S$. [*Hint:* Write $U = \text{span } S$ and use Theorem 6.]

 (b) Let $A_1, A_2, \ldots, A_k$ be $n \times n$ matrices. Show that the following are equivalent.

 (i) If $A_i\mathbf{b} = \mathbf{0}$ for all i (where $\mathbf{b}$ is a column in $\mathbb{R}^n$), then $\mathbf{b} = \mathbf{0}$.

 (ii) The set of all rows of the matrices A_i spans $\mathbb{R}^n$.

25. Let $[x_i] = (x_1, x_2, \dots)$ denote a sequence of real numbers x_i, and let
$$V = \{[x_i] \mid \text{only finitely many } x_i \neq 0\}.$$
Define componentwise addition and scalar multiplication on V as follows:
$$[x_i] + [y_i] = [x_i + y_i], \quad \text{and} \quad a[x_i] = [ax_i]$$
for a in $\mathbb{R}$.
Given $[x_i]$ and $[y_i]$ in V, define
$\langle [x_i], [y_i] \rangle = \sum_{i=0}^{\infty} x_i y_i$. (Note that this makes

sense since only finitely many x_i and y_i are nonzero.) Finally define
$$U = \{[x_i] \text{ in } V \mid \sum_{i=0}^{\infty} x_i = 0\}.$$
(a) Show that V is a vector space and that U is a subspace.
(b) Show that $\langle \, , \, \rangle$ is an inner product on V.
(c) Show that $U^{\perp} = \{\mathbf{0}\}$.
(d) Hence show that $U \oplus U^{\perp} \neq V$ and $U \neq U^{\perp\perp}$.

SECTION 10.3   Orthogonal Diagonalization

There is a natural way to define a symmetric linear operator T on a finite dimensional inner product space V. If T is such an operator, it is shown in this section that V has an orthogonal basis consisting of eigenvectors of T. This yields another proof of the principal axis theorem in the context of inner product spaces.

Theorem 1

Let $T : V \to V$ be a linear operator on a finite dimensional space V. Then the following conditions are equivalent.
1. V has a basis consisting of eigenvectors of T.
2. There exists a basis B of V such that $M_B(T)$ is diagonal.

PROOF

We have $M_B(T) = [C_B[T(\mathbf{e}_1)] \;\; C_B[T(\mathbf{e}_2)] \;\; \cdots \;\; C_B[T(\mathbf{e}_n)]]$ where $B = \{\mathbf{e}_1, \mathbf{e}_2, \dots, \mathbf{e}_n\}$ is any basis of V. Hence

$$M_B(T) = \begin{bmatrix} \lambda_1 & 0 & \cdots & 0 \\ 0 & \lambda_2 & \cdots & 0 \\ \vdots & \vdots & & \vdots \\ 0 & 0 & \cdots & \lambda_n \end{bmatrix} \quad \text{if and only if} \quad T(\mathbf{e}_i) = \lambda_i \mathbf{e}_i \text{ for each } i$$

by comparing columns. Theorem 1 follows.

A linear operator T on a finite dimensional space V is called **diagonalizable** if V has a basis of eigenvectors of V.

Example 1

Let $T : \mathbf{P}_2 \to \mathbf{P}_2$ be given by
$$T(a + bx + cx^2) = (a + 4c) - 2bx + (3a + 2c)x^2$$
Find the eigenspaces of T and hence find a basis of eigenvectors.

Solution   If $B_0 = \{1, x, x^2\}$, then

$$M_{B_0}(T) = \begin{bmatrix} 1 & 0 & 4 \\ 0 & -2 & 0 \\ 3 & 0 & 2 \end{bmatrix}$$

so $c_T(x) = (x + 2)^2(x - 5)$, and the eigenvalues of T are $\lambda = -2$ and $\lambda = 5$.

Thus $\left\{ \begin{bmatrix} 0 \\ 1 \\ 0 \end{bmatrix}, \begin{bmatrix} 4 \\ 0 \\ -3 \end{bmatrix}, \begin{bmatrix} 1 \\ 0 \\ 1 \end{bmatrix} \right\}$ is a basis of eigenvectors of $M_{B_0}(T)$, so

$B = \{x, 4 - 3x^2, 1 + x^2\}$ is a basis of $\mathbf{P}_2$ consisting of eigenvectors of T.

If V is an inner product space, the expansion theorem gives a simple formula for the matrix of a linear operator with respect to an orthogonal basis.

Theorem 2

Let $T : V \to V$ be a linear operator on an inner product space V.
If $B = \{\mathbf{e}_1, \mathbf{e}_2, \dots, \mathbf{e}_n\}$ is an orthogonal basis of V, then

$$M_B(T) = \left[\frac{\langle \mathbf{e}_i, T(\mathbf{e}_j) \rangle}{\|\mathbf{e}_i\|^2} \right]$$

PROOF

Write $M_B(T) = [a_{ij}]$. The jth column of $M_B(T)$ is $C_B[T(\mathbf{e}_j)]$, so

$$T(\mathbf{e}_j) = a_{1j}\mathbf{e}_1 + \cdots + a_{ij}\mathbf{e}_i + \cdots + a_{nj}\mathbf{e}_n$$

On the other hand, the expansion theorem (Theorem 4 §10.2) gives

$$\mathbf{v} = \frac{\langle \mathbf{e}_1, \mathbf{v} \rangle}{\|\mathbf{e}_1\|^2}\mathbf{e}_1 + \cdots + \frac{\langle \mathbf{e}_i, \mathbf{v} \rangle}{\|\mathbf{e}_i\|^2}\mathbf{e}_i + \cdots + \frac{\langle \mathbf{e}_n, \mathbf{v} \rangle}{\|\mathbf{e}_n\|^2}\mathbf{e}_n$$

for any $\mathbf{v}$ in V. The result follows by taking $\mathbf{v} = T(\mathbf{e}_j)$.

Example 2

Let $T : \mathbb{R}^3 \to \mathbb{R}^3$ be given by

$$T(a, b, c) = (a + 2b - c, \ 2a + 3c, \ -a + 3b + 2c)$$

If the dot product in $\mathbb{R}^3$ is used, find the matrix of T with respect to the standard basis $B = \{\mathbf{e}_1, \mathbf{e}_2, \mathbf{e}_3\}$ where $\mathbf{e}_1 = (1, 0, 0)$, $\mathbf{e}_2 = (0, 1, 0)$, $\mathbf{e}_3 = (0, 0, 1)$.

Solution   The basis B is orthonormal, so Theorem 2 gives

$$M_B(T) = \begin{bmatrix} \mathbf{e}_1 \cdot T(\mathbf{e}_1) & \mathbf{e}_1 \cdot T(\mathbf{e}_2) & \mathbf{e}_1 \cdot T(\mathbf{e}_3) \\ \mathbf{e}_2 \cdot T(\mathbf{e}_1) & \mathbf{e}_2 \cdot T(\mathbf{e}_2) & \mathbf{e}_2 \cdot T(\mathbf{e}_3) \\ \mathbf{e}_3 \cdot T(\mathbf{e}_1) & \mathbf{e}_3 \cdot T(\mathbf{e}_2) & \mathbf{e}_3 \cdot T(\mathbf{e}_3) \end{bmatrix} = \begin{bmatrix} 1 & 2 & -1 \\ 2 & 0 & 3 \\ -1 & 3 & 2 \end{bmatrix}$$

Of course, this can be found in the usual way.

It is not difficult to verify that an $n \times n$ matrix A is symmetric if and only if $X \cdot (AY) = (AX) \cdot Y$ holds for all columns X and Y in $\mathbb{R}^n$. The analog for operators is as follows:

Theorem 3

Let V be a finite dimensional inner product space. The following conditions are equivalent for a linear operator $T : V \to V$.
1. $\langle \mathbf{v}, T(\mathbf{w}) \rangle = \langle T(\mathbf{v}), \mathbf{w} \rangle$ for all $\mathbf{v}$ and $\mathbf{w}$ in V.
2. The matrix of T is symmetric with respect to every orthonormal basis of V.
3. The matrix of T is symmetric with respect to some orthonormal basis of V.
4. There is an orthonormal basis $B = \{\mathbf{e}_1, \mathbf{e}_2, \dots, \mathbf{e}_n\}$ of V such that
   $\langle \mathbf{e}_i, T(\mathbf{e}_j) \rangle = \langle T(\mathbf{e}_i), \mathbf{e}_j \rangle$ holds for all i and j.

PROOF

It is clear that $(2) \Rightarrow (3)$.

$(1) \Rightarrow (2)$. Let $B = \{\mathbf{e}_1, \dots, \mathbf{e}_n\}$ be an orthonormal basis of V, and write $M_B(T) = [a_{ij}]$. Then $a_{ij} = \langle \mathbf{e}_i, T(\mathbf{e}_j) \rangle$ by Theorem 2. Hence (1) and axiom P2 give

$$a_{ij} = \langle \mathbf{e}_i, T(\mathbf{e}_j) \rangle = \langle T(\mathbf{e}_i), \mathbf{e}_j \rangle = \langle \mathbf{e}_j, T(\mathbf{e}_i) \rangle = a_{ji}$$

for all i and j. This shows that $M_B(T)$ is symmetric.

$(3) \Rightarrow (4)$. Let $B = \{\mathbf{e}_1, \dots, \mathbf{e}_n\}$ be an orthonormal basis of V such that $M_B(T)$ is symmetric. By (3) and Theorem 2, $\langle \mathbf{e}_i, T(\mathbf{e}_j) \rangle = \langle \mathbf{e}_j, T(\mathbf{e}_i) \rangle$ for all i and j, so (4) follows from axiom P2.

$(4) \Rightarrow (1)$. Let $\mathbf{v}$ and $\mathbf{w}$ be vectors in V and write them as $\mathbf{v} = \sum_{i=1}^{n} v_i \mathbf{e}_i$ and $\mathbf{w} = \sum_{j=1}^{n} w_j \mathbf{e}_j$. Then

$$\langle \mathbf{v}, T(\mathbf{w}) \rangle = \left\langle \sum_i v_i \mathbf{e}_i, \sum_j w_j T(\mathbf{e}_j) \right\rangle = \sum_i \sum_j v_i w_j \langle \mathbf{e}_i, T(\mathbf{e}_j) \rangle$$

$$= \sum_i \sum_j v_i w_j \langle T(\mathbf{e}_i), \mathbf{e}_j \rangle$$

$$= \left\langle \sum_i v_i T(\mathbf{e}_i), \sum_j w_j \mathbf{e}_j \right\rangle$$

$$= \langle T(\mathbf{v}), \mathbf{w} \rangle$$

where we used (4) at the third stage. This proves (1).

A linear operator T on an inner product space V is called **symmetric** if $\langle \mathbf{v}, T(\mathbf{w}) \rangle = \langle T(\mathbf{v}), \mathbf{w} \rangle$ holds for all $\mathbf{v}$ and $\mathbf{w}$ in V.

Example 3

If A is an $n \times n$ matrix, let $T_A : \mathbb{R}^n \to \mathbb{R}^n$ be the matrix operator given by $T_A(\mathbf{v}) = A\mathbf{v}$ for all columns $\mathbf{v}$. If the dot product is used in $\mathbb{R}^n$, then T_A is a symmetric operator if and only if A is a symmetric matrix.

Solution If B is the standard basis of $\mathbb{R}^n$, then B is orthonormal if the dot product is used. We have $M_B(T_A) = A$ (by Example 4 §9.1), so the result follows immediately from part (3) of Theorem 3.

It is important to note that whether an operator is symmetric depends on which inner product is being used (see Exercise 2).

If V is a finite dimensional inner product space, the eigenvalues of an operator $T : V \rightarrow V$ are the same as those of $M_B(T)$ for any orthonormal basis B (see Theorem 3 §9.3). If T is symmetric, $M_B(T)$ is a symmetric matrix and so has real eigenvalues by Theorem 7 §5.5. Hence we have the following:

Theorem 4

A symmetric linear operator on a finite dimensional inner product space has real eigenvalues.

If U is a subspace of an inner product space V, recall that its orthogonal complement is the subspace $U^\perp$ of V defined by

$$U^\perp = \{\mathbf{v} \text{ in } V \,|\, \langle \mathbf{v}, \mathbf{u} \rangle = 0 \text{ for all } \mathbf{u} \text{ in } U\}.$$

Theorem 5

Let $T : V \rightarrow V$ be a symmetric linear operator on an inner product space V, and let U be a T-invariant subspace of V. Then:
1. The restriction of T to U is a symmetric linear operator on U.
2. $U^\perp$ is also T-invariant.

PROOF

1. U is itself an inner product space using the same inner product, and condition 1 in Theorem 3 that T is symmetric is clearly preserved.
2. If $\mathbf{v}$ is in $U^\perp$, our task is to show that $T(\mathbf{v})$ is also in $U^\perp$; that is, $\langle T(\mathbf{v}), \mathbf{u} \rangle = 0$ for all $\mathbf{u}$ in U. But if $\mathbf{u}$ is in U, then $T(\mathbf{u})$ also lies in U because U is T-invariant, so

$$\langle T(\mathbf{v}), \mathbf{u} \rangle = \langle \mathbf{v}, T(\mathbf{u}) \rangle = 0$$

using the symmetry of T and the definition of $U^\perp$.

The principal axis theorem (Theorem 2 §8.2) asserts that an $n \times n$ matrix A is symmetric if and only if $\mathbb{R}^n$ has an orthogonal basis of eigenvectors of A. The following result not only extends this theorem to an arbitrary n-dimensional inner product space, but the proof is much more intuitive.

Theorem 6 Principal Axis Theorem

The following conditions are equivalent for a linear operator T on a finite dimensional inner product space V.
1. T is symmetric.
2. V has an orthogonal basis consisting of eigenvectors of T.

PROOF

(1) $\Rightarrow$ (2). Assume that T is symmetric and proceed by induction on $n = \dim V$. If $n = 1$, *every* nonzero vector in V is an eigenvector of T, so there is nothing to prove. If $n \geq 2$, assume inductively that the theorem holds for spaces of dimension less than n. Let λ_1 be a real eigenvalue of T (by Theorem 4) and

choose an eigenvector $\mathbf{e}_1$ corresponding to λ_1. Then $U = \mathbb{R}\mathbf{e}_1$ is T-invariant, so $U^\perp$ is also T-invariant by Theorem 5 (T is symmetric). Because dim $U^\perp = n - 1$ (Theorem 6 §10.2) and because the restriction of T to $U^\perp$ is a symmetric operator (Theorem 5), it follows by induction that $U^\perp$ has an orthogonal basis $\{\mathbf{e}_2, \dots, \mathbf{e}_n\}$ of eigenvectors of T. Hence $B = \{\mathbf{e}_1, \mathbf{e}_2, \dots, \mathbf{e}_n\}$ is an orthogonal basis of V, which proves (2).

(2) $\Rightarrow$ (1). If $B = \{\mathbf{e}_1, \dots, \mathbf{e}_n\}$ is a basis as in (2), then $M_B(T)$ is symmetric (indeed diagonal), so T is symmetric by Theorem 3.

The matrix version of the principal axis theorem is an immediate consequence of Theorem 6. If A is an $n \times n$ symmetric matrix, then $T_A : \mathbb{R}^n \to \mathbb{R}^n$ is a symmetric operator, so let B be an orthonormal basis of $\mathbb{R}^n$ consisting of eigenvectors of T_A (and hence of A). Then $P^T AP$ is diagonal where P is the orthogonal matrix whose columns are the vectors in B (see Theorem 4 §9.2).

Similarly, let $T : V \to V$ be a symmetric linear operator on the n-dimensional inner product space V and let B_0 be any convenient orthonormal basis of V. Then an orthonormal basis of eigenvectors of T can be computed from $M_{B_0}(T)$. In fact, if $P^T M_{B_0}(T)P$ is diagonal where P is orthogonal, let $B = \{\mathbf{e}_1, \dots, \mathbf{e}_n\}$ be the vectors in V such that $C_{B_0}(\mathbf{e}_j)$ is column j of P for each j. Then B consists of eigenvectors by Theorem 3 §9.3, and they are orthonormal because B_0 is orthonormal. Indeed

$$\langle \mathbf{e}_i, \mathbf{e}_j \rangle = C_{B_0}(\mathbf{e}_i) \cdot C_{B_0}(\mathbf{e}_j)$$

holds for all i and j, as the reader can verify. Here is an example.

Example 4

Let $T : \mathbf{P}_2 \to \mathbf{P}_2$ be given by

$$T(a + bx + cx^2) = (8a - 2b + 2c) + (-2a + 5b + 4c)x + (2a + 4b + 5c)x^2$$

Using the inner product $\langle a + bx + cx^2, a' + b'x + c'x^2 \rangle = aa' + bb' + cc'$, show that T is symmetric and find an orthonormal basis of $\mathbf{P}_2$ consisting of eigenvectors.

Solution   If $B_0 = \{1, x, x^2\}$, then $M_{B_0}(T) = \begin{bmatrix} 8 & -2 & 2 \\ -2 & 5 & 4 \\ 2 & 4 & 5 \end{bmatrix}$ is symmetric, so T is symmetric.

This matrix was analyzed in Example 5 §8.2, where it was found that an *orthonormal* basis of eigenvectors is $\{\frac{1}{3}[1 \ 2 \ -2]^T, \frac{1}{3}[2 \ 1 \ 2]^T, \frac{1}{3}[-2 \ 2 \ 1]^T\}$. Because B_0 is orthonormal, the corresponding orthonormal basis of $\mathbf{P}_2$ is

$$B = \{\tfrac{1}{3}(1 + 2x - 2x^2), \tfrac{1}{3}(2 + x + 2x^2), \tfrac{1}{3}(-2 + 2x + x^2)\}.$$

Exercises 10.3

1. In each case, show that T is symmetric by calculating $M_B(T)$ for some orthonormal basis B.

(a) $T : \mathbb{R}^3 \to \mathbb{R}^3$;
$T(a, b, c) = (a - 2b, -2a + 2b + 2c, 2b - c)$;
dot product

◆(b) $T : \mathbf{M}_{22} \to \mathbf{M}_{22}$; $T\begin{bmatrix} a & b \\ c & d \end{bmatrix} = \begin{bmatrix} c - a & d - b \\ a + 2c & b + 2d \end{bmatrix}$;
inner product

$$\left\langle \begin{bmatrix} x & y \\ z & w \end{bmatrix}, \begin{bmatrix} x' & y' \\ z' & w' \end{bmatrix} \right\rangle = xx' + yy' + zz' + ww'$$

(c) $T: \mathbf{P}_2 \to \mathbf{P}_2$; $T(a + bx + cx^2)$

$= (b + c) + (a + c)x + (a + b)x^2$; inner product

$\langle a + bx + cx^2, a' + b'x + c'x^2 \rangle = aa' + bb' + cc'$

2. Let $T: \mathbb{R}^2 \to \mathbb{R}^2$ be given by $T(a, b) = (2a + b, a - b)$.

   (a) Show that T is symmetric if the dot product is used.

   (b) Show that T is *not* symmetric if $\langle X, Y \rangle = XAY^T$, where $A = \begin{bmatrix} 1 & 1 \\ 1 & 2 \end{bmatrix}$. [*Hint:* Check that $B = \{(1, 0), (1, -1)\}$ is an orthonormal basis.]

3. Let $T: \mathbb{R}^2 \to \mathbb{R}^2$ be given by $T(a, b) = (a - b, b - a)$. Use the dot product in $\mathbb{R}^2$.

   (a) Show that T is symmetric.

   (b) Show that $M_B(T)$ is *not* symmetric if the orthogonal basis $B = \{(1, 0), (0, 2)\}$ is used. Why does this not contradict Theorem 3?

4. Let V be an n-dimensional inner product space, and let T and S denote symmetric linear operators on V. Show that:

   (a) The identity operator is symmetric.

   ◆(b) rT is symmetric for all r in $\mathbb{R}$.

   (c) $S + T$ is symmetric.

   ◆(d) If T is invertible, then T^{-1} is symmetric.

   (e) If $ST = TS$, then ST is symmetric.

5. In each case, show that T is symmetric and find an orthonormal basis of eigenvectors of T.

   (a) $T: \mathbb{R}^3 \to \mathbb{R}^3$; $T(a, b, c) = (2a + 2c, 3b, 2a + 5c)$; use the dot product

   ◆(b) $T: \mathbb{R}^3 \to \mathbb{R}^3$; $T(a, b, c) = (7a - b, -a + 7b, 2c)$; use the dot product

   (c) $T: \mathbf{P}_2 \to \mathbf{P}_2$; $T(a + bx + cx^2) = 3b + (3a + 4c)x + 4bx^2$; inner product $\langle a + bx + cx^2, a' + b'x + c'x^2 \rangle = aa' + bb' + cc'$

   ◆(d) $T: \mathbf{P}_2 \to \mathbf{P}_2$;

   $T(a + bx + cx^2) = (c - a) + 3bx + (a - c)x^2$; inner product as in part (c)

6. If A is any $n \times n$ matrix, let $T_A: \mathbb{R}^n \to \mathbb{R}^n$ be given by $T_A(X) = AX$. Suppose an inner product on $\mathbb{R}^n$ is given by $\langle X, Y \rangle = X^T PY$, where P is a positive definite matrix.

   (a) Show that T_A is symmetric if and only if $PA = A^T P$.

   (b) Use part (a) to deduce Example 3.

7. Let $T: \mathbf{M}_{22} \to \mathbf{M}_{22}$ be given by $T(X) = AX$, where A is a fixed 2×2 matrix.

   (a) Compute $M_B(T)$, where

   $B = \left\{ \begin{bmatrix} 1 & 0 \\ 0 & 0 \end{bmatrix}, \begin{bmatrix} 0 & 0 \\ 1 & 0 \end{bmatrix}, \begin{bmatrix} 0 & 1 \\ 0 & 0 \end{bmatrix}, \begin{bmatrix} 0 & 0 \\ 0 & 1 \end{bmatrix} \right\}$.

   Note the order!

   ◆(b) Show that $c_T(x) = [c_A(x)]^2$.

   (c) If the inner product on $\mathbf{M}_{22}$ is $\langle X, Y \rangle = \text{tr}(XY^T)$, show that T is symmetric if and only if A is a symmetric matrix.

8. Let $T: \mathbb{R}^2 \to \mathbb{R}^2$ be given by $T(a, b) = (b - a, a + 2b)$. Show that T is symmetric if the dot product is used in $\mathbb{R}^2$ but that it is not symmetric if the following inner product is used: $\langle X, Y \rangle = XAY^T$, $A = \begin{bmatrix} 1 & -1 \\ -1 & 2 \end{bmatrix}$.

9. If $T: V \to V$ is symmetric, show that $T(U)^\perp = T^{-1}(U^\perp)$ holds for every subspace U of V. Here $T^{-1}(W) = \{\mathbf{v} \mid T(\mathbf{v}) \text{ is in } W\}$.

10. Let $T: \mathbf{M}_{22} \to \mathbf{M}_{22}$ be defined by $T(X) = PXQ$, where P and Q are nonzero 2×2 matrices. Use the inner product $\langle X, Y \rangle = \text{tr}(XY^T)$. Show that T is symmetric if and only if either P and Q are both symmetric or both are scalar multiples of $\begin{bmatrix} 0 & 1 \\ -1 & 0 \end{bmatrix}$. [*Hint:* If B is as in part (a) of Exercise 7,

   then $M_B(T) = \begin{bmatrix} aP & cP \\ bP & dP \end{bmatrix}$ in block form,

   where $Q = \begin{bmatrix} a & b \\ c & d \end{bmatrix}$.

   If $B_0 = \left\{ \begin{bmatrix} 1 & 0 \\ 0 & 0 \end{bmatrix}, \begin{bmatrix} 0 & 1 \\ 0 & 0 \end{bmatrix}, \begin{bmatrix} 0 & 0 \\ 1 & 0 \end{bmatrix}, \begin{bmatrix} 0 & 0 \\ 0 & 1 \end{bmatrix} \right\}$,

   then $M_{B_0}(T) = \begin{bmatrix} pQ^T & qQ^T \\ rQ^T & sQ^T \end{bmatrix}$, where $P = \begin{bmatrix} p & q \\ r & s \end{bmatrix}$.

   Use the fact that $cP = bP^T \Rightarrow (c^2 - b^2)P = 0$.]

11. Let $T: V \to W$ be any linear transformation and let $B = \{\mathbf{b}_1, \ldots, \mathbf{b}_n\}$ and $D = \{\mathbf{d}_1, \ldots, \mathbf{d}_m\}$ be bases of V and W, respectively. If W is an inner product space and D is orthogonal, show that

   $$M_{DB}(T) = \left[\frac{\langle \mathbf{d}_i, T(\mathbf{b}_j) \rangle}{\|\mathbf{d}_i\|^2} \right]$$

   This is a generalization of Theorem 2.

12. Let $T: V \to V$ be a linear operator on an inner product space V of finite dimension. Show that the following are equivalent.

(1) $\langle \mathbf{v}, T(\mathbf{w}) \rangle = -\langle T(\mathbf{v}), \mathbf{w} \rangle$ for all $\mathbf{v}$ and $\mathbf{w}$ in V.

♦ (2) $M_B(T)$ is skew-symmetric for every orthonormal basis B.

(3) $M_B(T)$ is skew-symmetric for some orthonormal basis B.

Such operators T are called **skew-symmetric** operators.

13. Let $T : V \to V$ be a linear operator on an n-dimensional inner product space V.

   (a) Show that T is symmetric if and only if it satisfies the following two conditions.

      (i) $c_T(x)$ factors completely over $\mathbb{R}$.

      (ii) If U is a T-invariant subspace of V, then $U^{\perp}$ is also T-invariant.

   (b) Using the standard inner product on $\mathbb{R}^2$, show that $T : \mathbb{R}^2 \to \mathbb{R}^2$ with $T(a, b) = (a, a + b)$ satisfies condition (i) and that $S : \mathbb{R}^2 \to \mathbb{R}^2$ with $S(a, b) = (b, -a)$ satisfies condition (ii) but that neither is symmetric. (Example 4 §9.3 is useful for S.)

   [*Hint for part* (a): If conditions (i) and (ii) hold, proceed by induction on n. By condition (i), let $\mathbf{e}_1$ be an eigenvector of T. If $U = \mathbb{R}\mathbf{e}_1$, then $U^{\perp}$ is T-invariant by condition (ii), so show that the restriction of T to $U^{\perp}$ satisfies conditions (i) and (ii). (Theorem 1 §9.3 is helpful for part (i)). Then apply induction to show that V has an orthogonal basis of eigenvectors (as in Theorem 6)].

14. Let $B = \{\mathbf{e}_1, \mathbf{e}_2, \dots, \mathbf{e}_n\}$ be an orthonormal basis of an inner product space V. Given $T : V \to V$, define $T' : V \to V$ by

$$T'(\mathbf{v}) = \langle \mathbf{v}, T(\mathbf{e}_1) \rangle \mathbf{e}_1 + \langle \mathbf{v}, T(\mathbf{e}_2) \rangle \mathbf{e}_2 + \cdots +$$

$$\langle \mathbf{v}, T(\mathbf{e}_n) \rangle \mathbf{e}_n = \sum_{i=1}^{n} \langle \mathbf{v}, T(\mathbf{e}_i) \rangle \mathbf{e}_i$$

   (a) Show that $(aT)' = aT'$.

   (b) Show that $(S + T)' = S' + T'$.

♦ (c) Show that $M_B(T')$ is the transpose of $M_B(T)$.

   (d) Show that $(T')' = T$, using part (c).
      [*Hint:* $M_B(S) = M_B(T)$ implies that $S = T$.]

   (e) Show that $(ST)' = T'S'$, using part (c).

   (f) Show that T is symmetric if and only if $T = T'$. [*Hint:* Use the expansion theorem and Theorem 3.]

   (g) Show that $T + T'$ and TT' are symmetric, using parts (b) through (e).

   (h) Show that $T'(\mathbf{v})$ is independent of the choice of orthonormal basis B. [*Hint:* If $D = \{\mathbf{f}_1, \dots, \mathbf{f}_n\}$ is also orthonormal, use the fact that $\mathbf{e}_i = \sum_{j=1}^{n} \langle \mathbf{e}_i, \mathbf{f}_j \rangle \mathbf{f}_j$ for each i.]

15. Let V be a finite dimensional inner product space. Show that the following conditions are equivalent for a linear operator $T : V \to V$.

   (1) T is symmetric and $T^2 = T$.

   (2) $M_B(T) = \begin{bmatrix} I_r & 0 \\ 0 & 0 \end{bmatrix}$ for some orthonormal basis B of V.

   An operator is called a **projection** if it satisfies these conditions. [*Hint:* If $T^2 = T$ and $T(\mathbf{v}) = \lambda\mathbf{v}$, apply T to get $\lambda\mathbf{v} = \lambda^2\mathbf{v}$. Hence show that 0, 1 are the only eigenvalues of T.]

16. Let V denote a finite dimensional inner product space. Given a subspace U, define $\text{proj}_U : V \to V$ as in Theorem 7 §10.2.

   (a) Show that proj_U is a projection in the sense of Exercise 15.

   (b) If T is any projection, show that $T = \text{proj}_U$, where $U = \text{im } T$. [*Hint:* Use $T^2 = T$ to show that $V = \text{im } T \oplus \text{ker } T$ and $T(\mathbf{u}) = \mathbf{u}$ for all $\mathbf{u}$ in im T. Use the fact that T is symmetric to show that ker $T \subseteq (\text{im } T)^{\perp}$ and hence that these are equal because they have the same dimension.]

SECTION 10.4 Isometries

We saw in Chapter 2 that rotations about the origin and reflections in a line through the origin are linear operators on $\mathbb{R}^2$. Similar geometric arguments establish that, in $\mathbb{R}^3$, rotations about a line through the origin and reflections in a plane through the origin are linear. We are going to give an algebraic proof of these results that is valid in any inner product space. The key observation is that reflections and rotations are distance preserving in the following sense. If V is an inner product space, a transformation $T : V \to V$ (not necessarily linear) is said to be

distance preserving if the distance between $T(\mathbf{v})$ and $T(\mathbf{w})$ is the same as the distance between $\mathbf{v}$ and $\mathbf{w}$ for all vectors $\mathbf{v}$ and $\mathbf{w}$; more formally, if

$$\|T(\mathbf{v}) - T(\mathbf{w})\| = \|\mathbf{v} - \mathbf{w}\| \quad \text{for all } \mathbf{v} \text{ and } \mathbf{w} \text{ in } V. \tag{$*$}$$

Distance preserving maps need not be linear. For example, if $\mathbf{u}$ is any vector in V, the transformation $T_{\mathbf{u}} : V \to V$ defined by $T_{\mathbf{u}}(\mathbf{v}) = \mathbf{v} + \mathbf{u}$ for all $\mathbf{v}$ in V is called **translation** by $\mathbf{u}$, and it is routine to verify that $T_{\mathbf{u}}$ is distance preserving for any $\mathbf{u}$. However, $T_{\mathbf{u}}$ is linear only if $\mathbf{u} = \mathbf{0}$. Remarkably, distance preserving operators that fix the origin are necessarily linear.

Theorem 1

Let V be an inner product space of dimension n, and consider a distance preserving transformation $T : V \to V$. If $T(\mathbf{0}) = \mathbf{0}$, then T is linear.

PROOF

We have $\|T(\mathbf{v}) - T(\mathbf{w})\|^2 = \|\mathbf{v} - \mathbf{w}\|^2$ for all $\mathbf{v}$ and $\mathbf{w}$ in V by ($*$), which gives

$$\langle T(\mathbf{v}), T(\mathbf{w}) \rangle = \langle \mathbf{v}, \mathbf{w} \rangle \quad \text{for all } \mathbf{v} \text{ and } \mathbf{w} \text{ in } V. \tag{$**$}$$

Now let $\{\mathbf{f}_1, \mathbf{f}_2, \dots, \mathbf{f}_n\}$ be an orthonormal basis of V. Then $\{T(\mathbf{f}_1), T(\mathbf{f}_2), \dots, T(\mathbf{f}_n)\}$ is orthonormal by ($**$) and so is a basis because $\dim V = n$. Now compute:

$$\langle T(\mathbf{v} + \mathbf{w}) - T(\mathbf{v}) - T(\mathbf{w}), T(\mathbf{f}_i) \rangle = \langle T(\mathbf{v} + \mathbf{w}), T(\mathbf{f}_i) \rangle - \langle T(\mathbf{v}), T(\mathbf{f}_i) \rangle - \langle T(\mathbf{w}), T(\mathbf{f}_i) \rangle$$
$$= \langle \mathbf{v} + \mathbf{w}, \mathbf{f}_i \rangle - \langle \mathbf{v}, \mathbf{f}_i \rangle - \langle \mathbf{w}, \mathbf{f}_i \rangle$$
$$= 0$$

for each i. It follows from the expansion theorem (Theorem 4 §10.2) that $T(\mathbf{v} + \mathbf{w}) - T(\mathbf{v}) - T(\mathbf{w}) = \mathbf{0}$; that is, $T(\mathbf{v} + \mathbf{w}) = T(\mathbf{v}) + T(\mathbf{w})$. A similar argument shows that $T(a\mathbf{v}) = aT(\mathbf{v})$ holds for all a in $\mathbb{R}$ and $\mathbf{v}$ in V, so T is linear after all.

Distance-preserving linear operators are called **isometries**.

It is routine to verify that the composite of two distance-preserving transformations is again distance preserving. In particular the composite of a translation and an isometry is distance preserving. Surprisingly, the converse is true.

Corollary

If V is a finite dimensional inner product space, then every distance-preserving transformation $S : V \to V$ is the composite of a translation and an isometry.

PROOF

If $S : V \to V$ is distance preserving, write $S(\mathbf{0}) = \mathbf{u}$ and define $T : V \to V$ by $T(\mathbf{v}) = S(\mathbf{v}) - \mathbf{u}$ for all $\mathbf{v}$ in V. Then $\|T(\mathbf{v}) - T(\mathbf{w})\| = \|\mathbf{v} - \mathbf{w}\|$ for all vectors $\mathbf{v}$ and $\mathbf{w}$ in V as the reader can verify; that is, T is distance preserving. Clearly, $T(\mathbf{0}) = \mathbf{0}$, so it is an isometry by Theorem 1. Since $S(\mathbf{v}) = \mathbf{u} + T(\mathbf{v}) = (T_{\mathbf{u}} \circ T)(\mathbf{v})$ for all $\mathbf{v}$ in V, we have $S = T_{\mathbf{u}} \circ T$, and the corollary is proved.

The corollary focuses our attention on the isometries, and the next theorem shows that, while they preserve distance, they are characterized as those operators that preserve other properties.

Theorem 2

Let $T : V \to V$ be a linear operator on a finite dimensional inner product space V. The following conditions are equivalent:
1. T is an isometry.   (T preserves distance)
2. $\|T(\mathbf{v})\| = \|\mathbf{v}\|$ for all $\mathbf{v}$ in V.   (T preserves norms)
3. $\langle T(\mathbf{v}), T(\mathbf{w}) \rangle = \langle \mathbf{v}, \mathbf{w} \rangle$ for all $\mathbf{v}$ and $\mathbf{w}$ in V.   (T preserves inner products)
4. If $\{\mathbf{f}_1, \mathbf{f}_2, \ldots, \mathbf{f}_n\}$ is an orthonormal basis of V, then $\{T(\mathbf{f}_1), T(\mathbf{f}_2), \ldots, T(\mathbf{f}_n)\}$ is also an orthonormal basis.   (T preserves orthonormal bases)
5. T carries some orthonormal basis to an orthonormal basis.

PROOF

$(1) \Rightarrow (2)$. Take $\mathbf{w} = \mathbf{0}$ in $(*)$.

$(2) \Rightarrow (3)$. Since T is linear, (2) gives $\|T(\mathbf{v}) - T(\mathbf{w})\|^2 = \|T(\mathbf{v} - \mathbf{w})\|^2 = \|\mathbf{v} - \mathbf{w}\|^2$. Now (3) follows.

$(3) \Rightarrow (4)$. By (3), $\{T(\mathbf{f}_1), T(\mathbf{f}_2), \ldots, T(\mathbf{f}_n)\}$ is orthogonal and $\|T(\mathbf{f}_i)\|^2 = \|\mathbf{f}_i\|^2 = 1$. Hence it is a basis because $\dim V = n$.

$(4) \Rightarrow (5)$. This needs no proof.

$(5) \Rightarrow (1)$. By (5), let $\{\mathbf{f}_1, \ldots, \mathbf{f}_n\}$ be an orthonormal basis of V such that $\{T(\mathbf{f}_1), \ldots, T(\mathbf{f}_n)\}$ is also orthonormal. Given $\mathbf{v} = v_1\mathbf{f}_1 + \cdots + v_n\mathbf{f}_n$ in V, we have $T(\mathbf{v}) = v_1 T(\mathbf{f}_1) + \cdots + v_n T(\mathbf{f}_n)$ so Pythagoras' theorem gives $\|T(\mathbf{v})\|^2 = v_1^2 + \cdots + v_n^2 = \|\mathbf{v}\|^2$. Hence $\|T(\mathbf{v})\| = \|\mathbf{v}\|$, and (1) follows by replacing $\mathbf{v}$ by $\mathbf{v} - \mathbf{w}$.

Before giving examples, we note some consequences of Theorem 2.

Corollary

Let V be a finite dimensional inner product space.
1. Every isometry of V is an isomorphism.
2. (a) $1_V : V \to V$ is an isometry.
   (b) The composite of two isometries of V is an isometry.
   (c) The inverse of an isometry of V is an isometry.

PROOF

(1) is by (4) of Theorem 2 and Theorem 1 §7.3, (2a) is clear, and (2b) is left to the reader. If $T : V \to V$ is an isometry and $\{\mathbf{f}_1, \ldots, \mathbf{f}_n\}$ is an orthonormal basis of V, then (2c) follows because T^{-1} carries the orthonormal basis $\{T(\mathbf{f}_1), \ldots, T(\mathbf{f}_n)\}$ back to $\{\mathbf{f}_1, \ldots, \mathbf{f}_n\}$.

The conditions in part (2) of the corollary assert that the set of isometries of a finite dimensional inner product space forms an algebraic system called a **group**. The theory of groups is well developed, and groups of operators are important in geometry. In fact, geometry itself can be fruitfully viewed as the study of those properties of a vector space that are preserved by a group of invertible linear operators.

Example 1

Rotations of $\mathbb{R}^2$ about the origin are isometries, as are reflections in lines through the origin: They clearly preserve distance and so are linear by Theorem 1. Similarly, rotations about lines through the origin and reflections in planes through the origin are isometries of $\mathbb{R}^3$.

Example 2

Let $T : \mathbf{M}_{nn} \to \mathbf{M}_{nn}$ be the transposition operator: $T(A) = A^T$. Then T is an isometry if the inner product is $\langle A, B \rangle = \text{tr}(AB^T) = \sum_{i,j} a_{ij} b_{ij}$. In fact, T permutes the basis consisting of all matrices with one entry 1 and the other entries 0.

The proof of the next result requires the fact (verify) that, if B is an orthonormal basis, then $\langle \mathbf{v}, \mathbf{w} \rangle = C_B(\mathbf{v}) \cdot C_B(\mathbf{w})$ for all vectors $\mathbf{v}$ and $\mathbf{w}$.

Theorem 3

Let $T : V \to V$ be an operator where V is a finite dimensional inner product space. The following conditions are equivalent.
1. T is an isometry.
2. $M_B(T)$ is an orthogonal matrix for every orthonormal basis B.
3. $M_B(T)$ is an orthogonal matrix for some orthonormal basis B.

PROOF

(1) $\Rightarrow$ (2). Let $B = \{\mathbf{e}_1, \dots, \mathbf{e}_n\}$ be an orthonormal basis. Then the jth column of $M_B(T)$ is $C_B[T(\mathbf{e}_j)]$, and we have

$$C_B[T(\mathbf{e}_j)] \cdot C_B[T(\mathbf{e}_k)] = \langle T(\mathbf{e}_j), T(\mathbf{e}_k) \rangle = \langle \mathbf{e}_j, \mathbf{e}_k \rangle$$

using (1). Hence the columns of $M_B(T)$ are orthonormal in $\mathbb{R}^n$, which proves (2).
(2) $\Rightarrow$ (3). This is clear.
(3) $\Rightarrow$ (1). Let $B = \{\mathbf{e}_1, \dots, \mathbf{e}_n\}$ be as in (3). Then, as before,

$$\langle T(\mathbf{e}_j), T(\mathbf{e}_k) \rangle = C_B[T(\mathbf{e}_j)] \cdot C_B[T(\mathbf{e}_k)]$$

so $\{T(\mathbf{e}_1), \dots, T(\mathbf{e}_n)\}$ is orthonormal by (3) and Theorem 2 applies.

It is important that B is *orthonormal* in Theorem 3. For example, $T : V \to V$ given by $T(\mathbf{v}) = 2\mathbf{v}$ preserves *orthogonal* sets but is not an isometry, as is easily checked.
If P is an orthogonal square matrix, then $P^{-1} = P^T$. Taking determinants yields $(\det P)^2 = 1$, so $\det P = \pm 1$. Hence:

Corollary

If $T : V \to V$ is an isometry where V is a finite dimensional inner product space, then $\det T = \pm 1$.

Example 3

If A is any $n \times n$ matrix, the matrix operator $T_A : \mathbb{R}^n \to \mathbb{R}^n$ is an isometry if and only if A is orthogonal (use the dot product in $\mathbb{R}^n$). Indeed, if E is the standard basis of $\mathbb{R}^n$, then $M_E(T_A) = A$.

Rotations and reflections that fix the origin are isometries in $\mathbb{R}^2$ and $\mathbb{R}^3$ (Example 1); we are going to show that these isometries (and compositions of them in $\mathbb{R}^3$) are the only possibilities. In fact, this will follow from a general structure theorem for isometries. Surprisingly enough, much of the work involves the two-dimensional case.

Theorem 4

Let $T : V \to V$ be an isometry on the two dimensional inner product space V. Then there are two possibilities.

   Either (1) There is an orthonormal basis B of V such that

$$M_B(T) = \begin{bmatrix} \cos\theta & -\sin\theta \\ \sin\theta & \cos\theta \end{bmatrix}, \quad 0 \le \theta < 2\pi$$

or   (2) There is an orthonormal basis B of V such that

$$M_B(T) = \begin{bmatrix} 1 & 0 \\ 0 & -1 \end{bmatrix}$$

Furthermore, type (1) occurs if and only if $\det T = 1$, and type (2) occurs if and only if $\det T = -1$.

PROOF

The final statement follows from the rest because $\det T = \det[M_B(T)]$ for any basis B. Let $B_0 = \{\mathbf{e}_1, \mathbf{e}_2\}$ be any ordered orthonormal basis of V and write

$$A = M_{B_0}(T) = \begin{bmatrix} a & b \\ c & d \end{bmatrix}; \quad \text{that is,} \quad \begin{aligned} T(\mathbf{e}_1) &= a\mathbf{e}_1 + c\mathbf{e}_2 \\ T(\mathbf{e}_2) &= b\mathbf{e}_1 + d\mathbf{e}_2 \end{aligned}$$

Then A is orthogonal by Theorem 3, so its columns (and rows) are orthonormal. Hence $a^2 + c^2 = 1 = b^2 + d^2$, so (a, c) and (d, b) lie on the unit circle. Thus angles θ and φ exist such that

$$\begin{aligned} a &= \cos\theta, & c &= \sin\theta & 0 \le \theta < 2\pi \\ d &= \cos\varphi, & b &= \sin\varphi & 0 \le \varphi < 2\pi \end{aligned}$$

Then $\sin(\theta + \varphi) = cd + ab = 0$ because the columns of A are orthogonal, so $\theta + \varphi = k\pi$ for some integer k. This gives $d = \cos(k\pi - \theta) = (-1)^k \cos\theta$ and $b = \sin(k\pi - \theta) = (-1)^{k+1}\sin\theta$. Finally

$$A = \begin{bmatrix} \cos\theta & (-1)^{k+1}\sin\theta \\ \sin\theta & (-1)^k \cos\theta \end{bmatrix}$$

If k is even, we are in type (1) with $B = B_0$, so assume k is odd. Then $A = \begin{bmatrix} a & c \\ c & -a \end{bmatrix}$. If $a = 1$ and $c = 0$, we are in type (2) with $B = \{\mathbf{e}_2, \mathbf{e}_1\}$. Otherwise A has eigenvalues 1 and -1 with (nonzero) eigenvectors

$$\begin{bmatrix} 1+a \\ c \end{bmatrix} \text{ and } \begin{bmatrix} -c \\ 1+a \end{bmatrix} \text{ respectively. Hence}$$

$$\mathbf{f}_1 = (1+a)\mathbf{e}_1 + c\mathbf{e}_2 \quad \text{and} \quad \mathbf{f}_2 = -c\mathbf{e}_1 + (1+a)\mathbf{e}_2$$

are orthogonal eigenvectors of T (verify) and we are in type (2) with

$$B = \left\{ \frac{1}{\|\mathbf{f}_1\|}\mathbf{f}_1, \ \frac{1}{\|\mathbf{f}_2\|}\mathbf{f}_2 \right\}.$$

If $V = \mathbb{R}^2$, Theorem 4 has a satisfying geometrical interpretation: The type 1 and type 2 isometries are just rotations and reflections, respectively. In fact, if B_0 is the standard basis, then the clockwise rotation R_θ about the origin through an angle θ has matrix

$$M_{B_0}(R_\theta) = \begin{bmatrix} \cos\theta & -\sin\theta \\ \sin\theta & \cos\theta \end{bmatrix}$$

(see Example 5 §2.5). On the other hand, if $S : \mathbb{R}^2 \to \mathbb{R}^2$ is the reflection in a line through the origin (called the **fixed line** of the reflection), let $\mathbf{f}_1$ be a unit vector pointing along the fixed line and let $\mathbf{f}_2$ be a unit vector perpendicular to the fixed line. Then $B = \{\mathbf{f}_1, \mathbf{f}_2\}$ is an orthonormal basis, $S(\mathbf{f}_1) = \mathbf{f}_1$ and $S(\mathbf{f}_2) = -\mathbf{f}_2$, so

$$M_B(S) = \begin{bmatrix} 1 & 0 \\ 0 & -1 \end{bmatrix}$$

Thus S is of type 2. Note that, in this case, 1 is an eigenvalue of S, and any eigenvector corresponding to 1 is a direction vector for the fixed line.

Example 4

In each case, determine whether $T_A : \mathbb{R}^2 \to \mathbb{R}^2$ is a rotation or a reflection, and then find the angle or fixed line:

(a) $A = \frac{1}{2}\begin{bmatrix} 1 & \sqrt{3} \\ -\sqrt{3} & 1 \end{bmatrix}$ (b) $A = \frac{1}{5}\begin{bmatrix} -3 & 4 \\ 4 & 3 \end{bmatrix}$

Solution   Both matrices are orthogonal, so (because $M_{B_0}(T_A) = A$, where B_0 is the standard basis) T_A is an isometry in both cases. In the first case, $\det A = 1$, so T_A is a counterclockwise rotation through θ, where $\cos\theta = \frac{1}{2}$ and $\sin\theta = -\frac{\sqrt{3}}{2}$. Thus $\theta = \frac{5\pi}{3}$ (or 300°). In (b), $\det A = -1$, so T_A is a reflection in this case. We verify that $\mathbf{d} = \begin{bmatrix} 1 \\ 2 \end{bmatrix}$ is an eigenvector corresponding to the eigenvalue 1. Hence the fixed line $\mathbb{R}\mathbf{d}$ has equation $y = 2x$.

We now give a structure theorem for isometries. The proof requires three preliminary results, each of interest in its own right.

Lemma 1

Let $T : V \to V$ be an isometry of a finite dimensional inner product space V. If U is a T-invariant subspace of V, then $U^\perp$ is also T-invariant.

PROOF

> Let $\mathbf{w}$ lie in $U^{\perp}$. We are to prove that $T(\mathbf{w})$ is also in $U^{\perp}$; that is, $\langle T(\mathbf{w}), \mathbf{u} \rangle = 0$ for all $\mathbf{u}$ in U. At this point, observe that the restriction of T to U is an isometry $U \to U$ and so is an isomorphism by the corollary to Theorem 2. In particular, each $\mathbf{u}$ in U can be written in the form $\mathbf{u} = T(\mathbf{u}_1)$ for some $\mathbf{u}_1$ in U, so
>
> $$\langle T(\mathbf{w}), \mathbf{u} \rangle = \langle T(\mathbf{w}), T(\mathbf{u}_1) \rangle = \langle \mathbf{w}, \mathbf{u}_1 \rangle = 0$$
>
> because $\mathbf{w}$ is in $U^{\perp}$. This is what we wanted.

To employ Lemma 1 to analyze an isometry $T : V \to V$ when dim $V = n$, it is necessary to show that a T-invariant subspace U exists such that $U \neq 0$ and $U \neq V$. We will show, in fact, that such a subspace U can always be found of dimension 1 or 2. If T has a real eigenvalue λ, this is no problem because, if $\mathbf{u}$ is any λ-eigenvector, $U = \mathbb{R}\mathbf{u}$ is T-invariant. But, in case (1) of Theorem 4, the eigenvalues of T are $e^{i\theta}$ and $e^{-i\theta}$ (the reader should check this), and these are nonreal if $\theta \neq 0$ and $\theta \neq \pi$. It turns out that every complex eigenvalue λ of T has absolute value 1 (Lemma 2 below); and that U has a T-invariant subspace of dimension 2 if λ is not real (Lemma 3).

Lemma 2

Let $T : V \to V$ be an isometry of the finite dimensional inner product space V. If λ is a complex eigenvalue of T, then $|\lambda| = 1$.

PROOF

> Choose an orthonormal basis B of V, and let $A = M_B(T)$. Then A is a real orthogonal matrix so, using the standard inner product $\langle X, Y \rangle = X^T \overline{Y}$ in $\mathbb{C}^n$, we get
>
> $$\| AX \|^2 = (AX)^T \overline{(AX)} = X^T A^T A \overline{X} = \| X \|^2$$
>
> for all X in $\mathbb{C}^n$. But $AX = \lambda X$ for some $X \neq 0$, whence $\| X \|^2 = \| \lambda X \|^2 = |\lambda|^2 \| X \|^2$. This gives $|\lambda| = 1$, as required.

Lemma 3

Let $T : V \to V$ be an isometry of the n-dimensional inner product space V. If T has a nonreal eigenvalue, then V has a two-dimensional T-invariant subspace.

PROOF

> Let B be an orthonormal basis of V, let $A = M_B(T)$, and (using Lemma 2) let $\lambda = e^{i\alpha}$ be a nonreal eigenvalue of A, say $AX = \lambda X$ where $X \neq 0$ in $\mathbb{C}^n$. Because A is real, complex conjugation gives $A\overline{X} = \overline{\lambda}\overline{X}$, so $\overline{\lambda}$ is also an eigenvalue. Moreover $\lambda \neq \overline{\lambda}$ (λ is nonreal), so $\{X, \overline{X}\}$ is linearly independent in $\mathbb{C}^n$ (the argument in the proof of Theorem 4 §5.5 works). Now define
>
> $$E_1 = X + \overline{X} \quad \text{and} \quad E_2 = i(X - \overline{X})$$
>
> Then E_1 and E_2 lie in $\mathbb{R}^n$, and $\{E_1, E_2\}$ is linearly independent over $\mathbb{R}$ because $\{X, \overline{X}\}$ is linearly independent over $\mathbb{C}$. Moreover
>
> $$X = \tfrac{1}{2}(E_1 - iE_2) \quad \text{and} \quad \overline{X} = \tfrac{1}{2}(E_1 + iE_2)$$

Now $\lambda + \bar{\lambda} = 2\cos\alpha$ and $\lambda - \bar{\lambda} = 2i\sin\alpha$, and a routine computation gives

$$AE_1 = E_1\cos\alpha + E_2\sin\alpha$$
$$AE_2 = -E_1\sin\alpha + E_2\cos\alpha$$

Finally, let $\mathbf{e}_1$ and $\mathbf{e}_2$ in V be such that $E_1 = C_B(\mathbf{e}_1)$ and $E_2 = C_B(\mathbf{e}_2)$. Then

$$C_B[T(\mathbf{e}_1)] = AC_B(\mathbf{e}_1) = AE_1 = C_B(\mathbf{e}_1\cos\alpha + \mathbf{e}_2\sin\alpha)$$

using Theorem 2 §9.1. Because C_B is one-to-one, this gives the first of the following equations (the other is similar):

$$T(\mathbf{e}_1) = \mathbf{e}_1\cos\alpha + \mathbf{e}_2\sin\alpha$$
$$T(\mathbf{e}_2) = -\mathbf{e}_1\sin\alpha + \mathbf{e}_2\cos\alpha$$

Thus $U = \text{span}\{\mathbf{e}_1, \mathbf{e}_2\}$ is T-invariant and two-dimensional.

We can now prove the structure theorem for isometries.

Theorem 5

Let $T : V \to V$ be an isometry of the n-dimensional inner product space V. Given an angle θ, write $R(\theta) = \begin{bmatrix} \cos\theta & -\sin\theta \\ \sin\theta & \cos\theta \end{bmatrix}$. Then there exists an orthonormal basis B of V such that $M_B(T)$ has one of the following block diagonal forms, classified for convenience by whether n is even or odd:

$$n = 2k+1 \quad \begin{bmatrix} 1 & 0 & \cdots & 0 \\ 0 & R(\theta_1) & \cdots & 0 \\ \vdots & \vdots & \ddots & \vdots \\ 0 & 0 & \cdots & R(\theta_k) \end{bmatrix} \text{ or } \begin{bmatrix} -1 & 0 & \cdots & 0 \\ 0 & R(\theta_1) & \cdots & 0 \\ \vdots & \vdots & \ddots & \vdots \\ 0 & 0 & \cdots & R(\theta_k) \end{bmatrix}$$

$$n = 2k \quad \begin{bmatrix} R(\theta_1) & 0 & \cdots & 0 \\ 0 & R(\theta_2) & \cdots & 0 \\ \vdots & \vdots & \ddots & \vdots \\ 0 & 0 & \cdots & R(\theta_k) \end{bmatrix} \text{ or } \begin{bmatrix} -1 & 0 & 0 & \cdots & 0 \\ 0 & 1 & 0 & \cdots & 0 \\ 0 & 0 & R(\theta_1) & \cdots & 0 \\ \vdots & \vdots & \vdots & \ddots & \vdots \\ 0 & 0 & 0 & \cdots & R(\theta_{k-1}) \end{bmatrix}$$

PROOF

We show first, by induction on n, that an orthonormal basis B of V can be found such that $M_B(T)$ is a block diagonal matrix of the following form:

$$M_B(T) = \begin{bmatrix} I_r & 0 & 0 & \cdots & 0 \\ 0 & -I_s & 0 & \cdots & 0 \\ 0 & 0 & R(\theta_1) & \cdots & 0 \\ \vdots & \vdots & \vdots & \ddots & \vdots \\ 0 & 0 & 0 & \cdots & R(\theta_t) \end{bmatrix}$$

where the identity matrix I_r, the matrix $-I_s$, or the matrices $R(\theta_i)$ may be missing. If $n = 1$ and $V = \mathbb{R}\mathbf{v}$, this holds because $T(\mathbf{v}) = \lambda\mathbf{v}$ and $\lambda = \pm 1$ by Lemma 2. If $n = 2$, this follows from Theorem 4. If $n \geq 3$, either T has a real eigenvalue and therefore has a one-dimensional T-invariant subspace $U = \mathbb{R}\mathbf{u}$ for any eigenvector $\mathbf{u}$, or T has no real eigenvalue and therefore has a two-dimensional T-invariant subspace U by Lemma 3. In either case $U^{\perp}$ is T-invariant (Lemma 1) and $\dim U^{\perp} = n - \dim U < n$. Hence, by induction, let B_1 and B_2 be orthonormal bases of U and $U^{\perp}$ such that $M_{B_1}(T)$ and $M_{B_2}(T)$ have the form given. Then $B = B_1 \cup B_2$ is an orthonormal basis of V, and $M_B(T)$ has the desired form with a suitable ordering of the vectors in B.

Now observe that $R(0) = \begin{bmatrix} 1 & 0 \\ 0 & 1 \end{bmatrix}$ and $R(\pi) = \begin{bmatrix} -1 & 0 \\ 0 & -1 \end{bmatrix}$. It follows that an even number of 1's or -1's can be written as $R(\theta_1)$-blocks. Hence, with a suitable reordering of the basis B, the theorem follows.

As in the dimension-2 situation, these possibilities can be given a geometric interpretation when $V = \mathbb{R}^3$ is taken as euclidean space. As before, this entails looking carefully at reflections and rotations in $\mathbb{R}^3$. If $Q : \mathbb{R}^3 \to \mathbb{R}^3$ is any reflection in a plane through the origin (called the **fixed plane** of the reflection), take $\{E_2, E_3\}$ to be any orthonormal basis of the fixed plane and take E_1 to be a unit vector perpendicular to the fixed plane. Then $Q(E_1) = -E_1$, whereas $Q(E_2) = E_2$ and $Q(E_3) = E_3$. Hence $B = \{E_1, E_2, E_3\}$ is an orthonormal basis such that

$$M_B(Q) = \begin{bmatrix} -1 & 0 & 0 \\ 0 & 1 & 0 \\ 0 & 0 & 1 \end{bmatrix}$$

Similarly, suppose that $R : \mathbb{R}^3 \to \mathbb{R}^3$ is any rotation about a line through the origin (called the **axis** of the rotation), and let E_1 be a unit vector pointing along the axis, so $R(E_1) = E_1$. Now the plane through the origin perpendicular to the axis is a R-invariant subspace of $\mathbb{R}^2$ of dimension 2, and the restriction of R to this plane is a rotation. Hence, by Theorem 4, there is an orthonormal basis $B_1 = \{E_2, E_3\}$ of this plane such that $M_{B_1}(R) = \begin{bmatrix} \cos\theta & -\sin\theta \\ \sin\theta & \cos\theta \end{bmatrix}$. But then $B = \{E_1, E_2, E_3\}$ is an ortho-normal basis of $\mathbb{R}^3$ such that the matrix of R is

$$M_B(R) = \begin{bmatrix} 1 & 0 & 0 \\ 0 & \cos\theta & -\sin\theta \\ 0 & \sin\theta & \cos\theta \end{bmatrix}$$

However, Theorem 5 shows that there are isometries T in $\mathbb{R}^3$ of a third type: those with a matrix of the form

$$M_B(T) = \begin{bmatrix} -1 & 0 & 0 \\ 0 & \cos\theta & -\sin\theta \\ 0 & \sin\theta & \cos\theta \end{bmatrix}$$

If $B = \{E_1, E_2, E_3\}$, let Q be the reflection in the plane spanned by E_2 and E_3, and let R be the rotation corresponding to θ about the line spanned by E_1. Then $M_B(Q)$ and $M_B(R)$ are as above, and $M_B(Q) \, M_B(R) = M_B(T)$ as the reader can verify. This means that $M_B(QR) = M_B(T)$ by Theorem 1 §9.1, and this in turn implies that $QR = T$ because M_B is one-to-one (see Exercise 26 §9.1). A similar argument shows that $RQ = T$, and we have Theorem 6.

Theorem 6

If $T : \mathbb{R}^3 \to \mathbb{R}^3$ is an isometry, there are three possibilities.

(a) T is a rotation, and $M_B(T) = \begin{bmatrix} 1 & 0 & 0 \\ 0 & \cos\theta & -\sin\theta \\ 0 & \sin\theta & \cos\theta \end{bmatrix}$ for some orthonormal basis B.

(b) T is a reflection, and $M_B(T) = \begin{bmatrix} -1 & 0 & 0 \\ 0 & 1 & 0 \\ 0 & 0 & 1 \end{bmatrix}$ for some orthonormal basis B.

(c) $T = QR = RQ$ where Q is a reflection, R is a rotation about an axis perpendicular to the fixed plane of Q and $M_B(T) = \begin{bmatrix} -1 & 0 & 0 \\ 0 & \cos\theta & -\sin\theta \\ 0 & \sin\theta & \cos\theta \end{bmatrix}$ for some orthonormal basis B.

Hence T is a rotation if and only if $\det T = 1$.

PROOF

It remains only to verify the final observation that T is a rotation if and only if $\det T = 1$. But clearly $\det T = -1$ in parts (b) and (c).

A useful way of analyzing a given isometry $T : \mathbb{R}^3 \to \mathbb{R}^3$ comes from computing the eigenvalues of T. Because the characteristic polynomial of T has degree 3, it must have a real root. Hence, there must be at least one real eigenvalue, and the only possible real eigenvalues are ± 1 by Lemma 2. Thus Table 1 includes all possibilities.

TABLE 1

Eigenvalues of T	Action of T
(1) 1, no other real eigenvalues	Rotation about the line $\mathbb{R}E$ where E is an eigenvector corresponding to 1. [Case (a) of Theorem 6.]
(2) -1, no other real eigenvalues	Rotation about the line $\mathbb{R}E$ followed by reflection in the plane $(\mathbb{R}E)^\perp$ where E is an eigenvector corresponding to -1. [Case (c) of Theorem 6.]
(3) $-1, 1, 1$	Reflection in the plane $(\mathbb{R}E)^\perp$ where E is an eigenvector corresponding to -1. [Case (b) of Theorem 6.]
(4) $1, -1, -1$	This is as in (1) with a rotation of π.
(5) $-1, -1, -1$	Here $T(X) = -X$ for all X. This is (2) with a rotation of π.
(6) $1, 1, 1$	Here T is the identity isometry.

Example 5

Analyze the isometry $T : \mathbb{R}^3 \to \mathbb{R}^3$ given by $T\begin{bmatrix} x \\ y \\ z \end{bmatrix} = \begin{bmatrix} y \\ z \\ -x \end{bmatrix}$.

Solution If B_0 is the standard basis of $\mathbb{R}^3$, then $M_{B_0}(T) = \begin{bmatrix} 0 & 1 & 0 \\ 0 & 0 & 1 \\ -1 & 0 & 0 \end{bmatrix}$, so

$c_T(x) = x^3 + 1 = (x+1)(x^2 - x + 1)$. This is (2) in Table 1. Write:

$$E_1 = \frac{1}{\sqrt{3}} \begin{bmatrix} 1 \\ -1 \\ 1 \end{bmatrix} \qquad E_2 = \frac{1}{\sqrt{6}} \begin{bmatrix} 1 \\ 2 \\ 1 \end{bmatrix} \qquad E_3 = \frac{1}{\sqrt{2}} \begin{bmatrix} 1 \\ 0 \\ -1 \end{bmatrix}$$

Here E_1 is a unit eigenvector corresponding to $\lambda_1 = -1$, so T is a rotation (through an angle θ) about the line $L = \mathbb{R}E_1$, followed by reflection in the plane U through the origin perpendicular to E_1 (with equation $x - y + z = 0$). Then, $\{E_1, E_2\}$ is chosen as an orthonormal basis of U, so $B = \{E_1, E_2, E_3\}$ is an orthonormal basis of $\mathbb{R}^3$ and

$$M_B(T) = \begin{bmatrix} -1 & 0 & 0 \\ 0 & \frac{1}{2} & -\frac{\sqrt{3}}{2} \\ 0 & \frac{\sqrt{3}}{2} & \frac{1}{2} \end{bmatrix}$$

Hence θ is given by $\cos \theta = \frac{1}{2}$, $\sin \theta = \frac{\sqrt{3}}{2}$, so $\theta = \frac{\pi}{3}$.

Let V be an n-dimensional inner product space. A subspace of V of dimension $n - 1$ is called a **hyperplane** in V. Thus the hyperplanes in $\mathbb{R}^3$ and $\mathbb{R}^2$ are, respectively, the planes and lines through the origin. Let $Q : V \to V$ be an isometry with matrix

$$M_B(Q) = \begin{bmatrix} -1 & 0 \\ 0 & I_{n-1} \end{bmatrix}$$

for some orthonormal basis $B = \{\mathbf{e}_1, \mathbf{e}_2, \dots, \mathbf{e}_n\}$. Then $Q(\mathbf{e}_1) = -\mathbf{e}_1$ whereas $Q(\mathbf{u}) = \mathbf{u}$ for each $\mathbf{u}$ in $U = \text{span}\{\mathbf{e}_2, \dots, \mathbf{e}_n\}$. Hence U is called the **fixed hyperplane** of Q, and Q is called **reflection** in U. Note that each hyperplane in V is the fixed hyperplane of a (unique) reflection of V. Clearly, reflections in $\mathbb{R}^2$ and $\mathbb{R}^3$ are reflections in this more general sense.

Continuing the analogy with $\mathbb{R}^2$ and $\mathbb{R}^3$, an isometry $T : V \to V$ is called a **rotation** if there exists an orthonormal basis $\{\mathbf{e}_1, \dots, \mathbf{e}_n\}$ such that

$$M_B(T) = \begin{bmatrix} I_r & 0 & 0 \\ 0 & R(\theta) & 0 \\ 0 & 0 & I_s \end{bmatrix}$$

in block form, where $R(\theta) = \begin{bmatrix} \cos \theta & -\sin \theta \\ \sin \theta & \cos \theta \end{bmatrix}$ and where either I_r or I_s (or both) may be missing. If $R(\theta)$ occupies columns i and $i + 1$ of $M_B(T)$, and if $W = \text{span}\{\mathbf{e}_i, \mathbf{e}_{i+1}\}$, then W is T-invariant and the matrix of $T : W \to W$ with respect to $\{\mathbf{e}_i, \mathbf{e}_{i+1}\}$ is $R(\theta)$. Clearly, if W is viewed as a copy of $\mathbb{R}^2$, then T is a rotation in W. Moreover, $T(\mathbf{u}) = \mathbf{u}$ holds for all vectors $\mathbf{u}$ in the $(n - 2)$-dimensional subspace $U = \text{span}\{\mathbf{e}_1, \dots, \mathbf{e}_{i-1}, \mathbf{e}_{i+2}, \dots, \mathbf{e}_n\}$, and U is called the **fixed axis** of the rotation T. In $\mathbb{R}^3$, the axis of any rotation is a line (one-dimensional), whereas in $\mathbb{R}^2$ the axis is $U = \{0\}$.

With these definitions, the following theorem is an immediate consequence of Theorem 5 (the details are left to the reader).

Theorem 7

Let $T : V \to V$ be an isometry of a finite dimensional inner product space V. Then there exist isometries $T_1, \ldots, T_k$ such that

$$T = T_k T_{k-1} \cdots T_2 T_1$$

where each T_i is either a rotation or a reflection, at most one is a reflection, and $T_i T_j = T_j T_i$ holds for all i and j. Furthermore, T is a composite of rotations if and only if $\det T = 1$.

Exercises 10.4

Throughout these exercises, V denotes a finite dimensional inner product space.

1. Show that the following linear operators are isometries.

   (a) $T : \mathbb{C} \to \mathbb{C}$; $T(z) = \bar{z}$; $\langle z, w \rangle = \text{re}(z\bar{w})$

   (b) $T : \mathbb{R}^n \to \mathbb{R}^n$; $T(a_1, a_2, \ldots, a_n)$
   $= (a_n, a_{n-1}, \ldots, a_2, a_1)$; dot product

   (c) $T : \mathbf{M}_{22} \to \mathbf{M}_{22}$, $T\begin{bmatrix} a & b \\ c & d \end{bmatrix} = \begin{bmatrix} c & d \\ b & a \end{bmatrix}$;

   $\langle A, B \rangle = \text{tr}(AB^T)$

   (d) $T : \mathbb{R}^3 \to \mathbb{R}^3$, $T[a, b, c] = \frac{1}{9}(2a + 2b - c,$
   $2a + 2c - b, \; 2b + 2c - a)$; dot product

2. In each case, show that T is an isometry of $\mathbb{R}^2$, determine whether it is a rotation or a reflection, and find the angle or the fixed line. Use the dot product.

   (a) $T\begin{bmatrix} a \\ b \end{bmatrix} = \begin{bmatrix} -a \\ b \end{bmatrix}$ ♦(b) $T\begin{bmatrix} a \\ b \end{bmatrix} = \begin{bmatrix} -a \\ -b \end{bmatrix}$

   (c) $T\begin{bmatrix} a \\ b \end{bmatrix} = \begin{bmatrix} b \\ -a \end{bmatrix}$ ♦(d) $T\begin{bmatrix} a \\ b \end{bmatrix} = \begin{bmatrix} -b \\ -a \end{bmatrix}$

   (e) $T\begin{bmatrix} a \\ b \end{bmatrix} = \frac{1}{\sqrt{2}}\begin{bmatrix} a + b \\ b - a \end{bmatrix}$

   ♦(f) $T\begin{bmatrix} a \\ b \end{bmatrix} = \frac{1}{\sqrt{2}}\begin{bmatrix} a - b \\ a + b \end{bmatrix}$

3. In each case, show that T is an isometry of $\mathbb{R}^3$, determine the type (Theorem 6), and find the axis of any rotations and the fixed plane of any reflections involved.

   (a) $T\begin{bmatrix} a \\ b \\ c \end{bmatrix} = \begin{bmatrix} a \\ -b \\ c \end{bmatrix}$ ♦(b) $T\begin{bmatrix} a \\ b \\ c \end{bmatrix} = \frac{1}{2}\begin{bmatrix} \sqrt{3}c - a \\ \sqrt{3}a + c \\ 2b \end{bmatrix}$

   (c) $T\begin{bmatrix} a \\ b \\ c \end{bmatrix} = \begin{bmatrix} b \\ c \\ a \end{bmatrix}$ ♦(d) $T\begin{bmatrix} a \\ b \\ c \end{bmatrix} = \begin{bmatrix} a \\ -b \\ -c \end{bmatrix}$

   (e) $T\begin{bmatrix} a \\ b \\ c \end{bmatrix} = \frac{1}{2}\begin{bmatrix} a + \sqrt{3}b \\ b - \sqrt{3}a \\ 2c \end{bmatrix}$

   ♦(f) $T\begin{bmatrix} a \\ b \\ c \end{bmatrix} = \frac{1}{\sqrt{2}}\begin{bmatrix} a + c \\ -\sqrt{2}b \\ c - a \end{bmatrix}$

4. Let $T : \mathbb{R}^2 \to \mathbb{R}^2$ be an isometry. A vector X in $\mathbb{R}^2$ is said to be **fixed** by T if $T(X) = X$. Let E_1 denote the set of all vectors in $\mathbb{R}^2$ fixed by T. Show that:

   (a) E_1 is a subspace of $\mathbb{R}^2$.

   (b) $E_1 = \mathbb{R}^2$ if and only if $T = 1$ is the identity map.

   (c) $\dim E_1 = 1$ if and only if T is a reflection (about the line E_1).

   (d) $E_1 = \{\vec{0}\}$ if and only if T is a rotation ($T \neq 1$).

5. Let $T : \mathbb{R}^3 \to \mathbb{R}^3$ be an isometry, and let E_1 be the subspace of all fixed vectors in $\mathbb{R}^3$ (see Exercise 4). Show that:

   (a) $E_1 = \mathbb{R}^3$ if and only if $T = 1$.

   (b) $\dim E_1 = 2$ if and only if T is a reflection (about the plane E_1).

   (c) $\dim E_1 = 1$ if and only if T is a rotation ($T \neq 1$) (about the line E_1).

   (d) $\dim E_1 = 0$ if and only if T is a reflection followed by a (nonidentity) rotation.

♦6. If T is an isometry, show that aT is an isometry if and only if $a = \pm 1$.

7. Show that every isometry preserves the angle between any pair of nonzero vectors

(see Exercise 31 §10.1). Must an angle-preserving isomorphism be an isometry? Support your answer.

8. If $T : V \to V$ is an isometry, show that $T^2 = 1_V$ if and only if the only complex eigenvalues of T are 1 and -1.

9. Let $T : V \to V$ be a linear operator. Show that any two of the following conditions implies the third:

(1) T is symmetric.
(2) T is an involution ($T^2 = 1_V$).
(3) T is an isometry.

[*Hint:* In all cases, use the definition $\langle \mathbf{v}, T(\mathbf{w}) \rangle = \langle T(\mathbf{v}), \mathbf{w} \rangle$ of a symmetric operator.

For (1) and (3) $\Rightarrow$ (2), use the fact that, if $\langle T^2(\mathbf{v}) - \mathbf{v}, \mathbf{w} \rangle = 0$ for all $\mathbf{w}$, then $T^2(\mathbf{v}) = \mathbf{v}$.]

10. If B and D are any orthonormal bases of V, show that there is an isometry $T : V \to V$ that carries B to D.

11. Show that the following are equivalent for a linear transformation $S : V \to V$ where V is finite dimensional:

(1) $\langle S(\mathbf{v}), S(\mathbf{w}) \rangle = 0$ whenever $\langle \mathbf{v}, \mathbf{w} \rangle = 0$;
(2) $S = aT$ for some isometry $T : V \to V$.

[*Hint:* Given (1), show that $\| S(\mathbf{e}) \| = \| S(\mathbf{f}) \|$ for all unit vectors $\mathbf{e}$ and $\mathbf{f}$ in V.]

SECTION 10.5 An Application to Fourier Approximation[4]

In this section we shall investigate an important orthogonal set in the space $\mathbf{C}[-\pi, \pi]$ of continuous functions on the interval $[-\pi, \pi]$, using the inner product.

$$\langle f, g \rangle = \int_{-\pi}^{\pi} f(x)g(x)\, dx$$

Of course, calculus will be needed. The orthogonal set in question is

$$\{1, \sin x, \cos x, \sin(2x), \cos(2x), \sin(3x), \cos(3x), \ldots\}$$

Standard techniques of integration give

$$\|1\|^2 = \int_{-\pi}^{\pi} 1^2\, dx = 2\pi$$

$$\|\sin kx\|^2 = \int_{-\pi}^{\pi} \sin^2(kx)\, dx = \pi \quad \text{for any } k = 1,\, 2,\, 3, \ldots$$

$$\|\cos kx\|^2 = \int_{-\pi}^{\pi} \cos^2(kx)\, dx = \pi \quad \text{for any } k = 1,\, 2,\, 3, \ldots$$

We leave the verifications to the reader, together with the task of showing that these functions are orthogonal:

$$\langle \sin(kx), \sin(mx) \rangle = 0 = \langle \cos(kx), \cos(mx) \rangle \quad \text{if } k \neq m$$

and

$$\langle \sin(kx), \cos(mx) \rangle = 0 \quad \text{for all } k \geq 0 \text{ and } m \geq 0$$

(Note that $1 = \cos(0x)$, so the function 1 is included.)

Now define the following subspace of $\mathbf{C}[-\pi, \pi]$:

$$F_n = \text{span}\{1, \sin x, \cos x, \sin(2x), \cos(2x), \ldots, \sin(nx), \cos(nx)\}$$

The aim is to use the approximation theorem (Theorem 8 §10.2); so, given a function f in $\mathbf{C}[-\pi, \pi]$, define the **Fourier coefficients** of f by

---

4 The name honours the French mathematician J. B. J. Fourier (1768–1830) who used these techniques in 1822 to investigate heat conduction in solids.

$$a_0 = \frac{\langle f(x), 1 \rangle}{\|1\|^2} = \frac{1}{2\pi}\int_{-\pi}^{\pi} f(x)\, dx$$

$$a_k = \frac{\langle f(x), \cos(kx) \rangle}{\|\cos(kx)\|^2} = \frac{1}{\pi}\int_{-\pi}^{\pi} f(x)\cos(kx)\, dx \quad k = 1, 2, \ldots$$

$$b_k = \frac{\langle f(x), \sin(kx) \rangle}{\|\sin(kx)\|^2} = \frac{1}{\pi}\int_{-\pi}^{\pi} f(x)\sin(kx)\, dx \quad k = 1, 2, \ldots$$

Then the approximation theorem (Theorem 8 §10.2) gives Theorem 1.

Theorem 1

Let f be any continuous real-valued function defined on the interval $[-\pi, \pi]$. If $a_0, a_1, \ldots,$ and $b_0, b_1, \ldots$ are the Fourier coefficients of f, then given $n \geq 0$,

$$f_n(x) = a_0 + a_1\cos x + b_1\sin x + a_2\cos(2x) + b_2\sin(2x) + \cdots$$
$$+ a_n\cos(nx) + b_n\sin(nx)$$

is a function in F_n that is closest to f in the sense that

$$\|f - f_n\| \leq \|f - g\|$$

holds for all functions g in F_n.

The function f_n is called the nth **Fourier approximation** to the function f.

Example 1

Find the fifth Fourier approximation to the function $f(x)$ defined on $[-\pi, \pi]$ as follows:

$$f(x) = \begin{cases} \pi + x & -\pi \leq x < 0 \\ \pi - x & 0 \leq x \leq \pi \end{cases}$$

Solution   The graph of $y = f(x)$ appears in the diagram. The Fourier coefficients are computed as follows. The details of the integrations (usually by parts) are omitted.

$$a_0 = \frac{1}{2\pi}\int_{-\pi}^{\pi} f(x)\, dx = \frac{\pi}{2}$$

$$a_k = \frac{1}{\pi}\int_{-\pi}^{\pi} f(x)\cos(kx)\, dx = \frac{2}{\pi k^2}[1 - \cos(k\pi)] = \begin{cases} 0 & \text{if } k \text{ is even} \\ \frac{4}{\pi k^2} & \text{if } k \text{ is odd} \end{cases}$$

$$b_k = \frac{1}{\pi}\int_{-\pi}^{\pi} f(x)\sin(kx)\, dx = 0 \qquad \text{for all } k = 1, 2, \ldots$$

Hence the fifth Fourier approximation is

$$f_5(x) = \frac{\pi}{2} + \frac{4}{\pi}\{\cos x + \frac{1}{3^2}\cos(3x) + \frac{1}{5^2}\cos(5x)\}$$

We say that a function f is an **even function** if $f(x) = f(-x)$ holds for all x; f is called an **odd function** if $f(-x) = -f(x)$ holds for all x. Examples of even functions are constant functions, the even powers $x^2, x^4, \ldots,$ and $\cos(kx)$; these functions are characterized by the fact that the graph of $y = f(x)$ is symmetric about the Y axis.

Examples of odd functions are the odd powers x, $x^3, \ldots$ and $\sin(kx)$ where $k > 0$, and the graph of $y = f(x)$ is symmetric about the origin if f is odd. The usefulness of these functions stems from the fact that

$$\int_{-\pi}^{\pi} f(x)\, dx = 0 \qquad\qquad \text{if } f \text{ is odd}$$

$$\int_{-\pi}^{\pi} f(x)\, dx = 2\int_{0}^{\pi} f(x)\, dx \quad \text{if } f \text{ is even}$$

These facts often simplify the computations of the Fourier coefficients. For example:

1. The Fourier sine coefficients b_k all vanish if f is even.
2. The Fourier cosine coefficients a_k all vanish if f is odd.

This is because $f(x)\sin(kx)$ is odd in the first case and $f(x)\cos(kx)$ is odd in the second case.

The functions 1, $\cos(kx)$, and $\sin(kx)$ that occur in the Fourier approximation for $f(x)$ are all easy to generate as an electrical voltage (when x is time). By summing these signals (with the amplitudes given by the Fourier coefficients), it is possible to produce an electrical signal with (the approximation to) $f(x)$ as the voltage. Hence these Fourier approximations play a fundamental role in electronics.

Finally, the Fourier approximations $f_1, f_2, \ldots$ of a function f get better and better as n increases. The reason is that the subspaces F_n increase:

$$F_1 \subseteq F_2 \subseteq F_3 \subseteq \cdots \subseteq F_n \subseteq \cdots$$

So, because $f_n = \operatorname{proj}_{F_n}(f)$, we get (see the discussion following Example 6 §10.2)

$$\|f - f_1\| \geq \|f - f_2\| \geq \cdots \geq \|f - f_n\| \geq \cdots$$

These numbers $\|f - f_n\|$ approach zero; in fact, we have the following fundamental theorem.[5]

Theorem 2

Let f be any continuous function in $\mathbf{C}[-\pi, \pi]$. Then

$$f_n(x) \text{ approaches } f(x) \text{ for all } x \text{ such that } -\pi < x < \pi.[6]$$

It shows that f has a representation as an infinite series, called the **Fourier series** of f:

$$f(x) = a_0 + a_1 \cos x + b_1 \sin x + a_2 \cos(2x) + b_2 \sin(2x) + \cdots$$

whenever $-\pi < x < \pi$. A full discussion of Theorem 2 is beyond the scope of this book. This subject had great historical impact on the development of mathematics, and has become one of the standard tools in science and engineering.

Thus the Fourier series for the function f in Example 1 is

$$f(x) = \frac{\pi}{2} + \frac{4}{\pi}\{\cos x + \frac{1}{3^2}\cos(3x) + \frac{1}{5^2}\cos(5x) + \frac{1}{7^2}\cos(7x) + \cdots\}$$

Since $f(0) = \pi$ and $\cos(0) = 1$, taking $x = 0$ leads to the series

$$\frac{\pi^2}{8} = 1 + \frac{1}{3^2} + \frac{1}{5^2} + \frac{1}{7^2} + \cdots$$

---

5   See, for example, R. V. Churchill and J. W. Brown, *Fourier Series and Boundary Value Problems*, 4th ed., (New York: McGraw-Hill, 1987).

6   We have to be careful at the end points $x = \pi$ or $x = -\pi$ because $\sin(k\pi) = \sin(-k\pi)$ and $\cos(k\pi) = \cos(-k\pi)$.

Example 2

Expand $f(x) = x$ on the interval $[-\pi, \pi]$ in a Fourier series, and so obtain a series expansion of $\frac{\pi}{4}$.

Solution Here f is an odd function so all the Fourier cosine coefficients a_k are zero. As to the sine coefficients:

$$b_k = \frac{1}{\pi}\int_{-\pi}^{\pi} x \sin(kx)\, dx = \frac{2}{k}(-1)^{k+1} \quad \text{for } k \geq 1$$

where we omit the details of the integration by parts. Hence the Fourier series for x is

$$x = 2[\sin x - \tfrac{1}{2}\sin(2x) + \tfrac{1}{3}\sin(3x) - \tfrac{1}{4}\sin(4x) + \cdots]$$

for $-\pi < x < \pi$. In particular, taking $x = \frac{\pi}{2}$ gives an infinite series for $\frac{\pi}{4}$:

$$\tfrac{\pi}{4} = 1 - \tfrac{1}{3} + \tfrac{1}{5} - \tfrac{1}{7} + \tfrac{1}{9} - \cdots$$

Many other such formulas can be proved using Theorem 2.

Exercises 10.5

1. In each case, find the Fourier approximation f_5 of the given function in $\mathbf{C}[-\pi, \pi]$.

   (a) $f(x) = \pi - x$

   ◆(b) $f(x) = |x| = \begin{cases} x & \text{if } 0 \leq x \leq \pi \\ -x & \text{if } -\pi \leq x < 0 \end{cases}$

   (c) $f(x) = x^2$

   ◆(d) $f(x) = \begin{cases} 0 & \text{if } -\pi \leq x < 0 \\ x & \text{if } 0 \leq x \leq \pi \end{cases}$

2. (a) Find f_5 for the even function f on $[-\pi, \pi]$ satisfying $f(x) = x$ for $0 \leq x \leq \pi$.

   ◆(b) Find f_6 for the even function f on $[-\pi, \pi]$ satisfying $f(x) = \sin x$ for $0 \leq x \leq \pi$.
   [*Hint:* If $k > 1$, $\int \sin x \cos(kx)$
   $= \frac{1}{2}\left[\frac{\cos[(k-1)x]}{k-1} - \frac{\cos[(k+1)x]}{k+1}\right]$.]

3. (a) Prove that $\int_{-\pi}^{\pi} f(x)\, dx = 0$ if f is odd and that $\int_{-\pi}^{\pi} f(x)\, dx = 2\int_0^{\pi} f(x)\, dx$ if f is even.

   (b) Prove that $\frac{1}{2}[f(x) + f(-x)]$ is even and that $\frac{1}{2}[f(x) - f(-x)]$ is odd for any function f. Note that they sum to $f(x)$.

◆4. Show that $\{1, \cos x, \cos(2x), \cos(3x), \ldots\}$ is an orthogonal set in $\mathbf{C}[0, \pi]$ with respect to the inner product $\langle f, g \rangle = \int_0^{\pi} f(x)g(x)\, dx$.

5. (a) Show that $\frac{\pi^2}{8} = 1 + \frac{1}{3^2} + \frac{1}{5^2} + \cdots$ using Exercise 1(b).

   (b) Show that $\frac{\pi^2}{12} = 1 - \frac{1}{2^2} + \frac{1}{3^2} - \frac{1}{4^2} + \cdots$ using Exercise 1(c).

Appendix A: Complex Numbers

The fact that the square of every real number is positive shows that the equation $x^2 + 1 = 0$ has no real root; in other words, there is no real number u such that $u^2 = -1$. So the set of real numbers is inadequate for finding all roots of all polynomials. This kind of problem arises with other number systems as well. The set of integers contains no solution of the equation $3x + 2 = 0$, and the rational numbers had to be invented to solve such equations. But the set of rational numbers is also incomplete because, for example, it contains no root of the polynomial $x^2 - 2$. Hence the real numbers were invented. In the same way, the set of complex numbers was invented, which contains all real numbers together with a root of the equation $x^2 + 1 = 0$. However, the process ends here: the complex numbers have the property that every polynomial with complex coefficients has a (complex) root. This fact is known as the fundamental theorem of algebra.

One pleasant aspect of the complex numbers is that, whereas describing the real numbers in terms of the rationals is a rather complicated business, the complex numbers are quite easy to describe in terms of real numbers. Every **complex number** has the form

$$a + bi$$

where a and b are real numbers, and i is a root of the polynomial $x^2 + 1$. Here a and b are called the **real part** and the **imaginary part** of the complex number, respectively. The real numbers are now regarded as special complex numbers of the form $a + 0i = a$, with zero imaginary part. The complex numbers of the form $0 + bi = bi$ with zero real part are called **pure imaginary** numbers. The complex number i itself is called the **imaginary unit** and is distinguished by the fact that

$$i^2 = -1$$

As the terms *complex* and *imaginary* suggest, these numbers met with some resistance when they were first used. This has changed; now they are essential in science and engineering as well as mathematics, and they are used extensively. The names persist, however, and continue to be a bit misleading: These numbers are no more complex than the real numbers, and the number i is no more imaginary than -1.

Much as for polynomials, two complex numbers are declared to be **equal** if and only if they have the same real parts and the same imaginary parts. In symbols,

$$a + bi = a' + b'i \quad \text{if and only if} \quad a = a' \text{ and } b = b'$$

The addition and subtraction of complex numbers is accomplished by adding and subtracting real and imaginary parts:

$$(a + bi) + (a' + b'i) = (a + a') + (b + b')i$$
$$(a + bi) - (a' + b'i) = (a - a') + (b - b')i$$

This is analogous to these operations for linear polynomials $a + bx$ and $a' + b'x$, and the multiplication of complex numbers is also analogous with one difference: $i^2 = -1$. The definition is

$$(a + bi)(a' + b'i) = (aa' - bb') + (ab' + ba')i$$

With these definitions of equality, addition, and multiplication, the complex numbers satisfy all the basic arithmetical axioms adhered to by the real numbers (the verifications are omitted). One consequence of this is that they can be manipulated in the obvious fashion, except that i^2 is replaced by -1 wherever it occurs, and the rule for equality must be observed.

Example 1

If $z = 2 - 3i$ and $w = -1 + i$, write each of the following in the form $a + bi$: $z + w$, $z - w$, zw, $\frac{1}{3}z$, and z^2.

Solution

$$z + w = (2 - 3i) + (-1 + i) = (2 - 1) + (-3 + 1)i = 1 - 2i$$
$$z - w = (2 - 3i) - (-1 + i) = (2 + 1) + (-3 - 1)i = 3 - 4i$$
$$zw = (2 - 3i)(-1 + i) = (-2 - 3i^2) + (2 + 3)i = 1 + 5i$$
$$\tfrac{1}{3}z = \tfrac{1}{3}(2 - 3i) = \tfrac{2}{3} - i$$
$$z^2 = (2 - 3i)(2 - 3i) = (4 + 9i^2) + (-6 - 6)i = -5 - 12i$$

Example 2

Find all complex numbers z such as that $z^2 = i$.

Solution

Write $z = a + bi$; we must determine a and b. Now $z^2 = (a^2 - b^2) + (2ab)i$, so the condition $z^2 = i$ becomes

$$(a^2 - b^2) + (2ab)i = 0 + i$$

Equating real and imaginary parts, we find that $a^2 = b^2$ and $2ab = 1$. The solution is $a = b = \pm\frac{1}{\sqrt{2}}$, so the complex numbers required are $z = \frac{1}{\sqrt{2}} + \frac{1}{\sqrt{2}}i$ and $z = -\frac{1}{\sqrt{2}} - \frac{1}{\sqrt{2}}i$.

As for real numbers, it is possible to divide by every nonzero complex number z. That is, there exists a complex number w such that $wz = 1$. As in the real case, this number w is called the **inverse** of z and is denoted by z^{-1} or $\frac{1}{z}$. Moreover, if $z = a + bi$, the fact that $z \neq 0$ means that $a \neq 0$ or $b \neq 0$. Hence $a^2 + b^2 \neq 0$, and an explicit formula for the inverse is

$$\frac{1}{z} = \frac{a}{a^2 + b^2} - \frac{b}{a^2 + b^2} \cdot i$$

In actual calculation, the work is facilitated by two useful notions: the conjugate and the absolute value of a complex number. The next example illustrates the technique.

Example 3

Solution

Write $\dfrac{3+2i}{2+5i}$ in the form $a + bi$.

Multiply top and bottom by the complex number $2 - 5i$ (obtained from the denominator by negating the imaginary part). The result is

$$\frac{3+2i}{2+5i} = \frac{(2-5i)(3+2i)}{(2-5i)(2+5i)} = \frac{(6+10)+(4-15)i}{(4+25)+0i} = \frac{16}{29} - \frac{11}{29}i$$

Hence the simplified form is $\frac{16}{29} - \frac{11}{29}i$, as required.

The key to this technique is that the product $(2 - 5i)(2 + 5i) = 29$ in the denominator turned out to be a *real* number. The situation in general leads to the following notation: If $z = a + bi$ is a complex number, the **conjugate** of z is the complex number, denoted $\bar{z}$, given by

$$\bar{z} = a - bi \quad \text{where } z = a + bi$$

Hence $\bar{z}$ is obtained from z by negating the imaginary part. For example, $\overline{(2+3i)} = 2 - 3i$ and $\overline{(1-i)} = 1 + i$. If we multiply z by $\bar{z}$, we obtain

$$z\bar{z} = a^2 + b^2 \quad \text{where } z = a + bi$$

The real number $a^2 + b^2$ is always nonnegative, so we can state the following definition: The **absolute value** or **modulus** of a complex number $z = a + bi$, denoted by $|z|$, is the positive square root $\sqrt{a^2 + b^2}$; that is,

$$|z| = \sqrt{a^2 + b^2} \quad \text{where } z = a + bi$$

For example, $|2 - 3i| = \sqrt{2^2 + (-3)^2} = \sqrt{13}$ and $|1 + i| = \sqrt{1^2 + 1^2} = \sqrt{2}$.

Note that if a real number a is viewed as the complex number $a + 0i$, its absolute value (as a complex number) is $|a| = \sqrt{a^2}$, which agrees with its absolute value as a *real* number.

With these notions in hand, we can describe the technique applied in Example 3 as follows: When converting a quotient $\frac{z}{w}$ of complex numbers to the form $a + bi$, multiply top and bottom by the conjugate $\bar{w}$ of the denominator.

The following list contains the most important properties of conjugates and absolute values. Throughout, z and w denote complex numbers.

C1. $z \pm w = \bar{z} \pm \bar{w}$

C2. $\overline{zw} = \bar{z}\,\bar{w}$

C3. $\overline{\left(\dfrac{z}{w}\right)} = \dfrac{\bar{z}}{\bar{w}}$

C4. $\overline{(\bar{z})} = z$

C5. z is real if and only if $\bar{z} = z$

C6. $z\bar{z} = |z|^2$

C7. $\dfrac{1}{z} = \dfrac{\bar{z}}{|z|^2}$

C8. $|z| \geq 0$ for all complex numbers z

C9. $|z| = 0$ if and only if $z = 0$

C10. $|zw| = |z|\,|w|$

C11. $\left|\dfrac{z}{w}\right| = \dfrac{|z|}{|w|}$

C12. $|z + w| \leq |z| + |w|$ (**triangle inequality**)

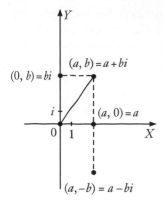

Figure A.1

All these properties (except property C12) can (and should) be verified by the reader for arbitrary complex numbers $z = a + bi$ and $w = c + di$. They are not independent; for example, property C10 follows from properties C2 and C6.

The triangle inequality, as its name suggests, comes from a geometric representation of the complex numbers analogous to identification of the real numbers with the points of a line. The representation is achieved as follows: Introduce a rectangular coordinate system in the plane (Figure A.1), and identify the complex number $a + bi$ with the point (a, b). When this is done, the plane is called the **complex plane**. Note that the point $(a, 0)$ on the X axis now represents the *real* number $a = a + 0i$, and for this reason, the X axis is called the **real axis**. Similarly, the Y axis is called the **imaginary axis**. The identification $(a, b) = a + bi$ of the geometric point (a, b) and the complex number $a + bi$ will be used in what follows without comment.[1] For example, the origin will be referred to as 0.

This representation of the complex numbers in the complex plane gives a useful way of describing the absolute value and conjugate of a complex number $z = a + bi$. The absolute value $|z| = \sqrt{a^2 + b^2}$ is just the distance from z to the origin. This makes properties C8 and C9 quite obvious. The conjugate $\bar{z} = a - bi$ of z is just the reflection of z in the real axis (X axis), a fact that makes properties C4 and C5 clear.

Given two complex numbers $z_1 = a_1 + b_1 i = (a_1, b_1)$ and $z_2 = a_2 + b_2 i = (a_2, b_2)$, the absolute value of their difference

$$|z_1 - z_2| = \sqrt{(a_1 - a_2)^2 + (b_1 - b_2)^2}$$

is just the distance between them. This gives the **complex distance formula**:

$$|z_1 - z_2| \text{ is the distance between } z_1 \text{ and } z_2$$

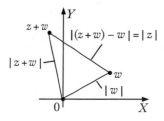

Figure A.2

This useful fact yields a simple verification of the triangle inequality, property C12. Suppose z and w are given complex numbers. Consider the triangle in Figure A.2 whose vertices are 0, w, and $z + w$. The three sides have lengths $|z|, |w|,$ and $|z + w|$ by the complex distance formula, so the inequality

$$|z + w| \leq |z| + |w|$$

expresses the obvious geometric fact that the sum of the lengths of two sides of a triangle is at least as great as the length of the third side.

The representation of complex numbers as points in the complex plane has another very useful property: It enables us to give a geometric description of the sum and product of two complex numbers. To obtain the description for the sum, let

$$z = a + bi = (a, b)$$
$$w = c + di = (c, d)$$

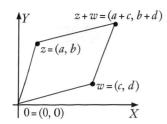

Figure A.3

denote two complex numbers. We claim that the four points 0, z, w, and $z + w$ form the vertices of a parallelogram. In fact, in Figure A.3 the lines from 0 to z and from w to $z + w$ have slopes

$$\frac{b - 0}{a - 0} = \frac{b}{a} \quad \text{and} \quad \frac{(b + d) - d}{(a + c) - c} = \frac{b}{a}$$

respectively, so these lines are parallel. (If it happens that $a = 0$, then both these lines are vertical.) Similarly, the lines from z to $z + w$ and from 0 to w are also parallel, so the figure with vertices 0, z, w, and $z + w$ is indeed a parallelogram. Hence, the

---

1 This is analogous to thinking of the real numbers as the points on a line.

complex number $z + w$ can be obtained geometrically from z and w by *completing* the parallelogram. This is sometimes called the **parallelogram law** of complex addition. Readers who have studied mechanics will recall that velocities and accelerations add in the same way; in fact, these are all special cases of *vector* addition.

The geometric description of what happens when two complex numbers are multiplied is at least as elegant as the parallelogram law of addition, but it requires that the complex numbers be represented in polar form. Before discussing this, we pause to recall the general definition of the trigonometric functions sine and cosine. An angle θ in the complex plane is in **standard position** if it is measured counterclockwise from the positive real axis as indicated in Figure A.4. Rather than using degrees to measure angles, it is more natural to use radian measure. This is defined as follows: The circle with its centre at the origin and radius 1 (called the **unit circle**) is drawn in Figure A.4. It has circumference 2π, and the **radian measure** of θ is the length of the arc on the unit circle counterclockwise from 1 to the point P on the unit circle determined by θ. Hence $90° = \frac{\pi}{2}$, $45° = \frac{\pi}{4}$, $180° = \pi$, and a full circle has the angle $360° = 2\pi$. Angles measured clockwise from 1 are negative; for example, $-i$ corresponds to $-\frac{\pi}{2}$ (or to $\frac{2\pi}{3}$).

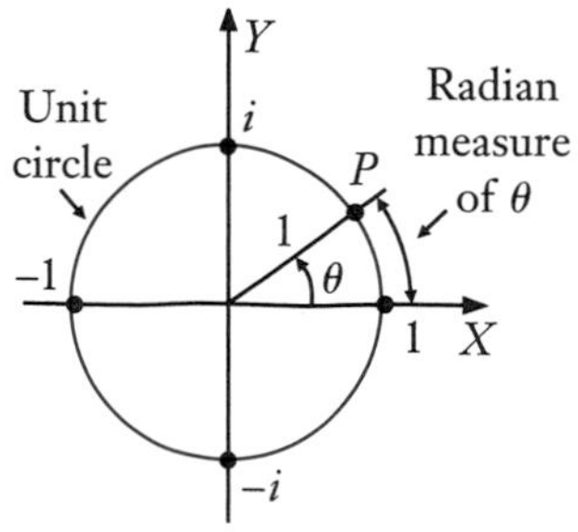

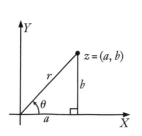

Figure A.4

If an acute angle θ (that is, $0 \leq \theta \leq \frac{\pi}{2}$) is plotted in standard position as in Figure A.4, it determines a unique point P on the unit circle, and P has coordinates $(\cos\theta, \sin\theta)$ by elementary trigonometry. However, *any* angle θ (acute or not) determines a unique point on the unit circle, so we *define* the **cosine** and **sine** of θ (written $\cos\theta$ and $\sin\theta$) to be the X and Y coordinates of this point. For example, the points

$$1 = (1,\ 0) \qquad i = (0,\ 1) \qquad -1 = (-1,\ 0) \qquad -i = (0,\ -1)$$

plotted in Figure A.4 are determined by the angles $0, \frac{\pi}{2}, \pi, \frac{3\pi}{2}$, respectively. Hence

$$\cos 0 = 1 \qquad \cos \tfrac{\pi}{2} = 0 \qquad \cos \pi = -1 \qquad \cos \tfrac{3\pi}{2} = 0$$
$$\sin 0 = 0 \qquad \sin \tfrac{\pi}{2} = 1 \qquad \sin \pi = 0 \qquad \sin \tfrac{3\pi}{2} = -1$$

Now we can describe the polar form of a complex number. Let $z = a + bi$ be a complex number, and write the absolute value of z as

$$r = |z| = \sqrt{a^2 + b^2}$$

If $z \neq 0$, the angle θ shown in Figure A.5 is called an **argument** of z and is denoted

$$\theta = \arg z$$

This angle is not unique ($\theta + 2\pi k$ would do as well for any $k = 0, \pm 1, \pm 2, \dots$). However, there is only one argument θ in the range $-\pi < \theta \leq \pi$, and this is sometimes called the **principal argument** of z.

Returning to Figure A.5, we find that the real and imaginary parts a and b of z are related to r and θ by

$$a = r\cos\theta$$
$$b = r\sin\theta$$

Hence the complex number $z = a + bi$ has the form

$$z = r(\cos\theta + i\sin\theta) \qquad r = |z|,\ \theta = \arg(z)$$

The combination $\cos\theta + i\sin\theta$ is so important that a special notation is used. This is

$$e^{i\theta} = \cos\theta + i\sin\theta$$

Figure A.5

With this notation, z is written

$$z = re^{i\theta} \qquad r = |z|, \ \theta = \arg(z)$$

This is a **polar form** of the complex number z. Of course it is not unique, because the argument can be changed by adding a multiple of 2π.

Example 4

Write $z_1 = -2 + 2i$ and $z_2 = -i$ in polar form.

Solution

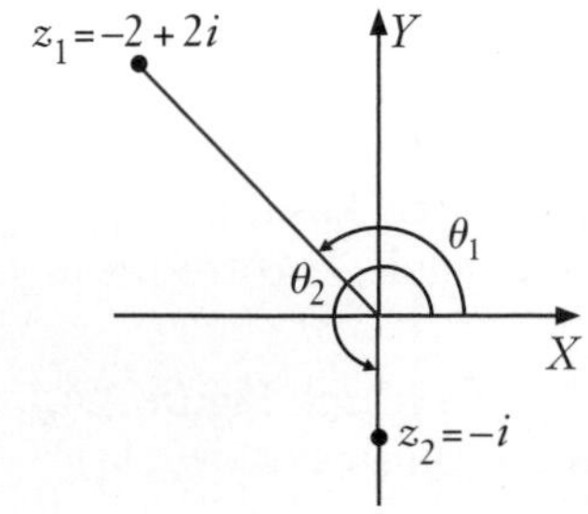

$z_1 = -2 + 2i$

$z_2 = -i$

Figure A.6

The two numbers are plotted in the complex plane in Figure A.6. The absolute values are

$$r_1 = |-2 + 2i| = \sqrt{(-2)^2 + 2^2} = 2\sqrt{2}$$
$$r_2 = |-i| = \sqrt{0^2 + (-1)^2} = 1$$

By inspection of Figure A.6, arguments of z_1 and z_2 are

$$\theta_1 = \arg(-2 + 2i) = \tfrac{3\pi}{4}$$
$$\theta_2 = \arg(-i) = \tfrac{3\pi}{2}$$

The corresponding polar forms are $z_1 = -2 + 2i = 2\sqrt{2}e^{3\pi i/4}$ and $z_2 = -i = e^{3\pi i/2}$. Of course, we could have taken the argument $-\tfrac{\pi}{2}$ for z_2 and obtained the polar form $z_2 = e^{-\pi i/2}$.

In the notation $e^{i\theta} = \cos\theta + i\sin\theta$, the number e is the familiar constant $e = 2.71828\dots$ from calculus. The reason for using e will not be given here; the reason why $\cos\theta + i\sin\theta$ is written as an *exponential* function of θ is that the **law of exponents** holds:

$$e^{i\theta} \cdot e^{i\phi} = e^{i(\theta+\phi)}$$

where θ and ϕ are any two angles. In fact, this is an immediate consequence of the addition identities for $\sin(\theta + \phi)$ and $\cos(\theta + \phi)$:

$$\begin{aligned}
e^{i\theta}e^{i\phi} &= (\cos\theta + i\sin\theta)(\cos\phi + i\sin\phi) \\
&= (\cos\theta\cos\phi - \sin\theta\sin\phi) + i(\cos\theta\sin\phi + \sin\theta\cos\phi) \\
&= \cos(\theta + \phi) + i\sin(\theta + \phi) \\
&= e^{i(\theta + \phi)}
\end{aligned}$$

This is analogous to the rule $e^a e^b = e^{a+b}$, which holds for real numbers a and b, so it is not unnatural to use the exponential notation $e^{i\theta}$ for the expression $\cos\theta + i\sin\theta$.

In fact, a whole theory exists wherein functions such as e^z, $\sin z$, and $\cos z$ are studied, where z is a *complex* variable. Many deep and beautiful theorems can be proved in this theory, one of which is the so-called fundamental theorem of algebra mentioned later (Theorem 5). We shall not pursue this here.

The geometric description of the multiplication of two complex numbers follows from the law of exponents.

Theorem 1 Multiplication Rule

If $z_1 = r_1 e^{i\theta_1}$ and $z_2 = r_2 e^{i\theta_2}$ are complex numbers in polar form, then

$$z_1 z_2 = r_1 r_2 e^{i(\theta_1 + \theta_2)}$$

In other words, to multiply two complex numbers, simply multiply the absolute values and add the arguments. This simplifies calculations considerably, particularly when we observe that it is valid for *any* arguments θ_1 and θ_2.

Example 5

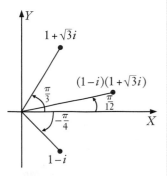

Figure A.7

Solution

Multiply $(1-i)(1+\sqrt{3}i)$ in two ways.

We have $|1-i| = \sqrt{2}$ and $|1+\sqrt{3}i| = 2$ so, from Figure A.7,

$$1 - i = \sqrt{2}e^{-i\pi/4}$$
$$1 + \sqrt{3}i = 2e^{i\pi/3}$$

Hence, by the multiplication rule,

$$
\begin{aligned}
(1-i)(1+\sqrt{3}i) &= (\sqrt{2}e^{-i\pi/4})(2e^{i\pi/3}) \\
&= 2\sqrt{2}e^{i(-\pi/4 + \pi/3)} \\
&= 2\sqrt{2}e^{i\pi/12}
\end{aligned}
$$

This gives the required product in polar form. Of course, direct multiplication gives $(1-i)(1+\sqrt{3}i) = (\sqrt{3}+1) + (\sqrt{3}-1)i$. Hence, equating real and imaginary parts gives $\cos(\frac{\pi}{12}) = \frac{\sqrt{3}+1}{2\sqrt{2}}$ and $\sin(\frac{\pi}{12}) = \frac{\sqrt{3}-1}{2\sqrt{2}}$.

If a complex number $z = re^{i\theta}$ is given in polar form, the powers assume a particularly simple form. In fact, $z^2 = (re^{i\theta})(re^{i\theta}) = r^2e^{2i\theta}$, $z^3 = z^2 \cdot z = (r^2e^{2i\theta})(re^{i\theta}) = r^3e^{3i\theta}$, and so on. Continuing in this way, it follows by induction that the following theorem holds for any positive integer n. The name honours Abraham De Moivre (1667–1754).

Theorem 2 De Moivre's Theorem

If θ is any angle, then $(e^{i\theta})^n = e^{in\theta}$ holds for all integers n.

PROOF

The case $n > 0$ has been discussed, and the reader can verify the result for $n = 0$. To derive it for $n < 0$, first observe that

$$\text{if} \quad z = re^{i\theta} \neq 0 \quad \text{then} \quad z^{-1} = \frac{1}{r}e^{-i\theta}$$

In fact, $(re^{i\theta})(\frac{1}{r}e^{-i\theta}) = 1e^{i0} = 1$ by the multiplication rule. Now assume that n is negative and write it as $n = -m$, $m > 0$. Then

$$(re^{i\theta})^n = [(re^{i\theta})^{-1}]^m = (\frac{1}{r}e^{-i\theta})^m = r^{-m}e^{i(-m\theta)} = r^n e^{in\theta}$$

If $r = 1$, this is De Moivre's theorem for negative n.

Example 6

Verify that $(-1+\sqrt{3}i)^3 = 8$.

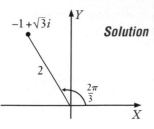

Figure A.8

Solution   We have $\left|-1+\sqrt{3}i\right| = 2$, so $-1+\sqrt{3}i = 2e^{2\pi i/3}$ (see Figure A.8). Hence De Moivre's theorem gives

$$(-1+\sqrt{3}i)^3 = (2e^{2\pi i/3})^3 = 8e^{3(2\pi/3)i} = 8e^{2\pi i} = 8$$

De Moivre's theorem can be used to find nth roots of complex numbers where n is positive. The next example illustrates this technique.

Example 7

Find the cube roots of unity; that is, find all complex numbers z such that $z^3 = 1$.

Solution   First write $z = re^{i\theta}$ and $1 = 1e^{i \cdot 0}$ in polar form. We must use the condition $z^3 = 1$ to determine r and θ. Because $z^3 = r^3 e^{3i\theta}$ by De Moivre's theorem, this requirement becomes

$$r^3 e^{3\theta i} = 1e^{0 \cdot i}$$

These two complex numbers are equal, so their absolute values must be equal and the arguments must either be equal or differ by an integral multiple of 2π:

$2k\pi$

$$r^3 = 1$$
$$3\theta = 0 + 2k\pi, \quad k \text{ some integer}$$

Because r is real and positive, the condition $r^3 = 1$ implies that $r = 1$. However,

$$\theta = \frac{2k\pi}{3}, \quad k \text{ some integer}$$

seems at first glance to yield infinitely many different angles for z. However, choosing $k = 0, 1, 2$ gives three possible arguments θ (where $0 \le \theta < 2\pi$), and the corresponding roots are

$$1e^{0i} = 1$$
$$1e^{2\pi i/3} = -\frac{1}{2} + \frac{\sqrt{3}}{2}i$$
$$1e^{4\pi i/3} = -\frac{1}{2} - \frac{\sqrt{3}}{2}i$$

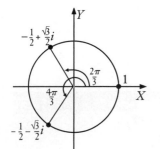

Figure A.9

These are displayed in Figure A.9. All other values of k yield values of θ that differ from one of these by a multiple of 2π—and so do not give new roots. Hence we have found all the roots.

The same type of calculation gives all complex **nth roots of unity**; that is, all complex numbers z such that $z^n = 1$. As before, write $1 = 1e^{0 \cdot i}$ and

$$z = re^{i\theta}$$

in polar form. Then $z^n = 1$ takes the form

$$r^n e^{n\theta i} = 1e^{0i}$$

using De Moivre's theorem. Comparing absolute values and arguments yields

$$r^n = 1$$
$$n\theta = 0 + 2k\pi, \quad k \text{ some integer}$$

Hence $r = 1$, and the n values

$$\theta = \frac{2k\pi}{n}, \quad k = 0, 1, 2, \ldots, n-1$$

of θ all lie in the range $0 \le \theta < 2\pi$. As before, *every* choice of k yields a value of θ that differs from one of these by a multiple of 2π, so these give the arguments of *all* the possible roots.

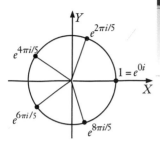

Figure A.10

Theorem 3 nth Roots of Unity

If $n \ge 1$ is an integer, the nth roots of unity (that is, the solutions to $z^n = 1$) are given by

$$z = e^{2\pi k i/n}, \quad k = 0, 1, 2, \ldots, n - 1$$

The nth roots of unity can be found geometrically as the points on the unit circle that cut the circle into n equal sectors, starting at 1. The case $n = 5$ is shown in Figure A.10, where the five fifth roots of unity are plotted.

The method just used to find the nth roots of unity works equally well to find the nth roots of any complex number in polar form. We give one example.

Example 8

Find the fourth roots of $\sqrt{2} + \sqrt{2}i$.

Solution

First write $\sqrt{2} + \sqrt{2}i = 2e^{\pi i/4}$ in polar form. If $z = re^{i\theta}$ satisfies $z^4 = \sqrt{2} + \sqrt{2}i$, then

$$r^4 e^{i(4\theta)} = 2e^{\pi i/4}$$

Hence $r^4 = 2$ and $4\theta = \frac{\pi}{4} + 2k\pi$, k an integer. We obtain four distinct roots (and hence all) by

$$r = \sqrt[4]{2}, \quad \theta = \frac{\pi}{16} + \frac{8k\pi}{16}, k = 0, 1, 2, 3$$

Thus the four roots are

$$\sqrt[4]{2}e^{\pi i/16} \qquad \sqrt[4]{2}e^{9\pi i/16} \qquad \sqrt[4]{2}e^{17\pi i/16} \qquad \sqrt[4]{2}e^{25\pi i/16}$$

Of course, reducing these roots to the form $a + bi$ would require the computation of $\sqrt[4]{2}$ and the sine and cosine of the various angles.

An expression of the form $ax^2 + bx + c$, where the coefficients $a \ne 0$, b, and c are real numbers, is called a **real quadratic**. A complex number u is called a **root** of the quadratic if $au^2 + bu + c = 0$. The roots are given by the famous **quadratic formula**:

$$u = \frac{-b \pm \sqrt{b^2 - 4ac}}{2a}$$

The quantity $d = b^2 - 4ac$ is called the **discriminant** of the quadratic $ax^2 + bx + c$, and there is no real root if and only if $d < 0$. In this case the quadratic is said to be **irreducible**. Moreover, the fact that $d < 0$ means that $\sqrt{d} = i\sqrt{|d|}$, so the two (complex) roots are conjugates of each other:

$$u = \tfrac{1}{2a}(-b + i\sqrt{|d|}) \quad \text{and} \quad \bar{u} = \tfrac{1}{2a}(-b - i\sqrt{|d|})$$

The converse of this is true too: Given any nonreal complex number u, then u and $\bar{u}$ are the roots of some real irreducible quadratic. Indeed, the quadratic

$$x^2 - (u + \bar{u})x + u\bar{u} = (x - u)(x - \bar{u})$$

has real coefficients ($u\bar{u} = |u|^2$ and $u + \bar{u}$ is twice the real part of u) and so is irreducible because its roots u and $\bar{u}$ are not real.

Example 9

Solution

Find a real irreducible quadratic with $u = 3 - 4i$ as a root.

We have $u + \bar{u} = 6$ and $|u|^2 = 25$, so $x^2 - 6x + 25$ is irreducible with u and $\bar{u} = 3 + 4i$ as roots.

The quadratic formula works for quadratics with complex coefficients. If $p \neq 0$ and v and w are complex numbers, the equation

$$px^2 + vx + w = 0$$

can be solved by an old technique called **completing the square** (it was known to the Arabian mathematician Al-Khowarizmi in the ninth century A.D.). The idea is to write the equation as $x^2 + \frac{v}{p}x = -\frac{w}{p}$ and then complete the square on the left by adding $\left(\frac{v}{2p}\right)^2$ to each side:

$$\left(x + \frac{v}{2p}\right)^2 = x^2 + \frac{v}{p}x + \left(\frac{v}{2p}\right)^2 = -\frac{w}{p} + \left(\frac{v}{2p}\right)^2 = \frac{v^2 - 4pw}{4p^2}$$

Taking square roots gives the complex version of the quadratic formula

$$x = \frac{-v \pm \sqrt{v^2 - 4pw}}{2p}$$

Of course, the discriminant $v^2 - 4pw$ is now a complex number, and we need the foregoing methods to find its square roots. Here is an example.

Example 10

Solution

Find all complex numbers z such that $z^2 - iz + (1 + 3i) = 0$.

The quadratic formula gives

$$z = \tfrac{1}{2}\left[i \pm \sqrt{i^2 - 4(1 + 3i)}\right] = \tfrac{1}{2}\left[i \pm \sqrt{-5 - 12i}\right] = \tfrac{1}{2}[i \pm w]$$

where $w = \sqrt{-5 - 12i}$. Hence $w^2 = -5 - 12i$; so if $w = a + bi$, equating real and imaginary parts gives $a^2 - b^2 = -5$ and $2ab = -12$. Hence $b = -6/a$, so $a^2 - 36/a^2 = -5$. This gives a quadratic in $a^2 : a^4 + 5a^2 - 36 = 0$, which factors as $(a^2 - 4)(a^2 + 9) = 0$. Thus $a = \pm 2$ and $b = -6/a = \mp 3$, so $w = \pm(2 - 3i)$. Finally,

$$z = \tfrac{1}{2}[i \pm w] = \tfrac{1}{2}[i \pm (2 - 3i)]$$

Hence the roots are $z = 1 - i$ and $-1 + 2i$.

If one root of a quadratic equation $px^2 + vx + w = 0$ is known, it is easy to find the other root. Because we can divide both sides of the equation by p, we state the result for quadratics with 1 as the coefficient of x^2.

Theorem 4

If u_1 and u_2 are the roots of the quadratic equation

$$x^2 + vx + w = 0$$

then $u_1 + u_2 = -v$ and $u_1 u_2 = w$.

PROOF

Because u_1 and u_2 are roots of $x^2 + vx + w = 0$ the factor theorem (see Section 6.5) asserts that the quadratic factors as

$$x^2 + vx + w = (x - u_1)(x - u_2)$$

The right side is $x^2 - (u_1 + u_2)x + u_1 u_2$, so the result follows because corresponding coefficients must be equal.

Example 11

Show that $u_1 = 1 + i$ is a root of $x^2 + (1 - 2i)x - (3 + i) = 0$ and then find the other root.

Solution $u_1^2 + (1 - 2i)u_1 - (3 + i) = (2i) + (3 - i) - (3 + i) = 0$, so u_1 is a root. If u_2 is the other root, then $u_1 + u_2 = -(1 - 2i)$ by Theorem 4, so $u_2 = -(1 - 2i) - u_1 = -2 + i$. Of course, this also follows from $u_1 u_2 = -(3 + i)$.

As we mentioned earlier, the complex numbers are the culmination of a long search by mathematicians to find a set of numbers large enough to contain a root of every polynomial. The fact that the complex numbers have this property was first proved by Gauss in 1797 when he was 20 years old. The proof is omitted.

Theorem 5 Fundamental Theorem of Algebra

Every polynomial of positive degree with complex coefficients has a complex root.

If $f(x)$ is a polynomial with complex coefficients, and if u_1 is a root, then the factor theorem (Section 6.5) asserts that

$$f(x) = (x - u_1)g(x)$$

where $g(x)$ is a polynomial with complex coefficients and with degree one less than the degree of $f(x)$. Suppose that u_2 is a root of $g(x)$, again by the fundamental theorem. Then $g(x) = (x - u_2)h(x)$, so

$$f(x) = (x - u_1)(x - u_2)h(x)$$

This process continues until the last polynomial to appear is linear. Thus $f(x)$ has been expressed as a product of linear factors. The last of these factors can be written in the form $u(x - u_n)$, where u and u_n are complex (verify this), so the fundamental theorem takes the following form.

Theorem 6

Every complex polynomial $f(x)$ of degree $n \geq 1$ has the form

$$f(x) = u(x - u_1)(x - u_2)\cdots(x - u_n)$$

where $u, u_1, \ldots, u_n$ are complex numbers and $u \neq 0$. The numbers $u_1, u_2, \ldots, u_n$ are the roots of $f(x)$ (and need not all be distinct), and u is the coefficient of x^n.

This form of the fundamental theorem, when applied to a polynomial $f(x)$ with *real* coefficients, can be used to deduce the following result.

Theorem 7

Every polynomial $f(x)$ of positive degree with real coefficients can be factored as a product of linear and irreducible quadratic factors.

In fact, suppose $f(x)$ has the form

$$f(x) = a_n x^n + a_{n-1}x^{n-1} + \cdots + a_1 x + a_0$$

where the coefficients a_i are real. If u is a complex root of $f(x)$, then we claim first that $\bar{u}$ is also a root. In fact, we have $f(u) = 0$, so

$$
\begin{aligned}
0 = \bar{0} = \overline{f(u)} &= \overline{a_n u^n + a_{n-1}u^{n-1} + \cdots + a_1 u + a_0} \\
&= \overline{a_n u^n} + \overline{a_{n-1}u^{n-1}} + \cdots + \overline{a_1 u} + \overline{a_0} \\
&= \bar{a}_n \bar{u}^n + \bar{a}_{n-1}\bar{u}^{n-1} + \cdots + \bar{a}_1\bar{u} + \bar{a}_0 \\
&= a_n \bar{u}^n + a_{n-1}\bar{u}^{n-1} + \cdots + a_1\bar{u} + a_0 \\
&= f(\bar{u})
\end{aligned}
$$

where $\bar{a}_i = a_i$ for each i because the coefficients a_i are real. Thus if u is a root of $f(x)$, so is its conjugate $\bar{u}$. Of course some of the roots of $f(x)$ may be real (and so equal their conjugates), but the nonreal roots come in pairs, u and $\bar{u}$. By Theorem 6, we can thus write $f(x)$ as a product:

$$f(x) = a_n(x - r_1)\cdots(x - r_k)(x - u_1)(x - \bar{u}_1)\cdots(x - u_m)(x - \bar{u}_m) \qquad (*)$$

where a_n is the coefficient of x^n in $f(x)$; $r_1, r_2, \ldots, r_k$ are the real roots; and $u_1, \bar{u}_1, u_2, \bar{u}_2, \ldots, u_m, \bar{u}_m$ are the nonreal roots. But the product

$$(x - u_j)(x - \bar{u}_j) = x^2 - (u_j + \bar{u}_j)x + u_j\bar{u}_j$$

is a real irreducible quadratic for each j (see the discussion preceding Example 9). Hence $(*)$ shows that $f(x)$ is a product of linear and irreducible quadratic factors, each with real coefficients. This is the conclusion in Theorem 7.

Exercises A

1. Solve each of the following for the real number x.
   (a) $x - 4i = (2 - i)^2$
   (b) $(2 + xi)(3 - 2i) = 12 + 5i$
   (c) $(2 + xi)^2 = 4$
   (d) $(2 + xi)(2 - xi) = 5$

2. Convert each of the following to the form $a + bi$.
   (a) $(2 - 3i) - 2(2 - 3i) + 9$
   (b) $(3 - 2i)(1 + i) + |3 + 4i|$
   (c) $\dfrac{1+i}{2-3i} + \dfrac{1-i}{-2+3i}$
   (d) $\dfrac{3-2i}{1-i} - \dfrac{3-7i}{2-3i}$
   (e) i^{131}
   (f) $(2 - i)^3$
   (g) $(1 + i)^4$
   (h) $(1 - i)^2(2 + i)^2$

3. In each case, find the complex number z.
   (a) $iz - (1 + i)^2 = 3 - i$
   (b) $(i + z) - 3i(2 - z) = iz + 1$
   (c) $z^2 = -i$
   (d) $z^2 = 3 - 4i$
   (e) $z(1 + i) = \bar{z} + (3 + 2i)$
   (f) $z(2 - i) = (\bar{z} + 1)(1 + i)$

4. In each case, find the roots of the real quadratic equation.
   (a) $x^2 - 2x + 3 = 0$
   (b) $x^2 - x + 1 = 0$
   (c) $3x^2 - 4x + 2 = 0$
   (d) $2x^2 - 5x + 2 = 0$

5. Find all numbers x in each case.
   (a) $x^3 = 8$
   (b) $x^3 = -8$
   (c) $x^4 = 16$
   (d) $x^4 = 64$

6. In each case, find a real quadratic with u as a root, and find the other root.
   (a) $u = 1 + i$ ◀(b) $u = 2 - 3i$
   (c) $u = -i$ ◆(d) $u = 3 - 4i$

7. Find the roots of $x^2 - 2\cos\theta\, x + 1 = 0$, θ any angle.

◆8. Find a real polynomial of degree 4 with $2 - i$ and $3 - 2i$ as roots.

9. Let re z and im z denote, respectively, the real and imaginary parts of z. Show that:
   (a) $\operatorname{im}(iz) = \operatorname{re} z$ (b) $\operatorname{re}(iz) = -\operatorname{im} z$
   (c) $z + \bar{z} = 2\operatorname{re} z$ (d) $z - \bar{z} = 2i\operatorname{im} z$
   (e) $\operatorname{re}(z + w) = \operatorname{re} z + \operatorname{re} w$, and $\operatorname{re}(tz) = t \cdot \operatorname{re} z$
      if t is real
   (f) $\operatorname{im}(z + w) = \operatorname{im} z + \operatorname{im} w$, and $\operatorname{im}(tz) = t \cdot \operatorname{im} z$
      if t is real

10. In each case, show that u is a root of the quadratic equation, and find the other root.
    (a) $x^2 - 3ix + (-3 + i) = 0$; $u = 1 + i$
    ◆(b) $x^2 + ix - (4 - 2i) = 0$; $u = -2$
    (c) $x^2 - (3 - 2i)x + (5 - i) = 0$; $u = 2 - 3i$
    ◆(d) $x^2 + 3(1 - i)x - 5i = 0$; $u = -2 + i$

11. Find the roots of each of the following complex quadratic equations.
    (a) $x^2 + 2x + (1 + i) = 0$
    ◆(b) $x^2 - x + (1 - i) = 0$
    (c) $x^2 - (2 - i)x + (3 - i) = 0$
    ◆(d) $x^2 - 3(1 - i)x - 5i = 0$

12. In each case, describe the graph of the equation (where z denotes a complex number).
    ⓐ $|z| = 1$ ◀(b) $|z - 1| = 2$
    (c) $z = i\bar{z}$ ◆(d) $z = -\bar{z}$
    (e) $z = |z|$ ◆(f) $\operatorname{im} z = m \cdot \operatorname{re} z$, m a real number

13. (a) Verify $|zw| = |z||w|$ directly for $z = a + bi$ and $w = c + di$.
    (b) Deduce (a) from properties C2 and C6.

14. Prove that $|w + z|^2 = |w|^2 + |z|^2 + w\bar{z} + \bar{w}z$ for all complex numbers w and z.

15. If zw is real and $z \neq 0$, show that $w = a\bar{z}$ for some real number a.

16. If $zw = \bar{z}v$ and $z \neq 0$, show that $w = uv$ for some u in $\mathbb{C}$ with $|u| = 1$.

17. Use property C5 to show that $(1 + i)^n + (1 - i)^n$ is real for all n.

18. Express each of the following in polar form (use the principal argument).
    (a) $3 - 3i$ ◀(b) $-4i$
    (c) $-\sqrt{3} + i$ ◉(d) $-4 + 4\sqrt{3}i$
    (e) $-7i$ ◆(f) $-6 + 6i$

19. Express each of the following in the form $a + bi$.
    (a) $3e^{\pi i}$ ◀(b) $e^{7\pi i/3}$
    (c) $2e^{3\pi i/4}$ ◀(d) $\sqrt{2}e^{-\pi i/4}$
    (e) $e^{5\pi i/4}$ ◉(f) $2\sqrt{3}e^{-2\pi i/6}$

20. Express each of the following in the form $a + bi$.
    (a) $(-1 + \sqrt{3}i)^2$ ◆(b) $(1 + \sqrt{3}i)^{-4}$
    (c) $(1 + i)^8$ ◆(d) $(1 - i)^{10}$
    (e) $(1 - i)^6(\sqrt{3} + i)^3$ ◆(f) $(\sqrt{3} - i)^9(2 - 2i)^5$

21. Use De Moivre's theorem to show that:
    (a) $\cos 2\theta = \cos^2\theta - \sin^2\theta$;
       $\sin 2\theta = 2\cos\theta\sin\theta$
    (b) $\cos 3\theta = \cos^3\theta - 3\cos\theta\sin^2\theta$;
       $\sin 3\theta = 3\cos^2\theta\sin\theta - \sin^3\theta$

22. ⓐ Find the fourth roots of unity.
    (b) Find the sixth roots of unity.

23. Find all complex numbers z such that:
    (a) $z^4 = -1$ ◀(b) $z^4 = 2(\sqrt{3}i - 1)$
    (c) $z^3 = -27i$ ◆(d) $z^6 = -64$

24. If $z = re^{i\theta}$ in polar form, show that:
    (a) $\bar{z} = re^{-i\theta}$ (b) $z^{-1} = \frac{1}{r}e^{-i\theta}$

25. Show that the sum of the nth roots of unity is zero. [*Hint:* $1 - z^n = (1 - z)(1 + z + z^2 + \cdots + z^{n-1})$ for any complex number z.]

26. (a) Suppose z_1, z_2, z_3, z_4, and z_5 are equally spaced around the unit circle. Show that
       $z_1 + z_2 + z_3 + z_4 + z_5 = 0$
       [*Hint:* $(1 - z)(1 + z + z^2 + z^3 + z^4) = 1 - z^5$ for any complex number z.]
    (b) Repeat (a) for any $n \geq 2$ points equally spaced around the unit circle.
    (c) If $|w| = 1$, show that the sum of the roots of $z^n = w$ is zero.

27. If $z = a + bi$, show that $|a| + |b| \le \sqrt{2} \cdot |z|$.
    [*Hint:* $(|a| - |b|)^2 \ge 0$.]

28. Let $z \ne 0$ be a complex number. If t is real, describe tz geometrically in terms of z if:
    (a) $t > 0$; (b) $t < 0$.

29. If z and w are nonzero complex numbers, show that $|z + w| = |z| + |w|$ if and only if one is a positive real multiple of the other.
    [*Hint:* Consider the parallelogram with vertices $0, w, z$, and $z + w$. Use the preceding exercise and the fact that if t is real, then $|1 + t| = 1 + |t|$ is impossible if $t < 0$.]

30. If a and b are *rational* numbers, let p and q denote numbers of the form $a + b\sqrt{2}$. If $p = a + b\sqrt{2}$ define $\tilde{p} = a - b\sqrt{2}$ and $[p] = a^2 - 2b^2$. Show that each of the following holds.
    (a) $a + b\sqrt{2} = a_1 + b_1\sqrt{2}$ only if $a = a_1$ and $b = b_1$
    (b) $\widetilde{p \pm q} = \tilde{p} \pm \tilde{q}$
    (c) $\widetilde{pq} = \tilde{p}\tilde{q}$
    (d) $[p] = p\tilde{p}$
    (e) $[pq] = [p][q]$
    (f) If $f(x)$ is a polynomial with rational coefficients and $p = a + b\sqrt{2}$ is a root of $f(x)$, then $\tilde{p}$ is also a root of $f(x)$.

Appendix B: Proofs

Logic plays a basic role in human affairs. Scientists use logic to draw conclusions from experiments, judges use it to deduce consequences of the law, and mathematicians use it to prove theorems. Logic arises in ordinary speech with assertions such as "If John studies hard, he will pass the course," or "If an integer n is divisible by 6, then n is divisible by 3."[1] In each case, the aim is to assert that if a certain statement is true, then another statement must also be true. In fact, if p and q denote statements, most theorems take the form of an **implication**: "If p is true, then q is true." We write this in symbols as

$$p \Rightarrow q$$

and read it as "p implies q." Here p is the **hypothesis** and q the **conclusion** of the implication. The verification that $p \Rightarrow q$ is valid is called the **proof** of the implication. In this section we examine the most common methods of proof[2] and illustrate each technique with some examples.

Method of Direct Proof

To prove that $p \Rightarrow q$, demonstrate directly that q is true whenever p is true.

Example 1

If n is an odd integer, show that n^2 is odd.

Solution If n is odd, it has the form $n = 2k + 1$ for some integer k. Then $n^2 = 4k^2 + 4k + 1 = 2(2k^2 + 2k) + 1$ also is odd because $2k^2 + 2k$ is an integer.

Note that the computation $n^2 = 4k^2 + 4k + 1$ in Example 1 involves some simple properties of arithmetic that we did not prove. These properties, in turn, can be proved from certain more basic properties of numbers (called axioms)—more about that later. Actually, a whole body of mathematical information lies behind nearly every proof of any complexity, although this fact usually is not stated explicitly. Here is a geometrical example.

---

1 By an *integer* we mean a "whole number"; that is, a number in the set $0, \pm 1, \pm 2, \pm 3, \dots$.

2 For a more detailed look at proof techniques see D. Solow, *How to Read and Do Proofs*, 2nd ed. (New York: Wiley, 1990); or J. F. Lucas, *Introduction to Abstract Mathematics*, Chapter 2 (Belmont, CA: Wadsworth, 1986).

Example 2

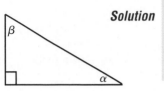

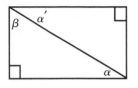

In a right triangle, show that the sum of the two acute angles is 90 degrees.

Solution   Construct the rectangle with the given triangle as the bottom half as shown. Then the top triangle in the rectangle is identical to the given triangle (the right angle and the sides adjacent to it are the same). Hence α' equals α, and so $\alpha + \beta = \alpha' + \beta$ is a right angle.

Geometry was one of the first subjects in which formal proofs were used—Euclid's *Elements* was published about 300 B.C. The *Elements* is the most successful textbook ever written, and contains many of the basic geometrical theorems that are taught in school today. In particular, Euclid included a proof of an earlier theorem (about 500 B.C.) due to Pythagoras. Recall that, in a right triangle, the side opposite the right angle is called the *hypotenuse* of the triangle.

Example 3

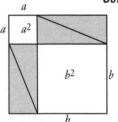

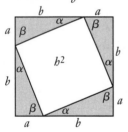

Pythagoras' Theorem
In a right-angled triangle, show that the square of the length of the hypotenuse equals the sum of the squares of the lengths of the other two sides.

Solution   Let the sides of the right triangle have lengths a, b, and h as shown, Consider two squares with sides of length $a + b$, and place four copies of the triangle in these squares as in the diagram. The central rectangle in the second square shown is a square because the angles α and β add to 90 degrees (using Example 2), so its area is h^2 as shown. Comparing areas shows that both $a^2 + b^2$ and h^2 equal the area of the large square minus four times the area of the original triangle, and hence are equal.

Sometimes it is convenient (or even necessary) to break a proof into parts, and deal with each case separately. We formulate the general method as follows:

Method of Reduction to Cases

To prove that $p \Rightarrow q$, show that p implies at least one of a list $p_1, p_2, \ldots, p_n$ of statements (the cases) and then show that $p_i \Rightarrow q$ for each i.

Example 4

Show that $n^2 \geq 0$ for every integer n.

Solution   This statement can be expressed as an implication: If n is an integer, then $n^2 \geq 0$. To prove it, consider the following three cases:

$$(1)\ n > 0; \qquad (2)\ n = 0; \qquad (3)\ n < 0.$$

Then $n^2 > 0$ in Cases (1) and (3) because the product of two positive (or two negative) integers is positive. In Case (2) $n^2 = 0^2 = 0$, so $n^2 \geq 0$ in every case.

Example 5

If n is an integer, show that $n^2 - n$ is even.

Solution   We consider two cases:

$$(1)\ n \text{ is even}; \qquad (2)\ n \text{ is odd}.$$

We have $n^2 - n = n(n - 1)$, so this is even in Case (1) because any multiple of an even number is again even. Similarly, $n - 1$ is even in Case (2) so $n(n - 1)$ is even for the same reason. Hence $n^2 - n$ is even in any case.

The statements used in mathematics are required to be either true or false. This leads to a proof technique which causes consternation in many beginning students. The method is a formal version of a debating strategy whereby the debater assumes the truth of an opponent's position and shows that it leads to an absurd conclusion.

Method of Proof by Contradiction

To prove that $p \Rightarrow q$, show that the assumption that both p is true and q is false leads to a contradiction.

Example 6

If r is a rational number (fraction), show that $r^2 \neq 2$.

Solution To argue by contradiction, we assume that r is a rational number and that $r^2 = 2$, and show that this assumption leads to a contradiction. Let m and n be integers such that $r = m/n$ is in lowest terms (so, in particular, m and n are not both even). Then $r^2 = 2$ gives $m^2 = 2n^2$, so m^2 is even. This means m is even (Example 1), say $m = 2k$. But then $2n^2 = m^2 = 4k^2$, so $n^2 = 2k^2$ is even, and hence n is even. This shows that n and m are both even, contrary to the choice of these numbers.

Example 7

Pigeonhole Principle
If $n + 1$ pigeons are placed in n holes, then some hole contains at least 2 pigeons.

Solution Assume the conclusion is false. Then each hole contains at most one pigeon and so, since there are n holes, there must be at most n pigeons, contrary to assumption.

The next example involves the notion of a *prime* number, that is an integer that is greater than 1 and that cannot be factored as the product of two smaller positive integers.

Example 8

If $2^n - 1$ is a prime number, show that n is a prime number.

Solution We must show that $p \Rightarrow q$ where p is the statement "$2^n - 1$ *is a prime*", and q is the statement "n *is a prime*." Suppose that q is false so that n is not a prime, say $n = ab$ where $a \geq 2$ and $b \geq 2$ are integers. If we write $2^a = x$, then $2^n = 2^{ab} = (2^a)^b = x^b$. Hence $2^n - 1$ factors:

$$2^n - 1 = x^b - 1 = (x - 1)(x^{b-1} + x^{b-2} + \cdots + x^2 + x + 1)$$

As $x \geq 4$, this expression is a factorization of $2^n - 1$ into smaller positive integers, contradicting the assumption that $2^n - 1$ is prime.

The next example exhibits one way to show that an implication is *not* valid.

Example 9

Show that the implication "*n is a prime* $\Rightarrow 2^n - 1$ *is a prime*" is false.

Solution

The first four primes are 2, 3, 5, and 7, and the corresponding values for $2^n - 1$ are 3, 7, 31, 127 (when $n = 2, 3, 5, 7$). These are all prime as the reader can verify. This result seems to be evidence that the implication is true. However, the next prime is 11 and $2^{11} - 1 = 2047$ or $23 \cdot 89$, which is clearly *not* a prime.

We say that $n = 11$ is a **counterexample** to the (proposed) implication in Example 9. Note that, if you can find even one example for which an implication is not valid, the implication is false. Thus disproving implications is in a sense easier than proving them.

The implications in Examples 8 and 9 are closely related: They have the form $p \Rightarrow q$ and $q \Rightarrow p$, where p and q are statements. Each is called the **converse** of the other and, as these examples show, an implication can be valid even though its converse is not valid. If *both* $p \Rightarrow q$ and $q \Rightarrow p$ are valid, the statements p and q are called **logically equivalent**. This is written in symbols as

$$p \Leftrightarrow q$$

and is read "p if and only if q". Many of the most satisfying theorems make the assertion that two statements, ostensibly quite different, are in fact logically equivalent.

Example 10

If n is an integer, show that "n is odd $\Leftrightarrow n^2$ is odd."

Solution

In Example 1 we proved the implication "n is odd $\Rightarrow n^2$ is odd." Here we prove the converse by contradiction. If n^2 is odd, we assume that n is not odd. Then n is even, say $n = 2k$, so $n^2 = 4k^2$, which is also even, a contradiction.

Many more examples of proofs can be found in this book and, although they are often more complex, most are based on one of these methods. In fact, linear algebra is one of the best topics on which the reader can sharpen his or her skill at constructing proofs. Part of the reason for this is that much of linear algebra is developed using the **axiomatic method**. That is, in the course of studying various examples it is observed that they all have certain properties in common. Then a general, abstract system is studied in which these properties are *assumed* to hold (and are called **axioms**). In this system, statements (called **theorems**) are deduced from the axioms using the methods presented in this section. These theorems will then be true in *all* the concrete examples, because the axioms hold in each case. But this procedure is more than just an efficient method for finding theorems in the examples. By reducing the proof to its essentials, we gain a better understanding of why the theorem is true and how it relates to analogous theorems in other abstract systems.

The axiomatic method is not new. Euclid first used it in about 300 B.C. to derive all the propositions of (euclidean) geometry from a list of 10 axioms. The method lends itself well to linear algebra. The axioms are simple and easy to understand, and there are only a few of them. For example, the theory of vector spaces contains a large number of theorems derived from only ten simple axioms.

Exercises B

1. In each case prove the result and either prove the converse or give a counterexample.

   (a) If n is an even integer, then n^2 is a multiple of 4.

   ◆(b) If m is an even integer and n is an odd integer, then $m + n$ is odd.

   (c) If $x = 2$ or $x = 3$, then $x^3 - 6x^2 + 11x - 6 = 0$.

   ◆(d) If $x^2 - 5x + 6 = 0$, then $x = 2$ or $x = 3$.

2. In each case either prove the result by splitting into cases, or give a counterexample.

   (a) If n is any integer, then $n^2 = 4k + 1$ for some integer k.

   ◆(b) If n is any odd integer, then $n^2 = 8k + 1$ for some integer k.

   (c) If n is any integer, $n^3 - n = 3k$ for some integer k. [*Hint:* Use the fact that each integer has one of the forms $3k$, $3k + 1$, or $3k + 2$, where k is an integer.]

3. In each case prove the result by contradiction and either prove the converse or give a counterexample.

   (a) If $n > 2$ is a prime integer, then n is odd.

   ◆(b) If $n + m = 25$ where n and m are integers, then one of n and m is greater than 12.

   (c) If a and b are positive numbers and $a \leq b$, then $\sqrt{a} \leq \sqrt{b}$.

   ◆(d) If m and n are integers and mn is even, then m is even or n is even.

4. Prove each implication by contradiction.

   (a) If x and y are positive numbers, then $\sqrt{x + y} \neq \sqrt{x} + \sqrt{y}$.

   ◆(b) If x is irrational and y is rational, then $x + y$ is irrational.

   (c) If 13 people are selected, at least 2 have birthdays in the same month.

5. Disprove each statement by giving a counterexample.

   (a) $n^2 + n + 11$ is a prime for all positive integers n.

   ◆(b) $n^3 \geq 2^n$ for all integers $n \geq 2$.

   (c) If $n \geq 2$ points are arranged on a circle in such a way that no three of the lines joining them have a common point, show that these lines divide the circle into 2^{n-1} regions. [The cases $n = 2$, 3, and 4 are shown in the diagram.]

$n = 2$ $\qquad$ $n = 3$ $\qquad$ $n = 4$

6. The number e from calculus has a series expansion

$$e = 1 + \frac{1}{1!} + \frac{1}{2!} + \frac{1}{3!} + \cdots$$

where $n! = n(n - 1) \cdots 3 \cdot 2 \cdot 1$ for each integer $n \geq 1$. Prove that e is irrational by contradiction. [*Hint:* If $e = m/n$, consider

$$k = n!\left(e - 1 - \frac{1}{1!} - \frac{1}{2!} - \frac{1}{3!} - \cdots - \frac{1}{n!} \right).$$

Show that k is a positive integer and that

$$k = \frac{1}{n + 1} + \frac{1}{(n + 1)(n + 2)} + \cdots < \frac{1}{n}.]$$

Appendix C:
Mathematical Induction

Suppose one is presented with the following sequence of equations:

$$1 = 1$$
$$1 + 3 = 4$$
$$1 + 3 + 5 = 9$$
$$1 + 3 + 5 + 7 = 16$$
$$1 + 3 + 5 + 7 + 9 = 25$$

It is clear that there is a pattern. The numbers on the right side of the equations are the squares 1^2, 2^2, 3^2, 4^2, and 5^2 and, in the equation with n^2 on the right side, the left side is the sum of the first n odd numbers. The odd numbers are

$$1 = 2 \cdot 1 - 1$$
$$3 = 2 \cdot 2 - 1$$
$$5 = 2 \cdot 3 - 1$$
$$7 = 2 \cdot 4 - 1$$
$$9 = 2 \cdot 5 - 1$$

and from this it is clear that the nth odd number is $2n - 1$. Hence, at least for $n = 1, 2, 3, 4$, or 5, the following is true:

$$1 + 3 + \cdots + (2n - 1) = n^2 \qquad (S_n)$$

The question arises whether the statement (S_n) is true for *every* n. There is no hope of separately verifying all these statements because there are infinitely many of them. A more subtle approach is required.

The idea is as follows: Suppose it is verified that the statement S_{n+1} will be true whenever S_n is true. That is, suppose we prove that, *if* S_n is true, then it necessarily follows that S_{n+1} is also true. Then, if we can show that S_1 is true, it follows that S_2 is true, and from this that S_3 is true, hence that S_4 is true, and so on and on. This is the principle of induction. To express it more compactly, it is useful to have a short way to express the assertion "If S_n is true, then S_{n+1} is true." As in Appendix B, we write this assertion as

$$S_n \Rightarrow S_{n+1}$$

and read it as "S_n implies S_{n+1}." We can now state the principle of mathematical induction.

The Principle of Mathematical Induction

Suppose S_n is a statement about the natural number n for each $n = 1, 2, 3, \ldots$.
Suppose further that:
1. S_1 is true.
2. $S_n \Rightarrow S_{n+1}$ for every $n \geq 1$.
Then S_n is true for every $n \geq 1$.

This is one of the most useful techniques in all of mathematics. It applies in a wide number of situations. The following examples illustrate this.

Example 1

Solution

Show that $1 + 2 + \cdots + n = \frac{1}{2}n(n+1)$ for $n \geq 1$.

Let S_n be the statement: $1 + 2 + \cdots + n = \frac{1}{2}n(n+1)$. We apply induction.

1. S_1 is true. The statement S_1 is $1 = \frac{1}{2}1(1+1)$, which is true.
2. $S_n \Rightarrow S_{n+1}$. We *assume* that S_n is true for some $n \geq 1$—that is, that

$$1 + 2 + \cdots + n = \frac{1}{2}n(n+1).$$

We must prove that the statement

$$S_{n+1}: \quad 1 + 2 + \cdots + (n + 1) = \frac{1}{2}(n + 1)(n + 2)$$

is also true, and we are entitled to use S_n to do so. Now the left side of S_{n+1} is the sum of the first $n + 1$ positive integers. Hence the second-to-last term is n, so we can write

$$1 + 2 + \cdots + (n + 1) = (1 + 2 + \cdots + n) + (n + 1)$$
$$= \tfrac{1}{2}n(n + 1) + (n + 1) \quad \text{using } S_n$$
$$= \tfrac{1}{2}(n + 1)[n + 2]$$

This shows that S_{n+1} is true and so completes the induction.

In the verification that $S_n \Rightarrow S_{n+1}$, we *assume* that S_n is true and use it to deduce that S_{n+1} is true. The assumption that S_n is true is sometimes called the **induction hypothesis**.

Example 2

Solution

If x is any number such that $x \neq 1$, show that $1 + x + x^2 + \cdots + x^n = \dfrac{x^{n+1} - 1}{x - 1}$ for $n \geq 1$.

Let S_n be the statement: $1 + x + x^2 + \cdots + x^n = \dfrac{x^{n+1} - 1}{x - 1}$.

1. S_1 is true. S_1 reads $1 + x = \dfrac{x^2 - 1}{x - 1}$, which is true because $x^2 - 1 = (x - 1)(x + 1)$.

2. $S_n \Rightarrow S_{n+1}$. Assume the truth of S_n: $1 + x + x^2 + \cdots + x^n = \dfrac{x^{n+1} - 1}{x - 1}$.

We must *deduce* from this the truth of S_{n+1}: $1 + x + x^2 + \cdots + x^{n+1} = \dfrac{x^{n+2} - 1}{x - 1}$.

Starting with the left side of S_{n+1} and using the induction hypothesis, we find

$$
\begin{aligned}
1 + x + x^2 + \cdots + x^{n+1} &= (1 + x + x^2 + \cdots + x^n) + x^{n+1} \\
&= \frac{x^{n+1} - 1}{x - 1} + x^{n+1} \\
&= \frac{x^{n+1} - 1 + x^{n+1}(x - 1)}{x - 1} \\
&= \frac{x^{n+2} - 1}{x - 1}
\end{aligned}
$$

This shows that S_{n+1} is true and so completes the induction.

Both of these examples involve formulas for a certain sum, and it is often convenient to use summation notation. For example, $\sum_{k=1}^{n}(2k-1)$ means that in the expression $(2k - 1)$, k is to be given the values $k = 1$, $k = 2$, $k = 3, \ldots, k = n$, and then the resulting n numbers are to be added. The same thing applies to other expressions involving k. For example,

$$\sum_{k=1}^{n} k^3 = 1^3 + 2^3 + \cdots + n^3$$

$$\sum_{k=1}^{5}(3k - 1) = (3 \cdot 1 - 1) + (3 \cdot 2 - 1) + (3 \cdot 3 - 1) + (3 \cdot 4 - 1) + (3 \cdot 5 - 1)$$

The next example involves this notation.

Example 3

Show that $\displaystyle\sum_{k=1}^{n}(3k^2 - k) = n^2(n+1)$ for each $n \geq 1$.

Solution   Let S_n be the statement: $\displaystyle\sum_{k=1}^{n}(3k^2 - k) = n^2(n+1)$.

1. S_1 is true. S_1 reads $(3 \cdot 1^2 - 1) = 1^2(1 + 1)$, which is true.
2. $S_n \Rightarrow S_{n+1}$. Assume that S_n is true. We must prove S_{n+1}:

$$
\begin{aligned}
\sum_{k=1}^{n+1}(3k^2 - k) &= \sum_{k=1}^{n}(3k^2 - k) + [3(n + 1)^2 - (n + 1)] \\
&= n^2(n + 1) + (n + 1)[3(n + 1) - 1] \quad \text{using } S_n \\
&= (n + 1)[n^2 + 3n + 2] \\
&= (n + 1)[(n + 1)(n + 2)] \\
&= (n + 1)^2(n + 2)
\end{aligned}
$$

This proves that S_{n+1} is true.

We now turn to examples wherein induction is used to prove propositions that do not involve sums.

Example 4

Show that $7^n + 2$ is a multiple of 3 for all $n \geq 1$.

Solution

1. S_1 is true. $7^1 + 2 = 9$ is a multiple of 3.
2. $S_n \Rightarrow S_{n+1}$. Assume that $7^n + 2$ is a multiple of 3 for some $n \geq 1$; say, $7^n + 2 = 3m$ for some integer m. Then
$$7^{n+1} + 2 = 7(7^n) + 2 = 7(3m - 2) + 2 = 21m - 12 = 3(7m - 4)$$
so $7^{n+1} + 2$ is also a multiple of 3. This proves that S_{n+1} is true.

In all the foregoing examples, we have used the principle of induction starting at 1; that is, we have verified that S_1 is true and that $S_n \Rightarrow S_{n+1}$ for each $n \geq 1$, and then we have concluded that S_n is true for every $n \geq 1$. But there is nothing special about 1 here. If m is some fixed integer and we verify that

1. S_m is true.
2. $S_n \Rightarrow S_{n+1}$ for every $n \geq m$.

then it follows that S_n is true for every $n \geq m$. This "extended" induction principle is just as plausible as the induction principle and can, in fact, be proved by induction. The next example will illustrate it. Recall that if n is a positive integer, the number $n!$ (which is read "n-factorial") is the product

$$n! = n(n - 1)(n - 2) \cdots 3 \cdot 2 \cdot 1$$

of all the numbers from n to 1. Thus $2! = 2$, $3! = 6$, and so on.

Example 5

Show that $2^n < n!$ for all $n \geq 4$.

Solution

Observe that $2^n < n!$ is actually false if $n = 1, 2, 3$.

1. S_4 is true. $2^4 = 16 < 24 = 4!$.
2. $S_n \Rightarrow S_{n+1}$ if $n \geq 4$. Assume that S_n is true; that is, $2^n < n!$. Then

$$
\begin{aligned}
2^{n+1} &= 2 \cdot 2^n \\
&< 2 \cdot n! && \text{because } 2^n < n! \\
&< (n + 1)n! && \text{because } 2 < n + 1 \\
&= (n + 1)!
\end{aligned}
$$

Hence S_{n+1} is true.

Exercises C

In Exercises 1–19, prove the given statement by induction for all n ≥ 1.

1. $1 + 3 + 5 + 7 + \cdots + (2n - 1) = n^2$

2. $1^2 + 2^2 + \cdots + n^2 = \frac{1}{6}n(n + 1)(2n + 1)$

3. $1^3 + 2^3 + \cdots + n^3 = (1 + 2 + \cdots + n)^2$

4. $1 \cdot 2 + 2 \cdot 3 + \cdots + n(n + 1) = \frac{1}{3}n(n + 1)(n + 2)$

5. $1 \cdot 2^2 + 2 \cdot 3^2 + \cdots + n(n + 1)^2$
$= \frac{1}{12}n(n + 1)(n + 2)(3n + 5)$

♦6. $\dfrac{1}{1 \cdot 2} + \dfrac{1}{2 \cdot 3} + \cdots + \dfrac{1}{n(n + 1)} = \dfrac{n}{n + 1}$

7. $1^2 + 3^2 + \cdots + (2n - 1)^2 = \frac{n}{3}(4n^2 - 1)$

8. $\dfrac{1}{1\cdot 2\cdot 3}+\dfrac{1}{2\cdot 3\cdot 4}+\cdots+\dfrac{1}{n(n+1)(n+2)}$

$$=\dfrac{n(n+3)}{4(n+1)(n+2)}$$

9. $1+2+2^2+\cdots+2^{n-1}=2^n-1$

10. $3+3^3+3^5+\cdots+3^{2n-1}=\frac{3}{8}(9^n-1)$

11. $\dfrac{1}{1^2}+\dfrac{1}{2^2}+\cdots+\dfrac{1}{n^2}\le 2-\dfrac{1}{n}$

12. $n<2^n$

13. For any integer $m>0$, $m!n!<(m+n)!$

♦14. $\dfrac{1}{\sqrt{1}}+\dfrac{1}{\sqrt{2}}+\cdots+\dfrac{1}{\sqrt{n}}\le 2\sqrt{n}-1$

15. $\dfrac{1}{\sqrt{1}}+\dfrac{1}{\sqrt{2}}+\cdots+\dfrac{1}{\sqrt{n}}\ge \sqrt{n}$

16. $n^3+(n+1)^3+(n+2)^3$ is a multiple of 9.

17. 5^n+3 is a multiple of 4.

18. n^3-n is a multiple of 3.

19. $3^{2n+1}+2^{n+2}$ is a multiple of 7.

♦20. Let $B_n=1\cdot 1!+2\cdot 2!+3\cdot 3!+\cdots+n\cdot n!$
Find a formula for B_n and prove it.

21. Let $A_n=(1-\frac{1}{2})(1-\frac{1}{3})(1-\frac{1}{4})\cdots(1-\frac{1}{n})$. Find a formula for A_n and prove it.

22. Suppose S_n is a statement about n for each $n\ge 1$. Explain what must be done to prove that S_n is true for all $n\ge 1$ if it is known that

(a) $S_n\Rightarrow S_{n+2}$ for each $n\ge 1$.

♦(b) $S_n\Rightarrow S_{n+8}$ for each $n\ge 1$.

(c) $S_n\Rightarrow S_{n+1}$ for each $n\ge 10$.

(d) Both S_n and $S_{n+1}\Rightarrow S_{n+2}$ for each $n\ge 1$.

23. If S_n is a statement for each $n\ge 1$, argue that S_n is true for all $n\ge 1$ if it is known that the following two conditions hold:

(1) $S_n\Rightarrow S_{n-1}$ for each $n\ge 2$.

(2) S_n is true for infinitely many values of n.

24. Suppose a sequence $a_1, a_2, \ldots$ of numbers is given that satisfies:

(1) $a_1=2$.

(2) $a_{n+1}=2a_n$ for each $n\ge 1$.

Formulate a theorem giving a_n in terms of n, and prove your result by induction.

25. Suppose a sequence $a_1, a_2, \ldots$ of numbers is given that satisfies:

(1) $a_1=b$.

(2) $a_{n+1}=ca_n+b$ for $n=1,2,3,\ldots$.

Formulate a theorem giving a_n in terms of n, and prove your result by induction.

26. (a) Show that $n^2\le 2^n$ for all $n\ge 4$.

(b) Show that $n^3\le 2^n$ for all $n\ge 10$.

Selected Answers

2. (b) $x = t$, $y = \frac{1}{3}(1 - 2t)$ or $x = \frac{1}{2}(1 - 3s)$, $y = s$ **(d)** $x = 1 + 2s - 5t$, $y = s$, $z = t$ or
$x = s$, $y = t$, $z = \frac{1}{5}(1 - s + 2t)$ **4.** $x = \frac{1}{4}(3 + 2s)$, $y = s$, $z = t$

5. (a) No solution if $b \neq 0$. If $b = 0$, *any* x is a solution.   **(b)** $x = \frac{b}{a}$ **7. (b)** $\begin{bmatrix} 1 & 2 & | & 0 \\ 0 & 1 & | & 1 \end{bmatrix}$ **(d)** $\begin{bmatrix} 1 & 1 & 0 & | & 1 \\ 0 & 1 & 1 & | & 0 \\ -1 & 0 & 1 & | & 2 \end{bmatrix}$

8. (b)
$$\begin{array}{rcrcrcl} 2x & - & y & & & = & -1 \\ -3x & + & 2y & + & z & = & 0 \\ & & y & + & z & = & 3 \end{array} \quad \text{or} \quad \begin{array}{rcrcrcl} 2x_1 & - & x_2 & & & = & -1 \\ -3x_1 & + & 2x_2 & + & x_3 & = & 0 \\ & & x_2 & + & x_3 & = & 3 \end{array}$$
9. (b) $x = -3$, $y = 2$ **(d)** $x = -17$, $y = 13$

10. (b) $x = \frac{1}{9}$, $y = \frac{10}{9}$, $z = \frac{-7}{3}$ **11. (b)** No solution   **14. (b)** F. $x + y = 0$, $x - y = 0$ has a unique solution.
(d) T. Theorem 1.   **16.** $x' = 5$, $y' = 1$, so $x = 23$, $y = -32$ **17.** $a = -\frac{1}{9}$, $b = -\frac{5}{9}$, $c = \frac{11}{9}$
19. \$4.59, \$5.20

1. (b) No, no **(d)** No, yes **(f)** No, no   **2. (b)** $\begin{bmatrix} 0 & 1 & -3 & 0 & 0 & 0 & 0 \\ 0 & 0 & 0 & 1 & 0 & 0 & -1 \\ 0 & 0 & 0 & 0 & 1 & 0 & 0 \\ 0 & 0 & 0 & 0 & 0 & 1 & 1 \end{bmatrix}$

3. (b) $x_1 = 2r - 2s - t + 1$, $x_2 = r$, $x_3 = -5s + 3t - 1$, $x_4 = s$, $x_5 = -6t + 1$, $x_6 = t$
(d) $x_1 = -4s - 5t - 4$, $x_2 = -2s + t - 2$, $x_3 = s$, $x_4 = 1$, $x_5 = t$ **4. (b)** $x = -\frac{1}{7}$, $y = -\frac{3}{7}$ **(d)** $x = \frac{1}{3}(t + 2)$, $y = t$
(f) No solution   **5. (b)** $x = -15t - 21$, $y = -11t - 17$, $z = t$ **(d)** No solution   **(f)** $x = -7$, $y = -9$, $z = 1$
(h) $x = 4$, $y = 3 + 2t$, $z = t$ **6. (b)** Denote the equations as E_1, E_2, and E_3. Apply gaussian elimination to
column 1 of the augmented matrix, and observe that $E_3 - E_1 = -4(E_2 - E_1)$. Hence $E_3 = 5E_1 - 4E_2$.
7. (b) $x_1 = 0$, $x_2 = -t$, $x_3 = 0$, $x_4 = t$ **(d)** $x_1 = 1$, $x_2 = 1 - t$, $x_3 = 1 + t$, $x_4 = t$
8. (b) If $ab \neq 2$, unique solution $x = \dfrac{-2 - 5b}{2 - ab}$, $y = \dfrac{a + 5}{2 - ab}$. If $ab = 2$: no solution if $a \neq -5$; if $a = -5$,
the solutions are $x = -1 + \frac{2}{5}t$, $y = t$. **(d)** If $a \neq 2$, unique solution $x = \dfrac{1 - b}{a - 2}$, $y = \dfrac{ab - 2}{a - 2}$.
If $a = 2$, no solution if $b \neq 1$; if $b = 1$, the solutions are $x = \frac{1}{2}(1 - t)$, $y = t$.
9. (b) Unique solution $x = -2a + b + 5c$, $y = 3a - b - 6c$, $z = -2a + b + 4c$, for any a, b, c.
(d) If $abc \neq -1$, unique solution $x = y = z = 0$; if $abc = -1$ the solutions are $x = abt$, $y = -bt$, $z = t$.

(f) If $a = 1$, solutions $x = -t$, $y = t$, $z = -1$. If $a = 0$, there is no solution. If $a \neq 1$ and $a \neq 0$, unique solution $x = \dfrac{a-1}{a}$, $y = 0$, $z = \dfrac{-1}{a}$. **10. (b)** 1 **(d)** 3 **(f)** 1 **11. (b)** 2 **(d)** 3

(f) 2 if $a = 0$ or $a = 2$; 3, otherwise. **12. (b)** False. $A = \begin{bmatrix} 1 & 0 & | & 1 \\ 0 & 1 & | & 1 \\ 0 & 0 & | & 0 \end{bmatrix}$ **(d)** False. $A = \begin{bmatrix} 1 & 0 & | & 1 \\ 0 & 1 & | & 0 \\ 0 & 0 & | & 0 \end{bmatrix}$

(f) False. $\begin{matrix} 2x - y = 0 \\ -4x + 2y = 0 \end{matrix}$ is consistent but $\begin{matrix} 2x - y = 1 \\ -4x + 2y = 1 \end{matrix}$ is not. **(h)** True, A has 3 rows, so there are at most 3 leading 1's. **16. (b)** $x^2 + y^2 - 2x + 6y - 6 = 0$ **18.** $\frac{5}{20}$ in A, $\frac{7}{20}$ in B, $\frac{8}{20}$ in C

Exercises 1.3 Homogeneous Equations (Page 19)

1. (b) False. $A = \begin{bmatrix} 1 & 0 & 1 & | & 0 \\ 0 & 1 & 1 & | & 0 \end{bmatrix}$ **(d)** False. $A = \begin{bmatrix} 1 & 0 & 1 & | & 1 \\ 0 & 1 & 1 & | & 0 \end{bmatrix}$ **(f)** False. $A = \begin{bmatrix} 1 & 0 & | & 0 \\ 0 & 1 & | & 0 \end{bmatrix}$ **(h)** False. $A = \begin{bmatrix} 1 & 0 & | & 0 \\ 0 & 1 & | & 0 \\ 0 & 0 & | & 0 \end{bmatrix}$

2. (b) $a = -3$, $x = 9t$, $y = -5t$, $z = t$ **(d)** $a = 1$, $x = -t$, $y = t$, $z = 0$; or $a = -1$, $x = t$, $y = 0$, $z = t$
4. (b) By Theorem 2 §1.2, there are $n - r = 6 - 1 = 5$ parameters and thus infinitely many solutions.
(d) If R is the row-echelon form of A, then R has a row of zeros and 4 rows in all. Hence R has $r = \text{rank } A = 1, 2,$ or 3. Thus there are $n - r = 6 - r = 5, 4,$ or 3 parameters and thus infinitely many solutions.

Exercises 1.4 An Application to Network Flow (Page 21)

1. (b)
$$f_1 = 85 - f_4 - f_7$$
$$f_2 = 60 - f_4 - f_7$$
$$f_3 = -75 + f_4 + f_6$$
$$f_5 = 40 - f_6 + f_7 \quad f_4, f_6, f_7 \text{ parameters}$$

2. (b) $f_5 = 15$
$25 \leq f_4 \leq 30$

3. (b) CD

Exercises 1.5 An Application to Electrical Networks (Page 23)

2. $I_1 = -\frac{1}{5}$, $I_2 = \frac{3}{5}$, $I_3 = \frac{4}{5}$ **4.** $I_1 = 2$, $I_2 = 1$, $I_3 = \frac{1}{2}$, $I_4 = \frac{3}{2}$, $I_5 = \frac{3}{2}$, $I_6 = \frac{1}{2}$

Exercises 1.6 An Application to Chemical Reactions (Page 25)

2. $2NH_3 + 3CuO \rightarrow N_2 + 3Cu + 3H_2O$ **4.** $15Pb(N_3)_2 + 44Cr(MnO_4)_2 \rightarrow 22Cr_2O_3 + 88MnO_2 + 5Pb_3O_4$
$+ 90NO$

Supplementary Exercises for Chapter 1 (Page 25)

1. (b) No. If the corresponding planes are parallel and distinct, there is no solution. Otherwise they either coincide or have a whole common line of solutions, that is, at least one parameter.
2. (b) $x_1 = \frac{1}{10}(-6s - 6t + 16)$, $x_2 = \frac{1}{10}(4s - t + 1)$, $x_3 = s$, $x_4 = t$
3. (b) If $a = 1$, no solution. If $a = 2$, $x = 2 - 2t$, $y = -t$, $z = t$. If $a \neq 1$ and $a \neq 2$, the unique solution is
$$x = \frac{8 - 5a}{3(a-1)}, \quad y = \frac{-2 - a}{3(a-1)}, \quad z = \frac{a+2}{3} \quad \textbf{4.} \quad \begin{bmatrix} R_1 \\ R_2 \end{bmatrix} \rightarrow \begin{bmatrix} R_1 + R_2 \\ R_2 \end{bmatrix} \rightarrow \begin{bmatrix} R_1 + R_2 \\ -R_1 \end{bmatrix} \rightarrow \begin{bmatrix} R_2 \\ -R_1 \end{bmatrix} \rightarrow \begin{bmatrix} R_2 \\ R_1 \end{bmatrix}$$
6. $a = 1$, $b = 2$, $c = -1$ **9. (b)** 5 of brand 1, 0 of brand 2, 3 of brand 3

Exercises 2.1 Matrix Addition, Scalar Multiplication, and Transposition (Page 34)

1. (b) $(a, b, c, d) = (-2, -4, -6, 0) + t(1, 1, 1, 1)$, t arbitrary **(d)** $a = b = c = d = t$, t arbitrary

2. (b) $\begin{bmatrix} -14 \\ -20 \end{bmatrix}$ **(d)** $(-12, 4, -12)$ **(f)** $\begin{bmatrix} 0 & 1 & -2 \\ -1 & 0 & 4 \\ 2 & -4 & 0 \end{bmatrix}$ **(h)** $\begin{bmatrix} 4 & -1 \\ -1 & -6 \end{bmatrix}$ **3. (b)** $\begin{bmatrix} 15 & -5 \\ 10 & 0 \end{bmatrix}$ **(d)** Impossible

(f) $\begin{bmatrix} 5 & 2 \\ 0 & -1 \end{bmatrix}$ **(h)** Impossible   **4. (b)** $\begin{bmatrix} 4 \\ \frac{1}{2} \end{bmatrix}$ **5. (b)** $A = -\frac{11}{3}B$ **6. (b)** $X = 4A - 3B,\ Y = 4B - 5A$

7. (b) $Y = (s,\ t),\ X = \frac{1}{2}(1+5s,\ 2+5t)$; s and t arbitrary   **8. (b)** $20A - 7B + 2C$

9. (b) If $A = \begin{bmatrix} a & b \\ c & d \end{bmatrix}$, then $(p,\ q,\ r,\ s) = \frac{1}{2}(2d,\ a+b-c-d,\ a-b+c-d,\ -a+b+c+d)$.

11. (b) If $A + A' = 0$, then $-A = -A + 0 = -A + (A + A') = (-A + A) + A' = 0 + A' = A'$.

14. (b) $s = 1$ or $t = 0$ **(d)** $s = 0$, and $t = 3$ **15. (b)** $\begin{bmatrix} 2 & 0 \\ 1 & -1 \end{bmatrix}$ **(d)** $\begin{bmatrix} 2 & 7 \\ -\frac{9}{2} & -5 \end{bmatrix}$

16. (b) $A = A^T$, so using Theorem 2 §2.1, $(kA)^T = kA^T = kA$.   **19. (b)** False. Take $B = -A$ for any $A \neq 0$.
   (d) True. Transposing fixes the main diagonal.
   (f) True. $(kA + mB)^T = (kA)^T + (mB)^T = kA^T + mB^T = kA + mB$
20. (c) Suppose $A = S + W$, where $S = S^T$ and $W = -W^T$. Then $A^T = S^T + W^T = S - W$, so $A + A^T = 2S$ and $A - A^T = 2W$. Hence $S = \frac{1}{2}(A + A^T)$ and $W = \frac{1}{2}(A - A^T)$ are uniquely determined by A.

Exercises 2.2 Matrix Multiplication (Page 47)

1. (b) $\begin{bmatrix} -1 & -6 & -2 \\ 0 & 6 & 10 \end{bmatrix}$ **(d)** $[-3\ \ -15]$ **(f)** $[-23]$ **(h)** $\begin{bmatrix} 1 & 0 \\ 0 & 1 \end{bmatrix}$ **(j)** $\begin{bmatrix} aa' & 0 & 0 \\ 0 & bb' & 0 \\ 0 & 0 & cc' \end{bmatrix}$

2. (b) $BA = \begin{bmatrix} -1 & 4 & -10 \\ 1 & 2 & 4 \end{bmatrix}$ $B^2 = \begin{bmatrix} 7 & -6 \\ -1 & 6 \end{bmatrix}$ $CB = \begin{bmatrix} -2 & 12 \\ 2 & -6 \\ 1 & 6 \end{bmatrix}$ $AC = \begin{bmatrix} 4 & 10 \\ -2 & -1 \end{bmatrix}$ $CA = \begin{bmatrix} 2 & 4 & 8 \\ -1 & -1 & -5 \\ 1 & 4 & 2 \end{bmatrix}$

3. (b) $(a,\ b,\ a_1,\ b_1) = (3,\ 0,\ 1,\ 2)$ **4. (b)** $A^2 - A - 6I = \begin{bmatrix} 8 & 2 \\ 2 & 5 \end{bmatrix} - \begin{bmatrix} 2 & 2 \\ 2 & -1 \end{bmatrix} - \begin{bmatrix} 6 & 0 \\ 0 & 6 \end{bmatrix} = \begin{bmatrix} 0 & 0 \\ 0 & 0 \end{bmatrix}$

5. (b) $A(BC) = \begin{bmatrix} 1 & -1 \\ 0 & 1 \end{bmatrix}\begin{bmatrix} -9 & -16 \\ 5 & 1 \end{bmatrix} = \begin{bmatrix} -14 & -17 \\ 5 & 1 \end{bmatrix} = \begin{bmatrix} -2 & -1 & -2 \\ 3 & 1 & 0 \end{bmatrix}\begin{bmatrix} 1 & 0 \\ 2 & 1 \\ 5 & 8 \end{bmatrix} = (AB)C$

7. (b) $AX = B$, where $A = \begin{bmatrix} -1 & 2 & -1 & 1 \\ 2 & 1 & -1 & 2 \\ 3 & -2 & 0 & 1 \end{bmatrix}$, $X = \begin{bmatrix} x_1 \\ x_2 \\ x_3 \\ x_4 \end{bmatrix}$, and $B = \begin{bmatrix} 6 \\ 1 \\ 0 \end{bmatrix}$ **8. (b)** $\begin{bmatrix} -2 \\ 2 \\ 0 \end{bmatrix} + t\begin{bmatrix} 1 \\ -3 \\ 1 \end{bmatrix}$ **(d)** $\begin{bmatrix} 3 \\ -9 \\ -2 \\ 0 \end{bmatrix} + t\begin{bmatrix} -1 \\ 4 \\ 1 \\ 1 \end{bmatrix}$

10. (b) $s\begin{bmatrix} -7 \\ 15 \\ 7 \\ 2 \\ 0 \end{bmatrix}$ **12. (b)** $m \times n$ and $n \times m$ for some m and n

13. (b) (i) $\begin{bmatrix} 1 & 0 \\ 0 & 1 \end{bmatrix}$, $\begin{bmatrix} 1 & 0 \\ 0 & -1 \end{bmatrix}$, $\begin{bmatrix} 1 & 1 \\ 0 & -1 \end{bmatrix}$ **(ii)** $\begin{bmatrix} 1 & 0 \\ 0 & 0 \end{bmatrix}$, $\begin{bmatrix} 1 & 0 \\ 0 & 1 \end{bmatrix}$, $\begin{bmatrix} 1 & 1 \\ 0 & 0 \end{bmatrix}$

16. (b) $AB = \begin{bmatrix} 2 & -1 & 3 & | & 1 \\ 1 & 0 & 1 & | & 2 \\ 0 & 0 & 1 & | & 0 \\ 0 & 0 & 0 & | & 1 \end{bmatrix}\begin{bmatrix} 1 & 2 & | & 0 \\ -1 & 0 & | & 0 \\ 0 & 5 & | & 1 \\ 1 & -1 & | & 0 \end{bmatrix} = \begin{bmatrix} \begin{bmatrix} 3 & 19 \\ 1 & 7 \\ 0 & 5 \end{bmatrix} + \begin{bmatrix} 1 & -1 \\ 2 & -2 \\ 0 & 0 \end{bmatrix} & | & \begin{bmatrix} 3 \\ 1 \\ 1 \end{bmatrix} + \begin{bmatrix} 0 \\ 0 \\ 0 \end{bmatrix} \\ (1-1) & | & 0 \end{bmatrix} = \begin{bmatrix} 4 & 18 & | & 3 \\ 3 & 5 & | & 1 \\ 0 & 5 & | & 1 \\ 1 & -1 & | & 0 \end{bmatrix}$

17. (b) $A^{2k} = \begin{bmatrix} 1 & -2k & | & 0 & 0 \\ 0 & 1 & | & 0 & 0 \\ \hline 0 & 0 & | & 1 & 0 \\ 0 & 0 & | & 0 & 1 \end{bmatrix}$ for $k = 0, 1, 2, \ldots$, $A^{2k+1} = A^{2k}A = \begin{bmatrix} 1 & -(2k+1) & | & 2 & -1 \\ 0 & 1 & | & 0 & 0 \\ \hline 0 & 0 & | & -1 & 1 \\ 0 & 0 & | & 0 & 1 \end{bmatrix}$ for $k = 0, 1, 2, \ldots$

18. (b) $\begin{bmatrix} I & 0 \\ 0 & I \end{bmatrix} = I_{2k}$ **(d)** 0_k **(f)** $\begin{bmatrix} X^m & 0 \\ 0 & X^m \end{bmatrix}$ if $n = 2m$; $\begin{bmatrix} 0 & X^{m+1} \\ X^m & 0 \end{bmatrix}$ if $n = 2m+1$.

20. (b) If Y is row i of the identity matrix I, then YA is row i of $IA = A$.   **22. (b)** $AB - BA$ **(d)** 0
24. (b) $(kA)C = k(AC) = k(CA) = C(kA)$
26. We have $A^T = A$ and $B^T = B$, so $(AB)^T = B^TA^T = BA$. Hence AB is symmetric if and only if $AB = BA$.

30. If $BC = I$, then $AB = 0$ gives $0 = 0C = (AB)C = A(BC) = AI = A$, contrary to the assumption that $A \neq 0$.

32. $\begin{matrix} 3 & v_1 \to v_4 \\ 0 & v_2 \to v_3 \end{matrix}$ 33. (b) False. $A = \begin{bmatrix} 1 & -1 & 1 \\ -1 & 1 & -1 \end{bmatrix}$ has a solution for $B = \begin{bmatrix} 0 \\ 0 \end{bmatrix}$ but not for $B = \begin{bmatrix} 1 \\ 0 \end{bmatrix}$

   (d) True. $A(X_1 - X_2) = AX_1 - AX_2 = B - B = 0$ (f) False. $A = \begin{bmatrix} 0 & 1 \\ 0 & 0 \end{bmatrix}$

   (h) True. $A(A + B) = (A + B)A$ gives $A^2 + AB = A^2 + BA$.

36. (b) If $A = [a_{ij}]$, then $\text{tr}(kA) = \text{tr}[ka_{ij}] = \sum_{i=1}^{n} ka_{ii} = k \sum_{i=1}^{n} a_{ii} = k \, \text{tr}(A)$.

   (e) Write $A^T = [a'_{ij}]$, where $a'_{ij} = a_{ji}$. Then $AA^T = \left(\sum_{k=1}^{n} a_{ik}a'_{kj} \right)$, so $\text{tr}(AA^T) = \sum_{i=1}^{n}\left[\sum_{k=1}^{n} a_{ik}a'_{ki} \right] = \sum_{i=1}^{n}\sum_{k=1}^{n} a_{ik}^2$.

38. (e) Observe that $PQ = P^2 + PAP - P^2AP = P$, so $Q^2 = PQ + APQ - PAPQ = P + AP - PAP = Q$.

40. (b) $(A+B)(A-B) = A^2 - AB + BA - B^2$, and $(A-B)(A+B) = A^2 + AB - BA - B^2$. These are equal if and only if $-AB + BA = AB - BA$; that is, $2BA = 2AB$; that is, $BA = AB$.

41. See V. Camillo, Communications in Algebra 25(6), (1997), 1767–1782; Theorem 2.

Exercises 2.3 Matrix Inverses (Page 59)

2. (b) $\frac{1}{5}\begin{bmatrix} 2 & -1 \\ -3 & 4 \end{bmatrix}$ (d) $\begin{bmatrix} 2 & -1 & 3 \\ 3 & 1 & -1 \\ 1 & 1 & -2 \end{bmatrix}$ (f) $\frac{1}{10}\begin{bmatrix} 1 & 4 & -1 \\ -2 & 2 & 2 \\ -9 & 14 & -1 \end{bmatrix}$ (h) $\frac{1}{4}\begin{bmatrix} 2 & 0 & -2 \\ -5 & 2 & 5 \\ -3 & 2 & -1 \end{bmatrix}$ (j) $\begin{bmatrix} 0 & 0 & 1 & -2 \\ -1 & -2 & -1 & -3 \\ 1 & 2 & 1 & 2 \\ 0 & -1 & 0 & 0 \end{bmatrix}$

(l) $\begin{bmatrix} 1 & -2 & 6 & -30 & 210 \\ 0 & 1 & -3 & 15 & -105 \\ 0 & 0 & 1 & -5 & 35 \\ 0 & 0 & 0 & 1 & -7 \\ 0 & 0 & 0 & 0 & 1 \end{bmatrix}$ 3. (b) $\begin{bmatrix} x \\ y \end{bmatrix} = \frac{1}{5}\begin{bmatrix} 4 & -3 \\ 1 & -2 \end{bmatrix}\begin{bmatrix} 0 \\ 1 \end{bmatrix} = \frac{1}{5}\begin{bmatrix} -3 \\ -2 \end{bmatrix}$

(d) $\begin{bmatrix} x \\ y \\ z \end{bmatrix} = \frac{1}{5}\begin{bmatrix} 9 & -14 & 6 \\ 4 & -4 & 1 \\ -10 & 15 & -5 \end{bmatrix}\begin{bmatrix} 1 \\ -1 \\ 0 \end{bmatrix} = \frac{1}{5}\begin{bmatrix} 23 \\ 8 \\ -25 \end{bmatrix}$ (f) $\begin{bmatrix} x \\ y \\ z \\ w \end{bmatrix} = \frac{1}{2}\begin{bmatrix} 0 & 1 & -1 & 1 \\ 0 & 1 & 1 & -1 \\ 2 & -1 & -1 & -1 \\ 0 & -1 & 1 & 1 \end{bmatrix}\begin{bmatrix} 1 \\ 0 \\ -1 \\ 2 \end{bmatrix} = \frac{1}{2}\begin{bmatrix} 3 \\ -3 \\ 1 \\ 1 \end{bmatrix}$

4. (b) $B = A^{-1}(AB) = \begin{bmatrix} 4 & -2 & 1 \\ 7 & -2 & 4 \\ -1 & 2 & -1 \end{bmatrix}$ 5. (b) $\frac{1}{10}\begin{bmatrix} 3 & -2 \\ 1 & 1 \end{bmatrix}$ (d) $\frac{1}{2}\begin{bmatrix} 0 & 1 \\ 1 & -1 \end{bmatrix}$ (f) $\frac{1}{2}\begin{bmatrix} 2 & 0 \\ -6 & 1 \end{bmatrix}$ (h) $-\frac{1}{2}\begin{bmatrix} 1 & 1 \\ 1 & 0 \end{bmatrix}$

6. (b) $A = \frac{1}{2}\begin{bmatrix} 2 & -1 & 3 \\ 0 & 1 & -1 \\ -2 & 1 & -1 \end{bmatrix}$ 8. (b) $AB = I$ 9. (b) False. $\begin{bmatrix} 1 & 0 \\ 0 & 1 \end{bmatrix} + \begin{bmatrix} 1 & 0 \\ 0 & -1 \end{bmatrix}$ (d) True. $A^{-1} = \frac{1}{3}A^3$

(f) False. $A = B = \begin{bmatrix} 1 & 0 \\ 0 & 0 \end{bmatrix}$ (h) True. If $(A^2)B = I$, then $A(AB) = I$; use Theorem 5.

10. (b) $(C^T)^{-1} = (C^{-1})^T = A^T$ because $C^{-1} = (A^{-1})^{-1} = A$.   11. (b) (i) Inconsistent. (ii) $\begin{bmatrix} x_1 \\ x_2 \end{bmatrix} = \begin{bmatrix} 2 \\ -1 \end{bmatrix}$

15. (b) $B^4 = I$, so $B^{-1} = B^3 = \begin{bmatrix} 0 & 1 \\ -1 & 0 \end{bmatrix}$ 16. $\begin{bmatrix} c^2 - 2 & -c & 1 \\ -c & 1 & 0 \\ 3 - c^2 & c & -1 \end{bmatrix}$ 18. $\begin{bmatrix} \sin\theta & -\cos\theta \\ \cos\theta & \sin\theta \end{bmatrix}$

19. (b) If column j of A is zero, $AY = 0$ where Y is column j of the identity matrix. Use Theorem 5.
    (d) If each column of A sums to 0, $XA = 0$ where X is the row of 1's. Hence $A^TX^T = 0$ so A has no inverse by Theorem 5 ($X^T \neq 0$).

20. (b) (ii) $(-1, 1, 1)A = 0$ 21. (d) If $PA = QA$, then right multiplication by A^{-1} gives $PAA^{-1} = QAA^{-1}$.

   Hence $PI = QI$, that is $P = Q$.   23. (d) (ii) $\begin{bmatrix} 2 & -1 & 0 \\ -5 & 3 & 0 \\ 0 & 0 & -1 \end{bmatrix}$ (iv) $\begin{bmatrix} 3 & -4 & 0 & 0 \\ -2 & 3 & 0 & 0 \\ 0 & 0 & 1 & 3 \\ 0 & 0 & 0 & -1 \end{bmatrix}$

24. (b) (ii) Verify that $=\begin{bmatrix} A & X \\ 0 & B \end{bmatrix}\begin{bmatrix} A^{-1} & -A^{-1}XB^{-1} \\ 0 & B^{-1} \end{bmatrix} = \begin{bmatrix} AA^{-1} & -XB^{-1}+XB^{-1} \\ 0 & BB^{-1} \end{bmatrix} = \begin{bmatrix} I & 0 \\ 0 & I \end{bmatrix}.$

Similarly: $\begin{bmatrix} A^{-1} & -A^{-1}XB^{-1} \\ 0 & B^{-1} \end{bmatrix}\begin{bmatrix} A & X \\ 0 & B \end{bmatrix} = \begin{bmatrix} I & 0 \\ 0 & I \end{bmatrix}.$

26. (b) $P = \begin{bmatrix} 1 & 0 \\ 0 & 0 \end{bmatrix}$, $Q = \begin{bmatrix} 1 & 0 \\ 0 & 0 \end{bmatrix}$, $R = \begin{bmatrix} 1 & 0 \\ 1 & 0 \end{bmatrix}$.   **27. (d)** If $A^n = 0$, $(I-A)^{-1} = I + A + \cdots + A^{n-1}$.

29. (b) $A[B(AB)^{-1}] = I = [(BA)^{-1}B]A$, so A is invertible by Exercise 10.

31. (a) Have $AC = CA$. Left-multiply by A^{-1} to get $C = A^{-1}CA$. Then right-multiply by A^{-1} to get $CA^{-1} = A^{-1}C$.

33. If $BX = 0$ where X is $n \times 1$, then $ABX = 0$ so $X = 0$ as AB is invertible. Hence B is invertible by

Theorem 5, so $A = (AB)B^{-1}$ is invertible.   **34. (b)** $B\begin{bmatrix} -1 \\ 3 \\ -1 \end{bmatrix} = 0$ so B is not invertible by Theorem 5.

38. (b) $(I - 2P)^2 = I - 4P + 4P^2$, and this equals I if and only if $P^2 = P$.   **40. (b)** $(A^{-1} + B^{-1})^{-1} = B(A+B)^{-1}A$

Exercises 2.4 Elementary Matrices (Page 68)

1. (b) Interchange rows 1 and 3 of I. $E^{-1} = E$.   **(d)** Add (-2) times row 1 of I to row 2. $E^{-1} = \begin{bmatrix} 1 & 0 & 0 \\ 2 & 1 & 0 \\ 0 & 0 & 1 \end{bmatrix}$

(f) Multiply row 3 of I by 5. $E^{-1} = \begin{bmatrix} 1 & 0 & 0 \\ 0 & 1 & 0 \\ 0 & 0 & \frac{1}{5} \end{bmatrix}$ **2. (b)** $\begin{bmatrix} -1 & 0 \\ 0 & 1 \end{bmatrix}$ **(d)** $\begin{bmatrix} 1 & -1 \\ 0 & 1 \end{bmatrix}$ **(f)** $\begin{bmatrix} 0 & 1 \\ 1 & 0 \end{bmatrix}$

3. (b) The only possibilities for E are $\begin{bmatrix} 0 & 1 \\ 1 & 0 \end{bmatrix}$, $\begin{bmatrix} k & 0 \\ 0 & 1 \end{bmatrix}$, $\begin{bmatrix} 1 & 0 \\ 0 & k \end{bmatrix}$, $\begin{bmatrix} 1 & k \\ 0 & 1 \end{bmatrix}$, and $\begin{bmatrix} 1 & 0 \\ k & 1 \end{bmatrix}$. In each case, EA has a

row different from C.   **5. (b)** No, 0 is not invertible.   **6. (b)** $\begin{bmatrix} 1 & -2 \\ 0 & 1 \end{bmatrix}\begin{bmatrix} 1 & 0 \\ 0 & \frac{1}{2} \end{bmatrix}\begin{bmatrix} 1 & 0 \\ -5 & 1 \end{bmatrix}A = \begin{bmatrix} 1 & 0 & 7 \\ 0 & 1 & -3 \end{bmatrix}.$

Alternatively, $\begin{bmatrix} 1 & 0 \\ 0 & \frac{1}{2} \end{bmatrix}\begin{bmatrix} 1 & -1 \\ 0 & 1 \end{bmatrix}\begin{bmatrix} 1 & 0 \\ -5 & 1 \end{bmatrix}A = \begin{bmatrix} 1 & 0 & 7 \\ 0 & 1 & -3 \end{bmatrix}.$

(d) $\begin{bmatrix} 1 & 2 & 0 \\ 0 & 1 & 0 \\ 0 & 0 & 1 \end{bmatrix}\begin{bmatrix} 1 & 0 & 0 \\ 0 & \frac{1}{5} & 0 \\ 0 & 0 & 1 \end{bmatrix}\begin{bmatrix} 1 & 0 & 0 \\ 0 & 1 & 0 \\ 0 & -1 & 1 \end{bmatrix}\begin{bmatrix} 1 & 0 & 0 \\ 0 & 1 & 0 \\ -2 & 0 & 1 \end{bmatrix}\begin{bmatrix} 1 & 0 & 0 \\ -3 & 1 & 0 \\ 0 & 0 & 1 \end{bmatrix}\begin{bmatrix} 0 & 0 & 1 \\ 0 & 1 & 0 \\ 1 & 0 & 0 \end{bmatrix}A = \begin{bmatrix} 1 & 0 & \frac{1}{5} & \frac{1}{5} \\ 0 & 1 & -\frac{7}{5} & -\frac{2}{5} \\ 0 & 0 & 0 & 0 \end{bmatrix}$

7. (b) $U = \begin{bmatrix} 1 & 1 \\ 1 & 0 \end{bmatrix} = \begin{bmatrix} 1 & 1 \\ 0 & 1 \end{bmatrix}\begin{bmatrix} 0 & 1 \\ 1 & 0 \end{bmatrix}$ **8. (b)** $A = \begin{bmatrix} 0 & 1 \\ 1 & 0 \end{bmatrix}\begin{bmatrix} 1 & 0 \\ 2 & 1 \end{bmatrix}\begin{bmatrix} 1 & 0 \\ 0 & -1 \end{bmatrix}\begin{bmatrix} 1 & 2 \\ 0 & 1 \end{bmatrix}$

(d) $A = \begin{bmatrix} 1 & 0 & 0 \\ 0 & 1 & 0 \\ -2 & 0 & 1 \end{bmatrix}\begin{bmatrix} 1 & 0 & 0 \\ 0 & 1 & 0 \\ 0 & 2 & 1 \end{bmatrix}\begin{bmatrix} 1 & 0 & -3 \\ 0 & 1 & 0 \\ 0 & 0 & 1 \end{bmatrix}\begin{bmatrix} 1 & 0 & 0 \\ 0 & 1 & 4 \\ 0 & 0 & 1 \end{bmatrix}$ **10.** $UA = R$ by Theorem 1, so $A = U^{-1}R$.

12. (b) $U = A^{-1}$, $V = I_2$; rank $A = 2$ **(d)** $U = \begin{bmatrix} -2 & 1 & 0 \\ 3 & -1 & 0 \\ 2 & -1 & 1 \end{bmatrix}$, $V = \begin{bmatrix} 1 & 0 & -1 & -3 \\ 0 & 1 & 1 & 4 \\ 0 & 0 & 1 & 0 \\ 0 & 0 & 0 & 1 \end{bmatrix}$; rank $A = 2$

17. (b) (i) $A \doteq A$ because $A = IA$ **(ii)** If $A \doteq B$, then $A = UB$, U invertible, so $B = U^{-1}A$. Thus $B \doteq A$.
(iii) If $A \doteq B$ and $B \doteq C$, then $A = UB$ and $B = VC$, U and V invertible. Hence $A = U(VC) = (UV)C$,
so $A \doteq C$.

19. (b) If $B \doteq A$, let $B = UA$, U invertible. If $U = \begin{bmatrix} d & b \\ -b & d \end{bmatrix}$, $B = UA = \begin{bmatrix} 0 & 0 & b \\ 0 & 0 & d \end{bmatrix}$ where b and d are not both zero (as U is invertible). Every such matrix B arises in this way: Use $U = \begin{bmatrix} a & b \\ -b & a \end{bmatrix}$—it is invertible by Example 4 §2.3.

22. (b) Multiply column i by $1/k$.

Exercises 2.5 Matrix Transformations (Page 80)

1. (b) $\begin{bmatrix} 0 & -1 \\ -1 & 0 \end{bmatrix}$ **(d)** $\begin{bmatrix} 0 & 1 \\ -1 & 0 \end{bmatrix}$ **(f)** $\begin{bmatrix} 1 & 0 & 0 \\ 0 & -1 & 0 \\ 0 & 0 & 1 \end{bmatrix}$ **(h)** $\begin{bmatrix} \cos\theta & 0 & -\sin\theta \\ 0 & 1 & 0 \\ \sin\theta & 0 & \cos\theta \end{bmatrix}$

2. (b) $T\left(2\begin{bmatrix} 0 \\ 1 \end{bmatrix}\right) \neq 2\begin{bmatrix} 0 \\ 0 \end{bmatrix}$ **3. (b)** $\begin{bmatrix} -1 \\ -1 \end{bmatrix}, \begin{bmatrix} 1 \\ -2 \end{bmatrix}$ **(d)** $\begin{bmatrix} 1 \\ -1 \end{bmatrix}, \begin{bmatrix} -1 \\ -2 \end{bmatrix}$

6. $T\left([x_1 \cdots x_n]^T + [y_1 \cdots y_n]^T\right) = T\left([x_1 + y_1 \cdots x_n + y_n]^T\right)$
$$= (x_1 + y_1)C_1 + \cdots + (x_n + y_n)C_n$$
$$= (x_1C_1 + \cdots + x_nC_n) + (y_1C_1 + \cdots + y_nC_n)$$
$$= T([x_1 \cdots x_n]^T) + T([y_1 \cdots y_n]^T)$$

Similarly $T\left(a[x_1 \cdots x_n]^T\right) = aT\left([x_1 \cdots x_n]^T\right)$ for all a in $\mathbb{R}$.

7. (b) aA **8. (b)** $T(-X) = T[(-1)X] = (-1)T(X) = -T(X)$.

9. (b) $[5 \ -1 \ 2 \ -4]^T = 2[1 \ 1 \ 1 \ 1]^T - 3[-1 \ 1 \ 0 \ 2]^T$, so
$$T\left([5 \ -1 \ 2 \ -4]^T\right) = 2[5 \ 1 \ -3]^T - 3[2 \ 0 \ 1]^T = [4 \ 2 \ -9]^T$$

12. (b) Reflection in the Y axis **(d)** Reflection in $y = -x$ **(f)** Rotation through $\frac{\pi}{2}$.

17. (b) If $B^2 = I$, then $T^2(X) = T[T(X)] = B[BX] = B^2X = IX = X = 1_{\mathbb{R}^n}(X)$ for all X in $\mathbb{R}^n$. Hence $T^2 = 1_{\mathbb{R}^n}$.
Conversely, if $T^2 = 1_{\mathbb{R}^n}$, then $B^2X = T^2(X) = 1_{\mathbb{R}^n}(X) = X = IX$ for all X in $\mathbb{R}^n$, so $B^2 = I$ by the uniqueness in Theorem 2.

Exercises 2.6 LU-Factorization (Page 89)

1. (b) $\begin{bmatrix} 2 & 0 & 0 \\ 1 & -3 & 0 \\ -1 & 9 & 1 \end{bmatrix}\begin{bmatrix} 1 & 2 & 1 \\ 0 & 1 & -\frac{2}{3} \\ 0 & 0 & 0 \end{bmatrix}$ **(d)** $\begin{bmatrix} -1 & 0 & 0 & 0 \\ 1 & 1 & 0 & 0 \\ 1 & -1 & 1 & 0 \\ 0 & -2 & 0 & 1 \end{bmatrix}\begin{bmatrix} 1 & 3 & -1 & 0 & 1 \\ 0 & 1 & 2 & 1 & 0 \\ 0 & 0 & 0 & 0 & 0 \\ 0 & 0 & 0 & 0 & 0 \end{bmatrix}$ **(f)** $\begin{bmatrix} 2 & 0 & 0 & 0 \\ 1 & -2 & 0 & 0 \\ 3 & -2 & 1 & 0 \\ 1 & 2 & 0 & 1 \end{bmatrix}\begin{bmatrix} 1 & 1 & -1 & 2 & 1 \\ 0 & 1 & -\frac{1}{2} & 0 & 0 \\ 0 & 0 & 0 & 0 & 0 \\ 0 & 0 & 0 & 0 & 0 \end{bmatrix}$

2. (b) $P = \begin{bmatrix} 0 & 0 & 1 \\ 1 & 0 & 0 \\ 0 & 1 & 0 \end{bmatrix}$ $PA = \begin{bmatrix} -1 & 2 & 1 \\ 0 & -1 & 2 \\ 0 & 0 & 4 \end{bmatrix} = \begin{bmatrix} -1 & 0 & 0 \\ 0 & -1 & 0 \\ 0 & 0 & 4 \end{bmatrix}\begin{bmatrix} 1 & -2 & -1 \\ 0 & 1 & -2 \\ 0 & 0 & 1 \end{bmatrix}$

(d) $P = \begin{bmatrix} 1 & 0 & 0 & 0 \\ 0 & 0 & 1 & 0 \\ 0 & 0 & 0 & 1 \\ 0 & 1 & 0 & 0 \end{bmatrix}$ $PA = \begin{bmatrix} -1 & -2 & 3 & 0 \\ 1 & 1 & -1 & 3 \\ 2 & 5 & -10 & 1 \\ 2 & 4 & -6 & 5 \end{bmatrix} = \begin{bmatrix} -1 & 0 & 0 & 0 \\ 1 & -1 & 0 & 0 \\ 2 & 1 & -2 & 0 \\ 2 & 0 & 0 & 5 \end{bmatrix}\begin{bmatrix} 1 & 2 & -3 & 0 \\ 0 & 1 & -2 & -3 \\ 0 & 0 & 1 & -2 \\ 0 & 0 & 0 & 1 \end{bmatrix}$

3. (b) $Y = \begin{bmatrix} -1 \\ 0 \\ 0 \end{bmatrix}$ $X = \begin{bmatrix} -1+2t \\ -t \\ s \\ t \end{bmatrix}$ s and t arbitrary **(d)** $Y = \begin{bmatrix} 2 \\ 8 \\ -1 \\ 0 \end{bmatrix}$ $X = \begin{bmatrix} 8-2t \\ 6-t \\ -1-t \\ t \end{bmatrix}$ t arbitrary

5. $\begin{bmatrix} R_1 \\ R_2 \end{bmatrix} \to \begin{bmatrix} R_1 + R_2 \\ R_2 \end{bmatrix} \to \begin{bmatrix} R_1 + R_2 \\ -R_1 \end{bmatrix} \to \begin{bmatrix} R_2 \\ -R_1 \end{bmatrix} \to \begin{bmatrix} R_2 \\ R_1 \end{bmatrix}$ **6. (b)** Let $A = LU = L_1U_1$ be LU-factorizations of the invertible matrix A. Then U and U_1 have no row of zeros and so (being row-echelon) are upper triangular with 1's on the main diagonal. Thus, using **(a)**, the diagonal matrix $D = UU_1^{-1}$ has 1's on the main diagonal. Thus $D = I$, $U = U_1$, and $L = L_1$.

7. If $A = \begin{bmatrix} a & 0 \\ X & A_1 \end{bmatrix}$ and $B = \begin{bmatrix} b & 0 \\ Y & B_1 \end{bmatrix}$ in block form, then $AB = \begin{bmatrix} ab & 0 \\ Xb + A_1Y & A_1B_1 \end{bmatrix}$, and A_1B_1 is lower triangular by induction.

Exercises 2.7 An Application to Input-Output Economic Models (Page 94)

1. (b) $\begin{bmatrix} t \\ 3t \\ t \end{bmatrix}$ **(d)** $\begin{bmatrix} 14t \\ 17t \\ 47t \\ 23t \end{bmatrix}$ **2.** $\begin{bmatrix} t \\ t \\ t \end{bmatrix}$ **4.** $P = \begin{bmatrix} bt \\ (1-a)t \end{bmatrix}$ is nonzero (for some t) unless $b = 0$ and $a = 1$.

In that case, $\begin{bmatrix} 1 \\ 1 \end{bmatrix}$ is a solution. If the entries of E are positive, then $P = \begin{bmatrix} b \\ 1-a \end{bmatrix}$ has positive entries.

7. (b) $\begin{bmatrix} 0.4 & 0.8 \\ 0.7 & 0.2 \end{bmatrix}$ **9. (b)** Use $P = \begin{bmatrix} 3 \\ 2 \\ 1 \end{bmatrix}$ in Theorem 2.

Exercises 2.8 An Application to Markov Chains (Page 101)

1. (b) Not regular   **2. (b)** $\frac{1}{3}\begin{bmatrix} 2 \\ 1 \end{bmatrix}$, $\frac{3}{8}$ **(d)** $\frac{1}{3}\begin{bmatrix} 1 \\ 1 \\ 1 \end{bmatrix}$, 0.312 **(f)** $\frac{1}{20}\begin{bmatrix} 5 \\ 7 \\ 8 \end{bmatrix}$, 0.306

4. (b) 50% middle, 25% upper, 25% lower   **6.** $\frac{7}{16}$, $\frac{9}{16}$ **8. (a)** $\frac{94}{450}$

(b) He spends equal time in compartments 3 and 4. The steady-state vector is $\frac{1}{18}\begin{bmatrix} 3 \\ 2 \\ 5 \\ 5 \\ 3 \end{bmatrix}$.

Supplementary Exercises for Chapter 2 (Page 103)

2. (b) $U^{-1} = \frac{1}{4}(U^2 - 5U + 11I)$.   **4. (b)** If $X_k = X_m$, then $Y + k(Y - Z) = Y + m(Y - Z)$. So $(k - m)(Y - Z) = 0$. But $Y - Z$ is not zero (because Y and Z are distinct), so $k - m = 0$ by Example 7 §2.1.

6. (d) Using parts **(c)** and **(b)** gives $I_{pq}AI_{rs} = \sum_{i=1}^{n}\sum_{j=1}^{n} a_{ij}I_{pq}I_{ij}I_{rs}$. The only nonzero term occurs when $i = q$ and $j = r$, so $I_{pq}AI_{rs} = a_{qr}I_{ps}$.

Exercises 3.1 The Cofactor Expansion (Page 114)

1. (b) 0   **(d)** -1 **(f)** -39 **(h)** 0   **(j)** $2abc$ **(l)** 0   **(n)** -56 **(p)** $abcd$ **5. (b)** -17 **(d)** 106

6. (b) 0   **7. (b)** 12   **9. (b)** F, $A = \begin{bmatrix} 1 & 1 \\ 2 & 2 \end{bmatrix}$ **(d)** F, $A = \begin{bmatrix} 2 & 0 \\ 0 & 1 \end{bmatrix} \rightarrow R = \begin{bmatrix} 1 & 0 \\ 0 & 1 \end{bmatrix}$

(f) F, $A = \begin{bmatrix} 1 & 1 \\ 0 & 1 \end{bmatrix}$ **(h)** F, $A = \begin{bmatrix} 1 & 1 \\ 0 & 1 \end{bmatrix}$ and $B = \begin{bmatrix} 1 & 0 \\ 1 & 1 \end{bmatrix}$ **10. (b)** 35   **11. (b)** -6 **(d)** -6

14. (b) $-(x - 2)(x^2 + 2x - 12)$ **15. (b)** -7 **16. (b)** $\pm\frac{\sqrt{6}}{2}$ **(d)** $x = \pm y$

21. (b) Let $X = [x_1 \ x_2 \cdots x_n]^T$, $Y = [y_1 \ y_2 \cdots y_n]^T$ and $A = [X + Y \ c_2 \cdots c_n]^T$. Expanding $\det A$ along column 1 we obtain:
$$
\begin{aligned}
T(X + Y) &= \det A \\
&= (x_1 + y_1)c_{11}(A) + (x_2 + y_2)c_{21}(A) + \cdots + (x_n + y_n)c_{n1}(A) \\
&= [x_1c_{11}(A) + x_2c_{21}(A) + \cdots + x_nc_{n1}(A)] + [y_1c_{11}(A) + y_2c_{21}(A) + \cdots + y_nc_{n1}(A)] \\
&= T(X) + T(Y).
\end{aligned}
$$

24. If A is $n \times n$, then $\det B = (-1)^{n(n-1)/2} \det A$.

Exercises 3.2 Determinants and Matrix Inverses (Page 126)

1. (b) $\begin{bmatrix} 1 & -1 & -2 \\ -3 & 1 & 6 \\ -3 & 1 & 4 \end{bmatrix}$ **(d)** $\frac{1}{3}\begin{bmatrix} -1 & 2 & 2 \\ 2 & -1 & 2 \\ 2 & 2 & -1 \end{bmatrix}$ **2. (b)** $c \neq 0$ **(d)** any c **(f)** $c \neq -1$ **3. (b)** -2

4. (b) 1 **6. (b)** $\frac{4}{9}$ **7. (b)** 16 **8. (b)** $\frac{1}{11}\begin{bmatrix} 5 \\ 21 \end{bmatrix}$ **(d)** $\frac{1}{79}\begin{bmatrix} 12 \\ -37 \\ -2 \end{bmatrix}$ **9. (b)** $\frac{4}{51}$

10. (b) $\det A = 1, -1$ **(d)** $\det A = 1$ **(f)** $\det A = 0$ if n is odd; nothing can be said if n is even

15. dA where $d = \det A$

19. (b) $\frac{1}{c^2}\begin{bmatrix} c & 0 & c \\ 0 & c^2 & c \\ -c & c^2 & c \end{bmatrix}, c \neq 0$ **(d)** $\frac{1}{2}\begin{bmatrix} 8-c^2 & -c & c^2-6 \\ c & 1 & -c \\ c^2-10 & c & 8-c^2 \end{bmatrix}$ **(f)** $\frac{1}{c^3+1}\begin{bmatrix} 1-c & c^2+1 & -c-1 \\ c^2 & -c & c+1 \\ -c & 1 & c^2-1 \end{bmatrix}, c \neq -1$

20. (b) T. $\det AB = \det A \det B = \det B \det A = \det BA$.

(d) T. $\det A \neq 0$ means A^{-1} exists, so $AB = AC$ implies that $B = C$.

(f) F. If $A = \begin{bmatrix} 1 & 1 & 1 \\ 1 & 1 & 1 \\ 1 & 1 & 1 \end{bmatrix}$, then adj $A = 0$. **(h)** F. If $A = \begin{bmatrix} 1 & 1 \\ 0 & 0 \end{bmatrix}$, then adj $A = \begin{bmatrix} 0 & -1 \\ 0 & 1 \end{bmatrix}$.

(j) F. If $A = \begin{bmatrix} -1 & 1 \\ 1 & -1 \end{bmatrix}$, then $\det(I + A) = -1$ but $1 + \det A = 1$.

22. (b) $5 - 4x + 2x^2$.

23. (b) $1 - \frac{5}{3}x + \frac{1}{2}x^2 + \frac{7}{6}x^3$.

24. (b) $1 - 0.51x + 2.1x^2 - 1.1x^3$; 1.25.

26. (b) Use induction on n where A is $n \times n$. It is clear if $n = 1$. If $n > 1$, write $A = \begin{bmatrix} a & X \\ 0 & B \end{bmatrix}$ in block form where

B is $(n-1) \times (n-1)$. Then $A^{-1} = \begin{bmatrix} a^{-1} & -a^{-1}XB^{-1} \\ 0 & B^{-1} \end{bmatrix}$, and this is upper triangular because B is upper

triangular by induction.

28. $-\frac{1}{21}\begin{bmatrix} 3 & 0 & 1 \\ 0 & 2 & 3 \\ 3 & 1 & -1 \end{bmatrix}$ **34. (b)** Have (adj $A)A = (\det A)I$; so taking inverses, $A^{-1} \cdot (\text{adj } A)^{-1} = \frac{1}{\det A}I$.

On the other hand, $A^{-1}\text{adj}(A^{-1}) = \det(A^{-1})I = \frac{1}{\det A}I$. Comparison yields $A^{-1}(\text{adj } A)^{-1} = A^{-1}\text{adj}(A^{-1})$,

and part **(b)** follows.

Exercises 3.3 Diagonalization and Eigenvalues (Page 141)

1. (b) $(x-3)(x+2)$; $3, -2$; $[4 \ -1]^T$, $[1 \ 1]^T$; $P = \begin{bmatrix} 4 & 1 \\ -1 & 1 \end{bmatrix}$; $P^{-1}AP = \begin{bmatrix} 3 & 0 \\ 0 & -2 \end{bmatrix}$.

(d) $(x-2)^3$; 2; $[1 \ 1 \ 0]^T$, $[-3 \ 0 \ 1]^T$; No such P; Not diagonalizable.

(f) $(x+1)^2(x-2)$; $-1, 2$; $[-1 \ 1 \ 0]^T$, $[-1 \ 0 \ 1]^T$, $[1 \ 1 \ 1]^T$; $P = \begin{bmatrix} -1 & -1 & 1 \\ 1 & 0 & 1 \\ 0 & 1 & 1 \end{bmatrix}$; $P^{-1}AP = \begin{bmatrix} -1 & 0 & 0 \\ 0 & -1 & 0 \\ 0 & 0 & 2 \end{bmatrix}$

(h) $(x-1)^2(x-3)$; $1, 3$; $[-1 \ 0 \ 1]^T$, $[1 \ 0 \ 1]^T$; No such P; Not diagonalizable. **2. (b)** $V_k = \frac{7}{3}2^k\begin{bmatrix} 2 \\ 1 \end{bmatrix}$.

(d) $V_k = \frac{3}{2}3^k\begin{bmatrix} 1 \\ 0 \\ 1 \end{bmatrix}$. **4.** $AX = \lambda X$ if and only if $(A - \alpha I)X = (\lambda - \alpha)X$. Same eigenvectors.

8. **(b)** $P^{-1}AP = \begin{bmatrix} 1 & 0 \\ 0 & 2 \end{bmatrix}$, so $A^n = P\begin{bmatrix} 1 & 0 \\ 0 & 2^n \end{bmatrix}P^{-1} = \begin{bmatrix} 9 - 8 \cdot 2^n & 12(1 - 2^n) \\ 6(2^n - 1) & 9 \cdot 2^n - 8 \end{bmatrix}$.

9. A has diagonalizing matrix $\begin{bmatrix} 1 & 3 \\ 1 & 0 \end{bmatrix}$ and B is already diagonal. However $AB = \begin{bmatrix} 2 & 3 \\ 0 & 2 \end{bmatrix}$ has only

one eigenvalue $\lambda = 2$ of multiplicity 2. But λ has only one basic eigenvector $\begin{bmatrix} 1 \\ 0 \end{bmatrix}$.

12. $\begin{bmatrix} 1 & 1 \\ 0 & 1 \end{bmatrix}$ is not diagonalizable by Example 8. But $\begin{bmatrix} 1 & 1 \\ 0 & 1 \end{bmatrix} = \begin{bmatrix} 2 & 1 \\ 0 & -1 \end{bmatrix} + \begin{bmatrix} -1 & 0 \\ 0 & 2 \end{bmatrix}$ and $\begin{bmatrix} 2 & 1 \\ 0 & -1 \end{bmatrix}$ has diagonalizing matrix

$P = \begin{bmatrix} 1 & -1 \\ 0 & 3 \end{bmatrix}$ and $\begin{bmatrix} -1 & 0 \\ 0 & 2 \end{bmatrix}$ is already diagonal. **14.** We have $\lambda^2 = \lambda$ for every eigenvalue λ (as $\lambda = 0$, 1) so

$D^2 = D$, and so $A^2 = A$ as in Example 9. **18. (b)** $c_{rA}(x) = \det[xI - rA] = r^n \det\left[\dfrac{x}{r}I - A\right] = r^n c_A\left(\dfrac{x}{r}\right)$

20. **(b)** If $\lambda \neq 0$, $AX = \lambda X$ if and only if $A^{-1}X = \frac{1}{\lambda}X$. The result follows.

21. **(b)** $(A^3 - 2A + 3I)X = A^3 X - 2AX + 3X = \lambda^3 X - 2\lambda X + 3X = (\lambda^3 - 2\lambda + 3)X$.

23. **(b)** If $A^m = 0$ and $AX = \lambda X$, $X \neq 0$, then $A^2 X = A(\lambda X) = \lambda AX = \lambda^2 X$. In general, $A^k X = \lambda^k X$ for all $k \geq 1$. Hence,
$\lambda^m X = A^m X = 0X = 0$, so $\lambda = 0$ (because $X \neq 0$).

25. **(a)** We have $P^{-1}AP = \lambda I$ by the diagonalization algorithm, so $A = P(\lambda I)P^{-1} = \lambda PP^{-1} = \lambda I$.
    (b) No. $\lambda = 1$ is the only eigenvalue.

Exercises 3.4 An Application to Linear Recurrences (Page 147)

1. **(b)** $x_k = \frac{1}{3}[4 - (-2)^k]$ **(d)** $x_k = \frac{1}{5}[2^{k+2} + (-3)^k]$. **2. (b)** $x_k = \frac{1}{2}[(-1)^k + 1]$

3. **(b)** $x_{k+4} = x_k + x_{k+2} + x_{k+3}$; $x_{10} = 169$

5. $\frac{1}{2\sqrt{5}}[(3 + \sqrt{5})\lambda_1^k + (-3 + \sqrt{5})\lambda_2^k]$ where $\lambda_1 = \frac{1}{2}(1 + \sqrt{5})$ and $\lambda_2 = \frac{1}{2}(1 - \sqrt{5})$.

7. $\frac{1}{2\sqrt{3}}[(2 + \sqrt{3})\lambda_1^k + (-2 + \sqrt{3})\lambda_2^k]$ where $\lambda_1 = 1 + \sqrt{3}$ and $\lambda_2 = 1 - \sqrt{3}$.

9. $\frac{34}{3} - \frac{4}{3}(-\frac{1}{2})^k$. Long term $11\frac{1}{3}$ million tons.

11. **(b)** $A\begin{bmatrix} 1 & \lambda & \lambda^2 \end{bmatrix}^T = \begin{bmatrix} \lambda & \lambda^2 & a + b\lambda + c\lambda^2 \end{bmatrix}^T = \begin{bmatrix} \lambda & \lambda^2 & \lambda^3 \end{bmatrix}^T = \lambda\begin{bmatrix} 1 & \lambda & \lambda^2 \end{bmatrix}^T$.

12. **(b)** $x_k = \frac{11}{30}3^k - \frac{11}{15}(-2)^k - \frac{5}{6}$

Exercises 3.5 An Application to Population Growth (Page 150)

1. **(b)** Dominant eigenvalue is 1. **(d)** The dominant eigenvalue is $\frac{1}{10}(3 + \sqrt{69}) = 1.13 > 1$ so it diverges.

4. Extinct if $\alpha < \frac{1}{5}$, Stable if $\alpha = \frac{1}{5}$, Diverges if $\alpha > \frac{1}{5}$

Exercises 3.6 Proof of the Cofactor Expansion Theorem (Page 154)

2. Consider the rows $R_p, R_{p+1}, \ldots, R_{q-1}, R_q$. In $q - p$ adjacent interchanges they can be put in the order
$R_{p+1}, \ldots, R_{q-1}, R_q, R_p$. Then in $q - p - 1$ adjacent interchanges we can obtain the order $R_q, R_{p+1}, \ldots, R_{q-1}, R_p$. This
uses $2(q - p) - 1$ adjacent interchanges in all.

Exercises 4.1 Vectors and Lines (Page 165)

1. **(b)** $\sqrt{6}$ **(d)** $\sqrt{5}$ **(f)** $3\sqrt{6}$.

2. **(b)** $\frac{1}{3}[-2 \ -1 \ 2]^T$. **4. (b)** $\sqrt{2}$ **(d)** 3. **6. (b)** $\overrightarrow{FE} = \overrightarrow{FC} + \overrightarrow{CE} = \frac{1}{2}\overrightarrow{AC} + \frac{1}{2}\overrightarrow{CB} = \frac{1}{2}(\overrightarrow{AC} + \overrightarrow{CB}) = \frac{1}{2}\overrightarrow{AB}$.

7. (b) Yes **(d)** Yes **8. (b)** $\vec{u}$ **(d)** $-(\vec{u}+\vec{v})$. **9. (b)** $[-1 \ -1 \ 5]^T$ **(d)** $[0 \ 0 \ 0]^T$ **(f)** $[-2 \ 2 \ -2]^T$.

10. (b) (i) $Q(5,-1,2)$ **(ii)** $Q(1,1,-4)$. **11. (b)** $\vec{x}=\vec{u}-6\vec{v}+5\vec{w}=[-16 \ 4 \ 9]^T$.

12. (b) $[a \ b \ c]^T=[-5 \ 8 \ 6]^T$. **14. (b)** $\frac{1}{4}[5 \ -5 \ -2]^T$. **17. (b)** $Q(0,7,3)$.

18. (b) $x'=\frac{1}{40}[-20 \ -13 \ 14]^T$. **20. (b)** $S(-3,5,0)$.

21. (b) T. $\|\vec{v}-\vec{w}\|=0$ implies that $\vec{v}-\vec{w}=\vec{0}$. **(d)** F. $\|\vec{v}\|=\|-\vec{v}\|$ for all $\vec{v}$ but $\vec{v}=-\vec{v}$ only holds if $\vec{v}=\vec{0}$.

   (f) F. If $t<0$ they have the *opposite* direction. **(h)** F. $\|-5\vec{v}\|=5\|\vec{v}\|$ for all $\vec{v}$, so it fails if $\vec{v}\neq\vec{0}$.

   (j) F. Take $\vec{w}=-\vec{v}$ where $\vec{v}\neq\vec{0}$.

22. (b) $(3,-1,4)+t(2,-1,5)$; $x=3+2t$, $y=-1-t$, $z=4+5t$

   (d) $(1,1,1)+t(1,1,1)$; $x=y=z=1+t$ **(f)** $(2,-1,1)+t(-1,0,1)$; $x=2-t$, $y=-1$, $z=1+t$

23. (b) P corresponds to $t=2$; Q corresponds to $t=5$. **24. (b)** No intersection

   (d) $P(2,-1,3)$; $t=-2, s=-3$ **29.** $P(3,1,0)$ or $P(\frac{5}{3},\frac{-1}{3},\frac{4}{3})$ **31. (b)** $\vec{CP}_k=-\vec{CP}_{n+k}$ if $1\leq k\leq n$, where

   there are $2n$ points.

33. $\vec{DA}=2\vec{EA}$ and $2\vec{AF}=\vec{FC}$, so $2\vec{EF}=2(\vec{EF}+\vec{AF})=\vec{DA}+\vec{FC}=\vec{CB}+\vec{FC}=\vec{FC}+\vec{CB}=\vec{FB}$.

   Hence $\vec{EF}=\frac{1}{2}\vec{FB}$.

Exercises 4.2 Projections and Planes (Page 177)

1. (b) 6 **(d)** 0 **(f)** 0 **2. (b)** π or 180° **(d)** $\frac{\pi}{3}$ or 60° **(f)** $\frac{2\pi}{3}$ or 120° **3. (b)** 1 or -17

4. (b) $t(-1,1,2)$ **(d)** $s(1,2,0)+t(0,3,1)$ **6. (b)** $29+57=86$ **8. (b)** $A=B=C=\frac{\pi}{3}$ or 60°

10. (b) $\frac{11}{18}\mathbf{v}$ **(d)** $-\frac{1}{2}\mathbf{v}$ **11. (b)** $\frac{5}{21}(2,-1,-4)+\frac{1}{21}(53,26,20)$ **(d)** $\frac{27}{53}(6,-4,1)+\frac{1}{53}(-3,2,26)$

12. (b) $\frac{1}{26}\sqrt{5642}$, $Q(\frac{71}{26},\frac{15}{26},\frac{34}{26})$ **13. (b)** $(0,0,0)$ **(d)** $(4,-15,8)$ **14. (b)** $-23x+32y+11z=11$

   (d) $2x-y+z=5$ **(f)** $2x+3y+2z=7$ **(h)** $2x-7y-3z=-1$ **(j)** $x-y-z=3$

15. (b) $(x,y,z)=(2,-1,3)+t(2,1,0)$ **(d)** $(x,y,z)=(1,1,-1)+t(1,1,1)$ **(f)** $(x,y,z)=(1,1,2)+t(4,1,-5)$

16. (b) $\frac{\sqrt{6}}{3}$, $Q(\frac{7}{3},\frac{2}{3},\frac{-2}{3})$ **17. (b)** Yes. The equation is $5x-3y-4z=0$. **19. (b)** $(-2,7,0)+t(3,-5,2)$

20. (b) None **(d)** $P(\frac{13}{19},\frac{-78}{19},\frac{65}{19})$

21. (b) $3x+2z=d$, d arbitrary **(d)** $a(x-3)+b(y-2)+c(z+4)=0$ $a, b,$ and c not all zero

   (f) $ax+by+(b-a)z=a$ a and b not both zero **(h)** $ax+by+(a+2b)z=5a+4b$ a and b not both zero

23. (b) $\sqrt{10}$ **24. (b)** $\frac{14}{2}$, $A(3,1,2)$, $B(\frac{7}{2},-\frac{1}{2},3)$ **(d)** $\frac{\sqrt{6}}{6}$, $A(\frac{19}{3},2,\frac{1}{3})$, $B(\frac{37}{6},\frac{13}{6},0)$

28. The four diagonals are (a, b, c), $(-a, b, c)$, $(a, -b, c)$ and $(a, b, -c)$ or their negatives. The dot products

   are $\pm(-a^2+b^2+c^2)$, $\pm(a^2-b^2+c^2)$, and $\pm(a^2+b^2-c^2)$.

34. (b) The sum of the squares of the lengths of the diagonals equals the sum of the squares of the lengths of the

   four sides.

38. (b) The angle θ between $\vec{u}$ and $(\vec{u}+\vec{v}+\vec{w})$ is given by $\cos\theta=\dfrac{\vec{u}\cdot(\vec{u}+\vec{v}+\vec{w})}{\|\vec{u}\|\|\vec{u}+\vec{v}+\vec{w}\|}=\dfrac{\|\vec{u}\|}{\sqrt{\|\vec{u}\|^2+\|\vec{v}\|^2+\|\vec{w}\|^2}}=\dfrac{1}{\sqrt{3}}$,

   because $\|\vec{u}\|=\|\vec{v}\|=\|\vec{w}\|$. Similar remarks apply to the other angles.

41. (b) This follows from **(a)** because $\|\vec{v}\|^2=a^2+b^2+c^2$.

44. (d) Take $(x_1, y_1, z_1)=(x, y, z)$ and $(x_2, y_2, z_2)=(y, z, x)$ in **(c)**.

Exercises 4.3 The Cross Product (Page 185)

3. (b) $\pm\frac{\sqrt{3}}{3}[1 \ -1 \ -1]^T$. **4. (b)** 0 **(d)** $\sqrt{5}$ **5. (b)** 7 **6. (b)** The distance is $\|\vec{v}-\vec{p}_0\|$; use **(a)**.

10. $\|\vec{AB}\times\vec{AC}\|$ is the area of the parallelogram determined by $A, B,$ and C.

12. Because $\vec{u}$ and $\vec{v} \times \vec{w}$ are parallel, the angle θ between them is 0 or π. Hence $\cos(\theta) = \pm 1$, so the volume is $|\vec{u} \bullet (\vec{v} \times \vec{w})| = \|\vec{u}\|\|\vec{v} \times \vec{w}\| |\cos(\theta)| = \|\vec{u}\| \|\vec{v} \times \vec{w}\|$. But the angle between $\vec{v}$ and $\vec{w}$ is $\frac{\pi}{2}$ so $\|\vec{v} \times \vec{w}\| = \|\vec{v}\|\|\vec{w}\| \cos(\frac{\pi}{2}) = \|\vec{v}\|\|\vec{w}\|$. The result follows.

15. (b) If $\vec{u} = [u_1 \ u_2 \ u_3]^T$, $\vec{v} = [v_1 \ v_2 \ v_3]^T$ and $\vec{w} = [w_1 \ w_2 \ w_3]^T$, then

$$\vec{u} \times (\vec{v} + \vec{w}) = \det \begin{bmatrix} \vec{i} & u_1 & v_1 + w_1 \\ \vec{j} & u_2 & v_2 + w_2 \\ \vec{k} & u_3 & v_3 + w_3 \end{bmatrix} = \det \begin{bmatrix} \vec{i} & u_1 & v_1 \\ \vec{j} & u_2 & v_2 \\ \vec{k} & u_3 & v_3 \end{bmatrix} + \det \begin{bmatrix} \vec{i} & u_1 & w_1 \\ \vec{j} & u_2 & w_2 \\ \vec{k} & u_3 & w_3 \end{bmatrix} = (\vec{u} \times \vec{v}) + (\vec{u} \times \vec{w})$$

where we used Exercise 21 §3.1.

16. (b) $(\vec{v} - \vec{w}) \bullet [(\vec{u} \times \vec{v}) + (\vec{v} \times \vec{w}) + (\vec{w} \times \vec{u})]$

$= (\vec{v} - \vec{w}) \bullet (\vec{u} \times \vec{v}) + (\vec{v} - \vec{w}) \bullet (\vec{v} \times \vec{w}) + (\vec{v} - \vec{w}) \bullet (\vec{w} \times \vec{u})$

$= -\vec{w} \bullet (\vec{u} \times \vec{v}) + 0 + \vec{v} \bullet (\vec{w} \times \vec{u})$

$= 0.$

22. Let $\vec{p}_1$ and $\vec{p}_2$ be position vectors of points in the planes, so $\vec{p}_1 \bullet \vec{n} = d_1$ and $\vec{p}_2 \bullet \vec{n} = d_2$. The distance is the length of the projection of $\vec{p}_2 - \vec{p}_1$ along $\vec{n}$; that is $\dfrac{|(\vec{p}_2 - \vec{p}_1) \bullet \vec{n}|}{\|\vec{n}\|} = \dfrac{|d_2 - d_1|}{\|\vec{n}\|}$.

Exercises 4.4 Matrix Transformations II (Page 191)

1. (b) $A = \begin{bmatrix} 1 & -1 \\ -1 & 1 \end{bmatrix}$, projection on $y = -x$.   **(d)** $A = \frac{1}{5}\begin{bmatrix} -3 & 4 \\ 4 & 3 \end{bmatrix}$, reflection in $y = 2x$.

(f) $A = \frac{1}{2}\begin{bmatrix} 1 & -\sqrt{3} \\ \sqrt{3} & 1 \end{bmatrix}$, rotation through $\frac{\pi}{3}$.

2. (b) The zero transformation.

3. (b) $\frac{1}{21}\begin{bmatrix} 17 & 2 & -8 \\ 2 & 20 & 4 \\ -8 & 4 & 5 \end{bmatrix}\begin{bmatrix} 0 \\ 1 \\ -3 \end{bmatrix}$ **(d)** $\frac{1}{30}\begin{bmatrix} 22 & -4 & 20 \\ -4 & 28 & 10 \\ 20 & 10 & -20 \end{bmatrix}\begin{bmatrix} 0 \\ 1 \\ -3 \end{bmatrix}$ **(f)** $\frac{1}{25}\begin{bmatrix} 9 & 0 & 12 \\ 0 & 0 & 0 \\ 12 & 0 & 16 \end{bmatrix}\begin{bmatrix} 1 \\ -1 \\ 7 \end{bmatrix}$ **(h)** $\frac{1}{11}\begin{bmatrix} -9 & 2 & -6 \\ 2 & -9 & -6 \\ -6 & -6 & 7 \end{bmatrix}\begin{bmatrix} 2 \\ -5 \\ 0 \end{bmatrix}$

4. (b) $\frac{1}{2}\begin{bmatrix} \sqrt{3} & -1 & 0 \\ 1 & \sqrt{3} & 0 \\ 0 & 0 & 1 \end{bmatrix}\begin{bmatrix} 1 \\ 0 \\ 3 \end{bmatrix}$ **6.** $\begin{bmatrix} \cos\theta & 0 & -\sin\theta \\ 0 & 1 & 0 \\ \sin\theta & 0 & \cos\theta \end{bmatrix}$

9. (b) Write $\vec{v} = [x \ y \ z]^T$ and $d = a^2 + b^2 + c^2$. Then:

$$Q_L(\vec{v}) = 2P_L(\vec{v}) - \vec{v} = \left\{ \frac{2}{d}\begin{bmatrix} a^2 & ab & ac \\ ab & b^2 & bc \\ ac & bc & c^2 \end{bmatrix} - \begin{bmatrix} 1 & 0 & 0 \\ 0 & 1 & 0 \\ 0 & 0 & 1 \end{bmatrix} \right\}\begin{bmatrix} x \\ y \\ z \end{bmatrix} = \frac{1}{d}\begin{bmatrix} a^2 - b^2 - c^2 & 2ab & 2ac \\ 2ab & b^2 - a^2 - c^2 & 2bc \\ 2ac & 2bc & c^2 - a^2 - b^2 \end{bmatrix}\begin{bmatrix} x \\ y \\ z \end{bmatrix}.$$

10. (b) Write $\vec{v} = [x \ y \ z]^T$ and $d = a^2 + b^2 + c^2$. Then:

$$Q_M(\vec{v}) = 2P_M(\vec{v}) - \vec{v} = \left\{ \frac{2}{d}\begin{bmatrix} b^2 + c^2 & -ab & -ac \\ -ab & a^2 + c^2 & -bc \\ -ac & -bc & a^2 + b^2 \end{bmatrix} - \begin{bmatrix} 1 & 0 & 0 \\ 0 & 1 & 0 \\ 0 & 0 & 1 \end{bmatrix} \right\}\begin{bmatrix} x \\ y \\ z \end{bmatrix}$$

$$= \frac{1}{d}\begin{bmatrix} b^2 + c^2 - a^2 & -2ab & -2ac \\ -2ab & a^2 + c^2 - b^2 & -2bc \\ -2ac & -2bc & a^2 + b^2 - c^2 \end{bmatrix}\begin{bmatrix} x \\ y \\ z \end{bmatrix}.$$

Exercises 4.5 An Application to Computer Graphics (Page 194)

1. (b) $\frac{1}{2}\begin{bmatrix} \sqrt{2}+2 & 7\sqrt{2}+2 & 3\sqrt{2}+2 & -\sqrt{2}+2 & -5\sqrt{2}+2 \\ -3\sqrt{2}+4 & 3\sqrt{2}+4 & 5\sqrt{2}+4 & \sqrt{2}+4 & 9\sqrt{2}+4 \\ 2 & 2 & 2 & 2 & 2 \end{bmatrix}$

5. (b) $P(\frac{9}{5}, \frac{18}{5})$

Supplementary Exercises for Chapter 4 (Page 195)

4. 125 knots in a direction θ degrees east of north, where $\cos\theta = 0.6$ ($\theta = 53°$ or 0.93 radians).
6. (12, 5). Actual speed 12 knots.

Exercises 5.1 Subspaces and Spanning (Page 202)

1. (b) Yes **(d)** No **(f)** No. **2. (b)** No **(d)** Yes, $X = 3Y + 4Z$. **3. (b)** No
10. $\text{span}\{a_1X_1, a_2X_2, \dots, a_kX_k\} \subseteq \text{span}\{X_1, X_2, \dots, X_k\}$ by Theorem 1 because, for each i, a_iX_i is in $\text{span}\{X_1, X_2, \dots, X_k\}$. Similarly, the fact that $X_i = a_i^{-1}(a_iX_i)$ is in $\text{span}\{a_1X_1, a_2X_2, \dots, a_kX_k\}$ for each i shows that $\text{span}\{X_1, X_2, \dots, X_k\} \subseteq \text{span}\{a_1X_1, a_2X_2, \dots, a_kX_k\}$, again by Theorem 1.
15. (b) $X = (X+Y) - Y = (X+Y) + (-Y)$ is in U because U is a subspace and both $X+Y$ and $-Y = (-1)Y$ are in U.
16. (b) True. $X = 1X$ is in U. **(d)** True. Always $\text{span}\{Y, Z\} \subseteq \text{span}\{X, Y, Z\}$ by Theorem 1. Since X is in $\text{span}\{X, Y\}$ we have $\text{span}\{X, Y, Z\} \subseteq \text{span}\{Y, Z\}$, again by Theorem 1.
20. If U is a subspace, then S2 and S3 certainly hold. Conversely, assume that S2 and S3 hold for U. Since U is nonempty, choose X in U. Then $0 = 0X$ is in U by S3, so S1 also holds. This means that U is a subspace.
22. (b) The zero vector 0 is in $U + W$ because $0 = 0 + 0$. Let P and Q be vectors in $U + W$, say $P = X_1 + Y_1$ and $Q = X_2 + Y_2$ where X_1 and X_2 are in U, and Y_1 and Y_2 are in W. Then $P + Q = (X_1 + X_2) + (Y_1 + Y_2)$ is in $U + W$ because $X_1 + X_2$ is in U and $Y_1 + Y_2$ is in W. Similarly, $a(P + Q) = aP + aQ$ is in $U + W$ for any scalar a because aP is in U and aQ is in W. Hence $U + W$ is indeed a subspace of $\mathbb{R}^n$.

Exercises 5.2 Independence and Dimension (Page 211)

1. (b) Yes. If $r[1\ 1\ 1]^T + s[1\ -1\ 1]^T + t[0\ 0\ 1]^T = [0\ 0\ 0]^T$, then $r + s = 0$, $r - s = 0$, and $r + s + t = 0$. These equations give $r = s = t = 0$.
(d) No. Indeed: $[1\ 1\ 0\ 0]^T - [1\ 0\ 1\ 0]^T + [0\ 0\ 1\ 1]^T - [0\ 1\ 0\ 1]^T = [0\ 0\ 0\ 0]^T$.
2. (b) Yes. If $r(X + Y) + s(Y + Z) + t(Z + X) = 0$, then $(r + t)X + (r + s)Y + (s + t)Z = 0$. Since $\{X, Y, Z\}$ is independent, this implies that $r + t = 0$, $r + s = 0$, and $s + t = 0$. The only solution is $r = s = t = 0$.
(d) No. In fact, $(X + Y) - (Y + Z) + (Z + W) - (W + X) = 0$.
3. (b) $\{[2\ 1\ 0\ -1]^T, [-1\ 1\ 1\ 1]^T\}$; dimension 2. **(d)** $\{[-2\ 0\ 3\ 1]^T, [1\ 2\ -1\ 0]^T\}$; dimension 2.
4. (b) $\{[1\ 1\ 0\ 1]^T, [1\ -1\ 1\ 0]^T\}$; dimension 2. **(d)** $\{[1\ 0\ 1\ 0]^T, [-1\ 1\ 0\ 1]^T, [0\ 1\ 0\ 1]^T\}$; dimension 3.
(f) $\{[-1\ 1\ 0\ 0]^T, [1\ 0\ 1\ 0]^T, [1\ 0\ 0\ 1]^T\}$; dimension 3.
5. (b) If $r(X + W) + s(Y + W) + t(Z + W) + u(W) = 0$, then $rX + sY + tZ + (r + s + t + u)W = 0$, so $r = 0$, $s = 0$, $t = 0$, and $r + s + t + u = 0$. The only solution is $r = s = t = u = 0$, so the set is independent. Since $\dim\mathbb{R}^4 = 4$, the set is a basis by Theorem 8.
6. (b) Yes **(d)** Yes **(f)** No.
7. (b) T. If $rY + sZ + 0$, then $0X + rY + sZ = 0$ so $r = s = 0$ because $\{X, Y, Z\}$ is independent.
(d) F. If $X \neq 0$, take $k = 2$, $X_1 = X$ and $X_2 = -X$.
(f) F. If $Y = -X$ and $Z = 0$, then $1X + 1Y + 1Z = 0$.
(h) T. This is a nontrivial, vanishing linear combination, so the X_i cannot be independent.
10. If $rX_2 + sX_3 + tX_5 = 0$ then $0X_1 + rX_2 + sX_3 + 0X_4 + tX_5 + 0X_6 = 0$ so $r = s = t = 0$.
12. If $t_1X_1 + t_2(X_1 + X_2) + \cdots + t_k(X_1 + X_2 + \cdots + X_k) = 0$, then
$(t_1 + t_2 + \cdots + t_k)X_1 + (t_2 + \cdots + t_k)X_2 + \cdots + (t_{k-1} + t_k)X_{k-1} + (t_k)X_k = 0$.
Hence all these coefficients are zero, so we obtain successively $t_k = 0$, $t_{k-1} = 0$, $\dots$, $t_2 = 0$, $t_1 = 0$.

14. (b) We show A^T is invertible (then A is invertible). Let $A^T X = 0$ where $X = [s\ \ t]^T$. This means $as + ct = 0$ and $bs + dt = 0$, so $s(aX + bY) + t(cX + dY) = (sa + tc)X + (sb + td)Y = 0$. Hence $s = t = 0$ by hypothesis.

15. (b) Each $V^{-1}X_i$ is in null(AV) because $AV(V^{-1}X_i) = AX_i = 0$. The set $\{V^{-1}X_1, \ldots, V^{-1}X_k\}$ is independent by Example 9. If Y is in null(AV), then VY is in null(A) so let $VY = t_1X_1 + \cdots + t_kX_k$ where each t_k is in $\mathbb{R}$. Thus $Y = t_1V^{-1}X_1 + \cdots + t_kV^{-1}X_k$ is in span$\{V^{-1}X_1, \ldots, V^{-1}X_k\}$.

18. We have $\{0\} \subseteq U \subseteq W$ where $\dim\{0\} = 0$ and $\dim W = 1$. Hence $\dim U = 0$ or $\dim U = 1$ by Theorem 8, that is $U = 0$ or $U = W$, again by Theorem 8.

Exercises 5.3 Orthogonality (Page 218)

1. (b) $\left\{ \frac{1}{\sqrt{3}}[1\ 1\ 1]^T,\ \frac{1}{\sqrt{42}}[4\ 1\ -5]^T, \frac{1}{\sqrt{14}}[2\ -3\ 1]^T \right\}$.

3. (b) $[a\ b\ c]^T = \frac{1}{2}(a - c)[1\ 0\ -1]^T + \frac{1}{18}(a + 4b + c)[1\ 4\ 1]^T + \frac{1}{9}(2a - b + 2c)[2\ -1\ 2]^T$.

(d) $[a\ b\ c]^T = \frac{1}{3}(a + b + c)[1\ 1\ 1]^T + \frac{1}{2}(a - b)\ [1\ -1\ 0]^T + \frac{1}{6}(a + b - 2c)[1\ 1\ -2]^T$.

4. (b) $[14\ 1\ -8\ 5]^T = 3[2\ -1\ 0\ 3]^T + 4[2\ 1\ -2\ -1]^T$.   **5. (b)** $t[-1\ 3\ 10\ 11]^T$, t in $\mathbb{R}$.

6. (b) $\sqrt{29}$ **(d)** 19

7. (b) F. $X = [1\ 0]^T$ and $Y = [0\ 1]^T$.

(d) T. Every $X_i \bullet Y_j = 0$ by assumption, every $X_i \bullet X_j = 0$ if $i \neq j$ because the X_i are orthogonal, and every $Y_i \bullet Y_j = 0$ if $i \neq j$ because the Y_i are orthogonal. As all the vectors are nonzero, this does it.

(f) T. Every pair of *distinct* vectors in the set $\{X\}$ has dot product zero (there are no such pairs).

10. (b) Take $n = 3$ in **(a)**, expand, and simplify.

11. (b) We have $(X + Y) \bullet (X - Y) = \|X\|^2 - \|Y\|^2$. Hence $(X + Y) \bullet (X - Y) = 0$ if and only if $\|X\|^2 = \|Y\|^2$; if and only if $\|X\| = \|Y\|$—where we used the fact that $\|Z\| \geq 0$ for every Z in $\mathbb{R}^n$.

Exercises 5.4 Rank of a Matrix (Page 226)

1. (b) $\{[2\ -1\ 1]^T, [0\ 0\ 1]^T\};$ $\left\{ \begin{bmatrix} 2 \\ -2 \\ 4 \\ -6 \end{bmatrix}, \begin{bmatrix} 1 \\ 1 \\ 3 \\ 0 \end{bmatrix} \right\};$ 2   **(d)** $\{[1\ 2\ -1\ 3]^T, [0\ 0\ 0\ 1]^T\};$ $\left\{ \begin{bmatrix} 1 \\ -3 \end{bmatrix}, \begin{bmatrix} 3 \\ -2 \end{bmatrix} \right\};$ 2

2. (b) $\{(1, 1, 0, 0, 0), (0, -2, 2, 5, 1), (0, 0, 2, -3, 6)\}$ **(d)** $\left\{ \begin{bmatrix} 1 \\ 5 \\ -6 \end{bmatrix}, \begin{bmatrix} 0 \\ 1 \\ -1 \end{bmatrix}, \begin{bmatrix} 0 \\ 0 \\ 1 \end{bmatrix} \right\}$ **3. (b)** No; no   **(d)** No

(f) Otherwise, if A is $m \times n$, we have $m = \dim(\text{row } A) = \text{rank } A = \dim(\text{col } A) = n$

7. (b) The basis is $\left\{ \begin{bmatrix} 6 \\ 0 \\ -4 \\ 1 \\ 0 \end{bmatrix}, \begin{bmatrix} 5 \\ 0 \\ -3 \\ 0 \\ 1 \end{bmatrix} \right\},$ so the dimension is 2. Have rank $A = 3$ and $n - 3 = 2$.

8. (b) $n - 1$ **10. (b)** Write $r = \text{rank } A$. Then **(a)** gives $r = \dim(\text{col } A) \leq \dim(\text{null } A) = n - r$.

12. If C_j is column j of A and $X = [x_1\ \cdots\ x_n]^T$, then $AX = x_1C_1 + \cdots + x_nC_n$. Thus $\{AX | X \text{ in } \mathbb{R}^n\} = \text{span}\{C_j\} = \text{col } A$.

15. (b) Let $\{U_1, \ldots, U_r\}$ be a basis of col(A). Then B is *not* in col(A), so $\{U_1, \ldots, U_r, B\}$ is linearly independent. Show that col$[A\ B] = \text{span}\{U_1, \ldots, U_r, B\}$.

Exercises 5.5 Similarity and Diagonalization (Page 236)

1. (b) traces $= 2$, ranks $= 2$, but $\det A = -5$, $\det B = -1$

(d) ranks $= 2$, determinants $= 7$, but tr $A = 5$, tr $B = 4$

(f) traces $= -5$, determinants $= 0$, but rank $A = 2$, rank $B = 1$

3. (b) If $B = P^{-1}AP$, then $B^{-1} = P^{-1}A^{-1}(P^{-1})^{-1} = P^{-1}A^{-1}P$.

4. (b) Yes. $P = \begin{bmatrix} -1 & 0 & 6 \\ 0 & 1 & 0 \\ 1 & 0 & 5 \end{bmatrix}$, $P^{-1}AP = \begin{bmatrix} -3 & 0 & 0 \\ 0 & -3 & 0 \\ 0 & 0 & 8 \end{bmatrix}$ (d) No, $c_A(x) = (x+1)(x-4)^2$

so $\lambda = 4$ has multiplicity 2. But $\dim(E_4) = 1$ so Theorem 6 applies.

8. (b) If $B = P^{-1}AP$ and $A^k = 0$, then $B^k = (P^{-1}AP)^k = P^{-1}A^kP = P^{-1}0P = 0$.

9. (b) The eigenvalues of A are all equal (they are the diagonal elements), so if $P^{-1}AP = D$ is diagonal, then $D = \lambda I$.

10. (b) A is similar to $D = \text{diag}(\lambda_1, \lambda_2, \ldots, \lambda_n)$ so (Theorem 1) $\text{tr}\,A = \text{tr}\,D = \lambda_1 + \lambda_2 + \cdots + \lambda_n$.

12. (b) $T_P(A)T_P(B) = (P^{-1}AP)(P^{-1}BP) = P^{-1}(AB)P = T_P(AB)$.

13. (b) If A is diagonalizable, so is A^T, and they have the same eigenvalues. Use (a).

17. (b) $c_B(X) = [x - (a + b + c)][x^2 - k]$ where $k = a^2 + b^2 + c^2 - [ab + ac + bc]$.

Exercises 5.6 An Application to Correlation and Variance (Page 242)

2. Let X denote the number of years of education, and let Y denote the yearly income (in 1000's). Then $\bar{x} = 15.3$, $s_x^2 = 9.12$ and $s_x = 3.02$, while $\bar{y} = 40.3$, $s_y^2 = 114.23$ and $s_y = 10.69$. The correlation is $r(X, Y) = 0.599$.

4. (b) Given the sample vector $X = [x_1\ x_2 \cdots x_n]^T$, let $Z = [z_1\ z_2 \cdots z_n]^T$ where $z_i = a + bx_i$ for each i. By (a) we have $\bar{z} = a + b\bar{x}$, so

$$s_z^2 = \tfrac{1}{n-1}\sum_i(z_i - \bar{z})^2 = \tfrac{1}{n-1}\sum_i[(a + bx_i)-(a + b\bar{x})]^2 = \tfrac{1}{n-1}\sum_i b^2[(x_i - \bar{x})^2 = b^2 s_x^2.$$

Now (b) follows because $\sqrt{b^2} = |b|$.

Exercises 5.7 An Application to Least Squares Approximation (Page 248)

1. (b) $y = \frac{64}{13} - \frac{6}{13}x$ (d) $y = -\frac{4}{10} - \frac{17}{10}x$

2. (b) $y = 0.127 - 0.024x + 0.194x^2$. Here $(M^TM)^{-1} = \dfrac{1}{4248}\begin{bmatrix} 3348 & 642 & -426 \\ 642 & 571 & -187 \\ -426 & -187 & 91 \end{bmatrix}$.

4. $s = 99.71 - 4.87t$, giving $g = 9.74$ (the true value of g is 9.81). If a quadratic in s is fitted, we get

$$s = 101 - \tfrac{3}{2}t - \tfrac{9}{2}t^2,\ \text{giving } g = 9.\ \text{Here } (M^TM)^{-1} = \frac{1}{4}\begin{bmatrix} 76 & -84 & 20 \\ -84 & 98 & -24 \\ 20 & -24 & 6 \end{bmatrix}.$$

6. (b) $f(x) = a_0$ here, so the sum of squares is $S = \sum_i(y_i - a_0)^2 = na_0^2 - 2a_0\sum_i y_i + \sum_i y_i^2$. Completing the square (as a function of a_0) gives $S = n[a_0 - \tfrac{1}{n}\sum_i y_i]^2 + [\sum_i y_i^2 - \tfrac{1}{n}(\sum_i y_i)^2]$. This is minimal when $a_0 = \tfrac{1}{n}\sum_i y_i$, the mean of the y_i.

Supplementary Exercises for Chapter 5 (Page 249)

(b) F   (d) T   (f) T   (h) F   (j) F   (l) T   (n) F   (p) F   (r) F

Exercises 6.1 Examples and Basic Properties (Page 257)

1. (b) No; S5 fails.   (d) No; S4 and S5 fail.

2. (b) No; only A1 fails.   (d) No   (f) Yes   (h) Yes   (j) No   (l) No; only S3 fails.   (n) No; only S4 and S5 fail.

4. The zero vector is $(0, -1)$; the negative of (x, y) is $(-x, -2 - y)$.   5. (b) $\mathbf{x} = \tfrac{1}{7}(5\mathbf{u} - 2\mathbf{v})$, $\mathbf{y} = \tfrac{1}{7}(4\mathbf{u} - 3\mathbf{v})$

6. (b) Equating entries gives $a + c = 0$, $b + c = 0$, $b + c = 0$, $a - c = 0$. The solution is $a = b = c = 0$.
   (d) If $a \sin x + b \cos y + c = 0$ in $\mathbf{F}[0, \pi]$, then this must hold for *every* x in $[0, \pi]$. Taking $x = 0, \frac{\pi}{2}$, and π, respectively, gives $b + c = 0$, $a + c = 0$, $-b + c = 0$ whence, $a = b = c = 0$.

7. (b) $4\mathbf{w}$ 11. If $\mathbf{v} + \mathbf{x} = \mathbf{0}$, then $\mathbf{x} = \mathbf{0} + \mathbf{x} = (-\mathbf{v} + \mathbf{v}) + \mathbf{x} = -\mathbf{v} + (\mathbf{v} + \mathbf{x}) = -\mathbf{v} + \mathbf{0} = -\mathbf{v}$.

12. (b) $(-a)\mathbf{v} + a\mathbf{v} = (-a+a)\mathbf{v} = 0\mathbf{v} = \mathbf{0}$ by Theorem 3. Because also $-(a\mathbf{v}) + a\mathbf{v} = \mathbf{0}$ (by the definition of $-(a\mathbf{v})$ in axiom A.5), this means that $(-a)\mathbf{v} = -(a\mathbf{v})$ by cancellation. Alternatively, use Theorem 3(4) to give $(-a)\mathbf{v} = [(-1)a]\mathbf{v} = (-1)(a\mathbf{v}) = -(a\mathbf{v})$.

15. (b) If $a\mathbf{v} = a\mathbf{w}$, then $\mathbf{v} = 1\mathbf{v} = (a^{-1}a)\mathbf{v} = a^{-1}(a\mathbf{v}) = a^{-1}(a\mathbf{w}) = (a^{-1}a)\mathbf{w} = 1\mathbf{w} = \mathbf{w}$.

Exercises 6.2 Subspaces and Spanning Sets (Page 263)

1. (b) Yes **(d)** Yes **(f)** No; not closed under addition or scalar multiplication, and 0 is not in the set.
2. (b) Yes **(d)** Yes **(f)** No; not closed under addition.
3. (b) No; not closed under addition. **(d)** No; not closed under scalar multiplication. **(f)** Yes
5. (b) If entry k of X is $x_k \neq 0$, and if Y is in $\mathbb{R}^n$, then $Y = AX$ where the only nonzero column of A is column $k = x_k^{-1}Y$. **6. (b)** $-3(x+1) + 0(x^2+x) + 2(x^2+2)$ **(d)** $\frac{2}{3}(x+1) + \frac{1}{3}(x^2+x) - \frac{1}{3}(x^2+2)$
7. (b) No **(d)** Yes; $\mathbf{v} = 3\mathbf{u} - \mathbf{w}$.
8. (b) Yes; $1 = \cos^2 x + \sin^2 x$ **(d)** No. If $1 + x^2 = a\cos^2 x + b\sin^2 x$, then taking $x = 0$ and $x = \pi$ gives $a = 1$ and $a = 1 + \pi^2$.
9. (b) Because $\mathbf{P}_2 = \text{span}\{1, x, x^2\}$, it suffices to show that $\{1, x, x^2\} \subseteq \text{span}\{1 + 2x^2, 3x, 1 + x\}$. But $x = \frac{1}{3}(3x)$; $1 = (1+x) - x$ and $x^2 = \frac{1}{2}[(1+2x^2)-1]$.
11. (b) $\mathbf{u} = (\mathbf{u}+\mathbf{w}) - \mathbf{w}$, $\mathbf{v} = -(\mathbf{u}-\mathbf{v}) + (\mathbf{u}+\mathbf{w}) - \mathbf{w}$, and $\mathbf{w} = \mathbf{w}$ **14.** No
18. $\mathbf{v}_1 = \frac{1}{a_1}\mathbf{u} - \frac{a_2}{a_1}\mathbf{v}_2 - \cdots - \frac{a_n}{a_1}\mathbf{v}_n$, so $V \subseteq \text{span}\{\mathbf{u}, \mathbf{v}_2, \ldots, \mathbf{v}_n\}$. **21. (b)** $\mathbf{v} = (\mathbf{u}+\mathbf{v}) - \mathbf{u}$ is in U.

Exercises 6.3 Linear Independence and Dimension (Page 270)

1. (b) If $ax^2 + b(x+1) + c(1-x-x^2) = 0$, then $a-c = 0$, $b-c = 0$, $b+c = 0$, so $a = b = c = 0$.
(d) If $a\begin{bmatrix}1&1\\1&0\end{bmatrix} + b\begin{bmatrix}0&1\\1&1\end{bmatrix} + c\begin{bmatrix}1&0\\1&1\end{bmatrix} + d\begin{bmatrix}1&1\\0&1\end{bmatrix} = \begin{bmatrix}0&0\\0&0\end{bmatrix}$, then $a+c+d = 0$, $a+b+d = 0$, $a+b+c = 0$, and $b+c+d = 0$, so $a = b = c = d = 0$.
2. (b) $3(x^2 - x + 3) - 2(2x^2 + x + 5) + (x^2 + 5x + 1) = 0$
(d) $2\begin{bmatrix}-1&0\\0&-1\end{bmatrix} + \begin{bmatrix}1&-1\\-1&1\end{bmatrix} + \begin{bmatrix}1&1\\1&1\end{bmatrix} = \begin{bmatrix}0&0\\0&0\end{bmatrix}$ **(f)** $\dfrac{5}{x^2+x-6} + \dfrac{1}{x^2-5x+6} - \dfrac{6}{x^2-9} = 0$
3. (b) Dependent: $1 - \sin^2 x - \cos^2 x = 0$ **4. (b)** $x \neq -\frac{1}{3}$
5. (b) If $r(-1, 1, 1) + s(1, -1, 1) + t(1, 1, -1) = (0, 0, 0)$, then $-r+s+t = 0$, $r-s+t = 0$ and $r-s-t = 0$, and this implies that $r = s = t = 0$. This proves independence. To prove that they span $\mathbb{R}^3$, observe that $(0, 0, 1) = \frac{1}{2}[(-1, 1, 1)+(1, -1, 1)]$ so $(0, 0, 1)$ lies in span $\{(-1, 1, 1), (1, -1, 1), (1, 1, -1)\}$. The proof is similar for $(0, 1, 0)$ and $(1, 0, 0)$.
(d) If $r(1+x) + s(x+x^2) + t(x^2+x^3) + ux^3 = 0$, then $r = 0$, $r+s = 0$, $s+t = 0$, and $t+u = 0$, so $r = s = t = u = 0$. This proves independence. To show that they span $\mathbf{P}_3$, observe that $x^2 = (x^2+x^3) - x^3$, $x = (x+x^2) - x^2$, and $1 = (1+x) - x$, so $\{1, x, x^2, x^3\} \subseteq \text{span}\{1+x, x+x^2, x^2+x^3, x^3\}$.
6. (b) $\{1, x+x^2\}$; dimension $= 2$ **(d)** $\{1, x^2\}$; dimension $= 2$
7. (b) $\left\{\begin{bmatrix}1&1\\-1&0\end{bmatrix}, \begin{bmatrix}1&0\\0&1\end{bmatrix}\right\}$; dimension $= 2$ **(d)** $\left\{\begin{bmatrix}1&0\\1&1\end{bmatrix}, \begin{bmatrix}0&1\\-1&0\end{bmatrix}\right\}$; dimension $= 2$ **8. (b)** $\left\{\begin{bmatrix}1&0\\0&0\end{bmatrix}, \begin{bmatrix}0&1\\0&0\end{bmatrix}\right\}$
10. (b) $\dim V = 7$ **(c)** $\dim V = n^2 - n + 1$ **11. (b)** $\{x^2 - x, x(x^2-x), x^2(x^2-x), x^3(x^2-x)\}$; $\dim V = 4$
12. (b) No. Any linear combination of such polynomials has $f(0) = 0$.
(d) No. $\left\{\begin{bmatrix}1&0\\0&1\end{bmatrix}, \begin{bmatrix}1&1\\0&1\end{bmatrix}, \begin{bmatrix}1&0\\1&1\end{bmatrix}, \begin{bmatrix}0&1\\1&1\end{bmatrix}\right\}$ consists of invertible matrices.
(f) Yes. $0\mathbf{u} + 0\mathbf{v} + 0\mathbf{w} = \mathbf{0}$ for every set $\{\mathbf{u}, \mathbf{v}, \mathbf{w}\}$.
(h) Yes. $s\mathbf{u} + t(\mathbf{u}+\mathbf{v}) = \mathbf{0}$ gives $(s+t)\mathbf{u} + t\mathbf{v} = \mathbf{0}$, whence $s+t = 0 = t$.
(j) Yes. If $r\mathbf{u} + s\mathbf{v} = \mathbf{0}$, then $r\mathbf{u} + s\mathbf{v} + 0\mathbf{w} = \mathbf{0}$, so $r = 0 = s$.
15. If a linear combination of the subset vanishes, it is a linear combination of the vectors in the larger set (coefficients outside the subset are zero) so it is trivial.
19. Because $\{\mathbf{u}, \mathbf{v}\}$ is linearly independent, $s\mathbf{u}' + t\mathbf{v}' = \mathbf{0}$ is equivalent to $\begin{bmatrix}a&c\\b&d\end{bmatrix}\begin{bmatrix}s\\t\end{bmatrix} = \begin{bmatrix}0\\0\end{bmatrix}$. Now apply Theorem 5 §2.3.

22. (b) Independent **(d)** Dependent. For example, $(\mathbf{u} + \mathbf{v}) - (\mathbf{v} + \mathbf{w}) + (\mathbf{w} + \mathbf{z}) - (\mathbf{z} + \mathbf{u}) = \mathbf{0}$.

25. If z is not real and $a^2 + bz^2 = 0$, then $a + bz = 0$ $(z \neq 0)$. Hence if $b \neq 0$, then $z = -ab^{-1}$ is real. So $b = 0$, whenever $a = 0$. Conversely, if z is real, say $z = a$, then $(-a)z + 1z^2 = 0$, contrary to the independence of $\{z, z^2\}$.

27. (b) If $UX = 0$, $X \neq 0$ in $\mathbb{R}^n$, then $RX = 0$ where $R \neq 0$ is row 1 of U. If $B \in \mathbf{M}_{mn}$ has each row equal to R, then $BX \neq 0$. But if $B = \sum r_i A_i U$, then $BX = \sum r_i A_i UX = 0$. So $\{A_i U\}$ cannot span $\mathbf{M}_{mn}$.

34. (b) $\dim 0_n = \frac{n}{2}$ if n is even and $\dim 0_n = \frac{n+1}{2}$ if n is odd.

Exercises 6.4 Finite Dimensional Spaces (Page 278)

1. (b) $\{(0, 1, 1), (1, 0, 0), (0, 1, 0)\}$ **(d)** $\{x^2 - x + 1, 1, x\}$
2. (b) Any three except $\{x^2 + 3, x + 2, x^2 - 2x - 1\}$
3. (b) Add $(0, 1, 0, 0)$ and $(0, 0, 1, 0)$. **(d)** Add 1 and x^3.
4. (b) If $z = a + bi$, then $a \neq 0$ and $b \neq 0$. If $rz + s\bar{z} = 0$, then $(r + s)a = 0$ and $(r - s)b = 0$. This means that $r + s = 0 = r - s$, so $r = s = 0$. Thus $\{z, \bar{z}\}$ is independent; it is a basis because $\dim \mathbb{C} = 2$.
7. (b) Not a basis **(d)** Not a basis **8. (b)** Yes; no
10. $\det A = 0$ if and only if A is not invertible; if and only if the rows of A are dependent (Theorem 3 §5.2); if and only if some row is a linear combination of the others (Lemma 2).
11. (b) No. $\{(0, 1), (1, 0)\} \subseteq \{(0, 1), (1, 0), (1, 1)\}$.
 (d) Yes. See Exercise 15 §6.3.
18. (b) Two distinct planes through the origin (U and W) meet in a line through the origin ($U \cap W$).
21. (b) The set $\{(1, 0, 0, 0, \ldots), (0, 1, 0, 0, 0, \ldots), (0, 0, 1, 0, 0, \ldots), \ldots\}$ contains independent subsets of arbitrary size.
23. (b) $\mathbb{R}\mathbf{u} + \mathbb{R}\mathbf{w} = \{r\mathbf{u} + s\mathbf{w} \mid r, s \text{ in } \mathbb{R}\} = \text{span}\{\mathbf{u}, \mathbf{w}\}$

Exercises 6.5 An Application to Polynomials (Page 283)

2. (b) $3 + 4(x - 1) + 3(x - 1)^2 + (x - 1)^3$ **(d)** $1 + (x - 1)^3$
6. (b) The polynomials are $(x - 1)(x - 2)$, $(x - 1)(x - 3)$, $(x - 2)(x - 3)$. Use $a_0 = 3$, $a_1 = 2$, and $a_2 = 1$.
7. (b) $f(x) = \frac{3}{2}(x - 2)(x - 3) - 7(x - 1)(x - 3) + \frac{13}{2}(x - 1)(x - 2)$.
10. (b) If $r(x - a)^2 + s(x - a)(x - b) + t(x - b)^2 = 0$, then evaluation at $x = a$ $(x = b)$ gives $t = 0$ $(r = 0)$. Thus $s(x - a)(x - b) = 0$, so $s = 0$. Use Theorem 4 §6.4.
11. (b) Suppose $\{p_0(x), p_1(x), \ldots, p_{n-2}(x)\}$ is a basis of $\mathbf{P}_{n-2}$. We show that
$$\{(x - a)(x - b)p_0(x), (x - a)(x - b)p_1(x), \ldots, (x - a)(x - b)p_{n-2}(x)\}$$
is a basis of U_n. It is a spanning set by part **(a)**, so assume that a linear combination vanishes with coefficients $r_0, r_1, \ldots, r_{n-2}$. Then $(x - a)(x - b)[r_0 p_0(x) + \cdots + r_{n-2}p_{n-2}(x)] = 0$, so $r_0 p_0(x) + \cdots + r_{n-2}p_{n-2}(x) = 0$. This implies that $r_0 = \cdots = r_{n-2} = 0$.

Exercises 6.6 An Application to Differential Equations (Page 289)

1. (b) e^{1-x} **(d)** $\dfrac{e^{2x} - e^{-3x}}{e^2 - e^{-3}}$ **(f)** $2e^{2x}(1 + x)$ **(h)** $\dfrac{e^{ax} - e^{a(2-x)}}{1 - e^{2a}}$ **(j)** $e^{\pi - 2x} \sin x$

5. (b) $ce^{-x} + 2$, c a constant **6. (b)** $ce^{-3x} + de^{2x} + \dfrac{x^3}{3}$ **7. (b)** $t = \dfrac{3\ln(\frac{1}{2})}{\ln(\frac{4}{5})} = 9.32$ hours

9. $k = \left(\dfrac{\pi}{15}\right)^2 = 0.044$

Supplementary Exercises for Chapter 6 (Page 290)

2. (b) If $YA = 0$, Y a row, we show that $Y = 0$; thus A^T (and hence A) is invertible. Given a column C in $\mathbb{R}^n$ write $C = \sum_i r_i(A\mathbf{v}_i)$ where each r_i is in $\mathbb{R}$. Then $YC = \sum_i r_i YA\mathbf{v}_i = 0$, so $Y = YI_n = Y[E_1 \ E_1 \ \cdots \ E_n] = [YE_1 \ YE_2 \ \cdots \ YE_n] = [0 \ 0 \cdots 0] = 0$, as required.
4. We have null $A \subseteq \text{null}(A^T A)$ because $AX = 0$ implies $(A^T A)X = 0$. Conversely, if $(A^T A)X = 0$, then $\|AX\|^2 = (AX)^T(AX) = X^T A^T AX = 0$. Thus $AX = 0$.

Exercises 7.1 Examples and Elementary Properties (Page 295)

1. **(b)** $T(\mathbf{v}) = \mathbf{v}A$ where $A = \begin{bmatrix} 1 & 0 & 0 \\ 0 & 1 & 0 \\ 0 & 0 & -1 \end{bmatrix}$

   (d) $T(A + B) = P(A + B)Q = PAQ + PBQ = T(A) + T(B)$; $T(rA) = P(rA)Q = rPAQ = rT(A)$

   (f) $T[(p + q)(x)] = (p + q)(0) = p(0) + q(0) = T[p(x)] + T[q(x)]$
   $T[(rp)(x)] = (rp)(0) = rp(0) = rT[p(x)]$

   (h) $T(X + Y) = (X + Y) \cdot Z = X \cdot Z + Y \cdot Z = T(X) + T(Y)$, and $T(rX) = (rX) \cdot Z = r(X \cdot Z) = rT(X)$

   (j) If $\mathbf{v} = r_1\mathbf{e}_1 + \cdots + r_n\mathbf{e}_n$ and $\mathbf{w} = s_1\mathbf{e}_1 + \cdots + s_n\mathbf{e}_n$, then $T(\mathbf{v} + \mathbf{w}) = r_1 + s_1 = T(\mathbf{v}) + T(\mathbf{w})$,
   and $T(r\mathbf{v}) = rr_1 = rT(\mathbf{v})$

2. **(b)** $\operatorname{rank}(A + B) \neq \operatorname{rank} A + \operatorname{rank} B$ in general. For example, $A = \begin{bmatrix} 1 & 0 \\ 0 & 1 \end{bmatrix}$ and $B = \begin{bmatrix} 1 & 0 \\ 0 & -1 \end{bmatrix}$.

   (d) $T(\mathbf{0}) = \mathbf{0} + \mathbf{u} = \mathbf{u} \neq \mathbf{0}$, so T is not linear by Theorem 1.

3. **(b)** $T(3\mathbf{v}_1 + 2\mathbf{v}_2) = 0$ **(d)** $T\begin{bmatrix} 1 \\ -7 \end{bmatrix} = \begin{bmatrix} -3 \\ 4 \end{bmatrix}$ **(f)** $T(2 - x + 3x^2) = 46$

4. **(b)** $T(x, y) = \frac{1}{3}(x - y, 3y, x - y)$; $T(-1, 2) = (-1, 2, -1)$ **(d)** $T\begin{bmatrix} a & b \\ c & d \end{bmatrix} = 3a - 3c + 2b$

5. **(b)** $T(\mathbf{v}) = \frac{1}{3}(7\mathbf{v} - 9\mathbf{w})$, $T(\mathbf{w}) = \frac{1}{3}(\mathbf{v} + 3\mathbf{w})$

8. **(b)** $T(\mathbf{v}) = (-1)\mathbf{v}$ for all $\mathbf{v}$ in V, so T is the scalar operator -1.

12. If $T(1) = \mathbf{v}$, then $T(r) = T(r \cdot 1) = rT(1) = r\mathbf{v}$ for all r in $\mathbb{R}$.

15. **(b)** $\mathbf{0}$ is in $T^{-1}(P)$ because $T(\mathbf{0}) = \mathbf{0}$ is in P. If $\mathbf{v}$ and $\mathbf{w}$ are in $T^{-1}(P)$, then $T(\mathbf{v})$ and $T(\mathbf{w})$ are in P. Hence $T(\mathbf{v} + \mathbf{w}) = T(\mathbf{v}) + T(\mathbf{w})$ is in P and $T(r\mathbf{v}) = rT(\mathbf{v})$ is in P, so $\mathbf{v} + \mathbf{w}$ and $r\mathbf{v}$ are in $T^{-1}(P)$.

18. Suppose $r\mathbf{v} + sT(\mathbf{v}) = \mathbf{0}$. If $s = 0$, then $r = 0$ (because $\mathbf{v} \neq \mathbf{0}$). If $s \neq 0$, then $T(\mathbf{v}) = a\mathbf{v}$ where $a = -s^{-1}r$. Thus $\mathbf{v} = T^2(\mathbf{v}) = T(a\mathbf{v}) = a^2\mathbf{v}$, so $a^2 = 1$, again because $\mathbf{v} \neq \mathbf{0}$. Hence $a = \pm 1$. Conversely, if $T(\mathbf{v}) = \pm\mathbf{v}$, then $\{\mathbf{v}, T(\mathbf{v})\}$ is certainly not independent.

21. **(b)** Given such a T, write $T(x) = a$. If $p = p(x) = \sum_{i=0}^{n} a_i x^i$, then

$$T(p) = \sum a_i T(x^i) = \sum a_i [T(x)]^i = \sum a_i a^i = p(a) = E_a(p). \text{ Hence } T = E_a.$$

Exercises 7.2 Kernel and Image of a Linear Transformation (Page 302)

1. **(b)** $\left\{ \begin{bmatrix} -3 \\ 7 \\ 1 \\ 0 \end{bmatrix}, \begin{bmatrix} 1 \\ 1 \\ 0 \\ -1 \end{bmatrix} \right\}$; $\left\{ \begin{bmatrix} 1 \\ 0 \\ 1 \end{bmatrix}, \begin{bmatrix} 0 \\ 1 \\ -1 \end{bmatrix} \right\}$; 2, 2   **(d)** $\left\{ \begin{bmatrix} -1 \\ 2 \\ 1 \end{bmatrix} \right\}$; $\left\{ \begin{bmatrix} 1 \\ 0 \\ 1 \\ 1 \end{bmatrix}, \begin{bmatrix} 0 \\ 1 \\ -1 \\ -2 \end{bmatrix} \right\}$; 2, 1

2. **(b)** $\{x^2 - x\}$; $\{(1, 0), (0, 1)\}$ **(d)** $\{(0, 0, 1)\}$; $\{(1, 1, 0, 0), (0, 0, 1, 1)\}$ **(f)** $\left\{ \begin{bmatrix} 1 & 0 \\ 0 & -1 \end{bmatrix}, \begin{bmatrix} 0 & 1 \\ 0 & 0 \end{bmatrix}, \begin{bmatrix} 0 & 0 \\ 1 & 0 \end{bmatrix} \right\}$; $\{1\}$

   (h) $\{(1, 0, 0, \ldots, 0, -1), (0, 1, 0, \ldots, 0, -1), \ldots, (0, 0, 0, \ldots, 1, -1)\}$; $\{1\}$

   (j) $\left\{ \begin{bmatrix} 0 & 1 \\ 0 & 0 \end{bmatrix}, \begin{bmatrix} 0 & 0 \\ 0 & 1 \end{bmatrix} \right\}$; $\left\{ \begin{bmatrix} 1 & 1 \\ 0 & 0 \end{bmatrix}, \begin{bmatrix} 0 & 0 \\ 1 & 1 \end{bmatrix} \right\}$

3. **(b)** $T(\mathbf{v}) = \mathbf{0} = [0, 0]$ if and only if $P(\mathbf{v}) = 0$ and $Q(\mathbf{v}) = 0$; that is, if and only if $\mathbf{v}$ is in $\ker P \cap \ker Q$.

4. **(b)** $\ker T = \operatorname{span}\{(-4, 1, 3)\}$; $B = \{(1, 0, 0), (0, 1, 0), (-4, 1, 3)\}$, $\operatorname{im} T = \operatorname{span}\{(1, 2, 0, 3), (1, -1, -3, 0)\}$

6. **(b)** Yes. $\dim(\operatorname{im} T) = 5 - \dim(\ker T) = 3$, so $\operatorname{im} T = W$ as $\dim W = 3$.   **(d)** No. $T = 0 : \mathbb{R}^2 \to \mathbb{R}^2$

   (f) No. $T : \mathbb{R}^2 \to \mathbb{R}^2$, $T(x, y) = (y, 0)$. Then $\ker T = \operatorname{im} T$.

   (h) Yes. $\dim V = \dim(\ker T) + \dim(\operatorname{im} T) \leq \dim W + \dim W = 2 \dim W$

   (j) No. Consider $T : \mathbb{R}^2 \to \mathbb{R}^2$ given by $T[x\ y]^T = [y\ 0]^T$.

   (l) No. Same example as **(j)**.

7. **(b)** Given $\mathbf{w}$ in W, let $\mathbf{w} = T(\mathbf{v})$, $\mathbf{v}$ in V, and write $\mathbf{v} = r_1\mathbf{v}_1 + \cdots + r_n\mathbf{v}_n$.
   Then $\mathbf{w} = T(\mathbf{v}) = r_1 T(\mathbf{v}_1) + \cdots + r_n T(\mathbf{v}_n)$.

8. **(b)** $\operatorname{im} T = \{\sum_i r_i \mathbf{v}_i \mid r_i \text{ in } \mathbb{R}\} = \operatorname{span}\{\mathbf{v}_i\}$.

10. T is linear and onto. Hence $1 = \dim \mathbb{R} = \dim(\text{im } T) = \dim(\mathbf{M}_{nn}) - \dim(\ker T) = n^2 - \dim(\ker T)$.

15. **(b)** $B = \{x - 1, \dots, x^n - 1\}$ is independent (distinct degrees) and contained in ker T. Hence B is a basis of ker T by **(a)**.

20. Define $T : \mathbf{M}_{nn} \to \mathbf{M}_{nn}$ by $T(A) = A - A^T$ for all A in $\mathbf{M}_{nn}$. Then ker $T = U$ and im $T = V$ by Example 3, so the dimension theorem gives $n^2 = \dim \mathbf{M}_{nn} = \dim(U) + \dim(V)$.

22. Define $T : \mathbf{M}_{nn} \to \mathbb{R}^n$ by $T(A) = AY$ for all A in $\mathbf{M}_{nn}$. Then T is linear with ker $T = U$, so it is enough to show that T is onto (then dim $U = n^2 - \dim(\text{im } T) = n^2 - n$). We have $T(0) = 0$. Let $Y = [y_1 \ y_2 \ \cdots \ y_n]^T \neq 0$ in $\mathbb{R}^n$. If $y_k \neq 0$ let $C_k = y_k^{-1} Y$, and let $C_j = 0$ if $j \neq k$. If $A = [C_1 \ C_2 \ \cdots \ C_n]$, then $T(A) = AY = y_1 C_1 + \cdots + y_k C_k + \cdots + y_n C_n = Y$. This shows that T is onto, as required.

28. **(b)** By Theorem 4 §6.4, let $\{\mathbf{u}_1, \dots, \mathbf{u}_m, \dots, \mathbf{u}_n\}$ be a basis of V where $\{\mathbf{u}_1, \dots, \mathbf{u}_m\}$ is a basis of U. By Theorem 3 §8.1 there is a linear transformation $S : V \to V$ such that $S(\mathbf{u}_i) = \mathbf{u}_i$ for $1 \leq i \leq m$, and $S(\mathbf{u}_i) = \mathbf{0}$ if $i > m$. Because each $\mathbf{u}_i$ is in im S, $U \subseteq$ im S. But if $S(\mathbf{v})$ is in im S, write
$\mathbf{v} = r_1 \mathbf{u}_1 + \cdots + r_m \mathbf{u}_m + \cdots + r_n \mathbf{u}_n$. Then $S(\mathbf{v}) = r_1 S(\mathbf{u}_1) + \cdots + r_m S(\mathbf{u}_m) = r_1 \mathbf{u}_1 + \cdots + r_m \mathbf{u}_m$ is in U. So im $S \subseteq U$.

Exercises 7.3 Isomorphisms and Composition (Page 310)

1. **(b)** T is onto because $T(1, -1, 0) = (1, 0, 0)$, $T(0, 1, -1) = (0, 1, 0)$, and $T(0, 0, 1) = (0, 0, 1)$. Use Theorem 3.

   (d) T is one-to-one because $0 = T(X) = UXV$ implies that $X = 0$ (U and V are invertible). Use Theorem 3.

   (f) T is one-to-one because $\mathbf{0} = T(\mathbf{v}) = k\mathbf{v}$ implies that $\mathbf{v} = \mathbf{0}$ (because $k \neq 0$). T is onto because $T(\frac{1}{k}\mathbf{v}) = \mathbf{v}$ for all $\mathbf{v}$. [Here Theorem 3 does not apply if dim V is not finite.]

   (h) T is one-to-one because $T(A) = 0$ implies $A^T = 0$, whence $A = 0$. Use Theorem 3.

4. **(b)** $ST(x, y, z) = (x + y, 0, y + z)$, $TS(x, y, z) = (x, 0, z)$ **(d)** $ST\begin{bmatrix} a & b \\ c & d \end{bmatrix} = \begin{bmatrix} c & 0 \\ 0 & b \end{bmatrix}$, $TS\begin{bmatrix} a & b \\ c & d \end{bmatrix} = \begin{bmatrix} 0 & a \\ d & 0 \end{bmatrix}$

5. **(b)** $T^2(x, y) = T(x + y, 0) = (x + y, 0) = T(x, y)$. Hence $T^2 = T$.

   (d) $T^2 \begin{bmatrix} a & b \\ c & d \end{bmatrix} = \frac{1}{2} T \begin{bmatrix} a+c & b+d \\ a+c & b+d \end{bmatrix} = \frac{1}{2} \begin{bmatrix} a+c & b+d \\ a+c & b+d \end{bmatrix}$

6. **(b)** No inverse; $(1, -1, 1, -1)$ is in ker T.   **(d)** $T^{-1} \begin{bmatrix} a & b \\ c & d \end{bmatrix} = \frac{1}{5} \begin{bmatrix} 3a - 2c & 3b - 2d \\ a+c & b+d \end{bmatrix}$

   (f) $T^{-1}(a, b, c) = \frac{1}{2}[2a + (b - c)x - (2a - b - c)x^2]$

7. **(b)** $T^2(x, y) = T(ky - x, y) = (ky - (ky - x), y) = (x, y)$ **(d)** $T^2(X) = A^2 X = IX = X$

8. **(b)** $T^3(x, y, z, w) = (x, y, z, -w)$ so $T^6(x, y, z, w) = T^3[T^3(x, y, z, w)] = (x, y, z, w)$. Hence $T^{-1} = T^5$. So $T^{-1}(x, y, z, w) = (y - x, -x, z, -w)$.

9. **(b)** $T^{-1}(A) = U^{-1} A$.

10. **(b)** Given $\mathbf{u}$ in U, write $\mathbf{u} = S(\mathbf{w})$, $\mathbf{w}$ in W (because S is onto). Then write $\mathbf{w} = T(\mathbf{v})$, $\mathbf{v}$ in V (T is onto). Hence $\mathbf{u} = ST(\mathbf{v})$, so ST is onto.

13. **(b)** Given $\mathbf{w}$ in W, write $\mathbf{w} = ST(\mathbf{v})$, $\mathbf{v}$ in V (ST is onto). Then $\mathbf{w} = S[T(\mathbf{v})]$, $T(\mathbf{v})$ in U, so S is onto. But then im $S = W$, so dim $U = \dim(\ker S) + \dim(\text{im } S) \geq \dim(\text{im } S) = \dim W$.

16. $\{T(\mathbf{e}_1), T(\mathbf{e}_2), \dots, T(\mathbf{e}_r)\}$ is a basis of im T by Theorem 5 §7.2. Hence $T : \text{span}\{\mathbf{e}_1, \dots, \mathbf{e}_r\} \to \text{im } T$ is an isomorphism by Theorem 1.

19. **(b)** $T(x, y) = (x, y + 1)$

26. **(b)** If $T(p) = 0$, then $p(x) = -xp'(x)$. We write $p(x) = a_0 + a_1 x + a_2 x^2 + \cdots + a_n x^n$, and this becomes
$$a_0 + a_1 x + a_2 x^2 + \cdots + a_n x^n = -a_1 x - 2a_2 x^2 - \cdots - na_n x^n$$
Equating coefficients yields $a_0 = 0$, $2a_1 = 0$, $3a_2 = 0$, $\dots$, $(n + 1)a_n = 0$, whence $p(x) = 0$. This means that ker $T = 0$, so T is one-to-one. But then T is an isomorphism by Theorem 3.

27. **(b)** If $ST = 1_V$ for some S, then T is onto by Exercise 13. If T is onto, let $\{\mathbf{e}_1, \dots, \mathbf{e}_r, \dots, \mathbf{e}_n\}$ be a basis of V such that $\{\mathbf{e}_{r+1}, \dots, \mathbf{e}_n\}$ is a basis of ker T. Since T is onto, $\{T(\mathbf{e}_1), \dots, T(\mathbf{e}_r)\}$ is a basis of im $T = W$ by Theorem 5 §7.2. Thus $S : W \to V$ given by $S\{T(\mathbf{e}_i)\} = \mathbf{e}_i$ for $i = 1, 2, \dots, r$ is an isomorphism. Hence $TS[T(\mathbf{e}_i)] = T(\mathbf{e}_i)$ for each i, that is $TS[T(\mathbf{e}_i)] = 1_W[T(\mathbf{e}_i)]$. This means that $TS = 1_W$ because they agree on the basis $\{T(\mathbf{e}_1), \dots, T(\mathbf{e}_r)\}$ of W.

29. Let $B = \{\mathbf{e}_1, \dots, \mathbf{e}_r, \mathbf{e}_{r+1}, \dots, \mathbf{e}_n\}$ be a basis of V with $\{\mathbf{e}_{r+1}, \dots, \mathbf{e}_n\}$ a basis of ker T. If $\{T(\mathbf{e}_1), \dots, T(\mathbf{e}_r), \mathbf{w}_{r+1}, \dots, \mathbf{w}_n\}$ is a basis of V, define S by $S[T(\mathbf{e}_i)] = \mathbf{e}_i$ for $1 \leq i \leq r$, and $S(\mathbf{w}_j) = \mathbf{e}_j$ for

$r + 1 \le j \le n$. Then S is an isomorphism by Theorem 1, and $TST(\mathbf{e}_i) = T(\mathbf{e}_i)$ clearly holds for $1 \le i \le r$. But if $i \ge r + 1$, then $T(\mathbf{e}_i) = \mathbf{0} = TST(\mathbf{e}_i)$, so $T = TST$ by Theorem 2 §8.1.

Exercises 7.4 More on Linear Recurrences (Page 319)

1. **(b)** $\{[1], [2^n], [(-3)^n]\}$; $x_n = \frac{1}{20}(15 + 2^{n+3} + (-3)^{n+1})$

2. **(b)** $\{[1], [n], [(-2)^n]\}$; $x_n = \frac{1}{9}(5 - 6n + (-2)^{n+2})$ **(d)** $\{[1], [n], [n^2]\}$; $x_n = 2(n-1)^2 - 1$

3. **(b)** $\{[a^n], [b^n]\}$ 4. **(b)** $[1, 0, 0, 0, 0, \ldots), [0, 1, 0, 0, 0, \ldots), [0, 0, 1, 1, 1, \ldots), [0, 0, 1, 2, 3, \ldots)$

7. By Remark 2,

$$[i^n + (-i)^n] = [2, 0, -2, 0, 2, 0, -2, 0, \ldots)$$

$$[i(i^n - (-i)^n)] = [0, -2, 0, 2, 0, -2, 0, 2, \ldots)$$

are solutions. They are linearly independent and so are a basis.

Exercises 8.1 Orthogonal Complements and Projections (Page 326)

1. **(b)** $\{[2 \ 1], \frac{3}{5}[-1 \ 2]\}$ **(d)** $\{[0 \ 1 \ 1], [1 \ 0 \ 0], \frac{1}{2}[0 \ -2 \ 2]\}$

2. **(b)** $X = \frac{1}{182}[271 \ -221 \ 1030] + \frac{1}{182}[93 \ 403 \ 62]$ **(d)** $X = \frac{1}{4}[1 \ 7 \ 11 \ 17] + \frac{1}{4}[7 \ -7 \ -7 \ 7]$

   (f) $X = \frac{1}{12}[5a - 5b + c - 3d, \ -5a + 5b - c + 3d, \ a - b + 11c + 3d, \ -3a + 3b + 3c + 3d]$
   $\qquad + \frac{1}{12}[7a + 5b - c + 3d, \ 5a + 7b + c - 3d, \ -a + b + c - 3d, \ 3a - 3b - 3c + 9d]$

3. **(a)** $\frac{1}{10}[-9 \ 3 \ -21 \ 33] = \frac{3}{10}[-3 \ 1 \ -7 \ 11]$ **(c)** $\frac{1}{70}[-63 \ 21 \ -147 \ 231] = \frac{3}{10}[-3 \ 1 \ -7 \ 11]$

4. **(b)** $\{[1 \ -1 \ 0], \frac{1}{2}[-1 \ -1 \ 2]\}$; $\text{proj}_U(X) = [1 \ 0 \ -1]$

   (d) $\{[1 \ -1 \ 0 \ 1], [1 \ 1 \ 0 \ 0], \frac{1}{3}[-1 \ 1 \ 0 \ 2]\}$; $\text{proj}_U(X) = [2 \ 0 \ 0 \ 1]$

5. **(b)** $U^{\perp} = \text{span}\{[1 \ 3 \ 1 \ 0], [-1 \ 0 \ 0 \ 1]\}$

6. If X is orthogonal to U, it is orthogonal to each X_i because X_i is in U. Conversely, if $X \cdot X_i = 0$ for each i and $Y = r_1 X_1 + \cdots + r_m X_m$ is in U, then $X \cdot Y = r_1(X \cdot X_1) + \cdots + r_m(X \cdot X_m) = 0$.

10. Let $\{E_1, E_2, \ldots, E_m\}$ be an orthonormal basis of U. If X is in U the expansion theorem gives $X = (X \cdot E_1)E_1 + (X \cdot E_2)E_2 + \cdots + (X \cdot E_m)E_m = \text{proj}_U(X)$.

14. Let $\{Y_1, Y_2, \ldots, Y_m\}$ be a basis of $U^{\perp}$, and let A be the $n \times n$ matrix with rows $Y_1^{\perp}, Y_2^{\perp}, \ldots, Y_m^{\perp}, 0, \ldots 0$. Then $AX = 0$ if and only if $Y_i \cdot X = 0$ for each $i = 1, 2, \ldots, m$; if and only if X is in $U^{\perp\perp} = U$.

16. **(d)** $E^T = A^T[(AA^T)^{-1}]^T(A^T)^T = A^T[(AA^T)^T]^{-1}A = A^T[AA^T]^{-1}A = E$
    $E^2 = A^T(AA^T)^{-1}AA^T(AA^T)^{-1}A = A^T(AA^T)^{-1}A = E$

Exercises 8.2 Orthogonal Diagonalization (Page 333)

1. **(b)** $\frac{1}{5}\begin{bmatrix} 3 & -4 \\ 4 & 3 \end{bmatrix}$ **(d)** $\frac{1}{\sqrt{a^2+b^2}}\begin{bmatrix} a & b \\ -b & a \end{bmatrix}$ **(f)** $\begin{bmatrix} \frac{2}{\sqrt{6}} & \frac{1}{\sqrt{6}} & -\frac{1}{\sqrt{6}} \\ \frac{1}{\sqrt{3}} & -\frac{1}{\sqrt{3}} & \frac{1}{\sqrt{3}} \\ 0 & \frac{1}{\sqrt{2}} & \frac{1}{\sqrt{2}} \end{bmatrix}$ **(h)** $\frac{1}{7}\begin{bmatrix} 2 & 6 & -3 \\ 3 & 2 & 6 \\ -6 & 3 & 2 \end{bmatrix}$

2. We have $P^T = P^{-1}$; this matrix is lower triangular (left side) and also upper triangular (right side—see Lemma 1 §2.6), and so is diagonal. But then $P = P^T = P^{-1}$, so $P^2 = I$. This implies that the diagonal entries of P are all ± 1.

5. **(b)** $\frac{1}{\sqrt{2}}\begin{bmatrix} 0 & 1 & 1 \\ \sqrt{2} & 0 & 0 \\ 0 & 1 & -1 \end{bmatrix}$ **(d)** $\frac{1}{3\sqrt{2}}\begin{bmatrix} 2\sqrt{2} & 3 & 1 \\ \sqrt{2} & 0 & -4 \\ 2\sqrt{2} & -3 & 1 \end{bmatrix}$ or $\frac{1}{3}\begin{bmatrix} 2 & -2 & 1 \\ 1 & 2 & 2 \\ 2 & 1 & -2 \end{bmatrix}$ **(f)** $\frac{1}{2}\begin{bmatrix} 1 & -1 & \sqrt{2} & 0 \\ -1 & 1 & \sqrt{2} & 0 \\ -1 & -1 & 0 & \sqrt{2} \\ 1 & 1 & 0 & \sqrt{2} \end{bmatrix}$

6. $P = \frac{1}{\sqrt{2k}}\begin{bmatrix} c\sqrt{2} & a & a \\ 0 & k & -k \\ -a\sqrt{2} & c & c \end{bmatrix}$

10. **(c)⇒(a).** By Theorem 1 let $P^{-1}AP = D = \text{diag}(\lambda_1, \ldots, \lambda_n)$ where the λ_i are the eigenvalues of A. By **(c)** we have $\lambda_i = \pm 1$ for each i, whence $D^2 = I$. But then $A^2 = (PDP^{-1})^2 = D^2 = I$. Since A is symmetric this is $AA^T = I$, proving **(a)**.

12. **(b)** If $B = P^T A P$, $P^T = P^{-1}$, then $B^2 = P^T A P P^T A P = P^T A^2 P$.

14. If X and Y are respectively columns i and j of I_n, then $X^T A^T Y = X^T A Y$ shows that the (i, j)-entries of A^T and A are equal.

17. **(b)** $\det \begin{bmatrix} \cos\theta & -\sin\theta \\ \sin\theta & \cos\theta \end{bmatrix} = 1$ and $\det \begin{bmatrix} \cos\theta & \sin\theta \\ \sin\theta & -\cos\theta \end{bmatrix} = -1$ [*Remark:* These are the *only* 2×2 examples.]

20. We have $AA^T = D$, where D is diagonal with main diagonal entries $\|R_1\|^2, \ldots, \|R_n\|^2$. Hence $A^{-1} = A^T D^{-1}$, and the result follows because D^{-1} has diagonal entries $1/\|R_1\|^2, \ldots, 1/\|R_n\|^2$.

22. **(b)** Because $I - A$ and $I + A$ commute,
$$PP^T = (I - A)(I + A)^{-1}[(I + A)^{-1}]^T(I - A)^T = (I - A)(I + A)^{-1}(I - A)^{-1}\,(I + A) = I.$$

Exercises 8.3 Positive Definite Matrices (Page 339)

1. **(b)** $U = \frac{\sqrt{2}}{2}\begin{bmatrix} 2 & -1 \\ 0 & 1 \end{bmatrix}$ **(d)** $U = \frac{1}{30}\begin{bmatrix} 60\sqrt{5} & 12\sqrt{5} & 15\sqrt{5} \\ 0 & 6\sqrt{30} & 10\sqrt{30} \\ 0 & 0 & 5\sqrt{15} \end{bmatrix}$

6. Let $X \neq 0$ in $\mathbb{R}^n$. Then $X^T(U^T A U)X = (UX)^T A(UX) > 0$ provided $UX \neq 0$. But if $U = [C_1 \; C_2 \; \cdots \; C_n]$ and $X = [x_1 \; x_2 \; \cdots \; x_n]^T$, then $UX = x_1 C_1 + x_2 C_2 + \cdots + x_n C_n \neq 0$ because $X \neq 0$ and the C_i are independent.

10. Let $P^T A P = D = \text{diag}(\lambda_1, \ldots, \lambda_n)$ where $P^T = P$. Since A is positive definite, each eigenvalue $\lambda_i > 0$. If $B = \text{diag}(\sqrt{\lambda_1}, \ldots, \sqrt{\lambda_n})$ then $B^2 = D$, so $A = PB^2P^T = (PBP^T)^2$. Take $C = PBP^T$. Since C has eigenvalues $\sqrt{\lambda_i} > 0$, it is positive definite.

12. **(b)** If A is positive definite, use Theorem 1 to write $A = U^T U$ where U is upper triangular with positive diagonal D. Then $A = (D^{-1}U)^T D^2 (D^{-1}U)$ so $A = L_1 D_1 U_1$ is such a factorization if $U_1 = D^{-1}U$, $D_1 = D^2$ and $L_1 = U_1^T$. Conversely, let $A^T = A = LDU$ be such a factorization. Then $U^T D^T L^T = A^T = A = LDU$, so $L = U^T$ by **(a)**. Hence $A = LDL^T = V^T V$ where $V = LD_0$ and D_0 is diagonal with $D_0^2 = D$ (the matrix D_0 exists because D has positive diagonal entries). Hence A is symmetric, and it is positive definite by Example 1.

Exercises 8.4 QR-Factorization (Page 343)

1. **(b)** $Q = \frac{1}{\sqrt{5}}\begin{bmatrix} 2 & -1 \\ 1 & 2 \end{bmatrix}$, $R = \frac{1}{\sqrt{5}}\begin{bmatrix} 5 & 3 \\ 0 & 1 \end{bmatrix}$ **(d)** $Q = \frac{1}{\sqrt{3}}\begin{bmatrix} 1 & 1 & 0 \\ -1 & 0 & 1 \\ 0 & 1 & 1 \\ 1 & -1 & 1 \end{bmatrix}$, $R = \frac{1}{\sqrt{3}}\begin{bmatrix} 3 & 0 & -1 \\ 0 & 3 & 1 \\ 0 & 0 & 2 \end{bmatrix}$

Exercises 8.5 Computing Eigenvalues (Page 346)

1. **(b)** Eigenvalues $4, -1$; eigenvectors $\begin{bmatrix} 2 \\ -1 \end{bmatrix}$, $\begin{bmatrix} 1 \\ -3 \end{bmatrix}$; $X_4 = \begin{bmatrix} 409 \\ -203 \end{bmatrix}$; $r_3 = 3.94$

   (d) Eigenvalues $\lambda_1 = \frac{1}{2}(3 + \sqrt{13})$, $\lambda_2 = \frac{1}{2}(3 - \sqrt{13})$; eigenvectors $\begin{bmatrix} \lambda_1 \\ 1 \end{bmatrix}$, $\begin{bmatrix} \lambda_2 \\ 1 \end{bmatrix}$; $X_4 = \begin{bmatrix} 142 \\ 43 \end{bmatrix}$; $r_3 = 3.3027750$

   (The true value is $\lambda_1 = 3.3027756$, to seven decimal places.)

2. **(b)** Eigenvalues $\lambda_1 = \frac{1}{2}(3 + \sqrt{13}) = 3.302776$, $\lambda_2 = \frac{1}{2}(3 - \sqrt{13}) = -0.302776$
$$A_1 = \begin{bmatrix} 3 & 1 \\ 1 & 0 \end{bmatrix}, \quad Q_1 = \frac{1}{\sqrt{10}}\begin{bmatrix} 3 & -1 \\ 1 & 3 \end{bmatrix}, \quad R_1 = \frac{1}{\sqrt{10}}\begin{bmatrix} 10 & 3 \\ 0 & -1 \end{bmatrix}$$
$$A_2 = \frac{1}{10}\begin{bmatrix} 33 & -1 \\ -1 & -3 \end{bmatrix}, \quad Q_2 = \frac{1}{\sqrt{1090}}\begin{bmatrix} 33 & 1 \\ -1 & 33 \end{bmatrix}, \quad R_2 = \frac{1}{\sqrt{1090}}\begin{bmatrix} 109 & -3 \\ 0 & -10 \end{bmatrix}$$
$$A_3 = \frac{1}{109}\begin{bmatrix} 360 & 1 \\ 1 & -33 \end{bmatrix} = \begin{bmatrix} 3.302775 & 0.009174 \\ 0.009174 & -0.302775 \end{bmatrix}$$

4. Use induction on k. If $k = 1$, $A_1 = A$. In general $A_{k+1} = Q_k^{-1}A_kQ_k = Q_k^TA_kQ_k$, so the fact that $A_k^T = A_k$
implies $A_{k+1}^T = A_{k+1}$. The eigenvalues of A are all real (Theorem 7 §5.3), so the A_k converge to an upper
triangular matrix T. But T must also be symmetric (it is the limit of symmetric matrices), so it is
diagonal.

Exercises 8.6   Complex Matrices (Page 355)

1. (b) $\sqrt{6}$ **(d)** $\sqrt{13}$ **2. (b)** Not orthogonal   **(d)** Orthogonal

3. (b) Not a subspace. For example, $i(0, 0, 1) = (0, 0, i)$ is not in U.   **(d)** This is a subspace.

4. (b) Basis $\{(i, 0, 2), (1, 0, -1)\}$; dimension 2   **(d)** Basis $\{(1, 0, -2i), (0, 1, 1, -i)\}$; dimension 2

5. (b) Normal only   **(d)** Hermitian (and normal), not unitary   **(f)** None   **(h)** Unitary (and normal); Hermitian
if z is real

8. (b) $U = \frac{1}{\sqrt{14}}\begin{bmatrix} -2 & 3-i \\ 3+i & 2 \end{bmatrix}$, $U^*ZU = \begin{bmatrix} -1 & 0 \\ 0 & 6 \end{bmatrix}$ **(d)** $U = \frac{1}{\sqrt{3}}\begin{bmatrix} 1+i & 1 \\ -1 & 1-i \end{bmatrix}$, $U^*ZU = \begin{bmatrix} 1 & 0 \\ 0 & 4 \end{bmatrix}$

(f) $U = \frac{1}{\sqrt{3}}\begin{bmatrix} \sqrt{3} & 0 & 0 \\ 0 & 1+i & 1 \\ 0 & -1 & 1-i \end{bmatrix}$, $U^*ZU = \begin{bmatrix} 1 & 0 & 0 \\ 0 & 0 & 0 \\ 0 & 0 & 3 \end{bmatrix}$

10. (b) $\|\lambda Z\|^2 = \langle \lambda Z, \lambda Z \rangle = \lambda\bar{\lambda}\langle Z, Z \rangle = |\lambda|^2\|Z\|^2$

11. (b) If the (k, k)-entry of Z is z_{kk}, then the (k, k)-entry of $\bar{Z}$ is $\bar{z}_{kk}$, so the (k, k)-entry of $Z^* = (\bar{Z})^T$ is $\bar{z}_{kk}$.
As $Z = Z^*$ we have $z_{kk} = \bar{z}_{kk}$, so z_{kk} is real.

14. (b) $(S^2)^* = S^*S^* = (-S)(-S) = S^2$; $(iS)^* = \bar{i}S^* = (-i)(-S) = iS$

(d) If $Z = H + S$, as given, then $Z^* = H^* + S^* = H - S$, so $H = \frac{1}{2}(Z + Z^*)$ and $S = \frac{1}{2}(Z - Z^*)$. Hence the
representation is unique if it exists. But, always, $Z = \frac{1}{2}(Z + Z^*) + \frac{1}{2}(Z - Z^*)$, and these are hermitian and
skew-hermitian, respectively.

16. (b) If $U^{-1} = U^*$, then $(U^{-1})^* = (U^*)^* = U = (U^{-1})^{-1}$.

18. (b) $H = \begin{bmatrix} 1 & i \\ -i & 0 \end{bmatrix}$ is hermitian but $iH = \begin{bmatrix} i & -1 \\ 1 & 0 \end{bmatrix}$ is not.

21. (b) Let $U = \begin{bmatrix} a & b \\ c & d \end{bmatrix}$ be real and invertible, and assume that $U^{-1}AU = \begin{bmatrix} \lambda & \mu \\ 0 & \nu \end{bmatrix}$. Then $AU = U\begin{bmatrix} \lambda & \mu \\ 0 & \nu \end{bmatrix}$, and first
column entries are $c = a\lambda$ and $-a = c\lambda$. Hence λ is real (c and a are both real and are not both 0), and
$(1 + \lambda^2)a = 0$. Thus $a = 0$, $c = a\lambda = 0$, a contradiction.

Exercises 8.7   Best Approximation and Least Squares (Page 363)

1. (b) $\frac{1}{36}\begin{bmatrix} -60 \\ 138 \\ 285 \end{bmatrix}$, $(A^TA)^{-1} = \frac{1}{36}\begin{bmatrix} 24 & -30 & -54 \\ -30 & 42 & 72 \\ -54 & 72 & 129 \end{bmatrix}$

2. (b) $\frac{1}{92}[-46x + 66x^2 + 60(2^x)]$, $(M^TM)^{-1} = \frac{1}{92}\begin{bmatrix} 230 & 0 & -92 \\ 0 & 34 & -36 \\ -92 & -36 & 76 \end{bmatrix}$

3. (b) $3 + \frac{1}{4}x^2 - \frac{7}{2}\sin(\frac{\pi x}{2})$, $(M^TM)^{-1} = \frac{1}{2}\begin{bmatrix} -24 & 7 & 35 \\ 7 & -2 & -10 \\ 35 & -10 & -49 \end{bmatrix}$

4. $y = -5.19 + 0.34x_1 + 0.51x_2 + 0.71x_3$ $(A^TA)^{-1} = \frac{1}{50160}\begin{bmatrix} 1035720 & -16032 & 10080 & -45300 \\ -16032 & 416 & -632 & 800 \\ 10080 & -632 & 2600 & -2180 \\ -45300 & 800 & -2180 & 3950 \end{bmatrix}$

6. (b) It suffices to show that the columns of $M = \begin{bmatrix} 1 & e^{x_1} \\ \vdots & \vdots \\ 1 & e^{x_n} \end{bmatrix}$ are independent. If $r_0\begin{bmatrix} 1 \\ \vdots \\ 1 \end{bmatrix} + r_1\begin{bmatrix} e^{x_1} \\ \vdots \\ e^{x_n} \end{bmatrix} = \begin{bmatrix} 0 \\ \vdots \\ 0 \end{bmatrix}$ then

$r_0 + r_1e^{x_i} - 0$ for each i. Thus $r_1(e^{x_i} - e^{x_j}) = 0$ for all i and j. Hence $r_1 = 0$ because two x_i are distinct. Then $r_0 = r_0 + r_1e^{x_i} = 0$ too.

Section 8.8 Finite Fields and Linear Codes (Page 378)

2. (b) If $2a = 0$ in $\mathbb{Z}_{10}$, then $2a = 10k$ for some integer k. Thus $a = 5k$.

6. (b) $\det A = 15 - 24 = 1 + 4 = 5 \neq 0$ in $\mathbb{Z}_7$, so A^{-1} exists. Since $5^{-1} = 3$ in $\mathbb{Z}_7$, we have

$$A^{-1} = 3\begin{bmatrix} 3 & -6 \\ 3 & 5 \end{bmatrix} = 3\begin{bmatrix} 3 & 1 \\ 3 & 5 \end{bmatrix} = \begin{bmatrix} 2 & 3 \\ 2 & 1 \end{bmatrix}.$$

7. (b) We have $5 \cdot 3 = 1$ in $\mathbb{Z}_7$ so the reduction of the augmented matrix is:

$$\begin{bmatrix} 3 & 1 & 4 & 3 \\ 4 & 3 & 1 & 1 \end{bmatrix} \rightarrow \begin{bmatrix} 1 & 5 & 6 & 1 \\ 4 & 3 & 1 & 1 \end{bmatrix} \rightarrow \begin{bmatrix} 1 & 5 & 6 & 1 \\ 0 & 4 & 5 & 4 \end{bmatrix} \rightarrow \begin{bmatrix} 1 & 5 & 6 & 1 \\ 0 & 1 & 3 & 1 \end{bmatrix} \rightarrow \begin{bmatrix} 1 & 0 & 5 & 3 \\ 0 & 1 & 3 & 1 \end{bmatrix}.$$

Hence $x = 3 + 2t$, $y = 1 + 4t$, $z = t$; t in $\mathbb{Z}_7$.

9. (b) $(1 + t)^{-1} = 2 + t$.

10. (b) The minimum weight of C is 5, so it detects 4 errors and corrects 2 errors.

11. (b) $\{00000, 01110, 10011, 11101\}$.

12. (b) The code is
$\{0000000000, 1001111000, 0101100110, 0011010111, 1100011110, 1010101111, 0110110001, 1111001001\}$.
This has minimum distance 5 and so corrects 2 errors.

13. (b) $\{00000, 10110, 01101, 11011\}$ is a $(5,2)$-code of minimal weight 3, so it corrects single errors.

14. (b) $G = [1 \; \mathbf{u}]$ where $\mathbf{u}$ is any nonzero vector in the code. $H = \begin{bmatrix} \mathbf{u} \\ I_{n-1} \end{bmatrix}$.

Exercises 8.9 An Application to Quadratic Forms (Page 387)

1. (b) $A = \begin{bmatrix} 1 & 0 \\ 0 & 2 \end{bmatrix}$ **(d)** $A = \begin{bmatrix} 1 & 3 & 2 \\ 3 & 1 & -1 \\ 2 & -1 & 3 \end{bmatrix}$ **2. (b)** $P = \frac{1}{\sqrt{2}}\begin{bmatrix} 1 & 1 \\ 1 & -1 \end{bmatrix}$; $Y = \frac{1}{\sqrt{2}}\begin{bmatrix} x_1 + x_2 \\ x_1 - x_2 \end{bmatrix}$; $q = 3y_1^2 - y_2^2$; 1, 2

(d) $P = \frac{1}{3}\begin{bmatrix} 2 & 2 & -1 \\ 2 & -1 & 2 \\ -1 & 2 & 2 \end{bmatrix}$; $Y = \frac{1}{3}\begin{bmatrix} 2x_1 + 2x_2 - x_3 \\ 2x_1 - x_2 + 2x_3 \\ -x_1 + 2x_2 + 2x_3 \end{bmatrix}$; $q = 9y_1^2 + 9y_2^2 - 9y_3^2$; 2, 3

(f) $P = \frac{1}{3}\begin{bmatrix} -2 & 1 & 2 \\ 2 & 2 & 1 \\ 1 & -2 & 2 \end{bmatrix}$; $Y = \frac{1}{3}\begin{bmatrix} -2x_1 + 2x_2 + x_3 \\ x_1 + 2x_2 - 2x_3 \\ 2x_1 + x_2 + 2x_3 \end{bmatrix}$; $q = 9y_1^2 + 9y_2^2$; 2, 2

(h) $P = \frac{1}{\sqrt{6}}\begin{bmatrix} -\sqrt{2} & \sqrt{3} & 1 \\ \sqrt{2} & 0 & 2 \\ \sqrt{2} & \sqrt{3} & -1 \end{bmatrix}$; $Y = \frac{1}{\sqrt{6}}\begin{bmatrix} -\sqrt{2}x_1 + \sqrt{2}x_2 + \sqrt{2}x_3 \\ \sqrt{3}x_1 + \sqrt{3}x_3 \\ x_1 + 2x_2 - x_3 \end{bmatrix}$; $q = 2y_1^2 + y_2^2 - y_3^2$; 2, 3

3. (b) $x_1 = \frac{1}{\sqrt{5}}(2x - y)$, $y_1 = \frac{1}{\sqrt{5}}(x + 2y)$; $4x_1^2 - y_1^2 = 2$; hyperbola

(d) $x_1 = \frac{1}{\sqrt{5}}(x + 2y)$, $y_1 = \frac{1}{\sqrt{5}}(2x - y)$; $6x_1^2 + y_1^2 = 1$; ellipse

7. (b) $3y_1^2 + 5y_2^2 - y_3^2 - 3\sqrt{2}y_1 + \frac{11}{3}\sqrt{3}y_2 + \frac{2}{3}\sqrt{6}y_3 = 7$

$y_1 = \frac{1}{\sqrt{2}}(x_2 + x_3)$, $y_2 = \frac{1}{\sqrt{3}}(x_1 + x_2 - x_3)$, $y_3 = \frac{1}{\sqrt{6}}(2x_1 - x_2 + x_3)$

Exercises 8.10 An Application to Systems of Differential Equations (Page 393)

1. (b) $c_1\begin{bmatrix} 1 \\ 1 \end{bmatrix}e^{4x} + c_2\begin{bmatrix} 5 \\ -1 \end{bmatrix}e^{-2x}$; $c_1 = -\frac{2}{3}$, $c_2 = \frac{1}{3}$

(d) $c_1\begin{bmatrix} -8 \\ 10 \\ 7 \end{bmatrix}e^{-x} + c_2\begin{bmatrix} 1 \\ -2 \\ 1 \end{bmatrix}e^{2x} + c_3\begin{bmatrix} 1 \\ 0 \\ 1 \end{bmatrix}e^{4x}$; $c_1 = 0$, $c_2 = -\frac{1}{2}$, $c_3 = \frac{3}{2}$

Exercises 9.1 The Matrix of a Linear Transformation (Page 400)

1. **(b)** $\begin{bmatrix} a \\ 2b-c \\ c-b \end{bmatrix}$ **(d)** $\frac{1}{2}\begin{bmatrix} a-b \\ a+b \\ -a+3b+2c \end{bmatrix}$

2. **(b)** Let $\mathbf{v} = a + bx + cx^2$. Then

$$C_D[T(\mathbf{v})] = M_{DB}(T)C_B(\mathbf{v}) = \begin{bmatrix} 2 & 1 & 3 \\ -1 & 0 & -2 \end{bmatrix}\begin{bmatrix} a \\ b \\ c \end{bmatrix} = \begin{bmatrix} 2a+b+3c \\ -a-2c \end{bmatrix}$$

Hence $T(\mathbf{v}) = (2a + b + 3c)(1, 1) + (-a - 2c)(0, 1) = (2a + b + 3c, a + b + c)$.

3. **(b)** $\begin{bmatrix} 1 & 0 & 0 & 0 \\ 0 & 0 & 1 & 0 \\ 0 & 1 & 0 & 0 \\ 0 & 0 & 0 & 1 \end{bmatrix}$ **(d)** $\begin{bmatrix} 1 & 1 & 1 \\ 0 & 1 & 2 \\ 0 & 0 & 1 \end{bmatrix}$ **4.** **(b)** $\begin{bmatrix} 1 & 2 \\ 5 & 3 \\ 4 & 0 \\ 1 & 1 \end{bmatrix}$; $C_D[T(a, b)] = \begin{bmatrix} 1 & 2 \\ 5 & 3 \\ 4 & 0 \\ 1 & 1 \end{bmatrix}\begin{bmatrix} b \\ a-b \end{bmatrix} = \begin{bmatrix} 2a-b \\ 3a+2b \\ 4b \\ a \end{bmatrix}$

   (d) $\frac{1}{2}\begin{bmatrix} 1 & 1 & -1 \\ 1 & 1 & 1 \end{bmatrix}$; $C_D[T(a + bx + cx^2)] = \frac{1}{2}\begin{bmatrix} 1 & 1 & -1 \\ 1 & 1 & 1 \end{bmatrix}\begin{bmatrix} a \\ b \\ c \end{bmatrix} = \frac{1}{2}\begin{bmatrix} a+b-c \\ a+b+c \end{bmatrix}$

   (f) $\begin{bmatrix} 1 & 0 & 0 & 0 \\ 0 & 1 & 1 & 0 \\ 0 & 1 & 1 & 0 \\ 0 & 0 & 0 & 1 \end{bmatrix}$; $C_D\left(T\begin{bmatrix} a & b \\ c & d \end{bmatrix}\right) = \begin{bmatrix} 1 & 0 & 0 & 0 \\ 0 & 1 & 1 & 0 \\ 0 & 1 & 1 & 0 \\ 0 & 0 & 0 & 1 \end{bmatrix}\begin{bmatrix} a \\ b \\ c \\ d \end{bmatrix} = \begin{bmatrix} a \\ b+c \\ b+c \\ d \end{bmatrix}$

5. **(b)** $M_{ED}(S)M_{DB}(T) = \begin{bmatrix} 1 & 1 & 0 & 0 \\ 0 & 0 & 1 & -1 \end{bmatrix}\begin{bmatrix} 1 & 1 & 0 \\ 0 & 1 & 1 \\ 1 & 0 & 1 \\ -1 & 1 & 0 \end{bmatrix} = \begin{bmatrix} 1 & 2 & 1 \\ 2 & -1 & 1 \end{bmatrix} = M_{EB}(ST)$

   (d) $M_{ED}(S)M_{DB}(T) = \begin{bmatrix} 1 & -1 & 0 \\ 0 & 0 & 1 \end{bmatrix}\begin{bmatrix} 1 & -1 & 0 \\ -1 & 0 & 1 \\ 0 & 1 & 0 \end{bmatrix} = \begin{bmatrix} 2 & -1 & -1 \\ 0 & 1 & 0 \end{bmatrix} = M_{EB}(ST)$

7. **(b)** $T^{-1}(a, b, c) = \frac{1}{2}(b + c - a, a + c - b, a + b - c)$ $M_{DB}(T) = \begin{bmatrix} 0 & 1 & 1 \\ 1 & 0 & 1 \\ 1 & 1 & 0 \end{bmatrix}$; $M_{BD}(T^{-1}) = \frac{1}{2}\begin{bmatrix} -1 & 1 & 1 \\ 1 & -1 & 1 \\ 1 & 1 & -1 \end{bmatrix}$

   (d) $T^{-1}(a, b, c) = (a - b) + (b - c)x + cx^2$ $M_{DB}(T) = \begin{bmatrix} 1 & 1 & 1 \\ 0 & 1 & 1 \\ 0 & 0 & 1 \end{bmatrix}$; $M_{BD}(T^{-1}) = \begin{bmatrix} 1 & -1 & 0 \\ 0 & 1 & -1 \\ 0 & 0 & 1 \end{bmatrix}$

8. **(b)** $M_{DB}(T^{-1}) = [M_{BD}(T)]^{-1} = \begin{bmatrix} 1 & 1 & 1 & 0 \\ 0 & 1 & 1 & 0 \\ 0 & 0 & 1 & 0 \\ 0 & 0 & 0 & 1 \end{bmatrix}^{-1} = \begin{bmatrix} 1 & -1 & 0 & 0 \\ 0 & 1 & -1 & 0 \\ 0 & 0 & 1 & 0 \\ 0 & 0 & 0 & 1 \end{bmatrix}$. Hence $C_B[T^{-1}(a, b, c, d)] =$

$M_{BD}(T^{-1})C_D(a, b, c, d) = \begin{bmatrix} 1 & -1 & 0 & 0 \\ 0 & 1 & -1 & 0 \\ 0 & 0 & 1 & 0 \\ 0 & 0 & 0 & 1 \end{bmatrix}\begin{bmatrix} a \\ b \\ c \\ d \end{bmatrix} = \begin{bmatrix} a-b \\ b-c \\ c \\ d \end{bmatrix}$, so $T^{-1}(a, b, c, d) = \begin{bmatrix} a-b & b-c \\ c & d \end{bmatrix}$.

12. Have $C_D[T(\mathbf{e}_j)] = $ column j of I_n. Hence $M_{DB}(T) = [C_D[T(\mathbf{e}_1)] \ \ C_D[T(\mathbf{e}_2)] \cdots C_D[T(\mathbf{e}_n)]] = I_n$.

16. **(b)** If D is the standard basis of $\mathbb{R}^{n+1}$ and $B = \{1, x, x^2, \ldots, x^n\}$, then

$$M_{DB}(T) = [C_D(T(1)) \ [C_D(T(x))\cdots[C_D(T(x^2))] = \begin{bmatrix} 1 & a_0 & a_0^2 & \cdots & a_0^n \\ 1 & a_1 & a_1^2 & \cdots & a_1^n \\ 1 & a_2 & a_2^2 & \cdots & a_2^n \\ \vdots & \vdots & \vdots & & \vdots \\ 1 & a_n & a_n^2 & \cdots & a_n^n \end{bmatrix}.$$

This matrix has nonzero determinant by Theorem 7 §3.2 (since the a_i are distinct), so T is an isomorphism.

20. (d) $[(S + T)R](\mathbf{v}) = (S + T)(R(\mathbf{v})) = S[(R(\mathbf{v}))] = T[(R(\mathbf{v}))] = SR(\mathbf{v}) + ST(\mathbf{v}) = [SR + ST](\mathbf{v})$ holds for all $\mathbf{v}$ in V.
Hence $(S + T)R = SR + TR$.

21. (b) If $\mathbf{w}$ lies in im$(S + T)$, then $\mathbf{w} = (S + T)(\mathbf{v})$ for some $\mathbf{v}$ in V. But then $\mathbf{w} = S(\mathbf{v}) + T(\mathbf{v})$,
so $\mathbf{w}$ lies in im S + im T.

22. (b) If $X \subseteq X_1$, let T lie in X_1^0. Then $T(\mathbf{v}) = \mathbf{0}$ for all $\mathbf{v}$ in X_1, whence $T(\mathbf{v}) = \mathbf{0}$ for all $\mathbf{v}$ in X. Thus T is in X^0 and
we have shown that $X_1^0 \subseteq X^0$.

25. (b) Given $T : \mathbb{R} \to V$, let $T(1) = a_1\mathbf{e}_1 + \cdots + a_n\mathbf{e}_n$, a_i in $\mathbb{R}$. For all r in $\mathbb{R}$, we have

$$(a_1S_1 + \cdots + a_nS_n)(r) = a_1S_1(r) + \cdots + a_nS_n(r) = a_1r\mathbf{e}_1 + \cdots + a_nr\mathbf{e}_n = rT(1) = T(r).$$

This shows that $a_1S_1 + \cdots + a_nS_n = T$.

27. (b) Write $\mathbf{v} = v_1\mathbf{e}_1 + \cdots + v_n\mathbf{e}_n$, v_j in $\mathbb{R}$. Apply E_i to get $E_i(\mathbf{v}) = v_1E_i(\mathbf{e}_1) + \cdots + v_nE_i(\mathbf{e}_n) = v_i$ by the definition
of the E_j.

Exercises 9.2 Operators and Similarity (Page 408)

1. (b) $\frac{1}{2}\begin{bmatrix} -3 & -2 & 1 \\ 2 & 2 & 0 \\ 0 & 0 & 2 \end{bmatrix}$

4. (b) $P_{B \leftarrow D} = \begin{bmatrix} 1 & 1 & -1 \\ 1 & -1 & 0 \\ 1 & 0 & 1 \end{bmatrix}$, $P_{D \leftarrow B} = \frac{1}{3}\begin{bmatrix} 1 & 1 & 1 \\ 1 & -2 & 1 \\ -1 & -1 & 2 \end{bmatrix}$, $P_{E \leftarrow D} = \begin{bmatrix} 1 & 0 & 1 \\ 1 & -1 & 0 \\ 1 & 0 & 0 \end{bmatrix}$, $P_{E \leftarrow B} = \begin{bmatrix} 0 & 0 & 1 \\ 0 & 1 & 0 \\ 1 & 0 & 0 \end{bmatrix}$

5. (b) $A = P_{D \leftarrow B}$, where $B = \{(1, 2, -1), (2, 3, 0), (1, 0, 2)\}$. Hence $A^{-1} = P_{B \leftarrow D} = \begin{bmatrix} 6 & -4 & -3 \\ -4 & 3 & 2 \\ 3 & -2 & -1 \end{bmatrix}$

7. (b) $P = \begin{bmatrix} 1 & 1 & 0 \\ 0 & 1 & 2 \\ -1 & 0 & 1 \end{bmatrix}$ **8. (b)** $B = \left\{ \begin{bmatrix} 3 \\ 7 \end{bmatrix}, \begin{bmatrix} 2 \\ 5 \end{bmatrix} \right\}$

9. (b) $c_T(x) = x^2 - 6x - 1$ **(d)** $c_T(x) = x^3 + x^2 - 8x - 3$ **(f)** $c_T(x) = x^4$

12. Define $T_A : \mathbb{R}^n \to \mathbb{R}^n$ by $T_A(X) - AX$ for all X in $\mathbb{R}^n$. If null $A =$ null B, then $\ker(T_A) =$ null $A =$ null $B = \ker(T_B)$ so,
by Exercise 28 §7.3, $T_A = ST_B$ for some isomorphism $S : \mathbb{R}^n \to \mathbb{R}^n$. If B_0 is the standard basis of $\mathbb{R}^n$, we have
$A = M_{B_0}(T_A) = M_{B_0}(ST_B) = M_{B_0}(S)M_{B_0}(T_B) = UB$ where $U = M_{B_0}(S)$ is invertible by Theorem 1 §9.2.
Conversely, if $A = UB$ with U invertible, then $AX = 0$ if and only $BX = 0$, so null $A =$ null B.

Exercises 9.3 Invariant Subspaces and Direct Sums (Page 418)

3. (b) If $\mathbf{v}$ is in $S(U)$, write $\mathbf{v} = S(\mathbf{u})$, $\mathbf{u}$ in U. Then

$$T(\mathbf{v}) = T[S(\mathbf{u})] = (TS)(\mathbf{u}) = (ST)(\mathbf{u}) = S[T(\mathbf{u})]$$

and this lies in $S(U)$ because $T(\mathbf{u})$ lies in U (U is T-invariant).

6. Suppose U is T-invariant for every T. If $U \neq 0$, choose $\mathbf{u} \neq \mathbf{0}$ in U. Choose a basis $B = \{\mathbf{u}, \mathbf{u}_2, \ldots, \mathbf{u}_n\}$ of V containing $\mathbf{u}$. Given any $\mathbf{v}$ in V, there is (by Theorem 3 §7.1) a linear transformation $T : V \to V$ such that $T(\mathbf{u}) = \mathbf{v}$, $T(\mathbf{u}_2) = \cdots = T(\mathbf{u}_n) = \mathbf{0}$. Then $\mathbf{v} = T(\mathbf{u})$ lies in U because U is T-invariant. This shows that $V = U$.

8. (b) $T(1 - 2x^2) = 3 + 3x - 3x^2 = 3(1 - 2x^2) + 3(x + x^2)$ and $T(x + x^2) = -(1 - 2x^2)$, so both are in U.

Hence U is T-invariant by Example 3. If $B = \{1 - 2x^2, x + x^2, x^2\}$ then $M_B(T) = \begin{bmatrix} 3 & -1 & 1 \\ 3 & 0 & 1 \\ 0 & 0 & 3 \end{bmatrix}$, so

$$c_T(x) = \det\begin{bmatrix} x-3 & 1 & -1 \\ -3 & x & -1 \\ 0 & 0 & x-3 \end{bmatrix} = (x - 3)\det\begin{bmatrix} x-3 & 1 \\ -3 & x \end{bmatrix} = (x - 3)(x^2 - 3x + 3)$$

9. (b) Suppose $\mathbb{R}\mathbf{u}$ is T_A-invariant where $\mathbf{u} \neq \mathbf{0}$. Then $T_A(\mathbf{u}) = r\mathbf{u}$ for some r in $\mathbb{R}$, so $(rI - A)\mathbf{u} = \mathbf{0}$.
But $\det(rI - A) = (r - \cos\theta)^2 + \sin^2\theta \neq 0$ because $0 < \theta < \pi$. Hence $\mathbf{u} = \mathbf{0}$, a contradiction.

10. (b) $U = \text{span }\{(1, 1, 0, 0), (0, 0, 1, 1)\}$ and $W = \text{span }\{(1, 0, 1, 0), (0, 1, 0, -1)\}$, and these four vectors form a basis
of $\mathbb{R}^4$. Use Example 9.

14. The fact that U and W are subspaces is easily verified using the subspace test. If A lies in $U \cap V$, then $A = AE = 0$; that is, $U \cap V = 0$. To show that $\mathbf{M}_{22} = U + V$, choose any A in $\mathbf{M}_{22}$. Then $A = AE + (A - AE)$, and AE lies in U [because $(AE)E = AE^2 = AE$], and $A - AE$ lies in W [because $(A - AE)E = AE - AE^2 = 0$].

18. **(b)** Let $U = \mathbb{R}\mathbf{p}$ be T_A-invariant. Then $T_A(\mathbf{p})$ is in U, say $T_A(\mathbf{p}) = \lambda\mathbf{p}$. Hence $A\mathbf{p} = \lambda\mathbf{p}$ so λ is an eigenvalue of A. This means that $\lambda = 0$ by **(a)**, so $\mathbf{p}$ is in $\ker(T_A)$. Thus $U \subseteq \ker(T_A)$. But $\dim[\ker(T_A)] \neq 2$ because $T_A \neq 0$, so $\dim[\ker(T_A)] = 1 = \dim(U)$. Hence $U = \ker(T_A)$.

20. Let B_1 be a basis of U and extend it to a basis B of V. Then $M_B(T) = \begin{bmatrix} M_{B_1}(T) & Y \\ 0 & Z \end{bmatrix}$ so
$$c_T(x) = \det[xI - M_B(T)] = \det[xI - M_{B_1}(T)]\det[xI - Z] = c_{T_1}(x)q(x).$$

22. **(b)** $T^2[p(x)] = p[-(-x)] = p(x)$, so $T^2 = 1$; $B = \{1, x^2; x, x^3\}$
    (d) $T^2(a, b, c) = T(-a + 2b + c, b + c, -c) = (a, b, c)$, so $T^2 = 1$; $B = \{(1, 1, 0); (1, 0, 0), (0, -1, 2)\}$

25. **(b)** $T^2(a, b, c) = T(a + 2b, 0, 4b + c) = (a + 2b, 0, 4b + c) = T(a, b, c)$, so $T^2 = T$; $B = \{(1, 0, 0), (0, 0, 1); (2, -1, 4)\}$

29. **(b)** $T_{f,\,\mathbf{z}}[T_{f,\,\mathbf{z}}(\mathbf{v})] = T_{f,\,\mathbf{z}}[f(\mathbf{v})\mathbf{z}] = f[f(\mathbf{v})\mathbf{z}]\mathbf{z} = f(\mathbf{v})\{f[\mathbf{z}]\mathbf{z}\} = f(\mathbf{v})f(\mathbf{z})\mathbf{z}$. This equals $T_{f,\,\mathbf{z}}(\mathbf{v}) = f(\mathbf{v})\mathbf{z}$ for all $\mathbf{v}$ if and only if $f(\mathbf{v})f(\mathbf{z}) = f(\mathbf{v})$ for all $\mathbf{v}$. Since $f \neq 0$, this holds if and only if $f(\mathbf{z}) = 1$.

30. **(b)** If $A = [\mathbf{p}_1 \ \mathbf{p}_2 \ \cdots \ \mathbf{p}_n]$ where $U\mathbf{p}_i = \lambda\mathbf{p}_i$ for each i, then $UA = \lambda A$. Conversely, $UA = \lambda A$ means that $U\mathbf{p} = \lambda\mathbf{p}$ for every column $\mathbf{p}$ of A.

Exercises 9.4   Block Triangular Form (Page 427)

1. **(b)** $c_A(x) = (x + 1)^3$; $P = \begin{bmatrix} 1 & 0 & 0 \\ 1 & 1 & 0 \\ 1 & -3 & 1 \end{bmatrix}$; $P^{-1}AP = \begin{bmatrix} -1 & 0 & 1 \\ 0 & -1 & 0 \\ 0 & 0 & -1 \end{bmatrix}$

    (d) $c_A(x) = (x - 1)^2 (x + 2)$; $P = \begin{bmatrix} -1 & 0 & -1 \\ 4 & 1 & 1 \\ 4 & 2 & 1 \end{bmatrix}$; $P^{-1}AP = \begin{bmatrix} 1 & 1 & 0 \\ 0 & 1 & 0 \\ 0 & 0 & -2 \end{bmatrix}$

    (f) $c_A(x) = (x + 1)^2(x - 1)^2$; $P = \begin{bmatrix} 1 & 1 & 5 & 1 \\ 0 & 0 & 2 & -1 \\ 0 & 1 & 2 & 0 \\ 1 & 0 & 1 & 0 \end{bmatrix}$; $P^{-1}AP = \begin{bmatrix} -1 & 1 & 0 & 0 \\ 0 & -1 & 0 & 0 \\ 0 & 0 & 1 & -2 \\ 0 & 0 & 0 & 1 \end{bmatrix}$

4. If B is any ordered basis of V, write $A = M_B(T)$. Then $c_T(x) = c_A(x) = a_0 + a_1x + \cdots + a_nx^n$ for scalars a_i in $\mathbb{R}$. Since M_B is linear and $M_B(T^k) = M_B(T)^k$, we have
$$M_B[c_T(T)] = M_V[a_0 + a_1T + \cdots + a_nT^n] = a_0I + a_1A + \cdots + a_nA^n = c_A(A) = 0$$
by the Cayley–Hamilton theorem. Hence $c_T(T) = 0$ because M_B is one-to-one.

Exercises 9.5   Jordan Canonical Form (Page 432)

2. $\begin{bmatrix} a & 1 & 0 \\ 0 & a & 0 \\ 0 & 0 & b \end{bmatrix}\begin{bmatrix} 0 & 1 & 0 \\ 0 & 0 & 1 \\ 1 & 0 & 0 \end{bmatrix} = \begin{bmatrix} 0 & 1 & 0 \\ 0 & 0 & 1 \\ 1 & 0 & 0 \end{bmatrix}\begin{bmatrix} b & 0 & 0 \\ 0 & a & 1 \\ 0 & 0 & a \end{bmatrix}$

Exercises 10.1   Inner Products and Norms (Page 439)

1. **(b)** P5 fails.   **(d)** P5 fails.   **(f)** P5 fails.   3. **(b)** $\frac{1}{\sqrt{\pi}}f$ **(d)** $\frac{1}{\sqrt{17}}\begin{bmatrix} 3 \\ -1 \end{bmatrix}$ 4. **(b)** $\sqrt{3}$ **(d)** $\sqrt{3\pi}$

8. P1 and P2 are clear since $f(i)$ and $g(i)$ are real numbers.

P3: $\langle f + g, h \rangle = \sum_i (f+g)(i) \cdot h(i) = \sum_i (f(i)+g(i)) \cdot h(i) = \sum_i [f(i)h(i) + g(i)h(i)]$
$$= \sum_i f(i)h(i) + \sum_i g(i)h(i) = \langle f, h \rangle + \langle g, h \rangle.$$

P4: $\langle rf, g \rangle = \sum_i (rf)(i) \cdot g(i) = \sum_i rf(i) \cdot g(i) = r\sum_i f(i) \cdot g(i) = r\langle f, g \rangle.$

P5: If $f \neq 0$, then $\langle f, f \rangle = \sum_i f(i)^2 > 0$ because some $f(i) \neq 0$.

12. (b) $\langle \mathbf{v}, \mathbf{v} \rangle = 5v_1^2 - 6v_1v_2 + 2v_2^2 = \frac{1}{5}[(5v_1 - 3v_2)^2 + v_2^2]$ **(d)** $\langle \mathbf{v}, \mathbf{v} \rangle = 3v_1^2 + 8v_1v_2 + 6v_2^2 = \frac{1}{3}[(3v_1 + 4v_2)^2 + 2v_2^2]$

13. (b) $\begin{bmatrix} 1 & -1 \\ -1 & 2 \end{bmatrix}$ **(d)** $\begin{bmatrix} 1 & 0 & -2 \\ 0 & 2 & 0 \\ -2 & 0 & 5 \end{bmatrix}$ **16. (b)** -15

20. 1. Using P2: $\langle \mathbf{u}, \mathbf{v} + \mathbf{w} \rangle = \langle \mathbf{v} + \mathbf{w}, \mathbf{u} \rangle = \langle \mathbf{v}, \mathbf{u} \rangle + \langle \mathbf{w}, \mathbf{u} \rangle = \langle \mathbf{u}, \mathbf{v} \rangle + \langle \mathbf{u}, \mathbf{w} \rangle.$
2. Using P2 and P4: $\langle \mathbf{v}, r\mathbf{w} \rangle = \langle r\mathbf{w}, \mathbf{v} \rangle = r\langle \mathbf{w}, \mathbf{v} \rangle = r\langle \mathbf{v}, \mathbf{w} \rangle$; the rest is P4.
3. Using P3: $\langle \mathbf{0}, \mathbf{v} \rangle = \langle \mathbf{0} + \mathbf{0}, \mathbf{v} \rangle = \langle \mathbf{0}, \mathbf{v} \rangle + \langle \mathbf{0}, \mathbf{v} \rangle$, so $\langle \mathbf{0}, \mathbf{v} \rangle = 0$. The rest is P2.
4. Assume that $\langle \mathbf{v}, \mathbf{v} \rangle = 0$. If $\mathbf{v} \neq \mathbf{0}$ this contradicts P5, so $\mathbf{v} = \mathbf{0}$. Conversely, if $\mathbf{v} = \mathbf{0}$, then $\langle \mathbf{v}, \mathbf{v} \rangle = 0$ by Part 3 of this theorem. **22. (b)** $15\|\mathbf{u}\|^2 - 17\langle \mathbf{u}, \mathbf{v} \rangle - 4\|\mathbf{v}\|^2$

26. (b) $\{(1, 1, 0), (0, 2, 1)\}$ **28.** $\langle \mathbf{v} - \mathbf{w}, \mathbf{v}_i \rangle = \langle \mathbf{v}, \mathbf{v}_i \rangle - \langle \mathbf{w}, \mathbf{v}_i \rangle = 0$ for each i, so $\mathbf{v} = \mathbf{w}$ by Exercise 27.
29. (b) If $\mathbf{u} = (\cos\theta, \sin\theta)$ in $\mathbb{R}^2$ (with the dot product) then $\|\mathbf{u}\| = 1$. Use **(a)** with $\mathbf{v} = (x, y)$.

Exercises 10.2 Orthogonal Sets of Vectors (Page 447)

1. (b) $\frac{1}{14}\left\{(6a + 2b + 6c)\begin{bmatrix} 1 \\ 1 \\ 1 \end{bmatrix} + (7c - 7a)\begin{bmatrix} -1 \\ 0 \\ 1 \end{bmatrix} + (a - 2b + c)\begin{bmatrix} 1 \\ -6 \\ 1 \end{bmatrix}\right\}$

(d) $\left(\frac{a+d}{2}\right)\begin{bmatrix} 1 & 0 \\ 0 & 1 \end{bmatrix} + \left(\frac{a-d}{2}\right)\begin{bmatrix} 1 & 0 \\ 0 & -1 \end{bmatrix} + \left(\frac{b+c}{2}\right)\begin{bmatrix} 0 & 1 \\ 1 & 0 \end{bmatrix} + \left(\frac{b-c}{2}\right)\begin{bmatrix} 0 & 1 \\ -1 & 0 \end{bmatrix}$

2. (b) $\{(1, 1, 1), (1, -5, 1), (3, 0, -2)\}$ **3. (b)** $\left\{\begin{bmatrix} 1 & 1 \\ 0 & 1 \end{bmatrix}, \begin{bmatrix} 1 & -2 \\ 3 & 1 \end{bmatrix}, \begin{bmatrix} 1 & -2 \\ -2 & 1 \end{bmatrix}, \begin{bmatrix} 1 & 0 \\ 0 & -1 \end{bmatrix}\right\}$

4. (b) $\{1, x-1, x^2 - 2x + \frac{2}{3}\}$ **6. (b)** $U^\perp = \text{span}\{[1\ -1\ 0\ 0], [0\ 0\ 1\ 0], [0\ 0\ 0\ 1]\}$, $\dim U^\perp = 3$, $\dim U = 1$

(d) $U^\perp = \text{span}\{2 - 3x, 1 - 2x^2\}$, $\dim U^\perp = 2$, $\dim U = 1$ **(f)** $U^\perp = \text{span}\left\{\begin{bmatrix} 1 & -1 \\ -1 & 0 \end{bmatrix}\right\}$, $\dim U^\perp = 1$, $\dim U = 3$

7. (b) $U = \text{span}\left\{\begin{bmatrix} 1 & 0 \\ 0 & 1 \end{bmatrix}, \begin{bmatrix} 1 & 1 \\ 1 & -1 \end{bmatrix}, \begin{bmatrix} 0 & 1 \\ -1 & 0 \end{bmatrix}\right\}$; $\text{proj}_U(A) = \begin{bmatrix} 3 & 0 \\ 2 & 1 \end{bmatrix}$ **8. (b)** $U = \text{span}\{1, 5 - 3x^2\}$; $\text{proj}_U(x) = \frac{3}{13}(1 + 2x^2)$

9. (b) $B = \{1, 2x - 1\}$ is an orthogonal basis of U because $\int_0^1 (2x - 1)\,dx = 0$.
Using it, we get $\text{proj}_U(x^2 + 1) = x + \frac{5}{6}$, so $x^2 + 1 = (x + \frac{5}{6}) + (x^2 - x + \frac{1}{6})$.

11. (b) This follows from $\langle \mathbf{v} + \mathbf{w}, \mathbf{v} - \mathbf{w} \rangle = \|\mathbf{v}\|^2 - \|\mathbf{w}\|^2$.

14. (b) $U^\perp \subseteq \{\mathbf{u}_1, \ldots, \mathbf{u}_m\}^\perp$ because each $\mathbf{u}_i$ is in U. Conversely, if $\langle \mathbf{v}, \mathbf{u}_i \rangle = 0$ for each i, and $\mathbf{u} = r_1\mathbf{u}_1 + \cdots + r_m\mathbf{u}_m$ is any vector in U, then $\langle \mathbf{v}, \mathbf{u} \rangle = r_1\langle \mathbf{v}, \mathbf{u}_1 \rangle + \cdots + r_m\langle \mathbf{v}, \mathbf{u}_m \rangle = 0$.

18. (b) $\text{proj}_U(-5, 4, -3) = (-5, 4, -3)$; $\text{proj}_U(-1, 0, 2) = \frac{1}{38}(-17, 24, 73)$

19. (b) The plane is $U = \{\vec{x} \mid \vec{x} \cdot \vec{n} = 0\}$ so $\text{span}\left\{\vec{n} \times \vec{w}, \vec{w} - \frac{\vec{n} \cdot \vec{w}}{\|\vec{n}\|^2}\vec{n}\right\} \subseteq U$. This is equality because both spaces have dimension 2 (using **(a)**).

20. (b) $C_E(\mathbf{b}_i)$ is column i of P. Since $C_E(\mathbf{b}_i) \cdot C_E(\mathbf{b}_j) = \langle \mathbf{b}_i, \mathbf{b}_j \rangle$ by **(a)**, the result follows.

Exercises 10.3 Orthogonal Diagonalization (Page 453)

1. (b) $B = \left\{\begin{bmatrix} 1 & 0 \\ 0 & 0 \end{bmatrix}, \begin{bmatrix} 0 & 1 \\ 0 & 0 \end{bmatrix}, \begin{bmatrix} 0 & 0 \\ 1 & 0 \end{bmatrix}, \begin{bmatrix} 0 & 0 \\ 0 & 1 \end{bmatrix}\right\}$; $M_B(T) = \begin{bmatrix} -1 & 0 & 1 & 0 \\ 0 & -1 & 0 & 1 \\ 1 & 0 & 2 & 0 \\ 0 & 1 & 0 & 2 \end{bmatrix}$

4. (b) $\langle \mathbf{v}, (rT)(\mathbf{w})\rangle = \langle \mathbf{v}, rT(\mathbf{w})\rangle = r\langle \mathbf{v}, T(\mathbf{w})\rangle = r\langle T(\mathbf{v}), \mathbf{w}\rangle = \langle rT(\mathbf{v}), \mathbf{w}\rangle = \langle (rT)(\mathbf{v}), \mathbf{w}\rangle$

(d) Given $\mathbf{v}$ and $\mathbf{w}$, write $T^{-1}(\mathbf{v}) = \mathbf{v}_1$ and $T^{-1}(\mathbf{w}) = \mathbf{w}_1$. Then
$$\langle T^{-1}(\mathbf{v}), \mathbf{w}\rangle = \langle \mathbf{v}_1, T(\mathbf{w}_1)\rangle = \langle T(\mathbf{v}_1), \mathbf{w}_1\rangle = \langle \mathbf{v}, T^{-1}(\mathbf{w})\rangle.$$

5. (b) If $B_0 = \{(1, 0, 0), (0, 1, 0), (0, 0, 1)\}$, then $M_{B_0}(T) = \begin{bmatrix} 7 & -1 & 0 \\ -1 & 7 & 0 \\ 0 & 0 & 2 \end{bmatrix}$ has an orthonormal basis of

eigenvectors $\left\{ \frac{1}{\sqrt{2}}\begin{bmatrix} 1 \\ 1 \\ 0 \end{bmatrix}, \frac{1}{\sqrt{2}}\begin{bmatrix} 1 \\ -1 \\ 0 \end{bmatrix}, \begin{bmatrix} 0 \\ 0 \\ 1 \end{bmatrix} \right\}$. Hence an orthonormal basis of eigenvectors of T is

$\left\{ \frac{1}{\sqrt{2}}(1, 1, 0), \frac{1}{\sqrt{2}}(1, -1, 0), (0, 0, 1) \right\}$.

(d) If $B_0 = \{1, x, x^2\}$, then $M_{B_0}(T) = \begin{bmatrix} -1 & 0 & 1 \\ 0 & 3 & 0 \\ 1 & 0 & -1 \end{bmatrix}$ has an orthonormal basis of eigenvectors

$\left\{ \begin{bmatrix} 0 \\ 1 \\ 0 \end{bmatrix}, \frac{1}{\sqrt{2}}\begin{bmatrix} 1 \\ 0 \\ 1 \end{bmatrix}, \frac{1}{\sqrt{2}}\begin{bmatrix} 1 \\ 0 \\ -1 \end{bmatrix} \right\}$. Hence an orthonormal basis of eigenvectors of T is

$\left\{ x, \frac{1}{\sqrt{2}}(1+x^2), \frac{1}{\sqrt{2}}(1-x^2) \right\}$.

7. (b) $M_B(T) = \begin{bmatrix} A & 0 \\ 0 & A \end{bmatrix}$, so $c_T(x) = \det\begin{bmatrix} xI_2 - A & 0 \\ 0 & xI_2 - A \end{bmatrix} = [c_A(x)]^2$.

12. $(1) \Rightarrow (2)$. If $B = \{\mathbf{e}_1, \ldots, \mathbf{e}_n\}$ is an orthonormal basis of V, then $M_B(T) = [a_{ij}]$ where $a_{ij} = \langle \mathbf{e}_i, T(\mathbf{e}_j)\rangle$ by Theorem 2. If (1) holds, then $a_{ji} = \langle \mathbf{e}_j, T(\mathbf{e}_i)\rangle = -\langle T(\mathbf{e}_j), \mathbf{e}_i\rangle = -\langle \mathbf{e}_i, T(\mathbf{e}_j)\rangle = -a_{ij}$. Hence $[M_V(T)]^T = -M_V(T)$, proving (2).

14. (c) The coefficients in the definition of $T'(\mathbf{e}_j) = \sum_{i=1}^{n} \langle \mathbf{e}_j, T(\mathbf{e}_i)\rangle \mathbf{e}_i$ are the entries in the jth column $C_B[T'(\mathbf{e}_j)]$ of $M_B(T')$. Hence $M_B(T') = [\langle \mathbf{e}_j, T(\mathbf{e}_i)\rangle]$, and this is the transpose of $M_B(T)$ by Theorem 2.

Exercises 10.4   Isometries (Page 466)

2. (b) Rotation through π **(d)** Reflection in the line $y = -x$ **(f)** Rotation through $\frac{\pi}{4}$

3. (b) $c_T(x) = (x-1)(x^2 + \frac{3}{2}x + 1)$. If $E = [1 \ \sqrt{3} \ \sqrt{3}]^T$, then T is a rotation about $\mathbb{R}E$.

(d) $c_T(x) = (x+1)(x+1)^2$. Rotation (of π) about the X axis.

(f) $c_T(x) = (x+1)(x^2 - \sqrt{2}x + 1)$. Rotation (of $-\frac{\pi}{4}$) about the Y axis followed by a reflection in the X-Z plane.

6. If $\|\mathbf{v}\| = \|(aT)(\mathbf{v})\| = |a|\|T(\mathbf{v})\| = |a|\,\|\mathbf{v}\|$ for some $\mathbf{v} \neq \mathbf{0}$, then $|a| = 1$ so $a = \pm 1$.

Exercises 10.5   An Application to Fourier Approximation (Page 470)

1. (b) $\dfrac{\pi}{2} - \dfrac{4}{\pi}\left[\cos x + \dfrac{\cos 3x}{3^2} + \dfrac{\cos 5x}{5^2} \right]$

(d) $\dfrac{\pi}{4} + \left[\sin x - \dfrac{\sin 2x}{2} + \dfrac{\sin 3x}{3} - \dfrac{\sin 4x}{4} + \dfrac{\sin 5x}{5} \right] - \dfrac{2}{\pi}\left[\cos x + \dfrac{\cos 3x}{3^2} + \dfrac{\cos 5x}{5^2} \right]$

2. (b) $\dfrac{2}{\pi} - \dfrac{8}{\pi}\left[\dfrac{\cos 2x}{2^2 - 1} + \dfrac{\cos 4x}{4^2 - 1} + \dfrac{\cos 6x}{6^2 - 1} \right]$

4. $\displaystyle\int \cos kx \cos lx \, dx = \frac{1}{2}\left[\dfrac{\sin[(k+l)x]}{k+l} - \dfrac{\sin[(k-l)x]}{k-l} \right]_0^\pi = 0$ provided that $k \neq l$

Appendix A Complex Numbers (Page 482)

1. **(b)** $x = 3$ **(d)** $x = \pm 1$ **2. (b)** $10 + i$ **(d)** $\frac{11}{26} + \frac{23}{26}i$ **(f)** $2 - 11i$ **(h)** $8 - 6i$

3. **(b)** $\frac{11}{5} + \frac{3}{5}i$ **(d)** $\pm(2 - i)$ **(f)** $1 + i$ **4. (b)** $\frac{1}{2} \pm \frac{\sqrt{3}}{2}i$ **(d)** $2, \frac{1}{2}$

5. **(b)** $-2, 1 \pm \sqrt{3}i$ **(d)** $\pm 2\sqrt{2}, \pm 2\sqrt{2}i$ **6. (b)** $x^2 - 4x + 13; 2 + 3i$ **(d)** $x^2 - 6x + 25; 3 + 4i$

8. $x^4 - 10x^3 + 42x^2 - 82x + 65$ **10. (b)** $(-2)^2 + 2i - (4 - 2i) = 0; 2 - i$

   (d) $(-2 + i)^2 + 3(1 - i)(-1 + 2i) - 5i = 0; -1 + 2i$ **11. (b)** $-i, 1 + i$ **(d)** $2 - i, 1 - 2i$

12. **(b)** Circle, centre at 1, radius 2 **(d)** Imaginary axis **(f)** Line $y = mx$

18. **(b)** $4e^{-\pi i/2}$ **(d)** $8e^{2\pi i/3}$ **(f)** $6\sqrt{2}e^{3\pi i/4}$ **19. (b)** $\frac{1}{2} + \frac{\sqrt{3}}{2}i$ **(d)** $1 - i$ **(f)** $\sqrt{3} - 3i$

20. **(b)** $-\frac{1}{32} + \frac{\sqrt{3}}{32}i$ **(d)** $-32i$ **(f)** $-2^{16}(1 + i)$

23. **(b)** $\pm\frac{\sqrt{2}}{2}(\sqrt{3} + i), \pm\frac{\sqrt{2}}{2}(-1 + \sqrt{3}i)$ **(d)** $\pm 2i, \pm(\sqrt{3} + i), \pm(\sqrt{3} - i)$

Appendix B Proofs (Page 489)

1. **(b)** If $m = 2p$ and $n = 2q + 1$ where p and q are integers, then $m + n = 2(p + q) + 1$ is odd. The converse is false: $m = 1$ and $n = 2$ is a counterexample.
   (d) $x^2 - 5x + 6 = (x - 2)(x - 3)$ so, if this is zero, then $x = 2$ or $x = 3$. The converse is true: each of 2 and 3 satisfies $x^2 - 5x + 6 = 0$.

2. **(b)** This implication is true. If $n = 2t + 1$ where t is an integer, then $n^2 = 4t^2 + 4t + 1 = 4t(t + 1) + 1$. Now t is either even or odd, say $t = 2m$ or $t = 2m + 1$. If $t = 2m$, then $n^2 = 8m(2m + 1) + 1$; if $t = 2m + 1$, then $n^2 = 8(2m + 1)(m + 1) + 1$. Either way, n^2 has the form $n^2 = 8k + 1$ for some integer k.

3. **(b)** Assume that the statement "one of m and n is greater than 12" is false. Then both $n \leq 12$ and $m \leq 12$, so $n + m \leq 24$, contradicting the hypothesis that $n + m = 25$. This proves the implication. The converse is false: $n = 13$ and $m = 13$ is a counterexample.
   (d) Assume that the statement "m is even or n is even" is false. Then both m and n are odd, so mn is odd, contradicting the hypothesis. The converse is true: If m or n is even, then mn is even.

4. **(b)** If x is irrational and y is rational, assume that $x + y$ is rational. Then $x = (x + y) - y$ is the difference of two rationals, and so is rational, contrary to the hypothesis.

5. **(b)** $n = 10$ is a counterexample because $10^3 = 1000$ while $2^{10} = 1024$, so the statement $n^3 \geq 2^n$ is false if $n = 10$. Note that $n^3 \geq 2^n$ *does* hold for $2 \leq n \leq 9$.

Appendix C Mathematical Induction (Page 494)

6. $\dfrac{n}{n+1} + \dfrac{1}{(n+1)(n+2)} = \dfrac{n(n+2)+1}{(n+1)(n+2)} = \dfrac{(n+1)^2}{(n+1)(n+2)} = \dfrac{1}{n+2}$

14. $(2\sqrt{n-1}) + \dfrac{1}{\sqrt{n+1}} = \dfrac{2\sqrt{n^2+n}+1}{\sqrt{n+1}} - 1 < \dfrac{2(n+1)}{\sqrt{n+1}} - 1 = 2\sqrt{n+1} - 1$

20. $B_n = (n + 1)! - 1$ **22. (b)** Verify each of $S_1, S_2, \ldots, S_8$.

Index